Methods in Enzymology

Volume 132
IMMUNOCHEMICAL TECHNIQUES
Part J
Phagocytosis and Cell-Mediated Cytotoxicity

METHODS IN ENZYMOLOGY

EDITORS-IN-CHIEF

Sidney P. Colowick   Nathan O. Kaplan

Methods in Enzymology

Volume 132

Immunochemical Techniques

Part J
Phagocytosis and Cell-Mediated Cytotoxicity

EDITED BY

Giovanni Di Sabato

DEPARTMENT OF MOLECULAR BIOLOGY
VANDERBILT UNIVERSITY
NASHVILLE, TENNESSEE

Johannes Everse

DEPARTMENT OF BIOCHEMISTRY
TEXAS TECH UNIVERSITY HEALTH SCIENCE CENTER
SCHOOL OF MEDICINE
LUBBOCK, TEXAS

1986

ACADEMIC PRESS, INC.
Harcourt Brace Jovanovich, Publishers
Orlando San Diego New York Austin
Boston London Sydney Tokyo Toronto

149007

ACADEMIC PRESS, INC.
Orlando, Florida 32887

United Kingdom Edition published by
ACADEMIC PRESS INC. (LONDON) LTD.
24–28 Oval Road, London NW1 7DX

LIBRARY OF CONGRESS CATALOG CARD NUMBER: 54-9110

ISBN 0–12–182032–7 (alk. paper)

PRINTED IN THE UNITED STATES OF AMERICA

86 87 88 89   9 8 7 6 5 4 3 2 1

Table of Contents

Section I. Phagocytosis: General Methodology

Section II. Special Methods for Measuring Phagocytosis

Section III. Specific Methods for the Isolation of Cells and Cellular Components

Section VI. Cell-Mediated Cytotoxicity

Section VII. Assays for Cytotoxic Activity

Section VIII. Cytotoxic Activity of Macrophages

Contributors to Volume 132

Article numbers are in parentheses following the names of contributors.
Affiliations listed are current.

DARRYL R. ABSOLOM (2, 3, 13), *Research Institute, The Hospital for Sick Children, Toronto, Ontario, Canada M5G 1X8; Departments of Surgery and Mechanical Engineering and Institute of Biomedical Engineering, University of Toronto, Toronto, Ontario, Canada M5S 1A4; and Department of Microbiology, State University of New York, Buffalo, New York 14214*

DOLPH O. ADAMS (39), *Departments of Pathology and Microbiology–Immunology, Duke University Medical Center, Durham, North Carolina 27710*

PATRICIA C. ANDREWS (20), *Angenics, Inc., Cambridge, Massachusetts 02139*

JOHN A. BADWEY (17, 19), *Department of Biological Chemistry, Harvard Medical School, Boston, Massachusetts 02115*

MARCO BAGGIOLINI (12, 22), *Theodor-Kocher-Institut, University of Bern, 3000 Bern 9, Switzerland*

ZVI BAR-SHAVIT (15), *The Hubert H. Humphrey Center for Experimental Medicine and Cancer Research, The Hebrew University–Hadassah Medical School, Jerusalem 91010, Israel*

FRANK A. BLUMENSTOCK (16), *Department of Physiology, Albany Medical College of Union University, Albany, New York 12208*

K. THEODOR BRUNNER (27), *Department of Immunology, Swiss Institute for Experimental Cancer Research, CH-1066 Épalinges, Switzerland*

PINA CARDARELLI (16), *La Jolla Cancer Research Foundation, Cancer Research Center (NCI), La Jolla, California 92037*

VINCENT CASTRANOVA (33), *Department of Physiology, West Virginia University, Morgantown, West Virginia 26506, and Biochemistry Section, Division of Respiratory Disease Studies, National Institute for Occupational Safety and Health, Morgantown, West Virginia 26505*

JAMES R. CATTERALL (44), *Department of Respiratory Medicine, City Hospital, Edinburgh EH10, 5SB, Scotland*

JEAN-CHARLES CEROTTINI (27), *Ludwig Institute for Cancer Research, Lausanne Branch, CH-1066 Épalinges, Switzerland*

K.-P. CHANG (43), *Department of Microbiology, University of Health Sciences/Chicago Medical School, North Chicago, Illinois 60064*

PHILIP H. COOPER (22), *Wander Research Institute, A Sandoz Research Unit, 3001 Bern, Switzerland*

ANDREW R. CROSS (21), *Department of Biochemistry, University of Bristol, Bristol BS8 1TD, England*

CLAIRE H. DAYTON (16), *Department of Physiology, Albany Medical College of Union University, Albany, New York 12208*

MARTIN DE BOER (8), *Central Laboratory of The Netherlands Red Cross Blood Transfusion Service, 1066 CX Amsterdam, The Netherlands*

WILLEM DEN OTTER (29, 36), *Pathology Institute, Department of Experimental Pathology, University of Utrecht, 3511 HX Utrecht, The Netherlands*

BEATRICE DEWALD (12), *Theodor-Kocher-Institut, University of Bern, 3000 Bern 9, Switzerland*

ROEL A. DE WEGER (29, 36), *Pathology Institute, Department of Experimental Pathology, University of Utrecht, 3511 HX Utrecht, The Netherlands*

BETH-ELLEN DRYSDALE (38), *Subdepart-*

ment of Immunology, Department of Molecular Biology and Genetics, The Johns Hopkins University School of Medicine, Baltimore, Maryland 21205

HOWARD D. ENGERS (27), *World Health Organization Centre for Research and Training in Immunology, Department of Pathology, University of Geneva, 1211 Geneva, Switzerland*

JOHANNES EVERSE (32, 37), *Department of Biochemistry, Texas Tech University Health Science Center, Lubbock, Texas 79430*

ISAIAH J. FIDLER (42), *Department of Cell Biology, M. D. Anderson Hospital and Tumor Institute, Houston, Texas 77030*

THEODORE G. GABIG (18), *Department of Medicine, Indiana University School of Medicine, Indianapolis, Indiana 46254*

RACHEL GOLDMAN (15), *Department of Membrane Research, Weizmann Institute of Science, Rehovot 76100, Israel*

MATTHEW B. GRISHAM (40, 41), *Department of Biochemistry, St. Jude Children's Research Hospital, Memphis, Tennessee 38101*

WERNER HAAS (30), *Basel Institute for Immunology, CH-4005 Basel, Switzerland*

ANGELA M. HARPER (21), *Department of Medicine, Faculty of Clinical Sciences, University College London, London WC1E 6JJ, England*

JAN HED (6), *Section of Clinical Immunology, The Blood Center, Linköping University, S-581 85 Linköping, Sweden, and Laboratory of Clinical Immunology, Karolinska Hospital, S-104 01 Stockholm, Sweden*

JOHN B. HIBBS, JR. (34), *Veterans Administration Medical Center, and Department of Medicine, Division of Infectious Diseases, University of Utah School of Medicine, Salt Lake City, Utah 84148*

JAMES R. HUMBERT (35), *Departments of Pediatrics and Microbiology, State University of New York at Buffalo and Children's Hospital of Buffalo, Buffalo, New York 14222*

MACHIKO IWAMOTO (31), *Department of Biochemistry, Tokyo Metropolitan Institute of Gerontology, 35-2 Sakae-cho, Itabashi-ku, Tokyo 173, Japan*

STEPHANIE L. JAMES (46), *Departments of Medicine and Microbiology, The George Washington University Medical Center, Washington, D.C. 20037*

M. MARGARET JEFFERSON (40, 41), *Department of Biochemistry, St. Jude Children's Research Hospital, Memphis, Tennessee 38101*

WILLIAM J. JOHNSON (39), *Department of Immunology, Smith Kline & French Laboratories, Philadelphia, Pennsylvania 19101*

OWEN T. G. JONES (21), *Department of Biochemistry, University of Bristol, Bristol BS8 1TD, England*

MANFRED L. KARNOVSKY (17, 19), *Department of Biological Chemistry, Harvard Medical School, Boston, Massachusetts 02115*

MARGARET KIELIAN (11), *Department of Cell Biology, Albert Einstein College of Medicine, Bronx, New York 10461*

DANUTA KONOPINSKA (14), *Institute of Chemistry, University of Wroclaw, 50-383 Wroclaw, Poland*

NORMAN I. KRINSKY (20), *Department of Biochemistry, Tufts University School of Medicine, Boston, Massachusetts 02111*

JAMES LEE (14), *ENSECO, Cambridge, Massachusetts 02138*

BRUCE A. LEFKER (18), *Department of Chemistry, Colorado State University, Fort Collins, Colorado 80521*

DORIS L. LEFKOWITZ (37), *Department of Biological Sciences, Texas Tech University, Lubbock, Texas 79409*

STANLEY S. LEFKOWITZ (37), *Department of Microbiology, Texas Tech University Health Science Center, Lubbock, Texas 79430*

BRUCE E. LEHNERT (4), *Life Sciences Division, Toxicology Group, Los Alamos National Laboratory, Los Alamos, New Mexico 87545*

TOSHIHARU MATSUMURA (31), *Department of Cancer Cell Research, Institute of Medical Science, University of Tokyo, 4-6-1 Shirokanedai, Minato-ku, Tokyo 108, Japan*

STEVE J. MCFAUL (32), *Letterman Army Institute of Research, Blood Research Division, Presidio of San Francisco, San Francisco, California 94129*

PAUL J. MILLER (9), *Clinical Investigations Section, Biological Therapeutics Branch, Biological Response Modifiers Program, National Cancer Institute–Frederick Cancer Research Facility, Frederick, Maryland 21701*

CHANDRA K. MITTAL (25, 26), *Laboratory of Biochemical Research, Department of Basic Sciences, The University of Illinois College of Medicine at Peoria, Peoria, Illinois 61656*

LINDA L. MOORE (35), *Departments of Microbiology and Otolaryngology, State University of New York at Buffalo and Children's Hospital of Buffalo, Buffalo, New York 14222*

C. A. NACY (43), *Department of Immunology, Walter Reed Army Institute of Research, Washington, D.C. 20307*

YOSHITAKA NAGAI (31), *Department of Biochemistry, Faculty of Medicine, University of Tokyo, 7-3-1 Hongo, Bunkyo-ku, Tokyo 113, Japan*

VICTOR A. NAJJAR (14), *Division of Protein Chemistry, Tufts University School of Medicine, Boston, Massachusetts 02111*

RAJIV NAYAR (42), *Department of Cell Biology, M. D. Anderson Hospital and Tumor Institute, Houston, Texas 77030*

CHRISTIAN F. OCKENHOUSE (45), *Medical College of Pennsylvania, Philadelphia, Pennsylvania 19129*

MASAKO OKAJIMA (38), *Subdepartment of Immunology, Department of Molecular Biology and Genetics, The Johns Hopkins University School of Medicine, Baltimore, Maryland 21205*

JOHN R. ORTALDO (28), *Laboratory of Experimental Immunology, Biological Response Modifiers Program, Division of Cancer Treatment, National Cancer Institute–Frederick Cancer Research Facility, Frederick, Maryland 21701*

R. D. PEARSON (43), *Department of Medicine, Division of Geographic Medicine, University of Virginia School of Medicine, Charlottesville, Virginia 22908*

EDGAR PICK (24), *Laboratory of Immunopharmacology, Department of Human Microbiology, Sackler School of Medicine, Tel Aviv University, Ramat Aviv 69978, Tel Aviv, Israel*

JACK S. REMINGTON (44), *Department of Immunology and Infectious Diseases, Research Institute, Palo Alto Medical Foundation, Palo Alto, California 94301, and Department of Medicine, Stanford University, Stanford, California 94305*

DIRK ROOS (8, 10), *Blood Cell Chemistry, Central Laboratory of The Netherlands Red Cross Blood Transfusion Service, 1066 CX Amsterdam, The Netherlands*

WALTER RUCH (22), *Wander Research Institute, A Sandoz Research Unit, 3001 Bern, Switzerland*

BERT A. RUNHAAR (29), *Pathology Institute, Department of Experimental Pathology, University of Utrecht, 3511 HX Utrecht, The Netherlands*

THOMAS M. SABA (16), *Department of Physiology, Albany Medical College of Union University, Albany, New York 12208*

ALAN J. SCHROIT (42), *Department of Cell Biology, M. D. Anderson Hospital and Tumor Institute, Houston, Texas 77030*

ANTHONY W. SEGAL (21), *Department of Medicine, Faculty of Clinical Sciences, University College London, London WC1E 6JJ, England*

SOMESH D. SHARMA (44), *Department of Immunology and Infectious Diseases, Research Institute, Palo Alto Medical Foundation, Palo Alto, California 94301*

HANNAH LUSTIG SHEAR (45), *Department of Medical and Molecular Parasitology, New York University Medical Center, New York, New York 10016*

HYUN S. SHIN (38), *Subdepartment of Immunology, Department of Molecular Biology and Genetics, The Johns Hopkins University School of Medicine, Baltimore, Maryland 21205*

JOHN A. STEINKAMP (4), *Life Sciences Division, Experimental Pathology Group, Los Alamos National Laboratory, Los Alamos, New Mexico 87545*

HENRY C. STEVENSON (9), *Clinical Investigations Section, Biological Therapeutics Branch, Biological Response Modifiers Program, National Cancer Institute–Frederick Cancer Research Facility, Frederick, Maryland 21701*

CARLETON C. STEWART (4), *Life Sciences Division, Experimental Pathology Group, Los Alamos National Laboratory, Los Alamos, New Mexico 87545*

THOMAS P. STOSSEL (5), *Hematology–Oncology Unit, Department of Medicine, Massachusetts General Hospital, Boston, Massachusetts 02114*

READ R. TAINTOR (34), *Veterans Administration Medical Center, Salt Lake City, Utah 84148*

SAMUEL T. TEST (23), *Institute of Cancer Research at Pacific Presbyterian Medical Center, San Francisco, California 94115*

EDWIN L. THOMAS (40, 41), *Department of Biochemistry, St. Jude Children's Research Hospital, Memphis, Tennessee 38101*

KNOX VAN DYKE (33), *Department of Pharmacology and Toxicology, West Virginia University, Morgantown, West Virginia 26506*

CAREL J. VAN OSS (1), *Department of Microbiology, State University of New York, School of Medicine, Buffalo, New York 14214*

MICHAEL R. VAN SCOTT (33), *Division of Pulmonary Diseases, Critical Care and Occupational Medicine, Department of Medicine, University of North Carolina, Chapel Hill, North Carolina 27514*

ALWIN A. VOETMAN (10), *Development and Quality Assurance, Central Laboratory of The Netherlands Red Cross Blood Transfusion Service, 1066 CX Amsterdam, The Netherlands*

HARALD VON BOEHMER (30), *Basel Institute for Immunology, CH-4005 Basel, Switzerland*

RU-QI WEI (37), *Department of Cell Biology and Anatomy, Texas Tech University Health Science Center, Lubbock, Texas 79430*

STEPHEN J. WEISS (23), *Simpson Memorial Research Institute, Division of Hematology/Oncology, Department of Internal Medicine, University of Michigan, Ann Arbor, Michigan 48105*

SAMUEL D. WRIGHT (7), *Laboratory of Cellular Physiology and Immunology, The Rockefeller University, New York, New York 10021*

CHARLES M. ZACHARCHUK (38), *Department of Medicine, Stanford University School of Medicine, Stanford, California 94305*

Preface

Increasingly, researchers and clinicians are recognizing the importance of granulocytic cells as part of our defense mechanisms against disease. Although the importance of the "white blood cells" in combating bacterial infections has been known for a number of decades, the role of macrophages and granulocytic lymphocytes in the body's defense against neoplasia is just beginning to be appreciated. Our understanding of the biochemical events associated with the biological functioning of these cells as well as those associated with their activation is in most cases still rudimentary at best. However, the rapidly increasing interest in the biochemistry of the reticuloendothelial system more than warrants a volume that covers the most current methodology in this field.

There is no doubt that much of the recent progress in understanding the biochemical basis of cytotoxicity is due to the pioneering work of Klebanoff and his discovery of the role of myeloperoxidase in this process about twenty years ago. Since that time a great deal of work has been done on the biochemical mechanism through which peroxidases exert their cytotoxic activity. New areas of research were subsequently indicated by the findings that peroxidases were able to stimulate mast cells into histamine release and that these enzymes promoted tumor regression *in vivo*. These results may form the basis for new multidisciplinary research areas relating to the biochemical bases for cell–cell interactions, cell–cell recognition, etc. It is our hope that this volume will stimulate new research in these exciting areas.

Several assays for cytotoxic activity, especially those involving complement, as well as antibody-dependent cell-mediated cytotoxicity, are covered in Volume 93 of this series. Moreover, assays for superoxide and other cytotoxic agents are described in Volume 105. In accordance with the policy of *Methods in Enzymology*, these methods have not been repeated in this volume. Readers wishing to use such methodology are referred to these and other previous volumes.

We wish to express our appreciation to Drs. Manfred L. Karnovsky, Seymour J. Klebanoff, Carel J. van Oss, and Zena Werb for their constructive criticisms during the preparation of this volume. Their comments, suggestions, and recommendations were of invaluable help to us in organizing this work.

It is with deep sorrow that we remember the help and advice we received from the Editors-in-Chief of *Methods in Enzymology*, Dr. Sidney P. Colowick and Dr. Nathan O. Kaplan. Both passed away during the preparation of this volume, Dr. Colowick on January 9, 1985, and Dr. Kaplan on April 15, 1986. By realizing the impact that *Methods in Enzymology* has had on the progress of biochemical research in the past thirty years we recognize the contribution made by Dr. Colowick and Dr. Kaplan to biochemistry. Individual volumes of this series can be found in almost every biological research laboratory, and there are few biochemists indeed whose research efforts have not at least at one time or another benefited from *Methods in Enzymology*. The contribution to biochemistry made by these two scientists is, therefore, truly without precedence, and rivals those made by Nobelists. The vacuum left by their passing will not easily be filled.

GIOVANNI DI SABATO
JOHANNES EVERSE

METHODS IN ENZYMOLOGY

EDITED BY

Sidney P. Colowick and Nathan O. Kaplan

VANDERBILT UNIVERSITY
SCHOOL OF MEDICINE
NASHVILLE, TENNESSEE

DEPARTMENT OF CHEMISTRY
UNIVERSITY OF CALIFORNIA
AT SAN DIEGO
LA JOLLA, CALIFORNIA

METHODS IN ENZYMOLOGY

EDITORS-IN-CHIEF

Sidney P. Colowick and Nathan O. Kaplan

VOLUME LX. Nucleic Acids and Protein Synthesis (Part H)
Edited by KIVIE MOLDAVE AND LAWRENCE GROSSMAN

VOLUME 61. Enzyme Structure (Part H)
Edited by C. H. W. HIRS AND SERGE N. TIMASHEFF

VOLUME 62. Vitamins and Coenzymes (Part D)
Edited by DONALD B. MCCORMICK AND LEMUEL D. WRIGHT

VOLUME 63. Enzyme Kinetics and Mechanism (Part A: Initial Rate and Inhibitor Methods)
Edited by DANIEL L. PURICH

VOLUME 64. Enzyme Kinetics and Mechanism (Part B: Isotopic Probes and Complex Enzyme Systems)
Edited by DANIEL L. PURICH

VOLUME 65. Nucleic Acids (Part I)
Edited by LAWRENCE GROSSMAN AND KIVIE MOLDAVE

VOLUME 66. Vitamins and Coenzymes (Part E)
Edited by DONALD B. MCCORMICK AND LEMUEL D. WRIGHT

VOLUME 67. Vitamins and Coenzymes (Part F)
Edited by DONALD B. MCCORMICK AND LEMUEL D. WRIGHT

VOLUME 68. Recombinant DNA
Edited by RAY WU

VOLUME 69. Photosynthesis and Nitrogen Fixation (Part C)
Edited by ANTHONY SAN PIETRO

VOLUME 70. Immunochemical Techniques (Part A)
Edited by HELEN VAN VUNAKIS AND JOHN J. LANGONE

VOLUME 71. Lipids (Part C)
Edited by JOHN M. LOWENSTEIN

VOLUME 72. Lipids (Part D)
Edited by JOHN M. LOWENSTEIN

VOLUME 85. Structural and Contractile Proteins (Part B: The Contractile Apparatus and the Cytoskeleton)
Edited by DIXIE W. FREDERIKSEN AND LEON W. CUNNINGHAM

VOLUME 86. Prostaglandins and Arachidonate Metabolites
Edited by WILLIAM E. M. LANDS AND WILLIAM L. SMITH

VOLUME 87. Enzyme Kinetics and Mechanism (Part C: Intermediates, Stereochemistry, and Rate Studies)
Edited by DANIEL L. PURICH

VOLUME 88. Biomembranes (Part I: Visual Pigments and Purple Membranes, II)
Edited by LESTER PACKER

VOLUME 89. Carbohydrate Metabolism (Part D)
Edited by WILLIS A. WOOD

VOLUME 90. Carbohydrate Metabolism (Part E)
Edited by WILLIS A. WOOD

VOLUME 91. Enzyme Structure (Part I)
Edited by C. H. W. HIRS AND SERGE N. TIMASHEFF

VOLUME 92. Immunochemical Techniques (Part E: Monoclonal Antibodies and General Immunoassay Methods)
Edited by JOHN J. LANGONE AND HELEN VAN VUNAKIS

VOLUME 93. Immunochemical Techniques (Part F: Conventional Antibodies, Fc Receptors, and Cytotoxicity)
Edited by JOHN J. LANGONE AND HELEN VAN VUNAKIS

VOLUME 94. Polyamines
Edited by HERBERT TABOR AND CELIA WHITE TABOR

VOLUME 95. Cumulative Subject Index Volumes 61–74 and 76–80
Edited by EDWARD A. DENNIS AND MARTHA G. DENNIS

VOLUME 96. Biomembranes [Part J: Membrane Biogenesis: Assembly and Targeting (General Methods; Eukaryotes)]
Edited by SIDNEY FLEISCHER AND BECCA FLEISCHER

Section I

Phagocytosis: General Methodology

[1] Phagocytosis: An Overview

By CAREL J. VAN OSS

Introduction

Phylogeny

Phagocytosis is the most important defense mechanism in all phyla of the animal kingdom. Among protozoa phagocytosis is not only the means of defense against foreign cells and particles,[1] but also the principal mode of ingesting food. Porifera, mollusca, annelids, and arthropods, as well as echinoderms and tunicates use ingestion by phagocytic cells to dispose of unwanted cells and small particles; their phagocytes and other, related cells also can combine together to encapsulate and isolate invading particles or organisms that are too large to be engulfed by one cell.[1-4] Notwithstanding the increasingly sophisticated humoral and cellular immune defense components that have evolved in the vertebrates, which culminated in the complex immune systems of birds and mammals, phagocytosis still remains the principal effector mechanism for the ultimate disposal of invading, foreign, effete, or otherwise unwanted cells or particles.

Phagocytosis in Antimicrobial Defense

The vast majority of the bacterial species to which man is exposed is not pathogenic. The inocuity of most of these bacteria is due to the fact that they spontaneously become phagocytized[5] as soon as they breach any of the natural protective barriers of the human organism. Only the relatively few species of bacteria that are endowed with mechanisms allowing them to evade phagocytic ingestion or avoid phagocytic digestion can properly be called pathogenic; to thwart the attacks of these few microorganisms, the whole complicated mammalian humoral and cellular

[1] M. J. Manning and R. J. Turner, "Comparative Immunobiology." Halsted/Wiley, New York, 1976.

[2] E. L. Cooper, "Comparative Immunology." Prentice-Hall, Englewood Cliffs, New Jersey, 1976.

[3] N. Cohen and M. M. Sigel, eds., "The Reticuloendothelial System," Vol. 3. Plenum, New York, 1982.

[4] G. Salt, "The Cellular Defense Reactions of Insects." Cambridge Univ. Press, London and New York, 1970.

[5] C. J. van Oss, C. F. Gillman, and A. W. Neumann, "Phagocytic Engulfment and Cell Adhesiveness," p. 33. Dekker, New York, 1975.

immunological selection and enhancement system developed, which ultimately again leads to the disposal of these pathogens, through phagocytosis.

Different immune deficiency diseases (involving, e.g., immunoglobulins, complement, or T cells) can shorten life to various degrees, but usually still leave life expectancies of several to many years. However, a drastic decrease in the number of circulating phagocytes can result in death within days or weeks through intractable bacterial infections, against which antibiotics are of little avail. In such infections, the bacteria involved often are of a variety that normally would not be considered pathogenic.

Thus the ultimate disposal of all infectious agents and other unwanted particles, which have been detected and processed with the help of other branches of the immune system, occurs through phagocytosis, so that phagocytosis represents the final and most indispensable step in our immunological defense system.

Phagocytes

Professional Phagocytes

Although the demarcation is not extremely sharp, we may make a distinction between professional and nonprofessional phagocytes, according to (1) whether or not the cells' principal known function is phagocytic, and/or (2) whether the cells can digest as well as ingest particles.

Polymorphonuclear Phagocytes

Polymorphonuclear leukocytes (PMNs, named after the variegated shapes of their usually multilobar nuclei),[5a] or granulocytes, are most conspicuous in the blood circulation; the peripheral blood of a normal adult contains about 2.5×10^{10} PMNs (which is approximately 0.1% of the circulating red cells). In the bone marrow, there is in addition a reserve of about 2.5×10^{12} PMNs and their immediate precursors. About 98% of the PMNs are neutrophils, 1–2% eosinophils, and 0.1 to 0.2% basophils (these distinctions are based on different staining reactions). The neutrophils are the most important circulating phagocytes; they form our first line of defense against invading particles, especially bacteria. Severe neutropenia (or granulocytopenia), i.e., a decrease of 90% or more in the normal numbers of circulating PMNs, incurs the likelihood of a fatal bacterial invasion. PMNs are end cells and usually live no longer than a few hours to a few days.

[5a] Abbreviations: PMN, polymorphonucleates; R, receptors.

Eosinophils also phagocytize (e.g., antigen–antibody complexes), but they have, in addition, other important functions in hypersensitivity and other antiparasitic reactions.[6]

Basophils probably do not phagocytize to any important degree; like mast cells (see below) they contain heparin and a large amount of histamine, and play an important role in IgE-mediated hypersensitivity. For morphological and physiological characteristics, see Bessis.[7]

Mononuclear Phagocytes

The various phagocytic mononuclear leukocytes that used to be taken to inhabit the "reticuloendothelial system" recently have been grouped together more aptly as the mononuclear phagocytes,[8] after the fact that the nuclei of these cells, although often lobar in shape, are essentially of one piece. One must not confound mononuclear phagocytes with "mononuclear cells," or "mononuclear leukocytes," which are terms used even to this day, to express an author's inability to distinguish between monocytes and lymphocytes under a given set of circumstances.

Mononuclear phagocytes comprise the promonocytes in the bone marrow, which give rise to the circulating monocytes, which ultimately mature into the mobile or fixed macrophages (or histiocytes, or reticulocytes[7]), found in the following tissues[7-9]:

Bone marrow: macrophages, monocytes, sinusoidal lining cells
Bone tissue: osteoclasts
Intestinal Peyer's patches: macrophages
Connective tissue (subcutaneous and in organs): histiocytes, macrophages, monocytes
Liver: Kupffer cells
Lung: alveolar macrophages; monocytes in the pleural fluid
Lymph nodes and lymphoid tissue: free and fixed macrophages; monocytes
Nervous system: microglial cells
Serous cavity: peritoneal macrophages
Spleen: free and fixed macrophages, sinusoidal lining cells, monocytes
Thymus: free and fixed macrophages, monocytes

[6] T. Yoshida and M. Torisu, eds., "Immunobiology of the Eosinophil." Elsevier Biomedical, New York, 1983.
[7] M. Bessis, "Living Blood Cells and Their Ultrastructure." Springer, New York, 1973.
[8] R. van Furth, *in* "Mononuclear Phagocytes" (R. van Furth, ed.), p. 1. Martinus Nijhoff, The Hague, 1980.
[9] C. J. van Oss, *in* "Principles of Immunology" (N. R. Rose, F. Milgrom, and C. J. van Oss, eds.), p. 155. Macmillan, New York, 1979.

Inflammation sites: macrophages; epithelioid or multinucleated giant cells, or Langerhans cells[7,8]

It should be noted that PMNs also can be found in all these sites, in various proportions to mononuclear phagocytes. Mononuclear phagocytes have a much longer life (up to 75 days[7]) than PMNs and can proliferate locally.[9,10] They essentially represent the second line of defense against invading particles. However, in the upper respiratory and digestive tracts, mononuclear cells can also assume a primary defense function.

Nonprofessional Phagocytes

Various types of circulating blood cells, as well as fixed tissue cells, that do not necessarily have phagocytosis as their main function and that do not digest ingested particles to a significant degree, nevertheless are capable of endocytosis. Examples of such nonprofessional phagocytes are lymphocytes,[7] platelets,[7,11] mast cells,[7] epithelial cells of various mucous membranes,[12] and also cultured fibroblasts and HeLa cells.[12] Even erythrocytes have been observed to ingest an occasional latex particle. In some important cases, endocytosis by nonprofessional phagocytes does not involve specific receptors (e.g., in the case of platelets,[11] and of HeLa cells[12]), which can facilitate the study of phagocytic ingestion from a purely surface-thermodynamic viewpoint.[11] Another interesting nonprofessional phagocytic cell model is that of *Arbacia* (sea urchin) eggs, which can ingest oil drops.[13]

Nonphagocytic Functions of Phagocytes

Professional phagocytes also may have important nonphagocytic functions. While the main role of neutrophilic PMNs is phagocytosis, eosinophils and basophils also play important roles in nonphagocytic antiparasite defenses and in other immediate hypersensitivity reactions (see above). Macrophages in particular are endowed with many nonphagocytic functions, e.g., (1) secretion of various compounds ("macrokines"), such as hydrolases, colony stimulating factor, transferrin, erythropoietin,

[10] W. T. Daems, *in* "The Reticuloendothelial System" (I. Carr and W. T. Daems, eds.), Vol. 1, p. 57. Plenum, New York, 1980.
[11] D. R. Absolom, D. W. Francis, W. Zingg, C. J. van Oss, and A. W. Neumann, *J. Colloid Interface Sci.* **85**, 168 (1982).
[12] E. Kihlström and G. Söderlund, *in* "Endocytosis and Exocytosis in Host Defense" (L. B. Edebo, L. Enerbäck, and O. I. Stendahl, eds.), p. 148. Karger, Basel, 1981.
[13] R. Chambers and M. J. Kopac, *J. Cell. Comp. Physiol.* **9**, 331 (1937).

prostaglandins, T and B cell stimulators, interferon,[14] and various complement components[15]; (2) extracellular killing of tumor cells and microorganisms[16,17]; and (3) antigen presentation in the antibody formation induction process.[14,16,17]

Phagocytosis

Chemotaxis and Diapedesis

Various relatively low-molecular-weight compounds, when forming a concentration gradient, can act as chemical attractants for phagocytic cells, causing them to migrate into the direction of increasing concentration of the attractant, by a process called chemotaxis. Apart from necrotactic substances, released by freshly lysed cells,[7] which give rise to phagocytic cell movement (over distances of up to 500 μm) to and ultimate disposal of cells that have recently been damaged (but not to any interaction with cells that have been dead for some time), by far the most important chemotactic agent is complement subfactor C5a, which has a molecular weight of 15,000.[18] Both via the classical and the alternate pathway C5 emits C5a (also known as C5a anaphylatoxin). In addition to the chemotaxin C5a, the presence of a cochemotaxin (or cocytotaxin) seems to be required; nucleotides such as ATP or cyclic AMP can function as cochemotaxins.[7] The chemotactic role of a complement factor that is released in the classical as well as in the alternate pathway guarantees the migration of PMNs and of monocytic phagocytes to virtually all microorganisms that interact with immunoglobulins and complement as well as to (e.g., gram-negative) microorganisms that can directly interact with complement from C3 on. An association product of C5b, C6, and C7 also has chemotactic activity.[18] Chemotaxis appears to be triggered through chemoattractant receptors on the leukocyte surface.[19]

The manner in which phagocytes move toward the chemotactic signal is as follows: first a pseudopod or a cytoplasmic veil (PMNs) or a frilly veil (monocytes and macrophages) is thrown forward and attaches to a

[14] I. Carr and W. T. Daems, *in* "The Reticuloendothelial System" (I. Carr and W. T. Daems, eds.), Vol. 1, p. 1. Plenum, New York, 1980.

[15] C. Bianco, O. Götze, and Z. A. Cohn, *in* "Mononuclear Phagocytes" (R. van Furth, ed.), p. 1443. Martinus Nijhoff, The Hague, 1980.

[16] S. J. Norman and E. Sorkin, "Macrophages and Natural Killer Cells." Plenum, New York, 1982.

[17] R. van Furth, ed., "Mononuclear Phagocytes," Part II. Martinus Nijhoff, The Hague, 1980.

[18] H. J. Müller-Eberhard, *in* "Textbook of Immunopathology" (P. A. Miescher and H. J. Müller-Eberhard, eds.), Vol. 1, p. 45. Grune & Stratton, New York, 1976.

[19] R. Snyderman and M. C. Pike, *Annu. Rev. Immunol.* **2,** 257 (1984).

solid surface or object.[7] (When freely suspended in a liquid, phagocytes are largely spherical; they cannot swim and they only start "crawling" upon making contact with a solid support.) Once the frontal pseudopod or veil is attached, it becomes rigid and the rest of the (still quite fluid) cytoplasm and nucleus streams toward it, leaving only a taillike extension (or uropod) behind.[7] Especially with PMNs, the moving phagocyte establishes a rather pinched waist, or ring, through which it seems to move,[7] which may be connected with its capacity to move through small openings between cells of vascular walls, by diapedesis (see below). PMNs move at the rate of 0.3 to 0.7 μm/sec,[7] the much slower monocytes at about one-third to one-half that speed.

When a chemotactic signal is received, the migration of phagocytes is not confined to the blood vessels and tissue structures that harbor them, but they can squeeze through the minuscule intracellular junctions between the cells that line the blood vessels or other tissues, by diapedesis,[7] much in the same manner in which they "crawl" through apparent ring-shaped strictures during ordinary locomotion. However, contrary to, e.g., lymphocytes, once phagocytes leave the peripheral blood circulation, they never return.[7]

Phagocytic Ingestion

The first step in the initiation of phagocytic ingestion is adhesion of the particle to the phagocyte's surface. The major physicochemical mechanism of that adhesion may be either a van der Waals-type attraction,[20] a hydrophobic attraction, and/or (more rarely) an electrostatic attraction[20,21] or a receptor-mediated bond (see below). Once adhesion has ensued, endocytosis of the adhering particle can take place, starting with an invagination of the phagocyte's outer cell membrane which then proceeds to enclose the particle completely in a phagocytic vacuole, or "phagosome." The phagosome then internalizes and fuses with one or more of the phagocyte's granules, or lysosomes, which contain a variety of hydrolytic enzymes.

Pinocytosis (or "drinking by cells") is the internalization of small amounts of extracellular liquid by the phagocyte, involving, as in endocytosis (described above), the invagination of the cell membrane and its progression into a vacuole, in the apparent absence of an initial specific or aspecific (i.e., a receptor-linked, or a purely physicochemical) adhesion step.

Exocytosis, i.e., the extrusion of a phagosome (which may contain a

[20] C. J. van Oss, D. R. Absolom, and A. W. Neumann, *Ann. N.Y. Acad. Sci.* **46,** 332 (1983).
[21] H. Nagura, J. Asai, and K. Kojima, *Cell Struct. Funct.* **2,** 21 (1977).

previously ingested particle and lysosomal enzymes), or a vacuole, can occasionally be observed. However, such a reversal of the endocytic process occurs relatively infrequently. Exocytosis by osteoclasts has been proposed as a mechanism for bone resorption.[22]

When measuring phagocytosis *in vitro,* it is useful to be able to distinguish between particles (e.g., bacteria) that are adhering to the outer surface of phagocytes, from those that are actually ingested. By light microscopy this is not always easy. When staphylococci or erythrocytes are used, the distinction can be made by lysing the adhering particles (staphylococci with lysostaphin; erythrocytes at low ionic strength); engulfed particles do not lyse: they are protected by the phagocyte. With the help of fluorescent antibacterial antibodies a distinction can also be made, as the fluorescence of such sensitized bacteria can be quenched with crystal violet when adhering to the outside of phagocytes, while ingested bacteria remain fluorescent.[23] With transmission electron microscopy, there usually is no difficulty in distinguishing between ingested and adhering particles (see this volume [6]).

Phagocytic Digestion (see this volume [11])

Once a phagosome, containing an ingested particle, has fused with a number of lysosomes (granules), the particle (e.g., a bacterium) is exposed to a large variety of hydrolases which rapidly destroy and digest most bacteria.[24] This lytic process does not affect the well-being of the phagocytic cell itself, because all the enzymatic reactions take place inside the phagosome, which has as its inner lining part of what used to be the outer membrane of the phagocyte, so that, topologically speaking, the digestion still takes place outside the phagocyte's cytoplasm.

Activation of Phagocytes

The most variegated agents can stimulate or "activate" phagocytes, e.g., they can cause phagocytic cells to diminish their migration and enhance their capacity to spread on surfaces (i.e., make them stay where they are), to increase their phagocytic avidity and activity, to become more hydrophilic,[5,25] and to accelerate their metabolic and digestive processes. Many such activating agents are produced by lymphocytes and

[22] U. Lucht, *in* "The Reticuloendothelial System" (I. Carr and W. T. Daems, eds.), Vol. 1, p. 705. Plenum, New York, 1980.
[23] J. Hed, *FEMS Lett.* **1,** 357 (1977).
[24] A. J. Sbarra and R. R. Strauss, eds., "The Reticuloendothelial System," Vol. 2. Plenum, New York, 1980.
[25] C. J. van Oss, *Annu. Rev. Immunol.* **32,** 19 (1978).

have become known under the general name of "lymphokines."[26] Although many lymphokines and other agents can activate PMNs, the activation of macrophages has, to date, been studied much more thoroughly.[14,16–18,27]

Even agents produced by macrophages themselves ("macrokines," see above), such as complement components,[15] can also activate them; especially complement components of the alternate pathway.[28]

Apart from the influence of the usual lymphokines,[17,18,26,27] phagocytic ingestion of very hydrophobic particles, such as mycobacteria, latex particles, or lipid droplets, also causes an increase in hydrophilicity and in phagocytic activity both in PMNs and in macrophages.[5,25] Activation of macrophages also causes an increase in the expression of various surface receptors.[27]

Surface Interactions between Phagocytes and Phagocytees

Cell–Particle Interactions (see also this volume [2])

Classically, the total interaction between particles and/or cells of various compositions and sizes is held to be the sum of their electrostatic repulsion (living cells and other biological particles generally carry a net negative charge), and their van der Waals attraction, also taking into account the interaction with and among the molecules of the liquid medium. That total interaction is described by the "DLVO theory," after the original Russian (Derjaguin and Landau) and Dutch (Verwey and Overbeek) authors of that theory.[29] The strong hydrophilicity of all blood cells, which are immersed in an aqueous medium, causes their mutual van der Waals attraction to be exceedingly slight, so that their negative surface potential normally keeps the cells about 100 Å apart, which gives rise to a very stable suspension.[20,30] In addition to the low van der Waals attraction and the sizable electrostatic repulsion, hydration effects[31] may play a role in the stability of blood cells.[30] In the normal course of events, bacteria also will be repelled by circulating blood cells (including phagocytes),[20]

[26] A. Khan and N. O. Hill, eds., "Human Lymphokines." Academic Press, New York, 1982.

[27] D. O. Adams and T. A. Hamilton, *Annu. Rev. Immunol.* **2,** 283 (1984).

[28] O. Götze, C. Bianco, J. S. Sundsmo, and Z. A. Cohn, *in* "Mononuclear Phagocytes" (R. van Furth, ed.), p. 1421. Martinus Nijhoff, The Hague, 1980.

[29] E. J. W. Verwey and J. Th. G. Overbeek, "Theory of the Stability of Lyophobic Colloids." Elsevier, Amsterdam, 1948.

[30] C. J. van Oss, *J. Dispersion Sci. Technol.* **6,** 131 (1985).

[31] D. M. LeNeveu, R. P. Rand, V. A. Parsegian, and D. Gingell, *Biophys. J.* **18,** 209 (1977).

but that repulsion is proportional to the radius of curvature of the particle or cell, so that the propensity of phagocytes to extrude pseudopodia with extremities of small radii of curvature allows them to overcome the electrostatic repulsion and thus to establish contact with most bacteria.[32] And as soon as an interparticle or intercellular distance of less than 8 Å (the "Debye" distance under physiological conditions)[20,29] is reached, electrostatic repulsions tend to become less dominant and van der Waals attractions may locally become preponderant and cause adhesion, or specific receptors may become sufficiently close to, e.g., Fc moieties, to attach to them, via both electrostatic and van der Waals attractions (see below).

ζ-Potentials

The respective negative surface potentials of both phagocytic and bacterial cells normally would prevent the establishment of contact between them, but owing to the small radius of curvature of bacteria and to the facility with which phagocytes can extend pseudopodia, that electrostatic repulsion usually can be overcome. Thus long-range repulsions due to negative ζ-potentials, rarely play an important role in the phagocytizability of bacteria, and, as already pointed out in the previous section, aspecific short-range interactions (i.e., interactions at distances <8 Å) are much less sensitive to repulsions due to ζ-potentials of the same sign of charge and the prevailing interactions tend to be of the van der Waals variety. Therefore, electrophoretic mobilities of bacteria do not afford a clear-cut correlation with the degree to which they become phagocytized.[33,34]

Surface Tensions and van der Waals Interactions

As soon as short-range (<8 Å) contact is established, van der Waals interactions become preponderant, and as van der Waals interactions are to an important extent proportional to interfacial tensions, nonspecific (i.e., nonreceptor mediated) phagocyte–phagocytee interactions can be accurately predicted once the surface tensions of the phagocytic cells, of the particles to be phagocytized, and of the liquid medicine are known.[5,9,25] Generally speaking, in aqueous media, nonopsonized particles will (and should) be phagocytized the more actively, the lower their surface tensions (or the more hydrophobic their surfaces).[5,9] This phenomenon can even be put to use to measure the surface tensions of

[32] C. J. van Oss, R. J. Good, and A. W. Neumann, *J. Electroanal. Chem.* **37**, 387 (1972).
[33] D. J. Wilkins and A. D. Bangham, *J. Reticuloendothel. Soc.* **1**, 233 (1964).
[34] T. R. Kozel, E. Reiss, and R. Cherniak, *Infect. Immun.* **29**, 295 (1980).

nonopsonized bacteria, as a function of their phagocytic ingestion by granulocytes[35] or platelets.[36]

Interactions in Aqueous Liquids vs Interactions in Blood and Other Biological Fluids

The van der Waals interaction between hydrophilic phagocytes and somewhat more hydrophobic bacteria and other particles becomes stronger with increasingly hydrophobic particles (see above) only when taking place in liquids such as water, which has a higher surface tension than both phagocytes and bacteria. On the other hand, in liquids with a lower surface tension than both phagocytes and bacteria, the strongest interaction occurs with the most hydrophilic particles.[37] And in liquids that have the same surface tension as phagocytes, the interaction is at a minimum, i.e., close to zero.[37] Now the effective surface tension of blood appears to be very close to that of most blood cells (including phagocytes),[38] so that short-range interactions between cells in blood tend to be quite feeble.[20] And as the average London–Lifshitz (LL) constants of blood cells also are very close[38a] to the LL constants of biological fluids,[38] the longer range LL attraction between cells immersed in biological fluids also is quite small.[38b]

Thus, interactions between phagocytes and nonopsonized particles in physiological saline solutions and in biological fluids are fundamentally different.

At first sight, such interactions should not take place at all in a biological fluid such as blood, but we know of course that they do occur all the same. Phagocytosis takes place quite actively in blood and in other biological media, and here also the more hydrophobic particles still are the ones to be most readily engulfed by phagocytic cells.[5,25] The reason for this may appear somewhat involved but is actually quite simple: in biological fluids containing plasma proteins, the most hydrophobic particles tend to adsorb the most protein. And IgG is one of the plasma proteins

[35] A. W. Neumann, D. R. Absolom, D. W. Francis, W. Zingg, and C. J. van Oss, *Cell Biophys.* **4,** 285 (1982).

[36] W. Zingg, D. R. Absolom, A. W. Neumann, D. W. Francis, and C. J. van Oss, *in* "Clinical Applications of Biomaterials" (A. J. C. Lee, T. Albrektsson, and P. I. Branemark, eds.), p. 255. Wiley, New York, 1982.

[37] A. W. Neumann, D. R. Absolom, C. J. van Oss, and W. Zingg, *Cell Biophys.* **1,** 79 (1979).

[38] C. J. van Oss, A. W. Neumann, R. J. Good, and D. R. Absolom, *Adv. Chem. Ser.* **188,** 107 (1980).

[38a] S. N. Omenyi, R. S. Snyder, C. J. van Oss, D. R. Absolom, and A. W. Neumann, *J. Colloid Interface Sci.* **81,** 401 (1981); S. N. Omenyi, R. S. Snyder, D. R. Absolom, C. J. van Oss, and A. W. Neumann, *J. Dispersion Sci. Technol.* **3,** 308 (1982).

[38b] C. J. van Oss, R. J. Good, and M. K. Chaudhury, *J. Colloid Interface Sci.* **111,** 378 (1986).

that is adsorbed most.[39] Thus (even in the absence of specific antibody to the bacterial surface), the more hydrophobic bacteria, when immersed in plasma, aspecifically adsorb IgG (to the extent of 3000 to 10,000 molecules per bacterium). A sufficient number of these adsorbed IgG molecules appear to have their Fc moieties protruding outward to allow them to bind specifically to the Fc receptors on PMNs and monocytic phagocytes. Phagocytes have Fc receptors for only IgG_1 and IgG_3. IgG_1 and IgG_3 also appear to adsorb more strongly to hydrophobic surfaces than the two other isotypes.[39] In the presence of complement, IgG_1 and IgG_3 adsorbed to bacterial (or other) particles also can trigger the complement cascade and cause additional interaction with phagocytes through their C3b receptors.

Thus, both in aqueous and in plasma protein-containing media, the most hydrophobic particles are the most phagocytized, in the first case because of direct hydrophobic and van der Waals interactions, and in the second case because of enhanced IgG adsorption which allows interactions with Fc receptors for adsorbed IgG_1 and IgG_3.

Importance of Receptors

As by far the most frequent encounters between bacteria (and other particles) and phagocytes take place in biological fluids that comprise immunoglobulins, the interaction between Fc and C3b and their receptors (R) on the phagocytes' surfaces is of preponderant importance in the initiation of phagocytosis. This holds for aspecific interactions (in which IgG only physically adsorbs onto hydrophobic bacteria) as well as for specific binding of antibodies, of the IgG_1 and IgG_3 classes, for interaction with Fc–R, and of the IgG_1, IgG_2, IgG_3, and IgM varieties plus complement for interaction with C3b–R. Especially with Fc/Fc–R interactions there is considerable competition between Fc–R interactions with ordinary circulating monomeric IgG and with IgG (specifically or aspecifically) bound to bacteria or other cells. Due to the vast amounts of freely circulating IgG, virtually all Fc–R on phagocytes are permanently occupied with monomeric IgG. For this reason there is, thermodynamically speaking, a greater probability for IgG-sensitized cell–phagocyte interaction in an organ such as the spleen (where both blood cell and phagocyte concentrations are considerably higher than in peripheral blood) than in circulating blood.[40] This agrees well with the experimental findings on the

[39] D. R. Absolom, C. J. van Oss, W. Zingg, and A. W. Neumann, *J. Reticuloendothel. Soc.* **31**, 59 (1982).

[40] C. J. van Oss, D. R. Absolom, and I. Michaeli, *Immunol. Invest.* **14**, 167 (1985).

comparative *in vivo* destruction of sensitized Rh_o (D)-positive erythro-cytes in the spleen and in the peripheral blood circulation.[41]

Many other receptors (especially on mononuclear phagocytes) have been described,[42] whose function in many cases is not directly related to phagocytosis. However, microbial lectins (in particular lectins with man-nose-binding activity), capable of attaching to phagocytic receptors, have recently been proposed as recognition determinants in the aspecific phagocytosis of a number of bacterial species.[43]

Phagocytic Recognition and Pathogenicity

Phagocytic Ingestion of Nonencapsulated Bacteria

Most if not all nonencapsulated bacteria are sufficiently hydrophobic to become spontaneously phagocytized, in aqueous fluids directly, be-cause of their hydrophobicity, and in plasma protein-containing liquids indirectly, because of their hydrophobicity, through adsorption of IgG and the subsequent phagocytosis via Fc and/or C3b receptors (see above). Thus most nonencapsulated bacteria are nonpathogenic because of the ease with which they are phagocytized (with the exception of those microorganisms that are readily ingested but which resist digestion, see below).

Phagocytic Ingestion of Encapsulated Bacteria

Bacterial capsules are very hydrophilic and as such resist attachment to and engulfment by phagocytes in aqueous liquids, on account of their hydrophilicity, and in plasma protein-containing liquids because any aspecifically adsorbed IgG attaches to the bacterial cell wall and remains well within the outer limits of the capsule,[44] which obviates any possibility of contact with phagocytic Fc–R. However, capsular proteins, glycopro-teins, or polysaccharides almost invariably are antigenic, so that in due course, i.e., within days after the first exposure to the capsular materials, the host starts producing specific antibodies to the capsule, which then attach to its outer periphery and trigger phagocytic ingestion,[44] via Fc–R and, in the presence of complement, also via C3b–R interactions.

[41] A. Fleer, F. W. van der Meulen, E. Linthout, A. E. G. Kr. von dem Borne, and C. P. Engelfriet, *Br. J. Haematol.* **39**, 425 (1978).

[42] P. E. McKeever and S. S. Spicer, *in* "The Reticuloendothelial System" (I. Carr and W. T. Daems, eds.), Vol. 1, p. 161. Plenum, New York, 1980.

[43] Z. Bar-Shavit, R. Goldman, I. Ofek, N. Sharon, and D. Mirelman, *Infect. Immun.* **29**, 417 (1980).

[44] M. W. Stinson and C. J. van Oss, *J. Reticuloendothel. Soc.* **9**, 503 (1971).

Antiphagocytic Stratagems of Microorganisms

Antiingestion stratagems may be divided into short-term and long-term mechanisms. Among the short-term mechanisms is the envelopment with a hydrophilic capsule, which however prevents phagocytic engulfment (see above) only until specific anticapsular antibodies are produced by the host.[44] Typical microorganisms that have adopted this approach are *Streptococcus pneumoniae*,[5] *Staphylococcus aureus* (S. Smith),[5,45] and *Streptococcus* spp. with M protein.[46] A longer term mechanism for the prevention of phagocytic ingestion consists of the attachment of IgG molecules to the bacterial cell wall by the Fc moiety, so that the Fab part protrudes. *Staphylococcus aureus* (type Cowan 1) does this by means of its cell wall "protein A," which will attach to the Fc ends of both nonspecific IgG and specific antibodies of the IgG class,[47] so that both aspecific adsorption of IgG and attachment of specific antibodies can be thwarted by turning their Fab groups outward, which are antiphagocytic.[48]

Antidigestion stratagems may consist of, e.g., simple outright indigestibility of bacterial cell walls[5] (e.g., *Mycobacteria* spp.), or of alterations brought about in the phagosomal membrane (e.g., by various parasitic protozoa[49]).

Phagocytotoxic stratagems depend mainly on lysing the phagocytes, or their granules. *Streptococcus* spp. can kill phagocytes by lysing their granules (and thus liberating their hydrolases intercellularly) with the streptolysins S and O[46]; by the same mechanism the staphylococcal Panton–Valentine leukocidin can lyse phagocytes. Staphylococcal α as well as β hemolysins also are leukocidal to phagocytes.[46]

[45] R. K. Cunningham, T. O. Soderström, C. F. Gillman, and C. J. van Oss, *Immunol. Commun.* **4**, 429 (1975).

[46] B. D. Davis, R. Dulbecco, H. N. Eiser, H. S. Ginsberg, and W. B. Wood, eds., "Microbiology," pp. 715–730. Harper & Row, New York, 1973.

[47] A. Forsgren and J. Sjöquist, *J. Immunol.* **97**, 822 (1966).

[48] C. J. van Oss, M. S. Woeppel, and S. E. Marquart, *J. Reticuloendothel. Soc.* **13**, 221 (1973).

[49] M. Aikawa and A. Kilejian, *in* "Lysosomes in Applied Biology and Therapeutics" (J. T. Dingle, P. J. Jacques, and I. H. Shaw, eds.), p. 31. Elsevier/North-Holland Publ., New York, 1979.

[2] Measurement of Surface Properties of Phagocytes, Bacteria, and Other Particles

By DARRYL R. ABSOLOM

1. Introduction

The role of surface properties in various diverse biological processes is now well established. Such phenomena include cell–cell and cell–surface interactions, opening and closing of the vasculature of the microcirculation, lubrication of joints, and antigen–antibody interactions. In addition, it has long been surmised that interfacial forces play an important role in the phagocytic engulfment of bacteria and other solid particles (e.g., carbon) by amoebas[1,2] or mammalian leukocytes.[3,4] In particular the determination of the degree of surface hydrophobicity of bacteria,[5–7] by means of various two-phase liquid systems showed, at least qualitatively, that the degree of hydrophobicity of bacteria played an important role in their phagocytosis. More recently similar qualitative studies on the relationship between particle surface hydrophobicity and the extent of phagocytic ingestion have been performed in several laboratories.[8–11] Virulent bacteria such as pneumococci, streptococci, *Klebsiella,* and others, which are generally encapsulated, are able to escape phagocytic ingestion by simply gliding away on smooth surfaces like glass, cellophane, and paraffin. These same bacteria are, however, readily engulfed when they are located on rough surfaces such as wet filter paper, cloth, and fiber glass. Nonvirulent bacteria, which generally lack capsular material, are

[1] R. A. Weisman and E. D. Korn, *Biochemistry* **6,** 485 (1967).
[2] E. D. Korn and R. A. Weisman, *J. Cell Biol.* **34,** 219 (1967).
[3] W. O. Fenn, *J. Gen. Physiol.* **3,** 465 (1921).
[4] D. Mouton, G. Biozzi, Y. Bouthillier, and C. Stiffel, *Nature (London)* **197,** 706 (1963).
[5] L. Edebo, F. Lindström, L. Sköldstam, O. Stendahl, and C. Tagesson, *Immunol. Commun.* **4,** 587 (1975).
[6] S. Mudd, M. McCutcheon, and B. Lucke, *Physiol. Rev.* **14,** 210 (1934).
[7] L. Perers, L. Andaker, L. Edebo, O. Stendahl, and C. Tagesson, *Acta Pathol. Microbiol. Scand., Sect. B: Microbiol.* **85B,** 308 (1977).
[8] C. Capo, P. Bongrand, A.-M. Benoliel, and R. Depieds, *Immunology* **36,** 501 (1979).
[9] M. Rabinovitch, *Proc. Soc. Exp. Biol. Med.* **124,** 386 (1967).
[10] B. Biozzy, C. Stiffel, B. N. Halpern, and D. Mouton, *Proc. Soc. Exp. Biol. Med.* **112,** 1017 (1963).
[11] D. F. Gerson, C. Capo, A.-M. Benoliel, and P. Bongrand, *Biochim. Biophys. Acta* **692,** 147 (1982).

fairly readily ingested. Because of the obvious importance of the role of the various surfaces (bacterial, phagocyte, and substrate), this sort of phagocytosis was termed "surface phagocytosis." This distinction was made in order to clearly differentiate between this mechanism of ingestion and ingestion mediated by specific opsonins such as capsular antibodies.

The aforementioned studies indicate that phagocytosis of bacteria can occur *in vitro* in the absence of opsonins (see also this volume [3]). A relevant question to be posed is whether surface phagocytosis is likely to occur *in vivo*. It is clear that phagocytosis can occur early in infection, long before the appearance of specific antibody in the serum. Encapsulated pneumococci injected into the footpad of rats are ingested within 30 min by granulocytes both in the popliteal nodes and in footpads. This rapidity of ingestion suggests that surface phagocytosis does in fact occur *in vivo*.

The observation that several types of microorganisms and colloidal particles are ingested in the complete absence of specific or nonspecific antibodies and other opsonins such as complement factors, tuftsin, etc., indicates that phagocytes do *not* require any special microbial surface structures in order for the process to occur. Furthermore, the ability of primitive protozoa such as amoeba to ingest a wide range of microorganisms and other materials (e.g., polystyrene particles[1,2]) suggests that phagocytosis is a primitive and very general process. Phagocytic cells have been shown, in the complete absence of any opsonins, to ingest particulate material as diverse as colloidal carbon,[3] oil droplets,[12] polystyrene latex,[13] starch beads,[3] liposomes,[14] and even chemically modified erythrocytes.[8,9,15,16] It does not seem likely that phagocytic cells would possess such a vast array of different membrane receptors allowing the specific discrimination among various kinds of particles that have never exerted any selective pressure during phylogenetic evolution. These observations must therefore be taken as additional evidence of the importance and role of surface forces in phagocytosis. The success and extent of the nonspecific ingestion process is largely determined by the physicochemical surface properties of the interacting species, i.e., bacteria and the phagocyte.

In pioneering studies the relationship between the extent of phagocytic ingestion and the contact angle (θ) which a drop of saline water

[12] T. P. Stossel, *Blood* **42**, 121 (1973).
[13] S. C. Silverstein, R. M. Steinman, and Z. A. Cohn, *Annu. Rev. Biochem.* **46**, 669 (1977).
[14] S. Batzri and E. D. Korn, *J. Cell Biol.* **66**, 621 (1975).
[15] C. Bona, A. Sulica, M. Dumitrescu, and D. Vranialici, *Nature (London)* **213**, 824 (1967).
[16] C. Capo, P. Bongrand, A. M. Benoliel, P. Pommier de Santi, and R. Depieds, *Ann. Immunol. (Paris)* **128C**, 23 (1977).

TABLE I
DEGREE OF PHAGOCYTOSIS AND CONTACT ANGLES OF A NUMBER OF BACTERIA

Organism	Contact angle ($\pm <1°$)[a]	Average number of bacteria phago- cytized per neutrophil[b]
Mycobacterium butyricum	70.0°	6.4 ± 1.1
Brucella abortus	27.0°	1.8 ± 0.3
Neisseria gonorrhoeae	26.7°	1.9 ± 0.3
Listeria monocytogenes (rough form)	26.5°	2.1 ± 0.4
Staphylococcus epidermidis	24.5°	2.5 ± 0.4
Escherichia coli, type 07	23.0°	2.2 ± 0.4
Enterobacter aerogenes	21.5°	2.0 ± 0.3
Streptococcus pyogenes	21.3°	2.1 ± 0.3
Salmonella typhimurium (rough form)	20.2°	1.4 ± 0.2
Streptococcus faecium	20.0°	2.1 ± 0.3
Salmonella arizonae	19.0°	1.6 ± 0.3
Staphylococcus aureus	18.7°	1.6 ± 0.3
Haemophilus influenzae (rough form)	18.6°	1.6 ± 0.3
Shigella flexneri	18.1°	1.0 ± 0.2
Human neutrophils	(17.8°–18.5°)[a]	
Haemophilus influenzae (group B)	17.6°	1.1 ± 0.1
Streptococcus pneumoniae, type unknown	17.3°	0.8 ± 0.2
Escherichia coli, type 0111	17.2°	0.6 ± 0.2
Streptococcus pneumoniae, type 1	17.0°	0.5 ± 0.1
Klebsiella pneumoniae	17.0°	0.4 ± 0.1
Salmonella typhimurium	17.0°	0.3 ± 0.1
Staphylococcus aureus, strain Smith	16.5°	0.2 ± 0.1

[a] The contact angles of peripheral human neutrophils from different donors tend to vary between these values.
[b] Plus or minus the standard error.

makes with layers of the bacteria was investigated. As is explained in Section 3.1., contact angles are a direct measure of surface hydrophobicity. The higher contact angle corresponds to a greater surface hydrophobicity (or, in quantitative terms, to a lower surface tension). The measured contact angles of a wide range of nonopsonized bacteria and human neutrophils is given in Table I. The relation between contact angle and the extent to which these same bacteria are ingested *in vitro* by human neutrophils is also given. Human neutrophils have a contact angle of 18°. As mentioned previously, what is important is the relative difference between the surface hydrophobicity of the bacteria and the phagocyte: those bacteria with a contact angle less than 18° tend to resist phagocytic ingestion [e.g., *Streptococcus pneumoniae* (type 1), *Haemophilus influenzae* (group B)]. On the other hand, bacteria with contact angles greater than 18° are readily ingested (e.g., *Brucella abortus, Neisseria gonorrhoeae*). As illustrated in Section 3.5, the extent of phagocytic ingestion of nonopsonized bacteria (or particles) can be predicted from a consideration of the relative surface hydrophobicities and the surface tension of the suspending liquid. A strong correlation emerges, and the bacteria clearly fall into two different classes: (1) The ones with contact angles $\theta > 18°$ readily become phagocytosed; (2) the ones with $\theta < 18°$, resist phagocytosis.

The most important aspect of the contact angles of bacteria (in connection with the degree to which they become phagocytized) is not the absolute value of θ, but its value as compared to the θ of phagocytic cells: bacteria more hydrophobic than the phagocytes easily become phagocytized; bacteria more hydrophilic than the phagocytic cells resist phagocytosis. It was also possible to demonstrate that the extent of adsorption of nonspecific antibodies onto bacteria is related to the surface tension of the bacterial surface.

Thus, in view of the importance of interfacial energetics in various biological processes such as phagocytosis, cell adhesion, and protein adsorption, we have investigated the development of various techniques, compatible with biological constraints, for determining the physicochemical surface properties of the interacting species. The main surface properties that are likely to be involved are surface hydrophobicity and charge of the two interacting membranes. The quantitative assessment of hydrophobicity, in terms of surface energies, is known as surface tension. The purpose of this article is to describe the strategies that have evolved for measuring cellular surface tensions. The techniques to be discussed here are limited to those methods which are applicable to the cellular elements involved in phagocytosis. A more general perspective has recently been published elsewhere.[17]

2. Theoretical Aspects of Surface Tension

Before discussing the various strategies some fundamentals are considered first. In analogy to bulk phases, various thermodynamic properties can be defined for a surface. For example, the surface energy $U^{(A)}$ of a flat surface can be defined as[18,19]

$$U^{(A)} = U^{(A)}(S^{(A)}, A, N_i^{(A)})$$ (1)

[17] A. W. Neumann, J. Visser, R. P. Smith, S. N. Omenyi, D. W. Francis, J. K. Spelt, E. B. Vargha-Butler, W. Zingg, C. J. van Oss, and D. R. Absolom, *Powder Technol.* **37,** 229 (1984).
[18] J. W. Gibbs, "The Scientific Papers," Vol. 1, p. 55. Dover, New York, 1961.
[19] L. Boruvka and A. W. Neumann, *J. Chem. Phys.* **66,** 5464 (1977).

where $S^{(A)}$ is the surface entropy, A is the surface area, $N_i^{(A)}$ is the number of moles of component i for the surface.

The surface tension γ is defined as

$$\gamma = \left(\frac{\partial U^{(A)}}{\partial A}\right)_{S^{(A)},\, N_i^{(A)}} \tag{2}$$

The surface tension can be defined instead in terms of derivatives of Legendre transforms of the surface energy.[20] The surface need not consist of one component, but can consist of several components. If these components are uniformly distributed, the surface can be considered homogeneous, with a surface tension as defined above. However, complications can arise when the surface of a multicomponent system consists of patches of different composition. Nevertheless, if these patches are below a critical size, of the order of 0.1 μm,[21] the surface can be considered homogeneous in certain respects, such as in contact angle and wetting phenomena. If the patches are above this critical size, the surface is heterogeneous, with different surface tensions as defined above for each patch. As the domains of different components on the membrane of a cell are typically of macromolecular size (i.e., $\ll$0.1 μm), the concepts of surface thermodynamics are applicable to cells.

The concept of surface tension as introduced above assumes that the phase under consideration is bounded by a vacuum. In cases where the phase under consideration borders on a second phase, the surface tension for that interface, called the interfacial tension, is determined by the nature of these phases. For example, for a solid–liquid interface the value of the interfacial tension, γ_{SL}, will depend on which solid and liquid are present. Since the value of the interfacial tension of, e.g., solids, depends on the phase with which the solid is in contact, some reference state must be chosen to be able to compare the surface properties of these solids.

That reference state is chosen as the solid in contact with its own vapor. Thus, the surface tension of a solid, γ_S, is defined as the surface tension for the interface between the solid and its vapor. This surface tension cannot be measured directly; it can, however, be related to the interfacial tension, γ_{SV}, between the solid and the vapor of some other material, such as a liquid. When the solid surface is brought into contact with such a vapor, some of the vapor may be adsorbed on the solid surface, lowering the apparent surface tension of that solid. The difference between γ_S and γ_{SV} is called the equilibrium spreading pressure.

[20] H. B. Callen, "Thermodynamics." Wiley, New York, 1960.
[21] A. W. Neumann, Adv. Colloid Interface Sci. 4, 105 (1974).

Very often, particularly for "low energy" surfaces, the adsorption of the vapor is negligible and the spreading pressure is small. The surface tension for the interface between the solid and the vapor, γ_{SV}, is then essentially the same as the solid surface tension, γ_S. Therefore, the value of the solid surface tension is usually taken to be the value of γ_{SV}.

For phagocytic cells and bacteria the situation is somewhat different, as they are usually in an aqueous medium, and bringing them to the reference state may cause changes in the structure and the properties of the membrane or the conformation of the protein molecules. Therefore, it may not always be possible to bring a cell or a protein physically into its reference state without changing or even destroying it. The surface tension of a cell should be looked upon as the surface tension which would exist in the reference state provided the surface structure, composition, etc., has not changed from its usual state. The purpose of this definition of the surface tension of a cell is to allow the comparison between the surface properties of different or transformed cells or microorganisms. Thus the surface tension of a biological entity is a reference surface tension, which reflects the surface properties of that cell or protein in its *usual* state.

When considering the experimental determination of the surface tension of a given material, the first question to be asked is whether the material is liquid or solid. While more or less trivial in the case of nonbiological systems, the question is important when considering biological materials. The question is crucial because there are usually several direct techniques for the determination of the surface tension of a liquid, but no *direct* techniques for measuring the surface tension of a solid. What matters with respect to the distinction between solids and liquids is whether the surface assumes a shape which is governed by surface tension alone and, of course, whether external fields such as gravity influence that shape. Such shapes are given by the Laplace equation of capillarity and are generally called Laplacian curves.[22] While cells often are indeed deformable, their shapes clearly are not solely determined by surface tensions. This is obvious, e.g., in the case of an activated platelet where the development of pseudopodia is due to nonsurface tension stimuli. In addition, many cells are known to possess an internal cytoskeleton, which makes them unlike liquids. It is therefore concluded that the only way of determining the surface tension of biological cells is by treating them as solids.

The determination of surface tensions of solids is possible only by indirect means and hence is inherently complex. Until a few years ago the

---

[22] Y. Rotenberg, L. Boruvka, and A. W. Neumann, *J. Colloid Interface Sci.* **93,** 169 (1983).

only general approach to determine surface tensions of solids utilized contact angle measurements. To that end, an empirical equation of state approach for determining interfacial tensions has been developed.[23,24] In essence, this approach uses the fact that if a liquid is in contact with a solid, the interfacial tension γ_{SL} between liquid and solid is given by[25]

$$\gamma_{SL} = f(\gamma_{SV}, \gamma_{LV}) \tag{3}$$

where γ_{SV} and γ_{LV} are the solid and the liquid surface tensions, respectively. Equation (3) can be used in conjunction with Young's equation

$$\gamma_{SV} - \gamma_{SL} = \gamma_{LV} \cos \theta \tag{4}$$

where θ is the contact angle between solid and liquid, to determine γ_{SV} and γ_{SL} from γ_{LV} and θ, the latter two quantities being readily amenable to direct experimental determination.

Equation (3) was formulated explicitly; computer programs[24,26] as well as tables[27] which allow one to determine γ_{SV} and γ_{SL} for a range of γ_{LV} and θ values are available.

3. Determination of Surface Tensions of Phagocytic Cells and Bacteria

3.1. Contact Angle Measurements

3.1.1. Liquid–Vapor Systems. Two major problems are encountered when attempting to obtain solid surface tensions from contact angles: (1) there is always the question of the magnitude of the equilibrium spreading pressure, and (2) the thermodynamic status of contact angles is a difficult problem.[21] This problem is highlighted by the phenomenon of contact angle hysteresis, which cannot be reconciled with Young's equation.[21] Without elaborating on this point, we simply wish to state that contact angles, measured in situations where contact angle hysteresis is due to surface heterogeneity, may be used in conjunction with Young's equation. Contact angles measured on rough surfaces may not be used in conjunction with Young's equation. The arguments supporting these contentions have been given in considerable detail elsewhere.[21] Since there were no independent means of checking either one of these concerns, the

[23] O. Driedger, A. W. Neumann, and P. J. Sell. *Kolloid-Z. Z. Polym.* **201**, 52 (1965).
[24] A. W. Neumann, R. J. Good, C. J. Hope, and M. Sejpal, *J. Colloid Interface Sci.* **49**, 291 (1974).
[25] C. A. Ward and A. W. Neumann, *J. Colloid Interface Sci.* **49**, 291 (1974).
[26] A. W. Neumann, O. S. Hum, D. W. Francis, W. Zingg, and C. J. van Oss, *J. Biomed. Mater. Res.* **14**, 499 (1980).
[27] A. W. Neumann, D. R. Absolom, D. W. Francis, and C. J. van Oss, *Sep. Purif. Methods* **9**, 69 (1980).

only possible approach available was to go ahead with suitable experiments and to assess subsequently the merit of the results by comparing them with other observations.

There is one further complication peculiar to biological systems, i.e., the question of the appropriate type of liquid to be used for the contact angle measurements. Thermodynamic considerations demand that for the approximation $\gamma_S \simeq \gamma_{SV}$ to hold, i.e., for the equilibrium spreading pressure to be negligible, the liquid surface tension has to be larger than the surface tension of the material on which the contact angle is measured. As it is anticipated that the surface tensions of cells are fairly large, water or buffer solutions seem to be the only safe choice. Physiological considerations lead of course to the same choice for the measuring liquid and we feel that contact angle measurements with any other liquids are probably useless due to various complications arising both from thermodynamic and physiological considerations as outlined above.

Contact angle measurements on layers of cells were first reported over 10 years ago.[28,29] The details of the experimental procedures have been reported several times since by various research groups.[30–36] In essence, layers of cells[28–35] or bacteria[37] are prepared by deposition from suspension or by ultrafiltration on anisotropic cellulose acetate membranes. Drops of saline or buffer of known surface tension γ_{LV} are deposited on the initially very wet layer and the contact angle is measured by one of the techniques described in Section 3.1.2. Initially the contact angle is close to zero, as expected. However, as water evaporates and *fresh* drops of saline are deposited on sites which had not been used for contact angle measurements previously, gradually increasing contact angles are observed. The importance of using fresh drops of saline for each measurement cannot be emphasized enough. With time, cells are likely to exude proteins, etc., into the liquid thereby (1) reducing its own effective surface

[28] C. J. van Oss and C. F. Gillman, *RES, J. Reticuloendothel. Soc.* **12**, 283 (1972).
[29] C. J. van Oss and C. F. Gillman, *RES, J. Reticuloendothel. Soc.* **12**, 497 (1972).
[30] C. J. van Oss, C. F. Gillman, and A. W. Neumann, "Phagocytic Engulfment and Cell Adhesiveness as Surface Phenomena." Dekker, New York, 1975.
[31] D. R. Absolom, C. J. van Oss, W. Zingg, and A. W. Neumann, *RES, J. Reticuloendothel. Soc.* **31**, 59 (1982).
[32] D. R. Absolom, D. W. Francis, W. Zingg, C. J. van Oss, and A. W. Neumann, *J. Colloid Interface Sci.* **85**, 168 (1982).
[33] C. J. van Oss, D. R. Absolom, A. W. Neumann, and W. Zingg, *Biochim. Biophys. Acta* **670**, 76 (1981).
[34] D. R. Absolom, A. W. Neumann, W. Zingg, and C. J. van Oss, *Trans. Am. Soc. Artif. Intern. Organs* **15**, 152 (1979).
[35] D. F. Gerson and D. Scheer, *Biochim. Biophys. Acta* **602**, 506 (1980).
[36] G. Adam and C. Schumann, *Cell Biophys.* **3**, 189 (1981).
[37] D. R. Absolom, F. V. Lambertii, Z. Policova, W. Zingg, C. J. van Oss, and A. W. Neumann, *J. Appl. Environ. Microbiol.* **46**, 90 (1983).

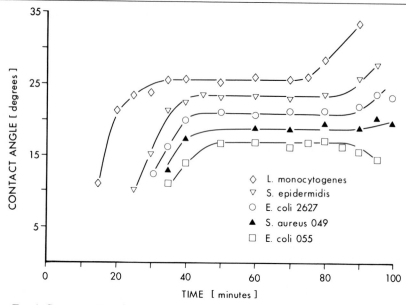

FIG. 1. Contact angles of saline on nonopsonized bacteria as a function of water evaporation from the wet biological substrate, measured in terms of time. Each point is the average of 10 individual contact angle readings on each of 4 different drops, at approximately the same time, on one and the same biological substrate. (Reproduced with permission from Academic Press, Inc.)

tension and (2) as a consequence changing the observed contact angle. Each measurement should be made as quickly as possible and certainly within 10 sec of deposition. After approximately half an hour to 1 hr, depending on circumstances such as the relative humidity, the observed contact angles of the fresh saline drops reach a plateau value, which persists for approximately another half-hour. Then the contact angles sometimes rapidly decrease or increase and in other cases simply become erratic.

Examples of such curves are given in Fig. 1 for bacteria and in Fig. 2 for various leukocytes. The interpretation of these curves is as follows: initially, the biological material, cells or proteins, are completely covered by water and the contact angle is equal to zero due to complete spreading as would be expected. As time passes and water evaporates, the water drops start reacting to the presence of the biological material. Finally, when the plateau is reached the observed advancing contact angle is a Young contact angle[21] and thus is determined exclusively by the properties of the layer of biological material. The final points beyond the plateau are attributed to conformational changes and denaturation of the speci-

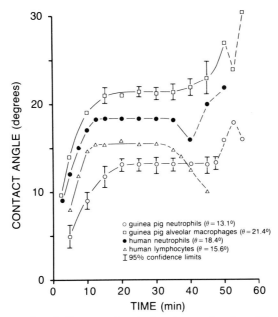

FIG. 2. Contact angles of saline on various leukocytes as a function of time. See legend to Fig. 1.

men. This interpretation is corroborated by the fact that changes in experimental conditions, such as relative humidity, change all features of the curves of Figs. 1 and 2, such as the initial slopes, or the length of the plateau, with the striking exception of the height of the plateau, which in our interpretation reflects solely the properties of the intact biological material. As an illustration of a plateau contact angle, shown in Fig. 3 is a drop of saline deposited on a layer of *Mycobacterium butyricum.*

The experiments of Figs. 1 and 2 were all performed at room temperature (22°). The surface tension of saline (0.15 M NaCl) at that temperature is 72.6 ergs/cm². From the contact angles (plateau values) of Figs. 1 and 2 and the surface tension of saline, γ_{LV}, the surface tensions of the biological materials were calculated from the equation of state approach as described earlier. The results are listed in the second column of Table II. Contact angle measurements have similarly been performed on layers of granulocytes,[31] platelets,[32] macrophages, lymphocytes,[30] and several serum proteins.[33] These results are also given in Table II.

In the case where cells are allowed to adhere to a substrate material (e.g., siliconized glass) the adherent cells should occupy *at least* 50% of the total surface area if reliable and repeatable contact angle measure-

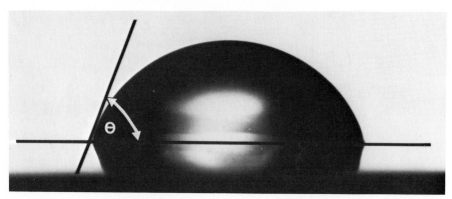

FIG. 3. Photomicrograph of a drop of saline on a layer of *Mycobacterium butyricum*. The measured plateau contact angle is 70.0°.

ments are to be obtained. This is not normally a problem but it is a caution which the investigator should bear in mind (cf. Section 6.1).

3.1.2. Techniques for Measuring Contact Angles. There are a wide variety of techniques available for the measurement of the contact angle in solid liquid–vapor systems.[38] The most accurate of these methods is based on the measurement of the capillary rise at a vertical plate, which has an uncertainty of less than ±0.1°.[21,38] Nevertheless, there are many situations in which it is desirable to utilize the more common sessile-drop configuration, because of its relative simplicity and the smaller quantities of liquid and solid surface area which are required. This technique is also more suitable for use with biological systems.

Sessile-drop contact angle measurement techniques may be considered in terms of three classes:

1. Those direct techniques which involve the placement of a tangent to the sessile-drop profile at the point of contact with the solid surface. This is most frequently done using a telescope equipped with a goniometer eyepiece, but it is perhaps more accurate to work with a photograph of the drop profile; see Fig. 3. In either case, the results are subjective and depend on the experience and skill of the operator, although certain training procedures have been proposed to improve both the accuracy and the precision,[38] which are usually stated to be within ±2°. As an aid in the placement of tangents to photographed drop profiles, Elliott and Riddiford[39] used a device they called a "tangentometer." This is simply a plane

[38] A. W. Neumann and R. J. Good, *in* "Surface and Colloid Science" (R. J. Good and R. R. Stromberg, eds.), Vol. 2, pp. 31–91. Plenum, New York, 1979.
[39] G. E. P. Elliott and A. C. Riddiford, *J. Colloid Interface Sci.* **23**, 389 (1967).

TABLE II
SURFACE TENSIONS OF BIOLOGICAL ENTITIES[a]

| System | Contact angle (via equation of state approach) | Engulfment | | | | | Suspension stability |
| | | Advancing solidification | Phagocytic ingestion | | Adhesion | Detachment | |
			Granulocytes	Platelets			
Granulocytes (human)	69.1	68.3	—	—	69.0	69.0	—
Lymphocytes (human)	70.1	70.6	—	—	—	—	—
Erythrocytes							
Human	—	64.9	—	—	64.6	—	64.3
Horse	—	65.1	—	—	—	—	65.4
Chicken	—	64.8	—	—	—	—	65.2
Turkey	—	65.1	—	—	—	—	65.7
Canine	—	63.9	—	—	—	—	64.4
Platelets (porcine)	67.6	—	—	67.9	—	—	—
Bacteria							
E. coli	69.7	—	69.6	69.3	69.6	—	—
S. aureus	69.1	—	68.7	68.8	69.3	—	—
S. epidermidis	67.1	—	66.9	67.3	66.0	—	—
L. monocytogenes	66.3	—	66.1	—	65.6	—	—
Proteins[b]							
Serum albumin (B)	70.2	—	—	—	70.2	—	—
Serum albumin (H)	70.3	—	—	—	67.7	—	—
Immunoglobulin G (H)	67.3	—	—	—	71.0	—	—
Immunoglobulin M (H)	69.4	—	—	—	71.0	—	—
α_2-Macroglobulin (H)	71.0	—	—	—	—	—	—
Transferrin (H)	66.8	—	—	—	—	—	—

[a] In ergs/cm^2; temperature, 22°.
[b] B, Bovine; H, human.

mirror mounted vertically at right angles to a straight-edge and positioned so that it is normal and flush with the photograph on which it is placed. The device is rotated until the profile is judged to form a smooth, continuous curve with its mirror image, thereby causing the straight-edge to be tangent at that point. This method has been carefully evaluated and was not found to offer any particular advantage. Indeed, it was difficult to avoid underestimating the contact angle using this approach.

2. Interference microscopy and the specular reflection of light from the liquid–vapor interface are two rather specialized methods of sessile-drop contact angle measurement which are discussed in Neumann and Good.[38] Both of these approaches have the advantage of being objective techniques which are independent of the experience of the operator. They are, however, restricted to situations where the contact angle is less than 90°, and are relatively cumbersome to use.

3. There are quite a number of measurement techniques based on the shape of sessile-drop profiles. (For a discussion of some of these, and for further references see, for example, Refs. 22, 38, 40.) Many of these are restricted to contact angles greater than 90°, and are difficult to use because they depend on the accurate location of specific points corresponding to rather ambiguous features of the drop profile, such as the equatorial radius and the apex. The determination of these selected points introduces a degree of operator subjectivity into the measurement with a concomitant loss of accuracy.

The axisymmetric drop shape analysis technique (ADSA),[22,40a] is an exception, however, and is unique in its ability to fit the Laplace equation of capillarity to an essentially arbitrary array of coordinate points representing a drop profile. This permits the simultaneous determination of the contact angle, the surface or interfacial tension, and the drop surface area and volume, all from a single drop photograph. The use of ADSA to measure surface tensions, particularly very low interfacial tensions (0.001 to 0.0001 mJ/m²) in liquid–liquid systems, has been described in Boyce et al.[41,42]

[40] C. Huh and R. L. Reed, J. Colloid Interface Sci. 91, 472 (1983).

[40a] Abbreviations: ADSA, axisymmetric drop shape analysis; HBSS, Hanks' balanced salt solution; DMSO, dimethyl sulfoxide; MABF, microaggregate blood filters; DMAC, dimethylacetamide; PEG, polyethylene glycol; BATH, bacterial adhesion to hydrocarbons; SPS, sulfonated polystyrene; FEP, fluorinated ethylene propylene; LDPE, low density polyethylene.

[41] J. F. Boyce, F. Schürch, Y. Rotenberg, and A. W. Neumann, Colloids Surf. 9, 307 (1984).

[42] Y. Rotenberg, S. Schürch, J. F. Boyce, and A. W. Neumann, in "Surfactants in Solution" (K. L. Mittal and B. Lindman, eds.), Vol. 3, pp. 2113–2120. Plenum, New York, 1984.

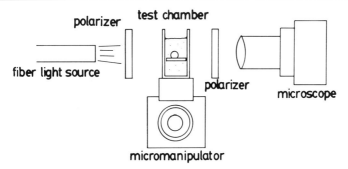

FIG. 4. Schematic representation of the equipment required for obtaining the most accurate contact angle measurements by means of the axisymmetric drop shape analysis (ADSA) technique.

The success of all drop-shape techniques depends on their ability to accurately model the complex shapes of liquid–vapor interfaces under the influence of gravity.

Accurate contact angles can only be obtained from drop profiles through the consideration of the entire profile shape, and not merely segments of it. A simple application of this concept is found in the work of Riley and Khoshaim,[43] in which drop shapes are approximated by elliptical curves, defined by three arbitrarily located profile coordinate points. As illustrated in the following section, this approximation can yield reasonable answers in certain cases, although the standard deviation of replicated results is quite large.

In general, the most accurate contact angle results will, of course, derive from the group of methods which fit the entire drop profile to the Laplace equation of capillarity. The optimum technique for achieving this curve fit is ADSA,[22] since all other approaches in this group require the accurate determination of a number of specific profile points (for example the apex, equatorial radius), as discussed above. By eliminating this critical dependence on a few preselected points, the accuracy of ADSA is limited only by the present technology of locating cartesian coordinates along an optical edge (the drop profile).

Apparatus: The main elements of the experimental apparatus are illustrated in the schematic in Fig. 4. A syringe-drive mechanism above the stage deposits sessile drops of the test liquid on the substrate material and ensures that the contact angles are in an advancing state. Photographs of the image are taken with a horizontally mounted stereomicroscope (Wild Heerbrug M7S) fitted with a 35-mm automatic camera. In most cases the

[43] D. J. Riley and B. H. Khoshaim, *J. Colloid Interface Sci.* **59**, 243 (1977).

overall magnification on the film plane is set at 4×. The microscope is mounted on a heavy, pivoting stand, and is supported at one end by an extendible laboratory jack. This arrangement permits the microscope to be tilted downward, slightly out of the horizontal, so that a small amount of drop reflection can be seen in the solid substrate. Regardless of the photographic system employed, if ADSA is to be used to measure surface tensions, it is important that the depth of field (focus) be kept to a minimum in order to permit the accurate linear calibration of the image scale.

The drops are illuminated from behind with a 150-W fiber optic light source shining through a heavily frosted diffuser. This minimizes the heat input to the drop and provides a uniformly bright background of white light. This results in the production of photographs of high contrast using Kodak Technical Pan 2415 film exposed at 100 ASA.

Drop-profile coordinates are acquired in one of two ways: The 35-mm negatives are either projected onto a translucent (rear projection) manual digitizing tablet, or they are processed with an automatic digital image analysis system. The former method is the simplest in terms of equipment and computer software, while the latter offers the advantages of speed, a somewhat greater precision, and the possibility of the direct *in situ* digitization of drop profiles, bypassing the need for intermediate, 35-mm negatives.

In our laboratory this is achieved using a Summagraphics 2020 Supergrid rear projection digitizing tablet (0.025-mm resolution), and a Leitz Prado Universal slide projector with an *f*2.8, 60-mm lens and film strip accessory. The projector is aligned so that the axis of the lens is normal to the vertically mounted screen, and at a distance of 60 cm the overall magnification from the negative is 42×. The digitizing tablet is connected directly to a PDP 11/23 PLUS (Digital Equipment Corporation) minicomputer, which is used to execute the ADSA algorithm, fitting the Laplace equation of capillarity to the arbitrarily selected drop-profile points.

Principles of ADSA use: The contact angle and surface tension are insensitive to either the locations, or the exact number of the digitized drop-profile coordinates. Usually, approximately 40 points are acquired over the length of a given profile. However, since the locations are arbitrary, one may omit sections of the profile image which may be poorly focused or of insufficient contrast. This capability is also useful if a pipet tip is left within a drop in order to smoothly add liquid, and thus sustain an advancing contact angle (see below).

In addition to an array of roughly 40 profile points, the use of ADSA to measure contact angles requires the relative location of the solid substrate, and some reference to the direction of the gravity vector. The surface position is, however, irrelevant to the calculation of liquid surface tension, which entails only a knowledge of the liquid density. To date, two methods have been employed to locate the position of the solid surface relative to the profile coordinates. The most convenient approach has been to tilt the camera downward, out of the horizontal a very small amount (less than 1°), thereby including in the photograph a small segment of the surface reflection of the drop. The complete image is, thus, football shaped, with the surface location defined by a line joining the two tips. The gravity vector is, then, perpendicular to this line.

Since all smooth surfaces are specularly reflective, this method is effective in every situation where the contact angle is less than, about 80°, and greater than roughly 100°. In between these limits the drop "tips," and for biological systems where the surface line is not well defined, it is necessary to adopt the second approach for surface location. A micromanipulator may be used to vertically position a small pin beside the drop, and to move it independently into focus so that its reflection serves to identify the substrate in the same manner as the drop tips (Fig. 5A). Here, again, the depth of field of the camera must be small in order to ensure that the pin, when it is in focus, lies in the same plane as the section of the drop being photographed. The use of only one pin means that a vertical (gravity vector) reference must also be included in the photograph. This is most easily accomplished by carefully aligning the edge of the photographic field with a plumb line or a bubble level. The edge of each negative then becomes a gravity reference to be used during digitization. Figure 5B shows the digitized drop profile points representing this same drop, and the Laplacian curve which was fitted to these points by the ADSA algorithm.

Both of these techniques compromise the accuracy of the profile image in the sense that they do not present a truly maximal meridonal section of the drop. Tests have shown, however, that camera inclination is not a significant source of error, provided that it is kept to a minimum, say 1° or less.

The success of any method which determines surface or interfacial tension from the shapes of fluid menisci, depends on the drop profile having some significant departure from spherical geometry.[42] The same is true, although to a lesser extent, for ADSA applied to the measurement of contact angles. For a given digitizing accuracy (the accuracy of locating a coordinate point along the profile of a drop image), the best results are achieved for drop profiles which are appreciably noncircular, but which, at the same time, are not so large that they are completely flat at the apex. This condition is not difficult to meet since there is considerable latitude in the shape and volume between the two undesirable extremes: the circular profile and the flattened "pancake" shape.

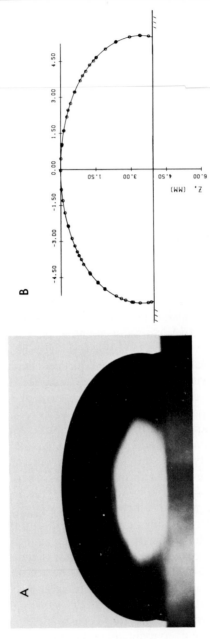

FIG. 5. (A) Photomicrograph of a drop of water on a siliconized (dimethyldichlorosilane) glass surface. (B) Axisymmetric drop shape analysis of the liquid profile in (A). Derived computational results: Measuring liquid surface tension (ergs/cm²), 72.08. Contact angle (degrees), 106.22.

TABLE III
COMPARISON OF SESSILE DROP ADVANCING CONTACT ANGLES MEASURED
BY SEVERAL DIFFERENT TECHNIQUES[a]

System	ADSA	Ellipse[b]	Goniometer	Tangentometer[c]
Hexadecane/FEP	53.0	53.6	53	49
	±0.2	±8.4	±1	±2
Methyl salicylate/FEP	69.3	68.4	69	64
	±0.5	±1.2	±1	±2
Water/siliconized	106.2	105.6	106	—
glass	±0.7	±3.3	±1	

[a] Contact angles (in degrees) are averages of six replications having a standard deviation reported as (±). FEP, Fluorinated ethylene propylene copolymer.
[b] Ellipse, elliptical curve fit of Riley and Khoshaim.[43]
[c] Tangentometer, device used by Elliott and Riddiford[39] to measure the contact angle of a sessile drop from a photographic print.

Table III compares the performance of ADSA with the elliptical fit of Riley and Khoshaim,[43] direct goniometer measurements, and the tangentometer technique of Elliott and Riddiford.[39] Each replication of ADSA consisted of digitizing a separately focused photograph of a given drop at a fixed projector focus. In the case of the elliptical fit, as recommended in Ref. 43, each replication was the average result of three sets of three coordinate points; i.e., each replication was itself the mean of three elliptical curve fits.

The ADSA contact angles agree well with the average values measured directly with the goniometer, while the tangentometer consistently underestimates the angle, although, being a subjective technique, it may be possible to train an operator to correct for this tendency. The precision of ADSA is better than that of the other methods.

The primary causes of uncertainty in ADSA stem from the limitations involved in sharply focusing a drop during photography and then accurately identifying coordinate points along the profile with either the manual digitizing tablet or the automatic digital image analyzer. In practice, these sources of random error are accommodated by averaging the results of three separately focused drop photographs, digitized at a constant projector focus. Table IV lists the average standard deviations encountered in such groups of three replications. The values are grouped according to the contact angle ranges within the data.

The precision of ADSA contact angle measurements, as reflected by the average standard deviations, is good, especially in comparison to other sessile-drop techniques.

TABLE IV
PRECISION ESTIMATES OF ADSA CONTACT ANGLE MEASUREMENTS[a]

Contact angle range (°)	Number of groups of three	Average standard deviation (°)	95% Student t confidence limits (°)	Average standard error of mean (°)[b]
50–55	25	0.40	0.14	0.23
70–75	35	0.42	0.10	0.24
104–106	4	0.57	0.18	0.33

[a] Average standard deviations encountered in groups of three replications of the measurement.

[b] Average standard error of mean equals $3^{-0.5}x$ (average standard deviation).

Conclusions: ADSA is the best available means of measuring the contact angles of sessile drops. Its chief advantages are as follows:

1. The method is objective, being independent of the skill and experience of the operator.

2. ADSA is applicable to all sessile drops, regardless of contact angle. It is not restricted to angles which are either less than, or greater than, 90°.

3. It is unique as a drop-shape method in that it achieves a least-squares fit of the Laplace equation of capillarity, utilizing an essentially arbitrary array of profile points without reference to specific (and difficult to define) points or drop dimensions, such as the apex or the equatorial radius.

4. The method is relatively straightforward and convenient to employ, requiring only a photograph (or negative) of the drop.

5. The precision of ADSA is considerably better than that of other sessile-drop contact angle methods.

ADSA fills a gap which exists in contact angle measurement techniques, between the "quick and dirty" goniometer approach and the rather sophisticated method of the capillary rise at a vertical plate.[21,38] In addition, ADSA provides a unique approach to the problem of accurately measuring low contact angles (<30°) or of determining liquid–liquid interfacial tensions, particularly if they are very low and, therefore, difficult to measure otherwise.[42]

3.2. Adhesion Experiments

Thermodynamically, cell adhesion to a solid substrate is favored when the change in free energy for the process of cell attachment is negative.[43a]

[43a] A. W. Neumann, D. R. Absolom, and C. J. van Oss, *Cell Biophys.* **1,** 79 (1979).

Fig. 6. Schematic representation of the process of cell adhesion. P, Cell; L, suspending liquid; S, substrate material.

For a system in which the effect of electrical charges can be neglected, this change is, per unit surface area,

$$\Delta F^{adh} = \gamma_{PS} - \gamma_{PL} - \gamma_{SL} \qquad (5)$$

where γ_{PS} is the particle (cell)/solid interfacial tension, γ_{PL} is the particle/ liquid interfacial tension, and γ_{SL} is the substrate/liquid interfacial tension as illustrated schematically in Fig. 6.

Cell or bacterial adhesion onto several polymer substrates (covering a range of surface tensions) varies as a function of substrate surface tension, γ_{SV}, and the suspending liquid surface tension, γ_{LV}. As an illustration of the influence of surface properties in determining the extent of cell–substrate interactions, reproduced in Fig. 7 are photomicrographs of the extent of erythrocyte adhesion to two polymer surfaces under various well-defined experimental conditions. It is the primary aim in these experiments to assess to what extent the number of cells adhering to various substrates corresponds to the free energy of adhesion. To this end, it is necessary to know the interfacial tensions γ_{PS}, γ_{PL}, and γ_{SL}. Considering Eq. (3) as a generic equation that allows the calculation of the interfacial tensions between two phases from the respective surface tensions of these phases, the three quantities can be calculated from γ_{SV}, the surface tension of the substrate, γ_{LV}, the surface tension of the suspending liquid, and γ_{PV}, the surface tension of the particle (cell).

One is therefore in the position to calculate the free energy of adhesion and to predict the pattern of adhesive behavior of cells to various polymer surfaces. For a given cellular species of a given surface tension γ_{PV} there are two functional parameters, i.e., the surface tension of the substrate γ_{SV} and the surface tension of the suspending liquid γ_{LV}. In these experiments both the substrate surface tension, γ_{SV} (by changing the substrate) and the surface tension of the suspending aqueous medium, γ_{LV} [through the admixture of suitable low-molecular-weight additives such as dimethyl sulfoxide (DMSO)] were varied.

Typical theoretical calculations of ΔF^{adh} for *Escherichia coli* with $\gamma_{PV} = 67.8$ ergs/cm^2 are given in Fig. 8. It becomes apparent that it is

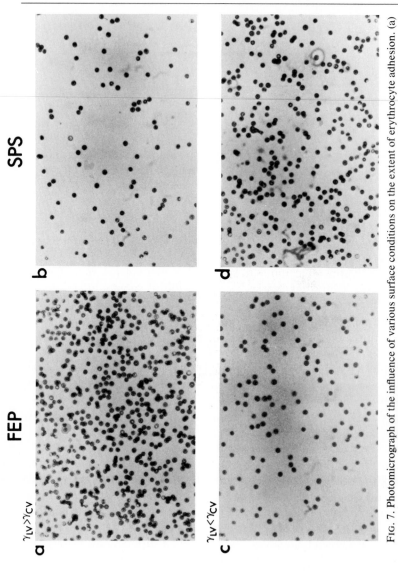

Fig. 7. Photomicrograph of the influence of various surface conditions on the extent of erythrocyte adhesion. (a) Fluorinated ethylene propylene copolymer (FEP, $\gamma_{SV} = 16.4$ ergs/cm^2); $\gamma_{LV} = 72.8$ ergs/cm^2. (b) Sulfonated polystyrene (SPS, $\gamma_{SV} = 66.7$ ergs/cm^2); $\gamma_{LV} = 72.8$ ergs/cm^2. (c) Fluorinated ethylene propylene copolymer (FEP, $\gamma_{SV} = 16.4$ ergs/cm^2); $\gamma_{LV} = 59.8$ ergs/cm^2. (d) Sulfonated polystyrene (SPS, $\gamma_{SV} = 66.7$ ergs/cm^2); $\gamma_{LV} = 59.8$ ergs/cm^2. The surface tension of the adhering erythrocytes is denoted by γ_{CV}. (Reproduced with permission from Academic Press, Inc.)

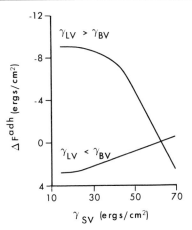

FIG. 8. A theoretical plot of the change in the free energy of adhesion (ΔF^{adh}) of a single bacterium species as a function of substrate surface tension, γ_{SV}. The bacterium considered is *E. coli* 2627 with a surface tension $\gamma_{PV} = 67.8$ ergs/cm². $\gamma_{LV} > \gamma_{PV}$; $\gamma_{LV} = 72.8$ ergs/cm². $\gamma_{LV} < \gamma_{PV}$; $\gamma_{LV} = 64.0$ ergs/cm².

necessary to distinguish between two cases. For

$$\gamma_{LV} > \gamma_{PV} \tag{6}$$

ΔF^{adh} increases (i.e., becomes more positive) with increasing γ_{SV}, predicting decreasing cell adhesion with increasing surface tension γ_{SV} of the substrate over a wide range of γ_{SV} values. On the other hand, when

$$\gamma_{LV} < \gamma_{PV} \tag{7}$$

the opposite behavior is predicted. For the case of the equality

$$\gamma_{LV} = \gamma_{PV} \tag{8}$$

ΔF^{adh} becomes equal to zero independently of the value of γ_{SV}. It is this latter fact which offers the possibility of an alternative method of determining the surface tension of cells: By determining the liquid surface tension γ_{LV} at which the level of cell adhesion becomes independent of the substrate surface tension, γ_{SV}.

As an example, reported here are the data for adhesion studies with human granulocytes. The substrates used, their preparation, contact angles with water, and surface tension are all available in Absolom et al.[34,43a] Adhesion experiments were performed with suspensions of granulocytes in Hanks' balanced salt solution (HBSS) to which varying amounts of DMSO were admixed, in order to change the surface tension of the suspending liquid. Each individual test was performed with 1 ml of granulocyte suspension, containing 1×10^6 cells, which was placed on the surfaces and was retained in wells formed from Teflon molds. The cells were

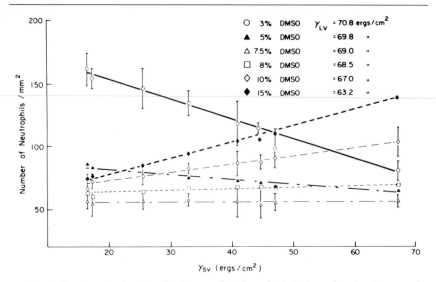

FIG. 9. Human granulocyte adhesion as a function of substrate surface tension, γ_{SV}, for various DMSO concentrations. Error limits: 95% confidence. For graphical reasons error limits are not given in all cases. (Reproduced with permission from Humana Press, Inc.)

then incubated at room temperature for 30 min. Thereafter the surfaces were carefully rinsed in a standardized procedure with HBSS to remove nonadhering cells. Next, the cells adhering to the surfaces were air dried and subsequently stained using Wright's stain. The number of cells adhering per unit surface area was then determined by means of a fully automatic digital image analysis system attached to a light microscope. Complete experimental details have been given elsewhere[34,43,43a,44] and are reiterated in Section 6.2. The results of these adhesion experiments are summarized in Fig. 9. The theoretical predictions are borne out by the experimental results. At the low DMSO concentrations, i.e., at high surface tensions γ_{LV} of the suspending medium, granulocyte adhesion decreases with increasing γ_{SV} of the substrate, whereas at low values of the surface tension γ_{LV} the number of adhering granulocytes increases with increasing γ_{SV}. At some intermediate value of γ_{LV}, adhesion does indeed become independent of substrate surface properties. As this is the case to which Eq. (8) refers, we have here a novel possibility of determining the surface tension γ_{PV} of cells. To this end, the slopes of the straight lines in Fig. 9 are evaluated by means of a computer least-squares fit and plotted versus γ_{LV} in Fig. 10. It is inferred that the slope becomes equal to zero at

[44] D. R. Absolom, W. Zingg, C. Thompson, Z. Policova, C. J. van Oss, and A. W. Neumann, *J. Colloid Interface Sci.* **104**, 51 (1985).

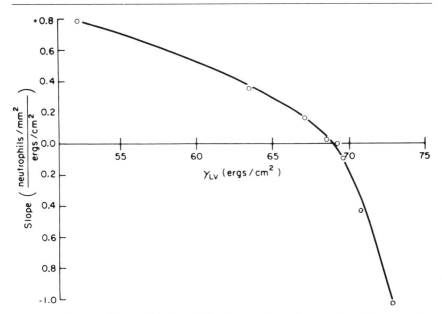

FIG. 10. Slopes of the straight lines of Fig. 9 versus the surface tension of the suspending liquid, $\gamma_{LV} = \gamma_{PV}$. The slope is zero at the point where $\gamma_{LV} = \gamma_{PV}$. (Reproduced with permission from Humana Press, Inc.)

$\gamma_{LV} = 69.0$ ergs/cm² implying in view of Eq. (8) that the surface tension of the granulocytes is also equal to 69.0 ergs/cm². This is in excellent agreement with the value of 69.1 ergs/cm² obtained from the contact angle measurements (cf. Table II).

Similar adhesion experiments were performed with various species of bacteria.[37] The number of bacteria adhering per unit surface area are plotted as a function of the substrate surface tension γ_{SV} for several values of γ_{LV} in Fig. 11. The overall appearance of these straight lines is quite similar to those of Fig. 9. Again, the surface tension γ_{PV} of these bacteria was determined by plotting the slopes of the straight lines of Fig. 11 versus γ_{LV} in Fig. 12. The γ_{PV} value, obtained as the intercept with the γ_{LV} axis (i.e., for the slope equal to zero) is reported in Table II. In all these systems the agreement with the results from the liquid–vapor contact angle measurements is striking.

The adhesion technique, although somewhat cumbersome, has considerable attraction since it has the advantage that it does not require, as does the contact angle method, the exposure of the biological material to an air interface.

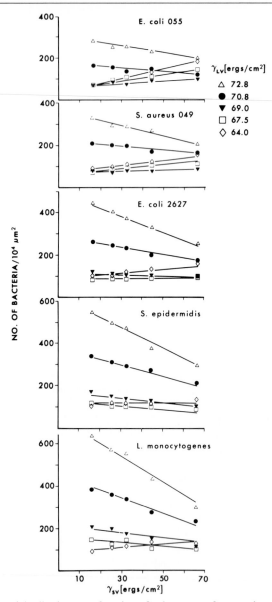

FIG. 11. Bacterial adhesion as a function of substrate surface tension, γ_{SV}, for various DMSO concentrations and hence for various liquid surface tensions, γ_{LV}. Error limits are similar in all cases to those indicated in Fig. 9. (Reproduced with permission from *Am. Soc. Microbiol.*)

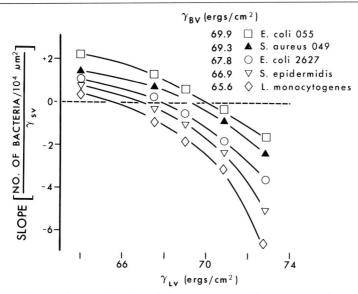

FIG. 12. Slopes of the straight lines of Fig. 11 versus γ_{LV}. The slope is zero for $\gamma_{LV} = \gamma_{PV}$. (Reproduced with permission from *Am. Soc. Microbiol.*)

There are numerous practical applications for the use of the type of information contained in plots of the type referred to above. For instance, we have been able to utilize such data in the design and development of more efficient microaggregate blood filters (MABF). Shown in Fig. 13 is the experimentally determined relation between MABF material surface tension and the extent of removal (i.e., efficiency) of cellular elements from whole blood. These data can be explained entirely in terms of the surface thermodynamic concepts outlined earlier in this section.

3.3. Elution of Cells from Substrates

In terms of the model underlying Eq. (5) it is clear that for negative values of ΔF^{adh} we expect cell attachment to a substrate. As the free energy of detachment, ΔF^{det}, is given by

$$\Delta F^{det} = -\Delta F^{adh} \tag{9}$$

it is clear that detachment of cells from substrates to which they are attached is not easily possible under the same conditions. However, it is possible to change the sign of ΔF^{adh} subsequent to cell adhesion and to make this ΔF^{det} negative, so that the process of cell detachment will be

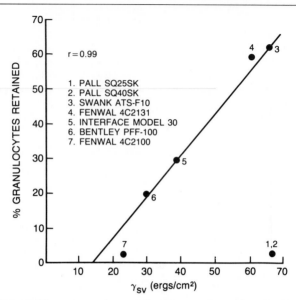

FIG. 13. Extent of human granulocyte adhesion to various microaggregate blood filters (MABF) as a function of MABF material surface tension, γ_{SV}. Correlation coefficient, r, is indicated. For complete experimental details see Ref. 121. (Reproduced with permission from Marcel Dekker, Inc.)

favored. Expressed in terms of surface tensions, there will be repulsion when[45]

$$\gamma_{SV} < \gamma_{LV} < \gamma_{PV} \tag{10a}$$

or

$$\gamma_{PV} < \gamma_{LV} < \gamma_{SV} \tag{10b}$$

On this basis we have reported recently on the development of a novel preparative method for the isolation of granulocytes from nylon fibers[46] and hence for determining the surface tension of these phagocytes. In that study the percentage yield of granulocytes was determined as a function of the surface tension of the elution liquid. The low-molecular-weight surface tension lowering additives used were DMSO and dimethyl acetamide (DMAC). The results are given in Fig. 14. It is noted that the points derived from the two additives essentially fall on the same curve, reinforcing the fact that it is the surface tensions of the interacting compo-

[45] A. W. Neumann, S. N., Omenyi, and C. J. van Oss, *J. Phys. Chem.* **86,** 1267 (1982).
[46] D. R. Absolom, C. J. van Oss, and A. W. Neumann, *Transfusion* **21,** 663 (1982).

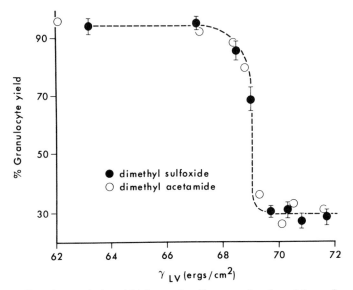

FIG. 14. Granulocyte elution yield from nylon fibers as a function of the surface tension of the eluting buffer. (Reproduced with permission from Lippincott, Ltd.)

nents which matter rather than the chemical composition of the fluid. The main feature of this plot is the dramatic rise in cell yield near a liquid surface tension of 69 ergs/cm² as the surface tension of the buffer is decreased by adding increasing amounts of DMSO and DMAC. As we infer from the inequalities, Eq. (10), we expect the change in the free energy $(\Delta F^{\mathrm{adh}})$ for the cell–substrate interactions to become positive as we decrease the liquid surface tension γ_{LV} from an initially high value to one just below γ_{PV}. Thus the results of Fig. 14 suggest that the surface tension of granulocytes is approximately equal to 69 ergs/cm². While cell elution as a means of determining the surface tension of cells may not have the same potential for accuracy as contact angles and adhesion strategies, it is certainly a very direct manifestation of interfacial tensions, essentially free from the thermodynamic complexities which are inherent in the contact angle technique. It also clearly illustrates the interrelation of interfacial tensions and van der Waals forces. It has recently been shown[45] that the net change in the free energy of adhesion as given by Eq. (5) may indeed be understood as an expression of the magnitude of the van der Waals interactions between the various phases.

It is interesting to compare the mechanisms of cell adhesion and cell elution. In cell adhesion, we observe generally a gradual increase of cell adhesion with decreasing free energy of adhesion, and in elution an on–

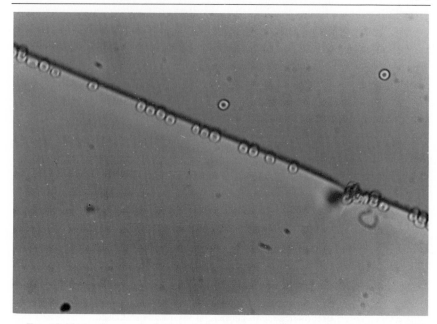

FIG. 15. Photomicrograph of aldehyde-fixed human erythrocytes at an ice/water solidification front. (Reproduced with permission from Humana Press, Inc.)

off mode of adhesion as we cross the borderline between attraction and repulsion. The reason for this difference is unclear at this time.

3.4. The Freezing Front Technique

Recently the advancing solidification front technique was used to explore its potential for the determination of the surface tension of biological cells.[47] This method is relatively new, but it has already been used to determine the surface tension of polymer particles,[48] protein-coated polymer particles,[49] and coal particles.[50] In the freezing front technique the cells are placed in the liquid phase of some matrix material that is then frozen in a carefully regulated manner.[47–50] The interactions between the advancing solidification front and the cells are then observed through a microscope, as illustrated by the photomicrograph in Fig. 15 for aldehyde-fixed human erythrocytes. The aim of these observations is the determi-

[47] J. K. Spelt, D. R. Absolom, W. Zingg, C. J. van Oss, and A. W. Neumann, *Cell Biophys.* **4,** 113 (1982).
[48] S. N. Omenyi, R. P. Smith, and A. W. Neumann, *J. Colloid Interface Sci.* **75,** 117 (1980).
[49] D. R. Absolom and Z. Policova, *J. Dispersion Sci. Technol.* **6,** 15 (1985).
[50] E. Vargha-Butler, M. R. Soulard, A. W. Neumann, and H. Hamza, *Can. Min. Metall. Bull.* **12,** 1 (1981).

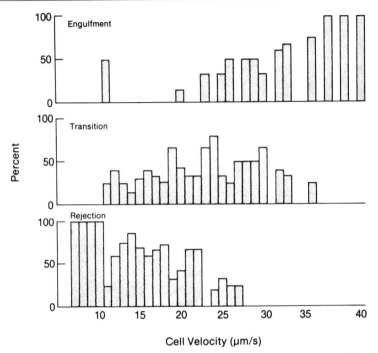

Cell Velocity (μm/s)

FIG. 16. Critical velocity, V_c, histogram for fresh human granulocytes. (Reproduced with permission from Humana Press, Inc.)

nation of the critical velocity of engulfment. This is the velocity of the advancing front at which cells are no longer pushed by the solid/melt interface, but are engulfed. A simple model of this phenomenon involves a balance between the viscous drag force acting on the cell and the repulsive force between the cell and the advancing freezing front arising from a positive free energy of adhesion.[51] At the critical velocity these forces are equal and engulfment is produced by any small disturbance.

As an illustration, Fig. 16 presents the data for an experiment using fresh human granulocytes. This histogram gives the percentage of observations at each velocity that were recorded as engulfment, transition, or rejection. The data are divided into these three regimes according to the *distance* of cell pushing. If the cell was displaced by the advancing solidification front by a distance greater than three times the major diameter of the cell, the event was classified as a "rejection." If the cell was pushed a distance equal to one to three times the cell's major diameter, the event

[51] S. N. Omenyi, A. W. Neumann, W. W. Martin, G. M. Lespinard, and R. P. Smith, *J. Appl. Phys.* **52,** 796 (1981).

was recorded as a "transition" and if engulfment of the cell by the advancing solidification front took place in a distance of less than the particle's major dimension, the event was considered as an "engulfment." These are somewhat arbitrary classifications but it should be noted that determination of the "critical velocity" is not significantly dependent on a single, individual observation (cf. Fig. 16).[47] On average, over 200 separate observations of cell-advancing front interactions were recorded to determine the critical velocity of each cell type. In all cases, at very low velocities there is 100% rejection (pushing) and at very high velocities there is 100% engulfment. Between these two extremes must lie the critical velocity (V_c) where the process of cell rejection gives way to that of cell engulfment. Thus, it is to be expected that the critical velocity will give rise to transition observations and that these will occur within a very narrow velocity range. The transition observations correspond to velocities just above that of the highest rejection velocity. Pushing distances at V_c are extremely short since the slightest disturbance (e.g., through turbulence, friction, thermophoretic gradient) in the balance of forces will produce immediate engulfment. The free energy of adhesion [see Eq. (5)] can be related to the critical velocity, V_c, through a scheme of dimensional analysis (γ_{PS} is the cell–solid interfacial tension, γ_{PL} is the cell–liquid interfacial tension, and γ_{SL} is the solid–liquid interfacial tension). The dimensional analysis interpretation has been described in detail elsewhere.[51] For particles with a major diameter less than 200 μm, the relationship between ΔF^{adh} and V_c is given by the following equation:

$$\Delta F_{\text{adh}} = 2.64 \times 10^5 \left(\frac{\rho_L^{0.847} T^{0.280} k_p^{0.720}}{\mu^{0.127} (\rho_P \cdot C_P)^{0.441}} \right) D^{0.407} V_c^{0.847} \tag{11}$$

where the individual terms are in SI units and are

ρ_P: cell density
ρ_L: liquid matrix density
C_P: heat capacity of the cell per unit mass
k_p: thermal conductivity of the cell
μ: viscosity of the liquid matrix
T: melting temperature of the matrix material
D: cell diameter
V_c: critical velocity

Computation of ΔF^{adh} from V_c measurements requires a knowledge of the physical properties of the cells and of the solid and liquid state of the matrix. This information is generally available from the literature. It is thus possible, using the experimentally determined V_c data, to calculate ΔF^{adh} through the use of Eq. (11). Having obtained a value for ΔF^{adh}, it is possible, through the use of the equation of state approach,[24] to obtain the various relevant interfacial tensions contained in Eq. (5) and in particular to determine the interfacial tension of the cells. As a matter of convention, the cell–liquid interfacial tension, γ_{PL}, obtained as described above, is generally expressed as a surface tension, γ_{PV}, with respect to a vapor phase. The relationship between interfacial tension and surface tension is shown below:

$$\Delta F^{adh} = \gamma_{PS} - \gamma_{PL} - \gamma_{SL}$$
$$= f(\gamma_{PV}, \gamma_{SV}) - f(\gamma_{PV}, \gamma_{LV}) - f(\gamma_{SV}, \gamma_{LV}) \qquad (12)$$

where γ_{PV}, γ_{SV}, and γ_{LV} are the cell (P)–, solid (S)–, and liquid (L)–vapor surface tensions, respectively.[25] The liquid surface tension, γ_{LV}, is usually easy to measure, e.g., by the Wilhelmy Plate method,[21] while the surface tension, γ_{SV}, of the solid matrix may be determined in a number of classical ways.[21] Thus, the calculations required to obtain the surface tension, γ_{PV}, of the engulfed cells can be summarized as follows:

$V_c \rightarrow$ dimensional analysis $\rightarrow \Delta F^{adh} \rightarrow$ equation of state "f" $\rightarrow \gamma_{PV}$
through Eq. (3)

To date, with respect to biological systems, the surface tensions of aldehyde-fixed erythrocytes of several species and of fresh human lymphocytes and granulocytes have been determined with the freezing front technique.[47] Water was used as the matrix material in all cases. The surface tensions obtained with the freezing front technique are those at the melting point of the matrix materials, i.e., 0° in the case of water. The values listed in Table I have been referred to room temperature by assuming that the temperature dependence of surface tension is $d\gamma_{PV}/dT = -0.1$ ergs/(cm²-°C).[21] The agreement of these results with those obtained from contact angles also from the adhesion protocol is good (cf. Table II). The practical aspects of this technique are described in Section 6.3.

The critical velocities of engulfment on which the above results are based are 13.5 μm/sec for the fresh human lymphocytes and 31.0 μm/sec for fresh human granulocytes, as illustrated in Fig. 17. The average diameter of granulocytes and lymphocytes is 13.0 and 11.0 μm, respectively. As the freezing front technique is not sensitive to such small differences in size, the considerable difference in critical velocity must be attributed essentially to the relatively small difference in surface tension between granulocytes and lymphocytes. This implies that the sensitivity of the technique to small differences or changes in surface properties is quite high.

3.5. Phagocytic Ingestion

The process of ingestion of a particle, such as a bacterium B by a phagocytic cell P, generates a cell–bacterium interface PB and annihilates

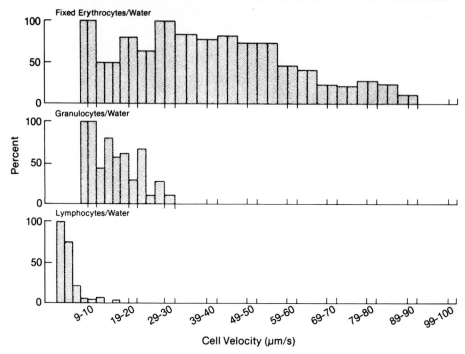

FIG. 17. A comparison of the critical velocity rejection histograms of human granulocytes, lymphocytes, and aldehyde-fixed erythrocytes in water.

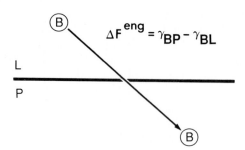

FIG. 18. Schematic representation of the process of phagocytic ingestion of a bacterium. B, Bacteria; L, suspending liquid; P, phagocyte.

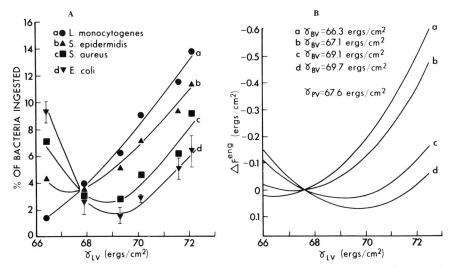

FIG. 19. Comparison of the theoretical free energy of engulfment (ΔF^{eng}) as a function of the surface tension, γ_{LV}, of the suspending aqueous media for four species of nonopsonized bacteria (A) and the experimentally determined extent of the same bacteria ingested by pig platelets in the same aqueous media (B). The errors indicated are 95% confidence limits. For graphical reasons they are given here in selected cases only. The error limits are similar in all cases. (Reproduced with permission from Academic Press, Inc.)

a bacterium–liquid interface BL as illustrated schematically in Fig. 18. This process may be considered from a surface thermodynamic perspective. If this is done, the net free energy for the process of engulfment is given by

$$\Delta F^{eng} = \gamma_{PB} - \gamma_{BL} \qquad (13)$$

where γ_{PB} is the phagocyte–bacterium interfacial tension and γ_{BL} the bacterium–liquid interfacial tension.

Since both γ_{PV} and γ_{BV} are fixed in an experiment with a certain bacterial species and a given type of phagocyte, the only means of influencing the outcome of the experiment by thermodynamic means is to vary the surface tension γ_{LV} of the liquid medium. Our first step then is to consider the thermodynamic prediction for the free energy of engulfing ΔF^{eng} as a function of the liquid surface tension, γ_{LV}. In Fig. 19B, the theoretical ΔF^{eng} is given for the phagocytic ingestion of four bacterial species by platelets.[32] The input information used are the values of the surface tensions of the nonopsonized bacteria and of platelets as determined from contact angles. The main feature of each individual curve is a minimum which occurs at $\gamma_{LV} = \gamma_{BV}$. Thus, if the experimental results follow this predicted pattern of ingestion yet another strategy of determining the surface tension of the ingested bacteria is available. This strategy consists simply in establishing the surface tension of the suspending liquid at which phagocytic ingestion is minimized. The second major feature is that the curves for the ingestion of the various bacteria have a common intersection point. This intersection occurs when $\gamma_{LV} = \gamma_{PV}$, i.e., when the surface tension of the suspending liquid is equal to the surface tension of the phagocytic cells. Thus phagocytic ingestion can be used in principle to determine both the surface tension of the ingested particles as well as the surface tension of the phagocytic cells.

Phagocytosis experiments were performed from this point of view with various bacteria and with both porcine platelets and human granulocytes as the phagocytic cells. The experimental procedures as well as the results have been described in detail elsewhere.[31,32] Reproduced in Fig. 19A are the experimental results of phagocytosis of nonopsonized bacteria by platelets. Comparison with the theoretical curves in Fig. 19B shows that there is a very close correspondence between the thermodynamic predictions and the experimental observations. The minima were identified as the surface tension of the bacteria and entered in Table II, together with the common intersection, which represents the surface tension of the platelets. The agreement with the results obtained from contact angle measurements, and in the case of the bacteria also with the results from the adhesion experiments, is good. These experiments were also performed with opsonized bacteria.[32] Contact angle measurements showed that the more hydrophobic bacteria remained virtually unchanged, whereas the surface tension of the more hydrophilic bacteria was slightly reduced by opsonization. The experimental curves corresponding to the theoretical curves given in Fig. 20 shows only minor differences, reflecting nothing more than the slightly lower surface tensions of the opsonized bacteria as confirmed directly by contact angle measurements. This implies that Fc receptors do not play a role in the phagocytosis of bacteria by platelets (see this volume [3]). A comparison between the extent of phagocytic ingestion of opsonized versus nonopsonized *E. coli* is given in Fig. 21.

Phagocytosis of bacteria by granulocytes is a different story.[31] Theoretically, i.e., from the point of view of the thermodynamic model, curves like those in Fig. 22 are expected. Typical experimental curves for the number of bacteria engulfed per granulocyte as a function of the surface tension of the suspending liquid are given in Fig. 23. The experimental curves for the ingestion by granulocytes, even of the nonopsonized bacteria, have a somewhat different appearance, the reasons for which are not completely understood at the present time. However, one feature in common with the theoretical curves is preserved, i.e., the minimum at $\gamma_{LV} = \gamma_{BV}$, so that phagocytosis by granulocytes can indeed be used for determi-

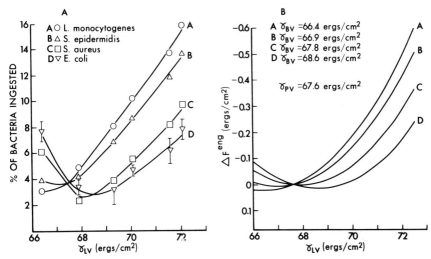

FIG. 20. The data for the four species of opsonized bacteria, presented in the format of Fig. 19. (Reproduced with permission from Academic Press, Inc.)

nations of the surface tension of ingested particles other than opsonized bacteria. Results obtained in this way are summarized in Table II.

In the case of opsonized bacteria it is inferred from Fig. 24 that there is no dependence at all of the extent of phagocytic ingestion on the surface tension of the suspending liquid. This indicates clearly that in this case ingestion does not follow the thermodynamic model. This finding is in

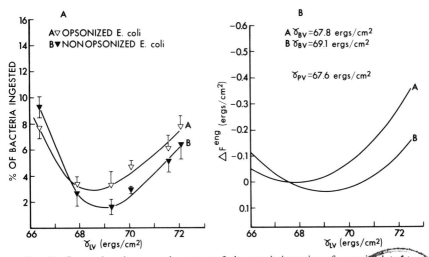

FIG. 21. Comparison between the extent of phagocytic ingestion of opsonized and non-opsonized E. coli in the same format as Figs. 19 and 20.

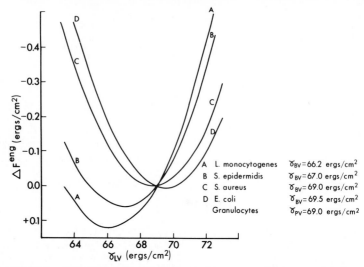

FIG. 22. Theoretical plots of the change in the free energy of human granulocyte engulfment (ΔF^{eng}) as a function of surface tension, γ_{LV}, of the suspending aqueous medium for four species of nonopsonized bacteria. The bacterial surface tension (γ_{BV}) values used are those obtained via contact angle measurements. The common intersection occurs at $\gamma_{LV} =$ 69 ergs/cm² which is equal to the surface tension, γ_{PV}, of the phagocyte; the minimum in each curve occurs when $\gamma_{LV} = \gamma_{BV}$. (Reproduced with permission from Humana Press, Inc.)

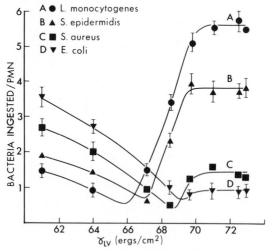

FIG. 23. Extent of ingestion of nonopsonized bacteria by human granulocytes as a function of the surface tension, γ_{LV}, of the suspending aqueous medium. The minima of the curves conform closely to the bacterial surface tension, γ_{BV}, values of the strains used. The principal deviations from the theoretical curves, shown in Fig. 22, are the plateaus each curve exhibits at high γ_{LV} values, and the lack of a common intersection point for all of the curves. PMN: Granulocyte. [Reproduced with permission from Alan R. Liss (Pty), Ltd.]

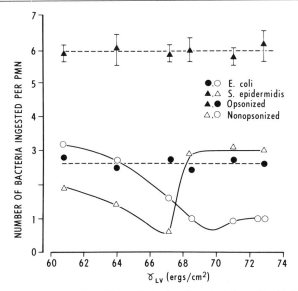

FIG. 24. Effect of opsonization of bacteria on the phagocytic ingestion by human granulocytes (solid line: nonopsonized bacteria; broken line: opsonized bacteria). For graphical reasons, 95% confidence limits are shown for only one case. Error limits are similar in all cases. (Reproduced with permission from Humana Press, Inc.)

agreement with the well-known fact that phagocytic ingestion of opsonized bacteria by granulocytes is Fc receptor mediated and hence specific and not nonspecific, as assumed in the thermodynamic model.[52] By implication it is clear that the thermodynamic model also is an excellent tool to help distinguish between specific and nonspecific interactions.

3.6. Suspension Stability Techniques

While the van der Waals interactions between unlike phases embedded in a liquid can be attractive as well as repulsive, these interactions between identical particles in a liquid can only be attractive or, at a minimum, zero. This implies that the van der Waals forces between cells of one type in suspension can only be attractive or zero in the limiting case. As attraction between cells enhances agglomeration and aggregation, maximum suspension stability is expected when the van der Waals attraction is zero. This occurs when the surface tension of the cells, γ_{CV}, is equal to the surface tension, γ_{LV}, of the suspending liquid medium, thereby providing a relatively simple method for determining γ_{CV}.

[52] A. W. Neumann, D. R. Absolom, W. Zingg, D. W. Francis, and C. J. van Oss, *Cell Biophys.* **4,** 285 (1983).

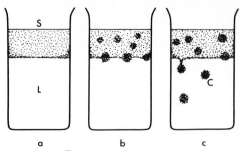

a   b   c

FIG. 25. Schematic representation of the sequence of events in the agglomeration of cells in droplet sedimentation. (a) Initial, well-dispersed state; (b) commencement of agglomeration; (c) fully developed agglomeration, which is manifested by the sedimentation of the agglomerated droplets into the lower phase (heavy water).

3.6.1. Droplet Sedimentation. We recently reported sedimentation experiments with aldehyde-fixed erythrocytes[53,54] from the same species that we had previously investigated using the advancing solidification front method (cf. Section 3.4.). The experimental protocol was based on the aforementioned reasoning. In essence, the aqueous suspension of fixed erythrocytes was layered on a cushion of heavy water, as indicated schematically in Fig. 25. Due to the attractive van der Waals forces between the erythrocytes, cell aggregates will form in the suspension. With time the aggregates will reach a certain size so that, under the influence of gravity, they will break through the H_2O-D_2O interface and start to sink into the D_2O cushion. Various experimental protocols are possible with this technique, including the determination of the time elapsed until the first aggregates form as a function of the surface tension of the suspending liquid medium.[53,54] In Fig. 26 the results of a somewhat different protocol are shown. Here the maximum cell concentration that did not give rise to aggregation for a preselected length of time was determined. The maxima in these curves correspond to the surface tension of the various species of fixed erythrocytes and occurs when $\gamma_{LV} = \gamma_{CV}$. The experimental data are summarized in Table II. A good agreement is obtained with the results obtained from the solidification front technique and also, in the case of fixed human erythrocytes, with the adhesion technique.[46] The practical aspects of this technique are discussed in detail in Section 6.4.

3.6.2. Sedimentation Volume. In this case simple sedimentation volume experiments are performed in which the sedimentation volumes are

[53] S. N. Omenyi, R. S. Snyder, C. J. van Oss, D. R. Absolom, and A. W. Neumann, *J. Colloid Interface Sci.* **81,** 402 (1981).

[54] S. N. Omenyi, R. S. Snyder, D. R. Absolom, C. J. van Oss, and A. W. Neumann, *J. Dispersion Sci. Technol.* **3,** 303 (1982).

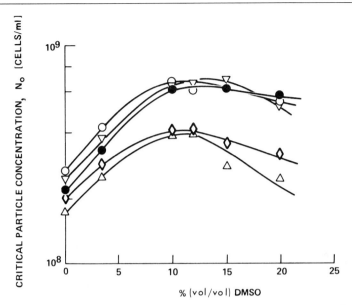

FIG. 26. Suspension stability measured by the "critical particle concentration" as a function of dimethyl sulfoxide concentration. Maximum stability occurs where $\gamma_{LV} = \gamma_{PV}$. $\triangle$, Turkey; $\diamondsuit$, chicken; $\bigcirc$, horse; $\triangledown$, canine; $\bullet$, human. (Reproduced with permission from Marcel Dekker, Ltd.)

measured as a function of the surface tension of the suspending liquid medium.[55] A fixed concentration (or mass) of cells (or particles) is suspended in a standard volume of various binary liquid mixtures. (The purpose of using such binary mixtures is to produce suspending liquid media with a range of surface tensions, γ_{LV}). Following thorough mixing, the suspensions were allowed to settle under the influence of gravity. The sedimentation volume is then recorded until no further change occurs. In aqueous mixtures (as required for biological systems) *maximum* sedimentation volume occurs when the surface tension of the suspending liquid media, γ_{LV}, is equal to the surface tension, γ_{CV}, of the suspended cells. Under such conditions, the van der Waals attraction between the cells is zero and hence the degree of compaction is reduced to a minimum, giving rise to maximum interparticle electrostatic and steric repulsion and hence maximum sedimentation volumes. As an example, we show in Fig. 27 a photograph of the sedimentation volume of polystyrene particles op-

[55] E. I. Vargha-Butler, T. K. Zubovitz, H. A. Hamza, and A. W. Neumann, *J. Dispersion Sci. Technol.* **6**, 357 (1985).

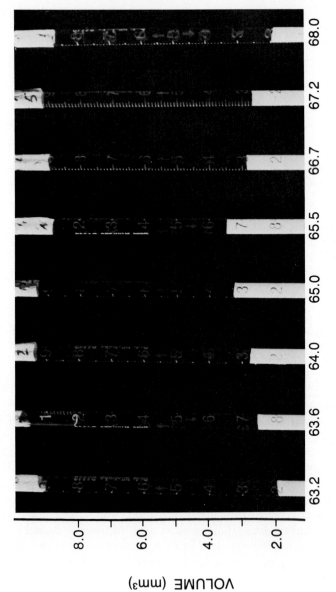

FIG. 27. Photograph of the sedimentation volume of a fixed mass of polystyrene particles preopsonized with human IgG (5 mg/ml) as a function of the surface tension, γ_{LV}, of the suspending liquid media.

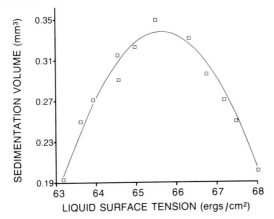

FIG. 28. Second-order polynomial computer curve fit of the data contained in Fig. 27. Sedimentation volume is plotted as a function of γ_{LV}. Maximum, 65.6 ergs/cm², occurs where $\gamma_{LV} = \gamma_{PV}$.

sonized with human immunoglobulin G, i.e., IgG.[56] In Fig. 28 the experimentally determined sedimentation volumes are plotted against the surface tension, γ_{LV}, of the suspending liquid mixture. In principle, this technique could be used to determine the surface tension of any type of blood cell or bacteria. The technique can be used to assess the effect of opsonization, or even pharmacological agents, on the surface properties of cells. Shown in Fig. 29 are the comparative data for fresh human erythrocytes obtained from cardiac arrhythmia patients (who had undergone clinical treatment with chlorpromazine) contrasted with erythrocytes obtained from untreated controls. It is clear from this figure that there is a marked lowering in the surface tension, γ_{PV}, of erythrocytes obtained from the patients presumably as a result of clinical treatment.

For biological systems the choice of binary mixtures is limited due to physiological constraints. The most practical systems include various buffers into which small volumes of surface tension lowering agents, such as propanol, DMSO, or DMAC are incorporated. The most appropriate methodology for this technique is described in detail in Section 6.5.

3.7. Qualitative Methods

3.7.1. Liquid–Liquid Contact Angles. More recently a new technique has been developed.[41,42] This involves liquid–liquid contact angles. For this purpose 8% (w/v) solutions of polyethylene glycol (PEG) 6000 and

[56] D. R. Absolom, W. Zingg, Z. Policova, T. Bruck, and A. W. Neumann, *J. Colloid Interface Sci.* (in press).

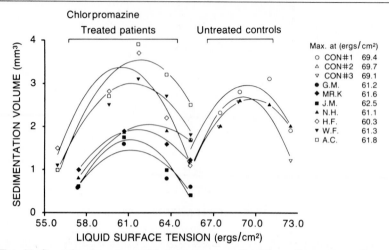

FIG. 29. Second-order polynomial computer curve fit of the sedimentation volumes of erythrocytes obtained from normal versus cardiac arrhythmia patients treated with chlorpromazine. Sedimentation volume is plotted as a function of the surface tension of the suspending liquids. Maximum sedimentation volume occurs where $\gamma_{LV} = \gamma_{PV}$ (see text for explanation).

Dextran T 500 are formed by dissolving the appropriate mass of the polymers in tissue culture media (e.g., RPMI 1640 or HBSS). Equal volumes of these two solutions are mixed together and then allowed to phase separate. The less dense upper phase is PEG rich and the more dense lower phase is Dextran rich. Layers of cells or bacteria are then deposited on some substrate material, e.g., anisotropic cellulose acetate by ultrafiltration, or through adhesion to siliconized glass slides (or tissue culture plastic) by adhesion. Following rinsing to remove serum proteins these cells are then immersed in a bath of PEG-rich fluid and a drop of a fixed volume (≈ 2 μl) of the Dextran-rich phase is deposited on the biological material. The contact angle which this droplet makes with the surface is then monitored as a function of time. This is achieved through the use of a stereomicroscope attached to a 35-mm camera which permits photographic recording of the image as illustrated in Fig. 30 for human granulocytes. Initially the contact angle is close to 180°, as expected. However, as contact is made and the Dextran droplet begins to interact with the cellular material, the contact angle is considerably reduced. After approximately 5 min the observed contact angles reach a stable plateau value which is taken as the effective contact angle measurement representative of the specimen. An example of such a curve is given in Fig. 31. In many respects this technique is similar to the well-known two-phase partition

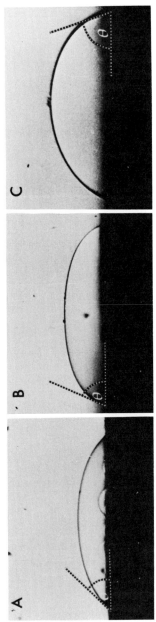

FIG. 30. A stereo photomicrograph of a liquid–liquid contact angle on Ficoll-Hypaque isolated human granulocytes before (A) and after (B,C) stimulation with the chemotactic peptide formyl-Met-Leu-Phe. In (B) the peptide concentration is 2×10^{-7} M and in (C) it is 2×10^{-5} M. The change in contact angle measurement between the untreated (A) control granulocytes ($\theta = 46°$) and the treated granulocytes ($\theta = 68°$ and $73°$, respectively) is approximately $25°$. [D. L. McIver and S. Schürch, University of Western Ontario, London, Ontario, private communication; see also S. Schürch, D. F. Gerson, and D. J. L. McIver, *Biochim. Biophys. Acta* **640**, 557–571 (1981).]

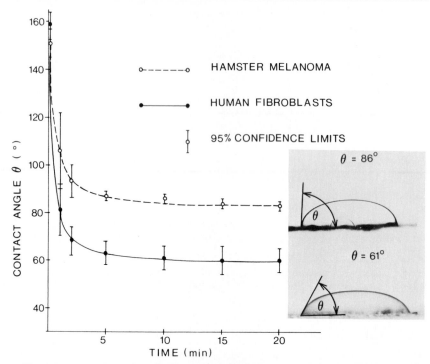

FIG. 31. A plot of liquid–solid–liquid contact angles measured as a function of time. A drop of tissue culture media (RPMI 1640) containing Dextran is placed on a monolayer of cells. The cells are adherent to tissue culture plastic and are completely immersed in a bathing fluid of RPMI 1640 containing polyethylene glycol.

technique developed many years ago.[35–37,57] Indeed, the present technique suffers from the same disadvantage: At this time it is still not possible to interpret and quantitate the solid–liquid–liquid contact angle in terms of the surface tension of the biological material. It is thus necessary to simply rank the material in terms of increasing or decreasing contact angle measurements.

Despite this apparent disadvantage, the technique has been successfully employed to document major differences in the surface properties of various cell lines including platelets, granulocytes, macrophages, fibroblasts, hepatocytes and melanoma cells, and vascular endothelium.[41,42] The major advantage of this technique is its extraordinary sensitivity to even small differences in cellular surface properties which are manifested as large contact angle changes.

[57] P. A. Albertsson, "Partition of Cells, Particles and Macromolecules." Wiley (Interscience), New York, 1971.

3.7.2. Two-Phase Partition Methods. The partition of microorganisms between a hydrophobic liquid and water was first utilized for the (qualitative) classification of bacterial surfaces according to their surface hydrophobicity, in the pioneering efforts of Mudd and Mudd.[58] All these techniques suffer from the drawback that they cannot be employed to provide quantitative information about the surface properties of the biological entities being investigated.

3.7.2.1. Two-phase polymer separation: Albertsson[57] introduced the use of immiscible aqueous Dextran and polyethylene glycol solution in partition separations. Stendahl *et al.*[59] used that method to demonstrate that the more hydrophobic (rough forms) of salmonellae are the ones that become phagocytized to the greatest extent; these results were later correlated with the more quantitative contact angle method (see Section 3.1) by Cunningham *et al.*[60] Stendahl *et al.*[61,62] also showed via partition studies that salmonellae opsonized with IgG antibodies become more hydrophobic, and that this increased hydrophobicity appeared to correlate with an enhanced phagocytic uptake. This method has the advantage that it is very sensitive to even small changes in surface properties. The sensitivity of the technique can be considerably enhanced through the use of either Dextran or polyethylene glycol polymers onto which various chemical groups have been substituted.[63]

3.7.2.2. Bacterial adherence to hydrocarbons: Rosenberg and colleagues have recently introduced an interesting and novel approach for measuring cell surface hydrophobicity.[64–67] The technique has primarily been employed for the characterization of bacteria but can, in principle, be applied to other cell types as well. The technique relies on quantitating the extent of bacterial adhesion to hydrocarbons (BATH). The technique may be briefly summarized as follows: To a fixed volume of aqueous suspension, at a standard bacterial concentration, various volumes of hydrocarbon (e.g., *n*-hexadecane, *n*-octane, or *p*-xylene) are added. After incubation at 30° for 10 min the binary system is vortexed for 2 min to

[58] S. Mudd and E. B. H. Mudd, *J. Exp. Med.* **40**, 647 (1924).

[59] O. Stendahl, C. Tagesson, and L. Edebo, *Infect. Immun.* **8**, 36 (1973).

[60] R. K. Cunningham, T. O. Soderström, C. F. Gillman, and C. J. van Oss, *Immunol. Commun.* **4**, 429 (1975).

[61] O. Stendahl, C. Tagesson, and L. Edebo, *Infect. Immun.* **10**, 316 (1974).

[62] O. Stendahl, C. Tagesson, K.-E. Magnusson, and L. Edebo, *Immunology* **32**, 11 (1977).

[63] K.-E. Magnusson and L. Johannson, *FEMS Microbiol. Lett.* **2**, 225 (1977).

[64] M. Rosenberg, D. Gutnick, and E. Rosenberg, *FEMS Microbiol. Lett.* **9**, 29 (1980).

[65] M. Rosenberg, A. Perry, E. A. Bayer, D. L. Gutnick, E. Rosenberg, and I. Ofek, *Infect. Immun.* **33**, 29 (1981).

[66] M. Rosenberg and E. Rosenberg, *J. Bacteriol.* **148**, 51 (1981).

[67] E. Weiss, M. Rosenberg, H. Judes, and E. Rosenberg, *Curr. Microbiol.* **7**, 125 (1982).

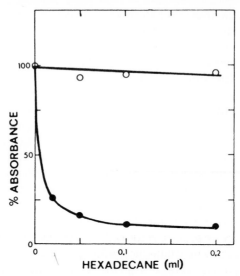

FIG. 32. A plot of the percentage adsorbance of the aqueous phase following phase separation as a function of the volume of hexadecane added to the bacterial suspension. Oral bacterial strains M2-14 (●) and D-26 (○). (Reproduced with permission from Elsevier, Inc.).

ensure complete mixing. The system is then allowed to phase separate and the aqueous phase carefully removed and its light adsorbance at 400 nm measured. When adherent cells are tested they attach to the hydrocarbon droplets and rise with these less dense drops following mixing. The adherent cells are therefore removed from the aqueous phase. The proportion of adherent cells is then determined by comparing the decrease in light adsorbance (following BATH) with the adsorbance of the aqueous suspension (prior to BATH) of known bacterial concentration. As an illustration, reproduced in Fig. 32, are the experimental data for the adherence of hydrophilic (D-26) and hydrophobic (M2-14) oral bacterial strains to hexadecane. This technique permits the ranking of bacterial surface hydrophobicity but it does not allow the quantitative assessment of surface tension.

3.7.3. Hydrophobic Chromatography. The chromatographic interaction of bacteria with various matrix materials, e.g., phenyl- and octyl-Sepharose, can also be employed to qualitatively rank the surface hydrophobicity of bacteria.[68] Rough (R) strains (hydrophobic) of *Salmonella,* for example, have been shown to adhere strongly to columns of phenyl-Sepharose whereas the encapsulated smooth strains were recovered al-

[68] C. J. Smyth, P. Jonsson, E. Olsson, O. Söderlind, J. Rosengren, S. Hjertén, and T. Wadström, *Infect. Immun.* **22**, 462 (1978).

most entirely in the void volume of the column.[69,70] The retention times can be correlated extremely well with data obtained by other qualitative as well as quantitative methods.[60]

3.7.4. Other Methods. 3.7.4.1. Toroid ring: An elegant method for indirectly measuring interfacial tension has been developed by Gingell and Todd.[71] They used a toroid ring with cells adsorbed to the surface. Following immersion of the ring at an oil–water interface, a completely flat meniscus can be obtained at one given level of the interface; the angle which this depth of the meniscal interface subtends with the surface of the torus can be used as a measure of the cellular interfacial tension.[71]

3.7.4.2. Cell adhesiveness: Van den Tempel[72] obtained information about the surface properties of particles by centrifugally detaching them after adsorption onto well-defined surfaces.[50] Earlier, Weiss measured the force of adhesion by subjecting the adherent cells to a shearing force in a rotating device.[73] McKeever measured the degree of macrophage adherence to flat surfaces by determining the force it takes to detach these cells from the substrate.[74] For this purpose fine glass needles were used as levers. The needles were calibrated against known masses by plotting the angle of curvature (λ) in the needles against the mass. The λ required to peel the cells from the substrate could be related to the curvature of the same needles against the fixed calibration masses. In this way, the force of detachment could be calculated. The proportion of phagocytic cells that adhere to glass[30,74] or plastic surfaces,[75] or to nylon,[76] or glass fibers[77,78] has also been utilized as a measure of cell adhesiveness and thus, indirectly, of cellular interfacial tensions.

3.7.4.3. Multiliquid contact angle method: The multiliquid contact angle (θ) method, developed by Zisman,[79–82] has also been used for deter-

[69] I. Stjernström, K.-E. Magnusson, O. Stendahl, and C. Tagesson, *Infect. Immun.* **18**, 261 (1977).

[70] K.-E. Magnusson, O. Stendahl, I. Stjernström, and L. Edebo, *Immunology* **36**, 439 (1979).

[71] D. Gingell and I. Todd, *J. Cell Sci.* **18**, 227 (1975).

[72] M. Van den Tempel, *Adv. Colloid Interface Sci.* **3**, 137 (1972).

[73] L. Weiss, *Exp. Cell Res.* **8**, 141 (1961).

[74] P. E. McKeever and J. B. L. Gee, *J. Reticuloendothel. Soc.* **18**, 221 (1975).

[75] C. J. van Oss, *Infect. Immun.* **4**, 54 (1971).

[76] M. Dy, A. Dimitriu, N. Thomson, and J. Hamburger, *Ann. Immunol. Inst. (Paris)* **125C**, 451 (1974).

[77] T. J. Greenwalt, M. Gajewski, and J. L. McKenna, *Transfusion* **2**, 221 (1962).

[78] H. E. Broxmeyer, G. van Zant, E. F. Schultz, L. A. Koltun, J. LoBue, and A. S. Gordon, *J. Reticuloendothel. Soc.* **18**, 118 (1975).

[79] W. A. Zisman, in "Adhesion and Cohesion" (P. Weiss, ed.), p. 176. Am. Elsevier, New York, 1962.

mining the so-called "critical surface tension, γ_c," of biological cells. This technique has the considerable drawback that it requires the exposure of the biological material to nonphysiological liquids (e.g., diiodomethane) that can interact both chemically and physically with them in numerous ways. A major advantage of the technique is that by plotting the cosine of the contact angle (θ) that the various liquids make with the cells against the experimentally determined surface tension of the liquids, it is possible to obtain, by extrapolation, cos $\theta = 1$, a Zisman plot to yield directly the γ_c of the surface. Instead of organic solvents, various high concentrations of salts (e.g., $NaNO_3$) dissolved in water can also be used. Baier and Zisman have reported on the hydrophobicity of collagen and gelatin surfaces through the use of 12 different organic liquids.[82] Baier and co-workers[83] have also reported on the use of the Zisman method for determining the surface properties of blood vessel walls as well as of potential cardiovascular implant materials.

4. Influence of Cell Surface Charge and Shape

The electrokinetic, or ζ-potential (as measured by means of cell microelectrophoresis[84–89]), of human blood cells, under physiological conditions is -11, -12, -16, and -18 mV for platelets, granulocytes, lymphocytes, and erythrocytes, respectively.[90] That potential, due to simple electrostatic repulsion, suffices to prevent any of these cells from approaching one another more closely than to within ≈ 50–80 Å, as long as they are reasonably smooth, and spherical or discoid in shape.[91] That

[80] W. A. Zisman, *Ind. Eng. Chem.* **55**, 18 (1963).
[81] W. A. Zisman, *Adv. Chem. Ser.* **43**, 1 (1964).
[82] R. E. Baier and W. A. Zisman, *Adv. Chem. Ser.* **145**, 155 (1975).
[83] R. E. Baier, R. C. Dutton, and V. L. Gott, *in* "Surface Chemistry of Biological Systems" (M. Blank, ed.), p. 235. Plenum, New York, 1970.
[84] C. J. van Oss, *Sep. Purif. Methods* **8**, 119 (1979).
[85] C. J. van Oss and R. M. Fike, *in* "Electrokinetic Separation Methods" (P. G. Rhigetti, C. J. van Oss, and J. W. van der Hoff, eds.), p. 111. Elsevier, Amsterdam, 1979.
[86] G. V. F. Seaman and D. E. Brooks, *in* "Electrokinetic Separation Methods" (P. G. Rhigetti, C. J. van Oss, and J. W. Van der Hoff, eds.), p. 95. Elsevier, Amsterdam, 1979.
[87] H. Yamazaki, *Thromb. Haemostasis* **44**, 43 (1980).
[88] M. Shimizu and T. Iwaguchi, *Electrophoresis* **2**, 45 (1981).
[89] P. J. Goetz, *in* "Clinical Electrophoresis: Clinical Applications and Methodology" (A. W. Pearce and D. Sabolovic, eds.), p. 445. Elsevier, Amsterdam, 1980.
[90] C. J. van Oss, R. J. Good, and A. W. Neumann, *J. Electroanal. Chem.* **37**, 387 (1972).
[91] R. J. Good, *J. Theor. Biol.* **37**, 413 (1972).

distance is too great to permit cellular interaction.[90,92] However, as soon as the cells develop protruding pseudopodia, spicules, villi, or other projections, with a radius of curvature less than 500 Å, then the diffuse electrostatic repulsion layer, extending to the so-called "secondary minimum" of 50–80 Å, is easily pierced by such protrusions, upon which contact to within a few angstroms can occur, thereby establishing molecular contact and thus allowing van der Waals attractions to prevail. Granulocytes as well as macrophages are characterized by their various protrusions of small radius of curvature, so that their contact with (and subsequent adhesion to and engulfment of) bacterial and other particles generally ensues without being hindered by the negative surface charge of virtually all bacteria.[93] In addition, bacteria and many viruses possess thin appendages (e.g., pili on gonococci[94] and adenoviruses[95]) and these are likely to be instrumental in the attachment of these microorganisms. The above experimental observations are likely to account for the fact that attempts to correlate the extent of bacterial–phagocyte attachment on the basis of electrostatic interactions, from ζ-potential considerations, have not succeeded[96–98] (see also Section 4.2 below).

4.1. Measurement of Electrokinetic Surface Potential

Electrokinetic methods provide considerable insight into the surface structure of biological cells without producing significant alterations of the cellular organization. In addition, the kinetics, course, and final result of chemical, enzymatic, immunological, viral, and pathological changes which result in the surface membrane properties being altered can be followed and documented by such methods. This information is valuable since the cell membrane peripheral area or surface region is the site of many such biological processes.

Under normal circumstances, biological cells or particles (e.g., vesicles) carry a net negative surface charge. Such charged particles, when placed in a voltage gradient in an electrophoresis chamber, will migrate toward the electrode of opposite charge. The resultant electrophoretic mobility, u, of a cell or particle is defined as the electrophoretic mobility velocity (meters/second) divided by the voltage gradient (volts/meter),

[92] A. D. Bangham, *Ann. N.Y. Acad. Sci.* **123**, 119 (1964).
[93] G. V. Sherbet, "The Biophysical Characterization of the Cell Surface." Academic Press, London, 1978.
[94] H. Smith, *Bacteriol. Rev.* **41**, 475 (1977).
[95] C. J. van Oss, C. F. Gillman, and R. J. Good, *Immunol. Commun.* **1**, 627 (1972).
[96] E. W. Joffe and S. Mudd, *J. Gen. Physiol.* **18**, 599 (1935).
[97] S. L. Daniels, Ph.D. Dissertation, University of Michigan, Ann Arbor (1967).
[98] T. R. Kozel, E. Reiss, and R. Cherniak, *Infect. Immun.* **29**, 295 (1980).

i.e., the potential difference between the electrodes. Such electrophoretic mobility measurements provide a basis from which the electrokinetic (or ζ) potentials, and surface charge densities, of the particles can be calculated. The magnitude, sign, and distribution of charge on any biological surface, cell, or molecule will influence the extent of its interaction with other surfaces (natural or synthetic) or molecules. Changes in the net electrical charge of blood, blood elements, tissues, cells, vessel walls, or components of body fluids may produce metabolic changes or alterations in the rheological properties of blood. In addition, alteration in electrical surface properties produce changes in cellular or vessel wall deformability and cell membrane viscosity, the extent of cellular aggregation, and cell membrane surface adsorption of and interaction with proteins or drugs contained in the body fluids.

Traditional particle electrophoresis equipment consists in essence of a transparent chamber, fitted with electrodes, in which the migration of cells or particles can be observed by means of a microscope. The cell or particle migration in a defined electric field is observed with the microscope focused at a stationary layer in the electrophoresis chamber at which layer fluid flow induced by the electric field (electroosmosis) does not influence the electrophoretic migration. The migration of the particles is timed across a graticule in the microscope ocular which has been calibrated in terms of real distance at the focal plane of the microscope. Knowing the distance over which each cell or particle migrates in a measured time interval, the velocity and hence the electrophoretic mobility (velocity per unit electric field strength applied) can be calculated. Individual particle or cell mobilities are then accumulated from the given sample population and mean mobilities and statistical information are calculated. The accumulated data are generally plotted in the form of a histogram of cell or particle number as a function of electrophoretic mobility.

The classical techniques of particle and cell microelectrophoresis have been well described by Abramson et al.,[99] and recently by Seaman for cylindrical chambers,[100] and Fuhrmann and Ruheustroth-Bauer for rectangular chambers.[101] More recently an extensive overview of the entire area of electrokinetics has been published.[102]

[99] H. A. Abramson, L. S. Mayer, and M. H. Gorin, in "Electrophoresis of Proteins," p. 43. Hafner, New York, 1964.
[100] G. V. F. Seaman, in "Cell Electrophoresis" (E. J. Ambrose, ed.), p. 4. Little, Brown, Boston, Massachusetts, 1965.
[101] G. F. Fuhrmann and G. Ruhenstroth-Bauer, in "Cell Electrophoresis" (E. J. Ambrose, ed.), p. 22. Little, Brown, Boston, Massachusetts, 1965.
[102] C. J. van Oss, Sep. Purif. Methods 8, 119 (1979).

The influence of the size and shape of particles on their electrophoretic mobilities has been discussed elsewhere.[103] The calculations required for converting measured cell electrophoretic mobilities into the associated surface of ζ-potentials are relatively simple but often are the source of confusion and should be treated with caution. As an example: when electrophoretic mobilities of small particles are measured under low ionic strength conditions the appropriate equation to describe the electrophoretic mobility (u) is Hückel's equation[104]:

$$u = \zeta\varepsilon/6\pi\eta \qquad (14)$$

where η is the viscosity and ε the dielectric constant of the medium. On the other hand, when the mobility is measured under conditions of high ionic strength ($\mu = 0.15\ M$ NaCl) the most appropriate equation is the von Smoluchowski equation[105]:

$$u = \zeta\varepsilon/4\pi\eta \qquad (15)$$

It should be emphasized that ζ-potentials that can be obtained from measured electrophoretic mobilities, in the cgs system, are expressed in electrostatic units (esu) of potential difference.[106] For instance, in the Hückel equation [Eq. (14)] it should be kept in mind that the electrophoretic mobility

$$u = U/X \qquad (16)$$

where U is the electrophoretic velocity and X the electric field strength. Thus, from Eqs. (14) and (16):

$$\zeta = 6\pi\eta U/X\varepsilon \qquad (17)$$

and in order to convert esu potential differences to volts, ζ as well as X should be multiplied by 300, so that the right-hand side of Eq. (17) must be multiplied by 90,000, if one is to express ζ in volts and if the electrophoretic mobility is given in cgs units ($cm^2sec^{-1}V^{-1}$). As the viscosity η of water decreases much more strongly with an increase in temperature than the dielectric constant ε, the temperature effect is considerable, and to

[103] J. Th. G. Overbeek and B. H. Bijsterbosch, in "Electrokinetic Separation Methods" (P. G. Righetti, C. J. van Oss, and J. W. Vanderhoff, eds.). Elsevier/North-Holland, Amsterdam, 1980.
[104] E. Hückel, Phys. Z. 25, 204 (1924).
[105] M. Z. von Smoluchowski, Phys. Chem. 92, 129 (1918).
[106] D. J. Shaw, "Electrophoresis," p. 4. Academic Press, New York, 1969.

achieve the same electrophoretic mobility, a particle or molecule needs to have twice as high a ζ-potential at 0 as at 35°.[102,107] For a comprehensive and detailed explanation of the possible pitfalls associated with these conversions the interested reader is referred to van Oss.[102]

4.2. Relationship of Cell Surface Charge to the Extent of Phagocytic Ingestion

All human blood cells have a net negative surface charge at physiological pH and ionic strength. This is due, in part, to the ionized carboxyl group of N-acetylneuraminic acid (sialic acid) since the negative charge is reduced when the cells are treated with neuraminidase. Such treatment can reduce the electrophoretic mobility of human granulocytes by as much as 55%.[108] Mature polymorphs exhibit a lower neuraminidase-sensitive ζ-potential (i.e., surface charge) than do bone marrow granulocyte precursors.[109]

As early as 1928, it was suggested on purely theoretical grounds that surface charge should influence distance of closest approach between the bacteria and the phagocytic cell and thus would affect the extent of phagocytic ingestion.[110] According to this thesis, electrostatic repulsion between negatively charged surfaces would hinder ingestion whereas oppositely charged particles (giving rise to attractive coulombic interactions) would promote enhanced phagocytic ingestion. Experimental evidence, however, does not support this theory.[96–98,111] Neuraminidase treatment of guinea pig neutrophils reduces the total sialic acid content of the cells by approximately 50% (thereby very substantially reducing the magnitude of their overall surface charge) and yet this treatment has no effect on either the extent of phagocytic ingestion or subsequent glycolysis.[111] Similar studies with human neutrophils have also revealed that neuraminidase treatment does not influence the extent of S. aureus ingestion.[112] These investigations have revealed that superoxide and H_2O_2 production were reduced while hexosemonophosphate shunt activity was not altered.[112,113] It should be mentioned, however, that neuraminidase treatment of guinea pig neutrophils has been reported as preventing the adsorption of influ-

[107] C. J. van Oss, Sep. Purif. Methods 4, 167 (1975).
[108] M. A. Lichtman and R. I. Weed, Blood 35, 12 (1970).
[109] M. A. Lichtman and R. I. Weed, Blood 39, 301 (1972).
[110] E. Ponder, J. Gen. Physiol. 11, 757 (1928).
[111] J. Noseworthy, H. Korchak, and M. L. Karnovsky, J. Cell. Physiol. 79, 91 (1972).
[112] M.-F. Tsan, K. H. Douglass, and P. A. McIntyre, Blood 49, 437 (1977).
[113] M.-F. Tsan and P. A. McIntyre, J. Exp. Med. 143, 1308 (1976).

enza virus onto the phagocytic cells.[114,115] It is not possible to discern from these studies whether this reduction is due to a surface charge effect or whether the enzyme has destroyed in some way a specific receptor necessary for viral adhesion.

5. Summary and Conclusions

1. Surface properties of cells and bacteria determine to a significant extent the outcome of the phagocytic process. The relative surface hydrophobicity of a bacterial strain and a phagocyte will not only affect the extent of ingestion of nonopsonized bacteria but will also determine the degree of nonspecific immunoglobulin adsorption (opsonization) and complement activation. Both of the latter will of course also influence phagocytic ingestion.

2. Surface hydrophobicity can be assessed by a variety of quantitative as well as qualitative techniques. Of the quantitative techniques the most useful for biological systems are the liquid–vapor contact angle (Section 3.1.1) and the more tedious adhesion strategy (Section 3.2). Of the qualitative techniques the liquid–liquid, two-phase partition, and hydrocarbon adherence techniques appear to be the most useful (Section 3.7).

3. The results obtained from the various quantitative techniques are in good agreement, as illustrated in Table II. The surface tensions of blood cells and proteins are relatively high; that is, these materials tend to be hydrophilic in their natural state. The surface properties of bacteria tend to vary markedly and can be extremely hydrophobic (e.g., see Fig. 3) as well as hydrophilic (cf. Table I). These variations are likely to influence considerably the extent to which they become aspecifically opsonized by serum and thus the extent of phagocytic ingestion.

4. The interpretation of the quantitative surface energy experiments described in this article aimed at the determination of the surface tensions of biological material. These arguments can, of course, be inverted so that it may be concluded that surface tensions, in the absence of specific receptor–ligand interactions, govern to a large degree biological processes such as cell adhesion, protein adsorption, cellular suspension stability, and phagocytosis.

5. The theoretical models described herein, which are based on quantitative surface tension considerations, can be employed to assess to what extent a particular biological process is nonspecific in nature and is modulated by van der Waals and electrostatic interactions only.

[114] H. S. Ginsberg and J. R. Blackmon, *Virology* **2**, 618 (1956).
[115] T. N. Fisher and H. S. Ginsberg, *Virology* **2**, 637 (1956).

6. The information generated in these studies can be used for the design and development of novel preparative techniques. Examples of this are the complete dissociation of antigen–antibody complexes,[116,117] isolation of large quantities of granulocytes[48] or proteins[118] by filtration hydrophobic chromatography, affinity chromatography procedures for the isolation of the putative antigen,[119,120] optimum design of microaggregate blood filters.[121]

6. Methodology

The purpose of this section is to describe in detail some of the techniques for measuring surface properties which have been developed more recently. This section is in no way complete and the interested reader is referred to the original publications cited in this article for a discussion of other techniques.

Table V summarizes some of the more important advantages and/or disadvantages of the four most easily applied techniques.

6.1. Measurement of Contact Angles on Biological and Other Highly Hydrated Surfaces

In recent years there has been a marked increase in the number of investigations which have employed contact angle measurements as a means of characterizing the surface properties of various biological systems. Measurement of contact angles (θ) on biological surfaces, while conceptually similar, requires a very different procedure from that used conventionally in measuring θ on, for example, polymer surfaces.

The primary reason for this difference stems from the fact that biological entities are highly hydrated and normally can be maintained and manipulated only in aqueous media. Practically speaking, a surface of a biological system being prepared for contact angle measurement will initially carry a thick aqueous layer. On the other hand, as a consequence of physiological constraints, the liquid of choice for these measurements is 0.15 M purified sodium chloride dissolved in deionized, distilled water (saline). Thus, when a drop of saline is placed on such a surface a zero contact angle (i.e., complete spreading) is generally observed initially. As

[116] C. J. van Oss, D. R. Absolom, and A. W. Neumann, *Immunol. Commun.* **8**, 11 (1979).

[117] C. J. van Oss, D. Beckers, C. P. Engelfriet, D. R. Absolom, and A. W. Neumann, *Vox Sang.* **40**, 367 (1981).

[118] D. R. Absolom, C. J. van Oss, and A. W. Neumann, *Transfusion* **21**, 663 (1982).

[119] D. R. Absolom, *Sep. Purif. Methods* **10**, 239 (1981).

[120] D. R. Absolom, *Immunol. Commun.* (submitted for publication).

[121] D. R. Absolom and E. L. Snyder, *J. Dispersion Sci. Technol.* **6**, 37 (1985).

TABLE V
ADVANTAGES AND DISADVANTAGES OF THE VARIOUS TECHNIQUES USED FOR
DETERMINING THE SURFACE PROPERTIES OF BIOLOGICAL ENTITIES

Method	Advantage	Disadvantage
Direct contact angle	1. Easy to perform	1. Relatively large scatter in experimental data
	2. Sample preparation is straightforward	2. Often difficult to accurately determine line of contact between the measuring liquid and substrate
Particle adhesion	1. Readily applicable to small particles (<15 μm)	1. Can only be employed for a narrow particle size range
	2. Low cell concentration required	2. Time consuming
	3. High accuracy	3. Relatively sophisticated instrumentation required
Freezing front	1. Readily applicable for a large size range (from 5 to 300 μm)	1. Difficult to employ with particle sizes less than 5 μm due to thermophoretic effects
	2. Very low cell concentration	2. Time consuming
	3. Very good reproducibility	3. Sophisticated instrumentation required
	4. Extreme sensitivity	
Sedimentation volume	1. Applicable to particles with a size range from 50 to 1 μm	1. Narrow particle size range is necessary
	2. Relatively simple and inexpensive method	2. Relatively large amount of sample required
		3. Limited accuracy and sensitivity

time passes and excess surface water evaporates, the contact angle becomes first small and finite, then increases, and eventually reaches a plateau value. As the system is allowed to dehydrate further, the contact angle starts to change again, at times erratically.

These are the experimental facts observed over and over, and without exception. Typical examples are given in Figs. 1 and 2. Thus, the question arises regarding what we mean by "the" contact angle on such surfaces. Unfortunately, not all investigators follow the above time dependence of the contact angle measurements and often do not specify the condition under which the contact angle measurements were performed. Such results are therefore open to question as to their significance. Measurements as described above were first performed over 10 years ago,[28] and the contact angle corresponding to the plateau was identified as the thermodynamically significant contact angle.[30] Apparently, several researchers in the area have not become aware of this strategy and interpretation. Therefore it is desirable to describe the procedure of contact angle measurements on highly hydrated surfaces more completely here.

In essence, layers of the biological entity of interest are prepared by deposition from suspension onto a hydrophobic surface, such as siliconized glass, or by ultrafiltration on anisotropic cellulose acetate membranes. When the adhesion method is employed, the adherent moieties should cover a minimum of 50% of the substrate material in order to ensure reproducible contact angle measurements.[122,123] Using a sonicated, chromic acid cleaned Hamilton syringe 1 to 5 μl of saline are then deposited on the initially very wet layer and the contact angle, which the drop makes with the surface, is determined. A standard volume of saline should always be employed for comparative studies.

Initially the contact angle is, as expected, close to zero. However, as water evaporates from the surface and *fresh* drops of saline are deposited on sites of the material which have not been used for measurements previously, gradually increasing contact angles are observed. Each measurement is made as quickly as possible and certainly within 10 sec of deposition of the drop. Both the left and right hand sides of the drop are examined and the readings recorded. Thereafter, a fresh drop of saline is deposited on a *new* location of the substrate material at various time intervals. The actual contact angle measurement may be performed using a standard goniometer with a suitable eyepiece grid. Generally a 10× objective lens is sufficient for a 1-μl drop. For such biological materials, a green filter has been found to be useful since it enhances the contrast between the substrate material and the edge of the liquid drop profile, thereby enhancing the accuracy of the measurement (cf. Section 3.1.2).

Recently a more sophisticated and considerably more accurate technique for measuring contact angles has been developed[23,24] which removes the operator subjectivity involved in such measurements (cf. Section 3.1.2). This technique employs a stereomicroscope coupled to an automatic 35-mm camera to photograph the saline drop profile on the substrate. Following development of the high contrast film the drop profile is digitized and the coordinates fitted to a theoretical Laplace drop profile. This technique is particularly suitable for the low contact angles

[122] J. L. Mege, C. Capo, A. M. Benoliel, C. Foa, and P. Bongrand, *Immunol. Commun.* **13**, 211 (1984).

[123] A. M. Benoliel, C. Capo, P. Bongrand, A. Ryter, and R. Depieds, *Immunology* **41**, 547 (1980).

(<30°) often observed on biological material and in which operator subjectivity can introduce substantial errors. In this automated technique the use of a red filter in the primary photography is recommended in order to enhance edge clarity in the negatives. This method gives rise to considerably enhanced precision and accuracy.

It should also be pointed out, however, that other workers have developed alternative strategies for measuring solid–liquid contact angles.[122–124] Generally these rely on the use of a known volume of liquid and the subsequent measurement of the apical height or diameter of the drop. Accuracy and precision of these techniques have not been reported.

The importance of using fresh drops of saline for each measurement cannot be overemphasized. With time the biological surfaces are likely to exude proteins and other soluble biological moieties, into the saline drop, thereby lowering the saline surface tension and thus altering the observed contact angle. After approximately 15 min, depending on circumstances such as relative humidity and temperature, the observed contact angles of fresh saline drops reach a plateau value which remains constant for some time as illustrated for various types of bacterial and mammalian cells in Figs. 1 and 2, respectively. Thereafter, the observed contact angles sometimes rapidly increase or decrease and in other cases simply become erratic.

Not surprisingly, the contact angles observed on biological and other highly hydrated surfaces are often quite small, in the range from 10–30°. There are, however, also examples of much larger contact angles. *Mycobacterium butyricum*, with a contact angle of 70°[30] is one such case, cartilage with a contact angle of approximately 100°[125] is another. Hydrogel surfaces are typically at the other end of the scale,[126] with contact angles often below 10°.

Our interpretation of these curves in the figures is as follows: initially the biological material is completely covered by water and the contact angle is equal to zero due to complete spreading, as would be expected. As time passes and the excess surface water evaporates, the saline drops start reacting to the presence of the biological material. Finally, when the plateau is reached the observed advancing contact angle is determined exclusively by the properties of the layer of biological material. The final points beyond the plateau are attributed to conformational and structural changes and possible denaturation of the specimen. Scanning electron

[124] C. Dahlgren and T. Sunquist, *J. Immunol. Methods* **40**, 171 (1981).
[125] J. Chappuis, I. A. Sherman, and A. W. Neumann, *Ann. Biomed. Eng.* **11**, 435 (1983).
[126] D. R. Absolom, M. H. Foo, W. Zingg, and A. W. Neumann, in "Polymers as Biomaterials" (S. W. Shalaby, A. S. Hoffman, B. D. Ratner, and T. A. Horbett, eds.), p. 149. Plenum, New York, 1984.

microscope studies have been reported which confirm the maintenance of structural integrity during the plateau region of the contact angle measurements and the subsequent concomitant loss of integrity and the observed scatter in the experimental data.[127] This contention is further corroborated by the fact that changes in experimental conditions, such as relative humidity, change all features of the contact angle *versus* time curves such as the initial slope, the duration of the plateau, the end values, with the striking single exception of the contact angle value on the plateau. The observed plateau values are unique for each biological system and reflect solely the properties of the intact biological material.

6.2. The Adhesion Technique

This technique relies on determining the extent of the interaction of cells with polymer surfaces of varying surface energy (i.e., surface tension) when suspended in liquids of various surface tensions. For the duration for which the cells are being exposed to the polymer surfaces they are retained in specially designed Teflon molds.

I. Polymer Surfaces (See Table VI for Commercial Names)

A. Preparation of Surfaces

1. Glass slides

a. Wipe microscope slide with tissue paper to remove any particles. Soak slides overnight in fresh chromic acid. (When the solution turns green, after frequent use, it loses its effectiveness and should be discarded).

b. Remove slides from the chromic acid and rinse four times with running glass-distilled H_2O water.

c. Place the slides in a Pyrex dish and cover with fresh methanol (or acetone). Sonicate for 15 min. Discard the methanol. Add fresh methanol to cover the slides and sonicate for an additional 15 min.

d. The glass slides are dried in a vacuum oven set at 37° and pressure of −30 cm of mercury.

e. The clean glass slides are stored under vacuum in a glass desiccator.

2. Siliconized glass slides

a. Place clean glass slides in a Teflon slide holder. Place the holder in a desiccator, with approximately 2 ml of dimethyldichlorosilane (handle in fumehood) in a small beaker. Place the desiccator under 15 units of vacuum for 48 hr at a temperature of 22°.

b. After 48 hr, the slides are rinsed by placing them in toluene for 15

[127] I. A. Sherman, J. Grayson, and W. Zingg, *Bibl. Anat.* **20** (1980).

min. Repeat twice and finally in acetone. Before rinsing slides in acetone, check to ensure the liquid front moves quickly across the slide and down the center.

c. The slides are air dried and stored under vacuum in a desiccator. Contact angle measurements must be made to check for complete siliconization of the glass surface.

3. Polystyrene

a. Polystyrene sheets are cut into 6 × 2 cm strips. Do not touch the surface, use Teflon-coated forceps.

b. The strips are placed in a beaker containing fresh methanol for 5 min, making sure none of the strips stick together. Discard the methanol and replace with fresh methanol for an additional 5 min.

c. Blot with Kimwipes.

d. Tape the ends of the strips (use autoclave tape) onto clean glass slides. Place the strips into a slide container and store under vacuum in a glass desiccator.

4. Sulfonated polystyrene (SPS)

a. Cut SPS sheet into 6 × 2 cm strips using Teflon-coated forceps. Generally only one side of the sheet is sulfonated, thus it is necessary to keep track of this side which exhibits a larger degree of wettability (i.e., produces a smaller contact angle with water).

b. Using forceps, dip each strip individually in hexane for 10 sec. Blot with Kimwipes. (Do not use alcohols with SPS.)

c. Tape the ends of the strips onto clean glass slides, with the nonsulfonated side against the slide. Place the slides into a slide container and store under vacuum.

5. Fluorinated ethylene propylene (FEP)

a. Cut sheet into 6 × 2 cm strips.

b. Place the strips in methanol (ensuring complete coverage) and sonicate for 15 min. Discard the methanol and replace with fresh methanol, sonicate for an additional 15 min.

c. Blot dry with Kimwipes.

d. To heat press: Cover the two steel plates with aluminum foil. For 24 FEP surfaces, arrange 24 clean glass slides in a 3 × 8 arrangement on one of the plates. Place a clean FEP surface on top of each glass slide and then place another clean glass slide on top of the FEP. Place the other metal plate under the metal plate with the foil and another metal plate on top. Thus, a total of four metal plates are used. For this purpose we use a Wabash heat press (Wabash, WI) with stainless steel platters.

e. Place the plates on the stainless steel plattens and set the thermostat at 250°F. Bring the top and bottom plattens together but do not apply any pressure. After 30 min, increase the thermostat to 450°F. After another 30 min, increase the thermostat to 580°F. Wait an additional 30 min and then apply 3 tons of pressure for 5 min. After 5 min, turn the temperature off but do not release the pressure.

f. Release the pressure when the temperature drops to 450°F. Place all the surfaces in a large, clean beaker of cold distilled water for about 30 min, as soon as the pressure is released. The smooth FEP surface and glass slides are easily separated. Store under vacuum.

6. *Teflon (polytetrafluoroethylene)*

a. Procedures for cutting, cleaning, and heat pressing are the same as for FEP. Except, use 9 tons of pressure for 1 hr.

b. Release the pressure when the temperature drops to 450°F. The smooth PTFE surface and glass slides separate easily. Store under vacuum.

7. *Low-density polyethylene (LDPE)*

a. Cut and clean using the same procedure as for FEP. Also, use the same preparation for heat press as described for FEP.

b. Set heat press temperature at 250°F. In 30 min, apply 3 tons of pressure for 5 min. Turn the temperature off but do not release the pressure.

c. Release the pressure when the temperature drops to 150°F. Place in a beaker of distilled water. The smooth LDPE surface and glass slides are easily separated. Store under vacuum.

8. *Acetal*

a. Procedures for cutting, cleaning and heat pressing are the same as for FEP. Except, increase the temperature after the first 30 min to 350°F. After 30 min (at 350°F), place under 3 tons of pressure for 5 min. Release the pressure when the temperature is 310°F and place in a beaker of distilled water. The glass and polymer surfaces are easily separated. Store under vacuum.

B. *Characterization of Surfaces.* Following preparation all of the surfaces should be checked for surface defects, e.g., surface roughness, and their surface properties examined with respect to contact angle measurement (cf. Table VI). Any surface which exhibits roughness or incorrect contact angle values should be discarded.

II. Suspending Liquids

In order to produce liquids of various surface tensions the cells are suspended in mixtures of HBSS (or any other suitable buffer) containing different amounts of a low-molecular-weight additive which lowers the surface tension. For this purpose, propanol, DMSO, or DMAC have been used previously.[31,32,43,46,48,116,117] Propanol has the advantage that very small concentrations produce a marked lowering of liquid surface tension. In addition propanol has an advantage over other alcohols in that it has a

TABLE VI
SOLID SUBSTRATES USED IN THE CELL ADHESION EXPERIMENTS

Material	Source	Preparation	Contact angle with water, θ_{H_2O} (degrees $\pm$ SD)	Surface tension, γ_{SV} (ergs/cm^2)[a]
Polytetrafluoroethylene (PTFE)	Commercial Plastics (Toronto, Canada)	Heat press	$108 \pm 4°$	17.6
Polystyrene (PS)	Central Research Laboratories, Dow Chemical (Midland, MI)	Film	$95 \pm 2°$	25.6
Low-density polyethylene (LDPE)	Commercial Plastics (Toronto, Canada)	Heat press	$84 \pm 4°$	32.5
Acetal	Commercial Plastics (Toronto, Canada)	Heat press	$64 \pm 1°$	44.6
Polyethylene terephthalate (PET)	Celanese, Ltd. (Toronto, Canada)	Heat press	$60 \pm 2°$	47.0
Sulfonated polystyrene (SPS)	Central Research Laboratories, Dow Chemical (Midland, MI)	Film	$24 \pm 3°$	66.7

[a] Derived via the equation of state approach.[24,25,27]

lower vapor pressure. On the other hand DMSO and DMAC, while not as effective in terms of lowering surface tension, have been used extensively in biological research[128] and their toxicity limits have been clearly established. The concentrations required in cell adhesion assays are far below the reported toxicity levels.

a. *Measurement of liquid surface tensions:* The surface tensions of the binary liquid mixtures may be measured using a modification of the Wilhelmy Balance technique. The principle of the technique is described below.

When a thin measuring plate suspended from an automatic electrobalance dips into the solution, there will be a capillary rise of the liquid at this measuring plate. The mass of liquid ΔM raised above the unperturbed level of the liquid is related to its surface tension γ_{LV} by

$$\gamma_{LV} = \Delta Mg/P \qquad (18)$$

g: acceleration due to gravity
P: perimeter of the measuring plate

Equation (18) assumes complete wetting or, in other words, zero contact angle.

A schematic diagram of the experimental set-up is shown in Fig. 33. The T-shaped stainless steel measuring plate is about 1 cm wide. It is suspended from an electrobalance F by means of a thin glass fiber C. The hang-down arrangement is protected against air currents in the laboratory by means of the protective glass sleeve E. The solution B is contained in a double-walled glass cell A, the inner diameter of which is about 4 cm. In order to minimize evaporation, the measuring cell A is covered by a lid with small holes for the passage of the glass fiber C, as well as for the thermocouple D. The measuring cell A is mounted in an enclosure O, in order to reduce disturbances from the outside. The interior of the enclosure O and the protective sleeve E are heated in order to prevent condensation on the nonimmersed parts of the measuring plate, the glass fiber C, and the sleeve E.

The temperature control is effected by pumping the thermostatted liquid from the thermostat I through the volume enclosed by the double walls of the measuring cell A. The temperature of the thermostat I can be raised and lowered at constant rates by means of the temperature programmer J. If necessary it is possible to minimize oxidation since argon may be percolated slowly from the container M through a bottle containing the solvent (not shown) and then introduced into the vapor phase above the solution. The electrobalance output is recorded, as a function

[128] J. C. de la Torre, ed., *Ann. N.Y. Acad. Sci.* **411**, 1 (1983).

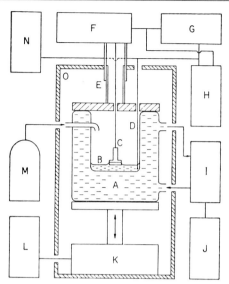

FIG. 33. Schematic representation of apparatus used to measure surface tension. A, Measuring cell; B, solution; C, glass fiber and measuring plate; D, thermocouple; E, protective sleeve for fiber; F, electrobalance; G, X–Y recorder; H, two-channel X–t recorder; I, circulating thermostat; J, temperature programmer; K, movable platform; L, drive for platform; M, argon bottle; N, temperature reference; O, enclosure.

of temperature, on the X–Y recorder G. For initial manipulations, including the observation of initial adsorption kinetic effects, a two-channel X–t recorder is also connected to facilitate separate observation of temperature and surface tension. After cleaning the measuring plate carefully in a methanol flame, it is suspended from the electrobalance. Contact between the measuring plate and the solution is established by raising the platform K by means of the motorized drive L. As stated above, the Wilhelmy technique assumes that the contact angle between the measuring plate and the solution is zero. Since on a relatively rough surface, such as the measuring plate, it is commonly expected that the receding contact angle is equal to zero, the measuring plate is at first immersed into the solution through a distance of approximately 2 mm and then retracted through exactly the same distance by first raising and subsequently lowering of the platform K. Thus, the depth of immersion will finally be zero under receding contact angle conditions.

The experimentally measured liquid surface tension γ_{LV} values of binary mixtures of phosphate-buffered saline containing varying amounts of propanol, DMSO, or DMAC are given in Table VII. It should be noted that since both DMSO and DMAC are extremely hygroscopic, the γ_{LV}

TABLE VII

EFFECT OF ADDING LOW-MOLECULAR-WEIGHT
SURFACTANTS ON THE SURFACE TENSION OF A
PHOSPHATE-BUFFERED SALINE[a]

Concentration (vol/vol)	Surface tension (ergs/cm^2) at 22°		
	n-Propanol	DMSO[b]	DMAC[c]
0.0	72.7	72.7	72.7
0.25	69.7	—	—
0.40	68.2	—	—
0.50	67.4	—	—
0.70	66.5	—	—
0.80	65.2	—	—
1.00	63.3	71.7	71.6
1.50	59.9	—	—
2.00	54.9	—	—
2.50	54.2	70.8	70.5
3.00	53.4	70.3	70.1
3.75	50.7	—	—
5.00	46.4	69.6	69.3
7.50	42.1	69.0	68.8
8.00	41.6	68.5	68.4
10.00	39.3	67.1	67.2
15.00	34.0	63.2	61.9

[a] Composition of buffer (in g/liter): NaCl (8.0); KCl (0.2); Na$_2$HPO$_4$ (1.15); KH$_2$PO$_4$ (0.2).
[b] DMSO, Dimethyl sulfoxide.
[c] DMAC, Dimethylacetamide.

values given should be taken only as guidelines. Any such mixture should always be measured individually immediately prior to use since the measured value will vary according to how long the bottle has been opened.

III. Preparation of Teflon Molds

In order to hold the cell suspension in place it is retained in wells bored into Teflon molds. The mold is clamped onto the polymer surface and leakage is prevented through the use of silastic rubber gaskets.

Before each experiment the Teflon molds are sonicated in methanol and then rinsed in deionized distilled water. (The mold should not be scratched through the use of cleaning brushes).

In our laboratory each mold has three wells, allowing each liquid condition to be repeated in triplicate. The wells are 1 cm in diameter and 0.5 cm in depth.

IV. Preparation of Cells

Following isolation of the cells of interest the cell suspension is adjusted to the desired concentration (usually 1×10^6 cells/ml) by means of dilution following Coulter counter analysis. Thereafter a known volume of the cell suspension is centrifuged gently to pellet the cells. The cell pellet is then resuspended in appropriate buffer to approximately two-thirds of the original volume. To this suspension DMSO, DMAC, or propanol is then added to yield the desired volume/volume concentration. This addition should be done carefully through a drop-by-drop addition of the reagent and continuous gentle mixing of the tube. Once the desired volume of surface tension lowering agent has been added, the volume of the cell suspension should be made up to the correct final volume through the addition of buffer. For liquid surface tension determination an identical protocol to that described above is carried out except that the cells are omitted. The pH, conductivity, osmolarity, and temperature of each solution should also be determined.

V. Protocol

A fixed volume of cell suspension (at a known concentration) is pipetted into the center of each well. The system is checked for leaks and sealed with parafilm to prevent evaporation.

The cell suspension is then left in place for a fixed length of time at the desired temperature. Thereafter the entire system is immersed in a large dish containing the same liquid mixture used to suspend the cells. While immersed the Teflon mold and silastic gaskets are removed. (Removal of the mold and gaskets while immersed is not a trivial step and should be performed in order to ensure reproducibility of experimental data.) Thereafter the surfaces are air dried and the number of cells adhering per unit surface area of polymer material for each liquid condition is determined.

For this purpose we employ a fully automatic image analysis system specifically programmed to assess (1) the pattern of adhesion, (2) the extent of cell adhesion, and (3) the degree of cell agglomeration. (These programs are available from the author upon request.)

The number of cells adhering per unit area can, of course, also be determined manually through the use of suitable objective magnifications and an ocular lens grid. In either case, manual or automatic, it should be noted however, that only the central two-thirds of the well area should be assessed for the extent of cell adhesion. This is because "wall effects" are considerable (as illustrated in Fig. 34) and vary in extent between polymer surfaces. Thus, for accurate comparisons these effects need to be

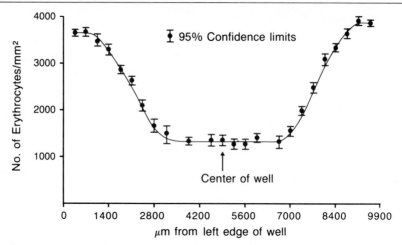

FIG. 34. Demonstration of the extent to which wall effects can influence experimental adhesion data. Substrate is FEP.

avoided. The experimental data are then plotted and analyzed as described in Section 3.2.

6.3. Advancing Solidification Front Technique

The basis of this technique has already been described in Section 3.4.

Apparatus: The problem of producing a planar solidification front and controlling its rate of advance is quite complex. In all cases the solidification front moves horizontally between two thin glass plates. This allows for the use of transmitted light optics and aids in the formation of vertically planar interfaces. Cooling is achieved by circulating either water or methanol from a thermostatted bath through copper tubes at the ends of the cell (cf. Fig. 35). Generally, the rate of advance of the freezing front is controlled by regulating this flow rate. The freezing chamber is mounted on a microscope stage between two specially designed brackets so that it can be moved horizontally in every direction. In addition, the freezing chamber used for the water experiments is placed in a small Plexiglas housing to prevent condensation on the cover glass.

Microscope and circulator/bath: The microscope used is a Leitz Orthoplan with an Orthomat automatic camera. Because of the Plexiglas condensation cover, it is necessary to use an objective lens with a long free working distance. A "heating stage" 32× objective (H 32×) is ideal and is used for all experiments in conjunction with a 1× tube factor and a 10× eyepiece. This gives an overall magnification of 320×. The eyepiece contains a graticule which is used to measure the front velocities. The

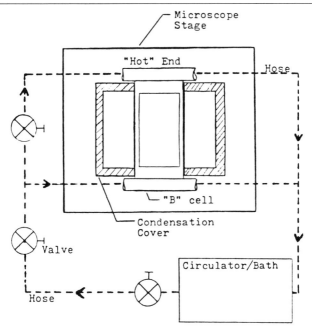

FIG. 35. Schematic of the apparatus used for advancing solidification front experiments with water as the matrix.

graticule is calibrated with a Leitz micrometer scale and found to have a grid division of 0.244 mm. The specimen length (S) represented by each of these small grid divisions is computed using the relation $S = 0.244/M_{obj}/$ TF. Here M_{obj} is the objective magnification and TF is the tube factor. Thus, for the 32× objective with the 1× tube factor the eyepiece graticule has fine divisions representing 7.6 μm.

In order to provide an estimate of the thickness of a given matrix layer, the microscope focusing knob should be calibrated. By alternately focusing on the upper and lower edges of a glass cover slip of known thickness it may be determined that each fine division of the knob represents 1.5 μm of vertical travel.

The thermostatted circulator/bath is a Super Lauda K2/R which has an operating temperature range of −25 to approximately 100°. Generally the bath temperature is fixed during an experiment; however, it is possible to vary this continuously by means of a low-speed dc motor attached to the thermostat. About 2.5 m of well-insulated, 5-mm i.d. plastic tubing connects the bath and the freezing cell on the microscope stage. A schematic of this arrangement is shown in Fig. 35.

The freezing chamber: The freezing cell has been designed expressly for use with water as the matrix. The wide bottom glass and narrow copper slides result in a very straight advancing solidification front. The flow rate through the "hot" end of the tube is regulated by means of an adjustable hose clamp. The "cold" end flow is controlled in one of two ways. Front velocities up to about 120 μm/sec are attainable if the flow is adjusted via the main valve attached to the water bath. Somewhat lower velocities, up to approximately 40 μm/sec, but also a finer control, are obtained if the flow is regulated by means of a needle valve located near the cell.

Freezing front experiments—water matrix: A new sample of water/ cell suspension is prepared for each set of freezing front experiments. The cells are washed three times by centrifugation in 0.15 M saline followed by three further washes in distilled, deionized water which serves as the matrix. The freezing chamber is installed on the microscope stage with the cooling hoses connected to both ends. Care is taken to ensure that both the bottom glass of the cell and the cover slip are clean. Several drops of the cell/water suspension are then put on the bottom glass near the "cold" end and covered with a 22 × 30-mm #2 cover slip. The liquid matrix layer is made very thin ($\simeq$100 μm) by gently depressing the cover slip and soaking up the excess liquid. Freezing is initiated by opening the main line valve and seeding the "cold" end of the cell with ice crystals. The initial velocity of the front is dependent on the flow rates through both ends of the cell. As it is not possible to maintain a constant front velocity for very long, the more general procedure is to let the advancing solidification front gradually decelerate and so make observations over a wide range of velocities. (In each new experiment a fresh volume of cell suspension is used.) This approach necessitates many repetitions of the experiment, since a given velocity is only observed once or twice each time. Videorecording of the front through the microscope as it decelerates greatly increases the information obtained in a single experiment.

To reduce the possibility of cell damage, the freezing front experiments are begun as soon as possible after the cells have been prepared. In order to produce planar ice fronts, it is necessary to remove the cells from HBSS and place them in distilled water. Even small quantities of residual salt serve to render the freezing fronts too dendritic and so the washing procedure needs to be thorough. This is complicated by the fact that lymphocytes and granulocytes are more fragile when placed in distilled water and cannot tolerate long periods of exposure to it. For this reason small fractions of the lymphocytes or granulocytes are washed separately in distilled water and then used for up to 5 min in the experiments, after which time a new fraction is prepared. In this way the possibility of cell damage due to extended exposure to distilled water is minimized. The cells are washed in the following way: About 0.5 ml of cell/saline solution is added to 0.5 ml of distilled water and centrifuged at 200 g for approximately 20 sec. The supernatant is removed and a further 0.5 ml of distilled water added. The cells are gently resuspended and then centrifuged for 15 sec at 200 g. After removal of the supernatant the cells are resuspended in about 0.5 ml of distilled water.

The freezing chamber is mounted in the condensation cover on the microscope stage and connected to the cooling hoses. A single drop of distilled water/cell suspension is placed on the bottom glass near the "cold" end. Again care must be taken to ensure that all glass surfaces are chromic acid cleaned prior to use. The drop is covered with a 22 × 30-mm, #2 cover slip so that roughly 5 mm of the cover slip lies over the copper ledge supporting the bottom glass. The cover slip is then depressed and any excess water removed to produce a thin matrix layer. The cooling flow is turned on, and after seeding the "cold" end of the matrix with ice, the condensation cover is closed and observations begins.

The ice front is then observed and videorecorded as it gradually decelerates. When the matrix is completely frozen, the valves are shut and the experiment repeated with a new drop of the water/cell suspension.

Freezing front data: The total number of observations made at a given velocity varies considerably because of the difficulties encountered trying to establish and maintain a specific rate of freezing front advance. Most of the observations are thus made on fronts which are continuously decelerating.

The pushing distance criteria for the classification of observations as engulfing, transition, and rejection are established through practical considerations and experience. It was found that when using the H32× objective (magnification of 320×) the shortest pushing distance which could be clearly distinguished is one fine graticule division or 8 μm. This then became the limiting distance between engulfing and transition. If the pushing distance is less than or equal to 8 μm engulfment is recorded. A transition observation should reflect the fact that the cell is indeed pushed or rejected by the solidification front, but is prematurely engulfed because of interfering factors (other than viscous drag). Using this definition, the critical velocity should produce transition-type observations. The cell is pushed along for a relatively short distance and is then engulfed because of some small disturbance which upsets the close balance of forces existing at or near the critical velocity. Such disturbances may be generated by friction with the walls of the freezing cell, by interactions with neighboring cells or particles, or by the geometry of the interface. On the other hand, if the pushing velocity is clearly subcritical, the viscous drag acting on the cell is much less than the repulsive van der Waals force and so the cell will not be as easily perturbed by these disturbances. For this reason, a rejection observation should be characterized by significantly greater pushing distances. After considerable experience with this system it was decided that 23 μm (or three graticule divisions) represents a reasonable limiting velocity between transition and rejection. If the cell is pushed at least this distance, the event is called a rejection. In terms of cell dimensions, 23 μm is approximately three times the diameter of an erythrocyte and roughly twice the diameter of a lymphocyte and granulocyte.

It may be noted that there are a significant number of engulfing observations at velocities which are quite low and clearly subcritical. Relative to this, is the wide distribution of transition observations, also extending into the low velocity regions. These apparent discrepancies are attributable to a number of factors which cause premature engulfment. First, as mentioned above, engulfing may be induced by friction with the walls of the freezing cell or by collision with adjacent cells or particles. Second, the geometry of the freezing front seems to play a significant role in cell behavior. If, for example, the front has a wedgelike vertical cross section cells may be forced to either the upper or lower glass surfaces and become trapped. This is often a problem at high rates of solidification and therefore necessitates many repetitions of the experiment.

In contrast to the observations of subcritical engulfing, the use of water as a matrix presents the possibility of supercritical pushing. Typically, an ice front will have a number of notches in it as the water freezes in wide "fingers." If a cell becomes lodged in such a notch, pushing can occur at velocities well above critical. This phenomenon, although not wholly understood, can be explained in terms of the increased area of contact between the cell and the ice which in turn raises the effective repulsive van der Waals force, and in terms of certain fluid effects resulting from the increased gap between the cell and the front.

Also related to the process of engulfing is the hydrodynamic stability of the cell as it is pushed through the matrix. The relevant example here is erythrocytes. These cells are disklike in shape. Since they usually float with the disk-plane horizontal, erythrocytes may be pushed in an intrinsically metastable orientation. Occasionally these cells will flip over and either become engulfed or continue to be pushed with the disk-plane vertical. Clearly, there will be a different critical velocity for each orientation. For this reason all observations of erythrocytes should be made with the cell in the horizontal plane (i.e., in the most stable orientation), so that when viewed from above the ring structure is apparent.

In spite of the scatter in the data (cf. Figs. 16 and 17) one important feature clearly stands out. In all cases, at high velocities 100% of the observations are rejections. This indicates that there must be some point, marked by the critical velocity, at which the mechanism of rejection gives way to that of engulfment. The data are analyzed as discussed in Section 3.4.

Figure 17 illustrates the large critical velocity differences which exist between three types of mammalian cells. This is of particular importance because it is a reflection of the tremendous sensitivity of the freezing front technique. What will be seen to be small differences in surface tension produce large changes in the critical velocity. This sensitivity lends great promise to the technique as a means of measuring the small surface tension differences which are believed to exist between healthy, normal cells, and pathological ones.

Thermophoresis of cells at a freezing front: Thermophoresis is the term used to describe the movement of small particles in a fluid down a temperature gradient. Although it has long been studied in gases, it was not until 1972 that McNab and Meisen[129] looked at the phenomenon in a liquid. During erythrocyte/water freezing front experiments, it has been noted that the cells began to move when the front was still anywhere from 8 to 76 μm away.[47,130] The cell motion is always away from the front and is at speeds as high as 5 to 10 μm/sec, although this varies considerably. Lymphocytes and granulocytes also experience this drifting motion but not to the same extent. It seems likely that these observations are the result of thermophoresis.

McNab and Meisen[129] developed the following expression for the thermophoretic velocity of a small particle in a liquid:

$$V_{th} = -0.26 \left(\frac{K}{2K + K_p} \right) \left(\frac{\mu}{pT} \right) \nabla T \tag{19}$$

Here, K, p, and μ are the liquid thermal conductivity, density, and absolute viscosity; K_p is the thermal conductivity of the particle and T and ∇T are the absolute temperature and temperature gradient present in the liquid at the particle location. Since the thermophoretic velocity is down the temperature gradient it might be expected that such a motion would be toward an advancing solidification front. This is contrary to what is observed; fixed erythrocytes begin to drift away from advancing fronts when they are still up to 80 μm distant. This suggests that the latent heat of fusion released by the solidification front raises the temperature of the liquid a short distance in front of it, thereby creating a local gradient

[129] G. S. McNab and A. Meisen, *J. Colloid Interface Sci.* **44**, 339 (1973).
[130] D. R. Absolom, Z. Policova, E. Moy, W. Zingg, and A. W. Neumann, *Cell Biophysics* **7**, 267 (1985).

which is opposite to the large-scale temperature gradient along the freezing cell.

6.4. Droplet Sedimentation Experiments

The fractionation of a mixture of biological particles or cells, e.g., in centrifugation, zone electrophoresis, or isoelectric focusing, frequently involves the layering of the particulate mixture on a solution of nonconducting gradient-forming molecules, e.g., sucrose or Ficoll, in an electrolyte. The particle suspension forms a zone of finite thickness, with a sharp interface between the sample and the liquid cushion. Under a field applied perpendicular to the layer, e.g., gravitational or electrical, different particle species within the mixture migrate to different levels, thus allowing their separation. However, at particle concentrations above a given minimum, the particles in the starting zone, and/or the particles in different, already-separated sample bands, aggregate together and separate out of the interface in clusters or droplets, which sediment toward the adjacent bands or into the liquid cushion. This leads to interface distortion and excessive band mixing, to the detriment of resolution. The above phenomenon is generally referred to as "streaming" or "droplet sedimentation."

The formation of droplets of a more concentrated particle suspension at the sample zone/liquid cushion interface is attributable to certain factors. The initial concentration of particles in the sample layer is a critical factor in initiating droplet formation. If the particle concentration is sufficiently high, the average distance between adjacent particles is small and the frequency of collision among particles will be high. If there is adhesion or fusion of particles upon collision, a build-up of aggregates of particles enhances droplet formation.

If the density of the aggregates in the sample zone exceeds that of the supporting liquid, the sample zone may become gravitationally unstable. A consequence of this instability is the sedimentation of droplets at the sample zone/liquid cushion interface. Ultimately, the interface becomes highly distorted and band mixing will occur.

Droplet sedimentation also depends on the relative difference between the diffusivities of the particles in the sample zone and the solute molecules in the liquid cushion. If the solute molecules of the liquid cushion diffuse into the sample zone, a local increase in the density of the sample zone close to the interface results. Convection will then set in and may lead to the formation of droplets of particles, which ultimately sediment. However, a slow and uniform diffusion of the particles into the liquid cushion does not create serious problems provided that the sample zone/liquid cushion interface remains sharp, i.e., no droplets of particles are formed. Under this condition, the sample zone may broaden somewhat without causing excessive band mixing.

Apart from achieving optimal separation, in most biological fractionation experiments one aims to handle maximum quantities of sample. It is possible to determine the maximum particle concentration that can remain stable on a given liquid cushion and to increase this particle concentration further by a modification of some of the operating conditions. An important factor that has to be considered is the effect of particle–particle interactions that cannot be avoided, especially when the particles are concentrated. Cell–cell and/or particle–particle interactions can be described in terms of the van der Waals, electrostatic, and hydrodynamic interaction forces. The van der Waals force between particles is always present, and for similar particles in a liquid medium this force is always attractive. It tends to bring close particles together to form aggregates, and therefore enhances the tendency for droplet formation and subsequent sedimentation. A reduction of this force to zero, however, will reduce droplet formation and thus sedimentation. With charged particles of the same sign of charge electrostatic forces play a role in keeping the particles dispersed, since these forces are repulsive. Therefore, the larger the electrostatic repulsive forces, the lower the probability of droplet formation and sedimentation. Hydrodynamic forces may arise due to particle or fluid motion. In such a case, the viscosity of the liquid may become an important parameter, influencing the stability of sample zones.

If we assume that only the attractive van der Waals and the repulsive electrostatic forces are required to determine whether aggregates of particles are formed within the sample zone, then the total potential energy for the system is the sum of the attractive and repulsive energies. If an additive to the liquid medium reduces the van der Waals forces sufficiently, so that electrostatic forces predominate, the particles will remain well dispersed, and formation of aggregates of particles and hence droplet sedimentation will be minimized. If, on the other hand, an additive is used to reduce the repulsive electrostatic forces by charge neutralization, allowing the particles to approach each other to a distance that is sufficiently small, allowing the attractive van der Waals forces to predominate, particle aggregates may be found. In such a case droplet sedimentation will be enhanced.

It is possible to illustrate the effects of van der Waals and electrostatic forces on droplet sedimentation, with glutaraldehyde-fixed human erythrocytes suspended in unbuffered 0.15 M NaCl aqueous solution. The liquid cushion on which the above particulate suspension is layered is D_2O made to ionic strength 0.15 with NaCl. DMSO may be used as an additive to reduce the surface tension of the liquid and thus modulate the attractive van der Waals forces between the particles (cells). Charge reversal is most easily accomplished by the addition of lanthanum nitrate. Initially, the critical concentration N_0 (suspension concentration above which suspension layer instability occurs) of particles for droplet sedimentation is determined experimentally in the absence of these additives. The experiment is then repeated in the presence of known amounts of these additives, and their effects on N_0 will indicate whether droplet sedimentation is enhanced or reduced.

Determination of critical cell concentration for droplet sedimentation: The experimental apparatus illustrated in Fig. 36 consists of a rectangular Perspex chamber (3 × 25 × 120 mm) which is two-thirds filled with the test liquid D_2O. Constant temperature (26°) is maintained in the tank by circulating thermostatted water through an external jacket. The sample (e.g., a suspension of human erythrocytes) is brought to the same temperature as the test liquid in the chamber. After 30 min, within which temperature equilibrium is attained, about 0.025 to 0.18 ml of the sample is slowly layered on the test liquid with a pipet, the tip of which is placed at the inner side of the Perspex chamber, about 3 mm above the meniscus of the test liquid. After the sample has spread on the surface of the test liquid, visual observation is made through a microscope at a magnification of 40×. A lamp placed behind the tank provides illumination for the microscopic observation. Timing commences immediately after the sample is layered, and the time at which interface distortion occurs (incipient droplet formation time) is recorded. If after a reasonable length of time, i.e., 200 to 240 sec, no droplets of particles are formed, the experiment is discontinued. To repeat the experiment, the used test liquid is siphoned

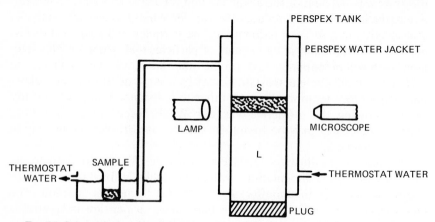

FIG. 36. Schematic drawing of the apparatus used for droplet sedimentation studies. The sample (S) is layered on the test liquid (L).

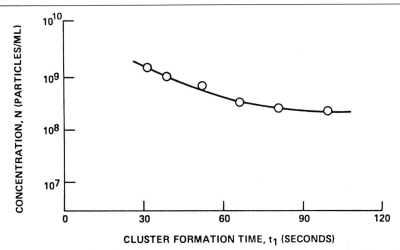

FIG. 37. Droplet formation times of glutaraldehyde-fixed human erythrocytes layered on a heavy-water (D_2O) cushion, as a function of cell concentration. (Reproduced with permission from Academic Press, Ltd.)

out of the chamber, the chamber dried, and refilled with the test liquid. Generally at least five experiments are performed for each given particle concentration, and the droplet formation time determined. The droplet formation times are found to vary by not more than ±4%. From these experiments a series of values of droplet formation times and the corresponding particle concentrations are obtained (Fig. 37). The lower limit of the curve, i.e., where no droplets of particles have been observed, was found to correspond to a particle concentration of 2.2×10^8 cells/ml. Since above this concentration under the prescribed conditions, droplets of particles may form, while below it, no droplets are formed within a reasonable length of time, it is designated the critical particle concentration for droplet formation.

From this figure it is clear that droplet formation and sedimentation strongly depend on particle concentration. The more concentrated the particle suspension, the sooner droplet formation occurs. For very dilute particle suspensions the sample zone remains stable for long periods of time. From Fig. 37 it is also possible to determine the maximum particle concentration that can be supported on a given liquid. In this manner the maximum sample concentration can be determined as a function of liquid surface tension, γ_{LV}, for use in optimized particle fractionation experiments.

Small amounts of DMSO are added to both the suspending physiological saline and the D_2O and for each amount of DMSO added, the critical particle concentration for droplet formation is determined experimentally. The results for several species of erythrocytes are shown in Fig. 26. As the DMSO concentration increases, the critical particle concentration rises sharply until a maximum is attained. This maximum corresponds to a critical particle concentration which is about three times larger than the value at 0% DMSO. This therefore shows very clearly that by adding appropriate amounts of an additive, it is possible to minimize droplet sedimentation and therefore increase the concentration of particles that can be supported on a given liquid. Beyond the maximum, the critical particle concentration decreases, but at a somewhat slower rate, probably because of the increase in density of the solution with the addition of more DMSO.

Erythrocyte suspension: Thus, these results show that at the lowest DMSO concentrations, i.e., at the highest surface tensions γ_{LV}, of the suspending medium (and hence large energy of attraction), suspension layer stability is lowest. As the DMSO concentration is increased and the surface tension of the suspending medium, γ_{LV}, correspondingly decreased, suspension layer stability is increased. At certain intermediate surface tensions where the surface tension of the liquid media equals that of the particulate material, i.e., $\gamma_{LV} = \gamma_{PV}$, maximum stability is achieved, and this condition is depicted by the maximum points in Fig. 26. The DMSO concentrations corresponding to the maximum points on Fig. 26 are determined by computer curve fitting and are seen to vary from 11.3 for turkey cells to 13.6% (v/v) DMSO for human cells. (Alternatively the data may be plotted as a function of the surface tension of the suspending liquids.) The maxima in both cases yield the surface tensions of the erythrocytes, since here $\gamma_{LV} = \gamma_{PV}$.

6.5. Sedimentation Volume Experiments

Sedimentation volume experiments are a well-established technique to study the stability of dispersions of powders in liquids. The behavior of such systems is believed to be governed by van der Waals and electrostatic interactions. It has been shown that there is a strong correlation between van der Waals-type interactions and surface tensions. The van der Waals interaction between two parallel, infinitely extended flat surfaces in a liquid medium was first calculated by Hamaker.[131] For the work done in bringing these surfaces from infinity to a distance d_0 he obtained

$$W = -(A_{132}/12\pi d_0^2) \tag{20}$$

where the coefficient A_{132} has subsequently been called the "Hamaker

[131] H. C. Hamaker, *Physica (Amsterdam)* **4**, 1058 (1937).

coefficient.'' If it is assumed that d_0 is so small that direct contact between the two solid phases occurs, then this work is the thermodynamic free energy of adhesion

$$\Delta F^{\text{adh}} = \gamma_{12} - \gamma_{13} - \gamma_{23} \tag{21}$$

where γ denotes interfacial tensions and the indices 1, 2, and 3 refer, respectively, to solid (1), solid (2), and liquid (3).

It has been shown that the free energy of adhesion can be positive, negative, or zero, implying that van der Waals interactions can be attractive as well as repulsive. While Eq. (20) can, strictly speaking, be expected to hold only for systems which interact by means of dispersion forces only, there are no such restrictions on Eq. (21). Since Eq. (21) describes fundamental patterns of the behavior of particles including macromolecules very well, independent of the type of molecular interactions present, it has been found convenient to define an "effective Hamaker coefficient" which reflects the free energy of adhesion.

While van der Waals interactions between unlike solids in a liquid may be attractive as well as repulsive, it is clear from the underlying thermodynamics that like particles can only attract each other, with zero as the minimum interaction in the limiting case where the surface tension of the suspending liquid, γ_{LV}, is equal to the particle surface tension, γ_{SV}.

It is interesting to speculate on what might happen to the sedimentation volume in the special case when $\gamma_{\text{LV}} = \gamma_{\text{SV}}$. There are at least two possible patterns of behavior, depending on whether or not agglomeration of the particles at the early stages of sedimentation is possible.

1. If there is no agglomeration at finite values of the van der Waals attraction least close packing of the sediment and hence a maximum of V_{sed} is expected. Under these circumstances the electrostatic double layer repulsion between like particles suspended in a polar (aqueous) medium dominate giving rise to a maximum in V_{sed}.

2. If there is agglomeration at finite values of the van der Waals attraction at early stages of sedimentation, then this agglomeration would cease when the van der Waals attraction approaches zero. Since the irregularly shaped aggregates resulting from agglomeration do not pack well, one would expect minimum sedimentation volume at zero van der Waals attraction.

In the case of a biological system, the cells suspended in a liquid may be expected to interact with each other according to the free energy of cohesion

$$\Delta F^{\text{coh}} = -2\gamma_{\text{PL}} \tag{22}$$

where γ_{PL} is the cell–liquid interfacial tension. This equation is simply a reduction of Eq. (21) since here we are dealing with identical particles interacting with each other. The larger the absolute value of the free energy of cohesion, the more pronounced will be the cell–cell interaction. The magnitude of the associated forces is expected to have an effect on aggregation and agglomeration and, as a consequence, on cell sedimentation rate and sedimentation volumes. The magnitude of the forces involved may be modulated by suspending the cells in liquids of various surface tensions, γ_{LV}. In practice the experiments are performed by suspending a known number of cells in a fixed volume of liquid. The liquids used are binary mixtures of buffer plus various increasing volumes of a surface tension lowering additive such as DMSO. Increasing additive concentration results in binary liquid mixtures with decreasing surface tensions (cf. Table VII). Since in these experiments only the final and stable V_{sed} value is measured the increase in liquid viscosity due to increasing concentrations of additives is of no concern and need not be considered (cf. comments relating to this problem in Section 6.4).

Thus this relatively simple concept gives rise to an easy method for determining the surface tension of a homogenous population of cells. For this purpose the following general strategy, described for erythrocytes, is followed:

1. Preparation of liquids: For comments relating to the choice of additives see Section 6.2.

2. Preparation of cells: The erythrocyte suspension is washed by centrifugation four times with 0.15 M NaCl to remove all unbound serum proteins. Thereafter, the cell concentration is adjusted by dilution to approximately 5×10^9 cells/ml following Coulter counter measurement.

3. Experimental strategy: Three hundred microliters of this well-dispersed suspension is then transferred into microcentrifuge polypropylene tubes with a capacity of 1 ml. These tubes are then filled to the top with the binary liquid mixture (see above). The cell suspension is mixed gently by inversion, recentrifuged, and the supernatant removed. This process is repeated three times.

Finally, 300 μl of the binary liquid mixture is added to the tubes. The cells are then *quantitatively* transferred to 1-ml calibrated Wintrobe tubes. These tubes are then filled to the top with the appropriate liquid and sealed with Parafilm to prevent evaporation. The tubes are inverted and the cell suspension mixed well. The cells are then allowed to sediment under the influence of gravity and sedimentation volume, V_{sed}, is determined as a function of time. A stable V_{sed} value generally occurs after approximately 4 hr. For convenience it is standard practice to read the

V_{sed} volumes after 16 hr. The stable V_{sed} value is then plotted as a function of the measured surface tension, γ_{LV}, of the various suspending liquids (cf. Figs. 27–29). For technical details as to how to measure the surface tension of the suspending liquids see Section 6.2.

Acknowledgments

Supported in part by research grants from the Medical Research (MT 5462, MT 8024, MA 9114), the Natural Science and Engineering Research (#UO-493) Councils of Canada, and the Ontario Heart Foundation (#4-12). In addition I should like to acknowledge the financial support of the Ontario Heart Foundation through the receipt of a Senior Research Fellowship.

[3] Basic Methods for the Study of Phagocytosis

By Darryl R. Absolom

1. Introduction

Interest in the measurement of phagocytic function has increased rapidly in recent years, with the recognition that neutrophils play a central role in determining the outcome of most bacterial and some fungal infections. As a consequence numerous assays for the investigation of human neutrophil function *in vitro* have been developed. A great deal of discussion has been devoted to the claims and counterclaims of various proponents of the desirability and advantages of a favored assay. In this chapter an attempt will be made to provide an overview of types of methods available for the study of phagocytosis. The encounter of neutrophils with a microbe is functionally divided into four phases: Attraction (directed and passive chemotaxis), binding of the microbe to the phagocyte membrane (attachment), ingestion (phagocytosis), and bactericidal activity (killing). As both chemotaxis and intracellular digestion have been discussed in detail elsewhere in this volume they will not be treated here. Instead focus will be directed toward the assessment of phagocytic ingestion. No attempt will be made to describe every assay that has been developed. Instead the various types of assays, including examples of those most frequently employed to study defective neutrophil function, will be described.

2. Preparation of Granulocytes and Monocytes from Blood

Depending on the type of phagocytic cell to be investigated, various and very different techniques have been described which can be employed for their isolation and purification. These techniques have been described elsewhere in this series and will not be duplicated here. For the separation of granulocytes and monocytes using density gradient media, see Vol. 108 [9] and this volume [8]; for the collection of peritoneal or alveolar phagocytic cells, see Vol. 108 [25]; for the collection of liver phagocytic cells, see Vol. 108 [27]; for the separation of phagocytic cells, see Vol. 108 [28], [29], [30], [31], and [32].

3. Assessment of Cell Viability

Before any functional studies are performed on the isolated and purified phagocytic cells the viability of the suspension should be established. In satisfactory preparations more than 95% of the cells will be viable. The most popular current viability assessment is by means of dye exclusion studies.

3.1. Dye Exclusion Tests

For this purpose three viability tests have been developed and differ based on the type of dye employed: trypan blue,[1] eosin-Y,[2] and nigrosin. By far the most popular (but not necessarily best) is the trypan blue dye exclusion test. The test procedures which have been developed for assessing viability by these three dyes are described in Protocol 1 together with comments regarding their relative efficacy and usefulness.

3.2. Fluorescein Diacetate Hydrolysis

Fluorescein diacetate is taken up by all cells but it is only hydrolyzed inside *viable* cells to yield a bright green fluorescence.[3] This is the most accurate test for cell viability but requires the use of a fluorescence microscope. The methodology associated with this preferred viability assay is described in detail in Protocol 1.

3.3. Acridine Orange–Ethidium Bromide Fluorescence

A solution (1 ppm) of both acridine orange and ethidium bromide in phosphate buffer can be used to stain viable versus nonviable cells differ-

---

[1] A. M. Pappenheimer, *J. Exp. Med.* **25,** 633 (1917).
[2] J. H. Hanks and J. H. Wallace, *Proc. Soc. Exp. Biol. Med.* **98,** 188 (1958).
[3] B. Rotman and B. W. Papermaster, *Proc. Natl. Acad. Sci. U.S.A.* **55,** 134 (1966).

entially.[4] Live cells take up acridine orange and exclude ethidium bromide, whereas nonviable cells take up ethidium bromide. When viewed through a fluorescent microscope the viable cells exhibit a bright green fluorescence whereas nonviable cells appear orange. The major disadvantage of this very sensitive system is the reported mutagenicity of both acridine orange and ethnidium bromide.[5] Also, acridine orange is known to be highly unstable. The methodology for this viability assay is described in Protocol 1.

4. General Considerations on Phagocytosis

Before describing the methodologies which have been developed to assay various phagocytic functions it may be worthwhile to consider some general principles first.

4.1. Environmental Factors

It is well known that the phagocytic process is influenced by changes in medium such as temperature, pH, osmolarity, and the presence or absence of various cations and anions. This needs to be considered when the suspending buffer media are selected for a particular investigation.

4.1.1. Temperature. A number of reports on the effect of temperature on phagocytic ingestion and bactericidal activity have appeared in the literature.[6–14] While the effect of temperature appears to vary with the particular aspect being investigated, in general optimum phagocytosis occurred within the physiological range. Efficiency was markedly reduced as the temperature was either raised beyond 40° or lowered below 33°.[6,9,12,14]

4.1.2. Ionic Composition. The presence of the divalent cations, Ca^{2+} and Mg^{2+}, is a necessary prerequisite for the process of phagocyto-

[4] D. R. Parks, V. M. Bryan, V. T. Oi, and L. A. Herzenberg, *Proc. Natl. Acad. Sci. U.S.A.* **76,** 1962 (1979).

[5] J. McCann, E. Choi, E. Yamasaki, and B. N. Ames, *Proc. Natl. Acad. Sci. U.S.A.* **72,** 5135 (1975).

[6] D. R. Absolom, C. J. van Oss, L. L. Moore, B. Park, and J. Humbert, *J. Reticuloendothel. Soc.* **27,** 561 (1980).

[7] R. M. Greendyke, R. E. Brierty, and S. N. Swisher, *Blood* **22,** 295 (1963).

[8] F. Allison, M. G. Lancaster, and J. L. Crosthwaite, *Am. J. Pathol.* **43,** 775 (1963).

[9] C. P. Craig and E. Suter, *J. Immunol.* **97,** 287 (1966).

[10] N. A. Cannarozi and S. E. Malawista, *Yale J. Biol. Med.* **46,** 177 (1973).

[11] B. Kvarstein, *Scand. J. Clin. Lab. Invest.* **24,** 271 (1969).

[12] G. L. Mandell, *Infect. Immun.* **12,** 221 (1975).

[13] P. K. Peterson, J. Verhoef, and P. G. Quie, *Infect. Immun.* **15,** 175 (1977).

[14] K. Staehr Johansen, E. M. Berger, and J. E. Repine, *Acta Pathol. Microbiol. Immunol. Scand.* **C91,** 355 (1983).

sis.[7,8,11,15-21] Maximum ingestion occurs at approximately 1–2 mM Ca^{2+} and Mg^{2+},[16] while higher concentrations tend to inhibit the process.[16,19] The role of cations in phagocytosis is complex and they may affect various stages in the following mechanisms.

1. Attachment: As discussed elsewhere in this volume ([13]), one of the major consequences of opsonization, and hence a major influence on phagocytic ingestion, is the activation of the complement system via either the classical or alternative pathways. Both pathways require the presence of divalent cations—Ca^{2+} being required for the classical pathway and Mg^{2+} for the alternative pathway.

2. Ingestion: Divalent cations are also required for the ingestion phase of phagocytosis. This was elegantly demonstrated by the requirement for presence of such ions in the phagocytic uptake of nonopsonized particles by neutrophils.[15,16,19] In studies related to the rate of phagocytic ingestion, Stossel reported that uptake of nonopsonized paraffin oil droplets (coated with albumin) was stimulated by various divalent cations.[19] In order of increasing effectiveness in promoting ingestion the cations were ranked as follows: Mn^{2+} > Co^{2+} > Mg^{2+} > Ca^{2+}. However, when opsonized particles were employed Mg^{2+} and Ca^{2+} were found to be more effective than either Mn^{2+} or Co^{2+}. In addition it was noted that lower concentrations of Ca^{2+} and Mg^{2+} were required in order to result in maximum ingestion rates when the particles were opsonized in comparison to nonopsonized particles.[16,19] The ingestion of polystyrene particles in the presence of serum (and therefore opsonized latex) by sheep neutrophils was found to be inhibited by Ba^{2+} concentrations ranging from 10^{-3} to 10^{-1} M.[16]

The mechanism by which Ca^{2+} and Mg^{2+} facilitate enhanced ingestion is not known. It has been suggested that they may play a role in facilitating the attachment of particle-associated opsonins to the corresponding receptor sites on the phagocytic cell membrane.[22,23] The necessity for the presence of cations to facilitate the attachment of IgG-opsonized particles has not been demonstrated.[24,25] However, C3b-coated particles definitely

[15] J. G. Hirsch and B. Strauss, *J. Immunol.* **92**, 145 (1964).
[16] D. J. Wilkins and A. D. Bangham, *J. Reticuloendothel. Soc.* **1**, 233 (1964).
[17] R. E. Bryant, *Yale J. Biol. Med.* **41**, 303 (1969).
[18] B. Kvarstein, *Scand. J. Clin. Lab. Invest.* **25**, 337 (1970).
[19] T. P. Stossel, *J. Cell. Biol.* **58**, 346 (1973).
[20] I. Talstad, *Scand. J. Haematol.* **9**, 516 (1972).
[21] R. I. Handin and T. P. Stossel, *N. Engl. J. Med.* **290**, 989 (1974).
[22] W. H. Lay and V. Nussenweig, *J. Immunol.* **102**, 1172 (1969).
[23] K. L. Banks and T. C. McGuire, *Immunology* **28**, 581 (1975).
[24] D. Roos, M. De Boer, and R. S. Weening, *in* "Movement, Metabolism and Bactericidal Mechanisms of Human Neutrophils," p. 223. Piccin Medical Books, Padua, Italy, 1977.
[25] D. R. Absolom, (submitted).

require the presence of Mg^{2+} to facilitate attachment and thus subsequent ingestion.[24]

Stossel has shown that the presence of monovalent cations is not a prerequisite for ingestion.[19,26] The ingestion rate of opsonized albumin-coated paraffin oil droplets is completely unaffected by either the absence or presence of K^+ in the extracellular fluid or by the addition of ouabain at concentrations that inhibit Na^+ and K^+ membrane transport.[19]

Thus, since many of the cell preparation protocols recommend the use of buffers which lack any divalent cations (in order to prevent cell clumping), the investigator should bear in mind the particular requirement of Ca^{2+} and Mg^{2+} in the phagocytic process. Optimum concentrations for phagocytosis are in the vicinity of 1.0 mM for Ca^{2+} and 1.5 mM for Mg^{2+}.[19] Cation complexing and precipitation should be borne in mind if other constituents in the medium are not compatible with these ions. For example, in an investigation of fluoride activation of the respiratory burst in human neutrophils it was necessary to lower the calcium concentration in order to prevent precipitation of the CaF_2 complex.[27]

4.1.3. Osmolarity. A related aspect of the aforementioned is the overall concentration of salts and sugars in the medium and their contribution to the final tonicity of the suspension.[28] High concentrations of salts[29,30] and sugars[31,32] have been shown to markedly inhibit phagocytosis. In a systematic study of the effect of osmolarity on the ingestion of starch particles by human neutrophils[31] it was shown that, in the absence of serum, the extent of phagocytic ingestion was not affected by variations in osmolarity from 205 to 348 mOsm/liter. Above 400 mOsm/liter, however, phagocytosis was markedly reduced. The mechanism by which hyperosmolarity affects phagocytosis is not completely understood but is thought to be related to an inhibition in pseudopod formation and shape changes in the phagocyte.[32] Under high osmolarity conditions the cells are likely to shrink[28] and to round up, thereby reducing the attachment phase in the process.[32] This effect *in vivo* is most likely to be noted in the urinary tract or pertinent areas of the kidney, where solute concentrations (e.g., of urea, sodium, and glucose in diabetics) are likely to produce hyperosmolarity conditions, thereby inhibiting phagocytosis.[30]

[26] T. P. Stossel, C. A. Alper, and F. S. Rosen, *J. Exp. Med.* **137,** 690 (1973).
[27] J. T. Curnutte, B. M. Babior, and M. L. Karnovsky, *J. Clin. Invest.* **63,** 637 (1979).
[28] A. J. Sbarra, W. Shirley, and J. S. Baumstark, *J. Bacteriol.* **85,** 306 (1963).
[29] A. J. Sbarra and W. Shirley, *J. Bacteriol.* **86,** 259 (1963).
[30] I. Chernew and I. Braude, *J. Clin. Invest.* **41,** 1945 (1962).
[31] T. D. Brogan, *Immunology* **10,** 137 (1966).
[32] C. J. van Oss, *Infect. Immun.* **4,** 54 (1971).

4.1.4. pH. The reported pH optimum for phagocytosis has varied considerably[8,33–35] and appears to be related to the particular phagocytic function being assessed. In general, however, a pH of 7.2 is favored. In pH conditions below pH 6.2, as is found in urine, phagocytosis is markedly reduced.[30]

4.2. Other Considerations

A distinction should be made between those particles that have actually been internalized and those which are simply attached to the surface of the phagocyte (see also this volume [6]). Thus, appropriate controls must be performed to establish that what is being observed is indeed true uptake and not merely accidental association. Such controls should include measurements of particle uptake in unincubated reaction mixtures (zero-time controls) as well as uptake measurements of reactions incubated at 4° (also see Section 5.1.1 and Protocol 3).

5. Measurement of Phagocytic Functions

A variety of methods have been developed for the measurement of phagocytosis. These techniques can be divided into two broad categories: (1) those which assess phagocytic ingestion per se; and (2) those assays which measure bactericidal activity of neutrophils. In the former the various techniques can themselves be grouped into two general classes: (1) The "microscopic category," in which the extent of particle uptake by individual cells is determined directly, and (2) the "macroscopic (or bulk) category," in which measurements of a certain parameter (e.g., oxygen consumption) are made on samples containing large numbers of cells. Such techniques can only indirectly measure the extent of phagocytic ingestion. Both classes of methods provide extremely useful information and enhanced insight into the phagocytic process. Examples of both types of methods will be described (see also Section 6).

5.1. Direct Assessment

In these techniques particles are added to a suspension or monolayer of cells and after incubation under a prescribed set of experimental conditions, the extent of phagocytosis is evaluated. Such assays directly measure the phagocytic process and as such are the easiest to correlate with

[33] W. O. Fenn, *J. Gen. Physiol.* **5**, 169 (1923).
[34] S. Mudd, M. McCutcheon, and B. Lucké, *Physiol. Rev.* **14**, 210 (1934).
[35] A. Beck, S. Bergner-Rabinowitz, and I. Ofek, *J. Bacteriol.* **100**, 1204 (1969).

neutrophil *in vivo* activities. Although conceptually simple the direct counting systems suffer from the following disadvantages: (1) It is often extremely difficult to separate the contribution to the measurement of membrane-bound (i.e., not ingested) from internalized particles.[36] (Recent advances have been made which make this a now less severe restriction; see below.) (2) They often fail to achieve reproducibility and therefore require the use of several suitable control experiments.

Four basic methods of counting have been developed for the purpose of quantitating phagocytic ingestion: (1) Microscopic methods: Direct visualization and quantitation of the ingested particles. (2) Isotopic labeling methods: Total radioactivity associated with a phagocytic cell suspension, following removal of the labeled particle, is determined. (3) Extraction methods: Extraction and solubilization of particles followed by quantitation, generally by spectrophotometric means. (4) Difference methods: The difference between the number of particles present in the bulk liquid pre- and postphagocytosis is measured.

5.1.1. Microscopic Methods. Following the addition of a predetermined number of particles (or bacteria) to a standard number of phagocytic cells (either in suspension or an adherent monolayer) for a defined length of time under fixed conditions of temperature, pH, etc., a smear of the cells is formed (if necessary) and the extent of ingestion is microscopically evaluated. Generally two types of data are collected: the mean number of particles ingested per cell (phagocytic index)[37–39] and the percentage of cells out of the total number viewed which have actually ingested particles (phagocytic activity index).[39,40] The technique has been applied to a number of different types of particles including bacteria, yeast,[41,42] India ink,[43] and rice starch.[44]

A number of difficulties exist with this kind of method:

1. As mentioned earlier it is often difficult to distinguish between an organism which has actually been internalized and one which is simply attached to the cell membrane but not ingested.[36] This problem is not

[36] T. P. Stossel, *Blood* **42**, 121 (1973).
[37] W. B. Leishman, *Br. Med. J.* **1**, 73 (1902).
[38] A. E. Wright and S. R. Douglas, *Proc. R. Soc. (London)* **72**, 357 (1903).
[39] C. J. van Oss and M. W. Stinson, *J. Reticuloendothel. Soc.* **8**, 397 (1970).
[40] H. J. Hamburger, *in* "Handbuch der Biologischen Arbeitsmethoden" (H. Abderhalden, ed.), Abt. IV, p. 953. Urban and Schwarzenberg, Berlin, 1927.
[41] R. I. Lehrer and M. J. Cline, *J. Bacteriol.* **98**, 996 (1969).
[42] M. E. Miller, *J. Pediatr.* **74**, 255 (1969).
[43] L. Gluck and W. A. Silverman, *Pediatrics* **20**, 951 (1957).
[44] Y. Matoth, *Pediatrics* **9**, 748 (1952).

peculiar to microscopic techniques but is also associated with isotopic labeling and extraction procedures. In well-stained smears, it is often possible to observe an area of degranulation surrounding the ingested particles, thereby enabling differentiation between adherent and ingested particles. Such differentiation is also made possible if the adherent particles can be selectively lysed without damage to the granulocyte or the ingested particles. For example, lysostaphin can be employed to enzymatically digest and thereby remove adherent staphylococci,[45-47] and by hemolysis in low ionic strength media in which only adhering, but not ingested, erythrocytes are lysed.[48,49] The extracellular, noningested particles can also be removed by treatment with various chemical agents.[50]

Recently, a more general method has been developed for distinguishing adherent from ingested bacteria.[51,52] This is based on the quenching of extracellular fluorescein-conjugated microorganisms by crystal violet. The ingested fluorescein-conjugated microorganisms are not quenched by crystal violet, which does penetrate the cell membrane of viable neutrophils. This technique is described in detail elsewhere in this volume [6].

Electron microscopy has also been used for the purpose of quantitating ingestion. It has the advantage in that complete internalization of the particle can be confirmed. In cases where this must be established beyond any doubt electron micrographs of serial sections must be examined to ensure complete enclosure within a phagocytic vacuole. However, it is not practical to routinely employ electron microscopy to quantitate the extent of phagocytosis.

2. With microscopic techniques phagocytosis is generally measured at a single time point. If that point is beyond the period of active phagocytosis then the measurement will reflect phagocytic capacity rather than the rate of ingestion. Thus phagocytic cells with an abnormally low ingestion *rate* may, after prolonged incubation period, have ingested the same number of particles as normal cells and thus an obvious defect (ingestion rate) will be missed by the investigator. It is for this reason that several re-

[45] D. E. Gardner, J. A. Graham, F. J. Miller, J. W. Illing, and D. L. Coffin, *Appl. Microbiol.* **25**, 471 (1973).

[46] M. L. J. van Schaik, R. S. Weening, A. Voetman, and D. Roos, *Annu. Rep. Karl Landsteiner Found.* Cntrl. Lab. Netherlands Red Cross Transfusion Service, p. 59, 1975.

[47] J. S. Tan, C. Watanakunakorn, and J. P. Phair, *J. Lab. Clin. Med.* **78**, 316 (1971).

[48] I. Gigli and R. A. Nelson, *Exp. Cell Res.* **51**, 45 (1968).

[49] A. Altman and T. P. Stossel, *Br. J. Haematol.* **27**, 241 (1974).

[50] P. Hersey, *Transplantation* **15**, 282 (1973).

[51] J. Hed, *FEMS Lett.* **1**, 357 (1977).

[52] J. Hed, in "Studies on the Phagocytosis by Human Polymorphonuclear Leukocytes Using a New Assay Which Allows Distinction between Attachment and Ingestion." M.D. Dissertation, Linköping University, Medical Microfilms #74, Linköping, Sweden, 1979.

searchers have emphasized the importance of taking readings at a number of time intervals following the period of initial contact between the cells and particles.[53,54] The same precautions of course also apply to other methods of assessment.

3. The phagocytic index and activity values obtained in the microscopic assays of phagocytosis vary with the type, number, and ratio of bacteria to phagocytic cells as well as the experimental conditions prevailing at the time of incubation. As different laboratories employ different conditions it is often difficult to compare the data obtained from two different laboratories.

Nevertheless, despite these apparent difficulties the direct microscopic technique using either suspensions or monolayers of the phagocytic cells[55–57] has provided a wealth of extremely useful information with regard to phagocytic processes and dysfunctions. Such information is particularly valuable when it is obtained in conjunction with other aspects of phagocytic ingestion, e.g., O_2 consumption.[6]

The number of different protocols which have been established is large. Since there is generally considerable overlap in the methods only two will be described in detail. The first allows the cells and particles to encounter one another in suspension[57a] (Protocol 3A). The second employs the monolayer technique,[55,57] in which the phagocytic cells are adherent to a siliconized glass surface (Protocol 3B). The monolayer technique requires less blood than the suspension method. The two techniques yield essentially the same quantitative results and there is little to choose between them except for personal preference.

5.1.2. Isotopic Labeling Methods. These methods involve the phagocytic ingestion of radiolabeled particles or macromolecules and the subsequent quantitation of radioactivity associated with the cells. Clearly this technique suffers from the same disadvantages as those described above in Section 5.1.1, in the sense that it is not possible to distinguish between surface-bound and ingested particles. Two variations of the method have been described. The first simply involves quantitation of the extent of phagocytic ingestion of radiolabeled particles, bacteria, or immune com-

[53] R. Michell, S. J. Pancake, J. Noseworthy, and M. L. Karnovsky, *J. Cell. Biol.* **40,** 216 (1969).

[54] T. P. Stossel, R. J. Mason, J. Hartwig, and M. Vaughan, *J. Clin. Invest.* **51,** 615 (1972).

[55] J. Newsome, *Nature (London)* **214,** 1092 (1967).

[56] H. Wehinger and M. Hofacker, *Eur. J. Pediatr.* **123,** 125 (1976).

[57] C. J. van Oss and C. F. Gillman, *J. Reticuloendothel. Soc.* **12,** 283 (1972).

[57a] R. C. Jandl, J. Andre-Schwartz, L. Borges-Dubois, R. S. Kipnes, B. J. McMurrich, and B. M. Babior, *J. Clin. Invest.* **61,** 1176 (1978).

plexes. The second employs the quantitation of the extent of uptake of radiolabeled macromolecules (e.g., albumin) accompanying the process of particle ingestion.

5.1.2.1. Radiolabeled particles: A wide variety of particles have been used, including bacteria and yeast,[58-64] platelets,[21] polystyrene latex spheres,[25,65] immune complexes,[66,67] and fungi[68] and starch particles.[53,69] In addition several different isotopes (^{3}H, ^{14}C, ^{32}P, ^{45}Ca, ^{51}Cr, ^{99}Tc) have been employed as the radioactive label. In the case of living organisms labeling was achieved by growing the cells either in or on media containing radioactive metabolites. The major advantage of this technique is that the tediousness associated with microscopic counting is avoided. One such experimental protocol for investigating the extent of phagocytic uptake of radiolabeled bacteria is described in Protocol 4.

5.1.2.2. Radiolabeled liquid medium: This technique is based on the observation that during phagocytic ingestion of a particle, a small volume of extracellular fluid is also taken up along with the particle.[70-72] Since the amount of extracellular fluid taken up in this way is small, the specific activity of radiolabeled macromolecules present in the fluid must be high. As one example, the use of ^{125}I-labeled human serum albumin has been employed to quantitate the extent of starch granule uptake by rabbit neutrophils.[73] Following appropriate incubation conditions the suspension is centrifuged and washed (to remove nonadherent and noningested ^{125}I-labeled albumin) and the radioactivity associated with the cellular elements determined. [It is also possible to measure the disappearance (through ingestion) of radiolabeled protein from the bulk liquid and thereby establish the extent of radioactive uptake by the phagocytic cells.[73]] A similar technique has been employed for use with human neu-

[58] W. Brzuchowska, *Nature (London)* **212,** 210 (1966).
[59] R. J. Downey and B. F. Diedrich, *Exp. Cell Res.* **50,** 483 (1968).
[60] T. Midtvedt and A. Trippestad, *Acta Pathol. Microbiol. Scand., Sect. B* **78,** 1 (1970).
[61] D. S. Miller and S. Beck, *J. Lab. Clin. Med.* **86,** 344 (1975).
[62] J. B. Suzuki, R. R. Booth, and N. Grecz, *J. Infect. Dis.* **123,** 93 (1971).
[63] J. Verhoef, P. K. Peterson, and P. G. Quie, *J. Immunol. Methods* **14,** 303 (1977).
[64] G. Dominque, K. Lloyd, and J. V. Schlegel, *Proc. Soc. Exp. Biol. Med.* **146,** 635 (1974).
[65] M. S. Al-Ibrahim, R. Chandra, R. Kishora, F. T. Valentine, and H. S. Lawrence, *J. Immunol. Methods* **10,** 207 (1976).
[66] P. A. Ward and N. J. Zvaifler, *J. Immunol.* **111,** 1771 (1973).
[67] R. Hällgren and G. Stalenheim, *Immunology* **30,** 755 (1976).
[68] E. Nordenfeld, *Acta Pathol. Microbiol. Scand. Sect. B,* **78,** 247 (1970).
[69] R. Baboolal and R. N. Powell, *Arch. Oral Biol.* **17,** 249 (1972).
[70] M. L. Karnovsky and A. J. Sbarra, *Am. J. Clin. Nutr.* **8,** 147 (1960).
[71] A. J. Sbarra, W. Shirley, and W. A. Bardawil, *Nature (London)* **194,** 255 (1962).
[72] R. R. Berger and M. L. Karnovsky, *Fed. Proc.* **25,** 840 (1966).
[73] Y. H. Chang, *Exp. Cell. Res.* **54,** 42 (1969).

trophils.[10] In general, however, the use of iodinated compounds should, in this type of technique, be avoided if possible, since neutrophils are known to be capable of catalyzing iodination, trans- and deiodination reactions,[74–79] in some cases extracellularly,[78,79] during the process of phagocytosis (also see Section 5.2.3.5).

5.1.3. Extraction Methods. In this group of methods particles which can be solubilized by various organic solvents are employed. When solubilized the particles have a distinctive adsorption coefficient and hence the concentration of the dissolved particles in the solvent can be readily determined. Both extraction of an easily dissolved solid particle (e.g., polystyrene latex) and of an easily stained lipid (oil red) have been employed.

5.1.3.1. Latex particles: The extent of ingestion of either polystyrene (PS)[79a] or polyvinyltoluene (PVT) latex spheres can be determined following washing of the phagocytic cell suspension (to remove adherent but not ingested particles) and subsequent extraction of PS or PVT through solubilization with dioxane.[11,80–83] The concentration of either component in the solvent can be measured by spectrophotometric means at the wavelength corresponding to their adsorption maxima. Through the use of standard concentration curves, or from the use of known extinction coefficients, it is possible to quantitate this method.

5.1.3.2. Oil droplets (oil red O): The oil red O method has been presented elsewhere in this volume [5] and will not be duplicated here. See Section 5.1.3.3. for a brief discussion of the significance of oil red O ingestion.

[74] S. J. Klebanoff, *J. Exp. Med.* **126**, 1063 (1967).
[75] J. Hakim, E. Cramer, P. Boivon, H. Troube, and J. Boucherot, *Eur. J. Clin. Invest.* **5**, 215 (1975).
[76] S. H. Pincus and S. J. Klebanoff, *N. Engl. J. Med.* **284**, 744 (1971).
[77] I. Olsson, T. Olofsson, and H. Odeberg, *Scand. J. Haematol.* **9**, 483 (1972).
[78] H. Odeberg, T. Olofsson, and I. Olsson, *Scand. J. Haematol.* **12**, 155 (1974).
[79] S. J. Klebanoff and C. B. Hamon, *J. Reticuloendothel. Soc.* **12**, 170 (1972).
[79a] Abbreviations: PS, polystyrene; PVT, polyvinyltoluene; NBT, nitroblue tetrazolium; PMA, phorbol myristate; MPO, myeloperoxidase; HMP, hexose monophosphate; 6-PGD, 6-phosphogluconate dehydrogenase; G-6-PD, glucose-6-phosphate dehydrogenase; 6PG, 6-phosphogluconate; CGD, chronic granulomatous disease; PAS, periodic acid–Schiff; PBS, phosphate-buffered saline; FCS, fetal calf serum; BSS, balanced salt solution; AO, acridine orange; EB, ethidium bromide; KRP, Krebs–Ringer phosphate; EDTA, ethylenediaminetetraacetic acid; HBSS, Hanks' balanced salt solution; HMDS, hexamethyldichlorosilane; MEM, minimum essential medium; LDH, lactate dehydrogenase; BSA, bovine serum albumin; LF, lactoferrin; CFU, colony-forming units.
[80] J. Roberts and J. H. Quastel, *Biochem. J.* **89**, 150 (1963).
[81] M.-F. Tsan and R. D. Berlin, *J. Exp. Med.* **134**, 1016 (1971).
[82] R. A. Weizman and E. D. Korn, *Biochemistry* **6**, 485 (1967).
[83] P. S. Sastry and L. E. Hokin, *J. Biol. Chem.* **241**, 3354 (1966).

5.1.3.3. Simultaneous oil red O–NBT reduction: Since frequently in the clinical evaluation of neutrophil samples nitroblue tetrazolium (NBT) is employed *simultaneously* with the oil red O test,[84] it has been included here for convenience. Oil red O content is indicative of phagocytic rate and/or capacity whereas NBT reduction is an *indirect* measure of oxygen metabolism reflecting the efficiency of the phagocyte suspension for "killing" bacteria; i.e., NBT reduction is an indicator of metabolic defects. The use of NBT reduction as an individual separate test, and the various precautions which should be kept in mind when using this assay, are discussed in more detail in Section 5.2.

The simultaneous oil red O–NBT reduction test is given in Protocol 6.

5.1.4. Removal of Test Particles from the Bulk Fluid Surrounding Phagocytic Cells. Most if not all assays of this type suffer from the drawback that it is extremely difficult to distinguish between bacteria attached (and their effect) to the phagocyte membrane from those (and their effect) which have actually been ingested. The most common approach in this type of protocol is to determine the difference between the number of particles (or bacteria) initially added to the system and the number recovered in the supernatant fluid following centrifugation of the phagocytic cells. The technique has been employed with a variety of test particles including bacteria,[85] C3b-sensitized sheep[48] or human erythrocytes,[86] and polystyrene particles.[80] In an interesting variation of this protocol, the decrease in uptake of radiolabeled metabolite has been employed to indirectly measure the removal of microorganisms from the extracellular fluid (and hence the extent of phagocytic ingestion).[87,88] Normally, the ingested organisms will no longer be able to utilize the metabolite and hence a quantitative decrease in the rate of metabolite uptake will be recorded. A metabolite which is *not* readily taken up by intracellular microbes or neutrophils must be employed.

5.2. Indirect Assessment of Ingestion

Phagocytic ingestion of particles by neutrophils is followed by a complex sequence of intracellular events. The quantitation of these events can be used as an indirect indicator of the extent and efficiency of the phagocytic process. These measurements themselves may either be direct, as for example, changes in oxygen consumption or indirect, as, for example,

[84] T. P. Stossel, R. K. Root, and M. Vaughan, *N. Engl. J. Med.* **286,** 120 (1972).
[85] Z. A. Cohn and S. I. Morse, *J. Exp. Med.* **110,** 419 (1959).
[86] M. E. Miller and U. R. Nilsson, *Clin. Immunol. Immunopathol.* **2,** 246 (1974).
[87] N. Foroozanfar, Z. Aghai, F. Ala, and J. R. Hobbs, *J. Immunol. Methods* **11,** 345 (1976).
[88] M. Yamamura, J. Boler, and H. Valdimarsson, *J. Immunol. Methods* **14,** 19 (1977).

measurement of metabolic change reflected by the reduction of nitroblue tetrazolium dye. The events include respiratory burst activity (e.g., oxygen consumption,[18,89–91] superoxide anion production,[91,92] glucose C-1 oxidation,[89,90,93,94] degranulation and enzyme release,[95,96] iodination,[97] chemiluminescence,[98–100] bactericidal,[15,100–102] Candicidal activity,[41,103] and NBT reduction.[21,104]

These indirect assessments of phagocytic activity provide valuable basic as well as clinical information. The following points should, however, be borne in mind:

1. Since these techniques quantitate phagocytic as well as postphagocytic events, they are a valid measure of phagocytosis *only* when the postphagocytic events are unimpaired. For example, as illustrated so elegantly by Klebanoff and Clark,[97] iodination of albumin is dependent not only on phagocytosis, but also on degranulation, myeloperoxidase activity, and the metabolic burst. Abnormalities in any of the latter may lead to decreased iodination without any decrease in phagocytic ingestion.

2. Any perturbation of the phagocyte membrane may affect metabolic activity and hence caution and appropriate controls should be included in studies employing these indirect tests. Such membrane perturbations, resulting in nonphagocytic ingestion respiratory burst activity, have been reported to occur as a result of agents as diverse as anti-leukocyte antibodies,[105] concanavalin A,[106] cytochalasin E,[107,108] the ionophore

[89] A. J. Sbarra and M. L. Karnovsky, *J. Biol. Chem.* **234**, 1355 (1959).
[90] H. Ohta, *Acta Haematol. (Japan)* **27**, 544 (1964).
[91] T. Mizuochi, Y. Nichimura, A. Sakai, O. Takenaka, and Y. Inada, *FEBS Lett.* **51**, 174 (1975).
[92] J. T. Curnutte, D. M. Whitten, and B. M. Babior, *N. Engl. J. Med.* **290**, 593 (1974).
[93] R. T. Skeel, R. A. Yankee, W. A. Spivak, L. Novikovs, and E. S. Henderson, *J. Lab. Clin. Med.* **73**, 327 (1969).
[94] I. D. Mickenberg, R. K. Root, and S. M. Wolff, *J. Clin. Invest.* **49**, 1528 (1970).
[95] G. Weissman and P. Dukor, *Adv. Immunol.* **12**, 283 (1970).
[96] I. M. Goldstein, *Prog. Allergy* **20**, 301 (1976).
[97] S. J. Klebanoff and R. A. Clark, *J. Lab. Clin. Med.* **89**, 675 (1977).
[98] V. E. Hemming, R. T. Hall, P. G. Rhodes, A. O. Shigeoka, and H. R. Hill, *J. Clin. Invest.* **58**, 1379 (1976).
[99] J. V. Grebner, E. L. Mills, B. H. Gray, and P. G. Quie, *J. Lab. Clin. Med.* **89**, 153 (1977).
[100] R. C. Allen, *Infect. Immun.* **15**, 828 (1977).
[101] J. W. Smith, J. A. Barnet, R. P. May, and J. P. Sanford, *J. Immunol.* **98**, 336 (1967).
[102] T. Laxdal, R. P. Messner, R. C. Williams, and P. G. Quie, *J. Lab. Clin. Med.* **71**, 638 (1968).
[103] R. I. Lehrer, *Infect. Immun.* **2**, 42 (1970).
[104] R. B. Johnston, M. R. Kemperer, C. A. Alper, and F. S. Rosen, *J. Exp. Med.* **129**, 1275 (1969).
[105] F. Rossi, M. Zatti, P. Patriarca, and R. Cramer, *J. Reticuloendothel. Soc.* **9**, 67 (1971).
[106] D. Romeo, G. Zabucchi, and F. Rossi, *Nature (London)* **243**, 111 (1973).

A23187,[109] as well as a number of agents such as surfactants, fatty acids, endotoxin, and phospholipase C.[110] (These metabolic changes are insensitive to rotenone, antimycin A, and cyanide. They also show an increase in KCN-insensitive NADPH oxidation.[105])

5.2.1. Glycolysis. Shortly after particles are added to a suspension of phagocytes, there is a marked increase in glucose consumption.[111-113] Metabolic inhibitors of anaerobic glycolysis, e.g., fluoride, 2-deoxyglucose, and iodoacetate, markedly decrease the uptake of particles by neutrophils, whereas inhibitors of oxidative phosphorylation, such as cyanide, antimycin, and dinitrophenol, have no such effect.[114-116]

In the resting state, normal neutrophils consume comparatively low volumes of oxygen. The primary energy source is through glycolysis even in the presence of unlimited concentrations of oxygen. The breakdown of glucose results in the formation of lactic acid and ATP. The lactic acid is released by the cell into the extracellular medium. Thus a relatively simple comparison can be made between neutrophil populations by assessing lactate production. The assessment can be performed in two ways: (1) the same neutrophil suspension can be monitored for lactate production before and after stimulation with a known concentration of particles; (2) a normal vs clinical population of neutrophils at the same cell concentration can be compared for lactate production when the same number of a single type of particles are added to the suspensions. Glycolytic activity is evaluated by measuring lactate excretion as described in Protocol 7.

5.2.2. Degranulation. Following attachment of a particle, either nonspecifically through adsorption or specifically through receptors on the cell membrane, the phagocyte forms pseudopods which surround and encase the particle, forming a vesicle—the so-called "phagosome"— lined with a portion of the cell membrane. The vesicles then break away

[107] A. Nakagawara, K. Takeshige, and S. Minakami, *Exp. Cell Res.* **87,** 392 (1974).
[108] A. Nakagawara, K. Takeshige, and S. Minakami, *J. Biochem. (Tokyo)* **77,** 567 (1975).
[109] E. Schell-Frederick, *FEBS Lett.* **48,** 37 (1974).
[110] F. Rossi, P. L. Patriarca, and D. Romeo, in "Future Trends in Inflammation" (G. P. Velo, D. A. Willoughby, and J. P. Giroud, eds.), p. 103. Piccin Medical Books, Rome, 1975.
[111] H. Stahelin, E. Suter, and M. L. Karnovsky, *J. Exp. Med.* **104,** 121 (1956).
[112] H. Stahelin, M. L. Karnovsky, and E. Suter, *J. Exp. Med.* **104,** 137 (1956).
[113] H. Stahelin, M. L. Karnovsky, A. E. Farnham, and E. Suter, *J. Exp. Med.* **105,** 265 (1957).
[114] R. J. Selvaraj and A. J. Sbarra, *Nature (London)* **211,** 1272 (1966).
[115] R. J. McRipley and A. J. Sbarra, *J. Bacteriol.* **94,** 1417 (1967).
[116] R. J. McRipley and A. J. Sbarra, *J. Bacteriol.* **94,** 1425 (1967).

from the cell membrane and move toward the interior of the cell. Simultaneously with this process, the phagocyte granules, containing the lytic enzymes, undergo violent movement adjacent to the phagosome followed by degranulation. Degranulation[95,96,117,118] occurs in tandem with ingestion, and the two processes cease simultaneously.[119,120] Degranulation results in the release of granules contents either into a phagosome or into the extracellular medium—referred to as intracellular and exocytic degranulation, respectively. Assessment of intracellular degranulation requires the isolation of the phagosomes and the quantitation of their enzyme content as described in Protocol 8A. In order to stimulate exocytic degranulation, the phagocytes are incubated with particles in the presence of an inhibitor of ingestion (e.g., cytochalasin B) so that granule contents are released into the extracellular fluid only. The experimental protocol for measuring exocytic degranulation is described in Protocol 8B.

Intracellular degranulation is calculated by determining the total concentration of enzyme in the phagosomes. Exocytic degranulation is determined by subtracting the amount of enzyme released in the presence of zymosan or phorbol myristate (PMA) from that released in their absence. The percentage degranulation is determined from this figure and the content of the granule constituents in whole cells. Typical values for exocytic degranulation are 10–20% with zymosan and 30–40% with PMA.

The neutrophil contains two types of granules: the azurophil (primary) granules which are peroxidase positive and the specific (secondary) granules which are peroxidase negative.[121–123] These two types of granules differ in their origin, their contents, and their behavior in response to stimuli. A list of the most important constituents of the granules is given in Table I. During phagocytosis, secondary (specific) granule release is directed primarily toward the outside of the cell, whereas primary (azurophil) granule degranulation is confined mainly to the internalized phagocytic vacuoles.[124,125] The granule contents include hydrolytic enzymes, bactericidal proteins, and myeloperoxidase (MPO). Myeloperoxidase, in combination with hydrogen peroxide and other oxygen metabo-

---

[117] E. L. Becker and P. M. Henson, *Adv. Immunol.* **17**, 93 (1973).
[118] L. J. Ignarro, *Agents Actions* **4**, 241 (1974).
[119] J. G. Hirsch, *J. Exp. Med.* **116**, 827 (1962).
[120] J. E. Trowell and D. B. Brewer, *J. Pathol.* **120**, 129 (1976).
[121] D. F. Bainton, J. L. Ullyot, and M. G. Farquhar, *J. Exp. Med.* **134**, 907 (1971).
[122] U. Bretz and M. Baggiolini, *J. Cell Biol.* **63**, 251 (1974).
[123] B. C. West, A. S. Rosenthal, N. A. Gelb, and H. R. Kimball, *Am. J. Pathol.* **77**, 41 (1974).
[124] M. S. Leffell and J. K. Spitznagel, *Infect. Immun.* **10**, 1241 (1974).
[125] M. S. Leffell and J. K. Sptiznagel, *Infect. Immun.* **12**, 813 (1975).

TABLE I
KNOWN CONSTITUENTS OF HUMAN NEUTROPHIL GRANULES[a]

Primary (azurophil) granules	Secondary (specific) granules
Acid hydrolases	Lysozyme
Acid β-glycerophosphatase	Lactoferrin
β-Glucuronidase	
N-Acetyl-β-glucosaminidase	Vitamin B_{12}-binding protein (cobalophilin)
α-Mannosidase	
Arylsulfatase	
β-Galactosidase	Collagenase[b]
5′-Nucleotidase	
α-Fucosidase	Acidic proteins
Acid protease (cathepsin)	
Neutral proteases	
Chymotrypsinlike protease	
Elastin	
Collagenase	
Cationic proteins	
Myeloperoxidase	
Lysozyme	
Acid mucopolysaccharide	

[a] Adapted from Klebanoff and Clark.[97]
[b] Secondary granule collagenase is released as a latent enzyme.

lites such as superoxide anions, is extremely active in killing ingested microorganisms.[97] The commonly measured enzymes, released from granules, include lysozyme, β-glucuronidase, and MPO. Specific enzyme assays are discussed in Section 5.2.3.8.

5.2.3. *Respiration Activity Burst.* Upon exposure to suitable stimuli, e.g., bacteria, the neutrophil undergoes a marked change in its pattern of oxygen metabolism. A sharp increase in oxygen consumption occurs[114,126–130]: under favorable circumstances the increase of oxygen consumption can be as much as 50 times that of resting levels.[131] Accompanying this increase in oxygen consumption is the formation of large quantities of D- and L-amino acid oxidases,[132,133] the superoxide anion

[126] R. J. Selvaraj and A. T. Sbarra, *Biochim. Biophys. Acta* **141** 243 (1967).
[127] R. J. Stjernholm and R. C. Manak, *J. Reticuloendothel. Soc.* **8,** 550 (1970).
[128] P. Patriarca, R. Cramer, S. Monocalvo, R. Rossi, and D. Romeo, *Arch. Biochem. Biophys.* **145,** 255 (1971).
[129] F. Rossi, D. Romeo, and P. Patriarca, *J. Reticuloendothel. Soc.* **12,** 150 (1972).
[130] B. B. Paul, R. R. Strauss, A. A. Jacobs, and A. J. Sbara, *Exp. Cell Res.* **73,** 456 (1972).
[131] R. K. Root, J. Metcalf, N. Oshino, and B. Chance, *J. Clin. Invest.* **55,** 945 (1975).
[132] M. J. Cline and R. J. Lehrer, *Proc. Natl. Acad. Sci. U.S.A.* **62,** 756 (1969).
[133] R. C. Skarnes, *Nature (London)* **225,** 1072 (1970).

(O_2^-),[134] as well as hydrogen peroxide.[135–137] In addition, there is also a substantial increase in the oxidation of glucose via the hexose monophosphate shunt.[136] Taken together these two events are termed the *respiratory activity burst*. Activation of the respiratory burst follows stimulus by about a 1-min period,[131] is a reversible process,[27] and requires neither internalization nor degranulation.[27,138]

The mechanism of the respiratory burst has been described in detail elsewhere[139–144] and needs no elaboration here. The stimulation of such an enhancement of activity which is not inhibited by the presence of cyanide distinguishes this process from the usual mitochondrial respiratory chain activity. The basis of the respiratory burst is the activation of a flavin-requiring enzyme,[131,132] dormant in resting cells, which catalyzes the reduction of oxygen to O_2^- at the expense of NADPH[92,145,146,146a] (see also Section 5.2.3.8.2.):

$$2O_2 + NADPH \rightarrow 2O_2^- + NADP^+ \tag{1}$$

Hydrogen peroxide is then formed by the reaction of O_2^- with itself[136] (also see Section 5.2.3.2):

$$HO_2^- + O_2 \rightarrow O_2 + H_2O_2 \tag{2}$$

The hexose monophosphate shunt provides a means by which the formed NADP$^+$ [cf. Eq. (1)] is converted back to NADPH.[133] Hydrogen peroxide (H_2O_2) and the superoxide anion (O_2^-) are, of course, extremely important in bactericidal killing.[116,143]

Several different techniques have been developed to measure the respiratory burst aspect of phagocytosis which, it should be emphasized, is determined by the extent of membrane stimulation. Some of these techniques measure general oxygen consumption whereas others measure the extent of by-product formation (e.g., H_2O_2 and O_2^-). Several of these

[134] J. T. Curnutte and B. M. Babior, *J. Clin. Invest.* **53**, 1662 (1974).
[135] G. Y. N. Iyer, M. F. Islam, and J. H. Quastel, *Nature (London)* **192**, 535 (1961).
[136] R. K. Root and J. A. Metcalf, *J. Clin. Invest.* **60**, 1266 (1977).
[137] J. Roberts and J. H. Quastel, *Nature (London)* **202**, 85 (1964).
[138] I. M. Goldstein, D. Roos, and H. B. Kaplan, *J. Clin. Invest.* **56**, 1155 (1975).
[139] L. R. DeChatelet, *J. Reticuloendothel. Soc.* **24**, 73 (1978).
[140] J. A. Badwey and M. L. Karnovsky, *Annu. Rev. Biochem.* **49**, 695 (1980).
[141] B. M. Babior, *N. Engl. J. Med.* **298**, 659 (1978).
[142] B. M. Babior, *N. Engl. J. Med.* **298**, 721 (1978).
[143] B. M. Babior, *Biochem. Biophys. Res. Commun.* **91**, 222 (1979).
[144] B. M. Babior, *in* "The Reticuloendothelial System: A Comprehensive Treatise. 2. Biochemistry and Metabolism" (A. J. Sbarra and R. R. Strauss, eds.), p. 339. Plenum, New York, 1980.
[145] B. M. Babior, J. T. Curnutte, and B. J. McMurrich, *J. Clin. Invest.* **58**, 989 (1976).
[146] B. M. Babior and R. S. Kipnes, *Blood* **50**, 517 (1977).
[146a] S. J. Klebanoff, *J. Bacteriol.* **95**, 2131 (1968).

methods, together with the underlying principle of their operation, are discussed individually below.

5.2.3.1. Oxygen consumption: Neutrophil consumption of oxygen can be measured manometrically or potentiometrically (with oxygen electrodes). Manometric measurement, described in Protocol 9A, is suitable when high concentrations of neutrophils are available. The potentiometric method, described in Protocol 9B, is considerably more sensitive. It suffers, however, from the disadvantage that it is considerably more difficult to use in long-term time studies.

5.2.3.2. Superoxide production (see also this volume [22], [23], [24]): One result of the respiratory burst, as indicated in Eq. (1), is the production of the superoxide anion as a consequence of the univalent reduction of oxygen. This highly reactive anionic radical can act either as a reductant (e.g., in the reduction of ferricytochrome *c*) or as an oxidant (e.g., in the oxidation of epinephrine). In addition, when two superoxide radicals interact, one is oxidized and the other is reduced, as shown in Eq. (2), resulting in the production of hydrogen peroxide (H_2O_2). This dismutation occurs spontaneously at low pH values[147,148] but is also catalyzed by the enzyme, superoxide dismutase.[149–151] Thus, the presence of superoxide dismutase will inhibit reactions dependent on the presence of the superoxide anion; indeed, the demonstration of such inhibition is itself used as evidence for the presence of the anion radical. Quantitation of either superoxide (O_2^-) or hydrogen peroxide (H_2O_2) production, as discussed below, can be used to document the extent of phagocytosis.

5.2.3.2.1. Cytochrome c reduction: The superoxide anion reduces cytochrome *c,* thereby providing the basis for comparatively simple, colorimetric assay for the quantitation of O_2^- production. As several electron donors can reduce cytochrome *c,* specificity of the reaction is of concern. However, of such potential electron donors only O_2^- is destroyed by the superoxide dismutase enzyme and, consequently, its production is quantitated by measuring superoxide dismutase inhibition of cytochrome *c* reduction. The use of the dismutase enzyme therefore confers specificity on the assay.

Cytochrome *c* does not penetrate into the neutrophil cytoplasm. As a consequence, only those superoxide anions which are released into the extracellular fluid are detected by the assays described below. For this

[147] D. Behar, G. Czapski, J. Rabani, L. M. Dorfman, and H. A. Schwarz, *J. Phys. Chem.* **81,** 1048 (1977).

[148] B. H. J. Bielski and A. O. Allen, *J. Phys. Chem.* **81,** 1048 (1977).

[149] I. Fridovich, *J. Biol. Chem.* **245,** 4053 (1970).

[150] J. Rabani, D. Klug, and I. Fridovich, *Isr. J. Chem.* **10,** 1095 (1972).

[151] I. Fridovich, *Photochem. Photobiol.* **28,** 733 (1978).

reason, it should be remembered that any change in the measured cytochrome c reduction (and by implication O_2^- concentration) may be due not only to alterations in superoxide production, but possibly also to alterations in the fraction released into the extracellular environment. Consequently any conclusions about neutrophil respiration activity based on O_2^- concentrations should be confirmed by quantitating oxygen uptake in order to distinguish between the two aforementioned possibilities.

Using cytochrome c, superoxide production can be measured by quantitating the amount of O_2^- generated over a selected arbitrary, but defined, time interval or alternatively by measuring its production continuously as a function of time. From a practical point of view, the former method is more suitable for a large number of samples (Protocol 10A), while the second method provides a better assessment of the kinetics of O_2^- production and particularly of the delay of onset, and rate of production in the early phase, of stimulation (Protocol 10B).

5.2.3.2.2. Nitroblue tetrazolium reduction: In addition to the above tests for quantitating extracellular O_2^- release, its production may also be approximated by the use of the NBT dye reduction test.[21,104,152–154] NBT is a soluble yellow redox dye which, when present in the extracellular fluid, is swept into the phagocytic vacuole together with the ingested particle. In the presence of superoxide anions the dye is chemically reduced,[155–157] to yield a dark purple insoluble compound (formazan) which can be extracted from the neutrophils with dioxane. Since formazan has an adsorption maximum at 580 nm, its concentration in the dioxane extract can be easily quantitated by spectrophotometric means.[158] Furthermore, since the reduced insoluble product is dark purple in color, and can be clearly discerned microscopically, it is easy to determine the percentage of cells containing the reduced NBT dye (formazan). It is thus not necessary to extract the product with dioxane if it is desired to assess oxygen metabolism *only*. In such instances the NBT dye is simply incubated with the phagocytes, and after a fixed period of time, the cells are microscopically examined for their purple content (slide NBT test). This test is described in detail in Protocol 11A.[159]

[152] A. W. Segal, *Lancet* **2,** 1248 (1974).
[153] R. Sher, R. Anderson, A. R. Rabson, and H. J. Kornhoff, *S. Afr. Med. J.* **48,** 209 (1974).
[154] B. H. Park, *J. Pediat.* **78,** 376 (1971).
[155] K. V. Rajagopalan and P. Handler, *J. Biol. Chem.* **239,** 2022 (1964).
[156] R. W. Miller, *Can. J. Biochem.* **48,** 935 (1970).
[157] C. Beauchamp and I. Fridovich, *Anal. Biochem.* **44,** 276 (1971).
[158] R. L. Baehner and D. G. Nathan, *N. Engl. J. Med.* **278,** 971 (1968).
[159] B. H. Park, S. M. Fikrig, and E. M. Smithwick, *Lancet* **2,** 532 (1968).

The sensitivity of this NBT-reduction test can be significantly enhanced through the inclusion of various types of agents into the system. These include polystyrene latex particles,[160] endotoxin,[161-164] immune complexes,[165-167] phorbol myristate acetate,[168,169] and bacterial culture filtrates.[162,170,171] These particles are ingested and thus facilitate (simultaneously) enhanced NBT uptake. This is known as the stimulated NBT test and is described in Protocol 11B for phorbol myristate acetate stimulation.

Nitroblue tetrazolium is toxic to neutrophils at high concentrations as assessed by trypan blue exclusion or the release of lactic acid dehydrogenase. Other tetrazolium salts are also reduced by human neutrophils to yield an insoluble colored precipitate and may be used as alternatives to NBT. Such salts include dimethylthiazolyldiphenyltetrazolium chloride, tetraphenylditetrazolium chloride, and triphenyl tetrazolium chloride (see also Section 8, Protocol 11A). Since several leukocyte, including neutrophil, enzyme systems are inactivated by sodium fluoride, the use of NaF–heparin should be avoided in NBT tests.

5.2.3.3. Hydrogen peroxide production (see also this volume [22], [23], [24]): The production of H_2O_2 by neutrophils during phagocytosis has been confirmed by a number of investigators using a variety of techniques. Hydrogen peroxide can be formed either from oxygen directly, through divalent reduction, e.g., by means of glucose oxidase,[172-174] or by a dismutation reaction between two O_2^- (or perhydroxyl radicals) as shown in Eq. (2). This can occur either spontaneously (below pH 5.0) or catalyzed (between pH 5.0 and 9.5) by a superoxide dismutase.[147-149] A number of distinct superoxide dismutases exist which differ in terms of

[160] S. K. Kim, E. Monz, and H. Wehunger, *Infektionen. Klun. Pediat.* **185,** 141 (1973).
[161] B. H. Park and R. A. Good, *Lancet* **2,** 616 (1970).
[162] G. Matula and P. Y. Paterson, *N. Engl. J. Med.* **285,** 311 (1971).
[163] H. D. Ochs and R. P. Igo, *J. Pediatr.* **83,** 77 (1973).
[164] D. Merzbach and N. Obendeanu, *J. Med. Microbiol.* **8,** 375 (1975).
[165] U. E. Nydegger, R. M. Anner, A. Gerebzotff, P. H. Lambert, and P. A. Miescher, *Eur. J. Immunol.* **3,** 465 (1973).
[166] L. M. Pachman, P. Jayanetra, and R. M. Rothbery, *Pediatrics* **52,** 823 (1973).
[167] C. Koch, N. Hoiby, and A. Wiik, *Acta Pathol. Microbiol. Scand.* **C83,** 144 (1975).
[168] J. E. Repine, J. E. White, C. C. Clawson, and B. M. Holmes, *J. Lab. Clin. Med.* **83,** 911 (1974).
[169] L. R. DeChatelet, P. S. Shirley, and R. B. Johnston, *Blood* **47,** 545 (1976).
[170] C. Koch, *Acta Pathol. Microbiol. Scand.* **B81,** 266 (1973).
[171] C. Koch and N. Hoiby, *Acta Pathol. Microbiol. Scand.* **B81,** 787 (1973).
[172] V. Massey, S. Strickland, S. G. Mayhew, L. G. Howell, P. C. Engel, R. G. Matthews, M. Schuman, and P. A. Sullivan, *Biochem. Biophys. Res. Commun.* **36,** 891 (1969).
[173] R. Nillson, F. M. Pick, and R. C. Bray, *Biochim. Biophys. Acta* **192,** 145 (1969).
[174] M. T. Stankovich, L. M. Schopfer, and V. Massey, *J. Biol. Chem.* **253,** 4971 (1978).

their metal component (copper, zinc, magnesium, or iron), their primary structure, and distribution in tissues and microorganisms. In neutrophils two distinct dismutases have been detected: one is cyanide sensitive and is found in the cytosol of phagocytic cells; the other is cyanide insensitive and is found in the particulate fraction.

Several techniques are available for measuring H_2O_2 concentration. These include the oxidation of formate by catalase,[135,175] the inhibition of catalase activity by aminotriazole,[176] the oxidation of leukodiacetyl-2,7-dichlorofluorescein[177–180] or scopoletin by horseradish peroxidase,[131,181,182] and the release of oxygen by catalase.[183] Of these methods, perhaps the most useful direct assay relies on the use of a fluorescent dye, scopoletin (7-hydroxy-6-methoxycoumarin), that undergoes fluorescence quenching when it is oxidized by horseradish peroxidase in the presence of H_2O_2. The assay may be performed either as a continuous (Protocol 12A) or as a discontinuous (Protocol 12B) assessment. The major problem associated with this technique is that since proteins are used, which do not penetrate the neutrophil membrane, only that H_2O_2 which is released extracellularly is measured (see Section 5.2.3.2.1 for a note of caution in this respect).

5.2.3.4. Hexose monophosphate shunt activity (see also this volume [17], [18]): As a result of the increase in metabolic activity accompanying the respiratory burst, there is an enhanced requirement for NADPH. This is obtained through the use of the alternative hexose monophosphate (HMP) shunt pathway. The presence of HMP shunt activity in neutrophils has been established by isotopic labeling techniques.[184] This route of glucose metabolism bypasses the normal glycolytic pathway entirely (cf. Section 5.2.1). Glucose is converted, via 6-P-gluconate, to pentose-P, which is oxidized via acetyl-P and triose-P as illustrated below:

$$\text{Glucose 6-phosphate} + \text{NADP}^+ \rightleftharpoons \text{6-phosphogluconate} + \text{NADPH} + \text{H}^+ \qquad (3)$$

A key enzyme in the hexose monophosphate shunt is glucose-6-phosphate dehydrogenase (G-6-PD). Phosphogluconate is then converted by a second enzyme, 6-phosphogluconate dehydrogenase (G-PGD), which cat-

[175] S. J. Klebanoff and S. M. Pincus, *J. Clin. Invest.* **50**, 2226 (1971).
[176] E. Margoliash and A. Novogrodsky, *Biochem. J.* **68**, 468 (1958).
[177] A. S. Keston and R. Brandt, *Anal. Biochem.* **11**, 1 (1965).
[178] M. J. Black and R. B. Brandt, *Anal. Biochem.* **58**, 246 (1974).
[179] B. B. Paul and A. J. Sbarra, *Biochim. Biophys. Acta* **156**, 168 (1968).
[180] J. W. T. Homan-Müller, R. S. Weening, and D. Roos, *J. Lab. Clin. Med.* **85**, 198 (1975).
[181] W. A. Andreae, *Nature (London)* **175**, 859 (1955).
[182] A. Boveris, E. Martino, and A. O. M. Stoppani, *Anal. Biochem.* **80**, 145 (1977).
[183] M. Zatti, F. Rossi, and P. Patriarca, *Experentia* **24**, 669 (1968).
[184] R. V. Coxon and R. J. Robinson, *Proc. R. Soc. (London)* **B145**, 232 (1956).

alyzes the reaction:

6-Phosphogluconate + NADP$^+$ $\rightleftharpoons$ D-ribulose phosphate + CO$_2$ + NADPH + H$^+$ (4)

The net end result of HMP shunt activity is enhanced NADPH yield and CO$_2$ production. One molecule of CO$_2$ is released for every molecule of glucose oxidized. The carbon released is in the 1 position; thus, the conversion of radioactive [1-^{14}C]glucose to ^{14}CO$_2$ can be employed as a measure of the oxidation of glucose via the HMP shunt as described in Protocol 13. In order to correct for mitochondrial oxidation (and hence CO$_2$ production), which occurs via the normal glycolytic pathway, [6-^{14}C]glucose oxidation is also monitored. In the resting neutrophil less than 5% of the metabolized glucose is converted to CO$_2$ through the HMP shunt.[89,127,185,186] However, following ingestion of particles there is a dramatic increase in glucose metabolism by the HMP pathway. Values as high as 40% have been reported.[89,113,127]

5.2.3.5. Iodination: The mechanisms by which neutrophils kill ingested bacteria and other microorganisms have been considerably clarified in recent years.[187-190] Klebanoff has shown that one of the most powerful antimicrobial systems involves the interaction between H$_2$O$_2$, a halide anion, and the myeloperoxidase system.[79,191,192] Hydrogen peroxide is produced, as discussed in Section 5.2.3.3, as a result of the metabolic burst associated with phagocytosis. The halide is known to be present as the chloride ion in the intracellular milieu, and the third component, myeloperoxidase (MPO), contained exclusively in the azurophilic granules, is released into the phagosomes on fusion. (For further discussion of MPO activity see Section 5.2.3.8.1.)

If radioactive iodine, ^{125}I, is added into the system, then a portion of this halide will be converted, by oxidation, into the protein-bound form through the action of H$_2$O$_2$ and myeloperoxidase. The iodination reaction can therefore be employed as a measure of neutrophil activity. It should be borne in mind however, that this test is dependent on both phagocytic ingestion as well as H$_2$O$_2$ production (during the metabolic burst) and myeloperoxidase activity. Thus a defect in any of these could result in diminished iodination and hence it is important to perform appropriate

[185] W. S. Beck, *J. Biol. Chem.* **232**, 271 (1958).
[186] W. S. Beck, *Ann. N.Y. Acad. Sci.* **75**, 4 (1958).
[187] R. C. Skarnes and D. W. Watson, *Bacteriol. Rev.* **21**, 273 (1957).
[188] J. G. Hirsch, *Bacteriol. Rev.* **24**, 133 (1960).
[189] S. J. Klebanoff, *Semin. Hematol.* **12**, 117 (1975).
[190] L. R. DeChatelet, *J. Infect. Dis.* **131**, 295 (1975).
[191] S. J. Klebanoff, *J. Biol. Chem.* **249**, 3724 (1974).
[192] S. J. Klebanoff, *in* "The Phagocytic Cell in Host Resistance" (J. A. Bellanti and D. H. Dayton, eds.), p. 45. Raven Press, New York, 1975.

individual, control experiments for ingestion, H_2O_2 production, and MPO activity, respectively.

The basis of the test is that the radiolabeled iodine halide, added to the extracellular medium, through the combined action of MPO and H_2O_2, is covalently incorporated into various protein substrates by oxidation. The amount of ^{125}I that becomes associated with trichloroacetic acid-precipitable material from neutrophils, incubated with opsonized zymosan, Na[125]I, and an acceptor protein such as albumin, is then measured as described in Protocol 14. Since albumin is used, only extracellular oxidation of the test protein is achieved and hence this technique only measures that H_2O_2 and MPO which is released into the extracellular fluid (also see Section 5.2.3.2.1).

5.2.3.6. Oxidation radicals: During phagocytosis a number of highly reactive oxidizing radicals are formed in addition to the superoxide anion. The most important of these are the hydroxyl radicals. The source of the $\cdot OH$ is thought to be the result of a trace metal acting as an oxidation–reduction catalyst of the interaction of O_2^- and H_2O_2, with the metal being alternatively reduced by O_2^- and oxidized by H_2O_2 as follows[193–195]:

$$O_2^- + M^{n+} \rightarrow O_2 + M^{(n-1)+} \qquad (5)$$
$$M^{(n-1)\pm} + H_2O_2 \rightarrow M^{n+} + OH^- + \cdot OH \qquad (6)$$
$$O_2^- + H_2O_2 \rightarrow O_2 + OH^- + \cdot OH \qquad (7)$$

where M^{n+} and M^{n-1} refer to the oxidized and reduced states of the metal, respectively. Of a large number of trace metals tested, only iron has been found to catalyze this reaction *in vivo*.

The formation of hydroxyl radicals can be monitored by a number of techniques[196] which rely on the ability of these radicals to release hydrocarbon gases from certain organosulfo compounds. This ability is not possessed by either superoxide (O_2^-) radicals or by H_2O_2. These techniques include the formation of ethylene from methional or 2-keto-4-thiomethylbutyric acid,[197–199] the formation of $^{14}CO_2$ from [^{14}C]benzoic acid,[200,201] the demonstration of an $\cdot OH$ spin-trap adduct by electron spin

[193] B. Halliwell, *FEBS Lett.* **92**, 321 (1978).
[194] B. Halliwell, *FEBS Lett.* **96**, 238 (1978).
[195] J. M. McCord and E. D. Day, *FEBS Lett.* **86**, 139 (1978).
[196] W. Bors, M. Saran, E. Lengfelder, C. Michel, C. Fichs, and C. Frenzel, *Photochem. Photobiol.* **28**, 629 (1978).
[197] S. F. Yang, *Arch. Biochem. Biophys.* **122**, 481 (1967).
[198] S. F. Yang, *J. Biol. Chem.* **244**, 4360 (1969).
[199] S. J. Klebanoff and H. Rosen, *J. Exp. Med.* **148**, 490 (1978).
[200] A. L. Sagone, M. A. Dekker, R. M. Wells, and C. DeMocko, *Biochim. Biophys. Acta* **628**, 90 (1980).
[201] A. L. Sagone, D. S. Mendelson, and E. N. Metz, *J. Lab. Clin. Med.* **89**, 1333 (1977).

resonance,[202-205] and the formation of methane from dimethyl sulfoxide.[206-208] Of these various techniques, the most favored for phagocytosis studies are (1) methane formation from dimethyl sulfoxide; methane formation can be readily detected by gas chromatography. (2) Ethylene formation from thioesters; for this purpose, the generally used substrate is 2-keto-4-thiomethylbutyric acid. The technical details of this particular assay are given in Protocol 15.

The data from these assays should be interpreted with caution. The oxidation radicals can be removed by various constituents of the assay fluid, including glucose, and by the neutrophils themselves. The amount of hydrocarbon released from the incubation medium, therefore, only reflects that fraction of radicals that react with the particular substrate material. Thus variations, e.g., in the composition of the fluid medium, can alter measured hydrocarbon release even though the rate and concentration of radical formation may not change.

5.2.3.7. Chemiluminescence: The use of chemiluminescence for the study of phagocytosis has been discussed elsewhere in this volume [33] and will not be duplicated here.

5.2.3.8. Specific enzyme assays: As mentioned previously, degranulation results in the release of several enzymes and proteins either intra- or extracellularly. Also, membrane stimulation during phagocytosis activates certain enzymes. The release and activity of these enzymes is often used as an indicator of the extent to which phagocytosis has occurred within a given population of neutrophils.

In assaying for such enzymes or proteins it is always important to establish that the measured activity is proportional to the amount of sample (or substrate) employed. If this caution is not exercised, it will not be possible to interpret comparisons between various samples. Also, absolute values calculated for any one sample will be meaningless. In determining enzyme activities, it is preferable to have the assay conditions such that the reaction rate is affected only by the concentration of the

[202] E. Finkelstein, G. M. Risen, and E. J. Rauckman, *Arch. Biochem. Biophys.* **200**, 1 (1980).
[203] E. G. Janzen, in "Free Radicals in Biology" (W. A. Pryor, ed.), Vol. 4, p. 115. Academic Press, New York, 1980.
[204] M. R. Green, H. A. O. Hill, M. J. Okelow-Zubkowska, and A. W. Jegal, *FEBS Lett.* **100**, 23 (1979).
[205] H. Rosen and S. J. Klebanoff, *J. Clin. Invest.* **64**, 1725 (1979).
[206] B. C. Gilbert, R. O. C. Norman, and R. C. Sealy, *J. Chem. Soc. (Perkin Trans.)* **2**, 303 (1975).
[207] A. P. Reuvers, C. L. Greenstock, J. Borsa, and J. D. Chapman, *Int. J. Radiat. Biol.* **24**, 533 (1973).
[208] J. E. Repine, J. W. Eaton, M. W. Anders, J. R. Hoidal, and R. B. Fox, *J. Clin. Invest.* **64**, 1642 (1979).

sample being investigated, and not by changes in substrate or cofactor levels.

In addition to the above, the effect of other constituents must also be taken into account. For example, the F^- anion significantly inhibits a number of leukocyte enzymes and hence the use of NaF–heparin should, if possible, be avoided when enzyme studies are to be performed on neutrophils isolated from the anticoagulated blood.

Despite the above concerns, properly performed enzyme assays have been extensively used in phagocytosis studies and continue to provide important information particularly with respect to neutrophil dysfunction.

5.2.3.8.1. Myeloperoxidase: This subject has been discussed elsewhere in this volume [20] and will not be duplicated here.

5.2.3.8.2. Superoxide anion oxidase (see also this volume [17]): Oxygen in its ground state contains two unpaired valence electrons. Each of these occupies a separate electronic orbital which can be filled through the addition of a second electron. When oxygen accepts a single electron it is converted into the superoxide anion (O_2^-) or its protonated form, the perhydroxyl radical (HO_2^-). A number of oxidases are known to reduce oxygen, in part or totally, to O_2^-. Initiation of the respiratory burst of phagocytes is due to one such oxidase which is responsible for catalyzing the reaction:

$$NADPH + 2O_2 \rightarrow 2O_2^- + NADP^+ + H^+ \qquad (8)$$

This enzyme is bound to the surface membrane of neutrophils. Assessment of oxidase activity is performed by measuring the quantity of superoxide anion produced. This is done colorimetrically since these anions are able to reduce cytochrome c as illustrated below:

$$O_2^- + Fe^{3+} Cyt\ c \rightarrow Fe^{2+} Cyt\ c + O_2 \qquad (9)$$

The reduced and oxidized forms of cytochrome c have different adsorption maxima. Since this reaction can be inhibited by the presence of superoxide dismutase (which converts the O_2^- radicals into H_2O_2), an easy method is available for establishing a background reference blank.

The most commonly employed method for measuring O_2^- by this enzymatic process is summarized in Protocol 16.

5.2.3.8.3. β-Glucuronidase: As mentioned previously, degranulation may be assessed by measuring the rate of appearance of granule-associated enzymes in the phagocytic vacuoles, in the extracellular medium, or in both areas.

During phagocytosis β-glucuronidase, a constituent of the primary (azurophil) granules, is released intracellularly since degranulation of these granules is confined primarily to the internalized phagocytic vacuoles. Despite this, however, a certain portion of this acid hydrolase is released into the extracellular fluid.

A colorimetric titration can be used to quantitate the extent of extracellular β-glucuronidase release as described in Protocol 17. The basis of the technique is the use of p-nitrophenyl β-D-glucuronide as the substrate for the enzyme. This results in the release, through enzymatic cleavage, of p-nitrophenol that has a chromophore which adsorbs strongly between 405 and 410 nm. Consequently, a relatively simple spectrophotometric assay is available for assessing β-glucuronidase activity and hence the extent of its release into the extracellular fluid (cf. Protocol 18).

5.2.3.8.4. Lysozyme: This enzyme has long been known to have antimicrobial activity.[209,210] Human neutrophils contain about 20 μg of lysozyme per 10^7 cells, which is found in both the azurophils and specific granules.[211,212] Lysozyme is a single polypeptide chain, cationic protein, with a molecular weight of 14,000. Its natural substrate is the bacterial wall. Lysozyme is an endoacetylmuramidase which specifically cleaves the β-(1,4) linkage between the repeating units of N-acetylmuramic acid and N-acetylglucosamine in the peptidoglycans of the bacterial wall.[213] Not all bacteria exhibit the same sensitivity to lysozyme. This is because the acetylmuramic acid residues often are replaced by tetrapeptides which crosslink the hexosamine chains. Such a substitution does not occur in *Micrococcus lysodeikticus,* rendering this bacterium particularly susceptible to N-acetylmuramic acid cleavage by lysozyme. Enzyme cleavage results in the rapid lysis of the bacteria, giving rise to an elegant, and yet simple, means for assaying for neutrophil release of lysozyme. This is performed by monitoring the change in the optical density (i.e., light adsorption) of an *M. lysodeikticus* suspension exposed to a degranulation extract or to a suspension of neutrophils as described in detail in Protocol 18. Lysis of the bacteria results in a decrease in light adsorption.

It should be mentioned that spontaneous secondary granule exocytosis, and particularly release of lysozyme and cobalophilin (vitamin B_{12} binding protein; see Section 5.2.3.9.2) into the extracellular fluid, is significantly enhanced when neutrophils make contact with surfaces such as unsiliconized glass or nylon *in vitro*.[214] Consequently, appropriate con-

[209] A. Fleming, *Proc. R. Soc. (London)* **B93**, 306 (1922).

[210] A. Fleming and V. D. Allison, *Br. J. Exp. Pathol.* **3**, 252 (1922).

[211] M. Baggiolini, U. Bretz, and B. Gusus, *Dtsch. Schweiz. Med. Wochenschr.* **104**, 129 (1974).

[212] M. S. Leffell and J. K. Spitznagel, *Infect. Immun.* **6**, 761 (1972).

[213] J. L. Strominger and D. J. Tipper, in "Lysozyme" (E. Osserman, W. Canfield, and C. Beckok, eds.), p. 169. Academic Press, New York, 1974.

[214] D. G. Wright and J. I. Gallin, *J. Immunol.* **123**, 285 (1979).

trols should be devised when quantitating release of these proteins as an indicator of phagocytosis.

5.2.3.8.5. Lactate dehydrogenase: The final step in anaerobic glycolysis (cf. Section 5.2.3.4) is the conversion of pyruvate into lactic acid, a reaction that is catalyzed by the enzyme, lactate dehydrogenase. This reaction requires the presence of NADPH and is as follows:

$$\text{Pyruvate} + \text{NADPH} + \text{H}^+ \rightleftharpoons \text{lactate} + \text{NADP}^+ \qquad (10)$$

Lactate dehydrogenase (LDH) has been demonstrated to occur in high concentrations in human neutrophils.[215,216] It is a soluble cytoplasmic enzyme. During granule isolation, however, as much as 45% of the enzyme adheres to azurophil granules. The granule-bound LDH is not active.[217] Extracellular release of LDH can be employed to reflect the extent of phagocytic ingestion as described in Protocol 19. The extent of enzyme activity is measured spectrophotometrically by monitoring the decrease in adsorbance at 340 nm. This occurs as NADPH is oxidized to yield NADP$^+$.

5.2.3.8.6. Glucose-6-phosphate dehydrogenase: During exocytosis neutrophil glucose-6-phosphate dehydrogenase (G-6-P-D), as well as 6-phosphogluconic dehydrogenase (6-PGD), is released into the extracellular fluid.[218–220] Enzyme activity measurements are performed measuring the rate of reduction of NADP. As the reduction product, NADPH, is formed there is an increase in adsorbance at 340 nm. Since the crude neutrophil extract contains G-6-PD as well as 6-PGD, both of which reduce NADP, correction should be made for 6-PGD activity. For this purpose, the difference in the reduction rate of NADP between the reaction mixture containing D-glucose 6-phosphate and 6-phosphogluconate (6-PG) and the reaction mixture containing only 6-PG is determined, and used to calculate G-6-PD activity. The results of the enzyme activities are expressed as nanomoles of NADP reduced per minute per unit mass of protein at 25°. The most common G-6-PD assay is described in Protocol 20.

5.2.3.8.7. Nucleotidase: Stimulated neutrophil suspensions have been shown to have the ability to cleave the phosphate moiety from ATP, AMP or *p*-nitrophenylphosphate molecules added to the solution. This was first

[215] W. S. Beck, *J. Biol. Chem.* **232**, 252 (1958).
[216] Y. Rabinowitz, *Blood* **27**, 470 (1966).
[217] S. John, N. Berger, M. J. Bonner, and J. Schultz, *Nature (London)* **215**, 1483 (1967).
[218] A. Kahn and J. C. Dreyfus, *Biochim. Biophys. Acta* **334**, 257 (1974).
[219] A. Kahn, O. Bertrand, D. Cottreau, P. Boivin, and J. C. Dreyfus, *Biochim. Biophys. Acta* **445**, 537 (1976).
[220] A. Kahn, A. Milani, J. Marie, D. Cottreau, and P. Bovin, *Biochimie* **57**, 325 (1975).

demonstrated for guinea pig neutrophils[221–223] and has subsequently been shown for rabbit[224] and human neutrophils[225] as well. Such hydrolytic ability is due to presence of ectoenzymes. Examples of these nucleotidases include the sodium/potassium—and magnesium—ATPases[224,225] and the 5'-nucleotidase.[221–223]

The basis of the radioassay described in Protocol 21 is the observation that activated charcoal or alumina suspension will bind (with high affinity) phosphorylated nucleotides but will not bind free phosphate molecules. The concentration of the aforementioned nucleotidases can therefore be determined by measuring the extent of reduction in the adsorption of the ^{32}P-labeled nucleotide to the activated charcoal or alumina matrix. (Following enzymatic hydrolysis the cleaved ^{32}P will not bind whereas the uncleaved ^{32}P-nucleotide will continue to bind.) The assay described (Protocol 21) is for the specific quantitation of Mg-ATPase concentration in human neutrophil suspensions. This method, however, can, with suitable modification of concentrations, substrates, and cofactors, be easily modified for assaying other nucleotidases.

It should be remembered, however, that the nucleotidases are ectoenzymes, i.e., enzymes located on the neutrophil membrane with their active sites directed externally toward the extracellular fluid. As a consequence of this location, during the ingestion of particles as much as 15 to 25% AMPase ectoenzyme activity disappears due to the internalization of specific portions of the plasmalemma during phagocytosis.[81,226,227] The enzyme that "disappears" from the cell surface is found in the phagolysosomes. The enzymes are inactive since internalization produces an inverted membrane which forms the phagocytic vacuole with the enzyme active sites directed inward and therefore latent. (Sonication results in the rupture of these vesicles and restoration of full enzyme activity.[121]) This reduction in nucleotidase activity, due to membrane internalization, should be borne in mind when interpreting assay data. Specifically, it should always be established that there is not a defect in phagocytic ingestion (cf. Section 5.1) in the neutrophil suspension being examined.

[221] J. W. DePierre and M. L. Karnovsky, *Science* **183**, 1096 (1976).

[222] J. W. DePierre and M. L. Karnovsky, *J. Biol. Chem.* **249**, 7111 (1974).

[223] J. W. DePierre and M. L. Karnovsky, *J. Biol. Chem.* **249**, 7121 (1974).

[224] E. L. Bekker, V. Talley, H. J. Showell, P. H. Naccache, and R. I. Sha'fi, *J. Cell Biol.* **77**, 329 (1978).

[225] J. Harlan, L. R. DeChatelet, D. B. Iverson, and C. E. McCall, *Infect. Immun.* **15**, 436 (1977).

[226] K. Takanaka and P. J. O'Brien, *Arch. Biochem.* **169**, 436 (1975).

[227] J. M. Oliver, T. E. Ukena, and R. D. Berlin, *Proc. Natl. Acad. Sci. U.S.A.* **71**, 394 (1974).

5.2.3.8.8. Acid and alkaline phosphatase: Acid phosphatase is a constituent of the primary (azurophil) granules and the demonstration of its presence in the extracellular fluid is indicative of degranulation. In contrast, alkaline phosphatase, in human neutrophils, is an ectoenzyme associated with the cell membrane. Thus, reduction in the activity of alkaline phosphatase can be taken as an indirect measure of the extent of phagocytic ingestion (cf. comments in Section 5.2.3.8.7).

Both enzymes can be detected by a calorimetric assay using *p*-nitrophenol phosphate as the substrate. The two phosphatases will hydrolyze this compound, cleaving the phosphate moiety, to yield *p*-nitrophenol which has a chromophore that absorbs strongly at 405–410 nm. The assay procedure is very similar to that described for β-glucuronidase (Section 5.2.3.8.3). However, since the optimum enzyme concentrations and pH conditions differ for the two enzyme assays, technical details for phosphatase analysis are given in Protocol 22 (also see Sections 7.2 and 7.3).

5.2.3.9. Nonenzyme markers: Degranulation of specific (secondary) granules results in the release of their contents into the extracellular fluid. Thus, the assay for these components can be used to assess the extent of degranulation due to phagocytic ingestion.

At the present time, the most specific biochemical markers for neutrophil secondary granules are lactoferrin and cobalophilin, the vitamin B_{12}-binding protein.[228–230]

5.2.3.9.1. Lactoferrin: This protein has a molecular weight of 77,000 and each molecule avidly binds two molecules of iron even though it has no heme groups. In the secondary granules it is found in the unsaturated form, apolactoferrin, which acts as a bacteristatic agent. This occurs through the ability of apolactoferrin to bind iron strongly and thus remove it from the environment, thereby interfering with iron (an essential nutrient) uptake by bacteria. The usual lactoferrin *release* from human neutrophils is approximately 4 μg/10^6 cells.[212,231] However, the lactoferrin *content* of the same neutrophils has been reported to be as high as 20 μg/10^6 cells.[232]

The usual assay for apolactoferrin is to measure the extent to which a

[228] M. Baggiolini, J. G. Hirsch, and C. de Duve, *J. Cell Biol.* **40**, 529 (1969).
[229] M. Baggiolini, C. de Duve, P. L. Masson, and J. F. Heremans, *J. Exp. Med.* **131**, 559 (1970).
[230] S. P. Kane and T. J. Peters, *Clin. Sci. Mol. Med.* **49**, 171 (1975).
[231] P. L. Masson, J. F. Heremans, and E. Schonne, *J. Exp. Med.* **130**, 643 (1969).
[232] R. M. Bennett and T. Kokocinski, *Br. J. Haematol.* **39**, 509 (1978).

solution of this protein inhibits growth of bacterial colonies.[233-236] As applied to neutrophil studies it is more common to use radiolabeled procedures such as those described in Protocol 22A and B.

5.2.3.9.2. Cobalophilin: vitamin B_{12}-binding protein: Secondary granule release results in the presence, in the extracellular fluid, of a vitamin B_{12}-binding protein also known as cobalophilin,[237,238] or cobalamin-binding protein.[238,239] This protein is immunologically identical to both transcobalamins (TC) I and III present in plasma.[239] It now seems likely that plasma TC III is derived from neutrophil exocytosis of cobalophilin. In addition, there is strong evidence that TC I, to which most circulating vitamin B_{12} (cobalamin) is bound, is also derived from neutrophil exocytosis of the secondary (specific) granules. The affinity of this protein for vitamin B_{12} provides a direct binding assay for determining its concentration in the extracellular fluid of neutrophil suspensions. The sensitivity of the assay can be markedly increased by using cyanocobalamin radiolabeled with cobalt (^{57}Co). The technical details of one such assay are described in Protocol 23.

5.3. Bactericidal Activity (Killing) (see also this volume [35])

Ten to 30 min after ingestion many pathogenic and nonpathogenic bacteria are inactivated: they fail to grow if placed on nutrient media and often exhibit structural alterations. The mechanisms by which leukocytes kill bacteria and other microorganisms have been considerably clarified, especially in the case of neutrophils. Accurate measurements of intracellular killing based on the recovery of live bacteria from phagocytic cells may be difficult because of the incomplete removal of noningested, surface-adherent bacteria. The most successful approach to circumvent these problems has included the use of obligate intracellular parasites,[240,241] the use of tissue culture medium which does not favor bacterial

[233] P. L. Masson and J. F. Heremans, *Protides Biol. Fluids* **14,** 115 (1966).
[234] P. L. Masson, J. F. Heremans, J. J. Prignot, and G. Wauters, *Thorax* **21,** 538 (1966).
[235] J. D. Oram and B. Reiter, *Biochim. Biophys. Acta* **170,** 351 (1968).
[236] J. G. Bishop, F. L. Schanbacher, L. C. Ferguson, and K. L. Smith, *Infect. Immun.* **14,** 911 (1976).
[237] J. D. Simon, W. E. Houck, and M. M. Albala, *Biochem. Biophys. Res. Commun.* **73,** 444 (1976).
[238] C. Gottlieb, K. S. Lau, L. R. Wasserman, and V. Herbert, *J. Haematol.* **25,** 875 (1965).
[239] E. Jacob, S. J. Baker, and V. Herbert, *Physiol. Rev.* **60,** 918 (1980).
[240] Y. T. Chang, *Appl. Microbiol.* **17,** 750 (1959).
[241] H. H. Hopps, E. B. Jackson, J. K. Danauskas, and J. E. Smadel, *J. Immunol.* **82,** 172 (1959).

growth,[242] or the use of antibiotics or enzymes such as lysostaphin[243,244] to destroy extracellular bacteria. Alternatively, bacterial killing may be assessed only during the lag period of growth before bacterial division resumes.[47,245,246] The efficiency of killing (defined as the percentage of ingested bacterial killed), which is independent of the efficiency of actual ingestion,[247] of various cell lines has been determined.

A number of methods have been developed for measuring the killing of microorganisms by neutrophils. One of the most common techniques relies on the washing of neutrophils following the ingestion phase (to remove adherent noningested bacteria), subsequent lysis of the phagocytes, and plating of the suspension on agar growth media. The number of bacterial colonies is taken as a measure of the bacterial concentration in the lysate. Clearly, the technique relies on the assumption of "one colony— one bacteria"—a relation which is often disputed. Such methods are generally referred to as "plating techniques." One such method is described in Protocol 24. The bactericidal assay measures the end point of several processes including opsonization, ingestion, and the metabolic alterations which are a prerequisite for bacterial killing. The limitations of this approach include those mentioned above relating to the separation of ingestion from nonspecific adherence (cf. Section 5.1.1 and Protocol 2). As a result of the number of diverse processes involved, the interpretation of the final killing "rate" is difficult. Furthermore, the test requires fairly long incubation times since the determination of viability is generally performed under nonsaturated conditions. During this time phagocyte aggregation (i.e., cell clumping) or bacterial agglutination can occur. A clump of bacteria will give rise to a single bacterial colony on an agar plate, thus leading to a gross underestimation of the number of viable organisms.

5.3.1. Candicidal Activity. An alternative method relies on the ability of viable organisms to exclude dye uptake. In one such technique the ingested particles, following lysis of the neutrophils, are exposed to methylene blue dye.[103] The dye is able to penetrate dead but not viable microorganisms. The most common microorganism employed in such studies is *Candida albicans.* In addition to the nonspecific methylene blue dye exclusion test, a specific staining method for assessing *Candida* viability has

[242] R. R. Martin, G. A. Warr, R. B. Couch, H. Yaeger, and V. Knight, *J. Infect. Dis.* **129,** 110 (1974).

[243] C. S. F. Easmon, *Br. J. Exp. Pathol.* **60,** 24 (1979).

[244] D. Adam, P. Phillip, and B. H. Belohradsky, *Aertzl. Forsch.* **25,** 181 (1971).

[245] D. Adam, W. Schaffert, and W. Marget, *Infect. Immun.* **9,** 811 (1974).

[246] D. Adam, F. Staber, B. H. Belohradsky, and W. Marget, *Infect. Immun.* **5,** 537 (1972).

[247] S. R. Simmons and M. L. Karnovsky, *J. Exp. Med.* **138,** 44 (1983).

been developed.[103] A plating method similar to that for bacteria has also been designed for use with *Candida*. These three variations for assessing the killing ability of human neutrophils with respect to *Candida* are described in Protocol 25A and B.

6. Protocol for Evaluating Suspected Phagocytic Dysfunction

The previous section outlined the principle of some of the assays which have been developed for assessing various aspects of the phagocytic process. More than a dozen of such tests have been developed.[248] They are listed in Table II together with comments regarding their principle of operation, the purpose of the test, and the range of normal values to be expected.

Phagocytic dysfunction may occur as a result of defects at one or more steps in the phagocytic process. Obviously it is generally not possible, nor even desirable, to perform all of these tests in order to screen for various clinical disorders. Neutrophil dysfunctions may be grouped into four broad categories: (1) margination, migration, and chemotactic disorders; (2) opsonization, receptor recognition, and ingestion disorders; (3) defective granule functions; and (4) impaired oxidative metabolism.

Listed in Table III are some selected tests which, when performed, will enable the investigator to rule out certain neutrophil dysfunction syndromes. Since chemotactic defects are fairly common, the neutrophils being studied should be evaluated in both the absence and presence of their homologous serum. Chemotaxis[249] and chemokinesis[250] and the various tests thereof have been described in detail elsewhere in this volume. If chemotaxis screening indicates a serum defect the fresh serum should be quantitated for total hemolytic complement activity and the serum concentration levels of the various immunoglobulins (types and subclasses).

Thereafter the following strategy is suggested.

1. The leukocyte adherence test, employing nylon wool fibers, is an easy and rapid test to perform.[251,252] It provides useful information about

[248] D. R. Miller, H. A. Pearson, R. L. Baehner, and C. W. McMillan in "Smith's Blood Diseases of Infancy and Childhood" (S. R. Simmons and M. L. Karnovsky, *et al.*, eds.), p. 540. The CV Mosby Co., St. Louis, Missouri, 1978.

[249] R. Snyderman, L. C. Altman, M. S. Hausman, and S. E. Mergenhagen, *J. Immunol.* **108,** 857 (1972).

[250] R. C. Wilkinson and R. B. Allan, *in* "Leukocyte Chemotaxis: Methods, Physiology and Clinical Implications" (J. I. Gallin and P. G. Quie, eds.), pp. 1–24. Raven Press, New York, 1978.

[251] R. R. MacGregor, P. J. Spagnuolo, and A. L. Lentnek, *N. Engl. J. Med.* **291,** 642 (1974).

[252] F. L. Rasp, C. C. Clawson, J. R. Hoidal, and J. E. Repine, *J. Reticuloendothel. Soc.* **25,** 101 (1979).

TABLE II
NEUTROPHIL FUNCTION TESTS[a]

Function	Principle of test	Purpose of test	Normal values
1. Microtubule			
Adherence	Neutrophils are passed over a microcolumn of nylon fiber packed in a Pasteur pipet and the percentage of cells adhering is calculated	Measures surface adherence of neutrophils or monocytes	60 to 80%
Chemotaxis	Neutrophils are sedimented on a Millipore filter, placed in Böyden chamber, and incubated for 3 hr at 37°, with a source of chemotactic factor placed on the opposite side of the filter. Chemotactic index is calculated by assessing the number of neutrophils migrating completely through the filter	Evaluates directed neutrophil movement in response to a variety of chemotactic factors. Test can be adapted to study cells, serum, or the effects of pharmacologic agents on chemotaxis	Chemotactic index: 13 ± 5 (without chemotactic factor), 67 ± 16 (with chemotactic factor)
Fluorescein–Concanavalin A capping	Neutrophils incubated with fluorescein–Con A; the movement of fluorescence into polar aggregates represents microtubule depolymerization	Tests for factors that control microtubule polymerization in neutrophils. Drugs (e.g., colchicine) and diseases (e.g., Chediak–Higashi syndrome) that interfere with microtubule assembly increase number of capped neutrophils	Fluorescence distribution: random >90%; capped <10%

(continued)

TABLE II (*continued*)

Function	Principle of test	Purpose of test	Normal values
Degranulation of lysosomal enzymes into phagosomes	Emulsified liquid phthalate oil, containing oil red O is phagocytized by neutrophils. Phagocytic vesicles are isolated from homogenized cells on a density gradient. Vesicles rise to the top of the gradient (due to the internalization of oil droplets and hence lower density), and the other cell components sediment	Assesses lysosomal movement in neutrophils, dependent in part on normal microtubular assembly	$14.5 \pm 2.2\%$ of total β-glucuronidase released into phagocytic vesicles within 45 min
Degranulation of lysosomal enzymes into extracellular fluid	Cytochalasin B inhibits ingestion and converts neutrophils to "secretory" cells releasing lysosomal enzymes during incubation with opsonized zymosan particles	Assesses lysosomal movement in neutrophils, dependent in part on normal microtubule assembly	20% of total β-glucuronidase released in 30 min
2. Microfilament			
Phagocytic uptake of oil red O particles	Liquid hydrocarbon, stained by oil red O dye and emulsified by sonication, is stabilized with *Escherichia coli* lipopolysaccharide. Requires opsonization by alternative C' pathway for efficient uptake by neutrophils	Measures the rate of uptake of particles by neutrophils: also measures functional alternative pathway opsonic capacity	0.138 $(0.121–0.157)$ mg phthalate oil taken up/min/10^7 cells

3. Oxidase

Oxygen consumption	Cyanide-insensitive O_2 consumption is associated with, but not required for, phagocytosis by neutrophils and monocytes	Phagocytic cells reduce oxygen to hydrogen peroxide (H_2O_2), superoxide anion (O_2^-), singlet oxygen ($^1O_2$), and hydroxyl radical ($\cdot$OH). Some or all of these reduction products are required for bacterial killing	Resting: 7.4 ± 3.8 μl O_2 consumed/hr/10^7 cells. Phagocytosis: 37.6 ± 22.5 μl O_2 consumed/hr/10^7 cells
Superoxide release	Univalent reduction product of oxygen is O_2^-; it and other substances reduce ferricytochrome c. The amount of ferricytochrome c reduction which is inhibited by purified superoxide dismutase (SOD) is due only to O_2^-	Measures release of O_2^- from neutrophils and monocytes at rest and during phagocytosis	Resting: 0.50 nmol O_2^-/15 min/10^7 cells. Phagocytosis: 1.0 nmol/15 min/10^7 cells
Nitroblue tetrazolium (NBT) reduction	Solubilized redox dyes of tetrazolium salts when incubated with neutrophils are reduced to insoluble formazan. Reduction occurs to a larger extent during phagocytosis. Addition of superoxide dismutase decreases amount of formazan formed, indicating that NBT reduction reflects O_2^- generation by neutrophils	Detects carriers of the X-linked form of the chronic granulomatous disease (CGD): detects all patients affected with CGD	Resting: 0.088 ± 0.040 OD/15 min/10^7 cells. Phagocytosis: 0.319 ± 0.112 OD/15 min/10^7 cells
$[1\text{-}^{14}C]$Glucose $\rightarrow$ $^{14}CO_2$	Oxidation of glucose through the hexose monophosphate shunt is stimulated in neutrophils and monocytes during phagocytosis	NADPH oxidation by O_2 and/or H_2O_2 occurs in neutrophils with an intact NADPH/NADH oxidase system activated during phagocytosis	Resting: 62.6 ± 10 nmol glucose oxidized/30 min/5 $\times$ 10^6 cells. Phagocytosis: 169 ± 28 nmol glucose oxidized/30 min/5 $\times$ 10^6 cells

(continued)

TABLE II (continued)

Function	Principle of test	Purpose of test	Normal values
[^{14}C]Formate → ^{14}CO$_2$	Oxidation of formic acid by neutrophils requires H$_2$O$_2$ and is catalyzed by catalase	This is an indirect quantitation of H$_2$O$_2$ produced during phagocytosis by neutrophils	Resting: 0.6 (0.2–1.1) nmol formate oxidized/hr/mg protein. Phagocytosis: 2.8 (1.1–5.9) nmol formate oxidized/hr/mg protein
Hydrogen peroxide release	Scopoletin fluorescence is extinguished when oxidized by H$_2$O$_2$ in presence of horseradish peroxidase	Provides sensitive, direct quantitation of release of H$_2$O$_2$ from neutrophils during phagocytosis	Resting: 0.012 ± 0.003 nmol H$_2$O$_2$ released/min/2.5 × 10^6 cells. Phagocytosis: 0.445 ± 0.064 nmol H$_2$O$_2$ released/min/2.5 × 10^6 cells
Iodination	Bacterial and other particulate proteins are iodinated following their ingestion by neutrophils and monocytes. Biochemical requirements are the presence of halide ions, peroxidase, and hydrogen peroxide	Measures capacity of specific granule peroxidase in phagocytes (neutrophils and monocytes) to be discharged into phagosomes; assesses available pool of H$_2$O$_2$ in region of phagolysosomes to iodinate ingested particles	Resting: 0.04 ± 0.03 nmol iodide consumed/hr/10^7 cells. Phagocytosis: 3.95 ± 0.82 nmol iodide consumed/hr/10^7 cells
Chemiluminescence	Measures light produced by neutrophils and monocytes during phagocytosis of zymosan	Measures capacity of neutrophils and monocytes to reduce O$_2$ to H$_2$O$_2$, O$_2^-$, and ^{1}O$_2$; all three are required for optimal chemiluminescence	142.5 ± 64 × 10^3 counts/min/13 min/10^7 cells
4. Bactericidal			
Bacterial killing	Opsonized live bacteria are ingested and killed by phagocytic blood cells	Measures capacity of phagocytes to kill opsonized bacteria; may also be used to measure opsonic activity of sera	Less than 10% of most bacterial species viable after 60 min at 37° incubation with an equal number of neutrophils

[a] Modified from D. R. Miller, H. A. Pearson, R. L. Baehner, and C. W. McMillan (eds.). "Smith's Blood Diseases of Infancy and Childhood," 4th Ed., p. 540. The CV Mosby Co., St. Louis, Missouri, 1978.

TABLE III
SELECTIVE LABORATORY TESTS FOR NEUTROPHIL
DYSFUNCTION ASSESSMENT

1. Direct microscopic evaluation (morphological abnormalities)
2. Boyden chamber chemotaxis
3. Nylon fiber adherence
4. Oil red O paraffin oil ingestion
5. Superoxide anion-dependent NBT reduction
6. Bactericidal assay

membrane surface properties which can be correlated with defects in the ability of the phagocyte to attach, or ingest, bacteria.[253,254]

2. The ability of neutrophils to recognize opsonized particles and to ingest them is best studied by employing the oil red O technique (see this volume [5]). This assay provides information about both membrane receptor recognition capability and ingestion capacity. The test performed with nonopsonized (i.e., albumin-stabilized) oil red O droplets will enable the investigator to evaluate defects due to membrane recognition as opposed to defects in the ability to actually ingest particles.

3. The NBT-reduction tests are an excellent means for screening for disorders of oxidative metabolism. The NBT test can be used to easily diagnose patients with chronic granulomatous disease (CGD) and related disorders. This dye reduction test can also be employed to identify the carrier state of CGD in the x-linked form.

4. The *in vitro* bactericidal activity assay, employing both normal serum and patient's serum (as separate opsonin sources), is an important assay when evaluating patients with recurrent infections.

5. Careful microscopic evaluation of a peripheral blood smear should always be performed. Specifically, the investigator should look for morphologic abnormalities in the leukocytes. Such abnormalities are readily observed in the case of Chediak–Higashi syndrome.

6. Histochemical stains are also useful; particularly when performed as an assessment of a specific enzyme marker (cf. Section 7), e.g., myeloperoxidase, or for the evaluation of certain hematologic malignan-

[253] D. R. Absolom, C. J. van Oss, R. J. Genco, D. W. Francis, and A. W. Neumann, *Cell Biophysics* **2**, 113 (1980).
[254] C. J. van Oss, D. R. Absolom, and A. W. Neumann, *in* "The Reticuloendothelial System" (S. M. Reichard and J. P. Filkins, eds.), Vol. 7A, p. 3. Plenum Press, New York, 1984.

TABLE IV
HISTOCHEMICAL DIFFERENTIATION BETWEEN
NEUTROPHILS AND MONOCYTES

Stain	Neutrophils	Monocytes
Peroxidase	$++^a$	$-/+$
Esterase		
α-Naphthyl acetate	$-/\pm$	$+++$
α-Naphthyl butyrate	$-$	$++/+++$
Naphthol chloroacetate	$+/++$	$-/\pm$
Periodic acid stain (PAS)	$+++$	$+$
Alkaline phosphatase	$+/++$	$-$
Acid phosphatase	$+/++$	$++$
Sudan black	$++$	$-/\pm$

^{a} Key: $-$, negative; $\pm$, weakly positive (faint staining or a few positive cells); $+$, moderate; $++$, strong; $+++$, strongly positive (virtually all cells).

cies, e.g., acute granulocytic and promyelocytic leukemia, etc. (cf. Section 9).

The aforementioned group of selected tests will enable the investigator to rule out one or more phagocytic dysfunction syndromes as the cause of any observed abnormality in phagocytic function.

7. Specific Stains for the Differentiation of Phagocytic Cell Type

Very often in phagocytic studies it is necessary to establish that the acquired data are due to ingestion by one specific phagocytic cell type as opposed to another. This is particularly important in direct microscopic assessment studies (cf. Section 5.1) in which often whole blood is employed. A number of histochemical stains exist which permit a clear differentiation between neutrophils and monocytes. The differential staining properties for these two cell types are illustrated in Table IV. For a complete discussion of histochemical staining techniques, as applied to the diagnosis and classification of hematological malignancies, the interested reader is referred to references.[255,256]

Discussed below are the principles of some of these staining techniques while technical details are summarized in Protocols 26–31.

[255] M. J. Cline, in "Leukocyte Function" (M. J. Cline, ed.), p. 130. Churchill Livingstone, New York, 1981.
[256] W. J. Williams, "Hematology." McGraw Hill, New York, 1977.

7.1. Leukocyte Peroxidase

As mentioned previously, peroxidase is a component of the primary granules and its activity in neutrophils and monocytes appears to be identical. This enzyme is important in the killing (or digestion) of microorganisms.

One technique for assaying for peroxidase utilizes the oxidation of benzidine dihydrochloride,[257] in the presence of hydrogen peroxide, to yield an insoluble brown product which is deposited at the site of enzyme activity (Protocol 26A). As a result of its reported carcinogenicity, care should be exercised when working with benzidine. In view of this danger, alternative staining procedures (Protocol 26B) have now been developed that rely on the use of 3-amino-9-ethylcarbazole.[258]

7.2. Alkaline Phosphatase

The histochemical reaction for alkaline phosphatase is positive for neutrophils and negative for monocytes.

The basis for the assay relies on the production of an insoluble hydrolysis reaction compound which is visible as brightly colored crystalloid specks in the cytoplasm.[259] The enzyme hydrolyzes naphthol AS-BI phosphate into the phosphate and reactive arylnaphtholamide moieties. The latter is readily coupled to a diazonium salt to yield the insoluble crystalline product. The precipitate is situated as discrete specks in the cytoplasm of neutrophils. The test is described in detail in Protocol 27.

7.3. Acid Phosphatase

Acid phosphatase hydrolyzes a number of hydroxynaphthoic acid anilide compounds to yield insoluble naphthols that readily complex, under acid pH conditions, with diazonium salts.[260,261] The insoluble precipitate is a colored compound and readily detected microscopically (Protocol 28). The diazonium salt used is hematoxylin, which in the presence of aluminum and ammonium sulfate is oxidized to the red-colored hematin precipitate. Thus, acid phosphatase activity is demonstrated by the presence of red granular precipitates. Moderate staining is seen in monocytes and relatively weak staining is present in mature neutrophils.

[257] L. S. Kaplow, *Blood* **26**, 215 (1965).
[258] L. S. Kaplow, *Am. J. Clin. Pathol.* **63**, 451 (1975).
[259] L. S. Kaplow, *Am. J. Clin. Pathol.* **39**, 943 (1963).
[260] M. S. Burstone, *J. Natl. Cancer Inst.* **21**, 523 (1958).
[261] T. Barka and P. J. Anderson, *J. Histochem. Cytochem.* **10**, 741 (1962).

7.4. Esterase Activity

Leukocytes contain both specific (substrate: chloroacetate) and nonspecific (substrate: α-naphthyl acetate or α-naphthyl butyrate) esterases. The reaction product of chloroacetate (specific esterase) is red, and that of both α-naphthyl acetate and α-naphthyl butyrate (nonspecific esterase) is orange–brown, permitting ready differentiation between specific and nonspecific enzyme content.

The specific esterase has a pH optimum between 7.0 and 7.6 and is insensitive to the presence of fluoride ions.[262] The specific esterases are found primarily in the azurophil granules of granulocytes.[263] The reaction with α-naphthol chloroacetate as the substrate is negative for monocytes and strongly positive for neutrophils (Protocol 29A).

The nonspecific esterases have a pH optimum between 6.0 and 6.3 and are inhibited by the presence of fluoride ions.[262] These nonspecific esterases generally stain monocytes the most strongly with neutrophils very seldom giving a positive reaction with α-naphthyl acetate. Staining with α-naphthyl butyrate is less sensitive than the acetate derivative, but it has the advantage of being more specific than the acetate esterase for detecting cells of the mononuclear phagocytic series. The esterase reaction, using either α-naphthyl acetate or α-naphthyl butyrate as the substrate, is strongly positive in mononuclear phagocytes and weak in neutrophils (Protocol 29B).

The assay relies on the ability of the leukocyte esterase to hydrolyze ester derivatives of naphthalene. This results in the production of a naphthol (or naphthyl) compound which is extremely reactive, and complexes with added diazonium salts to produce an insoluble colored precipitate at or near the site of enzyme activity within the cell cytoplasm.

7.5. Lipid Content

Sudan black B stains various lipids, including neutral fat, phospholipids, and sterols.[264] The extent of stain uptake corresponds to granule content. Hence, mature neutrophils (or neutrophils derived from patients with, e.g., acute granulocytic leukemia) contain many stained granules. Monocytes are generally unstained or, at most, contain only a few discrete stained granules (Protocol 30).

[262] C. Y. Li, K. W. Lam, and L. T. Yam, J. Histochem. Cytochem. **21,** 1 (1973).
[263] W. C. Moloney, K. McPherson, and L. Fliegelman, J. Histochem. Cytochem. **8,** 200 (1960).
[264] R. D. Lillie and H. J. Burtner, J. Histochem. Cytochem. **1,** 8 (1953).

7.6. Cellular Polysaccharides

The periodic acid–Schiff (PAS) stain oxidizes glycols and related compounds to aldehydes. The aldehydes react with leukofuchsin (Schiff reagent) to stain intracellular components, principally glycogen.[265,266]

Thus, PAS staining (Protocol 31) can be used to differentiate between neutrophils (in which the cytoplasm stains a deep pink or red in either a uniform or granular pattern) and monocytes (cytoplasm stains only faintly in a granular pattern).

8. Experimental Protocols

8.1. VIABILITY

A comparison of the efficiency of the dye exclusion tests for assessing cell viability has been published elsewhere.[267]

Protocol 1A: Determination of Viability by Trypan Blue Exclusion

The number or percentage of viable leukocytes in a cell suspension can be determined by staining cell populations with trypan blue.[1] Viable cells exclude the dye, while nonviable cells take up the dye, thereby enabling a direct visual distinction between unstained viable cells and blue-stained nonviable cells. After being stained with trypan blue, the cells must be counted within 3 min; after that time viable cells begin to take up the dye. Also, since trypan blue has a great affinity for proteins,[268] elimination of serum from the cell diluent will allow a more accurate determination of cell viability.

Materials and Reagents

Cell suspension at 2–5 × 10^6 cells/ml
Trypan blue, 0.2% (w/v) in water
5× Normal saline: 4.25% NaCl (w/v)

Procedure

1. On the day of use, mix four parts of 0.2% trypan blue with one part of 5× saline.
2. To one part of the trypan blue saline solution, add one part of the cell suspension (1 : 2 dilution).

[265] G. B. Wislocki, J. M. Rheingold, and E. W. Dempsey, *Blood* **4,** 562 (1949).
[266] D. Quangliano and F. G. J. Hayhoe, *Br. J. Haematol.* **6,** 26 (1960).
[267] W. E. Hathway, L. A. Newby, and J. H. Githens, *Blood* **23,** 517 (1964).
[268] P. F. Kruse and M. K. Paterson, *in* "Tissue Culture: Methods and Applications." Academic Press, New York, 1973.

3. Place cells in a hemacytometer and determine the number of unstained (viable) white blood cells and stained (dead) cells separately. For greater accuracy, count more than a combined total of
200 cells:

Number of viable cells/ml =
$$\text{(average number of viable cells in large square)}(10^4/\text{ml})(1/\text{dilution}) \quad (11)$$

$$\text{Percentage viable cells} = \left(\frac{\text{number of viable cells}}{\text{number of viable cells} + \text{number of dead cells}} \right) 100\% \quad (12)$$

Protocol 1B: Determination of Viability by Eosin Y Exclusion

The advantage of using eosin Y as a vital stain is that the time elapsed
before examining the cells is less critical than for trypan blue exclusion.[2]
The percentage of viable cells remains constant from 1–10 min after staining with eosin Y. However, some investigators find red (eosin Y)-stained
cells more difficult to recognize than blue (trypan blue)-stained cells.

Materials and Reagents

Cell suspension adjusted to 2×10^6 cells/ml
Saline, 0.85% NaCl (w/v)
Eosin Y, 0.2% (w/v) in saline. This solution is stable at room temperature. If a precipitate forms, pass the solution through Whatman #1
filter paper

Procedure

1. To one part of cell suspension, add one part of 0.2% eosin Y.
2. Determine the number of viable cells/milliliter and the percentage
of viable cells by the procedure described in the previous method.

Comment. Eosin Y may bind to proteins; therefore, elimination of
serum from the cell diluent promotes a more accurate determination of
cell viability.

Protocol 1C: Determination of Viability by Nigrosin Exclusion

This procedure is especially advantageous in tests where stained cells
must remain for several hours before examination, without disintegration
and where living cells are not killed by prolonged exposure to the dye.
Nigrosin exclusion may be the more accurate measure of macrophage
viability since viable macrophages on occasion take up trypan blue and
eosin Y. For routine determination of cell viability, nigrosin staining has
some disadvantages: (1) The uptake of nigrosin is slower than that of

either trypan blue or eosin Y; (2) inexperienced workers often confuse dead (brown–black) cells and live cells that are not in focus.

Materials and Reagents

Cell suspension adjusted to 2–20 × 10^6 cells/ml

Nigrosin, 1% (w/v) in water; filter stock solution through Whatman #1 paper

Balanced salt solution or saline (0.85% NaCl, w/v) containing 2.5–5% fetal calf serum

Procedure

1. Dilute the 1% nigrosin stock 1 : 10 in medium containing 2–5% fetal calf serum, just prior to use.
2. Mix the cell suspension with the above 0.1% nigrosin solution such that the cell suspension:nigrosin solution ratio is between 1 : 2 and 1 : 10.
3. Wait 5–10 min; then determine the number of viable cells/milliliter and the percentage of viable cells as described for trypan blue.

Protocol 1D: Determination of Viability with Fluorescein Diacetate

Fluorescein diacetate is taken in by all cells but is only hydrolyzed inside live cells to yield a strong green fluorescence.[3]

Materials and Reagents

Phosphate-buffered saline (PBS)

Fetal calf serum (FCS)

Balanced salt solution containing 5% fetal calf serum (BSS–5% FCS)

Fluorescence microscope

Cell suspension at approximately 10^6 cells/ml in BSS–5% FCS

Fluorescein diacetate, 5 mg/ml in acetone, stored in a tightly capped container at $-20°$

Procedure

1. Dilute fluorescein diacetate solution (Sigma, St. Louis, MO) (1 : 50) in PBS at room temperature (a fine suspension forms). Immediately, add 1 vol of the fluorescein diacetate suspension to 9 vol of the cell suspension.
2. Allow the mixture to stand at room temperature for 15 min. Examine cells with a fluorescence microscope. Viable cells will appear bright green.
3. Determine the percentage of viable cells as described previously (cf. Protocol 1A).

Protocol 1E: Determination of Viability with Acridine Orange–Ethidium Bromide

A one-part-per-million solution of acridine orange (AO) and ethidium bromide (EB) can be used to stain viable and nonviable cells differentially.[4]

Live cells take up acridine orange, whereas nonviable cells take up ethidium bromide. When viewed through a fluorescence microscope, viable cells appear green, while nonviable cells appear orange.

Materials and Reagents

Cell suspension at approximately 10^6 cells/ml
AO (Sigma or Aldrich Chemical). Caution: AO is hazardous
EB (Sigma or Aldrich Chemical). Caution: EB is carcinogenic
Phosphate-buffered saline (PBS)
Hemacytometer and coverslip
Fluorescence microscope; a tungsten–halogen illuminator is adequate
    and preferable to a mercury lamp
Solution of one-part-per-million AO/EB: Dissolve 0.1 mg of both AO
    and EB in 100 ml of PBS; divide into aliquots and freeze. Once
    thawed, store in the dark at 4°

Procedure

1. Mix one part cell suspension with one part AO/EB solution.
2. Place cells in a hemacytometer and determine the number of green
   (viable) and orange (nonviable) cells, using both UV and visible
   illumination at the same time. (The fluorescence of the cells is so
   strong that it is not necessary to turn off the visible illumination
   while observing the cells.)

8.2. DIFFERENTIATION BETWEEN ATTACHMENT AND INGESTION

Protocol 2: Fluorescence Quenching of FITC-Labeled Bacteria by Crystal Violet

This method can be used to differentiate clearly between attached and ingested microorganisms.[51,52] This method has been discussed elsewhere in this volume [6] and will not be duplicated here.

8.3. DIRECT MICROSCOPIC ASSESSMENT

Protocol 3A: Phagocytosis of Opsonized Zymosan Determined by Light Microscopy

Opsonization of Zymosan. Zymosan (Sigma, St. Louis, MO) suspended at a concentration of 5 mg/ml in autologous serum, is incubated at

37° for 20 min in a shaking water bath. Opsonization is stopped by placing the suspension in melting ice. The suspension is then centrifuged at 160 g for 5 min at 0°, the serum poured off, and the zymosan washed once with Krebs–Ringer phosphate (KRP) before it is suspended at a concentration of 18 mg/ml in KRP.

If desired, the zymosan may be boiled for 30 min in 0.15 M saline prior to opsonization. The incubation in serum should not be continued for longer than 15 to 20 min. Opsonization, through the attachment of C3b, is complete by that time, and further incubation will only decrease the level of opsonization through the action of the C3b inactivator present in the serum.[269] Zymosan should be stored frozen in a desiccator.

Phagocytosis. The neutrophil suspension (5 × 10⁷ cells/ml in PBS) and the suspension of opsonized zymosan are separately brought to 37°. Phagocytosis is begun by adding 0.50 ml of the zymosan suspension to 0.50 ml of neutrophils. Cells and zymosan are incubated together at 37° in a shaking water bath for various periods of time. At the desired time interval, a 0.1-ml portion of the incubation mixture is withdrawn and added to 0.1 ml of an ice-cold solution of *N*-ethylmaleimide (Sigma, St. Louis, MO) 2 mM in PBS. The remainder of the incubation mixture is discarded.

The cells and the zymosan should be mixed as soon as possible after both reach 37°. Every assay should include a sample in which the reaction is stopped immediately after the cells are mixed with the particles (zero time) and a sample in which cells and particles are incubated together on ice for the duration of the longest 37° incubation.

Measurement of Uptake. The remainder of the procedure is carried out at room temperature. To the *N*-ethylmaleimide-treated reaction mixture, add 2.0 ml of 0.25 M sucrose containing 0.1 mM disodium EDTA, 5 μM colchicine (Sigma, St. Louis, MO) and 1% bovine serum albumin. Disrupt neutrophil clumps by agitation, then deposit the cells, in 0.4 ml of the suspension, onto a glass slide with a cytocentrifuge (Shandon Southern). Dry the slide in air, fix for 10 min in absolute ethanol : formalin (9 : 1, v/v). Then stain the zymosan red using the PAS reagent, and counterstain the neutrophils with methyl green (2% aqueous solution for 5 min) followed by fast green (2% aqueous solution for 5 min). Finally, count the number of zymosan particles associated with each of at least 100 cells under oil immersion. The data may be presented as the average number of particles associated with each cell, or a distribution may be calculated in the form of a histogram showing the number of particles on the x-axis and the fraction of all cells associated with that number of particles on the y-axis.

The colchicine and EDTA treatments ensure uniform spreading of the cells during cytocentrifugation, whereas bovine serum albumin attaches

[269] J. Noseworthy and M. L. Karnovsky, *Enzyme* **13**, 110 (1972).

them firmly to the slide. The concentration of neutrophils and zymosan particles used in the incubation can present difficulties: too many cells or zymosan particles will give a smear too thick to interpret; if the numbers are too low, phagocytosis will not be optimal. This is a particular problem in experiments in which comparisons are made between phagocytosis and other neutrophil functions, because in such experiments it is best to use incubation conditions as similar as possible among the various assays, a consideration that limits the experimental design, possibly leading to problems with the cytocentrifuge smears. The conditions given above produce slides with a suitable number of neutrophils but more than an ideal number of zymosan particles.

Protocol 3B: In Vitro Monolayer Phagocytosis Test

Materials

Suspensions of *Staphylococcus epidermidis, Escherichia coli, Staphylococcus epidermidis* opsonized with pooled human IgG (Cohn fraction II)
Blood anticoagulated with EDTA or isolated neutrophil suspension (1×10^6 cells/ml)
Sterile HBSS containing 1% human albumin
Molten paraffin wax
Mackaness culture chambers
Cleaned round glass coverslips siliconized with hexamethyldichlorosilane (HMDS; Dow Corning)
Sterile 2.50-ml syringe fitted with a 26-gauge $\frac{1}{2}$-in. needle

Procedure

1. The bacteria are harvested by centrifugation, washed twice in saline, and resuspended in HBSS (pH 7.4) to give a final concentration of 2×10^8 bacteria/ml.
2. Whole blood or the neutrophil suspension is then placed in contact with the HMDS-treated coverslips and neutrophils allowed to adhere to them. The coverslips are placed in a moist chamber at 37° and incubated for 25 min.
3. Transfer the coverslips from the incubator to a Petri dish containing saline (37°); gently agitate until the clot floats clear of the glass surface. Continue this process until most of the residual red blood cells and fibrin are removed. The resulting neutrophil monolayer contains approximately 300 cells/mm².
4. Each coverslip is removed with as much saline as possible and inverted into its place on the Mackaness chamber prepared in ad-

vance. Dry the outer surface with a paper towel and quickly seal into position with molten paraffin.

5. Fill the resulting chamber with 0.8 ml HBSS (37°), leaving no air bubbles, using the 26-gauge needles. Invert the chamber and incubate at 37° for 15 min.

6. Drain the chamber and refill with 0.8 ml of a standardized suspension of bacteria or particles and incubate as indicated above.

7. Disassemble the chamber on a towel saturated with disinfectant. Gently wash the coverslips containing the monolayer in 37° HBSS and allow to air dry. Note: Do not blow dry! The cell side must face up.

8. Fix the coverslips to microscope slides with permanent mounting fluid, with the monolayer on the exposed surface, and gram stain. Note: Do not heat fix the film.

9. Examine under oil immersion and count a total of 100 neutrophils; record the number of neutrophils and the total number of cells engaged in phagocytosis, as well as the number of particles ingested per cell. These counts are taken on two transverses of the monolayer at right angles to each other through the center of the coverslip. The mean number of particles in the cells engaged in phagocytosis is the avidity index. The proportion of neutrophils that actually engulfed particles is the phagocytic index. The phagocytic activity is the mean number of particles engulfed per neutrophil (equal to the product of these two indices).

8.4. ISOTOPIC LABELING METHODS

Protocol 4: Ingestion of Radioactive Bacteria

Bacteria. Media for growing bacteria are obtained from Difco. *S. aureus* type 502A, a penicillin-sensitive strain routinely used in the study of phagocyte function, is purchased in a lyophilized form from the American Type Culture Collection, and is germinated in broth culture, according to the directions supplied by the vendor. The germinated bacteria are maintained by streaking from the broth onto slants of nutrient agar, which are incubated overnight at 37°, then stored in the refrigerator. The stored bacteria should be transferred weekly by growing them overnight in nutrient broth (Difco) and preparing fresh slants from this growth.

Radioactively labeled bacteria are prepared by growing *S. aureus* overnight in 10 ml of nutrient broth containing 10 μCi of radioactive amino acids ([14C]-labeled amino acids, New England Nuclear). The bacteria are then washed three times by centrifugation at 14,000 g for 10 min in

PBS. An aliquot is removed for determination of radioactivity and enumeration of bacterial colony-forming units, as described below. The remainder are killed by heating for 20 min in boiling water, suspended in PBS at 2×10^9 organisms/ml, and stored at $-20°$.

E. coli K12 can be maintained on trypticase-soy agar. For use, they are grown overnight in minimum essential medium (MEM) supplemented with 5 mM glucose containing 50 μCi uridine [^{14}C]glucose (New England Nuclear). The bacteria are then washed three times by centrifugation at 14,000 g for 10 min in PBS. An aliquot is removed for determination of radioactivity and enumeration of bacterial colony-forming units, as described below. The rest of the bacteria are killed by autoclaving and stored until needed at 10^9 bacteria/ml in PBS.

Measuring the Concentration of Bacteria. The concentration of bacteria is determined spectrophotometrically by measuring the optical density of the bacterial suspension with respect to its concentration of colony-forming units (see below; this calibration should be performed for each species and strain of microorganism employed). Radioactivity is determined by liquid scintillation counting, using a small volume of bacterial suspension that has been mixed with an equal volume of solubilizer [hydroxide of Hyamine (New England Nuclear) or NCS (Nuclear Chicago)] before adding the scintillation fluid in order to minimize quenching. With suspensions of *S. aureus* 502A, an optical density of 0.1 corresponds to roughly 5×10^7 organisms/ml. Staphylococci grown according to the above conditions should contain approximately 10,000 dpm/10^7 bacteria; the specific radioactivity can be varied within wide limits by changing the amount of radioactivity added to the broth culture during the bacterial growth phase.

Colony Counting. The relationship between optical density and the concentration of microorganisms in bacterial suspension is established by counting colony-forming units in a suspension of known optical density. For this purpose, 85-mm Petri dishes, containing a layer of nutrient agar (*S. aureus*) or trypticase-soy agar (*E. coli*), are prepared by placing 30 ml of molten 1.2% agar (melting point about 50°) in the dishes, and letting them cool to room temperature. Serial dilutions of the bacterial suspension are prepared by placing 10 μl of the suspension in 50 ml of sterile water, then using this diluted suspension to make 4 successive 10-fold dilutions in sterile water. These serial dilutions are made in duplicate or triplicate. The final volume used of each dilution is 0.25 ml. To each dilution is added 5 ml of liquid 0.6% nutrient agar (melting point: 40 to 42°), which has been cooled briefly from 45°. This mixture is immediately poured into one of the agar-containing Petri dishes prepared previously, held at room temperature for 30 min to permit the new layer of agar to solidify, then placed upside down in a 37° oven and incubated for 24 to 72 hr. Plates containing 50 to 500 colonies are selected and the colonies enumerated. From the number of colonies on the dish, the degree of dilution, and the optical density of the original suspension, the relationship between optical density and bacterial concentration can be calculated.

Colonies may grow poorly for several reasons. The bacteria may clump, so that a single colony-forming unit consists of more than one live microorganism; this does not usually occur with *Staphylococcus*, but if it presents a problem, it may be alleviated by including 1% serum albumin in the suspending medium. The bacteria may be killed if the agar in which they are finally suspended is too hot. Finally, the bacteria may be killed by a contaminating phage. This may be diagnosed by pouring an undiluted suspension of the bacteria directly onto an agar gel-containing Petri dish and incubating overnight; phage contamination will be indicated by the presence of many tiny clear areas (plaques) in what should normally be a uniform lawn of bacteria. Phage contamination is eliminated by careful sterilization of all constituents of the preparation: agar, buffers, glassware, and so on. Salts may precipitate during the sterilization of buffer, in which case fresh buffer must be prepared using autoclaved water.

A second problem is contamination. Contaminants may appear on the Petri dishes, a minor problem, or they may overgrow the desired bacteria in the broth culture. Contamination is indicated by a change in the appearance of the colonies on agar or an alteration in the properties of the bacteria during experimental manipulations. It is confirmed by bacteriologic identification of the microorganisms, a procedure that should be carried out from time to time as a routine precaution.

Measurement of Uptake. The heat-killed radiolabeled bacteria are thawed and sonicated at 30 cycles for 10 sec to disperse clumps. Bacteria (2×10^9) are then added to 1 ml of fresh serum and incubated for 20 min at 37° to ensure adequate opsonization. The bacteria are then centrifuged at 14,000 g for 10 min, washed once with KRP, and then resuspended in KRP at a final concentration of 2×10^8 (*E. coli*) or 2×10^9 (*S. aureus*) bacteria/ml. One-half milliliter of bacteria is then added to an equal volume of PBS containing 2.5×10^6 neutrophils. The suspension is shaken at 37° in a water bath for 20 min, after which 10 ml of ice-cold 0.15 M NaCl containing 1 mM N-ethylmaleimide is added. The neutrophils are then separated from the uningested bacteria by centrifugation at 280 g for 5 min. The cell pellet is then washed three times with 10 ml of ice-cold saline. The third wash should be checked in a scintillation counter to ensure that no radioactivity is present. The final pellet is resuspended in 0.5 ml normal saline and 0.1 ml fractions are added to 0.1 ml NCS (Nuclear Chicago) or Hyamine (New England Nuclear) and the radioactivity is determined. The radioactivity of a 0.04-ml portion of the opsonized bacteria only is also counted to determine the percentage of bacteria ingested by the neutrophils. Under normal circumstances, 10 to 20% of the bacteria are ingested.

As mentioned above, it is important to distinguish between attachment and ingestion. In this assay the appropriate control for attachment is to perform the incubation in the presence of 1 mM N-ethylmaleimide, which prevents ingestion.[269,270] This completely inhibits ingestion but does not affect nonspecific attachment of bacteria to the cells.

8.5. EXTRACTION METHODS

Protocol 5: Ingestion of Oil Droplets Containing Oil Red O

This method has been presented elsewhere in this volume [5] and will not be duplicated here.

Protocol 6: Simultaneous Oil Red O–NBT Reduction

NBT (Sigma Chemical Company, St. Louis, MO) is dispersed in KRP at a concentration of 2 mg/ml. The solubility of this material is variable and maximum solubility is approximately 0.6 mg/ml. The solution suspension therefore must be carefully filtered, either through 0.8-μm Millipore filters (Millipore Corp.) or Whatman #1 filter paper. If sterile equipment is used, the former technique has the advantage that the filtrate can be considered sterile and stored in lots for longer periods of time than nonsterile solutions. It is important that microbial contamination be minimized, since bacteria will enhance NBT reduction by the cells.

The opsonized oil droplets are prepared as described elsewhere in this volume [5]. Incubations are set up as follows:

	Tube 1	Tube 2
Opsonized oil droplets	0.2 ml	0.2 ml
Cell suspension	0.4 ml	0.4 ml
KRP	0.4 ml	—
NBT in balanced salt solution	None	0.4 ml

The reaction is started by adding the cells to the other premixed, prewarmed reagents. Tube 1 yields the ingestion rate, and tube 2 gives the NBT reduction rate. The rest of the assay is done as described elsewhere in this volume [5], except that the NBT–formazan is extracted with dioxane by heating the tube in a boiling water bath for 15 to 30 min. The adsorbance of NBT–formazan dioxane extract is read at 580 nm (A_{580})

[270] G. L. Mandell, *J. Reticuloendothel. Soc.* **11,** 129 (1972).

against a dioxane blank. This value is then converted to micrograms of formazan by multiplying the A_{580} value by 14.14, the extinction coefficient of formazan.

Calculation of Results

1. Ingested oil:

Phthalate oil (mg)/10^7 cells/min =

$$\frac{(A_{525})(\text{conversion factor})}{(5 \text{ min})(\text{neutrophils} \times 10^7)} = \text{initial ingestion rate} \quad (13)$$

where the value for the conversion factor (CF) is obtained as indicated in [5] and A_{525} is the absorption maxima of oil red O.

To determine NBT reduction, the A_{580} value of tube 1 at 580 nm must be subtracted from that of tube 2, since oil red O absorbs somewhat at 580 nm. Thus,

$$\frac{(A_{580} \text{ of tube } 2 - A_{580} \text{ of tube } 1)14.14}{(5 \text{ min})(\text{phagocytic cells} \times 10^7)} = \text{initial NBT reduction rate} \quad (14)$$

where the initial NBT reduction rate is expressed in micrograms of formazan per 10^7 phagocytes per minute.

8.6. INDIRECT ASSESSMENT

8.6.1. Glycolysis

Protocol 7: Lactate Excretion

Lactic acid secretion into the intracellular fluid may be measured enzymatically with a commercially available kit containing the necessary reagents [kit No. 826-UV (lactic acid, Sigma)]. Neutrophils are suspended in PBS at 10^7 cells/ml. One milliliter of the suspension is centrifuged at 250 g for 5 min, and the supernatant placed on ice. The remainder of the cell suspension is incubated for the desired time at 37°, then centrifuged to remove cells, and the supernatant placed on ice. At this point, the materials in the kit are employed. Glycine buffer (2.0 ml), 4.0 ml water, and 0.1 ml lactate dehydrogenase solution are placed in the commercial NAD-containing vial and mixed. From this vial, 1.4-ml portions are placed into each of two test tubes. Into these tubes are placed 0.1-ml portions of the two supernatants. These mixtures are then incubated at 37° for 30 min, after which their absorbance at 340 nm (A_{340}) against a water blank is determined. Lactate production is then calculated as follows:

$$\text{Lactate } (\mu\text{mol}/10^7 \text{ cells per time interval}) = 2.4(A^t_{340} - A^o_{340}) \quad (15)$$

where A_{340}^t is the adsorbance of the tube containing the second (reaction) supernatant and A_{340}^o is the absorbance of the other tube. A representative figure for normal human neutrophils is 0.15 μmol/10^7 cells/60 min.

8.6.2. *Degranulation*. Degranulation can be measured as the release of granule contents into a phagocytic vesicle or its release into extracellular fluid (exocytic degranulation). The assay of degranulation into phagocytic vesicles involves the isolation of these vesicles and the measurement of their enzyme content. For exocytic degranulation, cells are stimulated with a soluble agent or with a particulate agent in the presence of an inhibitor of ingestion (e.g., cytochalasin B) so that granule contents are released *only* into the external medium.

Protocol 8A: Intracellular Degranulation

Thirty milligrams of *E. coli* lipopolysaccharide (Difco) is mixed with 3 ml KRP. One milliliter of diisodecyl phthalate (Matheson, Coleman, and Bell) is layered on top. The mixture is sonicated, with the probe tip just below the oil–aqueous interface, for 90 sec at 40 amps. The emulsion is cooled and the oil droplet, stabilized with the lipopolysaccharide, is then opsonized with an equal volume of fresh serum at 37° for 20 min, and placed on ice. For assay, 1 ml of the opsonized emulsion is incubated for 5 min with 4 ml of KRP containing 7 to 8 $\times$ 10^7 granulocytes. The ingestion reaction is stopped by the addition of 10 ml ice-cold 0.15 M NaCl (saline) and the cells are centrifuged at 280 g for 5 min. The uningested oil is removed from the tube and the supernatant decanted. The cells are washed three times with ice-cold saline and finally resuspended in 4 ml of 0.34 M sucrose[270a] in 1 mM potassium phosphate, pH 7.4. Thereafter, the cells are homogenized over ice for 5 min, in a Potter-Elvehjem homogenizer, with a Teflon pestle driven at 1200 rpm. Part of the homogenate (0.5 ml) is removed for subsequent enzyme assay and the rest is layered over 2.5 ml of 0.45 M sucrose in a 10-ml tube which is then centrifuged at 100,000 g for 1 hr. The phagocytic vacuoles, because they have ingested oil droplets, are less dense, float to the top of the 0.45 M sucrose, and are removed using siliconized Pasteur pipets. Granule enzyme content is then measured, as described below, in both the homogenate and the phagocytic vacuoles.

Protocol 8B: Exocytic Degranulation

Zymosan (Nutritional Biochemicals or Sigma), approximately 100 mg, is washed twice with 10 ml 0.15 M NaCl (saline), centrifuged at 280 g for

[270a] Sucrose, 0.34 M, is brought to pH 7.4 with 1.0 M NaOH. Alternatively, sucrose is dissolved in 1 mM potassium phosphate buffer, pH 7.4, to a final concentration of 0.34 M.

10 min, and then incubated with fresh serum at 20 mg/ml for 20 min at 37°. The opsonized zymosan is separated from the serum by centrifugation at 280 g for 10 min, washed once with KRP, and resuspended in KRP at a concentration of 15 mg/ml. Cytochalasin B (Aldrich Chemical Company) is dissolved at 0.5 mg/ml in dimethyl sulfoxide. This may be stored at 0 to 4° indefinitely.

PMA (Sigma) may be used instead of zymosan. This is dissolved in dimethyl sulfoxide at 2 mg/ml and stored desiccated at −20°. When needed, PMA is diluted into PBS to a final concentration of 20 μg/ml.

To assay for exocytic degranulation, neutrophils, 0.5 to 2.0 × 10^7 in 0.5 ml KRP, are mixed with 3 mg zymosan plus 5 μg cytochalasin B in 0.5 ml KRP. (Cytochalasin B prevents ingestion.) The cells are shaken at 37° for 30 min and the reaction stopped by placing the mixture on ice and centrifuging at 280 g for 10 min. The supernatant is removed and assayed for enzyme contents. PMA (final concentration 1 μg/ml) can be used in place of opsonized zymosan plus cytochalasin B (which inhibits ingestion), if only the discharge of specific granules is being examined.[270b] As a control, incubations without zymosan or PMA (but including cytochalasin B) should be carried out. To measure release due to cell destruction, lactate dehydrogenase or G-6-PD activity is also determined (cf. Protocols 20 and 21, respectively).

Intracellular degranulation is calculated from the concentration of enzyme in the phagocytic vesicles. Extracellular degranulation is calculated by subtracting the amount of enzyme released in the presence of zymosan or PMA from that released in its absence. The percentage degranulation is determined from this figure and the content of the granule constituents in whole cells. Enzymes commonly measured are myeloperoxidase, lysozyme, and β-glucuronidase. Cobalamin (vitamin B$_{12}$)-binding protein can also be determined. Specific assays for granule contents are described below. Typical values for extracellular degranulation are 10 to 20% with zymosan and 30 to 40% with PMA.

8.6.3. Respiratory Burst Activity

8.6.3.1. Oxygen Consumption

Protocol 9A: Manometric Measurement of Oxygen Uptake

Equipment. The assay is done with a submersible differential respirometer (Gilson). A flask containing the reaction mixture fits on one arm of the respirometer, while the other arm leads to an air-filled chamber.

[270b] R. D. Estensen, J. G. White, and B. Holmes, *Nature (London)* **248**, 347 (1974).

The two arms are connected by a U-shaped tube, which is partly filled with colored mineral oil. The reaction side of the manometer is fitted with an airtight plunger that can be inserted or withdrawn by a micrometer screw. Uptake of oxygen by the reaction mixture will draw oil up the U tube. Oxygen consumption is determined from the volume the calibrated plunger must displace in order to return the oil to its original position. The entire apparatus is submerged in a constant temperature bath (with a precision of ±0.02°), the position of the oil column being read with a periscope. A valve connects the submerged chambers to the atmosphere, permitting temperature equilibration without dislodgment of the oil column from the manometer.

Determining Oxygen Uptake. The reaction vessel is a 25-ml Warburg flask with a sidearm and a center well. In the main compartment is placed 2×10^7 neutrophils in 1.5 ml of buffer. The activating agent, e.g., 8 mg of opsonized zymosan (cf. Protocol 8B) or PMA (2 μg), is placed in the sidearm of the flask in a volume of 0.5 ml. Into the center well is placed 0.1 to 0.2 ml of 1.0 M KOH followed by a pleated rectangle of Whatman 3-mm filter paper (4 to 5 pleats), long enough to extend about 0.5 cm above the walls of the center well; the pleats are spread to form a fan so that the KOH, which is drawn into the filter paper, will be able to absorb CO_2 effectively. The flask is then attached to the respirometer. Caution is exercised to ensure that the stopcock is turned first, ensuring that the respirometer chambers are open to the air. The apparatus is lowered into the water bath. After equilibrating for 5 min, the activating agent is poured into the main compartment, taking care not to contaminate the reaction mixture with KOH from the center well. After a further 5 min, the stopcock connecting the respirometer chambers with the atmosphere is closed, the meniscus of the oil column is brought into line with the index mark by means of the micrometer screw, and the micrometer reading is recorded. Ever 5 min thereafter, the column of oil is brought into line with the index mark, and the new micrometer reading is recorded. The differences between micrometer readings represents microliters of oxygen consumed during the corresponding interval of time. This value can be converted into micromoles of O_2 by multiplying by the factor:

$$O_2 \text{ (micromoles)} = 12.2/(273 + T) \tag{16}$$

where T is the temperature of the water bath in degrees centigrade.

Controls should include unstimulated cells and incubations without neutrophils. The latter will establish the time necessary to achieve temperature equilibration, since there will be no change in the level of the oil column in a completely equilibrated system. Because of the time required to achieve complete temperature equilibration, initial rates of oxygen uptake cannot be determined. Comparatively large numbers of cells are required for this method, because of the relatively low sensitivity of the assay. An advantage of the manometric method over the oxygen electrode technique, however, is that reactions can be followed over long periods of time.

Protocol 9B: Measuring Oxygen Uptake with the Oxygen Electrode

Equipment. Equipment needed for measuring oxygen uptake polarographically includes an incubation chamber, an oxygen electrode, and a strip chart recorder. The Gilson Oxygraph (Gilson Medical Electronics, Middleton, WI) is a suitable instrument that includes all of these elements. A constant temperature, circulating water bath is also needed to control the temperature of the incubation chamber.

Before use, the apparatus is calibrated to room air and 0% oxygen. The room air calibration is obtained using buffer equilibrated at the temperature to be used for the experiments. The anaerobic calibration is obtained by adding 1 to 2 mg of sodium dithionite (sodium hydrosulfite; $Na_2S_2O_4$) to the buffer after calibrating to room air. After calibrating to 0% oxygen, the incubation chamber must be thoroughly washed to remove all traces of dithionite.

Determining Oxygen Uptake. Into the incubation chamber place 1.3 ml of neutrophil suspension (concentration 4×10^6 cells/ml). Incubate for 5 min with vigorous stirring to permit temperature equilibration, measuring oxygen uptake during this interval to obtain a rate for the resting cells. Then add the prewarmed stimulus (4 mg of zymosan or 1 μg of PMA in a volume of 0.1 ml), and continue to measure oxygen uptake for an additional 5 to 10 min to obtain a rate for stimulated cells. Since oxygen uptake is likely not to be constant during the period of incubation, it is usually expressed by measuring the quantity of O_2 taken up over a given interval of time. This is preferable to attempting to estimate the maximum slope of the polarograph tracing, a maneuver that is subject to considerable error both with respect to the time at which the maximum occurs and the value of the slope itself. Nanomoles of oxygen consumed per milliliter of reaction mixture can be calculated from the following formula:

$$O_2 \text{ (nanomoles)} = [(\text{percentage uptake})(\text{solubility of oxygen})]/100 \quad (17)$$

where the solubility of oxygen in water at different temperatures is given in Table V.

In using the oxygen electrode, it is critical to exclude all air bubbles from the electrolyte between the membrane and the electrode itself, and

TABLE V
CONCENTRATION OF OXYGEN DISSOLVED IN
WATER WHICH IS IN EQUILIBRIUM WITH
ROOM AIR

Temperature (°)	$[O_2]$ (mM)
15	0.30
20	0.27
25	0.24
30	0.22
35	0.20
37	0.20
40	0.19

to make sure that the membrane is scrupulously clean. A major problem in measuring oxygen uptake by neutrophils with this technique is the tendency for the membrane to adsorb a layer of protein from the incubation mixture. This is suggested by a decrease in the measured rate of oxygen uptake during the course of an incubation, and is confirmed if oxygen uptake by a fresh incubation mixture of identical composition is slow compared with the previous measured rate. This problem can be minimized by using low concentrations of neutrophils, and is easily corrected for by changing the membrane of the electrode at frequent intervals.

A problem intrinsic to the oxygen electrode is a slow response time. This occurs since the oxygen concentration measured is actually that of the electrolyte behind the membrane, and it takes time for the oxygen in this compartment to equilibrate with the oxygen in the incubation mixture. The equilibration rate is maximum when the membrane is clean and the incubation mixture is stirred. However, even under ideal conditions several seconds are required for equilibration. For this reason, rapid measurements in rates of oxygen consumption (e.g., initial lag time) cannot be measured accurately, and initial rates are subject to uncertainty. In addition, the instrument may not be able to measure oxygen uptake accurately when the consumption rate is very rapid. Generally, the smaller the electrode, the shorter the response time. However, this increase in tracking ability is offset by a diminished sensitivity.

A discussion of the principles of the oxygen electrode has been reviewed by Estabrook.[271]

---

[271] R. W. Estabrook, in "Methods in Enzymology" Vol. 10, p. 41. Academic Press, New York, 1967.

8.6.3.2. Superoxide Production

Protocol 10A: Superoxide (O_2^-) Production Assessed by a Discontinuous (Fixed Time) Assay

Neutrophils are suspended in buffer at a concentration of $1–5 \times 10^6$ cells/ml. Into each of two test tubes, place 1.4 ml of neutrophil suspension. To one of the test tubes, add 10 μl superoxide dismutase solution (Sigma; 3 mg/ml in water). Incubate at 37° for 2 min. Then to each tube add 0.1 ml horse heart cytochrome c solution (Sigma type III of IV; 30 mg/ml in buffer) followed immediately by 1.5 ml stimulus (12 mg opsonized zymosan or 3 μg PMA) at 37°. Mix rapidly and place 1.5 ml of each mixture in ice (blank), and incubate the remainder at 37° for the desired time. Terminate the stimulation by placing the tubes in ice for 5 min. Then centrifuge at 1500 g for 20 min at 4° to remove cells and particulate stimuli. Finally, using a double-beam spectrophotometer, measure the absorbance of the supernatant by scanning between 530 and 570 nm, using the corresponding unincubated supernatant as a blank. Maximum absorbance (A) occurs at 550 nm and reflects the amount of cytochrome c reduced during the incubation. The difference between the amount reduced in the presence and absence of superoxide dismutase represents the amount of O_2^- generated during the incubation.

The change in absorbance due to the reduction of cytochrome c is calculated by subtracting the blank value from the value for the corresponding incubated supernatant. It is necessary to calculate absorbances by this procedure to correct for baseline differences between different samples.

The amount of O_2^- generated in the entire reaction mixture may be calculated from the following formula:

$$O_2^- \text{ (nanomoles)} = A[\text{volume of incubation mixture (ml)}]47.4 \quad (18)$$

where A is the peak height or maximum adsorbance at 550 nm. O_2^- generation is calculated on the basis that a change in adsorbance of 1.0, at 550 nm, corresponds to the presence of 47.4 nmol of O_2^-.

Typical values for human neutrophils are in the vicinity of 1 to 2 nmol/min per 10^6 cells.

In carrying out this assay, it is important that the cytochrome c concentration be high enough to trap all the superoxide. Incomplete trapping is especially likely to be a problem at low pH, because the rate of spontaneous dismutation of O_2^- increases rapidly with falling pH. Trapping effectiveness can be checked by repeating the assay at twice the original concentration of cytochrome c; if complete, there should be no change in the amount of O_2^--dependent cytochrome c reduction at the higher substrate concentration.

When following O_2^- production as a function of time, it is preferable to set up a separate assay for each time point rather than to draw aliquots from a single reaction mixture. This is because clumping of the neutrophils may lead to the withdrawal of nonrepresentative aliquots at late reaction times. Only a single blank is necessary for a time course obtained in this way.

The assay also may be performed by reading at fixed wavelengths if a scanning instrument is not available. Readings should be taken at 540, 550, and 560 nm, and the peak height (i.e., maximal absorbance, A) calculated by subtracting the average of the 540 and 560 nm readings from the reading at 550 nm.

Protocol 10B: Superoxide Production by a Continuous Assay

In this method, superoxide production is followed continuously by measuring the change in the absorbance of cytochrome c at 550 nm as a function of time. The reaction is performed in a spectrophotometer equipped with a thermostatted cell compartment. The reaction mixture is prepared by placing 50 μl of neutrophil suspension (5×10^7 cell/ml), 50 μl horse heart cytochrome c (Sigma type III or VI, 12.5 mg/ml), and 0.85 ml buffer in a 1-ml cuvette. The reaction mixture is then allowed to equilibrate to 37° (or the temperature being investigated) in the thermostatted cell compartment. The reference blank is prepared the same way, except that it contains, in addition, 10 μl of superoxide dismutase (1 mg/ml). After the reaction mixture has come to temperature, the reaction is started by adding 50 μl of stimulus [e.g., digitonin (0.2 mg/ml), PMA (20 μg/ml), or opsonized zymosan (15 mg/ml)] and the change in absorbance at 550 nm is monitored as a function of time. In a single-beam spectrophotometer, the reaction and the blank are measured consecutively, correcting for O_2^--independent cytochrome c reduction by subtracting the latter from the former. In a double-beam spectrophotometer, the blank is placed in the reference position, and the reactions in the two cuvettes are started simultaneously by adding the activating agent to both. This technique automatically corrects for O_2^--independent cytochrome c reduction.

O_2^- generation is calculated on the basis that a change in absorbance of 1.0 corresponds to 47.4 nmol of O_2^- [cf. Eq. (18) above].

Protocol 11A: NBT Reduction Test (see also this volume [24])

This test can be performed using anticoagulated whole blood or isolated neutrophils. NBT (Sigma) is dissolved as a 0.1% solution in 0.15 M

NaCl, filtered, and stored at between 2 and 8° in a dark bottle. Neutrophils (1×10^5) are added to KRP containing 5% albumin and 0.05% NBT in a total volume of 1 ml. This mixture is incubated for 30 min at 37°. The cells are then deposited onto a glass slide by cytocentrifuge (Shandon Southern), air dried, fixed with absolute methanol for 1 min, and stained for 1 min with safranin or freshly filtered Wright's stain.

For whole blood, a drop is placed on a clean glass/slide and incubated at 37° for 30 min in a moist chamber. The nonadherent cells are washed away with PBS. One milliliter of KRP containing 5% albumin and 0.05% NBT is then added to the slide, which is incubated for an additional 30 min at 37°. The solution is then washed away with PBS and the slide fixed with methanol and stained for 1 min with safaranin or Wright's stain.

The preparations are examined under oil immersion, and the fraction of cells containing precipitated formazan (purple) granules is determined, counting 100 to 200 cells.

The NBT test is useful for determining whether or not neutrophils are able to express a respiratory burst. Almost all normal neutrophils should reduce NBT. Failure of neutrophils to reduce NBT occurs in conditions in which the respiratory burst is not expressed.

Other tetrazolium salts may be employed as alternatives to NBT. These include dimethylthiazolyldiphenyltetrazolium chloride, tetraphenylditetrazolium chloride, and triphenyltetrazolium chloride.[272]

This test may yield higher values with heparinized blood than with blood anticoagulated with other agents. This is because heparin forms a precipitate with NBT, which is taken up by the neutrophils, activating the respiratory burst. If it is important to minimize NBT reduction by unstimulated neutrophils, EDTA or citrate should be used as anticoagulant.

Protocol 11B: Stimulated Nitroblue Tetrazolium Blue Reduction Test

As mentioned previously (cf. Section 5.2.3.2.2), it is possible to enhance the sensitivity of the NBT reduction test by incorporating into the assay system a phagocytic stimulus such as PMA or polystyrene latex particles. The technical procedure is very similar to that described above (Protocol 11A).

Polystyrene latex (0.81 μm, Difco Bacto-latex) is used at a volume : volume ratio of latex to neutrophil suspension of 1 : 100. The latex is added to the system immediately after the NBT dye.

PMA (Sigma) is used at a final concentration of 1 μg/ml of reaction mixture. This stimulant (PMA) is dissolved in dimethyl sulfoxide at 2 mg/ml and stored under desiccation at −20°. When required, the thawed

[272] B. Holmes and R. A. Good, *J. Reticuloendothel. Soc.* **12**, 216 (1972).

solution can be diluted into PBS to yield the desired concentration. Stimulated NBT test results are usually two to three times higher than the normal range obtained with the NBT test.

8.6.3.3. Hydrogen Peroxide Production (see also this volume [22], [23], [24])

Protocol 12A: Assessment of H_2O_2 Concentration by a Fluorimetric Assay

Preparation of Reagents. Scopoletin (7-hydroxy-6-methoxycoumarin) is dissolved in water to a concentration of 0.4 mM. Scopoletin (Sigma) is sparingly soluble in water and the solution should be mixed and warmed for about 1 hr to increase solubilization. The solution is stable in the dark and can be kept refrigerated for up to 1 year. Horseradish peroxidase (Sigma) is dissolved in PBS at 1 mg/ml and can be stored at $-20°$ until needed.

Assay. The excitation wavelength of a temperature-controlled spectrofluorometer is set at 380 nm and the emission wavelength at 460 nm. Neutrophils (about 1 to 2.5 $\times$ 10^6) are added to KRP containing 0.02 mg horseradish peroxidase, 8 nmol scopoletin, and stimulus, if desired (opsonized zymosan, 4 mg, or PMA, 1 μg), in a total reaction volume of 1.0 ml. The decrease in fluorescence is recorded as a function of time. Following an initial period, varying from 30 to 60 sec, in which no change occurs, a rapid decrease in fluorescence is noted. The rate of fluorescence decrease is constant for 2 to 5 min until 70% of the scopoletin is oxidized. The rate of H_2O_2 production is determined from this linear decrease in fluorescence. There is a baseline rate of scopoletin oxidation in the absence of phagocytic stimuli, which needs to be subtracted from the rate recorded in the presence of stimuli.

H_2O_2 Calibration Curve. To convert the measured changes in fluorescence to quantities of H_2O_2 produced, the change in fluorescence for a given concentration of H_2O_2 needs to be calibrated. This is done using a standard curve constructed as follows: Reference H_2O_2 is prepared by mixing 0.5 ml of concentrated H_2O_2 (Fisher) with 500 ml water, and determining the concentration of H_2O_2 in this solution spectrophotometrically at 235 nm (at this wavelength, the absorbance of 10 mM H_2O_2 is 0.89). This stock solution is then diluted in KRP by an additional factor of 50 to provide a standard solution containing approximately 0.25 mM H_2O_2. A standard curve is constructed by measuring the fluorescence of solutions containing 0.02 mg horseradish peroxidase, 8 nmol scopoletin, and 0, 5, 10, and 20 μl of the standard H_2O_2 solution (final volume 1.0 ml).

This method only records extracellular hydrogen peroxidase, and this needs to be kept in mind (cf. Section 5.2.3.2). A major disadvantage of this assay is the limitation in the amount of scopoletin that can be added to the assay. If too little is added, the rate is linear for only a very short time. If too much is added, the sensitivity of the fluorescence recorder needs to be decreased, making it difficult to accurately assess H_2O_2 production. A general problem with assays involving scopoletin is that they cannot be used with cell-free systems, due to interference by reduced pyridine nucleotides.

Protocol 12B: H_2O_2 Measurement by the Scopoletin–Horseradish Peroxidase Assay

A discontinuous assay for H_2O_2 production can also be employed, using the scopoletin–horseradish peroxidase assay. In this case, 0.1 mM sodium azide is added to resting and phagocytosing cells to inhibit endogenous peroxidase and catalase activity. (Scopoletin oxidation is not inhibited by this concentration of azide.) After the reaction is stopped by the addition of ice-cold 0.15 M NaCl, the cells are removed by centrifugation at 50 g for 10 min. To the supernatant is then added 0.02 mg horseradish peroxidase and 4–16 nmol of scopoletin (depending on the number of cells and the duration of the incubation), and the fluorescence is recorded.

The advantage to the discontinuous assay is that many reactions can be monitored simultaneously. The advantage of the continuous assay is the ease with which both the rate and the lag is response time can be measured.

8.6.4. Hexose Monophosphate Shunt Activity

Protocol 13: Glucose Oxidation

[1-^{14}C]Glucose and [6-^{14}C]glucose (New England Nuclear) are diluted with 50 mM glucose to a final specific activity of 25 μCi/ml. These stock solutions may be kept frozen at $-20°$ until needed. A working solution is prepared by diluting 200 μl of the stock solution with 800 ,1 of 50 mM glucose. Neutrophils (1 to 2.5 × 10^6) are added to KRP containing 1 mM [1-^{14}C]glucose, in the presence or absence of stimuli [e.g., 1 μg PMA, 4 mg opsonized zymosan, 0.1 ml polystyrene latex beads (0.81 μm)], in a final reaction volume of 1 ml. For most studies, the cells are added last. The reaction vial or tube is immediately covered with a rubber stopper through which a well, containing a filter paper wick and 0.2 ml of Hyamine hydroxide (New England Nuclear), has been inserted. The cell suspension is shaken at 37° for 20 to 30 min. The reaction is stopped, and the CO_2 is liberated from the buffer by injecting 1 ml 5 N H_2SO_4 through the rubber stopper. Care must be taken not to inject the H_2SO_4 into the well. The tubes are incubated for an additional 30 min at 37°, with shaking. The liberated CO_2 is adsorbed by the Hyamine hydroxide contained in the wells. The rubber stopper and well are then removed, and the wells placed into scintillation fluid, and the radioactivity determined.

Care must be taken to ensure that the assay mixture does not come in contact with the well, since most of the radioactivity remains in the assay mixture. Additions to the incubation mixture need to be made through the rubber stopper to prevent the loss of $^{14}CO_2$. Appropriate controls for measuring hexose monophosphate shunt activity include measurements made in the absence of phagocytic stimuli and the measurement of $^{14}CO_2$ release from [6-^{14}C]glucose (cf. Sections 5.2.3.4). Hexose monophosphate shunt activity is determined by subtracting the counts obtained from the oxidation of [6-^{14}C]glucose from those obtained from the oxidation of [1-^{14}C]glucose. Activity should be expressed in terms of counts per minute (cpm), not micromoles (μmol) or nanomoles (nmol), because the true intracellular specific activity of the hexose monophosphate shunt substrate is not known.

8.6.5. Iodination

When iodide is incubated with activated neutrophils, a portion of the free iodide is converted into a protein-bound form through the combined action of hydrogen peroxide and myeloperoxidase. The iodination reaction can be employed as a measure of neutrophil function. It is dependent on phagocytic ingestion as well as respiratory burst and myeloperoxidase activity, so a defect in any of these can result in diminished iodination.

The basis of the assay is the measurement of the amount of ^{125}I that becomes associated with trichloroacetic acid (TCA)-precipitatable material from cells incubated with opsonized zymosan, Na^{125}I, and an acceptor protein such as albumin.

Protocol 14: Incorporation of ^{125}I into TCA-Precipitable Material by Neutrophils

The reaction mixture consists of 40 nmol of sodium iodide containing 0.05 μCi Na^{125}I (New England Nuclear), 0.5 mg human serum albumin, 0.5 mg opsonized zymosan, and 2.5 × 10^6 neutrophils in a reaction volume of 0.5 ml, pH 7.4, containing 1 mM glucose. The reaction is started by adding the cells to the mixture which is then incubated at 37° for 20 min. It is stopped through the addition of 0.1 ml of 0.01 M sodium thiosulfate followed by 1.0 ml ice-cold 10% TCA. The precipitate is collected by centrifugation at 1900 g for 5 min and washed four times with 1-ml portions of TCA. The radioactivity in the washed precipitate is then determined by gamma counting. A blank, containing all components except neutrophils, is also run and the results subtracted from the experimental values. Activity is expressed as counts per minute.

8.6.6. Oxidizing Radical Production

Protocol 15: Detection of Oxidizing Radicals by Measuring the Release of Ethylene from 2-Keto-4-methylthiobutyric Acid

Into a 3.5-ml siliconized tube is placed 0.5 ml of neutrophil suspension (5×10^6 cells/ml buffer), followed by the addition of 0.1 ml of 2-keto-4-methylthiobutyric acid (Sigma), 10 mM in buffer. Following equilibration to 37°, 0.4 ml of a prewarmed solution of a stimulus (e.g., zymosan, PMA; for appropriate concentrations see Protocol 13) is added. The tube is immediately stoppered and then incubated for the desired length of time at 37°. The reaction is terminated by placing the tube in melting ice. A 0.5-ml sample of headspace gas is withdrawn with a Hamilton gas-tight syringe, and analyzed for ethylene content by gas chromatography. For this purpose a 6 in. $\times \frac{1}{4}$ in. glass column packed with Chromosorb 102, using an oven temperature of 100°, a carrier gas pressure of 40 psi, and a flame ionization detector are employed. The retention time for ethylene gas under these conditions is 45 sec. As a standard, pure ethylene, 18 ppm in nitrogen (Supelco, Inc.), is used. The height of the ethylene peak is proportional to the quantity of gas injected. The quantity of ethylene (in picomoles) released into the headspace gas during the incubation can therefore be calculated as follows:

$$\text{Ethylene (picomoles)} = 384RQ \qquad (19)$$

where R is the ratio of sample peak height to the height of the peak produced by 0.5 ml of 18 ppm ethylene standard; Q is the total gas volume (headspace plus syringe) divided by the volume of gas in the syringe. A typical yield for a 40-min incubation under the foregoing experimental conditions is 50 pmol of ethylene/5×10^6 cells.

Oxidizing radical assays must be interpreted with care. Oxidizing radicals can be scavenged by numerous constituents in the assay mixture, including components of the incubation medium (e.g., sucrose or glucose), and the neutrophils themselves. The amount of ethylene release from the incubation mixture reflects only that fraction of oxidizing radicals that is released extracellularly and which reacts with the trapping reagent. This reagent is in competition with all the other radical scavengers in the suspension. Variations in the composition of the reaction mixture can, therefore, alter ethylene production without altering the rate of radical formation.

8.6.7. Chemiluminescence. Methods for the study of phagocytosis by chemiluminescence have been presented elsewhere in this volume [33] and will not be duplicated here.

8.7. SPECIFIC ENZYME ASSAYS

8.7.1. Myeloperoxidase

The assay for myeloperoxidase has been presented elsewhere in this volume [20] and will not be duplicated here.

8.7.2. The O_2^--Forming Oxidase (see also this volume [24])

Protocol 16: Neutrophils are activated by incubating with opsonized zymosan. They are then homogenized and separated by centrifugation into particles and supernatant. The particles are then assayed for O_2^- production.

For activation, 2 ml neutrophil suspension (5×10^7 cells/ml PBS) is mixed with 2 ml opsonized zymosan (28 mg/ml in KRP or HBSS), containing 2 mM NaN$_3$). This mixture is then incubated with shaking for 7 min at 37°. The reaction mixture is chilled in ice and thereafter the cells are centrifuged at 160 g for 4 min at 0°. The cell pellet is washed once in 0.34 M sucrose, which had been adjusted to pH 7.4 with 0.2 M NaOH, and resuspended in the same solution. Next, the cells are disrupted by homogenization in a Potter-Elvehjem homogenizer and centrifuged at 160 g for 4 min at 0° to pellet unbroken cells, nuclei, and zymosan. The supernatant is then centrifuged at 27,000 g for 30 min at 0°. The pellet is washed once by resuspension in sucrose followed by centrifugation under the same conditions, and the resulting washed preparation of O_2^--forming particles is resuspended in 0.34 M sucrose (pH 7.4) and adjusted to a protein concentration of 1 mg/ml.

O_2^- production is measured continuously in a double-beam spectrophotometer as follows: The reaction mixture contains 0.2 ml of particle suspension, 0.2 ml cytochrome c solution (Sigma type III or VI; 10 mg/ml in water), 0.2 ml 1.0 mM NADPH in water, and 1.4 ml 0.1 M potassium phosphate buffer (pH 7.4). Exactly 1.0 ml of this suspension is placed in a 1-ml cuvette containing 3 μl of superoxide dismutase [bovine erythrocyte superoxide dismutase (Sigma), 10 mg/ml]. The remainder of the reaction mixture is placed in a second cuvette. The dismutase-containing cuvette is placed in the reference beam of the double-beam spectrophotometer, the other cuvette in the sample beam, and O_2^- production at room temperature is monitored by changes in absorbance at 550 nm. (Superoxide dismutase binds all of the free O_2^- radicals, thereby preventing them from reducing cytochrome c.)

Alternatively, O_2^- production can be measured at discrete time points. A reaction mixture is prepared as described above. One milliliter of the mixture is added to a tube containing 3 μl superoxide dismutase (10 mg/ml), which is incubated at room temperature for the desired time. The tube is then placed on ice (portion A). The remaining solution is incubated at room temperature for the desired time. Thereafter 3 μl superoxide dismutase (10 mg/ml) is added to this mixture which is then placed on ice (portion B). To measure O_2^--dependent cytochrome c reduction, a narrow spectrum (530 to 570 nm) is monitored with portion A in the reference beam and portion B in the sample beam. If a scanning instrument is not available, cytochrome c reduction can be measured by determining the absorbance of each sample at 540, 550, and 560 nm, respectively. The average of the 540 and 560 nm values is then subtracted from the 550 nm value to obtain a figure to be used in calculating the reduction of cytochrome c. (This method corrects for shifts in the baseline.) Optical density changes due to O_2^--dependent cytochrome c reduction are then calculated by subtracting the portion A value from the portion B value.

Rates of O_2^- production can be calculated as follows: For the continuous assay:

O_2^- production (nmol/min per mg protein)
$$= 47.4[\text{rate of change in adsorbance at 550 nm (units/min)}] \quad (20)$$

For the point assay:

O_2^- production (nmol/time interval per mg protein)
$$= 47.4[\Delta A_{550} \text{ (units)}] \quad (21)$$

O_2^- generation is calculated on the basis that a change in absorbance of 1.0, at 550 nm, corresponds to the presence of 47.4 nmol O_2^-. A typical activity for a good O_2^--forming particle preparation from human neutrophils is 10 nmol/min per mg protein.

O_2^--forming preparations have been produced using phorbol myristate[273] and digitonin[274] as the activating agent. For reasons that are not yet clear, little O_2^- is formed by particles isolated from cells activated with latex beads or fluoride. Cells may be disrupted by sonication instead of homogenization, though care must be taken not to inactivate the enzyme, which is susceptible to destruction by vigorous sonication.[275] The O_2^--forming activity is exceedingly sensitive to inactivation by salts (e.g., 0.1 M NaCl) and by even modest elevations of temperature (e.g., 37°). These restrictions must be borne in mind when considering modifications in the

[273] B. Dewald, M. Baggiolini, J. T. Curnutte, and B. M. Babior, *J. Clin. Invest.* **63**, 21 (1979).
[274] H. J. Cohen, M. E. Chovaniec, and W. A. Davies, *Blood* **55**, 355 (1980).
[275] T. G. Gabig and B. M. Babior, *J. Biol. Chem.* **254**, 9070 (1979).

method for preparing or assaying the particles. On the other hand, the particles are stable at 4° overnight or for longer periods at −70°, provided they are kept at low ionic strength and (if frozen) are thawed carefully.

8.7.3. β-Glucuronidase. β-Glucuronidase activity is assayed by measuring the release of *p*-nitrophenol from *p*-nitrophenyl-β-D-glucuronide by hydrolysis. *p*-Nitrophenol has an absorption maxima at 405 nm.

Protocol 17: Spectrophotometric Assay for β-Glucuronidase

p-Nitrophenyl-β-D-glucuronide (Sigma Chemical Company) is dissolved in water as a 0.01 *M* solution and stored at −70°. Reaction mixtures contain 0.5 ml sodium acetate buffer, pH 5.0, 0.1 ml of 1% Triton X-100, and 0.1 ml of 0.01 *M* *p*-nitrophenyl-β-D-glucuronide. The reaction is initiated by adding 0.3 ml of the supernatant from a degranulation preparation (cf. Section 8.6.2) or 5×10^5 neutrophils in a total volume of 0.3 ml. The assay mixture is incubated at 37° for 2 hr. The reaction is stopped with the addition of 1 ml of 0.2 *M* sodium hydroxide, and the change in absorbance at 405 nm (ΔA_{405}) is recorded. Reagent blanks, containing 0.3 ml of water instead of supernatant or cell suspension, are used to zero the spectrophotometer.

The time of incubation can be shorter if more enzyme activity is present. The activity rate is, however, linear for 2 hr.

The activity of the enzyme is calculated as follows:

Activity [(nmol/min per 0.3 ml of sample)] = $0.45[\Delta A_{405}$ (units/2 hr)]

A typical value is 0.1 nmol/min per 10^6 PMN.

8.7.4. Lysozyme. The assay for lysozyme is based on the change in light absorption of a suspension of *Micrococcus lysodeikticus,* when lysozyme acts on it.

Protocol 18: Assay for Lysozyme

Micrococcus lysodeikticus (5 mg) (Sigma) is mixed with 50 ml 0.15 *M* potassium phosphate buffer, pH 6.2. The absorbance of the suspension is measured at 450 nm and is adjusted to 0.6 or 0.7, if necessary, by adding more buffer or bacteria. The mixture (0.9 ml) is added to a cuvette, together with 0.05 ml of 1% Triton X-100. The reaction is started by adding 0.05 ml of supernatant from a neutrophil degranulation preparation or 5×10^5 neutrophils. The decrease in optical density is recorded as a function of time. The rate of this reaction is linear for only a short period of time. Consequently the change in absorbance for the first minute only is used to determine the rate of the reaction.

8.7.5. Lactate Dehydrogenase (LDH). The assay of this enzyme is based on the spectrophotometric measurement of NADPH consumption:

$$\text{Pyruvate} + \text{NADPH} \rightarrow \text{lactate} + \text{NADP}^+ \qquad (22)$$

Protocol 19

In a 1-ml quartz cuvette, 0.2 ml 0.15 M sodium phosphate buffer (pH 7.4), 0.1 ml 5 mM sodium pyruvate, 0.1 ml 1 mM NADPH, and 0.1 ml 1% Triton X-100 are mixed. The reaction is started by adding 0.5 ml of the material (cells or granule contents) to be assayed. The cuvette is immediately placed in a spectrophotometer and the decrease in absorbance at 340 nm (ΔA_{340}) determined. (The incubation is performed at room temperature.) A control without pyruvate, to measure NADPH oxidation, should be run. At 25°, 1 U of LDH activity oxidizes NADPH at the rate of 1 μmol/min.

The amount of LDH in the 0.5-ml sample is calculated as follows:

$$[\text{LDH activity (nmol lactate/min)}] = 161[\Delta A_{340} \text{ (units/min)}]$$

The LDH in 10^6 neutrophils will oxidize about 40 nmol lactate/min. Fresh solutions of NADPH should always be used.

8.7.6. Glucose-6-Phosphate Dehydrogenase (G-6-PD). The assay is based on the measurement of NADPH formed by the enzymatic reduction of NADP by glucose 6-phosphate. An increase in adsorption at 340 nm occurs due to the formation of NADPH.

Protocol 20: Assay for G-6-PD

Neutrophils (5 × 10^7/ml in PBS) are sonicated for 40 sec at 40 A to extract the enzyme. Fifty microliters of the sonicate is added to a 1-ml cuvette containing 0.1 ml of 1.0 M Tris–HCl, pH 8.0, containing 5 mM EDTA, 0.1 ml of 0.1 M MgCl$_2$, 0.1 ml of 2 mM NADP, and 0.55 ml water. The reaction is started with the addition of 0.1 ml of 6 mM glucose 6-phosphate, and the increase in absorbance at 340 nm is recorded at 25°. For degranulation studies, up to 0.3 ml of supernatant can be used in place of the neutrophils and the amount of water decreased accordingly.

A blank rate measurement is obtained in the absence of glucose 6-phosphate to control for nonspecific reduction of NADP. This assay actually measures the activity of both G-6-PD and 6-PGD, both of which reduce NADP to NADPH. If only the initial rate is examined, then G-6-PD activity alone is usually fairly accurately well estimated since it has a much higher turnover rate. If an exact determination of G-6-PD activity is required, the assay described above is performed in the presence of 0.1 ml of 6 mM glucose 6-phosphate and 0.1 ml of 6 mM 6-PG. This rate is subtracted from the rate obtained in the presence of only 0.1 ml of 6 mM 6-PG.

A typical value for G-6-PD activity, as measured in this assay, is 0.2 μmol/min per mg protein in the sample.

8.7.7. Nucleotidase. This class of enzymes catalyzes the cleavage of phosphate residues from nucleotides. The basis of the assay is that nucleotides, but not free phosphate molecules, bind to charcoal.

Protocol 21: Assay for Mg-ATPase Activity

Radioactive substrate is prepared by dissolving 250 μCi [γ-^{32}P]ATP (New England Nuclear) in 7.7 ml phosphate-buffered saline containing 1 mM ATP. The sample is divided into 1-ml aliquots and stored at $-70°$.

The substrate is prepared by adding to a solution of 0.4 mM ATP and 10 mM p-nitrophenyl phosphate (in PBS) sufficient radioactive material to yield a concentration (of ^{32}P) of 20,000 cpm/ml.

To measure Mg-ATPase, 0.5 ml of substrate is incubated for 20 min at 37° with 0.5 ml of the material to be assayed. The incubation mixture is then placed on ice and 1.0 ml 10% (w/v) charcoal in 10% trichloroacetic acid is added. This mixture is permitted to stand for a few minutes with occasional shaking. Thereafter the charcoal and precipitated protein are removed by Millipore ultrafiltration. The amount of radioactivity in a 1.0-ml portion of the filtrate (i.e., charcoal) is then determined by liquid scintillation counting. Controls include an incubation in which the material to be assayed is replaced by PBS. This enables the investigator to establish how much radioactivity can maximally be adsorbed onto the charcoal (control 1). The second control is a 0.25-ml sample of assay substrate, which is counted to determine the total amount of radioactivity originally contained in the assay substrate (control 2). The amount of ATP hydrolyzed by the 0.5-ml sample is calculated as follows:

[ATP hydrolyzed (nmol)] =
$$200[(\text{sample} - \text{control 1})/(\text{control 2} - \text{control 1})] \quad (23)$$

An unfractionated homogenate, originally containing 10^7 cells in 0.5 ml of buffer, catalyzes the hydrolysis of ~21 nmol ATP in 20 min.

The sensitivity of this assay can be varied by changing the concentration of unlabeled ATP in the assay substrate, as well as by varying the concentration of enzyme-containing material. Increasing the incubation time does not improve sensitivity. Results are most easily interpreted when 20 to 60% of the substrate is hydrolyzed during the incubation.

The purpose of the p-nitrophenyl phosphate included on the incubation mixture is to inhibit nonspecific phosphatases that might otherwise contribute to ATP hydrolysis, leading to erroneously high values for the level of Mg-ATPase.

8.7.8. Alkaline and Acid Phosphatase. Both enzymes may be measured through the use of the chromogenic substrate *p*-nitrophenyl phosphate, which is hydrolyzed by the enzyme at the appropriate pH to yield *p*-nitrophenol, a compound whose anion absorbs strongly in the blue region of the color spectrum, with an absorption maxima at 405 nm.

Protocol 22: Assay for Alkaline and Acid Phosphatase

The substrate solution consists of 4.5 mM *p*-nitrophenyl phosphate in 0.1 M diethanolamine–HCl buffer, pH 9.75, containing 0.05% Triton X-100 (alkaline phosphatase), or in 0.1 M acetic acid–sodium acetate buffer, pH 4.5, containing 0.05% Triton X-100 (acid phosphatase).

To measure phosphatase activity, 0.1 ml of sample is incubated for 10 to 30 min at room temperature with 1 ml of substrate solution. The reaction is stopped with 1 ml of 2 N NaOH, and the absorbance is measured spectrophotometrically at 405 nm (A_{405}). The activity of the phosphatase (micromoles phosphate per time interval per 0.1 ml of sample) is calculated as follows:

$$\text{Enzyme activity} = 1.15\, A_{405} \qquad (24)$$

The specific activity of unfractionated homogenates is typically around 100 μmol of phosphate/min per mg protein for acid phosphatase and 25 μmol/min per mg protein for alkaline phosphatase.[122]

Other substrates, including β-glycerol phosphate, 4-methylumbelliferyl phosphate, and other phosphate esters, have also been used for assaying neutrophil phosphatases. The catalytic activity of a given enzyme varies widely from substrate to substrate, so that comparisons between studies can only be made after specifying the substrate used in a particular assay.[276]

8.8. NONENZYME MARKERS

8.8.1. Lactoferrin (LF). This iron-binding protein is found only in neutrophils and is associated with the secondary (specific) granules only. Consequently, its extracellular release is a specific marker for studies of degranulation and ingestion in general. Although the total LF content of neutrophils is approximately 20 μg/10^6 cells,[232] only about 2–5 μg/10^6 cells is released during normal phagocytic stimulation.[212,231]

[276] U. Bretz and M. Baggiolini, *J. Cell Biol.* **59**, 696 (1973).

For assaying LF, radioimmunoassay,[277,278] enzyme-linked immunoadsorbant assay, and more recently radiometric assay[279] techniques have been developed. Only the RIA and radiometric assay will be described below. The radiometric assay has the advantage that it operates over a wide range of concentrations, has significantly shorter assay times, comparable sensitivity, and does not require the preparation of purified lactoferrin. In addition, the radiometric test appears to have the distinct advantage that it does not suffer from measurement inaccuracies due to protein or heparin: LF complexes.

Protocol 22A: Radioimmunoassay for LF

Reagents

1. Purified LF (from human breast milk) (Sigma)
2. Iron-depleted LF (apoLF) is prepared by dialysis of a 10 g/liter LF solution against 20 vol of 0.1 M citric acid
3. Radiolabeled LF is prepared using either ^{125}I or ^{59}Fe. Iodination is achieved using a modification of the chloramine-T method.[277] The protein is dissolved in a 0.4 M borate buffer, pH 8.0. Alternating LF may be labeled with ^{59}Fe. For this purpose, the chelating effect of nitrilotriacetate is used. One milliliter of 1 mM trisodium nitrotriacetate dissolved in 0.2 M phosphate buffer (pH 7.6) is added to 1 μM of ^{59}FeCl$_3$ (0.5 ml) containing 2 mCi of activity. Fifty microliters of this ^{59}Fe-chelate is then added to 1 mg of apoLF dissolved in 20 μl of 0.2 M phosphate buffer (pH 7.6). The reaction mixture is incubated at 37° for 3 hr. Unincorporated iron is removed by passage over a column of Sephadex G-25 equilibrated with 1 mM trisodium nitrotriacetate in 0.2 M phosphate (pH 7.6). This procedure results in the incorporation of about 95% of the ^{59}Fe label into the protein
4. IgG fraction of a goat anti-human LF antiserum (Atlantic Antibodies, Scarborough, ME)

Procedure[277]

Polypropylene counting vials are coated with high-affinity anti-LF antibodies. Following incubation at 37° for 30 min the vials are rinsed to remove nonadsorbed antibodies. Then purified LF, containing radiolabeled tracer (iodinated) ^{125}I-LF ($\approx$30 μCi/μg) in 0.9 ml of barbital–BSA

[277] R. M. Bennet and C. Mohla, *J. Lab. Clin. Med.* **88**, 156 (1976).
[278] F. M. Segars and J. M. Kincade, *J. Immunol. Methods* **14**, 1 (1977).
[279] K. L. Becker-Freeman, D. C. Anderson, and G. Buffone, *Clin. Chem.* **31**, 407 (1985).

diluent is added to the vials. Immediately add 100 μl of appropriately diluted test samples and LF standards. After incubation at 37° for 16 hr, the tubes are aspirated and rinsed four times with distilled water before counting the radioactivity associated with each tube.

From the standards of known LF concentration it is possible to construct a standard curve and hence determine the LF concentration in any unknown sample. It should be remembered, however, that this procedure is a competitive binding assay and hence the greater the amount of LF in the sample, the less radioactivity will be associated (bound) to the antibody-coated vials.

For an assessment of LF content of neutrophil granules see the end of Protocol 22B.

Protocol 22B: Radiometric Assay for LF

Reagents

1. Purified LF. A stock solution of LF purified from human milk (Sigma, St. Louis, MO) is dissolved in 60 mM barbital buffer (pH 8.6), containing 100 mM NaCl, 2 g/liter BSA, and 1 g/liter azide, and stored at −70°.

2. The IgG fraction of a high-affinity goat anti-human LF antiserum (Atlantic Antibodies, Scarborough, ME)

In order to confirm the specificity of the antiserum, an aliquot of ^{125}I-labeled anti-LF is solid phase adsorbed with LF coupled to Sepharose 4B. After adsorption, the antiserum should show no cross-reaction with either purified LF or neutrophil granular contents.

3. Radioiodination of Anti-LF Antibody. A volume equal to 2 mg of anti-LF antibody is radioiodinated using 3.5 mCi of (^{125}I) and 0.140 μl of lactoperoxidase immobilized on Sepharose beads (Worthington Biochemicals, Freehold, NJ). Following coupling, the bulk of free (^{125}I) is removed by passing the sample through a Sepharose G-25 column. The labeled antibody is subsequently dialyzed against PBS. The specific activity of the antibody preparation should average about 6 × 10^5 cpm/μg of protein.

4. Antibody Coating of Polystyrene Beads. Polystyrene beads (6 mm; Clifton Plastics, Inc., Clifton Heights, PA) are washed in a sintered glass funnel with freshly prepared 0.1 M carbonate buffer, pH 9.4. Fifteen micrograms of anti-LF antibody protein (IgG fraction) per bead, diluted to a final concentration of 100 μg/ml in carbonate buffer, is added to the beads very rapidly in order to achieve an even coating through nonspecific adsorption. This mixture is incubated at 4° overnight with gentle

shaking. After washing with carbonate buffer as above, any saturated reactive binding sites on the glass surface are quenched by immersing the beads in carbonate buffer containing FCS (10 ml/liter) and Tween 20 (5 ml/liter) and incubated for 1 hr at 4° with gentle shaking. The beads are then washed several times with PBS, alternating with PBS–Tween. The beads are now ready for use. Approximately 500 beads can be coated at once. The beads are stored at 4° in PBS containing 10 ml/liter FCS and 1 g/liter azide until needed.

Procedure[279]

To each tube, a single antibody-coated bead, 50 μl of sample (purified LF or neutrophil sonicates), and 200 μl of diluent (100 ml/liter FCS in PBS) are added and the tubes gently mixed by use of a rotating platform for 1 hr at ambient temperature. The beads are then washed with three 2-ml portions of Tween–PBS (0.5 ml/liter). The fluid is removed by decantation or aspiration. Subsequently 250 μl of radiolabeled antibody (diluted 1 : 500 in FCS–PBS) is added. The tubes are again rotated for 1 hr and washed as above. The beads are then transferred to clean tubes and the bound radioactivity determined. Nonspecific binding, which is determined by substituting diluent for the sample, is subtracted from all results. A standard curve is generated for every assay using a concentration range of 5–1500 ng/ml of LF. A sample containing 10,000 ng/ml LF is used to obtain maximum binding. The total time for the assay is 2.5–3 hr.

Nonspecific binding is less than 1% of total counts. It has been found that 100 g/liter BSA in PBS can be substituted for 100 ml/liter FCS without increasing nonspecific binding.

8.8.2. Total Neutrophil Granule LF Content.

Neutrophils are isolated from whole blood as described in Section 2.2.1. The neutrophil suspension is adjusted to a final concentration of 10^7 cells/ml. Portions of the cell suspension are then stimulated with appropriate phagocytic stimuli as dictated by the experimental protocol.

For determinations of total neutrophil LF content, 5×10^6 neutrophils in PBS containing 2 g/liter glucose are collected by centrifugation at 800 g for 5 min and the supernatants carefully removed and saved. LF is extracted by suspending the cell pellet in 0.5 ml of 1.0 ml/liter Triton X-100 in PBS, followed by sonication on ice for 60 sec. The neutrophil sonicates are diluted 1 : 50 with 2 mM phenylmethylsulfonyl fluoride (Sigma Chemical Company, St. Louis, MO) and stored at $-70°$. This reagent inhibits proteolytic degradation (by the proteases contained in the granules) of LF. Total content is defined to be equal to the LF concentration in the supernatant plus the concentration of LF in the cell pellet as determined by the radiometric assay described above (Protocol 22B).

8.8.3. Cobalamin. Cobalamin-binding (vitamin B_{12}-binding) protein of neutrophils is a good marker derived from specific granules. In the cell, it is largely unsaturated and can be measured by its ability to take up radioactive cobalamin (vitamin B_{12}) from the medium. Uptake is measured by a competition method using albumin-coated charcoal, which can remove free cobalamin from solution, but is unable to adsorb cobalamin that is bound to another protein.

Protocol 23: Determination of Cobalamin-Binding Protein

Suspensions containing 10^7 neutrophils per milliliter are used. Cobalamin-binding protein is released from the cells by sonication (total cobalamin-binding protein) or by exposure to an appropriate stimulus (see Section 8.6.2). Albumin-coated charcoal is prepared by mixing a 5% (w/v) suspension of charcoal (Norit "A," from Amend Drug and Chemical Company, New York, NY, or Darco G-60, from Atlas Chemical Industries, Wilmington, DE) in water with an equal volume of 1% (w/v) bovine serum albumin in water. To measure the binding of cobalamin to protein, 0.2 ml of the sample to be tested is mixed with 1.8 ml of 0.15 M NaCl (saline) and 1.0 ml cobalamin in saline (2.5 ng cyanocobalamin containing 20,000 cpm [^{57}Co]CN-Cbl per milliliter of saline). The mixture is then shaken at room temperature for 10 sec. Thereafter, 2 ml albumin-coated charcoal is added and the suspension centrifuged under conditions sufficient to pellet all of the added charcoal. The radioactivity is then determined in an aliquot (2.5 ml) of the supernatant as well as in 2.5 ml of an uncentrifuged sample in which the albumin-coated charcoal was replaced by an equal volume of 0.5% bovine serum albumin. Cobalamin binding, expressed as nanograms per 10^7 cell equivalents, is calculated as follows:

$$\text{Binding (ng/}10^7 \text{ cell equivalents)} = \frac{12.5[\text{cpm (charcoal-treated)}]}{[\text{cpm (untreated)}]} \quad (25)$$

A control should be performed in which the test sample is replaced by an equal volume of buffer, to ensure that the amount of charcoal employed can take up all the free cobalamin in the assay mixture.

If binding amounts to less than 10% or more than 80% of the added radioactivity, the experiment should be repeated, altering conditions to obtain a more intermediate value. If binding is low, the amount of sample may be increased or the concentration (but not the radioactivity) of cobalamin reduced. The opposite changes should be made if the binding is high. A typical range for total unsaturated cobalamin-binding capacity is 5 to 10 pg of cobalamin per 10^7 cell equivalents.

8.9. Bacterial Killing (see also this volume [35])

To measure bacterial killing, microorganisms and neutrophils are incubated together in suspension. At various intervals, a portion of the suspension is removed, the neutrophils are lysed, and the number of surviving microorganisms is determined.

Protocol 24: Pour Plate Method

Bacteria. Bacteria are grown, suspended in PBS, and their concentration adjusted as described previously (cf. Protocol 4), except that manipulations involving radioactivity are not performed, and in addition, the bacteria are not killed. Cultures may be obtained from clinical bacteriology laboratories. However, as different strains of the same species may vary in their susceptibility to neutrophils, reproducible results are more likely to be obtained with well-defined strains derived from the American Type Culture Collection. Reproducibility is also likely to be improved if the bacteria are always used at the same stage of growth, either log or stationary phase. Staphylococci and enterobacteriaceae grow satisfactorily on plain nutrient agar or broth and streptococci on Todd–Hewitt medium.

Opsonization. Bacteria are opsonized by the method described for zymosan (cf. Protocol 3A), except that centrifugation is carried out at 14,000 g for 20 min. Alternatively, nonopsonized bacteria may be used, provided that the killing assay is done in the presence of serum, as described below. Since some strains of enterobacteria are killed by serum alone, serum-resistant strains must be employed for this assay. Alternatively, serum-susceptible strains may be opsonized, provided the serum is heated for 30 min at 56° to inactivate complement. Such decomplemented serum will not opsonize, unless it contains antibody against the microorganism used in the assay.

Staphylococci and streptococci are ordinarily resistant to serum alone.

Killing. Into a sterile screw-cap tube (12 × 75 mm; Falcon) is placed 0.5 ml of neutrophil suspension containing 2×10^7 cells/ml. To this is added 0.5 ml of opsonized bacteria (2×10^6–2×10^8 organisms/ml) or 0.4 ml of unopsonized bacteria plus 0.1 ml autologous serum. A 10-μl portion of the suspension is immediately withdrawn and added to 10 ml distilled water (a zero time control). The tube is then capped and placed on an end-on-end shaker so that the fluid contents flow from one end to the other during each cycle. The shaker is placed in a 37° incubator. At the desired times, additional 10-μl portions are withdrawn and added to 10 ml of distilled water. The distilled water lyses the neutrophils and releases the ingested bacteria. The number of bacteria surviving in the diluted samples is then measured by the method of colony counting described earlier.

For some purposes, it is of interest to determine the survival of ingested bacteria only. This is accomplished by destroying surviving extracellular bacteria prior to dilution and colony counting. For the destruction of extracellular *S. aureus,* the usual test organism, 0.5 ml of the assay mixture, after a 10-min incubation, is added to 0.5 ml of 0.15 M NaCl containing 20 U/ml of lysostaphin (Sigma) and 15 μmol/ml NaF (solutions containing NaF must be prepared and stored in plastic vessels, since F⁻ etches glass). Sodium fluoride prevents further ingestion, while lysostaphin digests and kills the adherent *S. aureus,* but not other species of *Staphylococcus.* The incubation is continued for 20 min at 37°. Lysostaphin is then inactivated through the addition of 1.0 ml of freshly prepared 0.5% trypsin in PBS followed by incubation for a further 15 min at 37°. Then the neutrophils are centrifuged at 250 g for 5 min and the cell pellet lysed by resuspending it in 10 ml of water. Finally, a 50-μl portion of the incubation mixture is diluted and subjected to colony counting as described previously (cf. Protocol 4).

An alternative method for destroying extracellular bacteria employs antibiotics that do not penetrate the neutrophil.[243,280–283] In this procedure, 0.5 ml of the assay mixture is added to 0.5 ml of PBS containing 250 U/ml penicillin G and 250 U/ml streptomycin. These antibiotics will kill extracellular organisms, but since they do not penetrate the neutrophil membrane they have no effect on living intracellular bacteria. After 30 min at 37°, the neutrophils are centrifuged at 250 g for 5 min at 4°, washed twice in PBS to eliminate residual antibiotics, then lysed in distilled water and the numbers of surviving bacteria determined by colony counting, as described above. This method can be used with any bacteria that are sensitive to penicillin or streptomycin.

In measuring bacterial killing, a control should be included in which bacteria are incubated in the absence of neutrophils to determine whether bacteria are killed by components of the reaction mixture other than the cells. It is desirable to determine rates of killing by measuring killing at several time points, both early and late, rather than determining the extent of killing at a single time point. The latter approach may cause true differences in killing between different systems to be overlooked. Accu-

[280] C. O. Solberg, *Acta Med. Scand.* **191,** 383 (1972).
[281] J. W. Alexander and R. A. Good, *J. Lab. Clin. Med.* **71,** 971 (1968).
[282] K. N. Brown and A. Percival, *Scand. J. Infect. Dis. (Suppl)* **14,** 251 (1978).
[283] C. O. Solberg, *Scand. J. Infect. Dis. (Suppl.)* **14,** 246 (1978).

racy can also be improved by carrying out the killing assays at several concentrations of neutrophils and bacteria; minor abnormalities in killing can be reliably detected by this method. The possibility of contamination should be kept in mind.

8.10. KILLING OF *Candida albicans*

The killing of *Candida albicans* can be determined by a pour plate method similar to that used for bacteria, or it can be measured by a dye exclusion method.

Protocol 25A: Pour Plate Method

Candida. *Candida albicans* grown to stationary phase, in broth culture, are the organisms used for these experiments. The organisms are inoculated into 50 ml of Sabouraud–2% dextrose broth (Difco) and grown at 33° for 3 to 5 days. The culture at this time should consist of yeast forms of high viability, as determined by the exclusion of 0.2 mM methylene blue dye. The fungi are washed twice in distilled water containing 0.01% gelatin, then suspended in PBS. The concentration of organisms can be determined in a hemacytometer. The number of colony-forming units (CFU) can be measured by the pour plate method using Sabouraud agar of suitable concentration, as described in Protocol 24, except that with this organism the diluted suspension (1 ml) is not mixed with melted agar but is spread directly on the surface of the hardened agar in the Petri dish. Colonies may be counted after 48 to 72 hr incubation at 37°. Suitable preparations yield 1 CFU per one to two single organisms.

Opsonization. Opsonization is carried out as described for bacterial killing (cf. Protocol 26).

Killing. Mix 0.5 ml of neutrophil suspension (5 × 10⁶/ml) with 0.5 ml of *Candida* (5 × 10⁶/ml) or with 0.25 ml unopsonized *Candida* (10⁷/ml) plus 0.25 ml autologous serum. The suspension is brought to 37° prior to mixing and incubated for 1 to 3 hr on an end-to-end rocker. Survival is assayed by colony counting, as described above, after the neutrophils have been lysed in distilled water, by methylene blue exclusion, or by the specific staining method of Lehrer and Cline[41] discussed below.

To measure survival by methylene blue exclusion, leukocytes are lysed through the addition, to the reaction mixture, of 0.25 ml of 2.5% sodium deoxycholate. A 0.1-ml portion of the resulting mixture is added to an equal volume of 0.4 mM methylene blue in water, and uptake of the dye by the yeast is determined microscopically. At this concentration, the dye is able to penetrate dead, but not live, yeast organisms. Data are expressed as the percentage of cells stained by methylene blue.

Protocol 25B: Specific Staining Method

To measure survival by the specific staining method,[41] 0.1 ml of the assay mixture is mixed with 0.3 ml of PBS containing 15% FCS. Cells are deposited on a slide by means of cytocentrifuge (Shandon Southern). The slide is rapidly dried in air, fixed for 1 min in absolute methanol, then stained with Giemsa stain diluted 1 : 10 in distilled water, and used within 2 hr of preparation. The cytoplasm of viable *Candida* stains blue after this procedure. Nonviable *Candida* are easily recognized by the failure of their cytoplasm to take up stain and by the presence of other morphologic abnormalities.

In addition to their blue cytoplasm, some of the live *Candida* will have begun to develop germ tubes. Some investigators subclassify the live organisms as to those that have developed germ tubes and those that have not. *Candida albicans* is typically the *Candida* species most resistant to killing by neutrophils. Representative figures obtained by Lehrer[103] illustrate a normal level of ~40% killing in 90 min by human neutrophils.

9. Specific Staining Techniques for Differentiating between Phagocytic Cell Types

For a more complete discussion of the principles of the various staining techniques, see Section 7.

9.1. Leukocyte Peroxidase

Leukocyte peroxidases oxidize benzidine chloride and other substrates to yield a blue or brown oxidation product which is precipitated at the sight of enzyme activity.[257] This reaction only occurs in the presence of H_2O_2.

Protocol 26A: Benzidine Dihydrochloride[257]

Fixative: A mixture of 10 ml of 37% formaldehyde and 90 ml of absolute ethanol is used to fix the specimen.

Reagents

Ethanol, 30% (v/v) in water (100 ml)
Benzidine dihydrochloride (0.3 g)
$ZnSO_4 \cdot 7H_2O$, 0.132 M (1.0 ml)
Sodium acetate trihydrate ($NaC_2H_3O_2 \cdot 3H_2O$) (1.0 g)
Hydrogen peroxide, 3% (0.7 ml)
Sodium hydroxide, 1.0 N (1.5 ml)
Safranin O (0.2 g)

The reagents are mixed in the order given. Benzidine dihydrochloride may contain some insoluble material which should be removed by centrifugation. When zinc sulfate is added, a precipitate forms which dissolves when the remaining ingredients are added. The pH of the solution should be 6.0 ± 0.05. The solution is filtered and stored in a closed container at room temperature. It can be used repeatedly for months.

Procedure

1. Whole blood or isolated cell suspensions may be used. Anticoagulants (heparin, EDTA, oxalate) do not interfere with the test. Leukocyte peroxidase activity is unstable in light. Smears stored in the dark are satisfactory for as long as 3 weeks.

2. The smear is fixed by immersing it for 60 sec in fixative at room temperature.

3. Rinse for 15 to 30 sec in running tap water, drain, shaking off excess water. The slide is then immersed in the staining solution at room temperature for 30 sec.

4. The stained film is rinsed for a few seconds in running tap water, allowed to dry, and examined microscopically.

If better nuclear staining is desired at the expense of slightly less contrast, safranin may be eliminated from the staining mixture and the film stained in hematoxylin after the final rinse in step 4.

Interpretation. Peroxidase is a granule-associated enzyme. In the neutrophilic series peroxidase activity is demonstrated by blue granules within the cytoplasm. Peroxidase activity is localized to the azurophilic (nonspecific) granules at all stages of neutrophil development. Monocytes do not stain intensely as do granulocytes and show either tiny blue granules or faint, diffuse cytoplasmic staining.[284,285]

Protocol 26B: 3-Amino-9-ethylcarbazole Method[258]

Reagents

Fixative: A mixture of 25 ml of 37% formaldehyde solution, 45 ml of acetone, and 30 ml of water containing 20 mg Na_2HPO_4 and 100 mg KH_2PO_4. The final pH is about 6.6

The staining mixture is prepared as follows:

Acetic acid, 0.2 M, adjusted to pH 5.0–5.2 with 1 N NaOH (50 ml)
Dimethyl sulfoxide (reagent grade, Sigma) (6 ml)

[284] F. G. J. Hayhoe, D. Quagliano, and R. Doll, *in* "The Cytology and Cytochemistry of Acute Leukaemias." H. M. Stationery Office, London, England, 1964.
[285] F. G. J. Hayhoe and J. C. Cawley, *Clin. Haematol.* **1**, 49 (1972).

3-Amino-9-ethylcarbazole (Aldrich Chemical Company, WI) (10 mg)
Hydrogen peroxide, 0.3% (0.4 ml)

The solution is filtered before use. The final pH is 5.5. Mayer's hematoxylin is used as counterstain.

Procedure

1. Fix smears in the buffered formalin acetone solution at room temperature for 15 sec.
2. Rinse gently in distilled water.
3. Immerse fixed film in the staining mixture at room temperature for $2\frac{1}{2}$ min.
4. Wash in running tap water.
5. Stain in Mayer's hematoxylin for 8 min.
6. Wash in tap water, air dry, and mount in glycerine jelly.

Interpretation. Peroxidase activity is demonstrated by red–brown granules distributed identically to those observed with the benzidine method. Nuclei stain blue.

9.2. Alkaline Phosphatase

This leukocyte enzyme hydrolyzes naphthol AS-BI phosphate to yield an aryl naphtholamide derivative and a free phosphate molecule. The former is highly reactive and couples with diazonium salts to yield a colored insoluble precipitate which is visible microscopically.

Protocol 27

Reagents

Naphthol AS-BI phosphate (Sigma Chemical Company)
Fast red violet salt LB or fast violet B salt (5-benzamide-4-chloro-2-toluidine)
Dimethylformamide, reagent grade (Eastman Chemical Company)
2-Amino-2-methyl-1,3-propanediol (Matheson, Coleman, and Bell)
HCl, 0.1 M
Citric acid, 0.03 M
Sodium citrate, 0.03 M
Acetone
Hematoxylin powder
Sodium iodate
Aluminum potassium sulfate (J. T. Baker Chemical Company)
Fixative: To 32 ml of 0.03 M citric acid slowly add 300 ml of absolute acetone with stirring

Buffer: Dissolve 21 g of 2-amino-2-methyl-1,3-propanediol (Sigma) in distilled water and dilute to 1 liter. Add 50 ml of 0.1 M HCl to 250 ml of propanediol stock solution and dilute to 1000 ml with distilled water. The stock solution and the buffer are stored at 4°; warm the buffer to room temperature before using

Staining mixture: Dissolve 5 mg of naphthol AS-BI phosphate in 0.2 to 0.3 ml of dimethyl formamide in a 100-ml container. Add 60 ml of buffer and then approximately 50 mg of fast red violet LB. The mixture is filtered and should be used immediately

Counterstain: Add 1 g of hematoxylin to 500 ml of distilled water. Heat to boiling and add another 500 ml of water. Then add 200 mg of sodium iodate and 50 g of aluminum potassium sulfate, mix thoroughly, and store at room temperature in a dark bottle. Filter immediately before use

Procedure[259,286]

Smears prepared from whole blood or cell suspensions are used within 24 hr of preparation. The films are exposed to the fixative solution for 30 sec at room temperature, then washed gently in running tap water for 1 min and air dried. Fixed smears may be stored at −15 to −20° for 2 to 3 weeks, with loss of about 10% of activity.

The slides are placed in the staining mixture for exactly 10 min at room temperature. They are then washed in running tap water for 30 to 60 sec and counterstained for 5 to 8 min in the hematoxylin solution. The slides are air dried or blotted carefully after being washed in running tap water from 1 to 2 min.

Interpretation. At least 100 neutrophils are scored from 0 to 4+ on the basis of the intensity of precipitated dye. Criteria for scoring are shown in Table IV. The scores of the 100 cells are added. Neutrophils will stain positively and monocytes negatively.

9.3. Acid Phosphate

This enzyme hydrolyzes hydroxynaphthoic acid anilide substances to yield the naptholamide derivative and free phosphate. The derivative readily complexes at low pH with diazonium salts to form a colored, insoluble precipitate[260] localized at the site of enzyme activity.

---

[286] L. S. Kaplow and M. S. Burstone, *Nature (London)* **200,** 690 (1963).

Protocol 28

Reagents

Fixative: Ten milliliters of methanol is added to 90 ml of a mixture of 60 ml of acetone and 40 ml of 0.03 M citric acid which has been adjusted to pH 5.4 with 1 N NaOH

Substrate stock solution: Dissolve 10 mg naphthol AS-BIO phosphoric acid (Sigma Chemical Company) in 0.5 ml N,N-dimethylformamide (Eastman Kodak Company). Add 0.1 M acetate buffer, pH 5.0, to a volume of 100 ml. This solution is stable at 4° for 2 months

Substrate solution: Ten milliliters of substrate solution is mixed with 5 mg fast garnet GBC (Sigma). Filter immediately before use

Commercial Mayer's hemalum

Tartaric acid: Seventy-five milligrams of L(+)-tartaric acid is dissolved in 10 ml of 0.1 M acetate buffer, at pH 5.0, containing 1 mg of naphthol AS-BI phosphoric acid. The mixture is shaken well and the pH is adjusted to 5.0 with concentrated NaOH. The final concentration is adjusted to 0.05 M. Fast garnet GBC, 5 mg, is added and the mixture filtered before use

Procedure[261,287,288]

Smears are fixed at 4 to 10° for 30 sec, then washed three times with distilled water and dried at room temperature for 10 to 30 min.

Films are incubated in substrate-staining solution and tartaric acid solution for 45 min at 37°, then washed in tap water.

Films are counterstained with Mayer's hemalum for 1 to 3 min, mounted in glycerine jelly, and examined with an oil immersion lens.

Interpretation. The presence of acid phosphatase activity is revealed by the formation of a red granular precipitate. Intense staining is seen in macrophages; moderate staining in monocytes and weak staining in neutrophils.

9.4. Esterases

Histochemical distinction between specific and nonspecific esterases allows the clear distinction between neutrophils and monocytes or macrophages. Fixation is the same for all of the esterase reactions.

---

[287] C. Y. Li, L. T. Yam, and K. W. Lam, *J. Histochem. Cytochem.* **18,** 473 (1970).
[288] C. Y. Li, L. T. Yam, and K. W. Lam, *N. Engl. J. Med.* **284,** 357 (1972).

Protocol 29A: Specific Esterase

Reagents

Fixative: Buffered formalin–acetone mixture, pH about 6.6. Mix 20 mg Na_2HPO_4, 100 mg KH_2PO_4, 30 ml of distilled water, 45 ml of acetone, and 25 ml of 37% formaldehyde solution

Incubation mixture: Phosphate buffer (M/15, pH 7.6), 47.5 ml, and freshly prepared hexazotized new fuchsin, 0.25 ml: Dissolve 1 g new fuchsin (Allied Chemical Company, NJ) in 25 ml 2 N HCl. Dissolve 4 g sodium nitrite in 100 ml distilled water. Store at 4° and make a fresh solution each week. Equal volumes of the new fuchsin solution and the 4% sodium nitrite solution are allowed to react for 1 min before use. This is hexazotized new fuchsin.

Naphthol AS-D chloroacetate (Sigma Chemical Company), 5 mg in 2.5 ml N,N-dimethylformamide (Eastman Kodak Company)

The solutions of hexazotized new fuchsin and naphthol AS-D chloroacetate are added to the phosphate buffer. The pH is checked and adjusted to 7.4, if necessary. The mixture is used without filtration

Counterstain: Harris's hematoxylin (alum solution). Four milliliters of glacial acetate acid are added per 100 ml of stain solution

Rinse solution: Dilute NH_4OH. Mix 0.2 to 0.3 ml of concentrated NH_4OH (58%) with 100 ml tap water

Procedure[262,263]

Smears from the whole blood or cell suspensions may be freshly prepared or may be stored unfixed in the dark at room temperature for at least 2 weeks without loss of enzyme activity.

1. Place the films in the fixative for 30 sec at 4 to 10°.
2. Wash the films in three changes of distilled water and allow them to dry in air at room temperature for 10 to 30 min.
3. Immerse fixed blood or marrow films in the incubation mixture for 10 min at room temperature in a Coplin jar.
4. Wash in tap water.
5. Counterstain with Harris's hematoxylin for 10 min in a Coplin jar.
6. Rinse films in dilute NH_4OH until the color of the film changes from red to blue.
7. Wash in water, dry in air, mount in Permount, and examine with an oil immersion lens for a red precipitate.

Interpretation. Strongly positive in neutrophils and negative in monocytes and macrophages.

Protocol 29B: Nonspecific Esterase

   A. *α-Naphthyl Acetate*

   Reagents
   1. Phosphate buffer (M/15, pH 7.6), 44.5 ml
   2. Hexazotized pararosaniline, freshly prepared, 3.0 ml: Pararosaniline solution: Dissolve 1 g pararosaniline hydrochloride (Fischer Scientific Company) in 20 ml distilled water and 5 ml concentrated HCl with gentle warming. After the solution cools, it is filtered and the filtrate is stored at room temperature
   Mix equal volumes of the pararosaniline solution with 4% sodium nitrite solution and allow to react for 1 min. The sodium nitrite solution is kept at 4° and should be less than a week old
   3. α-Naphthyl acetate (Sigma Chemical Company), 50 mg in 2.5 ml ethylene glycol monomethyl ether
   4. The solution of hexazotized pararosaniline, the α-naphthyl acetate, and the phosphate buffer are combined. The pH is adjusted to 6.1 ± 0.3 with 1 *N* NaOH

   Procedure[289]

   1. Smears are placed in the incubation mixture for 45 min at room temperature in a Coplin jar.
   2. Wash in water.
   3. Counterstain with Harris's hematoxylin (as described above) for 10 min.
   4. Rinse slides in dilute NH_4OH until blue color appears, wash with water, dry in air, mount, and examine for the presence of the orange–brown reaction product.

   Interpretation. Strongly positive for monocytes and macrophages; weak staining for neutrophils.

   B. *α-Naphthyl Butyrate*

   Reagents
   1. Phosphate buffer (M/15, pH 6.3), 47.5 ml
   2. Hexazotized pararosaniline (freshly prepared, as above), 0.25 ml
   3. α-Naphthyl butyrate, 50 mg in a 2.6 ml ethylene glycol monomethyl ether (Sigma)
   4. The solutions of hexazotized pararosaniline and α-naphthyl butyrate are combined with the phosphate buffer. The mixture is filtered before use

[289] L. T. Yam, C. Y. Li, and W. H. Crosby, *Am. J. Clin. Pathol.* **55**, 283 (1971).

Procedure[262]

Immerse fixed blood or neutrophil suspension smears in the incubation mixture for 45 min at room temperature in a Coplin jar.

1. Wash in water.
2. Counterstain with Harris's hematoxylin for 10 min.
3. Rinse slide in dilute NH_4OH until blue color appears, wash with water, dry in air, mount in Permount, and examine with an oil immersion lens for the presence of the orange–brown reaction product.

Interpretation. Monocytes and macrophages stain strongly; neutrophils weakly.

9.5. Lipid Content

Neutrophils exhibit a strong positive reaction when stained with Sudan black B, which stains various lipids, phospholipids, and sterols. Monocytes exhibit only weak staining.

Protocol 30: Sudan Black

Reagents

1. Formalin: A 37% aqueous solution of formaldehyde
2. Buffer: Mix 16 g phenol with 30 ml absolute alcohol and add to a solution of 0.3 g $Na_2HPO_4 \cdot 12H_2O$ in 100 ml distilled water
3. Sudan black B stock solution: Dissolve 0.3 g Sudan black B powder in 100 ml absolute ethanol. This may require 1 or 2 days at room temperature
4. Staining solution: Mix 60 ml of the stain stock solution with 40 ml of buffer. This mixture is stable for several weeks at room temperature
5. Harris's hematoxylin: Dissolve 5 g of hematoxylin crystals in 50 ml absolute ethanol. Dissolve 100 g aluminum potassium sulfate (potassium alum) in 50 ml distilled water. Heat to aid dissolution. Mix the above two solutions and bring to a boil. Add 2.5 g mercuric oxide, add water, bringing the final volume to 1 liter and boil for 5 min. Place in melting ice to cool rapidly

Filter and add 40 ml glacial acetic acid. Solution is stable for 2 months

Procedure[264,290]

Air-dried smears are used.

Fix in formalin vapor for 10 min. The slides are placed in a sealed dish

[290] H. L. Sheehan and G. W. Storey, *Pathologic Bacteriol.* **59,** 336 (1947).

containing a small volume of 37% formalin. The slides are placed on a bed of glass beads to prevent direct contact with the formalin solution.

Place the fixed smear in the Sudan black B staining solution for 60 min. Rinse in 70% ethanol for 3 min to remove excess stain.

Wash in running tap water and counterstain with hematoxylin for 10 min. Wash in running tap water to remove excess stain. Air dry and examine microscopically with an out-immersion lens for the presence of black granules.

Interpretation

Neutrophils: strongly positive
Monocytes: weakly positive

9.6. Cellular Polysaccharides

Periodic acid oxidizes glycols and related compounds to form aldehydes which, in turn, can react with the Schiff reagent. This results in the release of fuchsin, which stains polysaccharides components within cells.[265,291]

Protocol 31: PAS Stain

Reagents

Alcoholic formalin: Mix 10 ml of 36% formaldehyde with 90 ml of 95% ethanol

Periodic acid: KIO_4 crystals, 0.69 g, are added to 100 ml of distilled water, followed by 0.3 ml of concentrated HNO_3. The mixture is slowly heated to dissolve the crystals and then is stored at room temperature

Amylase: 0.1 to 1.0 g of commercial malt diastase in 100 ml of sodium phosphate buffer, 0.02 M, at pH 6.0

Schiff reagent: Boil 1000 ml of distilled water, remove from heat, and immediately add 1 g of basic fuchsin. Add a small amount of the dye, stir into the water, and then add the remainder of the dye. Cool the mixture to 60° and filter. Add 2 g of $NaHSO_3$ or $Na_2S_2O_5$ and 20 ml of 1 N HCl to the filtrate. Store the solution at room temperature for 18 to 26 hr in a stoppered bottle. Add 300 ml of activated charcoal and shake the mixture for 1 min; filter and store at 4° in the dark. The solution may be used until it becomes slightly pink.

Sodium metabisulfite: Prepare a fresh solution by dissolving 0.5 g of sodium metabisulfite in 100 ml of water.

Hematoxylin counterstain (see Protocol 30).

[291] R. D. Hotchkiss, *Arch. Biochem.* **16**, 131 (1948).

Procedure[291]

Smears are dried in air, fixed in alcoholic formalin for 5 min, then washed in running tap water for 15 min.

The films are then treated with periodic acid for 10 min at room temperature, washed in running tap water for 5 min. Films previously stained with Romanovsky stains, such as Wright's or Giemsa, can be also stained with PAS reagents commencing at this point.

The films are immersed in the Schiff reagent for 10 min and are then transferred rapidly through three changes of sodium metabisulfite in separate containers. They are washed for 5 min in running water.

Cell nuclei are stained by immersing the films in Harris's hematoxylin for 10 min. They are then washed in running tap water for 5 min, air dried, and examined microscopically using an oil immersion lens.

Interpretation. Nuclei of all cells stain blue with the counterstain.

The cytoplasm of mature polymorphonuclear leukocytes stains deep pink or red. The stain may be uniform or granular in some cells. Immature neutrophils are less intensely stained. Monocyte cytoplasm stains faintly or moderately pink and may contain fine or sometimes coarse granules.

Acknowledgments

Some of the technical information contained in Sections 8 and 9 is based on the excellent text: "Leukocyte Function" (M. J. Cline, ed.). Churchill Livingstone, New York, 1981.

Supported in part by research grants from the Medical Research (#MT5462, #MT8024, #MA9114), the Natural Science and Engineering Research (#UO-493) Councils of Canada, and the Ontario Heart Foundation (#4-12, #AN-402). In addition, the financial support of the Ontario Heart Foundation through the receipt of a Senior Research Fellowship is acknowledged.

Section II

Special Methods for Measuring Phagocytosis

[4] *In Vitro* and *in Vivo* Measurement of Phagocytosis by Flow Cytometry

By CARLETON C. STEWART, BRUCE E. LEHNERT, and JOHN A. STEINKAMP

Introduction

Many *in vitro* assays have been developed to quantify the ingestion of opsonin-dependent and opsonin-independent phagocytosis as the overall physiologic importance of phagocytosis has been increasingly recognized. Often, phagocytosis is quantitated microscopically, but this approach is tedious and time consuming, and only a small proportion of the total mononuclear phagocyte population can be evaluated. In this chapter, we describe experimental approaches by which the phagocytic activity of virtually tens of thousands of cells in a population of macrophages can be rapidly measured *in vitro* on a cell-by-cell basis using flow cytometry.[1] Following a quick, quantitative microscopic confirmation that particles are engulfed and not merely adhered to the surfaces of the macrophages after a particle–cell coincubation period (see this volume [6]), phagocytosis measured by flow cytometry can be quantitatively described in two ways. First, the percentage of cells with engulfed particles can be determined. Second, individual particulate burdens within each cell can be expressed as frequency distributions showing the frequency of phagocytic cells with a given number of internalized particles. The percentage of cells with particles indexes the fraction of cells doing the work, whereas the particle–cell distribution data serve to index the extent of phagocytic work performed by each phagocytic macrophage.

Methods

In Vitro Assay

Cell Populations. Methodologies for obtaining mononuclear phagocytes from the bone marrow,[2] peripheral blood,[3] peritoneal cavity,[4] pleu-

---

[1] J. A. Steinkamp, J. S. Wilson, G. C. Saunders, and C. C. Stewart, *Science* **215**, 64 (1982).

[2] C. C. Stewart, *in* "Methods for Studying Mononuclear Phagocytes" (D. O. Adams, P. J. Edelson, and H. S. Koren, eds.), p. 5. Academic Press, New York, 1981.

[3] C. C. Stewart, *in* "Methods for Studying Mononuclear Phagocytes" (D. O. Adams, P. J. Edelson, and H. S. Koren, eds.), p. 21. Academic Press, New York, 1981.

[4] R. E. Conrad, *in* "Manual of Macrophage Methodology" (H. B. Herscowitz, H. T. Holden, J. A. Bellanti, and A. Ghaffar, eds.), p. 5. Dekker, New York, 1981.

ral space,[5] lymph nodes,[6] and the lung[7] have all been previously described in great detail. We refer to the cited publications for these procedures (see also this series, Vol. 108 [9], [25], [26], [27], [28], and [29]).

Media and Test Particles. Cell populations should be maintained in a complete medium that is buffered with an organic buffer and not with sodium bicarbonate. Macrophages are exquisitely sensitive to a pH exceeding 7.4. When a bicarbonate buffered medium is used, the appropriate pH cannot be maintained in the absence of a continuous atmosphere supplemented with 5 to 10% carbon dioxide. This problem can be circumvented by using an organic buffer such as HEPES or MOPS (10 to 20 mM). The medium we use is α-minimum essential medium supplemented with 20 mM 2-N-morpholinopropanesulfonic acid and 10% fetal bovine serum (α-10 MOPS). The osmolality of the medium is adjusted with sodium chloride to 300 mOsm and the pH is adjusted to 7.2. Another reagent used is the same medium containing 2% bovine serum albumin (fraction V, Sigma Chemical Company, St. Louis, MO). This medium is used to separate free test particles from the cells following the particle–macrophage coincubation phase of the assay. For test particles, we recommend using fluorescent microspheres from Polysciences, Inc. (Paul Valley Industrial Park, Warrington, PA). These polystyrene microspheres are available in different fluorescent colors over a range of sizes. Depending on the fluorescent color, the excitation wavelength will range from the ultraviolet to the red (see the table). For the phagocytic assay, it is important to use 1.5- to 2.0-μm-diameter microspheres because smaller particles may be pinocytosed by cells other than macrophages. For example, 0.8-μm-diameter microspheres are avidly pinocytosed by fibroblasts and some tumor cell types, but 1.5-μm-diameter microspheres are not. Also, it is important that the coefficient of variation (CV) [i.e., (standard deviation/mean)100%] of fluorescence of these spheres be less than 2% if the number of particles per cell is to be resolved by flow cytometry. With such a low CV, it is possible to resolve up to five particles per cell.

Assay System. The cell suspension is adjusted to 1×10^6 cells/ml and 3 ml is placed in a 50-ml polypropylene conical centrifuge tube (Falcon No. 2070, Becton Dickinson Labware, Oxnard, CA) and warmed to 37°. The large size of these tubes provides a high surface area to fluid volume ratio suitable for maintaining cells and microspheres in suspension during continuous mixing. Moreover, the polypropylene tubes provide a non-

[5] A. Zlotnik, A. Vatter, R. L. Hayes, E. Blumenthal, and A. J. Crowle, *RES: J. Reticuloendothel. Soc.* **31,** 207 (1982).
[6] D. E. Bice, D. L. Harris, C. T. Schnizlein, and J. L. Mauderly, *Drug Chem. Toxicol.* **2,** 35 (1979).
[7] B. E. Lehnert and J. Ferin, *RES: J. Reticuloendothel. Soc.* **33,** 293 (1983).

FLUORESCENT MICROSPHERES FOR USE IN FCM PHAGOCYTOSIS
EXPERIMENTS AND LASER EXCITATION

Polysciences catalog number	Absorbed dye excitation maximum (λ)	Laser/excitation wavelength (λ)	Typical fluorescence measurement barrier filter
17686	365 nm	HeCd/325 nm	GG400
17686	365 nm	Argon/333–363 nm	GG400
17686	365 nm	Krypton/337–356 nm	GG400
17687	458 nm	HeCd/441 nm	GG475
17687	458 nm	Argon/457 nm	GG495
17687	458 nm	Krypton/406–415 nm	GG455
17688[a]	554 nm	Argon/528 nm	OG570
17688[a]	554 nm	Krypton/530 nm	OG570
17689[a]	686 nm	Krypton/647 nm	RG695

[a] These microspheres also can be excited using an argon- or krypton-pumped CW dye laser.

wettable surface which further retards cell adherence; macrophages adhere to the surface of these tubes at a rate of only 10% per hour even when they are not agitated. The microspheres of a desired color are suspended in culture medium at a concentration of $1 \times 10^8/50 \ \mu l$ and sonicated (Ultrasonic Cleaner, Fischer Scientific, Medford, MA) at 37° for 30 sec in order to dissociate aggregates of the test particles. Immediately thereafter, 3×10^8 of the microspheres are added to the 3 ml of the cell suspension. With a 100 : 1 particle-to-cell ratio, phagocytosis is not limited by the availability of microspheres over the course of the study. The cells and particles are coincubated at 37° for 30–60 min, during which time a well-mixed system is maintained with a rotating wheel (Multi-Purpose Rotator, Scientific Industries, Inc., Springfield, MA). At desired times during this incubation period, cells can be sampled for kinetic studies of phagocytosis. During the incubation period, 2 ml of the medium containing 2% BSA is put in a 15-ml conical centrifuge tube (Corning No. 25310, Corning Glass Works, Corning, NY). One of these tubes is prepared for each of the suspensions that is being incubated. After the desired time of incubation, the cell–particle suspension is carefully layered on top of the BSA medium and each tube is centrifuged for 10 min at 150 g. The supernatants are aspirated and the cell pellet is resuspended in 3 ml culture medium. This latter centrifugation step separates free particles from the cells, as well as serving to strip loosely adherent microspheres from the cells' surfaces. After this procedure, the cells are ready for analysis by flow cytometry.

In Vivo Phagocytosis

It is also possible to quantitate *in vivo* phagocytosis. Microspheres can be injected intraperitoneally to measure phagocytosis by peritoneal macrophages, injected in the pleural space to measure phagocytosis by pleural macrophages, or administered intravenously to measure phagocytosis by blood monocytes and granulocytes. For all injections, a suspension of 1×10^8 microspheres in 0.25 ml is used. Microspheres can also be instilled into the lungs to measure phagocytosis by alveolar macrophages.

For studies of the phagocytic activity of alveolar macrophages *in vivo,* rats are anesthetized with Ethrane (enflurane, Airco, Madison, WI) and the animals are intrabronchially instilled with 2×10^8 particles in 0.5 ml normal saline using a blunt, 18-gauge needle tipped with a 20-gauge Teflon tube which is advanced just proximal to the carina. After the instillations, the animals are held vertically for 30 sec before being returned to their cages. At selected times thereafter, animal sacrifices are initiated with intraperitoneal injections of 50 mg pentobarbital sodium. Prior to the onset of apnea, the animals are exsanguinated via the carotid arteries. The trachea is then cannulated with an 18-gauge blunt needle secured with a ligature, the lungs and trachea are excised en bloc, and the heart and esophagus are removed. Bronchoalveolar lavage is initiated by instilling 6 ml of room-temperature, divalent cation-free, phosphate-buffered saline (pH 7.2) into the lungs and aspirating the lavage fluid during gentle lung massage. This procedure is repeated an additional seven times. The retrieved lavage fluids (generally >95% of the instilled volume) are pooled in a centrifuge tube maintained in ice. Aliquots of the cell suspensions are layered over newborn bovine serum and centrifuged at 360 *g* for 10 min (4°) to remove nonphagocytized microspheres. Following removal of the supernatants, the cells are vortexed and resuspended in 0° culture medium until analyzed in the flow cytometer.

By measuring phagocytosis at several sequential time points after instillation of the microspheres, the kinetics of particle phagocytosis following particle deposition in the lung and intrapulmonary, particle–macrophage relationships over the course of alveolar clearance can be studied. Concurrent with these studies, suspensions prepared from the thoracic lymph nodes can be analyzed for evidence of particle migration from the lungs to these sites.[8]

It should be pointed out that two different types of fluorescent microspheres may be used in the *in vitro* and *in vivo* phagocytic assays. We have applied this technique, for example, to determine if alveolar macro-

[8] S. Hyler, Y. E. Valdez, and B. E. Lehnert, *Toxicologist* **5**, 180 (1985).

phages that had not phagocytosed microspheres *in vivo* were also not phagocytic *in vitro,* or whether they simply did not encounter particles in the lung.[1] If microspheres of one color are instilled and after 2 hr the population is lavaged and then incubated with microspheres of a second color using the *in vitro* procedure, the maximum phagocytic activity of cells can be measured. With this approach, we found that macrophages which had not phagocytosed *in vivo* were capable of avid phagocytosis *in vitro.* Such results suggest the instilled particles were not evenly distributed in the alveoli and, consequently, the microspheres were not available to all members of the alveolar macrophage population *in situ.*

Flow Cytometry (see also this series, Vol. 108 [19])

The cells are analyzed for narrow angle light scatter (cell size) and for fluorescence (phagocytized spheres) as they flow through a flow cell[9] and intersect a laser beam of exciting light (e.g., 457 nm wavelength, argon-ion laser).[1] Optical sensors measure light scattering and fluorescence on a cell-by-cell basis and the signals are stored in the list mode data format in a computer.[10,11] The data are then reprocessed and displayed as single- or two-parameter cell-size and fluorescence frequency distribution histograms.

Typical Results

Typical flow cytometric data obtained with alveolar macrophages harvested 2 hr after an intrabronchial instillation of the microspheres are shown in Fig. 1A and B. Such results illustrate the types of data obtained with either the *in vitro* or *in vivo* phagocytic assay. Figure 1A shows the size distribution of lavage-free cells from a Sprague-Dawley rat instilled with 1.83-μm-diameter green fluorescent spheres. Region 1 represents polymorphonucleated leukocytes (PMN) and lymphocytes. Pulmonary alveolar macrophages that have or have not phagocytized microspheres are contained in region 2. Since the free nonphagocytized spheres are smaller than cells, they do not appear in the cell-size distribution and they are totally discriminated against. Thus, by requiring the fluorescence signals to be in coincidence with the light scattering signals from cells, only cells that have associated spheres are contained in the fluorescence distribution (see Fig. 1B).

Data from the fluorescence and size distributions can be used to determine the percentage of cells containing one or more spheres and the

[9] J. A. Steinkamp, *Rev. Sci. Instrum.* **55,** 1375 (1984).
[10] R. D. Hiebert, J. H. Jett, and G. C. Salzman, *Cytometry* **1,** 337 (1981).
[11] G. C. Salzman, S. F. Wilkins, and J. A. Whitfill, *Cytometry* **1,** 325 (1981).

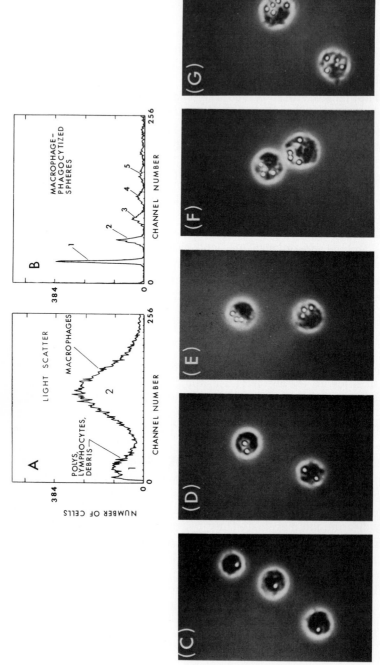

Fig. 1. (A) Cell-size distribution based on small-angle light scattering measurements of lavaged alveolar cells. (B) Fluorescence distribution of phagocytized spheres obtained by displaying only fluorescence signals associated with the light scatter signals from the macrophages. (C–G) Photomicrographs of rat alveolar macrophages containing spheres 1.83 μm in diameter that were sorted from peaks 1 to 5 of the fluorescence distribution of phagocytized spheres (B).

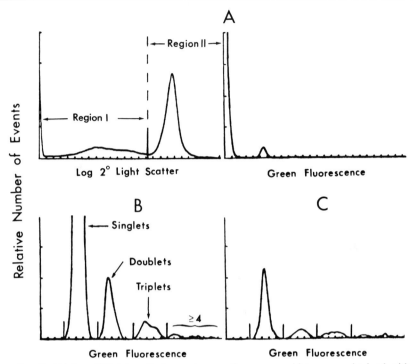

FIG. 2. (A) Log 2° light scattering and green fluorescence histograms obtained with a modal sample obtained 30 days after particle deposition in the lung. Left histogram: Region I of the light scatter histogram contains free particles, erythrocytes, and debris, while region II contains the nucleated cells. Right histogram: Fluorescence microspheres associated with the left light scatter distribution. (B) Distribution of single particles and aggregates of two or more microspheres is resolved when the correlated list mode data were reprocessed to obtain fluorescent events associated with events in region I of the light scatter histogram. The ordinate scale of the green fluorescent histogram has been expanded relative to that in (A) for the purpose of demonstration. (C) Distribution of microspheres that are associated with the cells in region II of the above light scatter histogram. The ordinate scale of the green fluorescent histogram has been expanded equivalently to that in (B).

percentage of cells having ingested one to five spheres, or more than five spheres. By dividing the total number of macrophages having phagocytized one or more spheres by the number of cells within region 2 (Fig. 1A—macrophages), the percentage of the cells that phagocytized the particles can be determined. This can be further expanded into the percentage of phagocytic alveolar macrophages containing one, two, three, four, five, and more than five spheres by dividing the total number of phagocytic cells into the number containing one, two, etc., spheres. Sorted cells containing one to five microspheres are shown in Fig. 1C–G.

Figure 2A–C illustrates the flow cytometric approach we recently used[8] to quantitate the migration of 1.9-μm-diameter fluorescent (green) microspheres to the thoracic lymph nodes after 4×10^8 of these particles were deposited in the lungs of rats. Cells and particles liberated from the nodes after dissection were prepared in suspension and they were excited by the 488-nm line of an argon laser. Log 2° light scattering and green fluorescence over a 515- to 545-nm wavelength range were measured. Signal processing electronics were set to trigger on any incoming event, i.e., no coincidence requirements. Data were acquired for 30,000 events per sample and stored in the correlated list mode fashion.[11]

In order to estimate the total number of particles associated with the nodal cells in each analyzed sample, the integrated sum of fluorescent events coincident with region II of the log 2° light scattering histogram, as illustrated in Fig. 2A and C, was determined. Those particulate events falling in the window range for ≥ 4 particles were assigned a minimum value of 4 for computational purposes. This sum was expressed as a fraction of the total number of cells analyzed in region II. This fraction, in conjunction with the total cell yield from the nodes of a given animal, was then used to determine the total numbers of particles associated with all the cells harvested. As with the alveolar macrophages, the particulate burdens in each nodal phagocyte were also determined.

The total number of free or non-cell-associated particles in the nodal samples was determined by analyzing region I of the log 2° light scatter histogram, Fig. 2A, for the fluorescent particulate events shown in Fig. 2B. The total number of fluorescent particles coincident with region I were expressed as a fraction of the total number of cellular events occurring in region II, and this fraction, in turn, was used in conjunction with nodal cell yields to determine the total number of "free" particles harvested from the lymph nodes.

Comments

We have described experimental techniques whereby phagocytosis by mononuclear phagocytes can be rapidly and accurately measured using multiparameter flow cytometry. Quantitative indices of phagocytosis obtained using the flow cytometric approach include the percentage of cells that actually phagocytose particles, and the distribution of particles in the cells that contain them. Therefore, the relative number of cells in a population that has performed the work of phagocytosis, as well as an index of the extent of the work performed by the phagocytic cells, is precisely obtained. Moreover, the *in vitro* suspension assay system outlined here allows one to obtain these quantitative indices of phagocytosis by evaluating the phagocytic activity of virtually every cell in an experimental population, unlike some conventional phagocytic assay systems that use monolayers of mononuclear phagocytes formed with adherent subpopulations only. As well, performance of the phagocytic assay with cells and particles in a well-mixed system negates the possibility that the phagocytic activities of the cells may be altered by adherence to a stationary substrate.[12] Furthermore, removal of adherent cells from a substrate is accompanied by death of a large proportion of the cells.

The phagocytic assay presented in this chapter, by the nature of the test particles used, measures "nonspecific" phagocytosis. However, since the fluorescent microspheres can be obtained with carboxyl groups, it may be possible to suitably coat them with appropriate opsonins for studying specific receptor-mediated phagocytosis. The finding by van Oss and co-workers that IgG adsorption onto polystyrene spheres augments their phagocytosis by granulocytes[13] lends tentative support for this possibility. Other investigators have used fluorescently labeled biotic particles such as bacteria in conjunction with flow cytometry to measure specific receptor-mediated phagocytosis.[14,15] The high coefficient of variation in the fluorescence from these biological test particles, however, restricts their usefulness. The percentage of cells that phagocytize can be determined, but individual cell burdens of particles cannot presently be resolved. Regardless, the microsphere assay that employs unopsonized particles may be especially appropriate for measuring the phagocytic activities of cells like the alveolar macrophages, inasmuch as these phagocytes routinely encounter particles deposited in the alveolar microenvironment, which is relatively deficient in proteins and opsonins.[16]

In summary, we know of no other automated method, except for the flow cytometric technique described here, that can measure the percentage of a macrophage population that has phagocytized, and also gives discrete information on particulate burdens in individual cells. The flow cytometric approach offers major advantages over existing, conventional techniques. Chiefly, data can be easily obtained from large sample sizes and, because of the rapidity of the method, multiple cell samples can be evaluated over a short period of time.

[12] B. E. Lehnert and P. E. Morrow, *Immunol. Commun.* **13**, 313 (1984).

[13] C. J. van Oss and M. W. Stenson, *RES: J. Reticuloendothel. Soc.* **8**, 397 (1970).

[14] P. Szejda, J. W. Parce, M. S. Seeds, and D. A. Boss, *J. Immunol.* **133**, 3303 (1984).

[15] C. F. Bassoe, O. D. Laerum, J. Glette, G. Hopen, B. Haneberg, and C. O. Solberg, *Cytometry* **4**, 254 (1983).

[16] H. Y. Reynolds and H. N. Newhall, *J. Lab. Clin. Med.* **84**, 559 (1974).

Acknowledgments

This work was supported in part by the Department of Energy, Office of Health and Environmental Research, and the Los Alamos National Flow Cytometry and Sorting Research Resource funded by the Division of Research Resources of NIH (Grant No. P41-RK 01315-02).

[5] Oil-Droplet Method for Measuring Phagocytosis

By THOMAS P. STOSSEL

Introduction

The basis of the oil-droplet method to assay phagocytosis is to stabilize particles containing a marker dye or radioactive compound, feed them to phagocytes, separate uningested particles from the cells by differential centrifugation, extract the marker from the cells, and measure it. The rate of association of the marker with the centrifuged cell pellet as a function of time reflects the rate of phagocytosis. Since its initial description[1] the oil-droplet method has been modified in many ways. Originally, the dye oil red O was incorporated as a marker into heavy paraffin oil. Subsequently, diisodecyl phthalate was found to have a more suitable density than paraffin oil. Other modifications have included the use of radioactive lipids as markers[2] and of spin-labeled cholestanone, which can be quenched by ascorbic acid, permitting assessment of the extent to which particles are either adherent to the cells or are incorporated within unsealed vacuoles. Extracellular particles or particles in unsealed vacuoles are quenched whereas particles within closed vacuoles are not.[3,4] Double-label experiments have been performed by feeding cells particles labeled with either oil red O or perylene, measuring the former spectrophotometrically and the latter fluorimetrically.[5]

The method has several advantages. The most important one is the fact that the density of suitable oil droplets is such that the efficiency of separation of uningested particles from phagocytes is very high; the ability to remove adherent particles markedly facilitates the precise quantitation of phagocytic rates. By varying the material utilized to stabilize the particles, the recognition mechanisms of phagocytosis can be studied.

[1] T. P. Stossel, R. J. Mason, J. H. Hartwig, and M. Vaughan, *J. Clin. Invest.* **51,** 615 (1972).
[2] A. Forsgren, D. Schmeling, and D. Zettervall, *Immunology* **32,** 491 (1977).
[3] M. J. Ueda, T. Ito, S.-I. Ohnishi, and T. S. Okada, *J. Cell Sci.* **51,** 173 (1981).
[4] T. Ito, M. J. Ueda, T. S. Okada, and S.-I. Ohnishi, *J. Cell Sci.* **51,** 189 (1981).
[5] R. D. Berlin, J. P. Fera, and J. R. Pfeiffer, *J. Clin. Invest.* **63,** 1137 (1979).

Finally, phagolysosomes containing ingested oil droplets can be isolated from the phagocytes in high yield.[6-14] One disadvantage of the assay is that the particles, unlike erythrocytes, zymosan, or latex, cannot be simply purchased as such but must be prepared by the investigator. Once made, however, most oil dispersions can be frozen and stored indefinitely and thawed prior to use. Another theoretical disadvantage is the fact that the particles are not uniform in size.

The assay has proved useful for measuring initial rates and extents of phagocytosis and for determining the effects of a variety of physiologic and pharmacologic influences on these rates.[1,15,16] It has brought out effects of additives, for example, colchicine, on phagocytosis which were missed with less quantitative assays.[1] The assay has been used to document impaired phagocytosis or the lack thereof by phagocytes of patients with disorders of phagocytic cells.[17-19] In certain cases oil droplets stabilized with selected emulsifiers are taken up by phagocytes only if first reacted with opsonins. In other circumstances opsonins increase a basal rate of ingestion of the particles. In both instances, the system allows the investigators to measure quantitively the effects of opsonins on the phagocytic process or else to test clinical samples for opsonic activity.[15,20-25] However, since most researchers of phagocytosis continue to

[6] T. P. Stossel, T. D. Pollard, R. J. Mason, and M. Vaughan, *J. Clin. Invest.* **50**, 1745 (1971).

[7] T. P. Stossel, R. J. Mason, T. D. Pollard, and M. Vaughan, *J. Clin. Invest.* **51**, 604 (1972).

[8] J. Oliver, T. E. Ukena, and R. D. Berlin, *Proc. Natl. Acad. Sci. U.S.A.* **71**, 394 (1974).

[9] Y. Rikihisa and D. Mizuno, *Exp. Cell Res.* **111**, 437 (1978).

[10] F. Amano and D. Mizuno, *FEBS Lett.* **115**, 193 (1980).

[11] P. D. Lew and T. P. Stossel, *J. Biol. Chem.* **255**, 5841 (1980).

[12] M. Volpi, P. H. Naccache, and R. I. Sha'afi, *J. Biol. Chem.* **258**, 4153 (1983).

[13] H. Lagast, P. D. Lew, and F. A. Waldvogel, *J. Clin. Invest.* **73**, 878 (1984).

[14] H. J. Cohen, M. Chovaniec, and W. A. Davies, *Blood* **55**, 355 (1980).

[15] T. P. Stossel, *J. Cell Biol.* **58**, 346 (1973).

[16] T. A. Lane and G. E. Lamkin, *Blood* **64**, 400 (1984).

[17] T. P. Stossel, R. K. Root, and M. Vaughan, *N. Engl. J. Med.* **286**, 120 (1972).

[18] L. A. Boxer, E. T. Hedley-Whyte, and T. P. Stossel, *N. Engl. J. Med.* **291**, 1093 (1974).

[19] M. A. Arnaout, J. Pitt, H. J. Cohen, J. Melamed, F. S. Rosen, and H. R. Colten, *N. Engl. J. Med.* **306**, 693 (1982).

[20] T. P. Stossel, C. A. Alper, and F. S. Rosen, *J. Exp. Med.* **137**, 690 (1973).

[21] T. P. Stossel, *Blood* **42**, 131 (1973).

[22] T. P. Stossel, R. J. Field, J. D. Gitlin, C. A. Alper, and F. S. Rosen, *J. Exp. Med.* **141**, 1329 (1975).

[23] T. P. Stossel, C. A. Alper, and F. S. Rosen, *Pediatrics* **52**, 134 (1973).

[24] S. S. Shurin and T. P. Stossel, *J. Immunol.* **120**, 1305 (1978).

[25] P. D. Lew, R. Zubler, P. Vaudaux, J. J. Farquet, F. A. Waldvogel, and P.-H. Lambert, *J. Clin. Invest.* **63**, 326 (1979).

count particles ingested by phagocytes observed in the light microscope, and since most of the results reported have not been checked by other methods, the extent to which this method can add substantively to the study of phagocytosis is unclear. In one case, however, the dose response for inhibition of phagocytosis by cytochalasin B was essentially identical as measured either by light microscopy[26] or the oil-droplet method.[27]

Procedure

Although a detailed procedure is described here that works, it must be emphasized that only the investigator's imagination and the behavior of the reagents limit the variations which could be attempted with this technique. The key elements are a stable oil emulsion, a quantifiable marker, and conditions in which phagocytes will ingest them. The conditions described were empirically derived, and by no means have alternatives or variations been exhaustively explored.

Preparation of Oil Droplets

A suitable oil is diisodecyl phthalate ($d = 0.95$) (Coleman, Matheson, and Bell, Cleveland, OH). To prepare a stock solution, approximately 2 g of oil red O (Allied Chemical Corp., Morristown, NJ) is added to 50 ml of diisodecyl phthalate in a mortar and the system is mixed with a pestle. The oil is then decanted into glass tubes and centrifuged to remove undissolved dye (which can be reutilized) at top speed in an intermediate velocity centrifuge (e.g., Sorvall RC 5). The colored oil solution can be stored indefinitely at room temperature.

If radioactive particles are to be prepared, a radioactive lipid or hydrocarbon (e.g., [14C]octadecane, Amersham-Searle) is added directly to the carrier oil. The radioactivity of the oil is determined in dioxane, and a conversion factor is calculated as described for oil red O (see p. 197). The quantity of radioactivity to be added depends on the number of cells to be tested and on the sensitivity of the radioactivity detection systems available.

To prepare the droplets, 1 ml of oil is layered over 3 ml of the emulsifying solution. The emulsifier is dissolved or suspended in the buffer solution to be used for the phagocytic assays which should therefore be some kind of physiologic salt solution (e.g., for mammalian phagocytes— Krebs–Ringer phosphate or bicarbonate solution or Hanks' balanced salt solution). Two substances which have been used most extensively as emulsifiers are bovine serum albumin (20 mg/ml) and *Escherichia coli*

[26] S. A. Axline and E. P. Reaven, *J. Cell Biol.* **62**, 647 (1974).
[27] J. H. Hartwig and T. P. Stossel, *J. Cell Biol.* **71**, 295 (1976).

lipopolysaccharide (40 mg/ml). The latter is particularly applicable for examining complement-dependent phagocytosis, since these particles are ingested only if first opsonized by serum containing the third component of complement (C3) and proteins of the alternative pathway of complement activation which deposit C3 on the particle surfaces. The lipopolysaccharide (serotype 026 : B6, Boivin preparation, Difco, Detroit, MI), should be dispersed in the salt solution by brief sonication. The serotype and method of preparation of the lipopolysaccharide are important details, because they influence both the stability of the emulsion and the ability of the opsonized particles to be ingested. For example, lipopolysaccharide prepared by the Westphal method does not produce as stable an emulsion; and if *E. coli* lipopolysaccharide serotype 011 : B8 is used to coat the oil droplets, they activate the complement system but are not ingested by phagocytes. The reasons for these particulars are not known.

To prepare the particles, the oil/buffer–emulsifier system is sonified. The sonicator probe should be just below the oil/buffer interface. The length of sonication should be about 90 sec, just until the tube is becoming too hot to handle, but the time may vary with the output of the particular sonicator. The emulsion can be used immediately after cooling or can be frozen. In the latter case, brief sonication after thawing is advisable.

When opsonization is indicated, as in the case of the lipopolysaccharide-coated droplets, the particles are suspended in an equal volume of fresh serum (from any source) and incubated 15–30 min at 37°. Proteolytic activity in serum will remove opsonically active C3 (the C3bi fragment of C3)[20,21] from the particles if the particles are left in serum too long. Therefore the particles should be cooled on ice, washed, or frozen after the opsonizing incubation. When proteins such as albumin are used as coating for the particles, the particles can also be opsonized with fresh serum. In this case the denatured protein on the oil surface appears to activate the alternative complement pathway. In addition, particles can be opsonized by incubating them with specific antibodies. The conditions of such incubations will depend on the titer and nature of the antibodies employed and must be designed to avoid agglutination.

To wash particles after opsonization, or to remove excess emulsifier, the suspension can be diluted with a 0.15 *M* NaCl solution and centrifuged. The particles float during centrifugation, and the infranatant solution can be removed by penetrating the pellicle with a needle and aspirating with a syringe. The ability of the particles to withstand repeated washing depends on the stability of the emulsion. Lipopolysaccharide-coated particles are particularly stable and can easily be washed in this manner.

Preparation of Phagocytic Cells

Methods for the preparation of test cells are described in Vol. 108 ([9], [25], [26], [27], [28], and [29]). Phagocytes from blood and exudate sources have been used with this assay. The only stipulations are that the cells be viable and not agglutinate during the procedure. In general, agglutination can be reduced by keeping the cells cold and in buffers of relatively low pH (7.0 or less) and without added divalent cations until ready to measure phagocytosis. Since optimal ingestion in some cases requires extracellular calcium and especially magnesium, these are added to final concentrations of 1 mM and the pH adjusted up to 7.4 just prior to the incubations for phagocytosis. The ability of a particular cell type to ingest the particle employed determines the volume of cells required to obtain a signal of reasonable magnitude for detection.

Assay of Ingestion

One approach is to incubate a relatively large volume of reaction mixture and remove samples at various times so as to obtain a time course of ingestion. For example, 4 ml of a 5% (v : v) cell suspension is incubated in siliconized or plastic flasks at 37°, and 1 ml of particles added at zero time. The flasks are gently shaken in a 37° water bath, and duplicate samples of 0.5 ml removed at 0, and other time points. The optimal times for measurement will depend upon the cells and particles selected. For albumin or lipopolysaccharide-coated droplets, polymorphonuclear leukocytes studied in the past ingest at a constant rate up to 10–15 min. Macrophages may internalize particles for up to 40 min.

The samples are immediately added to conical centrifuge tubes on ice containing 5 ml of 1 mM N-ethylmaleimide in 0.15 M NaCl. The cold and the SH-reacting compound are to stop ingestion. When the time course is completed, the tubes are centrifuged at 250 g for 10 min. The rim of uningested oil particles at the meniscus is dislodged by shaking the tube horizontally, and the supernatant fluid is decanted. The cell pellet is disaggregated by tapping on the bottom of the tube, and 5 ml of fresh, ice-cold 0.15 M NaCl solution is added. After a second centrifugation, the tubes are inverted to drain. Any oil droplets stuck where the meniscus was are removed with tissue. Then the cell pellets are disrupted and marker extracted by addition of 1 ml of *p*-dioxane. The tubes are tapped or vortexed to complete extraction. The tubes are centrifuged at 500 g for 15 min to sediment cell debris, and the optical density of the oil red O in the supernatant dioxane is then determined at 525 nm against a dioxane blank. If radioactivity was utilized instead of oil red O, the dioxane solution is suitable for direct addition to most commercial scintillation fluids for direct determination of radioactivity in the sample.

An alternative approach applicable for determining ingestion of multiple samples at a single time point is to add 0.2 ml of particles to 0.8 ml of cell suspension in conical centrifuge tubes and to incubate for the time and at the temperatures selected. The N-ethylmaleimide solution (6 ml) is then added to the tubes to stop the reaction, and the washing procedure follows as described above.

Although not extensively tested, the method can also be used with adherent cells. Cover slips or Petri dishes with adherent phagocytes are overlayed with the particle suspension which is decanted at suitable times. The surfaces with the cells are washed with cold salt solutions (it must be first determined whether N-ethylmaleimide causes the cells to detach) and then extracted with dioxane. Since fewer cells are incubated with this technique, it is probable that a radioactive marker will be more suitable than oil red O.

To determine the quantity of oil ingested per minute per cell, the following formula is used:

Initial ingestion rate
$$= [(OD)(\text{mg oil/OD unit})]/[(\text{cell protein/ml})(\text{time of incubation})] \quad (1)$$

(Cell numbers instead of cell protein, can also be used in the denominator.)

Where OD is the optical density of the dioxane extract and milligrams oil/OD unit is a calibration factor obtained by adding 10 μl of the stock oil red O in diisodecyl phthalate solution (see p. 194) to 10 ml of p-dioxane and reading the optical density at 525 nm. The calibration factor is computed by dividing 0.95 (the density of the oil) by the optical density.

Additional Comments

Although with the particles described the nonspecific adsorption of particles to cells tends to be low, it is advisable to perform control incubations at ice temperature or with metabolically paralyzed cells incapable of ingestion. While mixtures of cells, some of which may not be phagocytic, the uptake of particles can be visually monitored with Wright's-stained smears. The methanol in the stain removes the oil of the particles, and ingested particles appear as intracellular vacuoles.[28] External particles are completely removed and are therefore invisible.

The assay has been performed with human, guinea pig, and rabbit polymorphonuclear leukocytes and mononuclear phagocytes.[1–10] The ingestion response of different phagocytes to a given coating of oil droplets varies. For example, guinea pig peritoneal exudate polymorphonuclear leukocytes avidly ingest albumin-coated particles, whereas human pe-

---

[28] A. Altman and T. P. Stossel, *Br. J. Haematol.* **27**, 241 (1974).

ripheral blood neutrophils only do so when the particles are first coated with opsonically active C3. The density of different phagocytes also varies. For example, rabbit heterophils are less dense than guinea pig neutrophils. The guinea pig leukocytes fed paraffin oil-containing droplets sediment, but the rabbit cells tend to float. The use of diisodecyl phthalate, which has a higher density than paraffin oil, obviates this problem. However, these factors need to be kept in mind when designing experiments with this assay.

Acknowledgment

Supported by a grant from the Edwin S. Webster Foundation.

[6] Methods for Distinguishing Ingested from Adhering Particles

By JAN HED

Introduction

Most knowledge about the function of receptors of phagocytes is based on assays using opsonized erythrocytes. Hypotonic lysis is used to differentiate between attached and ingested erythrocytes. These assays are complicated by the fact that attached erythrocytes can be lysed by cytotoxic products released during the interaction.[1] In addition, scavengers from the erythrocytes can modulate the phagocyte function.[2,3] Since antibodies do not penetrate the plasma membrane of viable cells, immunofluorescence techniques[4] have been used to visualize attached particles. The advantage of this method is that nonmanipulated particles can be used. This is important when studying host–parasite interactions. This method, however, does not demonstrate the number of ingested particles which can only be calculated after fixation and staining of all the particles interacting with the phagocytes. Several other methods have been devised where different treatments release the attached particles. Lysostaphin has been used to digest noningested staphylococci,[5] and ex-

[1] G. Holm and S. Hammarström, *Clin. Exp. Immunol.* **13**, 29 (1973).
[2] W. L. Hand and N. L. King-Thompson, *Infect. Immun.* **40**, 917 (1983).
[3] J. Forslid, J. Hed, and O. Stendahl, *Immunology* **55**, 97 (1985).
[4] A. P. Punsalang, Jr., and W. D. Sawer, *Infect. Immun.* **8**, 255 (1973).
[5] J. Verhoef, P. K. Peterson, and P. G. Quie, *J. Immunol. Methods* **14**, 303 (1977).

cess of antigen or staphylococcal protein A (SpA)[5a] to dissociate attached immune complexes[6] and SpA–IgG[7] complexes, respectively. However, all these procedures necessitate further incubations which could influence the fate of the attached particles, making it difficult to evaluate the initial rate of phagocytosis.

A recently reported method[8] to differentiate between attached and ingested particles in phagocytosis is described in this chapter. The technique is based on the phenomenon known as fluorescence excitation transfer,[9,10] where a dye quenches the fluorescence of the attached particles leaving only the ingested particles fluorescent.

Methods

Assay Principle

In the phagocytic assay described in this chapter, fluorescent particles are used. After incubation, a dye is added which quenches the fluorescence of the attached particles, leaving the ingested particles fluorescent due to dye exclusion (Fig. 1). This simple technique called the fluorescence quenching method (FQ method) allows quantitation of attached and ingested particles in individual phagocytes. The quenching mechanism described is based on the phenomenon known as excitation energy transfer.[9,10] The efficiency of the transfer is dependent on (1) the integrated overlap between the excitation spectrum of the donor (fluorochrome) and the absorption spectrum of the acceptor (quenching dye), and (2) the distance between bound donor and acceptor molecules.

Problems that have to be considered when using this technique are the following:

1. Selection of a fluorochrome that can bind to the particle or selection of a particle that can bind the fluorochrome.

2. Selection of a quenching dye that can bind to the fluorescent particle but does not penetrate the plasma membranes of the phagocytes.

3. In quantitative cytofluorometric assays it is important that the fluorochrome should not be affected by the low pH in phagolysosomes

---

[5a] Abbreviations: SpA, staphylococcal protein A; FQ, fluorescence quenching; FITC, fluorescein isothiocyanate; Con A, concanavalin A; PBS, phosphate-buffered saline.

[6] P. A. Ward and N. J. Zvaifler, *J. Immunol.* **111**, 1771 (1973).

[7] R. Hällgren, L. Jansson, and P. Venge, *Immunology* **30**, 755 (1976).

[8] J. Hed, *FEMS Lett.* **1**, 357 (1977).

[9] T. Föster, *Radiat. Res., Suppl.* **2**, 326 (1960).

[10] T. Föster, *Discuss. Faraday Soc.* **27**, 7 (1960).

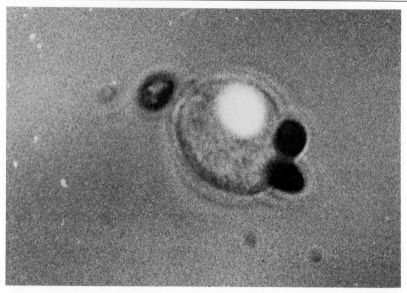

FIG. 1. A neutrophil with ingested and attached FITC-conjugated yeast particles after treatment with trypan blue, pH 4.4. Attached particles are quenched and dark whereas the ingested particle remains fluorescent.

and that the dye show a complete quenching of fluorescence and total dye exclusion.

Selection of Fluorochrome

Most studies concerning the FQ method have used fluorescein isothiocyanate (FITC),[8] fluorescamine[11] or glutaraldehyde-fixed erythrocytes.[12] We have primarily used FITC-conjugated heat-killed yeast particles (*Saccharomyces cerevisiae*), since our main interest is focused on the phagocytes and their receptor functions. The yeast particles are homogeneous and of suitable size, do not aggregate, are not lysed by phagocytes, are easy to prepare, and easy to coat with various substances such as IgG complement, and Con A. We have found zymosan, derived from *Saccharomyces cerevisiae* and often used in phagocytic experiments, less suitable primarily due to the variable size of particles and to the formation of aggregates which complicate quantitation of the interaction. Furthermore, the binding by zymosan particles of more immunoglobulin from normal serum than the heat-killed yeast particles[13] may activate the

[11] J. Hed, C. Dahlgren, and I. Rundquist, *Histochemistry* **79**, 105 (1983).
[12] J. D. Loik and S. C. Silverstein, *J. Immunol. Methods* **57**, 373 (1983).
[13] J. Hed and O. Stendahl, *Immunology* **45**, 727 (1982).

phagocytes and modulate their receptors. Here we describe the use of FITC-conjugated yeast particles, primarily to analyze the function of phagocytes. One important drawback of the FQ method is that the fluorochromes do not stain equally well all microorganisms since hydrophilic surface structures of the microbes block the fluorochrome from penetrating the surface and binding to the cell wall. The surface must be somehow modified before conjugation with the fluorochrome, making the particles less appropriate for these studies. In many cases, however, bacteria can be FITC conjugated without major alteration of the surface structures.[14]

Selection of Dye

Basic dyes, notably crystal violet, a constituent of the gram stain with high affinity for most microorganisms, show good quenching of FITC at neutral pH and were originally used in the development of the FQ method. Crystal violet has recently been used as a quenching agent in flow cytofluorometric techniques to quantitate ingestion and attachment.[15] The dye is difficult to use in cytofluorometry.[16,17] Crystal violet belongs to the lysosomotropic dyes and quenches the fluorescence of the ingested particles in neutrophils and macrophages by 45–70%.[16] Quenching agents with efficient dye exclusion should be sought among acid and direct dyes. Many blue and red anionic dyes are good quenchers at acid pH partly due to higher affinity for the FITC-conjugated particles at relatively low pH, and partly due to the quenching effect of the low pH per se. We have selected trypan blue (2 mg/ml) at pH 4.5 as a quenching agent since this dye is used to test cell viability. When trypan blue, pH 4.5, is used in flow cytofluorometric assay, a plateau is obtained when the concentration of the dye is 0.12 mg/ml or higher (Fig. 2), indicating quenching of only the attached particles. No such plateau is obtained with crystal violet, which shows a decrease in fluorescence intensity with increasing concentration, indicating an effect on ingested particles.[17]

Quantitative Assays

When the phagocytosis of particles like yeast is studied, it is easy to quantitate attached and ingested particles by the FQ method by counting. However, this is not possible when smaller particles or protein aggregates such as immune complexes[15] are phagocytized or when cytofluorometric assays are used. The sensitivity of FITC to changes in pH is a disadvan-

[14] L. Öhman, J. Hed, and O. Stendahl, *J. Infect. Dis.* **146**, 751 (1982).
[15] R. Bjerknes and C.-F. Bassoe, *Acta Pathol. Microbiol. Scand., Sect. C* **91C**, 341 (1983).
[16] S. Sahlin, J. Hed, and I. Rundquist, *J. Immunol. Methods* **60**, 115 (1983).
[17] J. Hed, S. G. O. Johansson, and P. Larsson, in preparation.

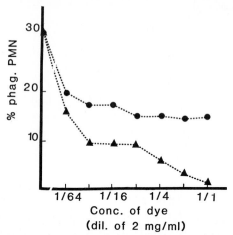

FIG. 2. Yeast-C3bi phagocytosis by human neutrophils measured by flow cytofluorometry and quenched with trypan blue (●) and crystal violet (▲).

tage in cytofluorometric assays. In order to obtain a measurement of the interaction of the particles with the phagocytes by using cytofluorometry (i.e., before adding the dye), the pH of the environment must be the same for the attached and ingested particles. We have found that the fluorescence of the FITC-conjugated particles is the same whether attached or ingested at pH 5.8 for human neutrophils and pH 4.4 for mouse peritoneal macrophages.[15] These values are recommended for measuring the total association in cytofluorometry.

We have recently overcome this quantitative problem in flow cytofluorometry[17] by calibrating the phagocytic system in such a way that only one particle interacts with each phagocytosing granulocyte. This system allows analysis of whole blood from patients without prior separation of the phagocytes since the percentage of phagocytosing neutrophils were not dependent on the concentration of cells.

Methodology

Preparation of FITC-Conjugated Heat-Killed Yeast Particles. Yeast particles (available from Sigma Chemical Company) are suspended in 0.9% NaCl to a thick suspension which is boiled in a water bath for 30 min. The heat-killed yeast particles are thoroughly washed five to seven times (120 g for 10 min) and resuspended in a suitable buffer at 10^9 particles/ml and stored at $-20°$ until used. To label with FITC (BBL Microbiology System, Cockeysville, MD), frozen heat-killed yeast particles are thawed and washed twice in 0.9% NaCl, incubated at a concentration of

10^9 particles/ml in a 0.1 M carbonate buffer, pH 9.5, containing 0.1 mg/ml FITC for 30 min at 37°. The particles are then washed five times by centrifugation at 120 g for 10 min and resuspended at an appropriate concentration in a suitable buffer and stored at $-20°$.

Opsonization with C3bi, Rabbit IgG, and Concanavalin A (Con A). The FITC-conjugated yeast particles (5 × 10^6 particles/ml) can be opsonized with C3bi, IgG, or Con A by incubation in 25% normal human serum, rabbit anti-yeast IgG (14 μg/ml), or 250 μg/ml Con A (Pharmacia, Uppsala, Sweden).[18] After incubation at 37° for 30 min the particles are washed in buffer and resuspended to appropriate concentration in the same buffer. The particles are designated yeast-C3bi, yeast-IgG, and yeast-Con A.

Preparation of the Quenching Dye. Trypan blue (2 mg/ml) (Merck, Darmstadt, West Germany) is dissolved in 0.15 M NaCl in 0.02 M citrate or acetate buffer, pH 4.4, containing crystal violet (2 mg/ml) (Merck, Darmstadt, West Germany) dissolved in PBS, pH 7.4. This dye is primarily used when efficient quenching of the attached particles is needed.

Phagocytic Assay (Fluorescence Quenching). Adherent phagocytes[13] are prepared by adding 0.1 ml of freshly drawn nonanticoagulated human blood or 0.1 ml of a granulocyte suspension (3 × 10^5 cells/ml) to the spots (diameter 13 mm) of multispot glass slides (Dynatech AG, Zug, Switzerland). The phagocytes are allowed to adhere to the glass surfaces for 30 min at 37° and nonadherent cells are washed away in warm buffer. The slides are then incubated with 0.1 ml of opsonized yeast particles (most often 2.5 × 10^6 particles/ml) for 15 min at 37°. The phagocytic process is stopped by decreasing the temperature to 4°. A few drops of the quenching dye are added and the slides examined in a incident UV-microscope by alternating between immunofluorescence and phase contrast.

In systems with phagocytes in suspension,[3] 0.2 ml of yeast particles (4 × 10^7 particles/ml) is mixed with an equal volume of neutrophils (10^7 cells/ml). The suspensions are incubated for 15 min at 37° with end-over-end rotation. Immediately after the incubation period, one drop of the suspension and one drop of trypan blue (2 mg/ml, pH 4.4) are put on a slide and examined in an incident UV-microscope as described above.

In the flow cytofluorometric assay,[17] 0.1-ml portions of EDTA–blood are lysed in 2 ml of NH_4Cl "Lysing Reagent" (Ortho Diagnostic Systems, Inc., Raritan, NJ) and centrifuged (1200 g, 5 min, 4°). The leukocyte pellets are washed once in Gey's medium (Gibco Laboratory) and resuspended in 0.2 ml of opsonized yeast–particle (5 × 10^6 particles/ml) suspensions. After incubation at 37° for 15 min the suspensions are diluted in

---

[18] J. Hed, O. Stendahl, and T. Sundqvist, *FEBS Lett.* **152**, 212 (1983).

cold PBS with and without trypan blue (2 mg/ml at pH 4.4) and examined in an Ortho spectrum III flow cytofluorometer (Ortho Diagnostic Systems, Inc., Raritan, NJ) (for more details, see Vol. 108 [19] and this volume [4]).

Calculation. Count the number of attached (blue-stained) and ingested (fluorescent) particles in individual phagocytes.

Total association: Total number of attached and ingested particles of 100 phagocytes

Total ingestion: Total number of ingested particles in 100 phagocytes

Percentage phagocytosis: The percentage of ingested particles of total interacted particles

Percentage phagocytic cells: The percentage of phagocytes ingesting particles

Furthermore, the assay allows detection of functional heterogeneity in the phagocyte population and identification of nonviable cells since each individual phagocyte is analyzed.

[7] Methods for the Study of Receptor-Mediated Phagocytosis

By SAMUEL D. WRIGHT

Introduction

The characterization of a receptor-mediated phagocytic event requires that two questions be answered. (1) Which receptors and ligands are involved in binding the particle to the phagocyte? (2) Which receptors signal the phagocyte to engulf the particle? The first step of phagocytosis, attachment of ligand-coated particles to phagocytes, requires only that there be sufficient crosslinks between phagocyte and particle to form a bridge. Metabolic energy is not needed for attachment, which can be mediated by any particle-bound ligand that has affinity for molecules on the surface of the phagocyte. Engulfment of particles, on the other hand, is an energy-dependent process in which pseudopods move around the particle in response to intracellular signals generated by binding of ligand to specific phagocytosis-promoting receptors on the cell surface. Some receptors, such as those for fibronectin (Fn),[1] mediate only the attachment of particles and do not generate intracellular signals for engulfment.[1a] Other receptors, such as those for IgG, promote both attachment and engulfment. A third type of receptor, exemplified by the receptors for C3, mediates attachment constitutively but promotes phagocytosis only after the phagocytes receive certain stimuli (lymphokines,[2] phorbol esters,[3] or Fn[4,5]).

The existence of these different categories of receptors poses special technical problems for the measurement of the attachment or engulfment mediated by an individual type of receptor. At the zone of contact between a phagocyte and a ligand-coated particle, both the receptors and the ligands are present in very high concentrations. Thus receptors or ligands that are in low abundance or receptor–ligand pairs that are of low affinity can still mediate phagocytosis if the particle bearing the ligands is attached to the phagocyte by a *different* ligand. This point is illustrated by the observation that erythrocytes bearing fewer than 600 IgG/erythrocyte are neither bound nor ingested by human monocytes. However, such lightly coated erythrocytes are avidly ingested if they also bear C3b.[6] Although the C3b receptors do not trigger ingestion, the contact between the erythrocyte and the cell surface mediated by C3 receptors enables the light coating of IgG to bind to Fc receptors, and the Fc receptors then promote phagocytosis.

The observation that phagocytosis may be promoted by ligands that do not by themselves cause attachment is of considerable interest because many particles commonly used as targets for phagocytosis have small amounts of phagocytosis-promoting ligands on their surfaces. For example, zymosan and rough strains of bacteria display a variety of carbohydrates that are recognized by one or more receptors on macrophages. These sparse ligands are not capable of causing phagocytosis unless the particle is attached to a macrophage by some other means. For example,

---

[1] Abbreviations: Fn, fibronectin; DGVB, dextrose, gelatin, veronal buffer; E, sheep erythrocytes; IgGαE and IgMαE, anti-sheep erythrocyte IgG and IgM; SA, *Staphylococcus aureus;* Con A, concanavalin A; C3b INA, C3b inactivator; DFP, diisopropyl phosphofluoridate; PLL, poly(L-lysine); PBS, phosphate-buffered saline; DNP, dinitrophenol; HSA, human serum albumin; PMA, phorbol myristate acetate; 2dG, 2-deoxyglucose.

[1a] M. P. Bevilacqua, D. Amrani, M. W. Mosesson, and C. Bianco, *J. Exp. Med.* 153, 42 (1981).

[2] J. A. Griffin and F. M. Griffin, Jr., *J. Exp. Med.* 150, 653 (1979).

[3] S. D. Wright and S. C. Silverstein, *J. Exp. Med.* 156, 1148 (1982).

[4] S. D. Wright, L. S. Craigmyle, and S. C. Silverstein, *J. Exp. Med.* 158, 1338 (1983).

[5] C. G. Pommier, S. Inada, L. F. Fries, T. Takahashi, M. M. Frank, and E. J. Brown, *J. Exp. Med.* 157, 1844 (1983).

[6] A. G. Ehlenberger and V. Nussenzweig, *J. Exp. Med.* 145, 357 (1977).

if C3 is attached to a zymosan particle, the zymosan will be readily ingested. An investigator who is unaware of the presence of endogenous ligands on the zymosan might conclude that the complement receptors signaled engulfment. For this reason it is essential to use particles that exhibit no ligands other than the ones placed on the particle by the investigator. Native erythrocytes are ideally suited to this purpose since their surfaces are not recognized by any of the known receptors on phagocytes, and they provide convenient surfaces to which antibodies and complement bind.

Another technical problem in the measurement of receptor function is a consequence of the geometry of phagocytosis. Ligand-coated particles are multivalent (10^3–10^5 ligands/particle), and thus they attach to phagocytes with an enormous net affinity. (If a monovalent ligand binds with an affinity of K, the net affinity for the interaction of dimers of the ligand with two receptors is roughly K^2, for trimers it is K^3, and so on.) So tight is the binding that monovalent ligands are ineffective as competitive inhibitors (i.e., they are usually unable to inhibit attachment or promote elution). For example, soluble IgG monomers are inefficient as inhibitors of the binding and ingestion of IgG-coated erythrocytes by macrophages.[7] Thus, it is difficult to determine the specificity of the particle–phagocyte interaction by studying competition with purified soluble ligands or with other competitors such as anti-receptor antibodies.

A partial solution is provided by culture surfaces covalently coated with ligand or antibody. When macrophages are plated on these substrates, receptors that are mobile in the plane of the membrane move to the substrate-adherent portion of the phagocyte and are trapped in this location by ligands or by anti-receptor antibodies bound to the substrate. The apical surface of the cell thus becomes depleted of a specific class of receptor. With this technique, the contribution of low-affinity receptor–ligand pairs to attachment and engulfment can be measured. Moreover, substrate-bound antibodies, even those that bind to but do not block the ligand-binding site of a receptor, will remove receptors from the apical portion of the cell. This property allows the assignment of the ligand specificity and function of the receptors recognized by the substrate-bound antibody. It also permits the identification of anti-receptor antibodies that bind to epitopes distant from the ligand binding site of the receptor, a class of antibodies not identifiable by most other techniques.

I will describe here methods designed to measure the function of individual types of receptors in attachment and engulfment of ligand-coated particles. One method involves the use of ligand- or antibody-coated sur-

[7] B. Diamond, B. R. Bloom, and M. D. Scharff, *J. Immunol.* **121**, 1329 (1978).

faces to selectively remove a single receptor type from the apical surface of the phagocyte. Such surfaces allow one to determine if an individual receptor is necessary for a given phagocytic event. The other principal method involves the use of erythrocytes that display a single type of ligand on their surface. These particles will usually bind to a single type of receptor on the surface of the phagocyte and permit one to determine if a given ligand and its receptor are sufficient to cause phagocytosis.

I will confine discussion to the principal opsonins of mammals, IgG and C3. Those interested in a broader discussion of phagocytosis should consult recent reviews[8,9] (see also this volume [2], [3], [12]).

General Methods

Purification and Culture of Mononuclear Phagocytes

Many procedures for the purification of mononuclear phagocytes exploit the capacity of the cells to adhere to glass or plastic surfaces. The resulting monolayers are difficult to bring into suspension and are thus unsuitable for experiments that require suspensions of phagocytes. Since all the procedures that involve ligand-coated substrates call for suspension of phagocytes, I will describe here a method for purifying human monocytes using centrifugation on gradients of Percoll[3] and a method for culturing the monocytes in suspension using Teflon beakers.[3] Alternative schemes for preparing monocytes and methods for preparing polymorphonuclear leukocytes are described elsewhere.[10] (See also this series, Vol. 108 [9].)

Purification of Human Monocytes

1. Percoll gradients are prepared as follows: The osmolarity of Percoll (Pharmacia) is first adjusted to isotonicity by adding one part 1.5 M saline to nine parts Percoll. Twenty-two milliliters of isotonic Percoll is then combined with 1 ml of normal human serum and 14.7 ml PBS in a 50-ml polycarbonate centrifuge tube. This mixture has a density intermediate between monocytes and lymphocytes (specific gravity ≈1.068). The Percoll is centrifuged in a Sorvall SS-34 rotor at 18,000 rpm for 25 min at 5° to form a gradient. One such gradient is sufficient to separate the cells from 80–100 ml of blood. Gradients are held on ice prior to use.

---

[8] S. C. Silverstein, R. M. Steinman, and Z. A. Cohn, *Annu. Rev. Biochem.* **46,** 669 (1977).

[9] S. D. Wright and S. C. Silverstein, *in* "Handbook of Experimental Immunology" (D. M. Weir, L. A. Herzenberg, C. C. Blackwell, and L. A. Herzenberg, eds.), 4th ed. Blackwell, Edinburgh, 1986 (in press).

[10] D. Roos, this volume [8].

2. Mononuclear cells (monocytes plus lymphocytes) are isolated on Ficoll-Hypaque (Pharmacia) gradients[11] using at least 80 ml of blood. Cell concentration is adjusted to 20–40 × 10^6 cells/ml in RPMI 1640 medium containing 10% normal human serum.

3. Five milliliters of mononuclear cells is layered onto the Percoll gradient, and the gradient is then promptly centrifuged in a swinging bucket rotor at 1500 g for 25 min at 5°. Two thick, hazy bands will be visible, a top band which contains monocytes and a bottom band which contains lymphocytes. The monocytes are pooled and washed twice in cold RPMI 1640 to remove the Percoll. The purity and yield of the monocytes should be determined by cytochemical staining of the fractions obtained (see this volume [8]).

Note: The whole blood should not be cooled before isolation of the mononuclear cells, but after the cells are separated from the plasma, they should be kept strictly cold. These measures minimize adhesion of cells to one another (monocytes with platelets and lymphocytes), which prevents the separation. Purified phagocytes should always be kept strictly cold during handling because they will otherwise adhere to the tubes and pipets.

Culturing Human Monocytes in Suspension (see also this volume [9]). Purified monocytes are suspended to 10^6 cells/ml in RPMI containing 12% normal human serum. The serum need not be autologous or blood type compatible and should not be heat inactivated. Ten milliliters of cells is added to each 60-ml Teflon beaker (Savillex Corp., Minnetonka, MN). This volume of fluid completely covers the hydrophobic bottom surface of the Teflon vessel but does not limit gas exchange. Beakers with screw caps may be used for culture since a cap that is screwed half-way on will allow sufficient gas exchange. When cultured in a 95% air/5% CO_2 incubator, cells will remain healthy for 12 days without a change of medium. Cultured cells can be removed from the beakers by cooling them on ice for 30 min, then pipetting vigorously with a Pasteur pipet. The tip of the pipet should be fire polished to prevent damage to the cells during pipetting.

Preparation of Ligand-Coated Erythrocytes

Materials

Buffer: $DGVB^{2+}$, 2.5 mM veronal buffer, pH 7.5, 75 mM NaCl, 2.5% dextrose, 0.05% gelatin, 0.15 mM $CaCl_2$, 0.5 mM $MgCl_2$. This buffer is used for handling erythrocytes. The dextrose prolongs the life of the erythrocyte, the gelatin prevents the erythrocytes from adhering

---

[11] A. Böyum, *Scand. J. Lab. Invest.* **97,** Suppl. 21, 77 (1968).

to tubes and pipets, and the low ionic strength favors the action of complement enzymes

Sheep erythrocytes (E): Sheep blood is drawn into an equal volume of Alserver's solution and should be held at 5° for at least 3 days and no longer than 3 weeks before use. Erythrocytes are washed several times in $DGVB^{2+}$, and the leukocytes are removed by aspirating the buffy coat at least twice. The washed cells are suspended at a final concentration of 10^9 cells/ml in $DGVB^{2+}$ (approximately 5% by volume). To determine the cell concentration, dilute 50 μl of cell suspension in 1.45 ml distilled water (to lyse the E) and read the optical density of the lysed cells at 541 nm. An OD of 0.350 corresponds to a cell density of 10^9 E/ml

IgGαE (raised in rabbits, Cordis Laboratories): Store frozen in small aliquots

IgMαE (raised in rabbits, Cordis Laboratories): Contaminating IgG should be removed by incubating the IgM with *Staphylococcus aureus* (SA). Wash 50 μl of 10% SA three times in PBS and resuspend the pellet of SA in 1 ml IgMαE. After 1 hr at 0° the SA is removed by centrifugation, and the IgM is frozen in small aliquots

Complement proteins (Cordis Laboratories): It is advantageous to use guinea pig C1, C4, and C2 rather than their human counterparts, but C3 and the C3b inactivator (C3bINA) should be of human origin. Currently, satisfactory C3 cannot be purchased, and must be purified from human plasma.[12] Human C3b and C3bi will bind, albeit weakly, to the corresponding receptors on murine phagocytes, but the interaction of these ligands with receptors of other species has not been characterized

Procedures

EIgG: To find the amount of IgGαE needed to cause optimal phagocytosis, first determine the concentration at which agglutination of E occurs. This is done by making serial 2-fold dilutions of IgGαE in 1 ml of $DGVB^{2+}$ and adding 1 ml of E to each. The mixtures are incubated at 37° for 30 min, then at 0° for 30 min with occasional agitation. Finally, the cells are washed 3 times in $DGVB^{2+}$ and resuspended to 10^8 cells/ml. Agglutination is obvious by visual inspection and usually occurs with 50–80,000 IgGs bound per E^3. Since optimal phagocytosis occurs with 30–50% that quantity of IgG bound, a dilution of antibody is chosen that is 2-fold below the minimum agglutinating dose. Record this dose and aliquot the antibody accordingly. EIgG are prepared as described for the titration above and can be stored on ice for no longer than 2 days.

[12] B. F. Tack and J. W. Prahl, *Biochemistry* **15**, 4513 (1976).

Note: The rabbit IgG called for in this procedure does not exhibit subclasses and appears to interact with all of the subclass-specific Fc receptors of human and murine phagocytes. Thus murine macrophages bind these EIgG via the receptor for IgG_{2a} ($FcR_{\gamma2a}$) and the receptor for IgG_1 and IgG_{2b} ($FcR_{\gamma1/2b}$). In order to observe the function of a single type of receptor, one should employ homologous IgG of a single subclass. For example, to observe the action of murine $FcR_{\gamma2a}$, one should coat the erythrocytes with murine $IgG_{2a}\alpha E$.[7]

ECon A: These erythrocytes are bound by phagocytes but are not ingested and thus are useful in control experiments. The Con A is titrated and used by the method used for IgG with one modification. PBS should be substituted for $DGVB^{2+}$, since the dextrose in the $DGVB^{2+}$ inhibits the binding of Con A.

EIgM: IgM activates the classical pathway of complement, yet it is not recognized by receptors on phagocytes. EIgM are thus used in the preparation of EC3b and can also be used as a control particle that will not be bound or ingested. The $IgM\alpha E$ is titrated and used exactly as described above for $IgG\alpha E$.

EC3b: These erythrocytes are prepared by sequentially treating EIgM with purified complement enzymes.[3]

EIgMC1: Combine 0.4 ml EIgM (10^9/ml in $DGVB^{2+}$), 2.0 ml $DGVB^{2+}$, and 200 μl guinea pig C1 (10,000 units/ml). Incubate at 0° for 40 min. Dilute with at least 10 ml $DGVB^{2+}$, sediment the cells, and resuspend them in 2 ml $DGVB^{2+}$.

EIgMC14: Add 400 μl guinea pig C4 (1000 units/ml). Incubate at 30° for 40 min. Dilute with at least 10 ml $DGVB^{2+}$, sediment the cells, and resuspend them in 2 ml $DGVB^{2+}$.

EIgMC142: Warm the EIgMC14 to 30°, then add 400 μl guinea pig C2 (1000 units/ml) (prewarmed to 30°). Incubate at 30° for 10.0 min. Without delay, dilute with at least 10 ml cold $DGVB^{2+}$, sediment the cells, and resuspend them in 2 ml $DGVB^{2+}$.

EC3b: Add 100 μl C3 (1 mg/ml in PBS). Incubate at 37° for 30 min. Wash three times and resuspend to 4 ml in $DGVB^{2+}$.

This preparation yields erythrocytes bearing fewer than 400 molecules of C1, C4, or C2 per cell but 30,000–100,000 molecules of C3 per cell (depending on the quality of the C3). The amount of C3 per E can be quantitated by using ^{125}I-labeled C3 during the preparation of the EC3b or by the binding of monoclonal ^{125}I-labeled $IgG\alpha C3$ to EC3b.[3] (For methods of iodination, see this series, Vol. 70, [11], [12], [13].) In order to observe optimal attachment of EC3b to phagocytes, at least 10,000–30,000 C3b/E are required, and optimal phagocytosis requires up to 130,000 C3b/E.

It is often desirable to construct a panel of erythrocytes that bear a variety of densities of C3/E. Such a panel allows the determination of the number of C3/E required to produce half-maximal binding or ingestion. The erythrocytes are made by incubating EIgMC142 intermediates with different dilutions of C3, and the resulting panel is then used in assays of binding or engulfment. The maximum amount of C3 that can be deposited appears to be limited by spatial constraints on the surface of the erythrocyte and is approximately 160,000/E.

EC3bi: These are prepared from EC3b by the action of the C3b inactivator (C3bINA). Before using C3bINA, contaminating proteases must be inactivated with diisopropyl phosphofluoridate (DFP, Sigma). (*Caution:* Use proper safety precautions with DFP. It is a volatile neurotoxin.)

Add 0.25 μl DFP to 400 μl C3bINA (500 units/ml). Incubate at 20° for 10 min, then add to 2 ml EC3b (10^8/ml). Incubate at 37° for 2.5 hr. Wash three times and resuspend in 2 ml of DGVB^{2+}.

It is important to verify that none of the C3 on the EC3b has been degraded to C3bi and that cleavage of C3b by C3bINA is complete. This is best done by examining the electrophoretic behavior of the cell-bound C3: the α chain of C3b runs with an apparent molecular weight of 130,000, while the α chain of C3bi runs with an apparent molecular weight of 60,000.[3,13] In order to observe these chains, EC3b and EC3bi are prepared using ^{125}I-labeled C3. The ^{125}I-labeled C3 is cleaved from the glycoproteins of the erythrocytes by suspending EC3b or EC3bi in 1% SDS, pH 11, and incubating at 37° for 30 min. The high pH is required to break the ester linkage between C3 and the components of the erythrocyte. The pH is adjusted to neutrality prior to SDS–PAGE, and the position of the bands is determined by autoradiography.

The EC3b and EC3bi should be stored at 10^8 cells/ml in DGVB^{2+} on ice and will last for 1–2 weeks. Antibiotics may be added to protect against bacterial growth.

Preparation of Ligand-Coated Culture Surfaces

Glass or plastic surfaces may be derivitized with ligands,[14] antibodies,[15] or cells[3,4] in order to specifically deplete the corresponding receptors from the apical surface of phagocytes spread on these surfaces. Surfaces may also be derivitized with components of the extracellular matrix. When macrophages adhere to certain of these molecules (Fn and serum

[13] S. K. Law, D. T. Fearon, and R. P. Levine, *J. Immunol.* **122,** 759 (1979).
[14] J. Michl, M. M. Pieczonka, J. C. Unkeless, and S. C. Silverstein, *J. Exp. Med.* **150,** 607 (1979).
[15] S. D. Wright, P. E. Rao, W. C. Van Voorhis, L. S. Craigmyle, K. Iida, M. A. Talle, E. F. Westberg, G. Goldstein, and S. C. Silverstein, *Proc. Natl. Acad. Sci. U.S.A.* **80,** 5699 (1983).

amyloid P component), the phagocytosis-promoting capacity of C3 receptors is dramatically enhanced.[4,5]

Many proteins will adsorb passively to tissue-culture plastic during a 1-hr incubation at 20°. For very sticky proteins such as Fn, passive adsorption provides an adequate coating of ligands.[4] In other cases, chemical means should be employed to obtain a dense, stable coating of ligand. A general method for coupling proteins with primary amino groups is described.[14]

Surfaces Coated with Proteins (see also this series, Vol. 108 [12])

1. Tissue-culture plastic surfaces or acid-washed glass surfaces are coated with poly(L-lysine) (PLL) by incubation with 0.1 mg/ml of PLL for 30 min at 20°. The surface is then washed in PBS taking care that it never goes dry.

2. The PLL surfaces are treated with 2% glutaraldehyde in PBS for 15 min at 20°. This allows the glutaraldehyde to react with amino groups of the PLL. The surfaces are again washed extensively with PBS.

3. The protein to be coupled is now added to the surface. The protein must be at least 25 μg/ml and should be dissolved in a buffer of neutral pH that bears no free amino groups. Allow coupling to proceed for 2–3 hr at 20° or overnight at 5°. Unreacted glutaraldehyde should then be blocked by adding 1 mg/ml serum albumin in 0.1 M glycine, pH 7, for 30 min at 20°.

Surfaces Coated with Antigen–Antibody Complexes. Several methods exist for coating surfaces with antigen–antibody complexes. Preformed complexes can be coupled by the method described above. Alternatively, an antigen such as serum albumin can be coupled to the surface, and then anti–serum albumin antibodies can be added to form the complex. However, the highest density and permanence is obtained by the following method[14]: PLL-coated surfaces are prepared as described above, and 0.05 M dinitrobenzene sulfonate dissolved in 0.15 M sodium carbonate buffer, pH 11.6, is added for 30 min at 20°. This procedure dinitrophenylates the amino groups of the PLL to yield a stable DNP-coated surface on which cells can attach and remain viable. Anti-DNP antibodies are next added and incubated for 1 hr at 20°. These combine with the DNP to form antigen–antibody complexes.

Because DNP is stable and nontoxic, the antibodies can be added after the phagocytes have spread on the DNP-coated surface, thus allowing one to synchronize the exposure of a cell population to ligand. This is especially valuable in measuring metabolic alterations that occur during phagocytosis. Commercially available anti-DNP antibodies (e.g., Accurate Chemical and Scientific Corp.) work well in this procedure, but better results are obtained using affinity-purified, high-affinity rabbit anti-DNP antibodies prepared by the method of Eisen.[16]

Surfaces Coated with Complement. These can be prepared by treating DNP-coated surfaces with IgMαDNP (Accurate Chemical and Scientific Corp., Westbury, NY) and then with complement enzymes as described above for EC3b and EC3bi. Alternatively, whole serum can be used as the source of complement. For this procedure, serum is diluted 2- to 5-fold with $DGVB^{2+}$, then incubated with the IgM-coated surface at 37° for 10–30 min. Short incubation times favor the deposition of C3b while longer incubation times cause the accumulation of its degradation product, C3bi. The serum deposits a very heavy coat of C3 on the surface, but other proteins such as C4, C4-binding protein, factor H, serum amyloid P component, and Fn may also be deposited. The presence of these other ligands limits the usefulness of this procedure.

Surfaces Coated with Cells. These surfaces serve two purposes: (1) They allow the construction of substrates which bear ligands that cannot be purified from the cells in question.[3] (2) They can *prevent* phagocytes from interacting with proteins on the underlying glass or plastic substrates.[4] For example, soluble Fn spontaneously adheres to the glass or plastic beneath spread macrophages, thereby making it difficult to expose cells to soluble Fn without also exposing them to substrate-bound Fn. Since Fn will not bind to erythrocytes, macrophages can be plated on a surface coated with erythrocytes to prevent attachment of Fn beneath the macrophages. A procedure for coupling sheep erythrocytes to a plastic surface follows.

Erythrocytes are washed into a buffer such as PBS that contains no protein. The cells should be washed just before use to prevent cell lysis from contributing protein to the solution. Sufficient cells are then added to form a double layer (as observed by microscopy) on a PLL-coated surface, and the surface is centrifuged at 400 g for 5 min. Unattached cells are removed by dipping the surfaces in PBS. When removing unattached cells, be certain not to direct a stream of buffer from a pipet onto an erythrocyte-coated surface: the bound cells will be ripped by shear forces. Hemoglobin may be removed from attached erythrocytes by subjecting the cells to five 1-sec dips in 5 mM sodium phosphate, pH 8. The resulting erythrocyte ghosts remain attached to the surface.

[16] H. N. Eisen, W. Gray, J. R. Little, and E. S. Simms, *Methods Immunol. Immunochem.* **1,** 351 (1967).

Use of Terasaki Plates for Microassays of Phagocytosis

In many instances the quantities of phagocytes, ligand-coated particles, or test substances is severely limited. This need not prevent measurements of phagocytosis since only 200–500 cells must be counted to obtain reliable data. Terasaki plates (Miles Laboratories) provide ideal microchambers for conducting phagocytosis experiments.[3] Each of the 60 wells on a Terasaki plate holds 5–10 μl of fluid, and the viewable surface of the plate accommodates 200–800 phagocytes, depending on their size. Because 60 wells can be loaded and washed at a time, it is easy to assay many conditions with duplicate or triplicate samples. Several special precautions, however, must be observed in using Terasaki plates:

1. The small quantity of buffer used readily evaporates and subjects the phagocytes to osmotic stress. To avoid evaporation, plates should be kept covered except when making additions, and manipulations should be performed in a cold room or with the plate resting on ice. All 37° incubations should be done in a humidified incubator.

2. The steep walls of each chamber slow the exchange of medium during washes. Therefore, the plate should be dipped at least five times in a wash buffer to assure a successful wash.

3. After the plates are dipped, surface tension holds approximately 10 μl of fluid in each well. In order to add reagents or particles to an individual well, the fluid must be removed by aspiration. Caution must be exercised so that the fluid level is not reduced so low that the cells in the middle of the well are damaged. This is readily accomplished with a fire-polished Pasteur pipet connected to a vacuum. When the tip of the pipet is placed on the ledge of each well, the current of air swirls fluid from the well into the aspirating pipet, leaving approximately 1 μl in the well. Do not extend the Pasteur pipet tip into the well or the cells will be damaged. Since 1 μl evaporates very quickly, additions should be made to the wells immediately after they are aspirated.

4. Dipping Terasaki plates allows the exchange of medium but does not provide enough mechanical force to suspend unbound particles or cells from the culture surface. Two methods can be used to wash off unbound particles. Method 1: Add sufficient buffer to the plate to cover the tops of all wells ($\approx$10 ml). With the plate resting on a smooth surface, shake the plate in fast tight circles for about 10 sec. The plate should trace circles of about 2 cm diameter at a rate of about five circles per second. This procedure sets up a mild current in the wells. Method 2: Cells can also be detached by a stream of buffer with a controlled flow rate. Set up a flask of PBS 50 cm above the bench top. Connect the flask by a length of tubing to a 10-μl capillary micropipet. The flow of buffer through the pipet

provides a stream that can be directed into the wells and will detach unbound cells but will not disturb receptor-mediated rosettes.

5. Terasaki plates must be viewed through an inverted microscope. The level of fluid in the plate is first raised above the walls of the wells so that light passing through the wells from the condenser is not refracted by a curved air/water interface. In order to obtain sufficient resolution to count phagocytosis, the best possible phase-contrast objective lens should be employed. The Zeiss 40× water-immersion lens has proved excellent in this capacity.

Measurement of Ligand-Dependent Binding and Phagocytosis

The procedures described have been used for measuring the binding and phagocytosis of ligand-coated sheep erythrocytes to human phagocytes using Terasaki microtest plates.[3] These procedures can be scaled up for use with 96-well plates or 24-well plates by increasing the volumes by a factor of about 10 or 50, respectively.

Assays of binding and phagocytosis are usually carried out in two steps. In the first, phagocytes are allowed to adhere and spread on a culture surface. When ligand-coated culture surfaces are employed, receptors for that ligand redistribute to the substrate-attached portion of the cell during this incubation. When Fn-coated surfaces are employed, ligation of Fn receptors causes C3 receptors to become capable of promoting phagocytosis. The first step also provides a period in which cells can be pretreated with drugs or antibodies before addition of the ligand-coated particles. In the second step, ligand-coated particles are added to the phagocytes, and the cultures are incubated at 37°. At the end of this step, the cells are washed, and binding or phagocytosis is assessed.

Preparation of Monolayers of Phagocytes

1. Preparation of culture surfaces: Phagocytes that spread on uncoated tissue culture surfaces exhibit depressed phagocytic activity, and thus the surfaces should be coated in advance of adding phagocytes. Directions for generating particular ligand-coated surfaces are given above. If no particular ligand is called for, the surface should be coated with a control protein such as human serum albumin (HSA) by incubating the well for 30 min at 20° with 1 mg/ml HSA in PBS.

2. Preparation of phagocytes: Phagocytes are suspended in a buffer that provides divalent cations, glucose, a source of protein, and in some cases, an inhibitor of proteolysis. A favorable buffer is as follows: PBS[17]

[17] R. Dulbecco and M. Vogt, *J. Exp. Med.* **99**, 167 (1954).

containing 3 mM glucose, 0.5 mg/ml HSA (Worthington), and 0.2 units/ml aprotinin (Sigma). The concentration of cells should be adjusted to account for the size of the phagocytes employed. Neutrophils are suspended at 2×10^6 cells/ml and macrophages at 0.5×10^6 cells/ml.

3. Plating phagocytes: Coated culture surfaces are washed with PBS, the overlying fluid is aspirated, then 5 μl of phagocytes are added per well. Terasaki wells hold 10 μl, thus leaving room for the addition of drugs or antibodies. Plates are next incubated for 30–90 min at 37° to allow the phagocytes to spread. Normally, 45 min is sufficient time to allow firm attachment of cells to the surface, redistribution of the receptors of phagocytes on ligand-coated surfaces, and activation of the receptors of phagocytes on Fn-coated surfaces. Plates are then washed by dipping in PBS and the overlying fluid is removed by aspiration.

Measurement of Ligand-Dependent Binding

Five microliters of ligand-coated erythrocytes (10^8 E/ml in DGVB^{2+}) is added to each well of phagocytes. Plates are then incubated at 37° for 20–30 min, a time by which ligand-dependent binding of erythrocytes is usually maximal. Attachment of most ligand-coated erythrocytes will also occur at 5° (EC3bi is an exception), but a longer incubation is necessary, usually 60 min. The monolayers are then washed to remove unattached erythrocytes (see above) and the attachment index is determined by phase-contrast microscopy as described below.

Measurement of Ligand-Dependent Phagocytosis

Ligand-coated erythrocytes are added to phagocytes as in assays of binding, and the preparation is incubated at 37° for 45–60 min, a time by which phagocytosis is usually maximal. Special steps must then be taken to distinguish ingested particles from those merely attached to the phagocyte. When erythrocytes are used, this is easily accomplished by dipping the monolayers into distilled water at 20°. Plates are given five consecutive dips of 1 sec in the distilled water and are then returned to isotonic buffer. Uningested erythrocytes are exposed to the distilled water and are lysed while ingested erythrocytes are protected from hemolysis.

Measurement of Attachment Index and Phagocytic Index

The attachment index is calculated by tallying the number of phagocytes with at least one bound erythrocyte, the number without a bound erythrocyte, and the total number of bound erythrocytes. At least 200 phagocytes should be counted in duplicate wells. From these three figures one may calculate the percentage of phagocytes binding at least one erythrocyte, the average number of erythrocytes per phagocyte, and the product of these two numbers, the attachment index. Assays of phagocytosis are quantitated in a similar fashion, except that only the ingested erythrocytes are counted.

It is desirable to describe the results of a binding or phagocytosis experiment with two numbers, the attachment or phagocytic index and the percentage of phagocytes binding or ingesting at least one erythrocyte. (The average number of erythrocytes per phagocyte can be back calculated from these two numbers.) The attachment or phagocytic index represents the number of erythrocytes bound or ingested by 100 phagocytes and constitutes a good overall estimate of the tendency of phagocytes to bind or engulf ligand-coated erythrocytes. The percentage of phagocytes binding or ingesting at least one erythrocyte provide a measure of the homogeneity of the response of the population of phagocytes. For example, an attachment index of 500 could be the result of all cells binding 5 erythrocytes per phagocyte or of one-half of the phagocytes binding no erythrocytes and the other half binding 10 erythrocytes per phagocyte.

The ability to distinguish heterogeneity in the response of a population of phagocytes is an important feature unique to cell-by-cell methods such as FACS analysis and visual counting. An additional benefit from the visual method is that the health and purity of the population can be observed for each condition chosen.

Special Techniques for Manipulating the Events of Phagocytosis

Monoclonal Anti-Receptor Antibodies

Monoclonal anti-receptor antibodies have been generated by immunizing animals with live phagocytes and then selecting antibodies with the desired specificity by any of several strategies.[17a] A large number of such antibodies now exist[17a] and they offer unique advantages for the study of receptor-mediated phagocytosis (for more details, see also this series, Vol. 108 [31]).

Often the specificity of monoclonal antibodies for receptors is greater than that of ligands. For example, while rabbit IgG binds to both $FcR_{\gamma2a}$ and $FcR_{\gamma1/2b}$ on murine phagocytes, the monoclonal antibody 2.4G2 recognizes exclusively $FcR_{\gamma1/2b}$. Antibodies such as this have established the

[17a] J. C. Unkeless and T. A. Springer, *in* "Handbook of Experimental Immunology" (D. M. Weir, L. A. Herzenberg, C. C. Blackwell, and C. A. Herzenberg, eds.), 4th ed. Blackwell, Edinburgh, 1986 (in press).

presence of multiple receptors for IgG[18] and C3[15] and also have been useful in functional studies of these receptors. Monoclonal antibodies can also be obtained that bind to a receptor but do not block its ligand-binding domain.[15] Such antibodies are useful tags since the presence of ligand will not obscure detection of the receptor. Several cautions, however, must be observed in the use of monoclonal antireceptor antibodies: (1) Monoclonal antibodies are not always specific for a single receptor. For example, the monoclonal antibody IB4 binds not only the C3bi receptor but two additional (and homologous) receptors on human leukocytes.[15] (2) Monoclonal antibodies against any cell surface component may block not only their target receptor but also Fc receptors. The reason for this appears to be that after binding to an antigen on the cell surface, the Fc domain can pivot and bind to nearby Fc receptors.[19] To avoid this type of Fc receptor blockade one should employ Fab or F(ab)$_2$ fragments or antibodies of the IgM class.

Activation of the Phagocytosis-Promoting Capacity of a Receptor by Ligation of a Different Type of Receptor

As mentioned above, the ability of C3 receptors to promote engulfment is regulated. In resting cells C3 receptors mediate only the binding of C3-coated particles, and ingestion does not occur unless the cells are simultaneously stimulated by another agent. Phorbol myristate acetate (PMA, 30 ng/ml) stimulates C3-mediated phagocytosis within 20 min of its addition.[3] In a typical experiment, macrophages are plated in the presence of PMA, they are washed, C3-coated erythrocytes are added, and phagocytosis is measured. Since PMA is hydrophobic it remains with the macrophages even after washing. Other phorbol esters such as phorbol dibutyrate are more water soluble and must be present along with the C3-coated erythrocytes in order to be effective.[20]

The phagocytosis-promoting capacity of the C3 receptors of human macrophages is also activated by Fn.[4,5] Soluble Fn is ineffective, and in order to observe activation, cells must be plated on Fn-coated surfaces.[4] As long as the phagocytes are in contact with the surface-bound Fn, the activated state of the C3 receptors remains stable.[20]

In murine phagocytes, C3 receptors can be activated by a unique lymphokine. The production and action of this lymphokine has been described by Griffin.[2]

---

[18] J. C. Unkeless, *J. Exp. Med.* **150,** 580 (1979).
[19] R. J. Kurlander, *J. Immunol.* **131,** 140 (1983).
[20] S. D. Wright, M. R. Licht, L. S. Craigmyle, and S. C. Silverstein, *J. Cell Biol.* **99,** 336 (1984).

Activation of C3 receptors involves a very subtle and transient alteration in the phagocyte. Activation by Fn or phorbol esters is not accompanied by a change in the number of C3 receptors per cell or in the ability of the receptors to bind ligand,[3,20] and activation is reversed within 20 min of the removal of the stimulus.[20] Thus activation affects only the ability of C3 receptors to generate the intracellular signals for engulfment. Since activation can augment the phagocytosis promoted by C3 receptors 100-fold, and activation can be caused by very small amounts of Fn on a culture surface, it is vital that Fn not contaminate any of the reagents used in the preparation of monolayers of phagocytes. Moreover, since macrophages secrete large amounts of Fn,[21] cells cultured for days on glass or plastic surfaces may cause the activation of their own receptors.

Methods for Halting Phagocytosis

It is often useful to halt phagocytosis during an experiment. When phagocytosis is stopped, antibodies, enzymes, or drugs can reach the receptors and ligands before they become sequestered in a phagosome. The simplest method to achieve this is to lower the temperature of the cells to 0°. Binding of particles occurs, but pseudopods cannot be extended.

Low temperature will halt nearly all biological processes, and thus a more selective means of inhibiting phagocytosis may be required. Phagocytosis can be inhibited by the metabolic poison, 2-deoxyglucose (2dG).[22] The cells must be preincubated with 50 mM 2dG for 1 hr in the absence of mannose or glucose, and 50 mM 2dG must be present throughout the course of the experiment. The reason for the action of 2dG on phagocytosis is not known, but it appears to act selectively on receptor-mediated phagocytosis, since pinocytosis and the uptake of latex beads are not affected.[22]

The movement of pseudopods requires the actin-based cytoskeleton, and phagocytosis can be inhibited with cytochalasin D (1 μg/ml),[23] a drug which binds to the barbed ends of actin filaments and prevents polymerization. Care must be taken in the use of cytochalasin D since it is quite toxic to phagocytes and even sublethal doses drastically change the morphology of the cells. Phagocytes round up, leaving thin arborized portions of the cells adherent to the substrate, and some of the rounded cells detach from the substrate.

Another means of halting the movement of pseudopods and closure of the phagosome is to mask the receptors before phagocytosis is com-

[21] K. Alitalo, T. Hovi, and A. Vaheri, *J. Exp. Med.* **151**, 602 (1980).
[22] J. Michl, D. J. Ohlbaum, and S. C. Silverstein, *J. Exp. Med.* **144**, 1465 (1976).
[23] S. G. Axline and E. P. Reaven, *J. Cell Biol.* **62**, 647 (1974).

plete.[24] Macrophages are incubated with ligand-coated erythrocytes for 1 hr at 5° to permit binding of the erythrocytes, and the remaining unligated receptors are then blocked by adding a saturating concentration of an appropriate anti-receptor antibody. When the cells are rewarmed to 37°, existing ligand–receptor bridges will remain, but no new receptor–ligand bonds can be made, and therefore the movement of pseudopods around the particle will stop. In this experiment, the anti-receptor antibody must be incubated with the cells for at least 30 min at 5° to block exposed receptors, and saturating levels of antibody should be maintained throughout the subsequent incubations to block receptors that move to the surface from intracellular stores. It is also important to employ an anti-receptor antibody that reacts with the ligand-binding portion of the relevant receptor.[9,15]

In some circumstances, it is possible to avoid the use of antibodies by removing ligands from one hemisphere of a particle before adding it to the phagocytes. Particles may be hemispherically decorated with IgG by incubating live cells such as lymphocytes with an anti-lymphocyte antibody and allowing the antibody to cap.[25] Pseudopods will advance as far as the IgG in the cap but will not extend beyond the cap, and phagocytosis will not be complete.

Perhaps the simplest method to prevent closure of the phagosome is to challenge the phagocyte with a particle too large to eat.[14] This is best done using the ligand-coated substrates described above. The "frustrated phagocytosis" of cells on such a substrate resembles conventional phagocytosis in all aspects except the radius of curvature of the target.[26]

Methods for Studying Alterations in the Metabolism of the Phagocyte Caused by Phagocytosis

Ligation of Fc receptors causes macrophages to manufacture and release large amounts of hydrogen peroxide[27] and metabolites of arachidonic acid.[28] Ligation of C3 receptors, on the other hand, causes the release of neither peroxide[29] nor arachidonic acid metabolites.[30] A deter-

[24] F. M. Griffin, Jr., J. A. Griffin, J. E. Leider, and S. C. Silverstein, *J. Exp. Med.* **142,** 1263 (1975).

[25] F. M. Griffin, Jr., J. A. Griffin, and S. C. Silverstein, *J. Exp. Med.* **144,** 788 (1976).

[26] S. D. Wright and S. C. Silverstein, *Nature (London)* **309,** 359 (1984).

[27] R. B. Johnston, Jr., J. E. Lehmeyer, and L. A. Guthrie, *J. Exp. Med.* **143,** 1551 (1976).

[28] C. A. Rouzer, W. A. Scott, J. Kempe, and Z. A. Cohn, *Proc. Natl. Acad. Sci. U.S.A.* **77,** 4279 (1980).

[29] S. D. Wright and S. C. Silverstein, *J. Exp. Med.* **158,** 2016 (1983).

[30] A. A. Aderem, S. D. Wright, S. C. Silverstein, and Z. A. Cohn, *J. Exp. Med.* **161,** 617 (1985).

mination of the role of any given phagocytosis-promoting receptor in these secretory phenomena is best studied with the aid of the ligand-coated substrates described earlier. For example, the release of hydrogen peroxide initiated by Fc receptors can be determined by comparing the release from cells spreading on IgG-coated surfaces with background release from cells spreading on surfaces coated with a control protein such as serum albumin. The advantages of the ligand-coated surfaces are that the investigator can be sure that gravity will bring all cells in contact with the surface, and that the resulting "frustrated phagocytes" are responding to a maximal stimulus. Further, such a surface obviates the need for particles (such as erythrocytes or bacteria) that contain enzymes that consume peroxide or possess membranes that can adsorb metabolites of arachidonic acid.

Acknowledgments

I wish to acknowledge the advice and wisdom of Dr. S. C. Silverstein in whose laboratory most of the methods described here were developed, and I wish to thank Dr. P. A. Detmers for critical reading of the manuscript. Supported by grant AI 22003 from the USPHS and JFRA-103 from the American Cancer Society.

Section III

Specific Methods for the Isolation of Cells and Cellular Components

[8] Purification and Cryopreservation of Phagocytes from Human Blood

By Dirk Roos and Martin de Boer

General Principles (see also this series, Vol. 108 [9])

Three types of phagocytic cell are found in human blood: monocytes, neutrophilic granulocytes, and eosinophilic granulocytes. For their separation, both biological and physical properties of the cells can be used. For instance, positive selection of phagocytes can be achieved by utilizing the adhesive or phagocytic properties of the cells. However, most methods nowadays utilize differences in specific gravity or size of the cells. Thus, either isopycnic centrifugation or velocity sedimentation can be used. Differences in electrophoretic mobility of blood cells have also been employed for their separation.[1]

Isopycnic Centrifugation

Isopycnic centrifugation separates cells on the basis of differences in specific gravities. According to Stokes' law, the velocity of a cell in a centrifugal field is proportional to the difference between the specific gravity of the cell and that of the medium. Thus, cells with a specific gravity higher than that of the medium will be sedimented, and those with a specific gravity lower than that of the medium will float to the surface. In this way, separation of cells is effected.

For cells with unknown specific gravities a gradient is made with material of increasing specific gravity. The cells are suspended in this gradient or layered on top of it, and the mixture is centrifuged. Each cell will then localize at a place where the specific gravity of the gradient material equals that of the cell. Afterward, the specific gravity of the medium is determined in each fraction, thus revealing the specific gravity of the cells in that fraction. Once this is known, routine separation from other cells may be carried out in the one-step procedure outlined in the previous paragraph.

Most commonly, blood cells are first separated by isopycnic centrifugation into a light fraction containing mononuclear leukocytes (monocytes + lymphocytes), basophils and a few early precursor cells, and a

---

[1] C. F. Gillman, P. E. Bigazzi, P. M. Bronson, and C. J. van Oss, *Prep. Biochem.* **4**, 457 (1974).

METHODS IN ENZYMOLOGY, VOL. 132

heavy fraction containing erythrocytes and granulocytes (neutrophils + eosinophils). Platelets are found mainly in the light fraction (see below). For this separation Bøyum[2] has used a mixture of Ficoll (a sucrose polymer) and Isopaque (an iodinated X-ray contrast medium) with a specific gravity of 1.077 g/cm.[3] Instead of Ficoll-Isopaque (or Ficoll-Hypaque), Percoll is now generally used. Percoll (a colloidal suspension of polyvinylpyrrolidone-coated silica particles) has a low viscosity, which permits cell separation at relatively low g forces in a short time. Moreover, Percoll can be easily made isoosmotic and brought to a physiologic pH over a large density range, shows a linear relation between concentration and density, and has no toxic effect on cells.[3–5]

Further separation of monocytes from lymphocytes and basophils, and of neutrophils and eosinophils from each other, can also be performed by isopycnic centrifugation, owing to the differences in specific gravity between the various types of blood cell (see the table). For this purpose, layers or gradients of Ficoll,[6,7] Ficoll-Isopaque,[8,9] Ficoll-Hypaque,[10] albumin,[10,11] sucrose,[12] Percoll,[4,5,13–19] and Nycodenz[20] have been used.

[2] A. Bøyum, Scand. J. Clin. Lab. Invest. **21,** Suppl. 97, 77 (1968).

[3] H. Pertoft, K. Rubin, L. Kjellén, T. C. Laurent, and B. Klingeborn, Exp. Cell Res. **110,** 449 (1977).

[4] S. D. Nathanson, P. L. Zamfirescu, S. I. Drew, and S. Wilbur, J. Immunol. Methods **18,** 225 (1977).

[5] A. J. Ulmer and H.-D. Flad, J. Immunol. Methods **30,** 1 (1979).

[6] R. M. Gorczynski, R. G. Miller, and R. A. Phillips, Immunology **19,** 817 (1970).

[7] F. Berthold, Blut **43,** 367 (1981).

[8] J. A. Loos, B. Blok-Schut, R. van Doorn, R. Hoksbergen, A. Brutel de la Rivière, and L. Meerhof, Blood **48,** 731 (1976).

[9] M. A. Vadas, J. R. David, A. Butterworth, N. T. Pisani, and T. A. Siongok, J. Immunol. **122,** 1228 (1979).

[10] W. D. Johnson, B. Mei, and Z. A. Cohn, J. Exp. Med. **146,** 1613 (1977).

[11] W. E. Bennet and Z. A. Cohn, J. Exp. Med. **123,** 145 (1966).

[12] H. I. Zeya, E. Keku, L. R. DeChatelet, M. R. Cooper, and C. L. Spurr, Am. J. Pathol. **90,** 33 (1978).

[13] G. L. Dettman and S. M. Wilbur, J. Immunol. Methods **27,** 205 (1979).

[14] F. Gmelig-Meyling and T. A. Waldmann, J. Immunol. Methods **33,** 1 (1980).

[15] H. Pertoft, A. Johnsson, B. Wärmegård, and R. Seljelid, J. Immunol. Methods **33,** 221 (1980).

[16] I. Gärtner, Immunology **40,** 133 (1980).

[17] A. J. Fluks, J. Immunol. Methods **41,** 225 (1981).

[18] I. Brandslund, J. Møller Rasmussen, D. Fisker, and S.-E. Svehag, J. Immunol. Methods **48,** 199 (1982).

[19] U. Feige, B. Overwien, and C. Sorg, J. Immunol. Methods **54,** 309 (1982).

[20] A. Bøyum, Scand. J. Immunol. **17,** 429 (1983).

Specific Gravity and Relative Volume of Human
Blood Cells[a]

Cells	Specific gravity (g/cm³)	Volume (channel number)
Platelets	1.054–1.062	<3
Monocytes	1.055–1.065	46–54
Lymphocytes	1.060–1.072	21–27
Basophils	1.065–1.075	30–33
Neutrophils	1.080–1.084	40–46
Eosinophils	1.082–1.090	45–52
Erythrocytes	1.090–1.110	6–9

[a] The cell volume is given as the channel number of the top fraction on a Coulter Counter Channelyzer (all measurements with the same setting of the instrument; no threshold setting). These numbers are roughly proportional to the cell volume.[42a]

Specific gravity and channel number were determined after equilibration of the cells in a medium of 290 mOsm and pH 7.4 at 25°. The results are given as the range of 10–20 experiments.

Changes in Cell Density

A problem with separation of phagocytes in this way is the uptake of gradient material by the cells, which changes their specific gravity during the separation.[8,21–23] Therefore, monocytes and granulocytes have to be incubated for some time in the absence of gradient material to recover their original specific gravity, before further purification is possible.

For the same reason we prefer to use isoosmotic gradients for cell separation, to avoid changes in cell density during centrifugation. Changes in pH too, may drastically change the physical properties of cells[24]; therefore, gradients should have the same (preferably physiological) pH throughout. Nevertheless, several investigators have reported good cell separations based on differences in specific gravity induced by hypo- or hyperosmotic media or by low pH.[9,17,19,20,24]

[21] T. A. W. Splinter, M. Beudeker, and A. van Beek, *Exp. Cell Res.* **111**, 245 (1978).

[22] J. S. J. Wakefield, J. S. Gale, M. V. Berridge, T. W. Jordan, and H. C. Ford, *Biochem. J.* **202**, 795 (1982).

[23] A. J. Ulmer, W. Scholz, M. Ernst, E. Brandt, and H.-D. Flad, *Immunobiology* **166**, 238 (1984).

[24] N. Williams and K. Shortman, *Aust. J. Exp. Biol. Med. Sci.* **50**, 133 (1972).

Streaming

Another general point to keep in mind with isopycnic centrifugation is to avoid overloading the gradients. Too many cells will cause "streaming" (bulk sedimentation) and gradient disturbance, due to the fact that at high concentrations the cells are influencing the specific gravity of the medium. Also passage of cells from one layer of gradient material into another (in discontinuous gradients) should be avoided, because the cells may first collect at the boundary between the two layers and then massively enter the denser layer, thus causing streaming. Preferably, the cells should be suspended in separation material of such density that the cells with the lowest specific gravity will float and those with higher specific gravity will sediment. In this way, no boundary has to be passed by the cells.

Removal of Platelets

Especially for the purification of monocytes, but also for granulocytes, care should be taken to avoid contamination of the cell preparations with platelets. Platelets have a specific gravity of about 1.058 g/cm^3; therefore, they will be present in the suspension of mononuclear leukocytes after centrifugation of the blood over a layer of 1.077 g/cm^3. Since the specific gravity of platelets and monocytes overlaps (see the table), it is impossible to separate these cells by isopycnic centrifugation. Instead, this should be done by velocity sedimentation since platelets are much smaller than monocytes.[24a]

Platelets can also be removed by defibrinating the blood during collection, but this procedure leads to loss of leukocytes. The use of heparin as anticoagulant causes the platelets to form microaggregates that tend to stick to leukocytes; thereafter, the platelets are very difficult to remove.[15] Conversely, when Dextran is used to aggregate and remove erythrocytes, the leukocytes to some extent also aggregate and trap platelets.[20] Therefore, the best way to avoid platelet contamination is to draw blood in citrate as the anticoagulant, purify monocytes by density centrifugation over Percoll, and remove the platelets by successive low-speed centrifugation steps (see protocol for monocyte purification).

To prevent platelet contamination of granulocyte preparations, the pellet fraction of the Percoll centrifugation step ($d = 1.077$ g/cm^3) must be carefully freed from supernatant gradient material by vacuum aspiration,

---

[24a] Velocity sedimentation separates cells on the basis of differences in sizes. According to Stokes' law, the velocity of a cell in a centrifugal field is proportional to the cell size. A short centrifugation at low speed will sediment leukocytes, but not platelets, and thus effectuate their separation.

because the platelets may have entered the gradient and stick to the reagent tube (see protocol for granulocyte purification).

Removal of Red Cells

For the initial separation of blood cells at 1.077 g/cm³, it is not necessary to remove the red cells first. After this separation the upper layer, consisting of monocytes + lymphocytes + basophils, usually contains hardly any red cells. The erythrocytes in the pellet can be removed later. In fact, the separation at 1.077 g/cm³ depends in part on the presence of the erythrocytes.[2] Since the red cells are heavier than most types of leukocyte (see the table), the red cells will tend to displace these leukocytes in a centrifugal field into a buffy coat at the top of the cell mixture. After entrance of the erythrocytes and the granulocytes into the separation material, the monocytes, lymphocytes, and basophils settle on top of this material. When the red cells are removed before separation of the leukocytes at 1.077 g/cm³, "streaming" causes incomplete separation in this step.

To remove red cells from leukocyte suspensions, several methods are employed. Dextran-induced clumping and accelerated sedimentation[25] of the red cells leads to incomplete removal of the red cells,[2] loss of leukocytes,[2,17,26] and contamination with platelets.[20] Hypotonic lysis of the erythrocytes (30 sec in distilled water) is effective, but not selective for red cells. Extreme care must be taken, therefore, to prevent damage to the leukocytes.

A better defined method is the specific lysis of erythrocytes by NH_4Cl in the presence of $KHCO_3$.[27] NH_3 and CO_2 penetrate into the cells; the resulting osmotic swelling is accelerated by chloride–bicarbonate exchange. Because the plasma membrane of leukocytes, in contrast to that of red cells, does not contain carbonic anhydrase, the lysis does not affect leukocytes as long as the temperature is kept at 4° to minimize diffusion. In a comparative study with granulocytes purified with and without the use of NH_4Cl, we have been unable to find any difference in functional capacity between these two preparations. Even the release of lysosomal enzymes, known to be inhibited by NH_4Cl,[28] was unaffected by the previous treatment of the cells with NH_4Cl. Treatment of the cells with NH_4Cl at room temperature leads to granulocyte damage, however.[29]

[25] W. A. Skoog and W. S. Beck, *Blood* **11**, 436 (1956).
[26] R. Hjorth, A.-K. Jonsson, and P. Vretblad, *J. Immunol. Methods* **43**, 95 (1981).
[27] D. Roos and J. A. Loos, *Biochim. Biophys. Acta* **22**, 565 (1970).
[28] M. S. Klempner and B. Styrt, *J. Clin. Invest.* **72**, 1793 (1983).
[29] H. Bolland, H. Pfisterer, W. Ruppelt, and G. Michlmayr, *Blut* **22**, 60 (1971).

Several reports have been published on one-step separations of whole blood into fractions containing either mononuclear leukocytes, granulocytes, or erythrocytes.[26,30–32] Complete separation of granulocytes and erythrocytes is seldom obtained, however. Moreover, for the purification of leukocytes from patients' blood, the method must be adapted,[33] but this problem may also be encountered with the classical separation at 1.077 g/cm³. For instance, the leukocytes from patients with infections, hypereosinophilia, or leukemia are often of lower specific gravity than those from healthy individuals.[12,34,35]

Purification of Monocytes

As indicated above, monocytes can be purified by successive isopycnic centrifugation steps (see protocol below). Other methods of monocyte purification employ differences in functional activity or cell size between monocytes and other cells. For optimal results, mononuclear phagocytes are first prepared, e.g., by centrifugation of the blood over a layer of Percoll (or other material) of 1.077 g/cm³. Thereafter, the monocytes are isolated by adherence, velocity sedimentation, or elutriation centrifugation.

Adherence of monocytes to glass or plastic, followed by washing away the nonadherent cells, carries the possibility of metabolic activation of the monocytes.[36] Moreover, detachment of the monocytes by mechanical means may easily lead to cell damage. This last problem is prevented by using EDTA or lidocaine for the detachment,[37–39] or by using a Teflon surface for the monocyte adherence (from which the monocytes will

[30] A. Ferrante and Y. H. Thong, J. Immunol. Methods 48, 81 (1982).
[31] J. Giudicelli, P. J. M. Philip, P. Delque, and P. Sudaka, J. Immunol. Methods 54, 43 (1982).
[32] R. J. Harbeck, A. A. Hoffman, S. Redecker, T. Biundo, and J. Kurnick, Clin. Immunol. Immunopathol. 23, 682 (1982).
[33] A. Ferrante, D. W. James, W. H. Betts, and L. G. Cleland, Clin. Exp. Immunol. 47, 749 (1982).
[34] I. Windqvist, T. Olofsson, I. Olsson, A.-M. Persson, and T. Hallberg, Immunology 47, 531 (1982).
[35] L. Prin, M. Capron, A.-B. Tonnel, O. Blétry, and A. Capron, Int. Arch. Allergy Appl. Immunol. 72, 336 (1983).
[36] P. T. Bodel, B. A. Nichols, and D. F. Bainton, J. Exp. Med. 145, 264 (1977).
[37] B. R. Bloom and B. Bennett, Science 153, 80 (1966).
[38] S. K. Ackerman and S. D. Douglas, J. Immunol. 120, 1372 (1978).
[39] J. J. Rinehart, B. J. Gormus, P. Lange, and M. E. Kaplan, J. Immunol. Methods 23, 207 (1978).

detach spontaneously after some time).[40] In general, the yield and purity of the monocyte suspensions prepared in this way are not very high.[39,41,42]

Since monocytes are larger than lymphocytes (see the table),[42a] these cells can be separated from each other by 1 g sedimentation or low-speed centrifugation in a gradient of low specific gravity.[43–45] Special devices have been constructed to minimize the gradient volume,[44] the separation time,[44,45] the "streaming" phenomenon,[44–46] and the loss of cells adhering to the walls of the gradient container.[46] In this way, large numbers of monocytes can be obtained in good yield and purity and in a relatively short time.

Excellent results have also been obtained by elutriation centrifugation. This method separates cells mainly on differences in cell size and to a much lesser extent also on differences in specific gravity.[47] In a conical sedimentation chamber within the rotor of a centrifuge, cells are held in equilibrium by the centrifugal force and the opposing centripetal flow of medium through the chamber. Large cells sediment with a relatively large velocity and equilibrate near the apex of the sedimentation chamber, where they encounter a high velocity of flow. The equilibrium of smaller cells is in the wider part of the chamber, where a lower medium flow exists. Cells with a low sedimentation constant cannot counteract the flow stream, not even in the widest part of the chamber, and are removed by the medium. In this way, cells can be separated by means of velocity sedimentation either by stepwise decreasing the speed of the rotor or by stepwise increasing the flow speed of the medium.

For optimal results, purification of monocytes by elutriation centrifugation also starts with mononuclear leukocytes. Improvements in speed

[40] W. E. Berdel, U. Fink, E. Thiel, K. Stünkel, E. Greiner, G. Schwarzkopf, A. Reichert, and J. Rastetter, *Immunobiology* **163**, 511 (1982).

[41] M. P. Brodersen and C. P. Burns, *Proc. Soc. Exp. Biol. Med.* **144**, 941 (1973).

[42] C. A. Koller, G. W. King, P. E. Hurtubise, A. L. Sagone, and A. F. LoBuglio, *J. Immunol.* **111**, 1610 (1973).

[42a] G. B. Segel, G. R. Cokelet, and M. A. Lichtman, *Blood* **57**, 894 (1981).

[43] R. D. Barr, J. Whang-Peng, and S. Perry, *Biomedicine* **26**, 112 (1977).

[44] W. S. Bont, J. E. de Vries, M. Geel, A. van Dongen, and J. A. Loos, *J. Immunol. Methods* **29**, 1 (1979).

[45] A. Tulp, J. G. Collard, A. A. M. Hart, and J. A. Aten, *Anal. Biochem.* **105**, 246 (1980).

[46] A. Tulp and M. G. Barnhoorn, *J. Immunol. Methods* **69**, 281 (1984).

[47] R. J. Sanderson, F. T. Shepperdson, A. E. Vatter, and D. W. Talmage, *J. Immunol.* **118**, 1409 (1977).

and flow control[48–50] and in cell sample introduction[51] have greatly improved the results (see protocol for monocyte purification by elutriation centrifugation). We have chosen to use the adaptions in speed control[48] and flow control[49] introduced by Bont et al., because this enables separation at a constant (high) flow rate by stepwise decreasing the rotor speed, leading to optimal and reproducible cell separations. The temperature and pH of the flow medium have proven to be of great importance for the purity of the monocytes in the final preparations.[52] Special equipment to monitor the composition of the cells in the outflow has facilitated the procedure.[50]

In our hands, elutriation of small numbers of mononuclear leukocytes ($<5 \times 10^7$) leads to relatively large losses of monocytes, but other investigators claim good results even then. We prefer to separate monocytes from small amounts of blood (≤100 ml) by isopycnic centrifugation over Percoll, and to use elutriation for larger volumes.

Purification of Neutrophils and Eosinophils

Since neutrophils and eosinophils (together called granulocytes) have a higher specific gravity than other leukocytes (see the table), the usual method of purification is by isopycnic centrifugation. In fact, the number of eosinophils in normal human blood is usually so small compared to the number of neutrophils ($\leq5\%$) that very few investigators remove the eosinophils from neutrophil preparations. The common method of neutrophil purification is therefore separation over a medium of 1.077 g/cm^3 and removal of the erythrocytes from the pellet fraction. Preparation of granulocyte suspensions by elutriation centrifugation has also been reported.[53,54]

For the purification of neutrophils from total granulocyte preparations, one can only use the slight difference between neutrophils and eosinophils in specific gravity, because there is too much overlap in cell volume (see the table). Since there is also considerable overlap in specific gravity, a certain selection of neutrophils with lower specific gravity will

[48] W. L. van Es and W. S. Bont, Anal. Biochem. 103, 295 (1980).
[49] C. G. Figdor, W. S. Bont, J. E. de Vries, and W. L. van Es, J. Immunol. Methods 40, 275 (1981).
[50] P. H. M. de Mulder, J. M. C. Wessels, D. A. Rosenbrand, J. B. J. M. Smeulders, D. J. T. Wagener, and C. Haanen, J. Immunol. Methods 47, 31 (1981).
[51] T. J. Contreras, J. F. Jemoniek, H. C. Stevenson, V. M. Hartwig, and A. S. Fauci, Cell. Immunol. 54, 215 (1980).
[52] R. S. Weiner and V. O. Shah, J. Immunol. Methods 36, 89 (1980).
[53] M. D. Persidsky and L. S. Olson, Proc. Soc. Exp. Biol. Med. 157, 599 (1978).
[54] J. F. Jemoniek, T. J. Contreras, J. E. French, and L. J. Shields, Transfusion 19, 120 (1979).

be the result of further purification. Due to the large excess of neutrophils, the eosinophils can be removed without losing more than about 10% of the neutrophils.

The large excess of neutrophils seriously hampers the purification of eosinophils. Therefore, most investigators have used blood from hypereosinophilic patients. However, it has become clear that eosinophils from such patients are functionally different from those of normal individuals.[34,35,55] Purification of eosinophils from the blood of normal individuals is difficult, but not impossible. Most reports describe purification by isopycnic centrifugation over Ficoll–Isopaque,[9,35,56] Ficoll–Hypaque,[55,57] or Percoll.[16,34] However, for high purity and recovery of the eosinophils, it is necessary to increase the difference in specific gravity between neutrophils and eosinophils. This may be achieved at hypertonic or slightly acidic conditions. Using a modification of Gärtner's method,[16] we have obtained good results when the granulocyte suspension was preincubated in a medium of pH 7.0 (instead of pH 7.4) before successive separations over Percoll of 1.082 g/cm^3 (see protocol).[58]

Another separation method has used the autofluorescence of eosinophils for enrichment by fluorescence-activated cell sorting.[59] This is a rather time-consuming process, however. The binding capacity of neutrophils, and the lack of binding capacity of eosinophils, for complexed rabbit IgG has been used to remove neutrophils, but this process proved to be rather inefficient.[60]

Recently, a new method has been described to separate eosinophils from other blood cells in one centrifugation step.[61] Whole normal blood anticoagulated with EDTA is incubated with 10^{-6} M formyl-methionyl-leucyl-phenylalanine (fMLP) and then centrifuged over a discontinuous Percoll gradient. From the intermediate cell layer, the erythrocytes are removed by hypotonic lysis. The resulting cell suspension contains almost pure eosinophils with a high yield. The principles of this separation procedure are the decrease in specific gravity of neutrophils caused by fMLP activation and the lack of reactivity of eosinophils with fMLP; this causes enough difference in specific gravities between the two cell types to permit purification of eosinophils without contamination with neutro-

[55] D. A. Bass, W. H. Grover, J. C. Lewis, P. Szejda, L. R. DeChatelet, and C. E. McCall, *J. Clin. Invest.* **66**, 1265 (1980).
[56] R. P. Day, *Immunology* **18**, 1955 (1970).
[57] J. E. Parillo and A. C. Fauci, *Blood* **51**, 457 (1978).
[58] M. Yazdanbakhsh, C. M. Eckmann, and D. Roos, *J. Immunol.* **135**, 1378 (1985).
[59] G. J. Weil and T. M. Chused, *Blood* **57**, 1099 (1981).
[60] P.-C. Tai and C. J. F. Spry, *Clin. Exp. Immunol.* **28**, 256 (1977).
[61] R. L. Roberts and J. I. Gallin, *Blood* **65**, 433 (1985).

phils. The advantages of this method are its ease, high yield, and lack of donor specificity. These enable one to purify eosinophils even from 10–20 ml of blood. However, a serious drawback is our finding that eosinophils purified in this way show lower metabolic reactivity than those purified by isopycnic centrifugation (M. Yazdanbakhsh and D. Roos, unpublished results). In contrast to the claim of Roberts and Gallin,[61] we found eosinophils to react with fMLP. Probably, eosinophils are metabolically "primed" by fMLP shortly after contact with this stimulus, and desensitized at a later stage. Therefore, we do not recommend this method when eosinophils are needed for metabolic or functional studies.

Cryopreservation of Phagocytes

Monocytes from human blood can easily be cryopreserved.[62] Only under extreme freezing conditions (e.g., without cryoprotectant or by immediate cooling to $-196°$) are these cells damaged. Under normal cryopreservation conditions (see protocol), monocytes are recovered in high yield after thawing, with at least 70% preservation of functional activities.

Cryopreservation of neutrophils has been attempted over the last 20 years with variable results.[63–65] Usually, the recovery of the neutrophils after thawing has been reported as low, and the viability of the cells has been found to decrease rapidly at 37°.[65,66] Especially the capacity to generate bactericidal oxygen products is strongly affected by freezing and thawing.[63,67–69] The only way to cryopreserve at least some properties of neutrophils, is to first prepare cytoplasts from these cells, which can subsequently be frozen and thawed with maintenance of antigenic, metabolic, and functional properties (see this volume [10]).

To our knowledge, no reports have been published on the cryopreservation of eosinophils. However, cytoplasts can also be prepared from eosinophils, and such structures can be cryopreserved in a similar way to neutrophil cytoplasts (M. Yazdanbakhsh and D. Roos, unpublished results).

---

[62] M. de Boer, R. Reijneke, R. J. van de Griend, J. A. Loos, and D. Roos, *J. Immunol. Methods* **43**, 225 (1981).
[63] S. C. Knight, J. A. O'Brien, and J. Farrant, *Cryobiology* **17**, 273 (1980).
[64] H. Bank, *Cryobiology* **17**, 187 (1980).
[65] J. Frim and P. Mazur, *Cryobiology* **17**, 282 (1980).
[66] F. J. Lionetti, S. M. Hunt, P. S. Lin, S. R. Kurtz, and C. R. Valeri, *Transfusion* **17**, 465 (1977).
[67] J. E. French, *Cryobiology* **17**, 252 (1980).
[68] P. Boonlayangoor, M. Telischi, S. Boonlayangoor, T. F. Sinclair, and E. W. Millhouse, *Blood* **56**, 237 (1980).
[69] A. Rowe and L. L. Lenny, *Cryobiology* **17**, 198 (1980).

Methods

Materials

a. Preparation of Percoll Suspensions. A solution of 1.54 M NaCl with 0.1 M NaH$_2$PO$_4$ is made in water and filter sterilized. Of this solution, 7.0 ml is added to 93 ml of sterile Percoll suspension (Pharmacia Fine Chemicals, Uppsala, Sweden). The resulting Percoll stock suspension has an osmolality of 290 mOsm and a pH of 7.45 at 25°. The specific gravity of this stock suspension will be around 1.1245 g/cm^3 at 25°, depending on the original Percoll batch. The procedure for measuring the osmolality and the specific gravity is given in the next section.

The Percoll stock suspension is further diluted to isotonic suspensions of desired densities that contain 5 mg of human albumin/ml and 13 mM trisodium citrate. For this purpose, the stock Percoll suspension is mixed with phosphate-buffered saline (PBS[69a]; 140 mM NaCl, 9.2 mM Na$_2$HPO$_4$, 1.3 mM NaH$_2$PO$_4$; pH 7.4), 5% human albumin (v/v) and trisodium citrate (130 mM) according to the following formula:

$$X(0.0056) + 0.1(0.0227) + 0.1(0.0219) + (0.8 - X)(0.1245)$$
$$= \text{desired specific gravity} - 1$$

in which X = ml of PBS/ml of final suspension
  0.0056 = specific gravity of PBS − 1
  0.1   = ml of albumin and citrate/ml of final suspension
  0.0227 = specific gravity of albumin (5%) − 1
  0.0219 = specific gravity of trisodium citrate (130 mM) − 1
  0.8 − X = ml of stock Percoll suspension/ml of final suspension
  0.1245 = specific gravity of stock Percoll suspension − 1

Thus, each ml of final Percoll suspension contains 0.8 ml of PBS + stock Percoll suspension, in a ratio $X:(0.8 - X)$ that varies with the desired specific gravity of the final Percoll suspension. For example, to prepare 100 ml of Percoll suspension with a specific gravity of 1.082 g/cm^3, one should mix 18.55 ml of PBS, 10 ml of albumin, 10 ml of citrate, and 61.45 ml of stock Percoll suspension.

Of course, each of the values of specific gravity used in this calculation must be checked. The pH of the final suspension should be between 7.2 and 7.4 at 25°, and the osmolality should be between 290 and 295 mOsm at 25°. All suspension and solutions are kept sterile.

[69a] Abbreviations: ACD, acid citrate dextrose (100 mM disodium citrate, 128 mM glucose; pH 5.0); DMSO, dimethyl sulfoxide; EDTA, ethylene diaminotetraacetate; FCS, fetal calf serum; fMLP, formyl-methyl-leucyl-phenylalanine; PBS, phosphate-buffered saline (140 mM NaCl, 9.2 mM Na$_2$HPO$_4$, 1.3 mM NaH$_2$PO$_4$; pH 7.4).

b. Measurement of Osmolality, Specific Gravity, and pH of Percoll Suspensions. A sample of 3–4 ml of Percoll is added to a tube, which is then closed with a stopper to prevent evaporation. The osmolality is measured in a 50-μl sample with an Osmomat 030 (Gonotec, Berlin, Federal Republic of Germany). Care must be taken to ensure that the freezing point depression is measured correctly: if freezing is started too early or too late, the real freezing point is not reached.

The specific gravity of the Percoll suspensions is measured in a 2-ml sample with a density meter (model DMA 46, Anton Paar, Graz, Austria). Care must be taken to prevent the formation of air bubbles in the viscous Percoll suspensions inside the apparatus.

If the pH of a Percoll suspension has to be corrected, this should be done with a buffer of pH 5.5–10, not with acid or base, because Percoll is not stable at extreme pH values.

Procedures

a. Blood Collection and Buffy Coat Preparation. For the purification of monocytes, blood (50 ml) is collected in 7.5 ml of sterile ACD (acid–citrate–dextrose; 100 mM disodium citrate, 128 mM glucose; pH 5.0). For purification of monocytes by elutriation centrifugation, a unit of blood (500 ml) is collected in 75 ml of sterile ACD in a plastic blood bag. This blood may be stored for not more than 4 hr at 25°. The blood is centrifuged (5 min, 4000 g, 25°) and the bag is placed in a plasma extractor. The plasma is then squeezed into a container. Thereafter, a surgical stomach clamp is placed over the entire breadth of the bag, about 2 cm below the plasma–cell interface. During this procedure, the bag remains in the plasma extractor. The leukocyte-rich buffy coat above the clamp (about 50–100 ml of cell suspension) is then transferred to a small vacuum bottle while the bag remains in an upright position in the plasma extractor. In this way, the leukocytes will enter the small bottle before the majority of the red cells. A buffy coat prepared this way contains about 70% of all leukocytes present in the original blood sample.

For the preparation of neutrophils, the blood can be handled as described above. Alternatively, the blood may be defibrinated by gentle shaking at room temperature for 10 min with sterile glass beads (0.5 cm in diameter, 2 beads/ml of blood) and separation of the blood from the clot. This procedure may lead to loss of some leukocytes. Also, heparinized blood (10 U/ml) may be used, but this will lead to platelet contamination of the leukocyte preparations.

For the purification of eosinophils from normal human donors, we have routinely used blood from donors with 3–10% eosinophils, without any clinical symptoms. When blood with less than 3% eosinophils is used,

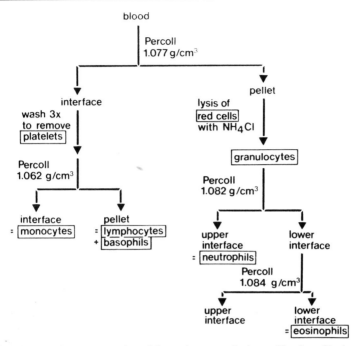

FIG. 1. Schematic representation of the various steps in the purification of leukocytes by isopycnic centrifugation.

the cell yield is usually very low. The blood (500 ml) should be collected in 50 ml of sterile trisodium citrate (138 mM; pH 7.0). Thereafter, a buffy coat can be prepared as described above.

All solutions needed for cell isolations are made and kept sterile. The isolation procedures themselves, however, can be carried out without special precautions, unless cells are purified that need to be incubated for more than 6 hr. Figure 1 shows a general scheme of leukocyte purification by isopycnic centrifugation.

b. Purification of Monocytes by Isopycnic Centrifugation. Citrated blood (see the previous section) is diluted with an equal volume of PBS containing 13 mM trisodium citrate. Buffy coat (50–100 ml) is diluted to 200 ml. Twenty-five milliliters of this mixture is carefully layered on 12 ml of Percoll (specific gravity 1.077 g/cm^3, 290 mOsm, pH 7.4, with albumin and citrate; see Materials) in 50-ml tubes. Care should be taken not to disturb the Percoll–blood interface. Alternatively, the Percoll may be injected with a long needle under the blood. The tubes are centrifuged at room temperature for 20 min at 1000 g. First, the plasma is collected in a separate tube. Thereafter, the cells from the interface between the Percoll and the plasma are collected with a Pasteur pipet, without taking too much Percoll suspension. The cells from two tubes are pooled in one tube, and the volume is made up to 50 ml with PBS with 13 mM trisodium citrate. The cell from the interface (lymphocytes, monocytes, and some basophils) are centrifuged (7 min, 400 g, room temperature), and the supernatant is carefully removed by vacuum aspiration. The cells are then resuspended to about 20 × 10^6 leukocytes/ml in PBS with 13 mM trisodium citrate and about 10% (v/v) autologous plasma. After centrifugation (5 min, 400 g, room temperature), the supernatant is removed and the cells are again suspended in the same medium. This washing procedure is repeated two times to remove platelets.

After the last washing step, the cells are suspended to no more than 20 × 10^6/ml in Earle's medium or minimal essential medium (GIBCO, Grand Island, NY) with 25 mM Tris–HCl, 13 mM trisodium citrate, and 10% autologous plasma (pH 7.4, at 25°, 290 mOsm) and incubated for 30 min at 37° to regain their original specific gravity. Thereafter, the suspension is centrifuged (5 min, 400 g, room temperature), the medium is aspirated, and the cells are resuspended to no more than 40 × 10^6/ml in Percoll (specific gravity 1.062 g/cm^3, 290 mOsm, pH 7.4, with albumin and citrate; see Materials). The cell suspension is carefully layered on top of a Percoll suspension of 1.077 g/cm^3 (height of each layer about 3 cm) in small test tubes. On top of the gradient about 0.2 ml PBS is layered.

After centrifugation (10 min, 1000 g, room temperature) the cells in the interface between medium and Percoll of 1.062 g/cm^3 are collected with a Pasteur pipet. Care should be taken during withdrawal of the pipet from the tube that a drop of cell suspension does not leave the outside of the pipet, thus contaminating the lower fraction of cells. At least three operations with the pipet are needed to remove all cells from an interface. The cells from the interface are collected into one tube, centrifuged (10 min, 500 g, room temperature) and resuspended in the desired medium. Lymphocytes can be harvested from the interface between the two Percoll layers. If monocytes have to be stored for longer than 30 min, the cells should be kept on ice, to prevent them from sticking to the tube.

With this procedure, monocytes from 50–500 ml of blood may be obtained in about 3 hr. The yield is about 70% of the monocytes in the blood or buffy coat. The purity is about 90%, with less than 10% lymphocytes or basophils. The viability is about 95%. For methods to evaluate these parameters, see point (g). The functional activity and the maintenance in culture of monocytes isolated in this way is also excellent.

    c. Purification of Monocytes by Elutriation Centrifugation (for more details see this series, Vol. 108 [20]). For purification of monocytes from larger volumes of blood (>100 ml), an elutriation centrifuge can be used.

First, a mononuclear leukocyte preparation is made, as described in the preceding section. The cells are then suspended in 10–20 ml of PBS with 13 mM trisodium citrate and 10% (v/v) autologous plasma. This cell suspension is injected into an elutriator centrifugation system (Beckman J2-21 centrifuge with a JE-6 elutriation rotor).

The elutriation medium is PBS with 13 mM trisodium citrate and 0.5% human albumin. To separate the cell populations, the flow rate is kept constant at 20 ml/min, while the rotor speed is diminished stepwise from 4000 rpm to 0, according to the methods described by Van Es and Bont[48] and Figdor *et al.*[49] The elutriation medium is cooled to 10° and the centrifuge to 15°.

After the cells have been injected, the rotor is started (4000 rpm). At the time of entrance of the cells into the elutriation chamber, collection of the outflow is started. After all the cells have entered the chamber, the rotor speed is diminished to 3700 rpm, and the bypass in which the cells were injected is closed. When 120 ml is collected in the first fraction (containing the platelets), the rotor speed is diminished to 2600 rpm, and 200 ml of outflow is collected. This fraction 2 contains most of the lymphocytes and erythrocytes. The third fraction (100 ml) is collected at 2500 rpm and contains some lymphocytes and monocytes. When cell clumps are visible in the chamber, the monocyte fraction (100 ml) is collected at about 1000 rpm. When no clumps are present, this fraction may be collected at 0 rpm.

The monocyte fraction is centrifuged at 400 g for 5 min at room temperature. If necessary, the basophils in this fraction are removed by isopycnic centrifugation. For this purpose, the cell pellet is suspended in 4 ml of Percoll with a specific gravity of 1.067 g/cm^3 and put in a small tube. About 0.2 ml of PBS is layered on top. The tube is centrifuged at room temperature for 10 min at 1000 g. The interface between the Percoll and the PBS contains the monocytes. These cells are collected with a Pasteur pipet [see point (b)], centrifuged (10 min, 500 g, room temperature), and resuspended in the desired medium. The basophils may be collected separately by including a Percoll layer of 1.075 g/cm^3 in the gradient; the lower interface then contains the basophils.[70]

This procedure takes about 3 hr. The yield, purity, and viability of the monocytes are comparable to those given in the preceding section.

d. Purification of Neutrophils. For the purification of neutrophils, the blood is first centrifuged over Percoll of 1.077 g/cm^3 as described under point (b). After removal of the plasma and the mononuclear leukocytes at the interface between plasma and Percoll, the remainder of the Percoll is

---

[70] M. de Boer and D. Roos, *J. Immunol.* **136**, 3447 (1986).

removed by vacuum aspiration to just above the cell pellet. The cell pellets are then thoroughly suspended in a 3-fold volume of ice-cold NH_4Cl medium to lyse the erythrocytes (155 mM NH_4Cl, 10 mM $KHCO_3$, and 0.1 mM EDTA, adjusted to pH 7.4 at 0° and filter sterilized). The tubes with the cell suspension in this medium are kept in ice. Every few minutes, the cell suspension is mixed again. After 7–8 min, the color of the cell suspension changes from red to black, indicating the hemolysis of the erythrocytes. After 10 min, the tubes are centrifuged for 5 min at 400 g and 4°. The supernatants are carefully aspirated and the pellets are resuspended in NH_4Cl medium of 0° with 0.5% human albumin. The cell suspensions are combined into one tube of 50 ml. After 15 min incubation in ice, the tubes are centrifuged (5 min at 400 g and 4°) and the supernatants are aspirated with a fine needle to remove as much red cell stroma and hemoglobin as possible without losing granulocytes. If the remaining pellets still have a somewhat red color, the NH_4Cl treatment can be repeated once more.

The remaining cells are suspended in one tube in PBS with 5 mg of human albumin per ml, centrifuged (5 min, 400 g, room temperature), and suspended in the desired medium. The resulting cell suspension contains 90–95% neutrophils, less than 2% lymphocytes, and 3–10% eosinophils. The contamination with platelets is negligible. The total procedure takes about 2 hr. The yield is about 90% of the original number of granulocytes. The viability is >98%. The cell functions are excellent.

Further purification of neutrophils may be achieved by isopycnic centrifugation (see Fig. 1).[58] First, the cells are suspended in Earle's medium or Hanks' balanced salt solution supplemented with 25 mM Tris–HCl and 5% (v/v) fetal calf serum (pH 7.0 at 37°). The cells are incubated in this medium for 30 min at 37°. The cells are then centrifuged (500 g, 7 min, room temperature) and the pellet is suspended to no more than 2 × 10⁷ cells/ml in Percoll with a specific gravity of 1.082 g/cm³ (290 mOsm, pH 7.4, containing citrate and albumin; see Materials). This suspension is layered in a small tube on top of half its volume of Percoll with a specific gravity of 1.100 g/cm³ (290 mOsm, pH 7.4, with citrate and albumin; see Materials). On top of the upper layer a small volume of PBS is added. The tubes are centrifuged at 1000 g for 20 min at room temperature. The cells from the *upper* interface are carefully collected [see point (b)], centrifuged (10 min, 500 g, room temperature), washed once with PBS containing 13 mM trisodium citrate and 0.5% human albumin, and resuspended in the desired medium.

With this procedure suspensions with >98% pure neutrophils are obtained. The yield is about 80%. The other parameters are comparable to those given above for granulocyte preparations.

e. Purification of Eosinophils. For the purification of eosinophils, the procedure is similar to that for purified neutrophils, except that blood is used from donors with 3–10% eosinophils and that the blood is drawn in trisodium citrate [see point (a)].[58] The cells from the *lower* Percoll interface of the neutrophil purification are collected after removal of the gradient material to just above this interface. The cells are resuspended in an equal volume of PBS containing 13 mM trisodium citrate and 0.5% human albumin, and centrifuged at 500 g for 10 min at room temperature. Sometimes this cell suspension contains already more than 85% eosinophils. If not, further purification is needed, as follows.

The cells are resuspended to no more than 2 × 10^7/ml in Percoll of 1.084 g/cm^3 (290 mOsm, pH 7.4, with citrate and albumin; see Materials). This suspension (4 ml) is layered in a small tube (Corning tissue culture tube, No. 25200, 13 ml) on top of 2 ml of Percoll with a specific gravity of 1.100 g/cm^3 (290 mOsm, pH 7.4, with citrate and albumin; see Materials). On top of the upper layer a small volume of PBS is added. The tubes are centrifuged at 1000 g for 20 min at room temperature. The cells at the interface between the two Percoll suspensions are carefully collected [see point (b)] after removal of the gradient material to just above this interface. The cells are then centrifuged (10 min, 500 g, room temperature), washed with PBS containing 13 mM trisodium citrate and 0.5% human albumin and suspended in the desired medium.

The resulting preparations contain 85–99% eosinophils, with a yield of about 60–70%. The total eosinophil purification takes about 4 hr. Neutrophils are the only contaminating cells. The viability is >98%. The cell functions are excellent.

f. Cryopreservation of Monocytes (for more details, see this series, Vol. 108 [36]). Monocytes are suspended to 10^7/ml in a physiological medium with 5% (v/v) FCS. The suspension is cooled to 0°, and an equal volume of ice-cold medium containing 35% (v/v) FCS and 20% (v/v) DMSO is added dropwise (about two drops/sec) with swirling. Portions of 2 ml of the resulting suspension are quickly transferred to 2.5-ml glass ampoules, which are then sealed in a flame. The ampoules are cooled with a controlled program[71] at a rate of 1.4° per min to −30°, with compensation for the heat of crystallization, followed by cooling to −100° at a rate of 6–7°/min. Thereafter, the ampoules are stored in liquid nitrogen.

Thawing of the monocyte suspensions is performed by gently shaking the ampoules in a water bath at 37°. When the ice in the ampoules has almost disappeared, the ampoules are placed in an ice bath. Thereafter,

[71] M. J. G. J. du Bois, P. T. A. Schellekens, J. J. F. M. de Wit, and V. P. Eijsvoogel, *Scand. J. Immunol.* **5**, Suppl. 5, 17 (1976).

the ampoules are opened and the content of two ampoules is transferred to a 50-ml polypropylene tube. The monocyte suspension is then very carefully diluted to 40 ml with ice-cold medium containing 0.5% human albumin. This dilution should be logarithmic: first 1 drop is added while the tube is shaken in ice water, then 2 drops, and so on, up to 16 drops; thereafter 1 dash of about 1 ml, 2 dashes, and so on. The whole procedure should take about 10 min. Immediately thereafter, the monocytes are centrifuged (5 min, 500 g, room temperature) and washed twice with physiological medium containing 0.5% human albumin.

With this procedure, the recovery of the monocytes is 95–98%, with 93% viable cells and about 70% of residual functional activities (chemotaxis, phagocytosis, respiratory burst, bacteria killing, cytotoxicity against red cells, and helper activity in lymphocyte activation).[62]

g. Evaluation of Cell Number, Purity, and Viability (see also this series, Vol. 108 [6]). The concentration of cells in a suspension is most easily counted electronically, e.g., with a Coulter counter (Coulter Electronics, Dunstable, UK). In the presence of 0.02% (w/v) saponin + 0.0001% (v/v) Triton X-100, only nuclei are counted. If erythrocytes have to be counted, the Triton–saponin should not be used. The counting accuracy is highest between 5 and 50 $\times$ 10^6 cells/ml (diluted 1000 times before counting). Cells that are not used immediately for experiments should be kept at a concentration of not more than 20 $\times$ 10^6/ml in a medium with some glucose and protein, in a plastic or siliconized glass tube. The metabolic functions of the cells are best preserved at 4°, but the receptor density at the cell surface may be drastically reduced by this procedure. Reexpression of the original number of surface receptors may take as long as 30 min at 37°.

A rough estimate of the purity of a leukocyte suspension may be obtained by electronic sizing, e.g., with a Coulter Channelyzer.[72] The proportion of monocytes and lymphocytes in a mononuclear leukocyte preparation can be estimated in this way, as well as the proportion of lymphocytes and granulocytes in a granulocyte preparation. However, monocytes, eosinophils, and neutrophils cannot be distinguished in this way, because the size of these cells are too close to each other (see the table).

A more accurate idea of the composition of a leukocyte preparation is obtained by microscopic examination after staining the cells with May-Grünwald/Giemsa. A sample of about 40,000 cells is centrifuged on a glass slide, pretreated with two drops of 0.5% human albumin (in PBS), with a cytocentrifuge (e.g., Shandon Elliot Cytospin centrifuge) for 10

---

[72] J. A. Loos, B. Blok-Schut, B. Kipp, R. van Doorn, and L. Meerhof, *Blood* **48**, 743 (1976).

min at 90 *g* and room temperature. The slide is dried in air for at least 5 min and then stained for 5 min with undiluted May-Grünwald solution (Merck, Darmstadt, Federal Republic of Germany), followed by staining for 30 min with Giemsa solution (Merck) diluted 20 times with 10 m*M* phosphate buffer, pH 7.0. The slide is then rinsed with tap water, air dried, and embedded in malinol (Chroma Gesellschaft, Schmid GmbH, Stuttgart-Untertürkheim, Federal Republic of Germany) for microscopic examination. At least 400 cells should be scored.

Cell integrity is usually evaluated by trypan blue exclusion. However, this is a notoriously inaccurate test. A much better test is the staining with fluorescein diacetate and ethidium bromide.[73–75] Intact cells hydrolyze fluorescein diacetate to fluorescein and acetate, and exclude ethidium bromide. Therefore, such cells stain bright green under light with a wavelength of 488 nm, whereas dead cells stain red, owing to coupling of ethidium bromide with DNA. Fluorescein diacetate is dissolved to 5 mg/ml in acetone; from this stock solution a dilution is made with PBS to 25 μg/ml. A cell suspension (4–6 × 10^5 cells in 100 μl of medium without protein) is mixed with fluorescein diacetate (2.5 μg in 100 μl of PBS) and ethidium bromide (100 μg in 100 μl of PBS). After 5 min at room temperature, the mixture is diluted with 1 ml of PBS. The number of red cells is detected by fluorescence microscopy; at least 400 cells should be scored.

[73] F. Celada and B. Rotman, *Proc. Natl. Acad. Sci. U.S.A.* **57**, 630 (1967).
[74] M. Edidin, *J. Immunol.* **104**, 1303 (1970).
[75] M. Takasugi, *Transplantation* **12**, 148 (1971).

[9] Suspension Culture of Human Monocytes

By PAUL J. MILLER and HENRY C. STEVENSON

Introduction

One of the most challenging tasks facing researchers involved with the *in vitro* handling of human monocytes is the development of culture techniques that most closely replicate the *in vivo* state of these cells. To this end, a number of guidelines concerning the culture of human monocytes have emerged: (1) It is advantageous to use the purest populations of monocytes possible to avoid the unintentional *in vitro* assessment of other possible contaminating leukocytes. (2) The cells should be harvested in a suspension state and cultured in a suspension state to replicate the *in vivo* suspension state of monocytes in the peripheral blood. (3) The cells should be cultured in the absence of potential activating substances (such as endotoxin and certain animal sera) by using only human serum (preferably autologous) with culture media or utilizing serum-free media. (4) Cultured monocytes should remain viable and functional for prolonged periods of time. (5) It would be desirable to develop the technology to culture large numbers of monocytes in suspension to allow for the execution of large experiments or in preparation for adoptive immunotherapy clinical trials. (6) The use of antibiotics and other protein synthesis inhibitors which could potentially alter the native state of these cells should be excluded.

Most of the currently used standard methodologies for culturing human monocytes are to a greater or lesser degree deficient in one or more of these criteria. It is still accepted practice to culture blood monocytes in polystyrene plates (to which they will adhere) followed by the removal of these cells by chelating agents, local anesthetics (such as lidocaine), or physically scraping them off with a rubber policeman. In addition, it is still standard practice to use media which have not been screened for endotoxin; similarly, the use of fetal calf serum and other nonhuman protein sources is quite common. Within recent years, several research groups in the United States and Europe have begun culturing monocytes in suspension.[1-4] This is only possible when the monocytes have been isolated by a technique such as Percoll gradients or counter-current centrifugal elutriation which purifies the cells in suspension.[5] The present chapter describes a monocyte suspension culture system, which when combined with counter-current centrifugal elutriation fulfills all of the requirements cited previously. Using this suspension culture system, up to 1×10^9 cells consisting of 95% pure monocytes can be cultured in suspension in serum-free media from a single individual for prolonged periods of time. The procedure takes approximately 3 hr from the time the leukocytes are harvested from the donor until the time that these cells are placed in *in vitro* culture. Specialized equipment is needed for this procedure, and the culture technique outlined below must be closely adhered to for optimal results.

[1] J. W. M. Van Der Meer, J. S. Van De Gevel, I. Elizenga-Claassen, and R. Van Furth, *Cell. Immunol.* **42**, 208 (1979).
[2] J. W. M. Van Der Meer, J. S. Van De Gevel, A. Blussé Van Oud Ablas, J. A. Kramps, T. L. Van Zwet, P. C. J. Leish, and R. Van Furth, *Immunology* **42**, 617 (1982).
[3] S. D. Wright and S. C. Silverstein, *J. Exp. Med.* **156**, 1149 (1982).
[4] H. C. Stevenson, E. Schlick, R. Griffith, M. A. Chirigos, R. Brown, J. Conlon, D. J. Kanapa, R. K. Oldham, and P. J. Miller, *J. Immunol. Methods* **70**, 245–255 (1984).
[5] H. C. Stevenson, this series [108], p. 242.

Methods

Materials

The basic materials required for successful suspension culture of human monocytes are (1) Teflon labware, (2) a source of serum-containing media that is compatible with human monocytes (or preferably a serum-free medium), and (3) a source of purified monocytes in suspension. Each of these will be described in detail:

A. Teflon labware: Several different sources of Teflon labware exist; some are obtained from commercial sources, others must be custom designed. Teflon labware suitable for monocyte culture is composed of three types of vessels: (1) Teflon bags, (2) Teflon culture bottles, and (3) specially tooled Teflon labware of the dimensions of standard polystyrene labware. Teflon bags have been extensively utilized in Europe (available from the DuPont de Nemours and Company, Geneva, Switzerland; 25-μm gauge supplied by Janssens' M. and L., St. Niklaas, Belgium), as described in detail elsewhere.[1,2] The disposable Teflon bags are sealed following the inoculation with monocytes by diathermic sealing apparatus (Super Sealboy 235; 210 W, audion Electro, Amsterdam, the Netherlands). The Teflon culture bottles are easily obtained from the Nalge Company, Rochester, NY; these culture vessels are available in sizes ranging from 10 ml up to 1 liter (as detailed in Wright and Silverstein).[3] Finally, we have also developed specially tooled Teflon labware to conform to the exact dimensions of standardly employed polystyrene labware. The Teflon sheets from which this labware was constructed were obtained from DuPont (duPont de Nemours, Wilmington, DE). The template for the production of each of these custom-made vessels was the Linbro 96-well flat-bottom plates (Flow Laboratories, McLean, VA) and the 24-well Costar plates (Costar Company, Cambridge, MA). Teflon labware is a very durable material capable of withstanding temperatures as low as −80°; it is also autoclavable. Since we have cultured monocytes in Teflon culture bottles and our custom-made Teflon labware, we will limit our discussion to these two systems in the subsequent parts of this chapter.

B. Culture media for monocytes: One has basically three choices for obtaining optimal media for culturing of human monocytes: (1) the use of screened, pooled human AB serum, (2) the use of autologous human serum, or (3) serum-free medium. There are several sources of pooled human AB serum. We use the serum provided by the Bio-Bee Company (Boston, MA); this firm and several others will allow the investigator to screen aliquots of different lots of human AB serum prior to purchase of a

COMPONENTS OF SERUM-FREE MEDIUM FOR HUMAN MONOCYTE
SUSPENSION CULTURE

Component	Final concentration
Iscove's modified Dulbecco's medium (IMDM)	
Human serum albumin (fatty acid free)	4 mg/ml
Cholesterol (greater than 99% pure)	20 μg/ml
L-α-Phosphatidylcholine	80 μg/ml
Human transferrin (98% pure)	1 μg/ml
Insulin (porcine)	0.128 U/ml
Ferrous chloride	7×10^{-11} M
2-Mercaptoethanol	10^{-7} M

single select lot. This precaution of prescreening lots of human AB serum is highly recommended. Most researchers use 5–20% human AB serum in RPMI 1640 medium (available from Flow Laboratories and many other sources). We use 10% human AB serum with RPMI 1640 when culturing monocytes in human serum. It is important, however, that the human serum as well as the RPMI media source be tested for absence of endotoxin [less than 0.025 ng/ml in the *Limulus* assay (Associates of Cape Cod, Inc., Woods Hole, MA)] prior to use. In addition, the serum should be heat inactivated and filtered prior to addition to the RPMI 1640.

The use of autologous serum is at least theoretically preferable to the use of pooled AB serum because it most closely replicates the environment from which the donors' monocytes are obtained. Autologous serum is harvested by collecting 50 ml of whole blood (which is allowed to clot) from the donor prior to removal of anticoagulated blood (used as the source of white cells). The tubes in which the whole blood is collected are then rimmed with a sterile wooden applicator (to remove adherent clots between the blood and the tube side wall), and the tubes are centrifuged at 1500 rpm for 10 min. The serum is removed, heat inactivated at 56° for 30 min, and then filtered prior to addition to the RPMI 1640.

We have developed a serum-free medium which successfully supports monocyte viability and functions in culture (see the table).[4] The Iscove's modified Dulbecco's medium is purchased from Hem Research, Inc., Rockville, MD. The serum human albumin, cholesterol, L-α-phosphatidylcholine, and human transferrin are purchased from Sigma Chemical Company, St. Louis, MO. The porcine insulin is obtained from Eli Lilly Company (Indianapolis, IN), the ferrous chloride is obtained from the M. Fisher Scientific Company, Fairlawn, NY, and the 2-mercaptoethanol is obtained from Eastman Kodak Company, Rochester, NY. The L-α-phosphatidylcholine and cholesterol are prepared together and sonicated with a Branson sonifier (Branson Sonic Power Company, Danbury, CT), with a microtip at a setting of 6, for 60 min at 4°. Following the addition of the other reagents, the entire serum-free medium is then filtered with a 0.45-μm Nalge filter (Nalge Company, Rochester, NY) and stored at −20° in 50-ml aliquots until used. Antibiotics are not used with any of the media mentioned above.

C. Source of purified monocytes in suspension: It is not recommended to attempt to culture monocytes in suspension if they have not been harvested and purified in suspension first (since they are likely to be agglutinated). Two basic methodologies exist to allow for the purification of monocytes in suspension: (1) Percoll gradients and (2) counter-current centrifugal elutriation. A number of established protocols for purifying human blood monocytes by Percoll gradients exist[6-8] (see also this series, Vol. 108 [9]). The potential disadvantages of utilizing Percoll gradients are that there have been conflicting reports as to the purity of cells obtained (usually in the 80% range) and the numbers of cells obtainable by gradients is usually quite low (usually <30 million cells). Counter-current centrifugal elutriation allows for the isolation of very large numbers of purified monocytes (95% pure; up to 1 billion cells harvested per normal donor). A detailed methods paper describing this technique has been published.[5] Regardless of the technique used for purifying blood monocytes in suspension, it is recommended that if they cannot be immediately added to the Teflon culture vessels they be stored in whichever culture medium will eventually be used for culture (10% autologous, 10% pooled AB serum, or serum-free media) and maintained in 50 ml polypropylene centrifuge tubes (Corning Plastics, Corning, NY) at 4°. It is also advised that prior to monocyte culture the monocyte cell substrate be analyzed for purity by the performance of three characterization techniques: esterase stain, latex bead ingestion, and morphology on Wright's stain, as described.[9]

Procedure

Basic Considerations. The suspension culture of human monocytes is basically a very straightforward procedure, providing that researchers have the proper purified cell suspensions and the proper reagents at their

[6] A. J. Ulmer and H. D. Flad, *J. Immunol. Methods* **30,** 1 (1979).
[7] K. Cochrum, D. Hanes, G. Fagan, J. Van Speybroeck, and F. K. Sturtevant, *Transplant. Proc.* **10,** 867 (1978).
[8] S. D. Nathanson, P. L. Zamfirescu, S. I. Drew, and S. Wilbur, *J. Immunol. Methods* **18,** 225 (1977).
[9] H. C. Stevenson, P. Katz, D. E. Wright, T. J. Contreras, J. F. Jemionek, V. M. Hartung, W. J. Flor, and A. S. Fauci, *Scand. J. Immunol.* **14,** 243 (1981).

disposal. Particular attention must be paid to sterility since we recommend culturing monocytes in the absence of antibiotics. For this reason, manipulations with the monocytes must be performed under a laminar flow hood. It is important to remember that all media and additives should be filtered through a 0.45-μm Nalge filter and sterile, disposable pipettes should be used. In addition, the Teflon labware can be easily sterilized by autoclaving or with ethylene oxide. One additional important consideration, however, is to ensure the absence of activating agents or toxins that might affect the function of the monocytes. To this end, all media should be endotoxin tested [to <0.1 ng of endotoxin/ml by a *Limulus* assay kit (Associates of Cape Cod, Inc., Woods Hole, MA)] and also tested for mycoplasma. In addition, although the Teflon labware can be sterilized by autoclaving or using ethylene oxide, this procedure does not ensure absence of chemical contaminants. All labware should be washed in an Alconox solution [1 tablespoon/4 liters of endoxin-free sterile distilled water (Alconox Company, New York, NY)], followed by soaking in sterile, endotoxin-free distilled water overnight. The next morning, the labware should be vigorously rinsed in sterile distilled water followed by immersion in 70% ethanol (diluted in sterile, endotoxin-free distilled water). The labware should then be removed from the alcohol solution, wrapped in an appropriate autoclavable gauze, and autoclaved. (The labware should be handled in a laminar flow hood with the use of sterile disposable gloves.)

Monocyte Suspension Culture Technique. Purified monocytes obtained from Percoll gradients or from counter-current centrifugal elutriation may be maintained in the medium in which they will be cultured in a 50-ml polypropylene tube at 4°, if more than a few minutes will elapse between the time of isolation and the actual culture of the cells. The cells ($1–5 \times 10^6$/ml) should be resuspended in the culture medium (10% autologous serum, 10% pooled AB serum, or serum-free medium). We generally incubate the cells in no more than 150 λ/well in a 96-well plate, 1 ml in a 24-well Costar plate, no more than 3 ml/30-ml Nalge Teflon bottle, and no more than 200 ml/1-liter Nalge bottle. This is to ensure an adequate surface-to-air ratio of the culture medium with the cells in suspension. The caps for the bottles should be loosely placed, as should the lids for the custom-made Teflon labware and the cultured monocytes, in a 37° incubator (5% CO_2) either stationary or, to ensure more complete distribution of the cells in suspension, placed on a rocker platform (7 cycles/min).

Refeeding and Harvesting of Monocytes Cultured in Suspension. Depending on the cell concentration and/or the use of activating signals, monocytes will metabolize varying amounts of the nutrients in the media. This can be most readily detected by noting a change of the pH of the media. Usually, regardless of the circumstances of culture, these cells are refed on a twice-a-week basis. Should the cells be seeded at high densities or should a powerful activation stimulus be included in the culture media, refeeding will be required more frequently. Harvesting of suspension culture monocytes is straightforward. Under a laminar flow hood, and using sterile gloves and a sterile pipet with a pipetman, the contents of the well are pipetted up and down. After approximately 15 sec of this dislodgement procedure, the contents of the well (or other culture vessel) can be aspirated and placed into a polypropylene centrifuge tube of appropriate size (to which monocytes do not adhere) and the tubes are centrifuged for 5 min at 1500 rpm. The pelleted cells are resuspended in an appropriate media. If the cells are to be refed and recultured, we generally resuspend them in 100% fresh media, perform a repeat cell count and viability prior to adding them back to a fresh Teflon culture vessel, and then incubate them at 37° (5% CO_2).

Additional Information. Suspension-cultured monocytes function similarly to adhered monocytes in a wide variety of immunologic assays. However, serum-free medium appears to enhance the secretion of certain cytokines (including interferon, colony-stimulating factor, and interleukin 1) from monocytes when compared to standard human serum medium formulations.[4] One notable exception in which monocytes appear to function distinctly when in suspension culture, as compared to adherent cultures, is antibody-dependent cellular cytotoxicity (ADCC) assays. Monocytes cultured in polystyrene plates maintained a higher degree of cytotoxicity in an ADCC assay against human red blood cells (as described in Poplack *et al.*[10]) when compared to monocytes cultured in serum-free media in Teflon culture vessels. The mechanism for the enhanced function in the ADCC assay of monocytes cultured in polystyrene is as yet unknown but may relate to known "spreading" of monocytes on polystyrene plates, thus exposing more membrane surface for participation in the ADCC reaction.

Suspension culture techniques of human monocytes may have clinical applications. As adoptive immunotherapy protocols are being developed for cancer patients,[11] the possibility of treating patients with their own activated tumoricidal monocytes has become a reality. In this regard it is quite clear that monocytes cultured on polystyrene labware are quite inadequate for two reasons: (1) since they are adhered to the polystyrene plastic they must be removed with noxious stimuli which could adversely affect their function and certainly make them too "clumpy" to consider

[10] D. G. Poplack, G. D. Bonnard, B. J. Holiman, and R. M. Blaese, *Blood* **48,** 809 (1976).
[11] H. C. Stevenson, K. A. Foon, D. J. Kanapa, T. Favilla, J. Beman, and R. K. Oldham, *Plasma Ther.* **5,** 237 (1984).

reinfusing into the patient. (2) The amount of time required to remove cells adhered onto plastic plates would make the management of the hundreds of millions of monocytes required for an *in vivo* clinical effect virtually impossible. In contrast, suspension culture of monocytes for clinical purposes is quite straightforward. The monocytes are not adherent and thus are not clumpy. They are cultured in human clinical grade reagents in the absence of antibiotics and thus do not contain sensitizing substances (potentially causing the development of allergic reactions) for patients. Finally, since the cells are cultured in suspension, it is very easy to harvest many hundreds of millions of monocytes in a period of a few minutes for preparation for reinfusion into patients.

Conclusion

The human monocyte suspension culture technique described is a precisely controlled *in vitro* culture system for maintaining monocytes in a state which most clearly mimics their *in vivo* condition. This system is the only technique currently available which meets the criteria of (1) high cell purity, (2) suspension culture state, (3) use of autologous serum or serum-free medium, (4) preservation of cell viability and function, (5) capacity to handle large numbers of cells, and (6) nonrequirement of antibiotics.

[10] Preparation and Cryopreservation of Cytoplasts from Human Phagocytes

By DIRK ROOS and ALWIN A. VOETMAN

Introduction

Intact phagocytes can be activated by soluble or particulate stimuli to generate reactive oxygen products and arachidonic acid metabolites. Although it is possible to prepare from activated phagocytes so-called podosomes (small plasma membrane vesicles) that continue to show the characteristics of activated cells,[1] it is impossible to activate podosomes prepared from resting cells. Apparently, podosomes made from resting phagocytes do not contain all the components necessary for the (activation of the) cell's synthesizing machinery. For this reason we have adapted to neutrophilic granulocytes a procedure described to enucleate

---

[1] H. J. Cohen, M. E. Chovaniec, and W. A. Davies, *Blood* **55**, 355 (1980).

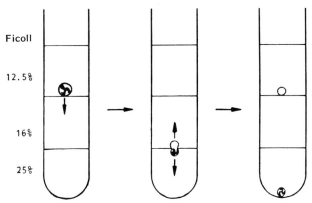

FIG. 1. Formation of cytoplasts and karyoplasts from neutrophils. The neutrophils are centrifuged to the surface of the 250 g/liter Ficoll layer. Since the specific gravity of the cells (1.073–1.084 g/cm³) is less than that of this Ficoll layer (1.0855 g/cm³), the cells cannot penetrate this layer. However, the nucleus and granules in the cell have specific gravities higher than 1.0855 g/cm³, and will therefore try to enter this Ficoll layer. At high temperatures (33°) and in the presence of cytochalasin B, the centrifugal force (81,000 g) will cause cell partitioning into cytoplasts (a plasma membrane vesicle filled with cytoplasm) and karyoplasts (a plasma membrane vesicle filled with nucleus, granules, and some cytoplasm). This process involves fusion of the plasma membrane to form a cytoplast and a karyoplast, without spilling of cell contents into the gradient. Due to the differences in specific densities, the cytoplasts float up to the interface between the 125 g/liter and the 160 g/liter Ficoll layer, whereas the karyoplasts are sedimented to the bottom of the tubes.

mouse L cells.[2] With this technique we have prepared cytoplasts that are devoid not only of nuclei but also of granules and mitochondria. Nevertheless, the cytoplasts can be stimulated to a number of functions that have until now only been observed with intact cells. Therefore, cytoplasts are ideal tools to study these functions in more detail, without interference of irrelevant cell structures. Moreover, neutrophil cytoplasts, in contrast to intact neutrophils, can be cryopreserved with complete maintenance of functional properties.

Methods

Principle of Enucleation

In a density gradient, cells will move until they reach a layer with a specific gravity higher than that of their own specific gravity. As depicted in Fig. 1, human neutrophils (1.073–1.084 g/cm³) in a Ficoll gradient with

[2] M. H. Wigler and I. B. Weinstein, *Biochem. Biophys. Res. Commun.* **63**, 669 (1975).

SPECIFIC GRAVITIES AND REFRACTIVE INDICES OF
FICOLL SOLUTIONS

Ficoll-70 solution (g/liter)	Specific gravity at 25° (g/cm³)	Refractive index at 25°
350	1.1198	1.385
250	1.0855	1.369
160	1.0578	1.356
125	1.0477	1.352

layers of 125, 160, and 250 g/liter will collect at the surface of the 250 g/liter layer (1.0855 g/cm³). However, the nucleus and granules in the neutrophils have specific gravities higher than 1.0855 g/cm³ and will therefore try to enter this Ficoll layer. At temperatures above the transition temperature of the lipids in the neutrophil plasma membrane (27°), and in the presence of the microfilament-destroying agent cytochalasin B, a high centrifugal force will cause partitioning of each neutrophil into a cytoplast (a plasma membrane vesicle filled with cytoplasm) and a karyoplast (a plasma membrane vesicle filled with nucleus, granules, and some cytoplasm). This process involves fusion of the plasma membrane to form a cytoplast and a karyoplast, without spilling of cell contents into the gradient. Due to the differences in specific densities, the cytoplasts float up to the interface between the 125 g/liter and the 160 g/liter Ficoll layer, whereas the karyoplasts are sedimented to the bottom of the tubes (Fig. 1).

Materials

Ficoll-70 (Pharmacia Fine Chemicals, Uppsala, Sweden) is dissolved to 350 g/liter in 25 mM Tris–HCl, pH 7.4. This solution is 400–500 mOsm. The solution is filter sterilized and stored at 4°. From this stock solution, dilutions are made just before use with phosphate-buffered saline (140 mM NaCl, 9.2 mM Na_2HPO_4, 1.3 mM NaH_2PO_4; pH 7.2) to Ficoll concentrations of 250 g/liter, 160 g/liter, and 125 g/liter. The osmolalities of these dilutions should be 280–300 mOsm. The specific gravities can be measured either directly with a densitometer or indirectly with a refractometer. These values should be as shown in the table.

To each of the dilutions, cytochalasin B (Sigma Chemical Company, St. Louis, MO) is added to a final concentration of 20 μM. For this purpose, a 20 mM stock solution of cytochalasin B is made by dissolving 10 mg of cytochalasin B in 1 ml of DMSO. Of this stock solution, 1 μl is added with a microsyringe for every milliliter of Ficoll solution.

Discontinuous Ficoll density gradients are made in polycarbonate ultracentrifuge tubes (2.5 × 8.9 cm; Beckman Instruments, Inc., Palo Alto, CA) by layering 4.5 ml of Ficoll (160 g/liter)–cytochalasin B carefully on top of 4.5 ml of Ficoll (250 g/liter)–cytochalasin B. These gradients are prewarmed for 2 hr at 37°.

Procedures

Cytoplast preparation: Neutrophilic granulocytes are purified from human blood as described in this volume [8]. The cell pellet is suspended to about 10^8 cells/ml in Ficoll (125 g/liter)–cytochalasin B. The cells are then preincubated for 5 min at 37°. Next, 4.5 ml of the cell suspension is layered carefully on top of each Ficoll density gradient (see Materials). Care should be taken not to overload the gradients: to each centrifuge tube not more than $6 × 10^8$ cells should be added. Immediately after loading the gradients, the tubes are centrifuged for 20 min at 81,000 g (middle of the tubes) and 33°, in a rotor prewarmed for 4 hr at 37°. If the temperature of the ultracentrifuge cannot be raised to 33° during the run, hot water may be added to the centrifuge tubes that do not contain gradients.

Immediately after centrifugation, the upper band of cellular material (between the Ficoll layer of 125 g/liter and that of 160 g/liter) is collected with a Pasteur pipet. Microscopically, this band consists for >99% of cytoplasts (Fig. 1). The cytoplast suspension is diluted 4-fold and washed several times with 0.5% human albumin in a physiologic medium to remove the cytochalasin B (centrifugation: 600 g, 10 min, room temperature). If the karyoplasts are needed for further experiments, the Ficoll gradients should contain an additional cushion of dense material (e.g., Ficoll of 350 g/liter) at the bottom of the centrifuge tubes, to prevent damage to the karyoplasts during centrifugation.

If cytoplasts are needed from cell types other than neutrophils, the Ficoll gradients should be adapted. The principal objective of the gradients is to provide a layer in which the intact cells cannot penetrate due to their lower specific density, whereas the karyoplasts can (see Fig. 1). For the preparation of cytoplasts from eosinophils, we have successfully used Ficoll layers of 125, 160, and 300 g/liter (unpublished observations). For lymphocytes a gradient of 125, 150, 160, 170, and 250 g of Ficoll/liter has been described.[3] For monocytes, layers of 100, 150, and 200 g/liter should be sufficient.

[3] J. Ohara, T. Sekiguchi, and T. Watanabe, *J. Immunol. Methods* **45,** 239 (1981).

Properties of Neutrophil Cytoplasts

Neutrophil cytoplasts consist of a closed vesicle of plasma membrane, filled with cytoplasm and some endoplasmic reticulum. The volume of the cytoplasts is about one-fourth, the surface area is about one-third, and the amount of cytoplasm is about one-half of that of the original neutrophils.[4]

Neutrophil cytoplasts phagocytize and kill bacteria.[4] Phagocytosis of *Staphylococcus aureus* is as high as that observed with intact neutrophils at low (<5) ratios of bacteria to cells, but each cytoplast does not ingest more than about five bacteria. The killing of *Escherichia coli* and *Staphylococcus aureus* by cytoplasts is low but significant. Neutrophil cytoplasts do not display chemotactic activity toward casein or zymosan-activated serum, in contrast to cytokineplasts prepared from neutrophils moving in a chemotactic gradient.[5,6]

Without stimuli, neutrophil cytoplasts show little oxygen metabolism. After addition of oxidase stimuli, however, they start consuming oxygen and generating superoxide and hydrogen peroxide, while the activity of the hexose monophosphate shunt increases. This activation has been observed with serum-opsonized zymosan particles, phorbol myristate acetate, formylmethionylleucylphenylalanine, concanavalin A, leukotriene B_4 and ionomycin, but not with complement component C5a or calcium ionophore A23187.[4,7-10] Calculated per unit of surface area, the activity of cytoplasts is comparable to that of intact neutrophils.[4] Cytoplasts also convert exogenous arachidonic acid via the lipoxygenase pathway.[11,12]

Neutrophil cytoplasts aggregate—as do intact neutrophils—in the presence of phorbol myristate acetate, formylmethionylleucylphenylalanine, concanavalin A, and leukotriene B_4.[7] Antibodies to neutrophil surface antigens react also with neutrophil cytoplasts.[4] This has been tested with a large number of different antibodies, both monoclonal and

[4] D. Roos, A. A. Voetman, and L. J. Meerhof, *J. Cell Biol.* **97**, 368 (1983).
[5] D. E. Dyett, S. E. Malawista, G. van Blaricom, D. A. Melnick, and H. L. Maleck, *J. Immunol.* **135**, 2090 (1985).
[6] D. E. Dyett, S. E. Malawista, P. H. Naccache, and R. I. Sha'afi, *J. Clin. Invest.* **77**, 34 (1986).
[7] H. M. Korchak, D. Roos, K. N. Giedd, E. M. Wynkoop, K. Vienne, L. E. Rutherford, J. P. Buyon, A. M. Rich, and G. Weissmann, *Proc. Natl. Acad. Sci. U.S.A.* **80**, 4968 (1983).
[8] A. A. Voetman, A. A. M. Bot, and D. Roos, *Blood* **63**, 234 (1984).
[9] R. Gennaro, T. Pozzan, and D. Romeo, *Proc. Natl. Acad. Sci. U.S.A.* **81**, 1416 (1984).
[10] Y. Ohno, B. E. Seligmann, and J. I. Gallin, *J. Biol. Chem.* **260**, 2409 (1985).
[11] S. C. Olson, J. T. O'Flaherty, and R. L. Wykle, *Fed. Am. Soc. Exp. Biol.* **43**, 1461A (1984).
[12] K. A. Haines, K. Giedd, and G. Weissmann, *Biochem. J.* **233**, 583 (1986).

polyclonal. Because neutrophil cytoplasts seem to contain more cytochrome *b*-558 and Mo-1 antigen in the plasma membrane than do intact neutrophils, it has been claimed that some degranulation may take place during cytoplast preparation.[13,14]

Cryopreservation of Neutrophil Cytoplasts

Intact neutrophils cannot be cryopreserved with maintenance of functional properties. For the cryopreservation of neutrophil cytoplasts, a simple procedure with DMSO can be followed.[8]

Neutrophil cytoplasts are suspended to 150×10^6/ml in a physiological medium with 10% (v/v) fetal calf serum (FCS). After the suspension has been cooled to 0°, an equal volume of ice-cold medium containing 10% (v/v) FCS and 20% (v/v) DMSO is added dropwise (about two drops/sec) to a cytoplast suspension with swirling. Portions of 2 ml of the resulting suspension are quickly transferred to 2.5-ml glass ampoules, which are then sealed in a flame and placed in a polystyrene insulating box with 10-mm-thick walls. The box is immediately placed in a freezer at $-70°$. After about 1 hr, the ampoules can be transferred to a storage tank with liquid nitrogen. Alternatively, the ampoules can also be cooled with nitrogen vapor in a programmed cooling device for the cryopreservation of lymphocytes.[15]

Thawing of the neutrophil cytoplasts is performed by gently shaking the ampoules in a water bath at 37°. When the ice in the ampoules has almost disappeared, the ampoules are placed in an ice bath. Directly thereafter, the ampoules are opened, and the content of two ampoules is transferred to a 50-ml polypropylene tube. The cytoplast suspension is then carefully diluted to 40 ml with 0.5% human albumin in ice-cold medium. This dilution should be logarithmic: first 1 drop is added while the tube is shaken in ice water, then 2 drops and so on, up to 16 drops; thereafter, 1 dash of about 1 ml, 2 dashes, and so on. The whole procedure should take about 10 min. Immediately thereafter, the cytoplasts are centrifuged (10 min, 600 *g*, 0°) and washed twice with physiological medium containing 0.5% human albumin.

With this procedure, cytoplast recoveries up to 100% are possible, with complete preservation of enzyme activities, antigenic properties, and functions (phagocytosis, oxygen consumption, hydrogen peroxide generation).[8]

[13] P. R. Petrequin, R. F. Todd, J. E. Smolen, and L. A. Boxer, *Clin. Res.* **33**, 385A (1985).
[14] F. K. Higson, L. Durbin, N. Pavlotsky, and A. I. Tauber, *J. Immunol.* **135**, 519 (1985).
[15] M. J. G. J. du Bois, P. Th. A. Schellekens, J. J. F. M. de Wit, and V. P. Eijsvoogel, *Scand. J. Immunol.* **5** (Suppl. 5), 17 (1976).

Applications of Neutrophil Cytoplasts

The unique properties of neutrophil cytoplasts open many new research areas. Thus, the lack of nucleus, granules, and mitochondria make cytoplasts excellent tools to study the importance of intracellular organelles in neutrophil functions. In this way, the possible role of lysosomal enzymes in neutrophil oxidase activity,[4,7,16,17] leukotriene B_4 production,[7,11,12] aggregation,[7,18] and cytotoxicity[19] has been examined. Because they constitute much simpler structures than whole neutrophils, cytoplasts have also been used to study the activation mechanism of neutrophils. For instance, the role of membrane depolarization,[7] changes in phospholipids,[20,21] free cytosolic calcium,[9,20,22–24] protein kinase C localization,[25] and protein phosphorylation[26] have been investigated in neutrophil cytoplasts. Also, modulation of receptor availability at the cell surface,[13,18,27] changes in the actin filament network,[28] and intracellular translocation of cytochrome b-558[10,14,29–31] have been studied with neutrophil cytoplasts. Neutrophil cytoplasts have even been used therapeutically in a rat model of neonatal sepsis.[32]

Finally, the possibility to cryopreserve cytoplasts with maintenance of functional properties enables one to study functional abnormalities of

[16] M. K. Wirtz and T. R. Green, *Clin. Res.* **33**, 357A (1985).

[17] F. Morel, J. Doussière, M.-J. Stasia, and P. V. Vignais, *Eur. J. Biochem.* **152**, 669 (1985).

[18] J. I. Gallin, J. A. Metcalf, D. Roos, B. Seligmann, and M. Friedman, *J. Immunol.* **133**, 415 (1984).

[19] G. M. Vercelotti, B. S. van Asbeck, and H. S. Jacob, *J. Clin. Invest.* **76**, 956 (1985).

[20] H. M. Korchak, K. Vienne, J. P. Buyon, C. Wilkenfeld, C. Roberts, M. C. Finkelstein, and D. Roos, *Clin. Res.* **32**, 558A (1984).

[21] E. M. Wynkoop, M. J. Broekman, H. M. Korchak, D. Roos, A. J. Marcus, and G. Weissmann, *Clin. Res.* **33**, 567A (1985).

[22] M. Torres and T. D. Coates, *Blood* **64**, 891 (1984).

[23] C. Wilkenfeld, C. Roberts, H. Korchak, K. Vienne, D. Roos, and G. Weissmann, *Clin. Res.* **32**, 326A (1984).

[24] J. E. Rickard and P. Sheterline, *Biochem. J.* **231**, 623 (1985).

[25] R. Gennaro, C. Florio, and D. Romeo, *Biochem. Biophys. Res. Commun.* **134**, 305 (1986).

[26] R. Gennaro, C. Florio, and D. Romeo, *FEBS Lett.* **180**, 185 (1985).

[27] J. O'Shea, B. Seligmann, J. I. Gallin, R. Chused, M. Berger, M. M. Frank, and E. Brown, *Fed. Am. Soc. Exp. Biol.* **43**, 1505A (1984).

[28] P. Scheterline, J. E. Rickard, and R. C. Richards, *Biochem. Soc. Trans.* **12**, 983 (1984).

[29] R. Lutter, R. van Zwieten, R. S. Weening, M. N. Hamers, and D. Roos, *J. Biol. Chem.* **259**, 9603 (1984).

[30] R. C. Garcia and A. W. Segal, *J. Biol. Chem.* **219**, 233 (1984).

[31] C. A. Parkos, C. G. Cochrane, M. Schmitt, and A. J. Jesaitis, *J. Biol. Chem.* **260**, 6541 (1985).

[32] G. Rothstein, R. D. Christensen, and T. E. Harper, *Clin. Res.* **33**, 418A (1985).

neutrophils from patients longitudinally or in one assay with those of other patients. Also, neutrophil cytoplasts can be frozen and serve as reference material or can be collected for studies that require large amounts of material from one particular donor.

[11] Assay of Phagosome–Lysosome Fusion

By MARGARET KIELIAN

Introduction

Particles ingested by cells are contained within a membrane-bounded vacuole or *phagosome*. In many instances this vacuole undergoes fusion with lysosomes, resulting in exposure of the particle to both the acidic pH[1] and hydrolytic enzymes[2] of the lysosome. The mechanisms which control the specificity, triggering, and kinetics of this and other intracellular fusion events are unclear. Also not understood is the mechanism by which a number of parasites inhibit fusion with lysosomes, in some cases resulting in their survival and multiplication within the host.[3,4] Better understanding of the effectors of the phagosome–lysosome (P–L)[4a] fusion reaction requires systems for its assay in cultured cells, and several types of assays have been described. Electron microscopy has been used to look for the appearance of lysosomal markers such as acid phosphatase or exogenously fed peroxidase, ferritin, colloidal gold, or thorium dioxide in the phagocytic vacuole. The degradation of radiolabeled particles by lysosomal hydrolases has been used as an indirect assay of P–L fusion.[5] Phagocytized particles of low density have been isolated by gradient centrifugation after ingestion, and the transfer of lysosomal enzymes into this "phagosome fraction" assayed.[6,7] An *in vitro* fusion system has also been described, in which light and electron microscopy were used to monitor the fusion of differentially labeled phagolysosome fractions.[8–10]

[1] S. Ohkuma and B. Poole, *Proc. Natl. Acad. Sci. U.S.A.* **75**, 3327 (1978).
[2] C. deDuve, *Science* **189**, 186 (1975).
[3] M. B. Goren, *Annu. Rev. Microbiol.* **31**, 507 (1977).
[4] M. A. Horwitz, *J. Exp. Med.* **158**, 2108 (1983).
[4a] Abbreviations: AO, acridine orange; FCS, fetal calf serum; HRP, horseradish peroxidase; MEM, Dulbecco's modified Eagle's medium; PBS, phosphate-buffered saline, pH 7.4; P–L fusion, phagosome–lysosome fusion.
[5] Z. A. Cohn, *J. Exp. Med.* **117**, 27 (1963).
[6] T. P. Stossel, R. J. Mason, T. D. Pollard, and M. Vaughan, *J. Clin. Invest.* **51**, 604 (1972).
[7] E. L. Pesanti and S. Axline, *J. Exp. Med.* **142**, 903 (1975).
[8] P. J. Oates and O. Touster, *J. Cell Biol.* **68**, 319 (1976).
[9] P. J. Oates and O. Touster, *J. Cell Biol.* **79**, 217 (1978).
[10] P. J. Oates and O. Touster, *J. Cell Biol.* **85**, 804 (1980).

In order to evaluate factors affecting P–L fusion in intact cells, we adapted the assay first described by Hart and Young.[11] Lysosomes are labeled with a fluorescent marker, and fluorescence microscopy is used to follow transfer of the marker into phagocytic vacuoles. This assay is simple, rapid, enables detailed and reproducible rate studies, and makes possible the analysis of P–L fusion using intact and viable cells as a test system. Cultured cells are prelabeled with acridine orange (AO), a fluorescent vial dye which by its weakly basic nature becomes concentrated primarily in the low pH environment of lysosomes.[12,13] AO is a metachromatic dye[14] which displays a brilliant orange fluorescence when concentrated and a green fluorescence when more dilute. After phagocytosis, the transfer of lysosomal fluorescence into the particle-containing vacuoles is evaluated by fluorescence microscopy, and orange-stained particles are counted as positive. Although it cannot be concluded that *no* fusion has occurred in green-stained vacuoles, the relative amount is less and is easily distinguished visually from that of the orange-stained vacuoles.

The cell used for these studies is the phagocytic macrophage, either primary cultures from mouse peritoneal cavity or a permanent macrophage-derived cell line. The choice of test particle is critical. A heat-killed, reduced, and alkylated yeast particle may be used. These particles are freely and uniformly permeable to AO (making the staining easy to evaluate), do not themselves concentrate the dye, and remain stably orange stained (the assay end point) for hours in viable cells. Other particles are less suitable for the fluorescence assay, including live yeast cells (which only show a "rim" of fluorescence and also are reported to inhibit fusion[11]), antibody-coated red blood cells (which are rapidly digested), and polystyrene latex beads (which bind AO nonspecifically). However, for some purposes these and other particles may prove useful. For example, modifications of this assay have been used to monitor the viability and characterize the fusion of various intracellular parasites.[3,15]

It is important to synchronize particle phagocytosis by the cells. The yeast particles are therefore opsonized with complement via the alternate pathway, see this volume [13], and this ligand is used to bind yeast to the macrophage complement receptor in the cold. Upon warming of the culture to 37°, rapid phagocytosis results. Under conditions of low particle/cell ratio (~2), more than 90% of the particles are ingested after 10 min at

[11] P. D. Hart and M. R. Young, *Nature (London)* **256**, 47 (1975).
[12] E. Robbins and P. I. Marcus, *J. Cell Biol.* **18**, 237 (1963).
[13] A. C. Allison and M. R. Young, *Life Sci.* **3**, 1407 (1964).
[14] M. E. Lamm and D. M. Neville, *J. Phys. Chem.* **69**, 3872 (1965).
[15] H. W. Murray and Z. A. Cohn, *J. Exp. Med.* **150**, 938 (1979).

$37°.$[16] Phagocytosis is thus essentially complete by the first time point of the fusion assay, and the rate of subsequent P–L fusion may be followed independent of the phagocytic rate.

Methods

Cells and Cell Culture

Resident peritoneal cells are prepared from female or male Nelson-Collins or Swiss Webster mice by lavaging the peritoneal cavity with 3–5 ml of sterile PBS without Ca^{2+} or Mg^{2+}.[16a] Cell suspensions are pooled, pelleted for 5 min at 1500 rpm, 4°, and resuspended in an appropriate volume of MEM containing 15–20% FCS (heat inactivated, 56°), 100 U/ml penicillin, and 100 μg/ml streptomycin. Cells are plated at densities of 6–10 × 10^6 peritoneal cells per 35-mm dish or 1.5–2 × 10^7 per 60-mm dish. One to two hours after plating, monolayers are washed with medium to remove nonadherent cells, resulting in a relatively pure primary macrophage culture.[17] For coverslip cultures to be used in the fluorescence assay, cells are resuspended to 4 × 10^6 peritoneal cells/ml and 100 μl layered onto a 12-mm glass coverslip previously sterilized by flaming in 90% ethanol. Ten coverslips can thus be conveniently placed in a 60-mm culture dish, and the cells cultured by the addition of 5 ml of medium after the 1-hr adherence and washing step. Cells are given fresh medium at least every other day, and for most experiments are used within 4 days of explant.

P388D$_1$, a macrophage-like permanent cell line[18] (obtainable from the American Type Culture Collection), is maintained in spinner culture in 10% FCS/MEM.[19] These cells are allowed to adhere on glass coverslips as described above for the fluorescence assay.

Particle Preparation for Fusion Studies

Preparation of Yeast Particles. Fresh baker's yeast is processed by the method of Lachman and Hobart.[20] Two packages of yeast (1/2 oz.) are resuspended in PBS and autoclaved for 30 min at 120°, washed with PBS until the supernatant is clear, then resuspended to 80 ml total volume in

[16] M. C. Kielian and Z. A. Cohn, *J. Cell Biol.* **85,** 754 (1980).

[16a] This series, Vol. 108 [25].

[17] Z. A. Cohn and B. Benson, *J. Exp. Med.* **121,** 153 (1965).

[18] H. S. Koren, B. S. Handwerger, and J. R. Wunderlick, *J. Immunol.* **114,** 894 (1975).

[19] J. C. Unkeless and H. N. Eisen, *J. Exp. Med.* **142,** 1520 (1975).

[20] P. J. Lachman and M. J. Hobart, *in* "Handbook of Experimental Immunology" (D. M. Weir, ed.), Vol. 1, p. 5A.1. Blackwell, Oxford, 1978.

0.1 M mercaptoethanol/PBS and stirred for 2 hr at 37° The yeast is then pelleted and washed once with PBS. It is next resuspended in 50 ml 0.85% NaCl containing 10 mM phosphate buffer, pH 7.2, and 30 mM iodoacetamide, and stirred at room temperature for 2 hr, keeping the pH adjusted to 7.2 if necessary. It is then washed three times with PBS, autoclaved in PBS 30 min at 120°, washed in PBS until the supernatant is clear, and made to a 50% (v/v) sterile stock in PBS and 0.02% NaN$_3$. This preparation can be stored indefinitely at 4°.

Opsonization of Yeast with Complement Components. Yeast is prepared as a 5% suspension in veronal buffer; 300 μl is mixed with 300 μl of mouse serum (fresh or stored at −70°) and incubated 30 min at 37°. The mixture is brought to 10 ml with cold veronal buffer and pelleted 10 min, 4°. The wash is repeated once and the opsonized yeast stored as a 1% (v/v) suspension in veronal buffer, and used within 3 days.

Veronal buffer: Made fresh daily; stocks are autoclaved separately and stored at 4°:

70 mM NaCl
2.5 mM Sodium 5,5-diethylbarbiturate, pH 7.35
2.5% Glucose
1 mM MgCl$_2$
0.15 mM CaCl$_2$

Fluorescence Assay of Phagosome–Lysosome Fusion

AO stock, 100 μg/ml in PBS, is filter sterilized and stored in the dark at 4°. It is made fresh monthly. It should be noted that AO binds to nucleic acids by intercalation and gloves should be worn when handling concentrated solutions.

1. The cells on coverslips are labeled by adding AO stock solution directly to the culture medium to give a final concentration of 5 μg/ml, and incubating for 20 min at 37°. To label a 60-mm dish containing 10 coverslips, for example, 250 μl AO stock is added to the 5 ml complete culture medium in the dish.

2. Each coverslip is then transferred with forceps to a well of a 24-well culture plate (Costar, Data Packaging, Cambridge, MA) containing 1 ml of 37° MEM/5% FCS, and incubated in the absence of AO for 10 min at 37°. This step helps to concentrate any free AO into the lysosomal compartment, decreasing background fluorescence and photodamage.

3. The medium is aspirated and 1 ml of a dilute suspension of opsonized yeast (0.004%, v/v) in PBS, 4°, is added to each well. The yeast suspension is centrifuged onto the cell monolayer by placing the culture dish in a centrifuge carrier designed for microtiter plates (Cooke Labora-

tory Products, Alexandria, VA) and centrifuging in an International centrifuge for 2 min at 1200 rpm, 4°. The opsonized yeast binds to the macrophage complement receptor; unbound particles are removed by washing each well with 1 ml of 4° MEM.

4. Each well is then given 1 ml 4° MEM/5% FCS and at time zero the entire plate is rapidly warmed to 37° by placing it in a 37° water bath inside a 37° incubator. Under these conditions, relatively synchronous phagocytosis of the bound yeast particles occurs. The average number of particles bound and ingested per cell is two.

5. Time points are taken by inverting coverslips over a drop of ice-cold PBS on a microscope slide, blotting, and rimming with nail polish. Slides are kept on ice and viewed immediately in a fluorescence microscope adjusted for fluorescein. A Zeiss photomicroscope III with a mercury lamp adjusted for epiilumination, a BG12 filter and fluorescein dichroic mirror for excitation, and a 53 barrier filter is appropriate for this purpose.

6. The intracellular yeast particles which are orange stained are counted as positive for P–L fusion; faintly green-stained particles are scored as negative. For each time point, at least 10 different microscope fields and duplicate coverslips should be examined, and a total of >200 particles counted. By using a diaphragm to control the intensity of the incident light, samples may be evaluated without extensive photodamage or cell death. A series of time points after the 10-min ingestion period enables the determination of the initial rate of P–L fusion as well as its final extent.

7. For the photography of the AO-labeled fusion samples the Zeiss photomicroscope III with an automatic exposure meter may be used. Kodak Tri-X film is used for black and white pictures and Kodak Ektachrome film, ASA 200, for color slides. Good, clear photographs require exposing the cells to the exciting light for as short a time as possible. Routinely, a field is located under reduced light, the picture is taken using the spot light meter of the microscope with a labeled cell centered in the spot, and the ASA setting of the camera is increased to 800 or even 1600 ASA. These factors contribute to decrease the exposure time of the film to less than 1 sec.

Figure 1 is a fluorescence picture of typical AO fusion assays in 1- and 4-day cultures. Punctate lysosomal staining occurs in the absence of extensive cytoplasmic or nuclear background. Both positive, brightly orange fluorescent yeast and negative, faintly green fluorescent yeast are observed. When counts of positive vs negative intracellular particles are made at varying times after ingestion, the results shown in Fig. 2 are typically obtained. The fusion rate is strongly affected by *in vitro* culture

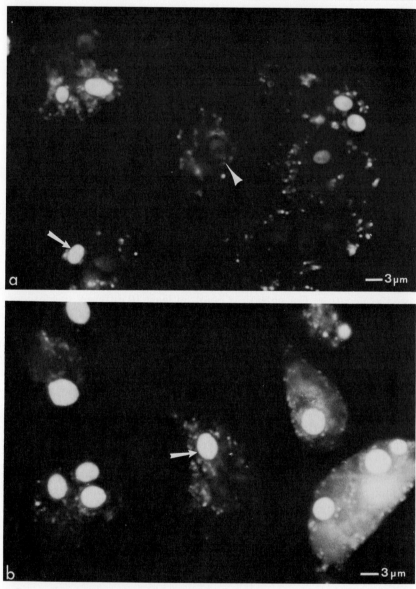

FIG. 1. Fluorescence micrographs of AO-fusion assays in 1- and 4-day-old cultures. (a) One-day cells 160 min after yeast ingestion. Both positively (arrow) and negatively (arrowhead) stained yeast are seen, as well as abundant punctate lysosomal staining. ×1500. (b) Four-day cells 40 min after ingestion. Abundant lysosomal staining is observed in cells containing several positively stained yeast. ×1598. From Kielian and Cohn.[16] Reproduced from *The Journal of Cell Biology*, by copyright permission of the Rockefeller University Press.

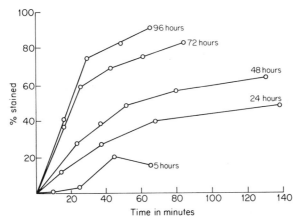

FIG. 2. Rate and extent of P–L fusion in mouse macrophages cultured for varying periods of time. The percent of intracellular particles positively stained is plotted vs time after particle ingestion. Length of cell culture is shown in hours. Average results from three separate experiments. From Kielian and Cohn.[16] Reproduced from *The Journal of Cell Biology*, by copyright permission of the Rockefeller University Press.

of mouse primary macrophages, being about 8-fold higher initially in 4-day cells than in cells cultured for 5 hr.[16] If 1-day cells with intracellular particles are cultured overnight, 90% of the particles are positively stained, implying that P–L fusion continues at this slow rate. When P–L fusion is examined in the P388D$_1$ macrophage line, the rate is similar to that of a 4-day primary culture.

Important Conditions for the Fluorescence Assay of P–L Fusion

Besides the considerations of choice of test particle, opsonization, etc., described above, it is important to note that the fluorescence assay can only be used under certain conditions. Since AO uptake is dependent on low intralysosomal pH, conditions which modify lysosomal pH or quench AO fluorescence interfere with the assay. Lysosomotropic drugs such as ammonium chloride and chloroquine, or metabolic inhibitors such as 2-deoxyglucose plus sodium azide all act to increase lysosomal pH[1] and also significantly decrease the lysosomal uptake of AO.[16]

When comparing P–L fusion in different cell populations in which the AO technique might be affected, it is important to quantitate the AO uptake by the cells to assure that comparable amounts of the fluorescent marker are concentrated. In addition, another P–L fusion assay should be used as a confirmatory assay when necessary. Electron microscopy, while less amenable to multiple time points, is unlikely to be affected by

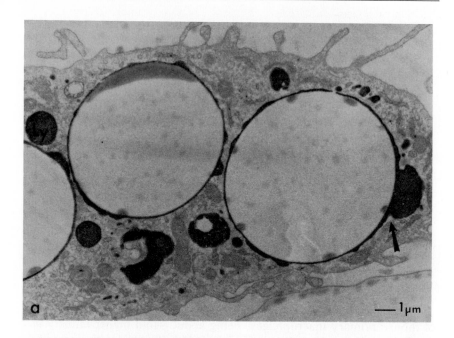

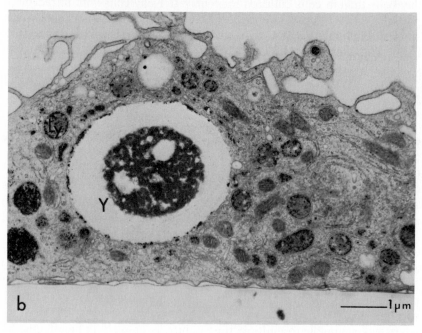

altered lysosomal pH. Lucifer yellow CH (Sigma Chemical Company), a recently described fluorescent marker which is taken up via pinocytosis,[21] might also be a useful control.

Quantitation of Acridine Orange Uptake. Coverslip cultures are labeled with AO as described for the fluorescence assay, washed in PBS, and extracted in 2 ml 95% ethanol. The fluorescence is read on an MPF-44 fluorometer (Perkin-Elmer Corp., Norwalk, CT) using an excitation wavelength of 490 nm, and recording emission at 520 nm. A standard curve of known AO concentrations in the same solvent is used. This curve is not affected by the addition of unlabeled cell extracts. AO concentrations are normalized for cell protein as determined with the fluorescamine protein assay[22] on parallel coverslips. Typical uptake by mouse macrophages is ~50 pmol AO/μg cell protein.[16]

Electron Microscopic Assay of P–L Fusion. To mark secondary lysosomes, cells are pulsed with either HRP or colloidal thorium dioxide (thorotrast, Fellows Testagar Div., Fellows Mfg. Co., Inc.) in 35-mm culture dishes, and the transfer of marker to phagosomes evaluated by electron microscopy.

1. HRP specimens: To label, cultures are allowed to pinocytose 2 mg/ml HRP in MEM for 2–3 hr at 37°, washed four times with MEM, and recultured in HRP-free medium for 20–60 min to allow transit of this fluid-phase marker to lysosomes. Cells can then be given yeast or latex particles and fixed 1 hr. after ingestion. After fixation, but before osmication, HRP is visualized by staining for 10 min at room temperature with freshly made DAB reagent [5 mg diaminobenzidine (Sigma Chemical Company), 10 ml 0.1 M Tris, pH 7.6, and 3.3 μl of 30% H_2O_2].[23] Specimens are processed for electron microscopy by standard techniques,[16] and thin sections examined without uranyl acetate and lead citrate staining. Under these conditions, positive fusion is indicated by a "rim" of electron-dense HRP

[21] D. K. Miller, E. Griffiths, J. Lenard, and R. A. Firestone, *J. Cell Biol.* **97**, 1841 (1983).
[22] P. Bohlen, S. Stein, W. Dairman, and S. Udenfriend, *Arch. Biochem. Biophys.* **155**, 213 (1973).
[23] R. C. Graham, Jr., and M. J. Karnovsky, *J. Histochem. Cytochem.* **14**, 291 (1966).

FIG. 3. Electron microscopic evaluation of P–L fusion using electron-dense lysosomal markers. (a) HRP assay in 2-day cultures given latex particles. Reaction product is seen in secondary lysosomes (Ly), and as a rim around latex vacoules. Much of the latex is dissolved by the propylene oxide used in specimen processing. ×4982. (b) Four-day cells prelabeled with thorotrast and given opsonized yeast. Thorotrast is seen in secondary lysosomes (Ly), and distributed linearly around an ingested yeast (Y). ×12,300. From Kielian and Cohn.[16] Reproduced from *The Journal of Cell Biology*, by copyright permission of the Rockefeller University Press.

MODULATORS OF P–L FUSION[a]

Fusion is unaffected by:	Number of particles phagocytosed
	Particle size
	Lysosome size
	Prior uptake of digestible or nondigestible solutes or particles[1]
	Enzyme pretreatment of the cell surface (trypsin, neuramini-dase, pronase, chymotrypsin)
	Cell surface cross-linking (concanavalin A)
	Particle surface cross-linking (concanavalin A or antibody)
	Cytoskeletal drugs (colchicine, cytochalasin B or D)[2]
	Increased lysosomal pH (via NH_4Cl or chloroquine)[3]
Fusion is enhanced by:	Increasing time of cell culture[1]
	In vivo macrophage activation[2]
	Phorbol myristate acetate pretreatment[4]
Fusion is inhibited by:	Incubation at temperatures $<37°$ ($Q_{10} = 2.5$, no fusion seen below 15°)[1]
	Lysosomal uptake of polyanions [dextran sulfate, suramin, poly(D-glutamate), heparin, chondroitin sulfate][3,5,6]

[a] Key to references: (1) M. C. Kielian and Z. A. Cohn, *J. Cell Biol.* **85,** 754 (1980); (2) M. C. Kielian and Z. A. Cohn, *J. Exp. Med.* **153,** 1015 (1981); (3) M. C. Kielian, R. M. Steinman, and Z. A. Cohn, *J. Cell Biol.* **93,** 866 (1982); (4) M. C. Kielian and Z. A. Cohn, *J. Exp. Med.* **154,** 101 (1981); (5) P. D. Hart and M. R. Young, *Nature (London)* **256,** 47 (1975); (6) P. D. Hart and M. R. Young, *Exp. Cell Res.* **118,** 365 (1979).

reaction product around the test particle (see Fig. 3a). This marker is easy to visualize and assays can be scored on the electron microscope without photographing. However, HRP is degraded by lysosomes and thus labeling must be performed just prior to particle ingestion.

2. Thorotrast specimens: Thorotrast, or colloidal thorium dioxide, is a particulate, electron-dense tracer which is also taken up by fluid-phase pinocytosis. Thorotrast is not degraded, and thus cells can be labeled at equivalent times in culture when pinocytic uptake should be comparable, and then further cultured under various experimental conditions. To label, cell monolayers are washed 2 hr after explant, allowed to pinocytose thorotrast in medium (1/100, v/v) for 12 hr, washed four times with MEM, and recultured for 1–4 days. Again, fusion is evaluated 1 hr after particle ingestion.

Thorotrast is present as discrete particles around the edge of the phagocytic vacuole, and within secondary lysosomes (Fig. 3b). The extent of fusion is determined by stereological analysis, to take the particulate nature of the marker into account. Micrographs at the same final magnification are projected through a 3× enlarger onto a grid of 1-cm squares. The number of times horizontal and vertical lines cross a phagocytic vacuole membrane vs the number of times lines cross thorotrast in the vacuole are scored. For each determination, from 25–60 vacuoles are scored, and total line crossing of 600–1500 are obtained. Results are expressed as the percentage of total possible crossings which are thorotrast positive. Colloidal gold of small particle size should in similar fashion be useable as a marker.

Conclusions

Using these assays, a number of treatments of the cell are found to affect P–L fusion using yeast as a phagocytic particle. The conditions which we have tested are summarized in the table. It appears that alterations in vacuole size, number, pH, or luminal membrane surface do not affect P–L fusion. Culture time, *in vivo* macrophage activation, and phorbol myristate acetate treatment all dramatically increase P–L fusion, while incubation temperatures <15° and lysosomal accumulation of polyanions such as dextran sulfate block fusion.

We found that our experimental results from the AO, HRP, and thorotrast assays were in relative agreement, in spite of the use of different markers, stereological analysis, and thin sections vs whole cell preparations.

Acknowledgment

The author would like to thank Dr. Zanvil Cohn of the Rockefeller University with whom this work was performed.

[12] Methods for Assessing Exocytosis by Neutrophil Leukocytes

By BEATRICE DEWALD and MARCO BAGGIOLINI

Introduction

Exocytosis

Exocytosis is the active release of preformed material, present in cytoplasmic storage organelles, into the extracellular space. It is a selective process which depends on the fusion of the membrane of the storage organelles with the plasma membrane, and which occurs without loss of cytosolic proteins. The release of storage macromolecules is of major importance for neutrophil function. Lytic enzymes, proteinases in particular, are required for intracellular digestion in phagosomes and, presumably, for diapedesis and chemotaxis. The same enzymes are involved in tissue damage at sites of inflammation. The assessment of exocytosis and of compounds or conditions that influence this process is therefore of interest to many investigators.

The methods described here apply to human neutrophils. Most of them, however, may be used, with minor adjustments, for neutrophils of other species and other granulocytes.

The Storage Compartments

Human neutrophils contain three types of storage organelles which discharge their content by exocytosis, the azurophil or primary granules, the specific or secondary granules, and smaller secretory vesicles.[1] Neutrophils from other sources are similarly equipped. Azurophil and specific granules have been identified in most species.[2] Smaller storage organelles have also been described, but so far secretory vesicles were demonstrated only in human neutrophils.[3] Ruminant neutrophils appear to constitute a special case; their major storage organelle is a large, peroxidase-negative granule distinct from the specific and azurophil granules.[4] For all neutrophils, the storage organelles are produced during the period of maturation in the bone marrow. The mature cells which are released into the circulation, and which usually are the object of experimental studies, have lost the capacity to form granules and are unable to replace the enzymes and other proteins which they release.

Markers of Exocytosis

Azurophil granules contain a wide variety of acid hydrolases, including N-acetyl-β-glucosaminidase, β-glucuronidase, α-mannosidase, cathepsin B and D, and β-glycerophosphatase, in addition to myeloperoxidase and two neutral proteinases, elastase and cathepsin G. For a number of reasons, e.g., easy assay conditions, stability, and no tendency to adhere to surfaces and cells, β-glucuronidase is the marker we prefer. Elastase can be used to estimate initial rates of azurophil granule marker release by the kinetic assay with methylsuccinylalanylalanylprolylva-

[1] M. Baggiolini and B. Dewald, *Contemp. Top. Immunobiol.* **14**, 221 (1984).
[2] M. Baggiolini, in "Handbook of Inflammation" (G. Weissmann, ed.), Vol. 2, p. 163. Elsevier, Amsterdam, 1980.
[3] B. Dewald, U. Bretz, and M. Baggiolini, *J. Clin. Invest.* **70**, 518 (1982).
[4] R. Gennaro, B. Dewald, U. Horisberger, H. U. Gubler, and M. Baggiolini, *J. Cell Biol.* **96**, 1651 (1983).

lylmethylcoumarin amide as substrate.[5] However, it has the disadvantage that the cell lysate contains an inhibitor which precludes the ready determination of the total cellular activity. Exocytosis from the specific granules can be best assessed by assaying for vitamin B_{12}-binding protein. Lactoferrin, another exclusive constituent of the specific granules has also been successfully used as a marker of release. It is assayed by immunodiffusion[6] or ELISA[7] methods, which in our opinion are less reliable than the assay of vitamin B_{12}-binding protein. Lysozyme is present in both azurophil and specific granules and is therefore not a useful marker for release from single compartments of the neutrophil. Gelatinase is exclusively contained in small storage organelles and is therefore the marker of choice for this compartment. Subcellular fractionation has shown that the fractions containing these vesicles are morphologically heterogenous and also contain minor amounts of acid hydrolases.

Methods

Materials

The suppliers used were Boehringer GmbH, Mannheim, FRG, for NADH; Cilag AG, Schaffhausen, Switzerland, for Ronpacon 440; Fluka AG, Buchs, Switzerland, for diisopropyl phosphofluoridate (DFP)[7a]; Merck AG, Darmstadt, FRG, for o-dianisidine, perhydrol (hydrogen peroxide), and sodium pyruvate; New England Nuclear, Dreieich, FRG, for [^{3}H]acetic anhydride, 50 mCi/mmol; Nyegaard and Company, Oslo, Norway, for Lympho-paque; Pharmacia AG, Uppsala, Sweden, for Ficoll, Ficoll-Paque, and Macrodex; The Radiochemical Centre, Amersham, GB, for cyano[^{57}Co]cobalamin, 10–20 μCi/μg; Serva GmbH, Heidelberg, FRG, for cytochalasin B and Norit A charcoal; Sigma Chemical Corp., St. Louis, MO, for A23187, aminophenylmercuric acetate, 4-methylumbelliferyl-β-D-glucuronide, 4-methylumbelliferyl-N-acetyl-β-D-glucosaminide, phorbol 12-myristate 13-acetate, and zymosan.

[5] L. A. Sklar, V. M. McNeil, A. J. Jesaitis, R. G. Painter, and C. G. Cochrane, J. Biol. Chem. 257, 5471 (1982).
[6] M. Baggiolini, C. de Duve, P. L. Masson, and J. F. Heremans, J. Exp. Med. 131, 559 (1970).
[7] H. S. Birgens, Scand. J. Haematol. 34, 326 (1985).
[7a] Abbreviations: ACD, acid–citrate–dextrose; EDTA, ethylenediaminetetraacetic acid; SDS, sodium dodecyl sulfate; DFP, diisopropyl phosphofluoridate; BSA, bovine serum albumin; APMA, 4-aminophenylmercuric acetate; TCA, trichloroacetic acid; PBS, phosphate-buffered saline; DMSO, dimethyl sulfoxide; fMLP, N-formyl-L-methionyl-L-leucyl-phenylalanine; PMA, phorbol myristate acetate.

Preparation of Neutrophils from Freshly Drawn Blood

The method described by Bøyum[8,9] is the preferred procedure for isolating neutrophils from freshly drawn blood. Blood (150 ml) is collected in a flask containing 30 ml of acid–citrate–dextrose (ACD) anticoagulant (0.20 M sodium citrate, 0.14 M citric acid, 0.22 M glucose), and is then gently mixed with 36 ml of 6% dextran 70 in 0.9% NaCl (e.g., Macrodex). Care should be taken to avoid foaming. The mixture is transferred into a narrow plastic or siliconized glass cylinder to allow sedimentation of the erythrocytes during 45–60 min at room temperature. The turbid upper portion of the suspension, which contains the leukocytes, is then aspirated and centrifuged at 400 g for 8 min at room temperature.[10] The cell pellet is resuspended with 24 ml of the supernatant containing 0.1% EDTA. Four gradients are prepared by layering 6 ml of this suspension on 3 ml of a Ficoll-Hypaque mixture consisting of 30 ml of 33.9% Hypaque (20 ml Ronpacon 440 diluted with 24 ml of water) and 72 ml of an aqueous solution of 9% Ficoll 400. Commercially available mixtures, e.g., Lympho-Paque or Ficoll-Paque, may be used instead. The gradients are centrifuged at 800 g for 20 min at 20° with the brake off. The upper layer and the mononuclear cells which collect at the interface between the two layers are removed. The pellet containing the neutrophils is taken up in saline. The four suspensions are combined, brought to 30 ml with saline, and centrifuged at 300 g for 5 min at 40°. Contaminating erythrocytes are removed by hypotonic lysis. The cell pellet is resuspended in 9 ml of ice-cold distilled water by gentle agitation on a Vortex mixer, and after 30 sec isotonicity is restored by adding 3 ml of ice-cold 0.6 M potassium chloride. The cells are centrifuged at 300 g for 5 min at 4°. The hypotonic treatment may be repeated once. The neutrophils are then washed in saline (300 g for 5 min at 4°) and finally resuspended in saline at a density of 10^8 cells/ml. The average purity is about 95%. The mean values ± SD obtained in 31 recent preparations were 94.5 ± 2.7% neutrophils and 4.7 ± 2.5% eosinophils.

Preparation of Neutrophils from Buffy Coats

Buffy coats are usually obtained from blood banks or transfusion centers. They are prepared by centrifuging anticoagulated blood (up to 500 ml) in glass bottles or plastic bags at 2600 g for 10 min at 20° with the brake off. The plasma supernatant is aspirated or pressed out of the bag and the buffy coat, i.e., the upper cell layer (50–60 ml) is collected. It contains the

[8] A. Bøyum, Scand. J. Clin. Lab. Invest. **21,** Suppl. 97, 31 (1968).
[9] A. Bøyum, Scand. J. Immunol. **5,** Suppl. 5 (1976).
[10] g_{max} values, calculated for the radial distance of the tube bottom, are given throughout.

white cells and platelets, many erythrocytes, and some residual plasma. The buffy coat is centrifuged at 800 g for 5 min at 20°. A platelet-containing supernatant, which is discarded, and a sediment are obtained. Eight milliliters of the top layer of this sediment is collected with a Pasteur pipet and diluted to 30 ml with Ca^{2+},Mg^{2+}-free PBS containing 300 U heparin/ml. Two 15-ml portions are transferred into 50-ml centrifuge tubes on top of a 10-ml layer of Lympho-Paque to which 0.58 ml of 3.2% sodium citrate had been added. The tubes are centrifuged at 500 g for 20 min at 20° with the brake off. The supernatant, the mononuclear cells at the interface, and the layer above the pellet are carefully withdrawn. Care is taken to remove the mononuclear cells as completely as possible including those sticking to the wall of the tube. The cell pellet containing neutrophils and erythrocytes is resuspended and washed in saline (300 g for 5 min at 4°). The erythrocytes are then lysed with ammonium chloride. The cells are resuspended in 40 ml of an ice-cold solution containing 155 mM NH_4Cl and 10 mM EDTA, pH 7.4. The tubes are kept on ice for 10 min with occasional mixing and are then centrifuged at 300 g for 5 min at 4°. The pellet is resuspended in saline and washed once (300 g for 5 min at 4°). If erythrocyte lysis appears incomplete, the treatment with ammonium chloride is repeated. The neutrophils are again washed and finally suspended in saline at a density of 10^8 cells/ml. The average purity of neutrophils prepared in this manner is 90% or higher.

Biochemical Assays

Gelatinase. Gelatinolytic activity is determined by a modification of the method of Harris and Krane,[11] using ^{3}H-acetylated gelatin as substrate.

Preparation of substrate: Gelatin is obtained by heat denaturation of rat skin collagen, and is acetylated with [^{3}H]acetic anhydride.

A typical preparation is described. Fifteen male rats, about 4 weeks old, are killed with ether. The skin is shaved, removed, freed from adhering subcutaneous tissue, cooled on ice, and finely minced with scissors. All subsequent steps are performed at 4–6°. The minced skin is suspended in 2.5 vol of 0.05 M Tris–HCl, pH 7.4, containing 1 M NaCl and 10 mM EDTA (buffer I) and kept for 24 hr under occasional stirring. A second portion of 2.5 vol of buffer I is then added and the suspension is kept for another 24 hr. The extract is subsequently filtered through cheesecloth, diluted 2-fold with buffer I, and cooled to 0° in an ice bath. Solid ammonium sulfate (32.1 g/100 ml yielding 45% saturation) is then slowly added while stirring with a glass rod. The mixture is allowed to stand at 4°

[11] E. D. Harris, Jr., and S. M. Krane, *Biochim. Biophys. Acta* **258**, 566 (1972).

overnight, and is then centrifuged at 100,000 g for 35 min. The supernatant is discarded and the pellet is resuspended in 150 ml 0.05 M Tris–HCl, pH 7.4, containing 0.5 M NaCl and 10 mM EDTA (buffer II), stirred overnight, and then dialyzed against four changes of 7 liters buffer II. A viscous yellowish suspension is obtained, which is diluted 1.5-fold with buffer II, and centrifuged at 100,000 g for 35 min. The supernatant is collected, and the soluble collagen precipitated with salt. To 250 ml of the supernatant 167 ml (two-thirds vol) of 4.25 M NaCl is slowly added under stirring. After standing overnight at 4°, the salt concentration is increased to 3 M by addition of 190 ml 4.25 M NaCl in small portions. After 24 hr at 4°, the mixture is centrifuged at 100,000 g for 35 min. The supernatant is discarded and the pellet resuspended in 80 ml of 0.05% acetic acid and stirred for 24 hr at 4°. The suspension is then centrifuged at 100,000 g for 35 min. The supernatant is dialyzed against two changes of 5 liters of 0.05 M Tris–HCl, pH 7.5, containing 0.5 M NaCl and 10 mM EDTA (buffer III), and the pellet is saved for reextraction. The volume of the dialyzed supernatant is brought to 140 ml and solid NaCl (14.6 g/100 ml) is slowly added under stirring to precipitate the collagen a second time. After 6 hr at 4° the mixture is centrifuged at 100,000 g for 35 min. The supernatant is discarded and the pellet dissolved in 80 ml 0.05% acetic acid. The solution is dialyzed against five changes of 5 liters of 0.05% acetic acid and finally lyophilized; 94.4 mg of collagen is obtained. The pellet saved for reextraction is stirred overnight with 200 ml 0.05% acetic acid at 4° and then centrifuged at 100,000 g for 35 min. The supernatant is dialyzed against three changes of 5 liters of buffer III. A slight precipitate that forms during dialysis is removed by centrifugation at 100,000 g for 35 min. To the clear solution solid NaCl (14.6 g/100 ml) is slowly added, the mixture is stirred for 6 hr at 4°, and then centrifuged at 100,000 g for 35 min. The pellet is dissolved in 120 ml 0.05% acetic acid. The solution is dialyzed against five changes of 5 liters of 0.05% acetic acid then lyophilized, yielding 547 mg collagen. Analysis by SDS-polyacrylamide gel electrophoresis shows that the preparation consists of type-I collagen.

To obtain gelatin, 100 mg of collagen is dissolved in 45 ml 0.05% acetic acid by stirring at 4° for 60 min. The solution is poured into two thin glass tubes (0.9 cm i.d., 50 cm long) and the collagen is denatured by incubating at 45° for 60 min. The gelatin preparation is then dialyzed against two changes of 5 liters of 0.05 M Tris–HCl, pH 7.6, containing 0.05 M NaCl. A slight precipitate formed during dialysis is removed by centrifugation at 100,000 g for 35 min. The gelatin solution is adjusted to 2 mg/ml and stored in portions of 1–2 ml at −65°.

To prepare ^{3}H-acetylated gelatin, 100 mg of collagen is dissolved, denatured as described above, and then dialyzed against two changes of 5 liters of 0.1 M sodium phosphate, pH 7.4, containing 0.14 M NaCl, 0.02 M KCl, and 0.05 M sodium acetate. Ten milligrams of [^{3}H]acetic anhydride (50 mCi/mmol) is dissolved in 0.5 ml acetonitrile and added immediately at 4° to the dialyzed gelatin solution. After careful mixing, the reaction mixture is kept at 4° for 60 min. It is then dialyzed against five changes of 10 liters of 0.05 M Tris–HCl, pH 7.6, containing 0.05 M NaCl. The final preparation contains 2 mg gelatin/ml and on average 1–2 × 10^6 cpm/ml. Small portions of 0.5–1.0 ml are stored at −65°.

DFP treatment of samples: Samples for the determination of gelatinase must be free of serine proteinase activities, contributed mainly by elastase and cathepsin G, which rapidly degrade gelatin. The method of choice is inhibition with diisopropyl phosphofluoridate (DFP). To 1 vol of sample are added 0.1 vol of 33 mM DFP in 0.55 M Tris–HCl, pH 7.6, containing 0.55 M NaCl and 0.55% Triton X-100, giving final concentrations of 3 mM, 0.05 M, 0.05 M, and 0.05%, respectively. After incubation at 37° for 60 min in a shaking water bath the reaction is stopped by the addition of 0.1 vol of 0.05 M Tris–HCl, pH 7.6, containing 0.05 M NaCl, 0.05% Triton X-100, and 1.2 mg BSA/ml. All manipulations with DFP are performed in a well-ventilated hood. The DFP-treated samples can be stored at −20°.

Assay of gelatinase activity: Human neutrophil gelatinase is latent. It is activated by 4-aminophenylmercuric acetate (APMA), which is included at the concentration of 2 mM in the assay mixture. ^{3}H-Acetylated gelatin is diluted 1 : 3 with unlabeled gelatin shortly before use. To 100 μl 0.05 M Tris–HCl, pH 7.6, containing 0.05 M NaCl, 22.5 mM CaCl$_2$, 0.2 mg BSA/ml, and 0.1% Triton X-100, is added 50 μl of diluted ^{3}H-acetylated gelatin and 50 μl of 9 mM APMA. The reaction is started by addition of 25 μl of the DFP-treated sample and is stopped after 60 min in a shaking water bath at 37° by the addition of 25 μl 0.05 M Tris–HCl, pH 7.6, containing 0.05 M NaCl and 100 mg BSA/ml, followed by 50 μl of 90% TCA (w/v). After standing on ice for 30 min, the samples are centrifuged at 1000 g for 15 min; 100 μl of the supernatant is mixed with 5 ml of scintillation fluid (Rialuma) and the radioactivity is determined.

Vitamin B$_{12}$-Binding Protein (see also this volume [3]). The method described by Kane *et al.*[12] is used with minor modifications. Before the assay, cyano[^{57}Co]cobalamin (15 μCi/μg) is diluted 1 : 500 in 0.1 M potassium phosphate buffer, pH 7.5, to obtain a solution containing 1.33 ng vitamin B$_{12}$/ml, and a suspension of albumin-coated charcoal is prepared by mixing equal volumes of a 1% aqueous solution of BSA and a 5% suspension of Norit A charcoal in water.

[12] S. P. Kane, A. V. Hoffbrand, and G. Neale, *Gut* **15,** 953 (1974).

Assay: 0.1 ml sample, 0.4 ml potassium phosphate (0.1 M), pH 7.5, containing 0.025% Triton X-100, and 1.0 ml of the diluted cyano[^{57}Co]cobalamin solution are mixed and incubated for 30 min in the dark at room temperature with occasional mixing on a Vortex. Two milliliters of the albumin-coated charcoal suspension is then added and after standing for another 15 min the mixtures are centrifuged at 2500 g for 15 min at 20°. The radioactivity of 2 ml of the supernatant is determined in a gamma-counter. The total activity of the diluted cyano[^{57}Co]cobalamin solution used is determined in a suitable aliquot. A sample blank is obtained by substituting PBS for the sample in the assay mixture and a total activity blank by omitting the cyano[^{57}Co]cobalamin solution. The amount of vitamin B_{12} bound by the sample is calculated by the formula

$$[1.33 \times \text{cpm (sample)}]/\text{cpm (total)} = \text{nanograms vitamin } B_{12}$$

The respective blank values are deducted.

β-Glucuronidase and N-Acetyl-β-Glucosaminidase (see also this volume [3]). The acid glycosidases are determined fluorimetrically using 4-methylumbelliferyl-glycosides as substrates.

Reagents

Buffer: 0.1 M sodium acetate, pH 4.0, containing 0.1% Triton X-100 for β-glucuronidase, and 0.1 M sodium citrate, pH 5.0, containing 0.1% Triton X-100 for N-acetyl-β-glucosaminidase

Substrate: 10 mM 4-methylumbelliferyl-β-D-glucuronide or 10 mM 4-methylumbelliferyl-N-acetyl-β-D-glucosaminide dissolved in their respective buffer. Fresh solutions are prepared daily

Standard: A stock solution containing 5 mM 4-methylumbelliferone in ethanol is diluted 1 : 100 in buffer before use. The stock solution can be stored at 4° in the dark for several weeks

Stop solution: A buffer containing 0.05 M glycine and 5 mM EDTA, adjusted to pH 10.4 with 1 M NaOH. This solution can be stored for weeks at room temperature

Assay. Sample (0.1 ml) and 0.1 ml substrate solution are mixed and incubated for 15 min at 37°. The reaction is stopped by addition of 3.0 ml stop solution. Liberated 4-methylumbelliferone is measured fluorimetrically (excitation: 365 nm, emission: 460 nm).

Myeloperoxidase (see also this volume [20])

Reagents

Substrate solution: To 10 parts of 0.1 M sodium citrate buffer, pH 5.5, are added 1 part of 0.1% o-dianisidine in ethanol, 1 part of a 1 mM

aqueous solution of H_2O_2, and Triton X-100, to yield a final concentration of 0.05%. This mixture is prepared shortly before use

35% Perchloric acid

Assay. Sample (0.1 ml) and 1.0 ml substrate solution are mixed and incubated for exactly 1 min at room temperature. The reaction is stopped with 1.0 ml of 35% perchloric acid, and the absorbance at 560 nm is measured.

Lactate Dehydrogenase (see also this volume [3])

Reagents

Sodium phosphate buffer, 0.1 M, pH 7.5
Sodium pyruvate, 4 mM, in water
NADH, 0.8 mM, in water
Triton X-100, 0.1% in water

Assay. Into a 1-ml cuvette are added 0.5 ml of buffer, 0.2 ml of NADH, 0.1 ml of Triton X-100, and 0.1 ml of sample. The reaction is started by addition of 0.1 ml of pyruvate and the change in absorbance at 340 nm is recorded continuously. Blanks are run in the absence of pyruvate.

Exocytosis Experiments (see also this volume [3])

Reagents

PBS contains 137 mM NaCl, 2.7 mM KCl, 8.1 mM Na_2HPO_4, 1.5 mM KH_2PO_4, 0.9 mM $CaCl_2$, and 0.49 mM $MgCl_2$. Ca^{2+},Mg^{2+}-free PBS has the same composition except that $CaCl_2$ and $MgCl_2$ are omitted

Opsonized zymosan is prepared by incubating zymosan (10–20 mg/ml) for 30 min at 37° in autologous serum. The opsonized particles are washed two times in PBS, and are finally suspended in PBS at a concentration of 25 mg/ml. Stock solutions of compounds used for stimulation or treatment of the cells, e.g., fMLP (10 mM), PMA (2 mg/ml), A 23187 (5 mM), and cytochalasin B (5 mg/ml) are made up in DMSO. Shortly before use, dilutions are prepared in PBS. The final concentration of DMSO in the incubation mixture should not exceed 0.2% (v/v)

Storage of the Cells. Following purification, the neutrophils are suspended in saline at a density of 10^8 cells/ml and are kept at 4–10° until use. Cell viability and responsiveness are less satisfactory upon storage on ice. The storage solution does not need buffering and should not contain calcium.

Procedure. Neutrophils (0.5–2.5 × 10^7) are suspended in 0.9 ml of PBS and are preincubated for 5–10 min at 37°. During this period, the cells

fMLP-Induced Exocytosis in Human Neutrophils[a]

	Gelatinase	Vitamin B$_{12}$-binding protein	β-Glucuronidase	Myeloperoxidase	Lactate dehydrogenase
Total	83,084 ± 22,015	653.5 ± 134.4	0.956 ± 0.269	266.1 ± 118.2	67.81 ± 22.2
Supernatant	24,601 ± 12,148	65.9 ± 8.4	0.026 ± 0.011	3.3 ± 1.8	2.98 ± 0.99
Pellet	57,782 ± 10,879	567.4 ± 118.7	0.925 ± 0.312	261.8 ± 117.3	62.46 ± 21.06
Percentage release[b]	28.6 ± 10.1	10.8 ± 2.9	2.9 ± 1.1	1.3 ± 0.3	5.0 ± 2.0
Percentage recovery[c]	102.5 ± 18.8	97.3 ± 6.7	98.3 ± 8.0	99.6 ± 3.2	97.0 ± 7.8

[a] Neutrophils (2.5×10^7/ml) stimulated for 10 min at 37° with 0.1 μM fMLP. In addition to the total amount of markers initially present in the cells (lysate of unstimulated cell suspension), the amount released (supernatant) and that remaining in the cells (pellet) are given. Absolute values are specific activities calculated per 10^6 cells and represent TCA soluble cpm/60 min for gelatinase, pg for vitamin B$_{12}$-binding protein, and nmol/min for β-glucuronidase, myeloperoxidase, and lactate dehydrogenase. The percentage release by unstimulated cells was 0.7 ± 0.7 for gelatinase, 1.4 ± 0.6 for vitamin B$_{12}$-binding protein, 1.0 ± 0.5 for β-glucuronidase, 0.8 ± 0.5 for myeloperoxidase, and 3.3 ± 1.0 for lactate dehydrogenase. The figures shown in the table were not corrected for these control values. Mean ± SD from seven separate experiments.

[b] Amounts recovered in the supernatant in percentage of the sum of the amounts in supernatant and pellet.

[c] Amounts in supernatant plus pellet in percentage of amount initially present in the cells (total).

may be exposed to cytochalasin B (5 μg/ml) or other compounds. The cells are then stimulated by addition of 0.1 ml of the appropriate stimulus in PBS, and the incubation is continued under mild agitation in a water bath at 37°. The reaction is stopped by rapid cooling in ice followed by centrifugation at 800 g for 10 min at 4°. Immediate rapid centrifugation may be adopted when the incubation times are very short. The supernatant is aspirated, diluted with an equal volume of PBS containing 0.1% Triton X-100, and set aside for assays. A suspension of the pellet is prepared with 1 ml of PBS containing 0.2% Triton X-100, kept in ice during 20 min with occasional mixing for extraction, diluted with 3 ml of PBS and centrifuged at 800 g for 10 min at 4°. The clear cell extract is used for assays. Similarly, an extract from untreated cells is prepared for the determination of the total cellular content of the markers.

Sample Storage. The samples for assays, i.e., the total cell lysate, the supernatant, and pellet, are prepared and stored in the cold. Lactate dehydrogenase and vitamin B_{12}-binding protein are rapidly inactivated and must therefore be assayed within a short time. The samples for the assay of gelatinase are treated with DFP on the same day and can then be stored overnight at 4°. Samples for the assay of β-glucuronidase, N-acetyl-β-glucosaminidase, and myeloperoxidase may be stored overnight at 4°.

Calculation of Results. A balance sheet of the results obtained in a series of experiments on exocytosis induced in human neutrophils by fMLP is shown in the table. In order to compare exocytosis from different subcellular compartments and under different conditions, release is conveniently expressed as percentage of the total amount of the respective marker. The validity of such figures depends on reliable assay conditions and good recoveries. It is therefore essential to calculate specific activities and percentage recoveries in each exocytosis experiment.

Controls. The accurate assessment of stimulus-dependent exocytosis requires comparison with exocytosis obtained under control conditions, i.e., in samples handled in the same way but without addition of a stimulus or samples exposed to the stimulus but incubated at 0°. Values obtained in the absence of stimulus are given in the table.

Finally, it is important to test for the specificity of the release observed. Exocytosis should occur without damage of the cells, i.e., it should not be accompanied by a loss of cytosolic components. The release of the cytosolic enzyme, lactate dehydrogenase, is the most common way to test for cell viability. It is important to recognize that some lactate dehydrogenase is always released under control conditions (see the table).

Section IV

Opsonization and Nonspecific Promoters of Phagocytosis

[13] Opsonins and Dysopsonins: An Overview

By DARRYL R. ABSOLOM

1. Introduction

Phagocytosis refers to the physical act of engulfment, i.e., the process by which a particle is recognized by the phagocyte as being foreign and is ingested by the cell and subsequently sequestered in an intracellular vacuole. This process is initiated by the migration (chemotaxis) of the phagocyte to the site of microbial infiltration. Chemotaxis is largely responsible for bringing the phagocyte into physical contact with the particle and is thus a critical primary step in the recognition phase of the phagocytic process.

Once contact between the particle and granulocyte has been achieved, a cohesive force prevents their separation despite extracellular fluid flow. This critical step in the phagocytic process is known as the adhesive phase. It occurs largely through the action of particle-associated opsonins (e.g., IgG and C3b) which bind in a specific receptor–ligand manner with complementary membrane receptors on the surface of the phagocyte. This adhesive phase can also be mediated, in the absence of such specific opsonins, by nonspecific opsonins, e.g., α_2-macroglobulin and fibronectin. These agents alter the surface properties of the particles, thereby either increasing or decreasing the extent of the particle–phagocyte interaction due to a modulation of the nonspecific (i.e., van der Waals and electrostatic) forces involved in the adhesion step. Following adhesion, cellular pseudopods form as a result of the gelation and contraction of phagocyte microfilaments[1-3] in an energy-dependent process.[4-6] The activation of microfilaments results in the production and movement of pseudopods around the particle until the particle is completely surrounded by membranous material. The resultant vacuole (containing the ingested particle) retains its connection to the cell surface by a membranous stalk for a short period after which the opposed membranes fuse and separation occurs from the outer membrane surface. The resulting vacuole, now

[1] A. C. Allison, P. Davies, and S. de Petris, *Nature (London), New Biol.* **232,** 153 (1971).
[2] L. A. Boxer, E. T. Hedley-White, and T. P. Stossel, *N. Engl. J. Med.* **291,** 1093 (1974).
[3] E. P. Reaven and S. G. Axline, *J. Cell Biol.* **59,** 12 (1973).
[4] P. Bodel and S. E. Malawista, *Exp. Cell Res.* **56,** 15 (1969).
[5] L. A. Boxer, R. L. Baehner, and J. Davis, *J. Cell. Physiol.* **91,** 89 (1977).
[6] K. Levin, *Scand. J. Clin. Lab. Invest.* **32,** 67 (1973).

referred to as a phagosome, is thus set free in the interior of the cell where the digestive process will occur.

The prerequisite for such ingestion and subsequent digestion is the formation of the critical adhesive bond between the particle to be ingested and the phagocytic cell. Several biological entities are known to either promote (opsonins) or retard (dysopsonins) this event. The term "opsonins," in the strictest sense, is generally taken to refer to some physical entity which adsorbs onto a surface, thereby rendering that surface more palatable to phagocytes. Opsonins (and dysopsonins) primarily affect the critical first step in phagocytosis, i.e., the *attachment* phase between the phagocyte and the particle. Opsonins are not generally regarded as being those factors which influence subsequent events in the process such as actual internalization of the particle and/or digestion. It should be emphasized, however, that even in the complete absence of any opsonins, that phagocytic ingestion of particles, including bacteria, can occur.[7,8] Such ingestion is mediated by the nonspecific recognition of the particle. The extent of such ingestion is determined by the relative surface energies of the interacting components, i.e., phagocyte, particle, and suspending liquid medium.[9] For example bacteria, latex beads, and carbon particles can be ingested by a mechanism that is independent of the presence of Fc–Fc receptor interaction. However, in the presence of an Fc receptor–ligand type of interaction the efficiency of the process is considerably enhanced. This aspect of phagocytosis has been discussed in greater detail in a separate section in this volume [2].

2. Opsonins

Following the pioneering studies of Metchnikoff[10] on the widespread role of phagocytosis as a primary defense mechanism in a variety of both uni and multicellular organisms, it was soon established that optimum phagocytic ingestion of some particles required either the addition of serum to the system or the pretreatment of the particles with serum.[11] Indeed, as early as 1903 Wright and Douglas concluded from their studies that "We have here conclusive proof that the blood fluids modify the

[7] C. J. van Oss, *Annu. Rev. Microbiol.* **32,** 19 (1978).

[8] C. J. van Oss, C. F. Gilman, and A. W. Neumann, "Phagocytic Engulfment and Cell Adhesiveness." Dekker, New York, 1975.

[9] C. J. van Oss, D. R. Absolom, and A. W. Neumann, in "The Reticuloendothelial System" (S. M. Richard and J. P. Filkins, eds.), Vol. 7A, Chapter 1. Plenum, New York, 1984.

[10] E. Metchnikoff, "Immunity in Infective Diseases." Cambridge Univ. Press, London and New York, 1905.

[11] A. E. Wright and S. R. Douglas, *Proc. R. Soc. London* **72,** 357 (1903).

bacteria in a manner which renders them a ready prey to the phagocyte. We may speak of this as an 'opsonic' effect (opsono—to cater for; to prepare victuals for) and we may employ the term 'opsonins' to designate the elements in the blood fluid which produce this effect."[11] The adsorption of certain proteins onto the surface of bacteria or other particles may influence the extent of phagocytic ingestion. Often, this adsorption greatly facilitates phagocytosis of the particles and these proteins (or other substances) are then referred to as *opsonins* (phagocytosis-promoting substances from the Greek το οψον: a relish, seasoning, or sauce) and the process of specific or aspecific adsorption of such a protein is referred to as *opsonization*.

It was clear from these early studies that two types of opsonins, one heat labile and the other heat stable, are present in serum. The opsonic properties of normal human serum were found to be considerably reduced by heat treatment at temperatures above 50°, with complete inactivation occurring in some experiments when serum was heated at 60–65° for 10–15 min.[11] In similar investigations Hektoen and Ruediger reported the loss of opsonic activity of human, rabbit, and guinea pig serum for a single strain of streptococcus on heating to 54–56° and of dog serum on heating to 58–60° for 30 min.[12] As controls in these experiments, fresh serum samples from nonimmunized animals or humans were employed and opsonization (as assessed by enhanced ingestion) of a wide variety of bacteria was noted.

During the course of these investigations, the effect of serum dilution on opsonic activity was noted: increased serum dilution resulting in a decreased bacterial ingestion. In contrast, Neufeld and Rimpau[13] documented the presence of opsonins in immune sera which were effective against the immunizing bacterial strains, even at high serum dilutions and which, in addition, were stable (i.e., retained their opsonic activity) at 56°. These latter opsonins were initially referred to as "bacteriotropins." This term, however, is no longer in general use. The term opsonin has been extended to include all substances, both heat stable and heat labile, which interact with particles to render them more susceptible to phagocytic ingestion.[14]

It is now the generally accepted view that the *specific* opsonins which are heat stable are immunoglobulins and that the specific heat-labile opsonins are components of the complement system. These two systems generally act synergistically to prepare the particle for ingestion. The heat-stable immunoglobulins serve both to identify (through immune com-

[12] L. Hektoen and G. F. Ruediger, *J. Infect. Dis.* **2**, 128 (1905).
[13] F. Neufeld and W. Rimpau, *Dtsch. Med. Wochenschr.* **30**, 458 (1904).
[14] H. K. Ward and J. F. Enders, *J. Exp. Med.* **57**, 527 (1933).

plex formation) the invasive foreign particle and to activate the complement cascade. It should be emphasized, however, that several other serum factors also can play a role in opsonization. These nonspecific opsonins are discussed separately below in Section 2.5.

2.1. Heat-Stable Opsonins

2.1.1. Immunoglobulins. 2.1.1.1. Specific antibodies: Early studies indicated that the heat-stable opsonins which appeared in the serum of immunized animals[13] were specific for the immunizing agent and were adsorbed from the serum by the strain used for immunization but not by other organisms.[15-17] This observation strongly suggested that these opsonins were specific antibody which, when present in sufficient concentration, could efficiently act as opsonins in the complete absence of heat-labile complement components.

The role of specific antibody in opsonization was first demonstrated clearly in studies using pneumococcus.[14,18] Encapsulated pneumococci, which normally easily resist phagocytosis,[14,19,20] are readily ingested when they are pretreated with type-specific antibody directed against the capsular polysaccharide antigens.[20,21] The importance of opsonization by capsular pneumococcal antibody was established early. The clinical response of pneumococcal pneumonia in man to type-specific antipneumococcal serum is often dramatic and animal investigations suggested that this was due to the opsonic activity of specific antibody. The intravenous injection of type-specific antipneumococcal serum into rats infected intrabronchially with virulent type-1 pneumococcus protected the rodents from an otherwise fatal lobar pneumonia.[22] The injected antibody was seen to cause bacterial agglutination of the microorganisms in the advancing edema zone of the lesion. Pneumococci immobilized in this manner were rapidly overtaken by the infiltrating polymorphs and ingested.[22] More recently, type-specific antibody directed against the antigenic M-protein of group A hemolytic streptococci has been shown to be opsonic.[23] Strain-specific heat-stable opsonins have also been shown to be present in the

[15] G. Dean, *Proc. R. Soc. London* **74**, 506 (1905).
[16] W. Bulloch and G. T. Western, *Proc. R. Soc. London* **76**, 531 (1906).
[17] G. Dean, *Proc. R. Soc. London* **79**, 399 (1907).
[18] R. Muir and W. B. M. Martin, *Proc. R. Soc. London* **79**, 187 (1907).
[19] B. D. Davis, R. Dubecco, H. N. Eisen, and W. B. Wood, "Microbiology." Harper & Row, Hagerstown, Maryland, 1973.
[20] W. B. Wood, *Bacteriol. Rev.* **24**, 41 (1960).
[21] L. Edebo, E. Kihlström, K.-E. Magnusson, and O. Stendahl, *Proc. Br. Soc. Cell Biol. (Cell Adhes. Motil.)* (1979).
[22] W. B. Wood, *J. Exp. Med.* **73**, 201 (1941).
[23] R. C. Lancefield, *J. Immunol.* **89**, 307 (1962).

sera of patients with severe staphylococcal infections.[24] Such opsonins can be adsorbed not only by the infecting strain but also by type-specific staphylococcal mucopeptides.[24,25] Patients immunized with purified lipopolysaccharides prepared from *Pseudomonas aeruginosa*[26] or with a polyclonal vaccine directed against *P. aeruginosa*[27-29] have been shown to develop type-specific heat-stable opsonins which can be removed from the patient's serum through adsorption by either the corresponding live or heat treated microorganism, or through the addition of the lipopolysaccharide of the immunizing stain. Serotype-specific heat-stable opsonins directed against the lipopolysaccharides of *P. aeruginosa* are frequently found in the sera of patients with pseudomonas infections but not in the sera from normal subjects.[30]

The sera of patients with subacute bacterial endocarditis are excellent sources of heat-stable opsonins in man. High dilutions of heated serum continue to exhibit potent opsonic activity directed against the infecting bacterial strain in infections due to *Streptococcus viridans*, *Staphylococcus aureus*, *Staphylococcus epidermidis*, and *Streptococcus pneumoniae*.[31] In contrast, many patients with staphylococcal osteomyelitis exhibit little or no type-specific heat-stable opsonic activity although heat-stable opsonins are generally present in osteomyelitis caused by gram-negative bacteria.[32]

Rabbits with experimental staphylococcal osteomyelitis, however, have a marked opsonic antibody response, as do rabbits with experimental staphylococcal endocarditis.[33] The aforementioned studies all point strongly toward the clearly established role and involvement in phagocytosis of type-specific antibody as opsonins. The nature of the opsonic immunoglobulins is discussed separately below in Section 2.1.3.

2.1.1.2. Aspecific adsorption of immunoglobulins: Although there are body fluids, serving as a liquid medium for phagocytosis, in which very little protein and extremely small amounts of IgG are present (e.g., in the urinary bladder), most body fluids contain sizeable amounts of protein,

[24] D. W. Humphreys, L. J. Wheat, and A. White, *J. Lab. Clin. Med.* **84**, 122 (1974).
[25] L. J. Wheat, D. W. Humphreys, and A. White, *J. Lab. Clin. Med.* **83**, 73 (1974).
[26] L. S. Young, *J. Infect. Dis.* **126**, 277 (1972).
[27] A. B. Bjornson and J. G. Michael, *J. Infect. Dis.* **128**, 182 (1973).
[28] A. B. Bjornson and J. G. Michael, *J. Infect. Dis.* **130**, 119 (1974).
[29] A. B. Bjornson and J. G. Michael, *J. Infect. Dis.* **130**, 127 (1974).
[30] J. G. Crowder, H. B. Devlin, M. Fisher, and A. White, *J. Lab. Clin. Med.* **83**, 853 (1974).
[31] T. Laxdal, R. P. Messner, R. C. Williams, and P. G. Quie, *J. Lab. Clin. Med.* **71**, 638 (1968).
[32] R. C. Williams, J. H. Dossett, and P. G. Quie, *Immunology* **17**, 249 (1969).
[33] D. S. Nickerson, J. A. Kazmierowski, J. H. Dossett, R. C. Williams, and P. G. Quie, *J. Immunol.* **102**, 1235 (1969).

TABLE I
ADSORPTION OF 1% HUMAN SERUM ALBUMIN (HSA) AND 1% HUMAN IgG
ONTO BACTERIA IN AN AQUEOUS MEDIUM (HBSS)[a,b]

Bacteria	Contact angle of bacteria with drops of saline water (degrees)	Bacterial surface tension (ergs/cm^2)	Protein adsorbed (fg)/bacterial particle ($\pm$SD)		
			HSA	IgG (FII)	IgG (monoclonal)[c]
Escherichia coli 055	16.7	69.6	11 $\pm$ 5	92 $\pm$ 23	56 $\pm$ 7
Staphylococcus aureus 49	18.8	68.8	16 $\pm$ 8	170 $\pm$ 23	92 $\pm$ 18
Staphyloccus epidermidis 47	23.4	67.0	28 $\pm$ 7	521 $\pm$ 63	201 $\pm$ 19
Listeria monocytogenes	25.3	66.2	58 $\pm$ 17	724 $\pm$ 81	259 $\pm$ 28

[a] The surface tensions of HBSS, HSA, and IgG are 72.8, 70.2, and 67.5 ergs/cm^2, respectively.
[b] Absolom et al.[36]
[c] Myeloma IgG #1024 (IgG$_3$).

usually including IgG as well as albumin. Apart from a general tendency of lowering the surface tension of the aqueous medium, of which the relevance to phagocytic engulfment has been discussed elsewhere in this volume ([2]), the degree to which various proteins become physically adsorbed onto bacterial surfaces must be taken into account in interpreting the interfacial interactions between bacteria and phagocytes under conditions approaching those prevailing *in vivo*. Especially important is the physical adsorption of IgG by bacteria, as that, in most (but not in all) cases, is tantamount to opsonization.[9,34-37] The nonspecific physical adsorption of two plasma proteins and its relation to subsequent phagocytic ingestion has been studied in detail.[36,37]

Human serum albumin (the protein present in plasma in by far the highest concentration) and human IgG (the principal heat-stable opsonin),[34,35] in the form of pooled polyclonal IgG (Cohn's fraction II) as well as purified human monoclonal IgG subclasses (IgG$_1$, IgG$_2$, IgG$_3$, and IgG$_4$) adsorption onto various bacteria and polystyrene latex particles has been studied extensively. The adsorption of all these proteins (in 1% concentration) is measured through the residual radioactivity (^{125}I was

[34] C. J. van Oss and M. W. Stinson, *RES, J. Reticuloendothel. Soc.* **8**, 397 (1970).
[35] M. W. Stinson and C. J. van Oss, *RES, J. Reticuloendothel. Soc.* **9**, 503 (1971).
[36] D. R. Absolom, C. J. van Oss, W. Zingg, and A. W. Neumann, *RES, J. Reticuloendothel. Soc.* **31**, 59 (1982).
[37] A. W. Neumann, D. R. Absolom, D. W. Francis, W. Zingg, and C. J. van Oss, *Cell Biophys.* **4**, 285 (1982).

TABLE II
ADSORPTION OF 1% HUMAN MONOCLONAL IgG[a] ONTO BACTERIA
IN WHOLE SERUM AND IN WHOLE PLASMA[b]

Bacteria	Contact angle of bacteria with drops of saline water (degrees)	Bacterial surface tension (ergs/cm²)	IgG adsorbed (fg)/bacterium (±SD)	
			In whole serum	In whole plasma
Escherichia coli 055	16.7	69.6	1.8 ± 0.7	0.6 ± 0.3
Staphylococcus aureus 49	18.8	68.8	3.8 ± 1.0	2.3 ± 1.1
Staphyloccus epidermidis 47	23.4	67.0	5.4 ± 1.2	3.6 ± 1.5
Listeria monocytogenes	25.3	66.2	8.3 ± 1.5	6.3 ± 1.1

[a] Myeloma #1024 (IgG_3).
[b] Absolom *et al.*[36]

used as a label) of the various bacteria and polystyrene, after extensive washing following the adsorption step.[36] Table I shows the extent of adsorption of HSA and IgG onto these particles. Clearly, the more hydrophobic (i.e., lowest surface tension) the particle (or bacteria), the more they adsorb both HSA and IgG, and any given bacterium adsorbs more IgG than HSA, in accordance with the fact that IgG is more hydrophobic than HSA.[8] All four bacteria tested adsorbed more pooled polyclonal (FII) than monoclonal IgG, since pooled polyclonal IgG still comprises various more or less specific antibodies against different bacterial antigenic determinants, which is not the case with monoclonal IgG. Thus, IgG does, as it should for thermodynamic reasons,[38] become physically adsorbed onto bacterial surfaces to a significantly greater amount than HSA; IgG becomes adsorbed most extensively to the more hydrophobic bacteria. Even in whole plasma or serum, monoclonal IgG still physically adsorbs to a significant extent to all bacteria tested, again the most strongly to the most hydrophobic particles (see Table II). Although less IgG becomes adsorbed to the bacteria in the presence of whole plasma than in HBSS,[38a] even under these conditions 3000 to 30,000 IgG molecules become adsorbed per bacterium, which still is more than ample for opsonization.[8,36] Experimental observations illustrate that immune complexes comprising three or more IgG molecules rapidly disappear from the circu-

[38] C. J. van Oss, D. R. Absolom, A. W. Neumann, and W. Zingg, *Biochim. Biophys. Acta* **670**, 64 (1981).
[38a] Abbreviations: HSA, human serum albumin; BSA, bovine serum albumin; EGTA, ethylene glycol bis(β-aminoethyl ether) N,N',N'-tetraacetic acid; HBSS, Hanks' balanced salt solution; sIgA, secretory IgA.

lation *in vivo*.[39–42] More recent work further confirms that in experimental immune complex disease, elicited in rabbits with bovine serum albumin (BSA) as the antigen, only complexes of molecular weight (MW) $\simeq$ 400,000 to 450,000 (IgG_2–BSA to IgG_2–BSA_2) and those of MW 600,000 to 670,000 (IgG_3–BSA_2 to IgG_3–BSA_3) remain in circulation, i.e., are not cleared by phagocytosis.[43,44]

2.1.1.3. Role of specific versus aspecific opsonization: The opsonizing role of IgG, specific as well as aspecific (FII) and nonspecific (monoclonal) is now well established.[34–36] The part of the IgG molecule endowed with opsonizing power is the Fc moiety (of IgG subclasses IgG_1 and IgG_3) while the Fab and $F(ab')_2$ fractions are totally devoid of such opsonic activity.[8,45] Enzymatic and chemical procedures,[46] as well as immunochemical modifications,[47,48] which affect the integrity of the Fc region of IgG molecules, remove all opsonic activity of the structurally compromised immunoglobulins (see also Section 2.1.3.1).

Studies with the strongly encapsulated *S. aureus* S. Smith indicated that IgG is active as an opsonin only when situated at the outermost surface of the bacterial cell, or, if present, of its capsule.[34] Opsonization with IgG is accompanied by an increase of hydrophobicity of the bacterial surface[8,36,49] which is to be expected to, and does, give rise to increased phagocytic ingestion *in vitro,* in aqueous media.

It is tempting to ascribe the increased phagocytic engulfment mainly to the greater surface hydrophobicity,[8,36] but in 1975 it was noted[7,8] that the normally already rather hydrophobic *Listeria monocytogenes* (contact angle $\theta \simeq 25°$), which therefore cannot and does not become more hydrophobic upon opsonization with IgG, nevertheless after incubation with IgG becomes noticeably more phagocytized.[36] This strengthens the suspicion that there is more to the function of IgG than simply to increase the hydrophobicity of bacteria (or to trigger the fixation of complement), and

[39] W. P. Arend and M. Mannik, *J. Immunol.* **107,** 63 (1971).
[40] M. Mannik and W. P. Arend, *J. Exp. Med.* **134,** 19 (1971).
[41] M. Mannik, W. P. Arend, A. P. Halland, and B. C. Gilland, *J. Exp. Med.* **133,** 713 (1971).
[42] C. J. van Oss, C. F. Gillman, and A. W. Neumann, *Immunol. Commun.* **3,** 77 (1974).
[43] F. C. Germuth, L. B. Sentrefit, and G. R. Dressman, *Johns Hopkins Med. J.* **130,** 344 (1972).
[44] A. M. Fagundus, Ph.D. Dissertation, p. 117. S.U.N.Y., Buffalo (1980).
[45] C. J. van Oss, M. S. Woeppel, and S. E. Marquart, *RES, J. Reticuloendothel. Soc.* **13,** 221 (1973).
[46] P. G. Quie, R. P. Messner, and R. C. Williams, *J. Exp. Med.* **128,** 553 (1968).
[47] R. P. Messner, T. Laxdal, P. G. Quie, and R. C. Williams, *J. Clin. Invest.* **47,** 1109 (1968).
[48] P. A. Ward and N. J. Zvaifler, *J. Immunol.* **111,** 1777 (1973).
[49] R. K. Cunningham, T. O. Söderström, C. F. Gilman, and C. J. van Oss, *Immunol. Commun.* **4,** 429 (1975).

TABLE III
COMPARISON BETWEEN THE PHAGOCYTIC INGESTION BY POLYMORPHONUCLEAR CELLS
OF NONOPSONIZED BACTERIA, AND BACTERIA OPSONIZED WITH 1% IgG (FII, AS WELL
AS MONOCLONAL), AND BACTERIA TREATED WITH HSA[a]

Treatment	Phagocytic activity $(N \pm SE)^b$			
	E. coli	*S. aureus*	*S. epidermidis*	*L. monocytogenes*
Nonopsonized	0.8 ± 0.2	1.5 ± 0.1	3.6 ± 0.1	5.4 ± 0.2
	(16.7°)[c]	(18.8°)[c]	(23.4°)[c]	(25.3°)[c]
Opsonized with	2.4 ± 0.2	3.6 ± 0.2	5.6 ± 0.1	8.2 ± 0.1
1% IgG (FII)	(19.4°)[c]	(21.7°)[c]	(23.6°)[c]	(25.5°)[c]
Opsonized with 1%	2.0 ± 0.1	3.8 ± 0.2	4.8 ± 0.2	7.8 ± 0.1
IgG (monoclonal)[d]	(18.9°)[c]	(21.5°)[c]	(22.6°)[c]	(24.8°)[c]
Treated with	0.6 ± 0.2	1.4 ± 0.1	3.2 ± 0.2	4.8 ± 0.2
1% HSA	(16.5°)[c]	(18.2°)[c]	(21.6°)[c]	(23.8°)[c]

[a] Absolom *et al.*[36]
[b] Average number of bacteria ingested per cell.[8]
[c] The contact angles with drops of saline water, in degrees, of each of the bacteria after the treatment indicated in the left-hand column are given in parentheses.
[c] Myeloma #1024 (IgG$_3$).

that interactions between IgG (via its Fc moiety) and Fc receptors on phagocytic cells play an important role in phagocytic ingestion.

Table III shows that in all cases opsonization with both polyclonal IgG (FII) and (essentially) monoclonal IgG$_3$ (myeloma) significantly enhances the extent of phagocytic ingestion.[9,36] With *E. coli* and *S. aureus,* that enhancement parallels an increase in hydrophobicity (as measured by an increase in contact angle), but with *S. epidermidis* and *L. monocytogenes,* no such increase in contact angle is noticeable, nor given the contact angle with drops of saline water for IgG of $\theta \simeq 23.5°$, could such an increase in contact angle be expected with the latter two, already hydrophobic, bacterial species.[8] Treatment with HSA in all cases causes a slight decrease in contact angle as well as in phagocytic engulfment, which is to be expected, in view of the marked hydrophilicity of HSA.[36] The data in Table II pertaining to *S. epidermidis* and *L. monocytogenes* clearly show that IgG can play an important opsonic role independent of changes in bacterial hydrophobicity.[36,50] IgG subclasses IgG$_1$ and IgG$_3$ are mainly implicated in opsonization, at least vis-à-vis monocytes and macrophages[51] and most likely also as far as neutrophils are concerned. It is interesting to note that IgG$_1$ and IgG$_3$ also appear to be the IgG forms that

[50] W. Davies, Ph.D. Dissertation, University of Sydney, Sydney, Australia (1975).
[51] A. Nisonoff, J. E. Hopper, and S. B. Spring, "The Antibody Molecule," p 96. Academic Press, New York, 1975.

are principally adsorbed onto polystyrene latex particles out of whole human serum.[52,53]

In conclusion, human IgG (principally IgG_1 and IgG_3), specific as well as aspecific, is a powerful opsonin, promoting enhanced phagocytic ingestion of bacteria, in part through an increase in bacterial surface hydrophobicity, and for a probably more important part through (strong and specific) interactions with Fc receptors on the surface of phagocytic cells (see Section 2.1.3). For opsonization, IgG is only active when situated at the periphery of the bacterial surface or of its capsule; the Fc moiety comprises the opsonizing determinant. Specific as well as aspecific IgG_1 and IgG_3 are the principal "heat-stable" opsonins.

2.1.2. *Phagocytic Cell Membrane Fc Receptors for Opsonic Immunoglobulins.* Phagocytic cells depend on cell surface receptors to "sense" changes in their microenvironment prior to the expression of their effector functions. One of these surface receptors, the Fc receptor, specifically recognizes a site on the Fc region of immunoglobulins and is intimately involved in the mediation of the process of phagocytosis and subsequent cytotoxic events and certain secretory responses.[54,55]

Antibody molecules, directed against antigens on the surface of the microorganism, bind with immunological specificity via the antibody active site in the Fab region. The carboxyl terminal Fc portion of the antibody molecule remains free to attach to receptor sites on the surface of the phagocyte. In pioneering studies, surface membrane Fc receptors for IgG were readily demonstrated on mononuclear phagocytes,[56–59] whereas their presence on neutrophils was much more difficult to prove. The existence of Fc receptors on human neutrophils could not be demonstrated through the use of opsonized erythrocytes coated with anti-Rh_0–IgG antibody.[56,58,60] However, neutrophil adherence studies with human, rabbit, and guinea pig heterophile antibodies to sheep erythrocytes coated with isologous IgG antibody were interpreted as indicating the presence

---

[52] R. S. Weening, D. Roos, and M. L. J. van Schaik, *in* "Inborn Errors of Immunity and Phagocytosis" (F. Gittler, J. W. T. Seakins, and R. A. Harkness, eds.), p. 291. University Park Press, Baltimore, Maryland, 1979.

[53] D. R. Absolom, in preparation.

[54] J. K. Spitznagel, *in* "The Granulocyte: Function and Clinical Utilization" (T. J. Greenwalt and G. A. Jamieson, eds.), p. 103. Alan R. Liss, Inc., New York, 1977.

[55] C. F. Nathan, N. W. Murray, and Z. A. Cohn, *N. Engl. J. Med.* **303,** 622 (1980).

[56] A. F. LoBuglio, R. S. Contran, and J. H. Jandl, *Science* **158,** 198 (1967).

[57] H. Huber and H. H. Fudenberg, *Int. Arch. Allergy Appl. Immunol.* **34,** 18 (1968).

[58] H. Huber, S. D. Douglas, and H. H. Fudenberg, *Immunology* **17,** 7 (1969).

[59] H. Huber, M. J. Polley, W. D. Linscott, H. H. Fudenberg, and H. J. Müller-Eberhard, *Science* **162,** 1281 (1968).

[60] M. J. Cline and R. I. Lehrer, *Blood* **32,** 423 (1968).

of membrane Fc receptors.[61-63] The presence of IgG Fc surface receptors on human neutrophils was also demonstrated through the use of a bactericidal assay[64] and directly by binding assays using radiolabel[65] and autographic techniques[66] using labeled anti-IgG antibody. Preformed antigen–antibody complexes bind to surface receptors of granulocytes more avidly than does uncomplexed antibody.[67] This may be related to a greater number of binding sites associated with a multimolecular complex and thus to an enhanced energy of interaction resulting in more stable binding.[67-69]

2.1.2.1. Detection and quantitation of Fc receptors: Before discussing Fc receptors in more detail, it is appropriate to define what is meant by the term "receptor." In the present context receptor is taken to refer to a distinct molecular entity present on the phagocytic cell membrane and which possesses the following properties[67]:

1. *Saturability.* There are a finite number of receptors per cell, all of which can be occupied at high free-ligand concentration.
2. *Specificity.* The receptor is capable of binding to only certain structural portions of only certain classes or subclasses of proteins.
3. *High affinity.* The specific counter-ligands are bound with a high energy, implying an intimate stereochemical relationship.
4. *Induction of cellular response.* The result of the high-energy association between the counter-ligand and the receptor is the initiation of some cellular response, e.g., ingestion, enzyme release. This property is important since it distinguishes the receptor interaction from the association due to carrier molecules such as serum albumin.

Several different techniques have been used to establish the presence of Fc receptors on various phagocytic cell lines.[70] These include the following:

1. *The rosette assay* (see this series, Vol. 108 [7]). This is a widely used semiquantitative (at best) method which has proved useful in

[61] P. M. Henson, *Immunology* **16,** 107 (1969).
[62] P. M. Henson, H. B. Johnson, and H. L. Spiegelberg, *J. Immunol.* **109,** 1182 (1972).
[63] P. M. Henson, *Immunol. Commun.* **5,** 757 (1976).
[64] R. P. Messner and J. Jelinek, *J. Clin. Invest.* **49,** 2165 (1970).
[65] J. M. Phillips-Quagliata, B. B. Levine, and J. W. Uhr, *Nature (London)* **222,** 1290 (1969).
[66] K. Ishizaka, H. Tomioka, and T. Ishizaka, *J. Immunol.* **105,** 1459 (1970).
[67] K. J. Dorrington, *in* "Immunology of Receptors" (B. Cinader, ed.), p. 183. Dekker, New York, 1977.
[68] J. C. Unkeless and H. N. Eisen, *J. Exp. Med.* **142,** 1520 (1975).
[69] J. C. Unkeless, H. Fleit, and I. S. Mellman, *Adv. Immunol.* **31,** 247 (1981).
[70] K. J. Dorrington, *in* "Molecular Immunology" (M. Z. Atassi, C. J. van Oss, and D. R. Absolom, eds.), Chapter 29. Dekker, New York, 1984.

determining immunoglobulin class and subclass specificity of the receptor. The indicator cell is an erythrocyte that has been coated with immunoglobulin.

2. *Radioligand binding assays* (see this series, Vol. 93 [10]). The use of radiolabeled (generally ^{125}I or ^{131}I) immunoglobulins has provided quantitative information regarding the number, intrinsic binding affinity, and heterogeneity of the Fc receptors. (Fluorescence labels can also be employed.)

Quantitative information has been obtained using binding assays involving radiolabeled monomeric IgG followed by Scatchard analysis of the data. The presently available data may be summarized as follows[67]: macrophages possess approximately 1×10^6 receptors which have an average intrinsic binding affinity (for monomeric IgG) of between 1×10^6 and 1×10^8 liter/M. The binding of monomeric IgG is, of course, *functionally* not important and such interactions exhibit a high rate of dissociation. Binding of polyvalent IgG ligands (e.g., opsonized erythrocytes, bacteria, or soluble immune complexes) is greatly favored because of their ability to bind simultaneously to several receptors. The avidity of such binding can reach very high levels.[71]

With respect to the heterogeneity of Fc receptors on human phagocytes the present evidence suggests that they possess only one class of Fc receptor for IgG. However, this Fc receptor is capable of binding all four human IgG subclasses albeit with markedly different affinities: $IgG_3 > IgG_1 \gg IgG_2 \geq IgG_4$[72] (see also Section 2.1.3.1).

Unkeless *et al.*[69] have recently written an excellent review article on the isolation, biochemical characterization, and physicochemical nature of Fc receptors.

2.1.3. Nature of the Opsonic Immunoglobulins. 2.1.3.1. IgG: The opsonizing role of IgG, specific, as well as aspecific (FII) and nonspecific (monoclonal, myeloma), is well established.[34,35] The part of the IgG endowed with opsonizing power is the Fc moiety (of IgG subclasses IgG_1 and IgG_3) while the Fab and $F(ab')_2$ fractions are totally devoid of it.[45] Integrity of the Fc region is a prerequisite. The opsonic activity of human IgG is lost when it is reacted with rheumatoid factor (19 S antibodies directed against the patient's own IgG) or rabbit anti-human IgG antiserum.[47,48] Furthermore, chemical treatment of IgG with reagents such as sodium metaperiodate, mercaptoethanol (reduction), or pepsin digestion, all of which affect the Fc tail portion of the antibody molecule, results in

[71] D. R. Absolom and C. J. van Oss, *CRC Crit. Rev. Immunol.* **6,** 1 (1986).
[72] H. Huber, S. D. Douglas, J. Nusbacher, S. Kowcha, and R. E. Rosenfield, *Nature (London)* **229,** 419 (1971).

the loss of IgG opsonizing power.[46] Binding of opsonic antibodies to neutrophils is inhibited by free intact IgG and by Fc fragments[64] presumably because of the competition for receptor sites. Human neutrophils initially exposed to intact polyclonal IgG, or to the purified Fc fragment, and subsequently to fluorescein-labeled anti-IgG, show binding of both competitive species to the cell membrane. In contrast the $F(ab')_2$ fragment was not detected on the cell surface by this technique,[73] indicating the absence of a membrane receptor to that moiety. Sheep and chicken Fc fragments (prepared from the homologous IgG) do not bind to human neutrophils, indicating some degree of species specificity in the Fc receptor.[73] The presence of Fc receptors has been detected on up to 90% of circulating human neutrophils by rosette formation with erythrocytes sensitized with blood group antibodies (EA rosettes),[74,75] whereas only 20–30% of ovine neutrophils, obtained from the mammary gland, bound immunoglobulin.[76]

In hyperimmune antisera the primary heat-stable opsonin is IgG. This was clearly demonstrated in series of elegant dilution studies using rabbits immunized with *P. mirabilis*,[77] in patients with subacute bacterial endocarditis due to a variety of organisms,[31] in patients suffering with osteomyelitis due to *S. marcescens*,[32] and in several *in vitro* systems.[34–36,45]

IgG subclasses IgG_1 and IgG_3 are mainly implicated in opsonization[78] and have been found to bind with the greatest efficiency to neutrophils. Purified IgG_1 and IgG_3 subclasses (derived from myeloma serum) were found to inhibit EA rosette formation whereas myeloma IgG_2 and IgG_4 subclasses did not inhibit rosette formation.[64,69] Radiolabeling studies have shown that ^{125}I-labeled myeloma IgG_1 and IgG_3 bind with a much larger energy (avidity) to monocytes in comparison with the avidity of IgG_2 and IgG_4 in which IgG_4 binds slightly better than IgG_2.[72,79,80] If the iodinated myeloma IgG subclasses were aggregated, e.g., through heat treatment, the extent of binding increases for all subclasses, although the relative order and magnitude of binding remains similar to that observed for the nonaggregated myeloma subclasses.[81] Henson *et al.* employed the quantitation of extracellular lysosomal secretion as a marker of the stimu-

[73] A. Sajnani, N. Ranadive, and H. Movat, *Life Sci.* **14**, 2427 (1974).

[74] L. Wong and J. D. Wilson, *J. Immunol. Methods* **7**, 69 (1975).

[75] S. H. Pross, J. A. Hallock, R. Armstrong, and C. W. Fishel, *Pediatr. Res.* **11**, 135 (1977).

[76] D. L. Watson, *Immunology* **31**, 159 (1976).

[77] J. W. Smith, J. A. Barnett, R. P. May, and J. P. Sanford, *J. Immunol.* **98**, 336 (1967).

[78] N. Haeffner-Cavaillon, M. Klein, and K. J. Dorrington, *J. Immunol.* **123**, 1905 (1979).

[79] M. D. Alexander, J. A. Andrews, R. G. Q. Leslie, and N. J. Wood, *Immunology* **35**, 115 (1978).

[80] F. C. Hay, G. Torrigiani, and I. M. Roitt, *Eur. J. Immunol.* **2**, 257 (1972).

[81] D. A. Lawrence, W. O. Weigel, and H. S. Spiegelberg, *J. Clin. Invest.* **55**, 368 (1975).

lation of enzyme production through the binding of IgG to the neutrophil surface membrane.[62] When aggregated, all four subclasses result in lysosyme secretion, although IgG_1 and IgG_3 are slightly more effective.[63] Degranulation of neutrophils was promoted through the ingestion of polystyrene latex particles precoated with myeloma IgG of all four subclasses. The degree of degranulation was related to the opsonizing subclass with the order of effectiveness being reported as $IgG_1 > IgG_3 \gg IgG_2 \geq IgG_4$.[82] This reported order is somewhat surprising in view of the fact that IgG_3 is known to be a more effective opsonin than IgG_1. However, since in these studies the extent of subclass adsorption onto the latex particles was not quantitated it is difficult to determine whether the extent of degranulation in these studies is not related more to the surface concentration of the adsorbed protein rather than to the nature of the IgG subclass. Weening et al. have shown that IgG_1 and IgG_3 are the IgG forms that are principally adsorbed onto polystyrene latex particles from whole human serum.[52] Absolom et al. have reported similar findings with respect to the nonspecific adsorption of myeloma IgG subclasses onto various strains of bacteria.[36] They have also noted that the most hydrophobic bacteria adsorbed the largest amounts of either IgG_1 or IgG_3. More recently Absolom[53] has demonstrated that the extent of particle (both latex and bacterial) ingestion, and the consequent rate of oxygen consumption, is related not only to the immunochemical nature of the opsonizing IgG subclass, but also depends on the surface concentration of the adsorbed protein.

As mentioned previously (Section 2.1.1.3), the integrity of the Fc portion of the IgG molecule is critical for conveying opsonic activity. Papain or pepsin-digested IgG, which lacks the antibody Fc region, binds to particles but lacks opsonic activity, i.e., it does not promote enhanced phagocytic ingestion.[46,83] Inhibition studies using human myeloma IgG subclasses demonstrated that IgG_1 and IgG_3 inhibited EAC-rosette formation more effectively than IgG_2 and IgG_4.[57] Furthermore, only Fc fragments derived from IgG_1 and IgG_3 were capable of inhibiting rosette formation between human monocytes and anti-D-coated human erythrocytes.[84]

2.1.3.2. IgM: The large majority of investigators has been unable to detect receptors for the IgM type antibody on neutrophils.[31,48,61] These results are in apparent contradiction with the earlier studies of Robbins et al., which suggested that particle-adsorbed IgM had a high degree of opsonic activity.[85] These workers reported that IgM opsonins to *S. typhi-*

[82] M. S. Leffell and J. K. Spitznagel, *Infect. Immun.* **12**, 813 (1975).
[83] R. I. Handin and T. P. Stossel, *N. Engl. J. Med.* **290**, 989 (1974).
[84] G. O. Okafor, M. N. Turner, and F. C. Hay, *Nature (London)* **248**, 228 (1974).
[85] J. B. Robbins, K. Kenny, and E. Suter, *J. Exp. Med.* **122**, 385 (1965).

murium were approximately 500 times as effective as the corresponding specific IgG opsonins. Removal of the preopsonized bacteria from the bloodstream of the experimental mice was used as a measure of the effectiveness of phagocytic ingestion. In apparent support for these conclusions, in a more substantive quantitative investigation, Rowley and Turner[86] reported that as few as eight molecules of IgM per bacterium were sufficient for opsonization of *S. adelaide*. In contrast, in order to produce the same rate of bacterial disappearance (from the peritoneal cavity of mice) 2200 molecules of IgG per bacterium were required. It should be remembered, however, that both of the aforementioned studies utilized an *in vivo* system in order to measure the effectiveness of phagocytic ingestion. In both cases it is thus possible, indeed likely, that the antibody-coated bacteria would interact with the complement system of the animal into which the preopsonized bacteria were injected, thereby explaining the apparently more effective opsonizing power of IgM.

It is unlikely that IgM alone, under conditions where the complement system is inactive, has any opsonic activity vis-à-vis neutrophils.[34] In this respect it should be noted that Pondman and his colleagues have shown that 19 S anti-Forssman IgM antibody molecules require the presence of complement in order to result in the ingestion of sheep erythrocytes whereas 7 S anti-Forssman IgG antibody do not.[87,88] In similar studies Dossett *et al.* found that the addition of complement was required in order to result in phagocytic ingestion of erythrocytes already sensitized with a 19 S immunoglobulin which alone exhibited no opsonic activity.[89] These workers concluded that while the opsonic activity of both IgG[75] and IgM (19S) antibodies is considerably amplified through the interaction with complement, only IgG exhibited opsonic activity in the absence of complement. In a series of double antibody studies, Reed was able to show *in vitro* that the heat-stable serum antibody which opsonized several *Shigella* strains was IgG-type antibody.[90] IgM was found to be ineffective unless complement was deliberately added to the system (see also Section 2.4). IgM has a rather pronounced hydrophilicity, with a surface tension of 69.4 mJ/m², and thus adsorption of this immunoglobulin, onto the surface of microorganisms, will not result in the production of conditions which are favorable for nonspecific phagocytic ingestion.[9,38] Finally, it is

[86] D. Rowley and K. J. Turner, *Nature (London)* **210,** 496 (1966).

[87] B. T. Gerlings-Petersen and K. W. Pondman, *Vox Sang.* **7,** 655 (1962).

[88] B. T. Gerlings-Petersen and K. W. Pondman, *Proc. Congr. Int. Soc. Blood Transfus. 10th, 1964,* p. 829 (1965).

[89] J. H. Dossett, R. C. Williams, and P. G. Quie, *Pediatrics* **44,** 49 (1969).

[90] W. P. Reed, *Immunology* **28,** 1051 (1975).

interesting to note that alveolar macrophages have been shown to have receptors for IgG and C3b but not for IgM.[91]

2.1.3.3. IgA: IgA [as well as secretory IgA (sIgA) present on the surface lymphocytes] may be regarded as a true dysopsonin, or a phagocytosis-inhibiting substance. As such it is discussed in further detail in Section 3.

2.1.3.4. IgD and IgE: The *opsonic* role of these two immunoglobulins has not been clearly established. Specifically, it would appear, from enzyme secretion studies, that neutrophils do not possess Fc receptors for IgD. In contrast, high density and affinity IgE Fc receptors are known to occur on a variety of phagocytic cell types including basophils and macrophages.[92-94]

2.2. Heat-Labile Opsonins

2.2.1. Evidence. A wide variety of particles, including many strains of bacteria and inert polymers, are rendered readily susceptible to enhanced phagocytic ingestion, i.e., they are opsonized, through incubation with nonimmune sera. The opsonic activity of these sera are almostly completely lost when the sera are heated at 56° for 30 min. The physicochemical and biochemical nature of these heat-labile opsonins have been investigated by numerous researchers. The evidence for the involvement of the complement system, and more specifically for the involvement of components of complement, in heat-labile opsonization is overwhelming. In 1969, Smith and Wood[95] summarized this evidence as follows: (1) The heat-labile opsonins have approximately the same heat lability as complement; (2) both heat-labile opsonins and the complement system are active at 37° but neither is active at 0°; (3) both exhibit the same degree of reduction in activity when immune aggregates are added to the sera; (4) both the heat-labile opsonins and complement activity are adsorbed from serum nonspecifically by bacteria at 37° but not at 0°; (5) both systems are largely dependent on the presence of Ca^{2+} and Mg^{2+} for their activity; (6) both exhibit an inhibition of activity in the presence of zymosan, purified cobra venom factor, and ammonium hydroxide. These conclusions have been confirmed in studies utilizing either individual complement components or clinical sera obtained from patients with various complement

---

[91] H. Y. Reynolds, J. P. Atkinson, H. H. Neuball, and M. M. Frank, *J. Immunol.* **114,** 1813 (1975).

[92] A. K. Sobotka, F. J. Malveaux, G. Marone, L. L. Thomas, and L. M. Lichtenstein, *Immunol. Rev.* **41,** 171 (1978).

[93] H. Metzger, *Immunol. Rev.* **41,** 186 (1978).

[94] K. Ishizaka and T. Ishizaka, *Immunol. Rev.* **41,** 109 (1978).

[95] M. R. Smith and W. B. Wood, *J. Exp. Med.* **130,** 1209 (1969).

component deficiencies. Such studies have clearly established that the complement system, if not the sole, is the primary source of heat-labile opsonins.

Evidence for the role of the opsonic properties of complement components acting independently of bound antibody mainly derives from studies involving the alternative pathway. This evidence may be summarized briefly as follows.

1. Serum can opsonize particles in the complete absence of specific immunoglobulins, including "natural" antibody.[96] Newborn serum which does not contain antibodies directed specifically against the test microorganisms is able to give rise to substantial opsonic activity.[97] Preheating of this sera for 30 min at 56° almost entirely destroys its opsonic activity. (Nonspecific antibody adsorption or cross-reacting immunoglobulins cannot, however, be completely ruled out in these investigations.)

2. Agammaglobulinemic sera, with only trace amounts of detectable immunoglobulin, exhibit considerable opsonizing power for a variety of bacterial strains.[98] Removal of trace quantities of immunoglobulin from agammaglobulinemic sera by means of affinity adsorption neither diminishes nor abolishes opsonic activity of these sera.[99] However, such activity is completely removed when the agammaglobulinemic sera are pretreated at 56° for 30 min and when preformed antigen–antibody complexes are added to the sera.[97] (Such immune complexes would bind and thus reduce and remove the available complement.)

3. The earlier components of complement (C1, C2, and C4) are not prerequisites for the heat-labile opsonization of particles. Thus, human serum devoid in C1q opsonizes *E. coli* and *S. aureus*.[99] The substitution of normal human serum by C4-deficient serum as a source of heat-labile opsonins does not alter the extent of phagocytic ingestion of pneumococci by guinea pig neutrophils.[100] Hydrocarbon droplets (paraffin oil) onto which a layer of human serum albumin is adsorbed are opsonized equally well by both C2-deficient sera and by a serum devoid of C2 and C4.[101]

4. Activation of the alternative pathway requires the presence of Mg^{2+} but not Ca^{2+} which, however, is required for activation of the classical pathway. A mixture (Mg-EGTA) of EGTA and $MgCl_2$ can therefore be employed to inhibit the classical pathway (by chelating the available

[96] J. A. Winkelstein, *J. Pediatr.* **82,** 747 (1973).
[97] J. Sterzl, *Folia Microbiol. (Prague)* **8,** 240 (1963).
[98] R. C. Williams and P. G. Quie, *J. Immunol.* **106,** 51 (1971).
[99] H. E. Jasin, *J. Immunol.* **109,** 26 (1972).
[100] J. A. Winkelstein, H. S. Shin, and W. B. Wood, *J. Immunol.* **108,** 1681 (1972).
[101] T. P. Stossel, *J. Cell Biol.* **58,** 346 (1973).

Ca^{2+}) without affecting the alternative pathway.[102] Exposure of several strains of bacteria to normal nonimmune human sera treated in this way with Mg-EGTA results in strong opsonic activity of the bacteria,[103,104] probably through the production of opsonically active C3.

  2.2.2. Antibody Activation of Opsonic Complement. Complement-derived opsonins may be produced through activation of the complement system either via the classical or alternative pathway. The heat-stable (antibody) and heat-labile (complement) components of serum generally interact in a manner which is synergistic rather than merely additive. A large number of investigators have thus reported that when fresh nonimmune sera is mixed with heat-inactivated hyperimmune sera, the opsonic activity of the resultant mixture is much greater than the opsonic activity of either serum alone.[16,28,29,86,105–107] The antibody which reacts with complement to produce the opsonic effect is most often an antibody population with a selective antigenic specificity and developed as a result of a previous exposure of the host to a particular invasive microorganism. However, antibodies with a broad specificity, which cross-react with certain common antigenic determinants on a variety of bacterial strains, will also give rise to complement activation. Such antibodies may develop as a result of the individual's exposure to the same or similar cross-reacting antigens in the intestinal flora or food.[108] In addition to the above, nonspecific adsorption of immunoglobulins, particularly IgG, onto the surface of bacteria will give rise to complement activation. Antibody activation of the complement system, and thus the production of opsonic complement components, can be achieved in either of two ways: (1) classical pathway activation through the $C\overline{142}$ pathway, or (2) activation of the alternative pathway through either the production of activated C3b or activation of the C1-bypass pathway. Sandberg *et al.* have suggested that only the Fc portion of antibodies is required for activation of the classical pathway, whereas the F(ab')$_2$ fragment is required for activation of the alternative pathway.[109,110] The studies cited in this section lead to the conclusion that

[102] D. P. Fine, S. R. Marney, D. G. Colley, J. S. Sergent, and R. M. Des Prez, *J. Immunol.* **109**, 807 (1972).

[103] A. Forsgren and P. G. Quie, *Immunology* **26**, 1251 (1974).

[104] A. Forsgren and P. G. Quie, *Infect. Immun.* **10**, 402 (1974).

[105] H. E. Eggers, *J. Infect. Dis.* **5**, 263 (1908).

[106] J. W. Shands, H. Stalder, and E. Suter, *J. Immunol.* **96**, 68 (1966).

[107] R. D. Diamond, J. E. May, M. A. Kane, M. M. Frank, and J. E. Bennett, *J. Immunol.* **112**, 2260 (1974).

[108] J. G. Michael, J. L. Whitby, and M. Landy, *J. Exp. Med.* **15**, 131 (1962).

[109] A. L. Sandberg, B. Oliveria, and A. G. Osler, *J. Immunol.* **106**, 282 (1971).

[110] A. L. Sandberg, O. Götze, H. J. Müller-Eberhard, and A. G. Osler, *J. Immunol.* **107**, 920 (1971).

maximal extent and rates of phagocytic ingestion require the presence of both heat-stable and heat-labile opsonins. Either system alone, however, may facilitate opsonization and result in enhanced phagocytic ingestion as compared to nonopsonized particles. The different mechanisms by which bacteria may be opsonized is best examined in the context of the different host–parasite relationships that exist. Since the heat-labile opsonizing system is present in nonimmune sera and, as has been shown above, is not dependent on the presence of specific antibody, it probably plays an important role in the early and critical pre-antibody stages of infections. Later, if the infection lasts, specific antibody is produced and, acting in concert with the complement system, also acts as a most effective opsonin. In the presence of large concentrations of specific IgG type antibody, the contribution by complement to the opsonizing process is likely to be less critical.

As mentioned previously, these different mechanisms of opsonization are not mutually exclusive but probably act concurrently, and in a synergistic manner, *in vivo*. However, one mechanism may be favored over another depending on the type of invasive microorganism, the stage of infection, and the presence or absence (and concentration) of specific antibody. It is worthwhile reiterating that even in the complete absence of opsonizing agents phagocytic ingestion of the particles is likely to occur although to a lesser extent.

2.2.3. Nature of Complement-Derived Components. The primary source of the heat-labile opsonins is derived from C3. In addition, however, C4 and C5 opsonic activity also has been described.

2.2.3.1. C3-derived opsonins: Early studies implied that only C1, C2, and C4 were required to produce opsonic activity via the classical complement pathway.[87,111] Subsequent studies have now unequivocally established that the presence of C3 is also a necessary prerequisite.[61,63,112–114] At the present time it is the generally accepted belief that the major opsonic component of complement, derived from activation either of the classical or alternative pathways, is a by-product of C3. This is supported by investigations using the sequential addition of isolated and purified complement (C) components to antibody-sensitized *E. coli,* which indicate that opsonization takes place at the *E. coli*–antibody–C1423 stage.[8,115,116]

[111] E. E. Ecker and G. Lopez-Castro, *J. Immunol.* **55,** 169 (1947).

[112] I. Gigli and R. A. Nelson, *Exp. Cell Res.* **51,** 45 (1968).

[113] R. B. Johnston, M. R. Klemperer, C. A. Alper, and F. S. Rosen, *J. Exp. Med.* **129,** 1275 (1969).

[114] H. S. Shin, M. R. Smith, and W. B. Wood, *J. Exp. Med.* **130,** 1229 (1969).

[115] C. J. van Oss and C. F. Gillman, *Res, J. Reticuloendothel. Soc.* **12,** 497 (1972).

[116] C. J. van Oss and C. F. Gillman, *Immunol. Commun.* **2,** 415 (1973).

Further sequential addition of C5, C6, C7, C8, and C9 do not result in any enhanced phagocytic ingestion. Thus, C1423 appears to be the principal "heat-labile" opsonin. Such conclusions are fully supported by clinical studies of the opsonic activity of C3-deficient sera.[117]

The dominant opsonin generated from complement is believed to be the high-molecular-weight C3b fragment of C3. Stossel et al. have suggested that opsonic activity per se does not reside in the C3b fragment as such but rather in a low-molecular-weight derivative.[118] It is well documented that C3b can bind to a wide variety of substances. The mechanism of this binding has yet to be elucidated. Particles (including bacteria) with surface-bound C3b attach with a strong affinity to the membrane of neutrophils. It appears that this is due to the presence of C3b receptors on neutrophil membranes (see also Section 2.3). Due to the presence of C3b inactivators[119,120] in serum, subsequent exposure of particle-bound C3b to sera results in the conversion (through cleavage) of the bound C3b into the antigenically distinct C3c and C3d fragments.[121] As a consequence such particles lose their ability to bind to neutrophil C3b receptors. Thus, C3b attachment to C3b receptors must occur reasonably rapidly if phagocytic ingestion is to be enhanced (C3b bound to surfaces that activate the alternative complement pathway appear to be relatively resistant to degradation by the inactivators).[121a] Cleavage results in the release of C3c but leaves C3d firmly attached to the particle. Neutrophils do not possess receptors for C3d[122] in contrast to monocytes[91] and lymphocytes.[123]

   2.2.3.2. C4-derived opsonins: Immune adherence studies with B lymphocytes[124,125] and neutrophils[126] using sheep erythrocytes sensitized with C1 and C4b suggest the presence on the cell surface of C4b receptors. A characteristic property of opsonins is their reactivity in immune adherence studies. The implication of this being that C4b is an opsonin.

[117] S. G. Osofsky, B. H. Thompson, T. F. Lint, and H. Gerwurz, *J. Pediatr.* **90,** 180 (1977).

[118] T. P. Stossel, R. J. Field, J. D. Gitlin, C. A. Alper, and F. S. Rosen, *J. Exp. Med.* **141,** 1329 (1975).

[119] N. Tamura and R. A. Nelson, *J. Immunol.* **99,** 582 (1967).

[120] P. J. Lachmann and H. J. Müller-Eberhard, *J. Immunol.* **100,** 691 (1968).

[121] H. J. Müller-Eberhard, in "The Phagocytic Cell in Host Resistance" (J. A. Bellanti and D. H. Dayton, eds.), p. 87. Raven Press, New York, 1975.

[121a] D. T. Fearon and K. F. Austen, *N. Engl. J. Med.* **303,** 259 (1980).

[122] G. D. Ross, M. J. Polley, E. M. Rabellina, and H. M. Grey, *J. Exp. Med.* **138,** 798 (1973).

[123] A. Eden, G. W. Miller, and V. Nusenzweig, *J. Clin. Invest.* **52,** 3239 (1973).

[124] V. A. Bokisch and A. T. Sobel, *J. Exp. Med.* **140,** 1336 (1974).

[125] A. T. Sobel and V. A. Bokisch, *Fed. Proc., Fed. Am. Soc. Exp. Biol.* **33,** 759 (1974).

[126] G. D. Ross and M. J. Polley, *J. Exp. Med.* **141,** 1163 (1975).

2.2.3.3. C5-derived opsonins: C5-deficient mouse serum exhibits a decreased opsonic activity of yeast *in vitro.*[127] *In vivo* studies also demonstrate that C5-deficient mice are more susceptible to experimentally induced candidiasis than are normal mice.[128,129] However, in view of the role of C5 in influencing the inflammatory response in general, e.g., production of a chemotactic factor, such *in vivo* and clinical[130,131] studies are difficult to interpret as direct evidence of the opsonic role of C5.

2.3. Neutrophil Membrane Complement Receptors

For an excellent comprehensive review of the nature, location, and function of various complement receptors, the interested reader is referred to Arnaout and Colten.[131a]

As mentioned previously, maximum phagocytic uptake will occur when both the heat-labile and heat-stable opsonins are operative. A striking synergistic enhancement of the phagocytic ingestion has been noted when both types of opsonins are present.[132] The effect of (particle) surface-bound C3b is that it very considerably reduces the amount of adsorbed (specific or nonspecific) IgG required to promote particle ingestion.[132] Several researchers have suggested that the primary function of the complement C3b receptor is to promote adherence of the C3b-coated particle to the neutrophil membrane, whereas the Fc–Fc receptor interaction is required in order to facilitate actual ingestion of the particles.[132–134] In this light it is interesting to note that van Oss and colleagues have shown that the formation of the *E. coli*–antibody–$C\overline{1423}$ complex is accompanied by a marked increase in the surface hydrophobicity of the complex.[8,115] This increase in hydrophobicity strongly enhances the propensity for the C3b complex to bind to the neutrophil membrane through nonspecific adsorption mechanisms[9] in addition to the C3b receptor-mediated binding.[132] Indeed, a number of nonimmunological agents which promote enhanced binding of the particle to the phagocyte membrane have been shown to readily mimic the contribution of C3 to the opsonization

[127] M. E. Miller and U. R. Nilsson, *Clin. Immunol. Immunopathol.* **2,** 246 (1974).

[128] R. Morelli and L. T. Rosenberg, *Infect. Immun.* **3,** 521 (1971).

[129] R. Morelli and L. T. Rosenberg, *J. Immunol.* **107,** 476 (1971).

[130] M. E. Miller and P. J. Koblenzer, *J. Pediatr.* **80,** 879 (1972).

[131] S. I. Rosenfeld, J. Baum, R. T. Steigbigel, and J. P. Leddy, *J. Clin. Invest.* **57,** 1635 (1976).

[131a] M. A. Arnaout and H. R. Colten, *Mol. Immunol.* **21,** 1191 (1984).

[132] A. G. Ehlenberger and V. Nussenzweig, *J. Exp. Med.* **145,** 357 (1977).

[133] B. Mantovani, *J. Immunol.* **115,** 15 (1975).

[134] D. J. Scribner and D. Fahrney, *J. Immunol.* **116,** 892 (1976).

TABLE IV
MODES OF OPSONIZATION

Mode	Heat-stable opsonization	Heat-labile opsonization
Aspecific opsonization Most pronounced with the most hydrophobic bacteria, which are therefore the least pathogenic[a]	Via aspecific adsorption of IgG$_1$ and/or IgG$_3$, through Fc γ receptors on phagocytes (neutrophils as well as macrophages)	Via aspecific adsorption of IgG$_1$ and/or IgG$_3$, in the presence of C$\overline{1423}$; mediated by C3b through C3b receptors on phagocytes (neutrophils as well as macrophages)
Specific opsonization Specific opsonization occurs only after specific antibodies have been formed especially against capsular antigens	Via specific binding of IgG$_1$ and/or IgG$_3$ through Fc γ receptors on phagocytes (neutrophils as well as macrophages)	Via specific binding of IgG$_1$, IgG$_3$, and/or IgG$_2$ and/or IgM, in the presence of C$\overline{1423}$; mediated by C3b through C3b receptors on phagocytes (neutrophils as well as macrophages)

[a] That is, in the sense of being the most prone to phagocytic engulfment.

process.[132] Such agents increase the surface hydrophobicity of the particles, thereby facilitating enhanced particle–phagocyte interactions.[135]

The physicochemical nature of the C3b membrane receptor has not been fully elucidated.[131a] Since the receptor is inactivated by trypsin but not by neuraminidase[61] it is likely that at least part of the moiety is proteinaceous.

C3d receptors have been demonstrated as being present on the membrane of both macrophages[91,132] and lymphocytes[122] but not on neutrophils.[123] A definitive distinction between C3b and C3d receptors has not yet been documented. Indeed, on some cells, the two receptors appear to be very closely related if not identical.[124,126]

2.4. Cooperativity of IgG and C3b in the Enhancement of Phagocytosis

The various modes of opsonization are summarized in Table IV. The interaction between C3b and the membrane C3b receptors represents an important aspect of phagocytic ingestion. In vitro erythrophagocytosis of

[135] C. J. van Oss, D. R. Absolom, and A. W. Neumann, Ann. N.Y. Acad. Sci. **416**, 332 (1983).

IgM-sensitized cells requires the presence of C3b.[112] In studies using human monocytes,[58] IgM-type antibody was shown to be ineffective as an opsonin for erythrocytes. However, in the presence of complement, and specifically when approximately 1000 molecules of C3b were bound to the IgM-coated erythrocytes, rapid ingestion occurred.[58] In contrast, erythrophagocytosis readily occurs in the absence of complement when the erythrocytes are sensitized with IgG-type antibodies alone[112] due to the Fc–Fc receptor interaction (cf. Section 2.1.2). In these *in vitro* studies, ingestion was completely inhibited by concentrations of free IgG far below that occurring *in vivo* in normal serum. Inhibition by free IgG, however, was overcome when approximately 100 molecules of C3b were bound to each erythrocyte.[58] These studies have provided considerable insight into the phagocytic process which may be summarized as follows:

1. IgG and C3b receptors present on the membrane of phagocytes probably function independently of each other in promoting ingestion.
2. The two kinds of receptors may exert a cooperative effect on the ingestion of particles opsonized with IgG in the presence of inhibitory concentrations of free IgG. This latter condition clearly pertains to *in vivo* conditions and underlines the importance of the role of C3b as a promoter of phagocytosis.

2.5. Other Serum Factors Which Facilitate Phagocytosis

A number of reports have appeared in the literature describing the presence in serum of natural phagocytosis-enhancing factors which are distinct from opsonic immunoglobulins and activated complement components. Generally these are substances that can aspecifically adsorb or adhere onto microorganisms thereby altering the microbe's surface properties.

2.5.1. Fibronectin (see this volume [16]). One extremely well-known example is fibronectin, a general opsonin and adhesion-inducing agent. This protein is also known under a wide variety of other names[136]: cold-insoluble globulin, antigelatin factor, microfibrillar protein, opsonic protein, fibroblast surface antigen, galactoprotein-a, cell surface protein, cell-spreading factor, zeta, large external transformation-sensitive protein, as well as opsonic surface-binding α_2-glycoprotein.[137,138,138a] This opsonin, first detected in rat serum, has molecular weight and electrophoretic mo-

[136] K. M. Yamada and K. Olden, *Nature (London)* **275**, 179 (1978).
[137] F. A. Blumenstock, T. M. Saba, P. Weber, and R. Laffin, *J. Biol. Chem.* **253**, 4287 (1978).
[138] T. M. Saba, *Immune Syst. Infect. Dis., Int. Convoc. Immunol., 4th, 1974*, p. 489 (1975).
[138a] C. J. van Oss, C. F. Gillman, P. M. Bronson, and J. R. Border, *Immunol. Commun.* **3**, 329 (1974).

bility characteristics similar to those of human α_2-macroglobulin.[136] Fibronectin stimulates rapid phagocytic uptake of particles such as lipid vesicles or gelatin-coated colloidal gold.[136]

2.5.2. Human Serum (HS)α_2-Glycoprotein. van Oss *et al.* were able to demonstrate that HSα_2-glycoprotein adsorbs strongly onto the surface of *E. coli* and *S. aureus.*[138a] In doing so, the microbial surface was rendered significantly more hydrophobic and phagocytic ingestion of the treated microorganisms was considerably enhanced.

2.5.3. Tuftsin (see this volume [14]). Tuftsin was discovered as a result of an observation by Najjar *et al.*[139–143] that a certain fraction (subsequently called leucokinin) of homologous IgG bound to autologous neutrophils. The active portion promoting such interaction was found to be a tetrapeptide, Thr-Lys-Pro-Arg, located in the Fc portion of the leucophilic IgG.[144–146] Two enzymes appear to be responsible for the production of tuftsin from leucokinin. The first is leukokinase located on the outer surface of neutrophil membranes and which cleaves at the amino terminal end of the L-threonine residue; the second is an enzyme produced in the spleen which cleaves at the carboxyl terminal end of the L-arginine residue.[141,145,146] Splenectomized animals have been shown not to produce tuftsin although their leucokinin levels are normal.[142,147]

Tuftsin enhances phagocytic ingestion 2- to 3-fold by neutrophils as well as macrophages.[142] The mechanism by which tuftsin stimulates phagocytosis has not been completely unravelled. The tetrapeptide, unlike most opsonins, binds to the phagocytic cell rather than to the ingested particle. Since tuftsin is a strongly basic (i.e., positively charged) peptide it is likely that it attaches, via electrostatic bonds, to the negative charge groups, already present in high density, on neutrophils. It may be speculated that negatively charged microorganisms (opsonized and nonopsonized) will then interact with the positively charged tuftsin, thereby promoting neutrophil–bacterium interactions and hence enhanced phagocytic ingestion (see this volume [2]).

[139] B. V. Fidalgo and V. A. Najjar, *Proc. Natl. Acad. Sci. U.S.A.* **57,** 957 (1967).
[140] B. V. Fidalgo and V. A. Najjar, *Biochemistry* **6,** 3386 (1967).
[141] V. A. Najjar and K. Nishioka, *Nature (London)* **228,** 672 (1970).
[142] V. A. Najjar and A.Constantopoulos, *Res, J. Reticuloendothel. Soc.* **12,** 197 (1972).
[143] V. M. Najjar and V. A. Najjar, *Mol. Cell. Biochem.* **13,** 45 (1976).
[144] V. A. Najjar, *Adv. Enzymol.* **41,** 129 (1974).
[145] K. Nishioka, A. Constantopoulos, P. S. Satoh, and V. A. Najjar, *Biochem. Biophys. Res. Commun.* **47,** 172 (1972).
[146] K. Nishioka, A. Constantopoulos, P. S. Satoh, W. Mitchell, and V. A. Najjar, *Biochim. Biophys. Acta* **310,** 217 (1973).
[147] V. A. Najjar, B. V. Fidalgo, and E. Stitt, *Biochemistry* **7,** 2376 (1968).

2.5.4. C-Reactive Protein. Also shown to significantly enhance the phagocytic ingestion of bacteria by human neutrophils is the so-called C-reactive protein. This is a group III-type acute-phase protein. A several hundredfold increase in the concentration of C-reactive protein in human serum occurs following a large variety of inflammatory or injurious stimuli, including many infections, surgical or other trauma, fractures, myocardial infarctions, and during chronic inflammatory states such as rheumatoid arthritis.[148,149] Following stimulus, the serum concentration level of C-reactive protein increases sharply, usually beginning within a few hours.

C-reactive protein owes its name to the circumstances which led to its discovery. It was originally detected in human serum because of its ability to precipitate with the somatic C-polysaccharide pneumococcus antigen.[150] This reaction has since been found to result from the binding of this protein to phosphocholine moieties in the polysaccharide.[151,152] It should be emphasized that C-reactive protein is not an immunoglobulin and that substrate: C-protein binding is not an antigen–antibody type of reaction. This protein is isolated from acute phase serum by means of 50 to 75% saturated sodium sulfate along with the serum albumin fraction.[153] It can be separated from serum albumin by dialysis against distilled water under which conditions the C-reactive protein precipitates.

While the precise biological role of C-reactive protein is unclear, a large number of recognition and activation functions are exhibited by the protein and which have the potential for affecting inflammation and other defence mechanisms.[154,155] Included among the recognition functions is binding to phosphocholine moieties (such as those found on leukocyte membranes), galactans, and polycations. In addition, it has the capacity to recognize, and the ability to bind to, a number of substances such as exogenous materials such as those derived from tissue breakdown. Once this complexing to "foreign" particles has occurred many cellular and biochemical components of both the inflammatory and immune responses are modulated. The mechanism by which complexed C-reactive protein enhances phagocytic ingestion is not completely understood. However, it is known, that surface-bound C-reactive protein activates the comple-

[148] I. Kushner, *Ann. N.Y. Acad. Sci.* **389**, 39 (1982).
[149] J. J. Morley and I. Kushner, *Ann. N.Y. Acad. Sci.* **389**, 406 (1982).
[150] T. J. Abernathy and O. T. Avery, *J. Exp. Med.* **73**, 173 (1941).
[151] M. McCarty, *Ann. N.Y. Acad. Sci.* **389**, 1 (1982).
[152] J. E. Volanakis and M. H. Kaplan, *Proc. Soc. Exp. Biol. Med.* **136**, 612 (1971).
[153] C. M. MacLeod and O. T. Avery, *J. Exp. Med.* **73**, 183 (1941).
[154] H. Gewurz, C. Mold, J. Siegel, and B. Fiedel, *Adv. Intern. Med.* **27**, 345 (1982).
[155] H. Gewurz, *Hosp. Pract.* **17**, 67 (1982).

ment system[154] and further that the complexed protein can bind with a high affinity, to the membrane of various phagocytes.[149] Thus it would appear that C-reactive protein mediated opsonization results in complement as well as C-receptor modulated phagocytosis.

2.5.5. *Others.* Several other serum components have, over the years, been implicated as stimulators of phagocytosis and thus have been called opsonins. These include the following heat-labile factors: α_1-globulin and β-globulin reported by Tullis and Surgenor in 1956[156]; the so-called "coopsonin" originally described by Stollerman *et al.*[157] and which was shown to enhance ingestion of the normally resistant group A streptococci which possess the M-protein dysopsonin on their outer surface.[158] Purified Hageman factor, or clotting factor XII, causes a decrease in the surface hydrophobicity of neutrophils and a concomitant increase in their phagocytic activity[8] and adhesiveness.[159]

Heparin reduces the surface hydrophobicity of macrophages and an increase in their phagocytic activity.[8,160] Levamisole (the L-isomer of 2,3,5,6-tetrahydro-6-phenylimidazo[2,1-*b*]thiazole hydrochloride), which has been shown to have immunostimulating[161,162] as well as tumor inhibiting[162] properties, also reduces the surface hydrophobicity of guinea pig macrophages and causes an increase in their phagocytic activity.[8]

3. Dysopsonins

A number of naturally occurring substances are known to inhibit phagocytic ingestion.

3.1. IgA

IgA (as well as secretory IgA, sIgA) may be regarded as a true dysopsonin. It prevents the adherence of microorganisms to epithelial surfaces, as well as their ingestion by neutrophils and macrophages.[163] The aforementioned observations have been clearly demonstrated despite the fact that subpopulations of both neutrophils and monocytes possess Fc recep-

[156] J. L. Tullis and D. M. Surgenor, *Ann. N.Y. Acad. Sci.* **66,** 386 (1956).

[157] G. H. Stollerman, M. Rytel, and J. Ortiz, *J. Exp. Med.* **117,** 1 (1963).

[158] G. J. Domingue and W. A. Pierce, *J. Bacteriol.* **89,** 583 (1965).

[159] J. W. Alexander and R. A. Good, *in* "Immunobiology for Surgeons," p. 10. Saunders, Philadelphia, Pennsylvania, 1970.

[160] A. G. Kitchen and R. Megirian, *J. Biol. Chem.* **246,** 3760 (1971).

[161] G. Renoux and M. Renoux, *C. R. Hebd. Seances Acad. Sci., Ser. D* **272,** 349 (1972).

[162] G. Renoux and M. Renoux, *Nature (London), New Biol.* **240,** 217 (1972).

[163] H. Y. Reynolds, W. M. Merrill, E. D. Amento, and G. P. Nagel, *in* "Secretory Immunity and Infection" (J. R. McGhee, J. Mestecky, and J. L. Babb, eds.), p. 533. Plenum, New York, 1978.

tors specific for IgA.[164] Purified colostral IgA (11 S) has no opsonizing effect on a wide variety of gram-negative and gram-positive microorganisms.[46] Colostral IgA also has no opsonic activity for several *Shigella* strains.[90] Human 11 S secretory IgA (sIgA) exhibits no opsonic activity with respect to human erythrocytes in a system containing human neutrophils or monocytes as the phagocytic cells.[165] Similar negative opsonic activity results were found for sIgA and *S. aureus*.[166]

The marked hydrophilicity of IgA makes it uniquely suited for preventing viruses from penetrating cells.[38] The dysopsonic function of IgA and sIgA may be especially important in preventing infection (particularly viral infections) in the upper respiratory and lower digestive mammalian anatomy where exclusion and subsequent removal of microorganisms (e.g., by coughing) is feasible without the assistance of phagocytosis.[167]

3.2. Others

3.2.1. Pharmacological. Several pharmacological agents have been reported to affect the extent of phagocytic ingestion. Generally these agents act either by altering the surface properties of the phagocyte or particle (or both), thereby affecting nonspecific antibody adsorption and hence complement activation, or by altering the metabolic activity of the phagocyte. The latter mechanism is not regarded as an opsonic activity and therefore will not be considered here.

Antibiotics have the potential to modulate the opsonic process at several levels. They may alter the deposition of opsonins on the bacterial surface by direct interaction with the opsonic proteins; by a modification of the attachment of opsonins to the bacterial surface; by removal (or exposure) of opsonin-binding (or inhibiting) molecules on the bacterial surface; by altering neutrophil Fc or C3b receptors. The interactions of antibiotics with bacteria and host defense mechanisms are multifaceted and complex. Overall, however, it appears that relatively brief periods of treatment of bacteria with low concentrations of cell-wall active antibiotics enhance susceptibility to neutrophil killing. It seems likely that this is a direct consequence of antibiotic alteration of the bacterial cell wall. The nature and mechanism by which those structural changes are induced have not yet been clarified.

[164] M. W. Fanger, L. Shen, J. Pugh, and G. M. Bernier, *Proc. Natl. Acad. Sci. U.S.A.* **79,** 3640 (1980).

[165] A. Zipursky, E. J. Brown, and J. Bienenstock, *Proc. Soc. Exp. Biol. Med.* **142,** 181 (1973).

[166] I. D. Wilson, *J. Immunol.* **108,** 726 (1972).

[167] J. F. Heremans, *Immune Syst. Infect. Dis., Int. Convoc. Immunol., 4th, 1974,* p. 376 (1975).

A wide variety of *in vitro* data are available regarding the role of antibiotics in various phagocytic systems. Some of those data are summarized in Table V.[167a] Altered phagocytosis may reflect either altered opsonization or a modified interaction of antibiotic-treated organisms with the phagocytic cell. For an outstanding overview of antibiotic–neutrophil interactions the interested reader is referred to Yourtee and Root.[167a]

The mechanisms by which the attachment or ingestion phase is altered have not been clearly defined. Several antimicrobial agents have been shown to be capable of inhibiting activation of the complement cascade which is known to be a major source of opsonins (cf. Section 2.2). For example, sulfadiazine, tetracycline, and gentamicin have all been shown to inhibit the alternative pathway *in vitro*. Such inhibition may well significantly impair the ability of the host to generate chemotactic and opsonic factors *in vivo*, although this remains to be proved.

In addition, significant alterations in the surface hydrophobicity of either the phagocyte or the bacteria have been implicated. (For a discussion of the relation between the extent of phagocytic ingestion and membrane surface hydrophobicity see this volume [2].) A number of antibiotics, in subinhibitory doses, depress the phagocytic ingestion of bacteria by neutrophils, e.g., penicillin, chloromycetin, polymyxin, bacitracin,[8] aureomycin,[168] and oral phenethicillin.[169] In the first mentioned group the effect of the antibiotic has been to produce a change in the surface properties of the bacteria which become markedly more hydrophilic.[8] On the other hand, lectins, such as phytohemagglutinin, concanavalin A, and ampicillin cause a decrease in surface hydrophobicity of the phagocytes themselves[8] and consequently a marked decrease in the extent of phagocytic ingestion.[8,170] Subinhibitory doses of ampicillin, chloramphenicol, and tetracycline result in an enhanced phagocytic uptake of *L. monocytogenes* by human macrophages.[171] The mechanism by which this occurs has not been clearly established but it is likely to be related to surface hydrophobicity changes in the membranes of either the macrophages or the bacteria.

Sugars have been reported to decrease phagocytic activity of neutro-

[167a] E. L. Yourtee and R. K. Root, *in* "Advances in Host Defence Mechanisms" (J. I. Gallin and A. S. Fauci, eds.), p. 187. Raven Press, New York, 1982.

[168] J. Muñoz and R. Geister, *Proc. Soc. Exp. Biol. Med.* **75**, 367 (1950).

[169] D. B. Louria, *in* "Opportunistic Pathogens" (J. E. Prier and H. Friedman, eds.), p. 1. University Park Press, Baltimore, Maryland, 1974.

[170] E. S. Golub, "The Cellular Basis of the Immune Response." Sinauer Assoc., Sunderland, Massachusetts, 1977.

[171] D. Adam, W. Schaffert, and W. Marget, *Infect. Immun.* **9**, 811 (1974).

TABLE V
MODIFICATION OF SPECIFIC MICROBE–PHAGOCYTE INTERACTIONS BY ANTIBIOTICS[a]

Function	Increase	Decrease	No effect
Phagocytosis			
Opsonization	Group A streptococci		Group A streptococci
	Lincomycin		Penicillin
	Clindamycin		Chloramphenicol
	Erythromycin		
	Group B streptococci		
	β-Lactams		
	Vancomycin		
Ingestion	Group A. streptococci,		Group A streptococci,
	S. pneumoniae		S. pneumoniae
	Lincomycin		Penicillin
	Clindamycin		Erythromycin
	S. aureus		Chloramphenicol
	Nafcillin		
	E. coli		S. aureus
	Gentamicin		β-Lactams
	Streptomycin		Vancomycin
	Polymyxin B		P. aeruginosa
	P. aeruginosa		Gentamicin
	Carbenicillin		
	L. monocytogenes		
	Ampicillin		
	Chloramphenicol		
	Tetracycline		
	B. recurrentis		
	Penicillin		
Killing	S. aureus	S. aureus	S. aureus
	β-Lactams	Oxacillin	Tetracycline
	Vancomycin		Chloramphenicol
	Group A streptococci		Erythromycin
	Penicillin		Clindamycin
	S. pneumoniae,		Gentamicin
	S. sanguis		E. coli
	β-Lactams		β-Lactams
	E. coli		
	Chloramphenicol[94,96]		
	Gentamicin		
	B. fragilis		
	Chloramphenicol		

(continued)

TABLE V (*continued*)

Function	Increase	Decrease	No effect
Susceptibility to lysozyme	Group A streptococci Penicillin *S. pneumoniae, S. fecalis, S. aureus* Nafcillin *N. catarrhalis* Polymyxin B *K. pneumoniae, S. typhimurium* Polymyxin B		*S. aureus* Oxacillin Cephalothin Vancomycin *N. catarrhalis* Nafcillin

[a] Those antibiotics which have been found to alter certain interactions of specific bacteria with leukocytes are listed next to the organism studied.[167a]

phils both *in vitro* and *in vivo*.[172] High glucose levels (0.2 to 0.8%) cause a decrease in phagocytic uptake by both neutrophils and macrophages.[173] No alteration was observed in the surface hydrophobicity of the phagocytes and the cause of the decrease in phagocytic ingestion appears to be due to a suppression of pseudopod formation and the increased tendency for the neutrophils to assume a spherocytic shape. Such shape changes, and particularly the reduced ability to form pseudopods, will give rise to an enhanced electrostatic repulsion between the both negatively charged phagocyte and bacteria,[174] thereby preventing the first critical step of attachment and thus resulting in decreased phagocytic ingestion (see this volume [2]). These shape changes may be responsible for the increased incidence of bacterial infections among uncontrolled diabetics.[171] In galactosemia the high galactose concentration (30 mM) may similarly cause shape changes in the phagocytes accounting, in part, for the increased susceptibility to infection in this clinical condition.[175,176]

   3.2.2. Microbial Components. In order to prevent their phagocytic ingestion several microorganisms have evolved mechanisms which are antiphagocytic in nature. These mechanisms may be classified broadly into two groups: (1) The release of soluble components (toxins) which inhibit various metabolic functions of the phagocytes; and (2) the develop-

[172] A. Sanchez, J. L. Reezer, H. S. Lau, P. Y. Yahiku, R. E. Willard, P. J. McMillan, S. Y. Cho, A. R. Magie, and U. D. Register, *Am. J. Clin. Nutr.* **26,** 1180 (1973).
[173] C. J. van Oss, *Infect. Immun.* **4,** 54 (1971).
[174] C. J. van Oss, R. J. Good, and A. W. Neumann, *J. Electroanal. Chem.* **37,** 387 (1972).
[175] W. J. Litchfield and W. W. Wells, *Infect. Immun.* **13,** 728 (1976).
[176] W. J. Litchfield and W. W. Wells, *Infect. Immun.* **16,** 189 (1977).

ment of special surface structures which enable them to escape phago-cytic ingestion.

3.2.2.1. Toxins: These products do not really constitute true dysop-sonins since generally they are agents which readily penetrate the neutro-phil plasma membranes and cause intracellular damage. Consequently they will be discussed only briefly here. Examples of such toxins are the S and O streptolysins formed by streptococci which cause lysis of the intra-cytoplasmic granules with the release of toxic agents and hence damage to the neutrophil itself.[177-179] Another example is staphylococcal leucocidin whose mechanisms of destruction is similar to that described above.[180] A similar leucocidin from *Pseudomonas* strains has also been reported.[181,182] The antiphagocytic role of such toxins *in vivo* has not been clearly estab-lished.

3.2.2.2. Microbial surface structures: One of the principal means of protection against opsonization and ingestion is the bacterial capsule. This is a thick "slimy" layer of extremely hydrophilic polysaccharide, protein, or glycoprotein surrounding the bacterial cell. Aspecific opsoni-zation through the adsorption of nonspecific IgG (cf. Section 2.1.1.2) is rendered ineffective by the capsule, since adsorption of IgG only takes place onto the bacterial wall, which lies well inside the capsular material, so that no Fc tails protrude.[20,21,115] Thus Fc–Fc receptor interactions can-not occur.

At the same time, aspecific ingestion is thwarted by the pronounced hydrophilicity of the capsular surface. Thus, only those antibodies specifi-cally formed to bind onto the capsule at its outer periphery are capable of opsonization.[21,115] A now classic illustration of this property is the pneumococci, which when fully encapsulated (smooth or S strain) com-pletely resists phagocytic ingestion and the clinically significant virulence of this microorganism is due to encapsulation.[19] Enzymatic removal of the capsular material (which increases surface hydrophobicity) or the attach-ment of specific anticapsular antibodies (Fc–Fc receptor reaction) pro-motes rapid phagocytic ingestion of the S strain, thereby greatly reducing the pathogenicity of this microorganism. In striking contrast the nonen-capsulated rough or R pneumococci strains are readily ingested and gen-erally avirulent.

[177] A. W. Bernheimer and L. L. Schwartz, *J. Pathol. Bacteriol.* **79**, 37 (1960).

[178] J. G. Hirsch, *J. Exp. Med.* **116**, 827 (1962).

[179] G. Weissman, B. Becher, and L. Thomas, *J. Cell Biol.* **22**, 115 (1964).

[180] A. M. Woodin, *in* "Lysosomes in Biology and Pathology" (A. Durgle, ed.), p. 395. North-Holland Publ., Amsterdam, 1973.

[181] W. Scharmann, *Infect. Immun.* **13**, 836 (1976).

[182] W. Scharmann, *Infect. Immun.* **13**, 1046 (1976).

Another mechanism for thwarting opsonization (and thus phagocytic ingestion) is displayed by various strains of staphylococci which generally possess protein A as a cell wall component.[183] Protein A binds IgG (IgG$_1$ IgG$_2$, and IgG$_4$) molecules at the Fc tail, so that only the Fab moieties protrude from the bacterial surface.[183] This effectively prevents the Fc portion from interacting with the membrane Fc receptors on the phagocyte, while at the same time conferring an overall hydrophilic surface on the bacteria by means of the thus exposed (hydrophilic) Fab domains.[8] In this manner both specific (Fc–Fc receptor) and nonspecific (surface hydrophobicity) modes of phagocytic ingestion are overcome. Thus, in the presence of IgG, protein A-rich strains of staphylococci are ingested to a significantly less extent than are protein A-poor strains.[184] In the absence of IgG (or in the presence of IgG-deficient sera) protein A-rich strains are ingested to a greater extent than are protein A-poor strains.[184]

Edebo and Norman[185] have shown that the virulence of *S. typhimurium* 395 MS and the *R* mutants MRO-MRIO was dependent on the length of the associated basal core oligosaccharide and the S-antigen (a repeating carbohydrate unit). These bacterial strains were ingested to different degrees by human neutrophils. Using hydrophobic interaction chromatography as a means of assessing bacterial surface hydrophobicity, they were able to demonstrate a striking correlation between bacterial hydrophobicity and the extent of neutrophil ingestion.[186] In the case of the rough mutants (R strains) the quantity of outer membrane proteins and total membrane lipopolysaccharide is reduced whereas the phospholipid content is increased.[187] These *R* mutants, which exhibit a marked increase in surface hydrophobicity, are readily ingested by both rabbit and human neutrophils and were the least virulent.[188] In contrast, the *S* mutants, which are very much more hydrophilic, are ingested (in both the absence and presence of nonimmune sera) to a very much lesser extent and tend to be much more virulent.[189] Similar studies have been employed to correlate the *in vivo* pathogenicity (i.e., resistance to phagocytosis) of different *E.*

[183] A. Nisonoff, J. E. Hopper, and S. B. Spring, "The Antibody Molecule," p. 69. Academic Press, New York, 1975.
[184] P. K. Peterson, J. Verhoef, L. D. Sabath, and P. G. Quie, *Infect. Immun.* **15**, 760 (1977).
[185] L. Edebo and B. Norman, *Acta Pathol. Microbiol. Scand., Sect. B: Microbiol. Immunol.* **78B**, 75 (1970).
[186] K.-E. Magnusson, O. Stendahl, I. Stjernström, and L. Edebo, *Immunology* **36**, 439 (1979).
[187] G. F. Ames, E. N. Spudich, and N. Nikaido, *J. Bacteriol.* **11**, 406 (1974).
[188] O. Stendahl, C. Tagesson, and M. Edebo, *Infect. Immun.* **8**, 36 (1973).
[189] O. Stendahl, Ph.D. Dissertation, Linköping University, Sweden (1973).

coli strains with their surface hydrophobicity.[190] Bacterial hydrophobicity was shown to be dependent on the relative presence of the K (acidic polysaccharide) and O (lipopolysaccharide) antigens on the bacterial surface.[191]

4. Clinical Consequences of Defective Opsonization

The focus of this discussion is restricted to the concepts outlined earlier in this chapter. No attempt has been made to comprehensively survey the literature.

Phagocytic disorders may be divided into extrinsic and intrinsic defects. Included in the extrinsic category are antibody and complement abnormalities and deficiencies in the "secondary" opsonins. Susceptibility to infection resulting from phagocytic dysfunctions range from mild recurrent skin infections to severe overwhelming fatal systemic infection. As a rule, such patients exhibit a marked susceptibility to bacterial infection and generally have little difficulty with viral or protozoan infections. Numerous tests can be performed to evaluate phagocytic dysfunction as described elsewhere in this volume ([3]). Disorders of ingestion and opsonic function can be evaluated with a phagocytic index study which involves allowing the neutrophils to ingest microorganisms under standard conditions and then determining the average number of particles ingested per phagocytic cell. A number of experimental systems have been devised to permit rapid quantitation by physical or chemical means. These techniques, generally based on the use of labeled particles, include oil red O,[192] iodinated bacteria,[193,194] starch,[193] and immune complexes[48] (see this volume [3]). The purpose of this section is to briefly allude to the clinical consequences of opsonic dysfunction. Excellent comprehensive reviews of general neutrophil dysfunction[195–198] and impaired opsonization/phagocytic ingestion[199] have been published elsewhere.

[190] O. Stendahl, B. Normann, and L. Edebo, *Acta Pathol. Microbiol. Scand., Sect. 190B,* **87B,** 85 (1979).

[191] I. Orskov, F. Orskov, B. Jann, and K. Jann, *Bacteriol. Rev.* **41,** 667 (1977).

[192] T. P. Stossel, *Semin. Hematol.* **12,** 83 (1975).

[193] R. H. Michell, S. J. Pancake, and J. Noseworthy, *J. Cell Biol.* **40,** 216 (1969).

[194] F. Ulrich, *Am. J. Physiol.* **220,** 958 (1971).

[195] S. J. Klebanoff, *in* "Textbook of Medicina" (P.B. Beeson and W. McDermott, eds.), pp. 1476–1480. Saunders, Philadelphia, Pennsylvania, 1975.

[196] R. B. Johnston, A. R. Lawton, and M. D. Cooper, *Med. Clin. North Am.* **57,** 421 (1973).

[197] P. G. Quie, *Medicine (Baltimore)* **52,** 411 (1973).

[198] J. A. Winkelstein and R. H. Drachman, *Pediatr. Clin. North Am.* **21,** 551 (1974).

[199] J. F. Soothill and B. A. M. Harvey, *Arch. Dis. Child.* **51,** 91 (1976).

Opsonization abnormalities may be divided into two categories, depending on the cause of the defect: (1) Humoral: decreased opsonic activity of serum or the natural opsonins; and (2) cellular: decreased ability of the phagocytic cells to ingest properly opsonized particles.

4.1. Humoral Defects

Improper (generally decreased) opsonization of particles has been documented in patients with clinically defective immunoglobulin, complement component, and natural tuftsin disorders.[200] Sera from patients with these diseases fail to opsonize microorganisms resulting in an enhanced susceptibility to infection. For example, some infants lack specific antibodies against the capsule of *Haemophilus influenzae;* their sera are deficient in opsonic power against the organisms, and they are prone to infection by this microbe.[201] Additional examples are cited in Table VI.[202–207]

4.1.1. Immunoglobulin Deficiencies. 4.1.1.1. Hypogammaglobulinemia: In view of the role of immunoglobulin molecules in the opsonization of particles (cf. Section 2) it is reasonable to suspect that this would be reduced in patients with hypogammaglobulinemia. The clinical evidence is mixed. Forssman reported normal phagocytic ingestion by neutrophils of two hypogammaglobulinemic patients[208] whereas Mickenberg *et al.* reported that the sera from three patients with severe hypogammaglobulinemia results in a marked reduction in opsonic activity against a range of bacteria.[209] Since the reduction in phagocytic ingestion was only partial it was suggested that the observed opsonic response is due to the presence of normal levels of complement components in their sera. In these studies, defective opsonization could be corrected through the addition of purified, concentrated nonspecific IgG to the patient's sera,[209] reflecting the importance and role of nonspecific antibody adsorption to the microbe surface.[36]

4.1.2. Complement Deficiencies. 4.1.2.1. C3 deficiency: A variety of complement deficiencies and abnormalities of complement function

[200] C. F. Alper, T. P. Stossel, and F. A. Rosen, *in* "The Phagocytic Cell and Host Resistance" (D. Dayton and G. Bellanti, eds.), p. 121. Raven Press, New York, 1979.

[201] P. Anderson, R. B. Johnston, and D. H. Smith, *J. Clin. Invest.* **51,** 31 (1972).

[202] A. Constantopoulos and V. A. Najjar, *Acta Paediatr. Scand.* **62,** 645 (1973).

[203] R. K. Root, L. Ellman, and M. M. Frank, *J. Immunol.* **109,** 477 (1972).

[204] T. P. Stossel, C. A. Alper, and F. S. Rosen, *Pediatrics* **52,** 134 (1973).

[205] H. E. Jasin, J. H. Orozco, and M. Ziff, *J. Clin. Invest.* **53,** 343 (1974).

[206] F. C. McDuffie and M. W. Brumfield, *J. Clin. Invest.* **51,** 3007 (1972).

[207] R. B. Johnston, S. L. Newman, and A. G. Struth, *N. Engl. J. Med.* **288,** 803 (1973).

[208] O. Forssman, *Acta Pathol. Microbiol. Scand.* **65,** 366 (1965).

[209] I. D. Mickenberg, R. K. Root, and S. M. Wolff, *J. Clin. Invest.* **49,** 1528 (1970).

TABLE VI

HUMORAL ABNORMALITIES ASSOCIATED WITH DEFECTIVE OPSONIZATION[a]

Abnormality	Etiology	Effect on neutrophil function	Clinical consequences
Antibody deficiency syndromes	Early infancy Genetic Acquired idiopathic Secondary to lymphatic neoplasms	Deficiency of serum opsonic activity for encapsulated pathogens[202]	Recurrent pyogenic infections
Disorders of complement C4 deficiency	Genetic absence	Subtle alterations in serum opsonic activity[203]	Possible increase in susceptibility to infection
C3 deficiency	Genetic absence Genetic hypercatabolism	Profoundly deficient serum chemotactic and opsonic activity[202]	Recurrent pyogenic infections
Properdin factor B deficiency	Some neonates	Deficient serum opsonic activity[204]	(?) A cause of susceptibility to infections
Depletion of multiple factors, inhibition of activation	Rheumatoid arthritis, systemic lupus erythematosus, sickle cell anemia, other inflammatory states	Deficient serum opsonic activity[205–207]	Recurrent infections

[a] T. P. Stossel, in "The Granulocyte: Function and Clinical Utilization" (T. J. Greenwalt and G. A. Jamieson, eds.), p. 87. Alan R. Liss, Inc., New York, 1977.

have been associated with increased susceptibility to infection. Activation of C3 and deposition of C3b on the particle to be phagocytized is of prime importance in opsonization (see Sections 2.2 and 2.3). Thus, it is not surprising that low serum levels of C3 and failure to activate C3 are associated with infections. Defective serum opsonic activity has been demonstrated in neonates and this decrease in phagocytic ingestion correlates well with their low C3 and C5 levels.[209] Similar data have been reported for adults[210,211] and the defect could be corrected through the addition of purified C3.

4.1.2.2. C3 hypercatabolism: Increased susceptibility to infection has been described in patients with C3 hypercatabolism.[211,212] Two families have been described in which an infant had infections with increased susceptibility to *S. aureus* and various gram-negative bacilli.[213,214] Studies using ^{125}I-labeled C3 have shown that the low serum C3 level is due to an increased catabolism, decreased synthesis, and an abnormally high ratio of extravascular to plasma C3 activity. It is believed that these patients contained a protease that has C3 as the substrate and that this enzyme may be identical to the so-called "nephritic factor" which has been shown to deplete C3 by activating alternative pathway activation.[215] In *in vitro* studies, the opsonization abnormalities of these sera could not be corrected though the addition of small volumes of normal serum.[216]

4.1.2.3. C5 deficiency: Generalized seborrheic dermatitis and intractable diarrhea have been associated with functional abnormalities of C5.[213,214] Family studies have shown a decreased opsonization of baker's yeast by the patient's sera which could be restored to normal through the addition of purified C5[217] (however, also see Section 2.2.3.3).

4.2. Cellular Defects

4.2.1. Tuftsin Deficiency. Severe and recurrent systemic infections occur as a result of the absence of this phagocytosis-stimulating tetrapep-

[210] G. J. McCracken and H. F. Eichenwald, *Am. J. Dis.Child.* **121**, 120 (1971).

[211] C. A. Alper, N. Abramson, R. B. Johnston, K. J. Jandl, and F. R. Rosen, *N. Engl. J. Med.* **282**, 349 (1970).

[212] C. A. Alper, K. J. Bloch, and F. S. Rosen, *N. Engl. J. Med.* **288**, 601 (1973).

[213] M. E. Miller and U. R. Nilsson, *N. Engl. J. Med.* **282**, 354 (1970).

[214] J. C. Jacobs and M. E. Miller, *Pediatrics* **49**, 535 (1972).

[215] L. G. Hunsicker, S. Ruddy, C. B. Carpenter, P. H. Schur, J. P. Merrill, H. J. Müller-Eberhard, and K. F. Austen, *N. Engl. J. Med.* **287**, 835 (1972).

[216] J. A. Charlesworth, D. G. Williams, E. Sherington, P. J. Lachman, and D. K. Peters, *J. Clin. Invest.* **53**, 1578 (1974).

[217] M. E. Miller and P. J. Koblenzer, *J. Pediatr.* **80**, 879 (1972).

tide. Tuftsin deficiency has been reported as a familial disorder[202,218,219] and also has been reported in the case of splenectomized patients[142,219] and experimental animals.[147]

4.2.2. General. In contrast to the catalog of well-defined serum disorders, there are only a few clear-cut instances of defects in the *ingestion* process where a cellular abnormality per se has been unequivocally identified. Such cellular defects include (1) abnormal neutrophil responses associated with energy metabolism defects; (2) neutrophil actin dysfunction; (3) impaired chemotactic responses; and (4) combinations of the above.

Patients with rheumatoid arthritis,[220] multiple myeloma,[221,222] systemic lupus erythematosus,[223] leukemia,[224] hepatic cirrhosis,[225] and severe bacterial or viral infections[226,227] have been said to have diminished neutrophil adhesiveness or ability to ingest, in addition to serum abnormalities. It should be mentioned, however, that the mechanism by which the reduction in ingestion occurs is largely unknown. As defective leukotaxis, intracellular killing and enzyme production are not directly related to opsonization per se such conditions will not be discussed further here.

Corticosteroid therapy, especially with high and sustained doses, has been shown to inhibit neutrophil adhesiveness and ability of various phagocytes to ingest.[228–230] Since corticosteroids have long been known to diminish the clearance of sensitized cells by intact animals, an attractive theory is that the drug acts in part by decreasing the cell ingestion capacity of mononuclear phagocytes.

[218] A. Constantopoulos, V. A. Najjar, and J. W.Smith, *J. Pediatr.* **80,** 564 (1972).

[219] V. A. Najjar, *J. Pediatr.* **87,** 1121 (1975).

[220] R. A. Turner, H. R. Schumacher, and A. R. Myers, *J. Clin. Invest.* **52,** 1632 (1973).

[221] R. Penny and D. A. G. Galton, *Br. J. Haematol.* **12,** 633 (1966).

[222] L. E. Spitler, P. Spath, L. Petz, N. Cooper, and H. H. Fundenberg, *Br. J. Haematol.* **29,** 279 (1975).

[223] L. Brandt and H. Hedberg, *Scand. J. Haematol.* **6,** 348 (1969).

[224] L. Brandt, *Scand. J. Haematol.* **2,** 1 (1967).

[225] J. S. Tan, R. G. Strauss, J. Akabutu, C. A. Kauffman, A. M. Maurer, and J. P. Phair, *Am. J. Med.* **57,** 251 (1974).

[226] C. E. McCall, J. Caves, M. R. Cooper, and L. R. DeChateler, *J. Infect. Dis.* **124,** 68 (1971).

[227] C. S. Hosking, M. G. Fitzgerald, and M. J. Shelton, *Aust. Paediatr. J.* **14,** 1 (1978).

[228] R. R. MacGregor, P. J. Spagnuolo, and A. L. Lentnek, *N. Engl. J. Med.* **291,** 642 (1974).

[229] R. I. Handin and T. P. Stossel, *Blood* **46,** 1016 (1975).

[230] A. D. Schreiber, J. Parsons, P. McDermott, and R. A. Cooper, *J. Clin. Invest.* **56,** 1189 (1975).

5. Measurement of Effective Opsonization

The extent and effectiveness of opsonization is generally monitored indirectly by the increase in the phagocytic response which it induces. Such activity is assessed by various direct and indirect techniques which have been described in detail elsewhere in this volume [3]. These techniques include microscopic, isotopic, fluorescence, extraction, and enzymatic methods. Isotopic and fluorescence techniques have the advantage that they permit quantitation of the extent of adsorbed opsonin per unit particle surface area. They suffer, however, from the drawback that such methods cannot themselves be used to assess the consequence of effective opsonization, i.e., they do not provide direct information as to how such adsorption will influence the extent of phagocytic ingestion. Indirect correlations are made between the extent of opsonin adsorption and the cellular response (e.g., ingestion, O_2 consumption, enzyme release) it invokes.

Acknowledgments

Supported in part by research grants from the Medical (#MT-5462, #MT-8024, #MA-9114) and Natural Science and Engineering (#UO-493) Research Council of Canada and the Ontario Heart Foundation (#4-12, #AN-402). In addition the author should like to acknowledge financial support of the Ontario Heart Foundation through the receipt of a Senior Research Fellowship.

[14] Tuftsin

By Victor A. Najjar, Danuta Konopinska, and James Lee

Introduction

Tuftsin, L-Thr-L-Lys-L-Pro-L-Arg, is a biologically active oligopeptide which is designed to stimulate all known functions of phagocytic cells, in particular granulocytes and macrophages/monocytes: phagocytosis, pinocytosis, motility,[1] processing of the antigen with immunogenic stimu-

[1] V. A. Najjar and N. Bump, in "Immune Modulation Agents and Their Mechanisms" (R. L. Fennichel and M. A. Chirigos, eds.), p. 229. Dekker, New York, 1984.

lation,[2,3] and bactericidal[4-6] and tumoricidal activities.[7-13] In addition, it rejuvenates age-depressed macrophage and T cell cytolysis[2,3,7,14] as well as disease-depressed monocyte chemotaxis, as it occurs in Hodgkin's disease and systemic lupus erythematosus.[15-17] It augments the formation of superoxide anion O_2^-),[18,19] which is dismutated to yield H_2O_2 and together with O_2^- give rise to the hydroxyl radical. All these compounds take part in the killing of bacteria and cancer cells. Furthermore, tuftsin stimulation raises the level of cyclic guanylate acid and depresses that of cyclic adenylic acid.[20]

Like many active oligopeptides, it is a part of a carrier molecule, leukokinin, a specific γ-globulin that binds to phagocytic cells. It comprises residues 289–292 and is liberated in the active form by two enzymes.

[2] E. Tzehoval, S. Segal, Y. Stabinsky, M. Fridkin, Z. Spirer, and M. Feldman, *Proc. Natl. Acad. Sci. U.S.A.* **75,** 3400 (1978).

[3] I. Florentin, M. Bruley-Rosset, N. Kiger, J. L. Imbach, F. Winternitz, and G. Mathé, *Cancer Immunol. Immunother.* **5,** 211 (1978).

[4] D. Blok-Perkowska, F. Muzalewski, and D. Konopińska, *Antimicrob. Agents Chemother.* **25,** 134 (1984).

[5] J. Martinez and F. Winternitz, *Ann. N.Y. Acad. Sci.* **419,** 23 (1983).

[6] J. Martinez, F. Winternitz, and J. Vindel, *Eur. J. Med. Chem.* **12,** 511 (1977).

[7] M. Bruley-Rosset, T. Hercend, J. Martinez, H. Rappaport, and G. Mathé, *JNCI, J. Natl. Cancer Inst.* **66,** 1113 (1981).

[8] R. Catane, S. Schlanger, P. Gottlieb, J. Halpern, A. J. Treves, Z. Fuks, and M. Fridkin, *Proc. Am. Assoc. Cancer Res. Am. Soc. Clin. Oncol.* **22,** 371 (1981).

[9] R. Catane, S. Schlanger, L. Weiss, S. Penchas, Z. Fuks, and A. J. Treves, *Ann. N.Y. Acad. Sci.* **419,** 251 (1983).

[10] V. A. Najjar, L. Linehan, and D. Konopińska, *Ann. N.Y. Acad. Sci.* **419,** 261 (1983).

[11] V. A. Najjar, D. Konopińska, M. K. Chaudhuri, D. E. Schmidt, and L. Linehan, *Mol. Cell. Biochem.* **41,** 3 (1981).

[12] K. Nishioka, *Br. J. Cancer* **39,** 342 (1979).

[13] K. Nishioka, G. F. Babcock, J. H. Phillips, R. A. Banks, and A. A. Amoscato, *Ann. N.Y. Acad. Sci.* **419,** 234 (1983).

[14] M. Bruley-Rosset, I. Florentin, and G. Mathé, *Mol. Cell. Biochem.* **41,** 113 (1981).

[15] M. Kavai, K. Lukacs, G. Szegedi, M. Szekerke, and J. Erchegyi, *Immunol. Lett.* **2,** 219 (1981).

[16] K. Lukacs, E. Berenyi, M. Kavai, G. Szegedi, and M. Szekerke, *Cancer Immunol. Immunother.* **15,** 162 (1983).

[17] K. Lukacs, G. Szabo, I. Sonkoly, E. Vegh, J. Gacs, M. Szekerke, and G. Szegedi, *Immunopharmacology* **7,** 171 (1984).

[18] G. L. Tritsch and P. W. Niswander, *Mol. Cell. Biochem.* **49,** 49 (1982).

[19] H. P. Hartung and K. V. Toyka, *Agents Actions* **15,** 38 (1984).

[20] Y. Stabinsky, Z. Bar-Shavit, M. Fridkin, and R. Goldman, *Mol. Cell. Biochem.* **30,** 71 (1980).

A human mutation of tuftsin has been found in the United States,[21-29] and Japan.[30] This has been designated as Congenital Tuftsin Deficiency, which manifests itself in very low phagocytosis activity and clinically with repeated severe infections which can be fatal in infancy and which become greatly ameliorated, but still quite evident, in the adult.[23-26] In one such patient the mutated peptide (Thr-Glu-Pro-Arg) was isolated, identified, and synthesized. In this peptide, the lysine triplets AAA or AAG become GAA or GAG coding for glutamic acid. An acquired deficiency of tuftsin may be found in electively splenectomized patients or where the spleen function is severely compromised as in sickle cell disease.[25-29] This deficiency results from the fact that one of the two enzymes required for its liberation as the active tetrapeptide is a spleen enzyme.

Synthesis of Radioactive Tuftsin

Tuftsin is commercially available from several companies but that produced by Vega (Tucson, AZ) is consistently the purest. Radioactive tuftsin has been used in the isolation and characterization of tuftsin receptors,[31-33] as well as in radioimmunoassay of serum in cases of suspected tuftsin deficiency, both congenital and acquired.[1,20,31,34]

The method has been developed in our laboratory and represents a significant modification of that described by Stabinsky *et al.*[31] In short, N^{α}-BOC-N^{ε}-Z-Lys-N-hydrosuccinimide ester (OSu) is coupled to proline

[21] V. A. Najjar, *Lymphology* **3**, 23 (1970).

[22] V. A. Najjar, *in* "Advances in Enzymology" (A. Meister, ed.), p. 129. Wiley, New York, 1974.

[23] V. A. Najjar, *J. Pediatr.* **87**, 1121 (1975).

[24] V. A. Najjar, *in* "Biological Membranes" (D. Chapman and D. F. H. Wallach, eds.), p. 191. Academic Press, London, 1976.

[25] V. A. Najjar, *Exp. Cell Biol.* **46**, 114 (1978).

[26] V. A. Najjar, *Klin. Wochenschr.* **57**, 751 (1979).

[27] V. A. Najjar, *in* "The Reticuloendothelial System" (A. J. Sbarra and R. Strauss, eds.), p. 45. Plenum, New York, 1980.

[28] V. A. Najjar, *Med. Biol.* **59**, 134 (1981).

[29] V. A. Najjar and A. Constantopoulos, *RES, J. Reticuloendothel. Soc.* **12**, 197 (1972).

[30] K. Inada, N. Nemoto, A. Nishijima, S. Wada, M. Hirata, and M. Yoshrida, *in* "Phagocytosis: Its Philosophy and Pathology" (Y. Kobobun and N. Kobayashi, eds.), p. 101. University Park Press, Baltimore, Maryland, 1979.

[31] Y. Stabinsky, P. Gottlieb, V. Zakuth, Z. Spirer, and M. Fridkin, *Biochem. Biophys. Res. Commun.* **83**, 599 (1978).

[32] N. Bump and V. A. Najjar, *Mol. Cell. Biochem.* **63**, 137 (1984).

[33] V. A. Najjar and N. J. Bump, *Int. Forum Pept.*, pp. 17–21 (1984).

[34] V. A. Najjar, M. K. Chaudhuri, D. Konopińska, B. D. Beck, P. P. Layne, and L. Linehan, *in* "Augmenting Agents in Cancer Therapy" (E. M. Hersh, M. A. Chirigos, and M. J. Mastrangelo, eds.), p. 459. Raven Press, New York, 1981.

to form the dipeptide N^α-BOC-N^ε-Lys-Pro-OH. This is then coupled to N^α-BOC-Thr-OSu after neutralization with triethylamine (TEA).[34a] This protected tripeptide is reacted with N-hydroxysuccinimide in the presence of the coupling reagent dicyclohexylcarbodiimide (DCC). The protected and activated tripeptide OSu is now coupled to radioactive arginine. This is followed by reduction with hydrogen to remove the benzyloxycarbonyl (Z) group. It is then deprotected with 50% trifluoroacetic acid (TFA) in dichloromethane (CH_2Cl_2) (v/v) to remove the butyloxycarbonyl (BOC) group.

Procedure

H-N^α-Z-Lys-Pro-OH Trifluoroacetate (TFA) Salt (I). L-Proline, 3.65 g (31.7 mmol), dissolved in 32 ml 1 M NaOH to pH 7.5, is added to N^α-BOC-N^ε-Z-Lys-N-hydroxysuccinimide ester (OSu) (Sigma), 13.85 g (30 mmol), in dioxane, 32 ml. It is stirred at room temperature for 20 hr and evaporated to dryness *in vacuo*. It is then dissolved in 30 ml water and washed 3× with ethyl ether, 10 ml. The aqueous layer is carefully acidified to pH 2–3 with 1 N HCl. The precipitate of the cloudy aqueous product is extracted 3× with ethyl acetate, 30 ml. The pooled extract is washed 3× with water, 10 ml, dried over anhydrous Na_2SO_4, evaporated *in vacuo* down to approximately 15 ml. Petroleum ether (35–40 ml, kept at 4–6° overnight) is added slowly to turbidity. An oily product separates. The ether layer is decanted and the oily product is desiccated under vacuum to constant weight, 10.65 g (76% yield). The material is then deprotected in 15 ml of 50% TFA and dichloromethane (v/v), evaporated approximately to half volume, and 30 ml of ethyl ether is added. A white precipitate is produced. The ether layer is decanted. The ether precipitation is repeated two more times. Finally the material is desiccated *in vacuo* to constant weight, 7.0 g (63% yield).

N^α-BOC-Thr-N^ε-Z-Lys-Pro-OH (II). The dipeptide, TFA salt (I), 6.9 g (14.4 mmol), is dissolved in a solution composed of 20 ml dimethylformamide (DMF), water, 20 ml, and triethylamine (TEA), 3.92 ml. This is added to a solution containing N^α-BOC-Thr-OSu, 4.36 g (13.8 mmol), in 15 ml of DMF. The pH of this mixture is adjusted to about 7.5 with TEA. This is stirred at room temperature for 20 hr and the solvents evaporated. To the remaining product (oil) approximately an equivalent of 1 N HCl is added to pH 2–3 and extracted 3× with ethyl acetate, 10 ml. The ethyl acetate is quickly washed once with 1 N citric acid followed by three washes with water, 10 ml each, dried over anhydrous Na_2SO_4, and evapo-

[34a] Abbreviations: TEA, triethylamine; DCC, dicyclohexylcarbodiimide; TFA, trifluoroacetic acid; OSU, hydroxysuccinimide ester; BOC, butyloxycarbonyl; DMF, dimethylformamide; PBS, phosphate-buffered saline; BSA, bovine serum albumin.

rated *in vacuo*. The product is dissolved in 15 ml of ethyl acetate and 45 ml of petroleum ether, that was added slowly at room temperature to slight turbidity and kept at 4–6° overnight. The precipitate is filtered and dried to constant weight, 6.9 g (11.9 mmol) (86% yield).

N^α-BOC-Thr-N^ε-Z-Lys-Pro-N-hydroxysuccinimide Ester (III). The tripeptide (II), 7.7 g (13.3 mmol), is reacted with 1.5 g (13.0 mmol) of N-hydroxysuccinimide (Sigma), with 2.9 g of DCC (14.0 mmol) in 20 ml of dry dioxane overnight, filtered, and evaporated *in vacuo*. The resulting oily product is dissolved in 30 ml of ethyl acetate, washed quickly with 0.5 N HCl (2×, 10 ml), saturated solution of sodium bicarbonate (2×, 15 ml), with water (2×, 15 ml), dried over anhydrous sodium sulfate, and evaporated to yield an oily precipitate. It is then triturated with dry ether to a white gummy solid to yield 2.5 g of material, melting point 148–150°. A single spot on thin-layer chromatography is obtained with the following solvent systems: acetonitrile : water (9 : 1), R_f 0.85, benzene : water (3 : 1), R_f 0.2, CHCl$_3$: methanol : acetic acid : water (6 : 4 : 5 : 2), R_f 0.91, *n*-butanol : acetic acid : water (4 : 1 : 1), R_f 0.71.

[³H]Tuftsin. The tripeptide (III), 120 mg (17 μmol) is mixed with 1 mCi of [2,3,4,5-³H]Arg (specific activity 70 Ci/mmol), in 4 ml of dioxane : water (1 : 1) (v/v), and NaHCO$_3$ is added to a pH of approximately 8.4. The incubation is continued for 20 hr at room temperature. The material is then dried *in vacuo* and extracted twice with methanol and again dried. The residue is dissolved in 10% acetic acid in methanol and subjected to catalytic hydrogenation with 140 mg of 5% palladium on barium sulfate for 4 hr, followed by 60 mg of the same overnight. The material is then filtered. This treatment removes the Z protecting group at N^ε of lysine. The N^α-BOC is then removed by treatment with 5 ml of 50% TFA in CH$_2$Cl$_2$ (v/v) for 20 min. The material is then dried *in vacuo,* washed twice with CH$_2$Cl$_2$, and dried again. The material is then purified by chromatography on a column (1 × 32 cm) of Dowex AG 50W-X4 with a linear gradient between 150 ml of 1.2 M pyridine acetate, pH 4, and 150 ml of 2.5 M pyridine acetate, pH 6, at a flow rate of 14 ml/hr.[35] The purification is monitored by radioactivity. The peak of [³H]tuftsin that coincides with that of nonradioactive tuftsin is harvested as pure radioactive tuftsin.

The yield does not exceed 40%. The major component is an analog of tuftsin. This arises when the amino group of arginine attacks the succinimidyl component at the carbonyl–nitrogen bond instead of the carboxyl ester of proline. The result is the incorporation of the linear succinimide into the tetrapeptide which on thin layer gives a blue color with

[35] K. Nishioka, P. S. Satoh, A. Constantopoulos, and V. A. Najjar, *Biochim. Biophys. Acta* **310,** 230 (1973).

ninhydrin spray. This reaction was discovered by Savrda.[36] It is inherent in the reaction of proline-OSu structure.

Another good method for the preparation of radioactive tuftsin is through the synthesis of H-Thr-Lys-3,4-dehydro-Pro-Arg-OH followed by catalytic reduction with tritium gas. This is done simply by using N^α-BOC-3,4-dehydroproline (Bachem) in place of N^α-BOC proline during solid phase synthesis with a modified method[37] of Merrifield.[38]

In brief, N^α-BOC-N^G-Tos-arginine is esterified to a chloromethyl resin, a copolymer of polystyrene, and 1% divinylbenzene containing 1–2 mmol Cl/g. The BOC group is then removed with 50% TFA. After neutralization with triethylamine, N^α-BOC-3,4-dehydroproline is then coupled to H-Tos-Arg-resin with DCC. The removal of N^α-BOC of the dipeptide and neutralization are again repeated to yield H-3,4-dehydro-Pro-N^G-Tos-Arg-resin. N^α-BOC-N^ε-Z-Lys is now coupled to the αNH_2 group of the dipeptide with DCC and processed as before to yield H-N^ε-Z-Lys-3,4-dehydro-Pro-Arg-resin. Similarly, N^α-BOC-O-Bzl-threonine is also coupled with DCC to the αNH_2 group of the tripeptide and again processed. Finally, with one single reagent, trifluoromethanesulfonic acid, the tetrapeptide is removed from the resin along with all protecting groups N^α-BOC, N^ε-Z, and N^G-Tos, to yield H-Thr-Lys-3,4-dehydro-Pro-Arg-OH. The dehydroproline residue is then subjected to catalytic reduction with tritium gas. This latter can be done by firms such as New England Nuclear.

Radioimmunoassay of Tuftsin

Radioactive tuftsin, prepared as described above, and antibody to tuftsin are required for the radioimmunoassay.

Preparation of Antisera to Tuftsin

1. Tuftsin is too small an oligopeptide to be immunogenic. To render it antigenic, it should be coupled to a large protein. To that end, the hapten α-aminophenylacetyltuftsin[31,39] or p-aminophenylacetylglycylglycyltuftsin were prepared.[29] The purpose of the glycyl–glycyl residue is to allow the antibody to recognize all four residues if possible. The hapten is coupled to bovine serum albumin (BSA) through the diazo reaction.[39] Several other proteins can be used. Rabbits are immunized with 200–400

---

[36] J. Savrda, *J. Org. Chem.* **42**, 3199 (1977).
[37] M. K. Chaudhuri and V. A. Najjar, *Anal. Biochem.* **95**, 305 (1979).
[38] R. B. Merrifield, *Biochemistry* **3**, 1385 (1964).
[39] Z. Spirer, V. Zakuth, N. Bogair, and M. Fridkin, *Eur. J. Immunol.* **7**, 69 (1977).

μg of the antigen in complete Freund adjuvant by subcutaneous injections in several areas of the back. This is repeated in 2 weeks and then in 4 more weeks. Two weeks hence, serum can be prepared. Such antisera are capable of binding the [^{3}H]tuftsin or the hapten p-aminophenylacetyl-Gly-Gly-tuftsin iodinated at the phenyl ring with ^{125}I. In order to evaluate the antibody level at various times, blood is withdrawn immediately before the immunization injection with the BSA–hapten complex. Withdrawing a blood sample just before the injection merely avoids lowering of the antibody blood titer by the injected antigen.

2. Rabbit antiserum to human γ-globulin, as well as goat antiserum to human γ-globulin, as obtained from several commercial suppliers, possesses strong anti-tuftsin activity. It has been known for some time that the heavy chain of γ-globulin contains tuftsin as part of its structure (residues 289–292). In leukokinin, a leukophilic γ-globulin, the tetrapeptide, is cleaved at the carboxy-terminal arginine but is covalently still attached at its amino end to the heavy chain. When unfractionated γ-globulin, which always contains leukokinin, is used for immunization of rabbits or goats, a fraction of the antibody produced would be directed against the free carboxy terminal. The orientation of tuftsin on its natural carrier would be similar to its orientation on its artificial BSA carrier. In both cases the tetrapeptide is anchored at the amino-terminal threonine with the carboxy-terminal arginine oriented outward. This antiserum displays very good affinity to [^{3}H]tuftsin with a titer equal to, and occasionally surpassing, that of the antiserum generated by BSA-bound tuftsin. We have therefore adopted the procedure of screening such antisera from various suppliers for high titers sufficient for radioimmunoassay. Such assays with commercial antibody to human γ-globulin duplicate quantitatively assays with anti-BSA-tuftsin.

Radioimmunoassay of tuftsin has been carried out in serum of patients suspected of tuftsin deficiency, either congenital or acquired after splenectomy. Here γ-globulin from patients' serum is precipitated in 33% saturated ammonium sulfate and dialyzed against 2 liters of 0.1 M phosphate buffer, pH 8.1. Ten milligams of this preparation is incubated at 37° with 0.5 mg of trypsin for 1 hr in a final volume of 2.5 ml. Four volumes of 95% ethanol are then added to terminate the reaction. The mixture is then placed in an 80° bath for 10 min or more. It is then cooled in an ice bath and centrifuged at 3000 g for 10–15 min. The supernatant is dried *in vacuo* and dissolved in 0.25 ml of Krebs–Ringer solution. An aliquot of 100 μl can be used for radioimmunoassay or phagocytic stimulation assay.[25–30,34,39] Spirer *et al.*[39] have simplified this procedure by omitting the γ-globulin preparation and treating serum directly with trypsin.

Procedure

Various known amounts of tuftsin or samples to be tested are assayed (2–2000 μg) in triplicates in phosphate-buffered saline (PBS), pH 7.0 (0.5 ml final volume). Control tubes contain PBS only. Radioactive tuftsin is prepared in PBS to yield a radioactivity of 15,000 cpm/1.0 μl. This is added to all tubes followed by 0.2 ml of antiserum in appropriate dilutions in PBS with ethylenediaminetetraacetic acid, 0.05 M, and 1% normal rabbit serum. The samples are incubated for 2 hr at 37° and overnight at 4–6°. Three milliliters of PBS is then added to each tube and centrifuged at 2000–3000 g for 20 min. The supernatant is discarded and any adherent fluid is wiped out with a strip of filter paper, and the radioactivity of the precipitate is determined. A semilogarithmic plot is made with log concentration on the abscissa and the percentage of maximum specific binding on the ordinate.

Various dilutions of the trypsin-treated serum or its γ-globulin are added to the standard set and the deviation in the tuftsin curve can be used to estimate the amount of tuftsin present in the unknown sample. In order to determine nonspecific binding, 1 μmol of unlabeled tuftsin is added to each tube in a parallel duplicate run.

Normal values per milliliter of serum that we obtained with γ-globulin digestion are slightly higher than those obtained with serum digestion. The range for normal is 255–500 ng/ml.[1,39] For electively splenectomized patients, levels are found to be lower (118 ng/ml). By contrast, patients with tuftsin deficiency syndrome yield very high false values of 2500–3500 ng/ml. This is due to the fact that in place of the natural tuftsin peptide (Thr-Lys-Pro-Arg), these patients possess a mutant peptide (Thr-Glu-Pro-Arg) that binds to the antibody approximately 10 times more strongly than tuftsin.

Details of classical procedures of RIA should be consulted and followed.[40]

Acknowledgments

This work was supported by the Public Health Service Grant # AI09116, The March of Dimes Birth Defects Foundation Grant # 1-556, and the American Cancer Society Grant # RDP-32E.

[40] J. J. Langone and H. Van Vunakis, eds. "Immunochemical Techniques" *in* "Methods in Enzymology" (S. P. Colowick and N. O. Kaplan, eds.), Vol. 74. Academic Press, Inc., New York, 1981.

[15] Substance P and Neurotensin

By Zvi Bar-Shavit and Rachel Goldman

Introduction

The basic tetrapeptide tuftsin (see the table) was described by Najjar and collaborators as the entity responsible for the phagocytosis-enhancing activity of leukophilic γ-globulins (see this volume [14]). Since then, this peptide was shown to affect many of the known functions of phagocytic cells, such as phagocytosis, bactericidal activity, motility, cytotoxicity toward tumor target cells, and immunogenic functions of macrophages.[1] During our studies on the mechanism of action of the peptide, we searched for natural peptides with sequences resembling tuftsin. We have found two neuropeptides [substance P (SP), and neurotensin (NT)][1a] which contained sequences analogous to tuftsin (see the table). These peptides and their partial sequences containing the tuftsin-related domains displace labeled tuftsin from phagocytes and enhance their phagocytic capability.[2-4]

Both peptides exhibit a broad spectrum of pharmacological effects in the central nervous system, as well as at the periphery. SP is widespread in both central and peripheral nervous systems (i.e., nerve endings in the skin, fibers terminating around blood vessels, intestinal plexuses, peripheral nerves, central spinal canal, and a variety of brain regions). At the periphery SP is recognized for its hypotensive, vasodilatory, and smooth muscle-contracting properties. In the central nervous system, SP is suggested to play a role in sensory nerve transmission. NT is distributed in various parts of the brain, gastrointestinal tissues, plasma, and in synovial tissue and fluid. As a result of its action on the central nervous system, the

---

[1] V. A. Najjar and M. Fridkin, eds., "Antineoplastic, Immunogenic and Other Effects of the Tetrapeptide Tuftsin: A Natural Macrophages Activator," Ann. N.Y. Acad. Sci., Vol. 419. N.Y. Acad. Sci., New York, 1983.

[1a] Abbreviations: SP, substance P; NT, neurotension; Z, zymosan; PBS, phosphate-buffered saline.

[2] Z. Bar-Shavit, R. Goldman, Y. Stabinsky, P. Gottlieb, M. Fridkin, V. I. Teichberg, and S. Blumberg, *Biochem. Biophys. Res. Commun.* **94**, 1445 (1980).

[3] Z. Bar-Shavit, S. Terry, S. Blumberg, and R. Goldman, *Neuropeptides (Edinburgh)* **2**, 325 (1982).

[4] R. Goldman, Z. Bar-Shavit, E. Shezen, S. Terry, and S. Blumberg, *in* "Macrophages and Natural Killer Cells: Regulation and Function" (S. J. Normann and E. Sorkin), p. 133. Plenum, New York, 1982.

Substance	Sequence
Tuftsin	Thr-Lys-Pro-Arg
SP (1–4)	Arg-Pro-Lys-Pro
SP[7]	Arg-Pro-Lys-Pro-Gln-Gln-Phe-Phe-Gly-Leu-Met(NH$_2$)
NT (8–13)	Arg-Arg-Pro-Tyr-Ile-Leu
NT (6–13)	Lys-Pro-Arg-Arg-Pro-Tyr-Ile-Leu
NT (1–10)	pGlu-Leu-Tyr-Glu-Asn-Lys-Pro-Arg-Arg-Pro
NT[7]	pGlu-Leu-Tyr-Glu-Asn-Lys-Pro-Arg-Arg-Pro-Tyr-Ile-Leu

[a] Peptides sharing the capacity to augment the phagocytic capability of phagocytic cells. All amino acids are in the L-configuration. The homologous sequences are underlined.

peptide causes hypothermia, analgesia, diminished locomotion activity, and muscle relaxation. In addition, NT has vasodilatory hypotensive effects. *In vitro* the peptide causes contraction of guinea pig ileum, and relaxation of rat duodenum. (For reviews see Refs. 5–8.)

The physiological relevance of the interaction of SP and NT with macrophages and granulocytes is not entirely clear. Both were shown to interact with mast cells[9,10] and to be strong vasodilators.[5,6] Both affect the mobilization of human granulocytes[11,12] and enhance the phagocytic capability of these cells as well as macrophages.[2,3,11] SP is released from sensory nerves following injury, axon reflex, and antidromic stimulation and thus is thought to be involved in neurogenic inflammation.[9,13] The physiological levels of NT in the blood are in the range that have been shown to be effective *in vitro* in the augmentation of phagocytosis.[14] Thus, SP and NT have properties and are located or released at sites that will enable them to act as mediators at various stages of inflammatory reactions.

[5] S. E. Leeman, E. A. Mroz, and R. E. Carraway, *in* "Peptides in Neurobiology" (H. Gainer, ed.), p. 99. Plenum, New York, 1977.

[6] A. Rokaeus, *Acta Physiol. Scand., Suppl.* **501**, 1 (1981).

[7] S. Watson, *Life Sci.* **25**, 797 (1984).

[8] C. B. Nemeroff, D. Luttinger, and A. J. Prange, Jr., *Trends Neurosci.* **3**, 212 (1980).

[9] F. Lembeck and P. Holzer, *Naunym-Schmiedeberg's Arch. Pharmacol.* **310**, 175 (1979).

[10] L. H. Lazarus, M. H. Perrin, and M. R. Brown, *J. Biol. Chem.* **252**, 7174 (1977).

[11] R. Goldman, Z. Bar-Shavit, and D. Romeo, *FEBS Lett.* **159**, 63 (1983).

[12] W. A. Marasco, H. J. Showell, and E. L. Becker, *Biochem. Biophys. Res. Commun.* **99**, 1065 (1981).

[13] F. Lembeck, R. Gamse, and J. Jaun, *in* "Substance P" (U. S. von Euler and B. Pernow, eds.), p. 171. Raven Press, New York, 1977.

[14] R. J. Miller, *Med. Biol.* **59**, 65 (1981).

Synthetic SP and NT and some shorter peptides containing their partial sequences are commercially available (Sigma, St. Louis, MO; Vega, Tucson, AZ). Using the method described below we were able to detect activity in 10^{-12} M NT and 10^{-8} M SP.

Assay for SP and NT

The assay for phagocytosis, which we have used for peptides, is applicable also to vitamin D_3.[15] For phagocytosis assays, advantage is taken of the fact that zymosan, a cell wall preparation of *Saccharomyces cerevisiae,* contains high amounts of mannan,[16] a polymer containing mannose residues. Oxidation by sodium periodate and reduction of the generated aldehyde groups by tritiated sodium borohydride has been suggested as a method of glycoprotein labeling.[17] We have used this method to label zymosan particles and obtained particles of a specific activity that have enabled us to assess phagocytosis using relatively small amounts of both, phagocytes and particles. Usually specific activities of $3-8 \times 10^3$ dpm/10^5 particles are obtained.

The labeling procedure does not affect the gross morphology of the particles (assessed by microscopic examination) and the uptake of ^{3}H-Z, as assessed by enumeration, is the same as that of unlabeled zymosan particles. The radioactivity of the particles is stable for at least a few months and no leakage of label is detectable in the supernatant of the suspended particles.

Preparation of ^{3}H-Labeled Zymosan Particles (^{3}H-Z)

Zymosan particles (Immunological Reagent, ICN Pharmaceuticals, Inc., Cleveland, OH) (50 mg) are washed twice in 0.9% NaCl (saline) and suspended in 0.2 M sodium acetate buffer (5 ml, pH 4.5). Sodium metaperiodate (NaIO$_4$, BDH Chemicals, Ltd., Poole, England) (0.01 M, 5 ml) is added and the suspension is stirred for 60 min in the dark at 4°. The oxidation step is terminated by five washes with phosphate buffer (0.1 M, pH 8.0). The particles are then resuspended in 3 ml of the above phosphate buffer to which 10 mCi of sodium boro[^{3}H]hydride (156 mCi/mg, the Radiochemical Center, Amersham, England) is added. The reaction is allowed to proceed overnight at 4° in the dark. The excess of the borohydride is decomposed by lowering the pH to pH 5.0 with acetic acid. The

---

[15] Z. Bar-Shavit, D. Noff, S. Edelstein, M. Meyer, S. Shibolet, and R. Goldman, *Calcif. Tissue Int.* **33,** 673 (1981).
[16] T. S. Stewart and C. E. Ballou, *Biochemistry* **7,** 1855 (1968).
[17] This series, Vol. 108 [25] and [26].

particle suspension is washed five times with phosphate-buffered saline (PBS) and incubated in PBS containing 0.1 M hydroxylamine for 30 min at room temperature. Following five washes in PBS, the particles are resuspended in PBS and stored at 4° until use.

Phagocytosis

Cell Plating. Thioglycolate-elicited mouse peritoneal macrophages are collected with standard methods.[18] Cells are plated (in the amounts specified in the legends to the figures) onto either 24-well (16 mm diameter) tissue culture plates (Costar, Cambridge, MA) or onto flat-bottom 96-well Linbro plates (IS-FB-96; Flow Laboratories, Inc., Irvine, Scotland).

For microscopic enumeration of phagocytosed particles, macrophages are plated onto 13-mm cover glasses placed in wells of the 24-well plates. Macrophages are allowed to adhere for 2 hr, washed, and cultured in medium containing serum for 24 hr.

Assays. Cell monolayers are washed twice in PBS and incubated in quadruplicate with the particles suspended in PBS, at the specified experimental conditions at 37°. The phagocytic assay is carried out for 30 min. The monolayers are washed five times with PBS and dissolved in 5% SDS overnight. The radioactivity present in the dissolved monolayers is determined in toluene–Triton X-100-based liquid scintillation solution in a counter equipped with an internal correction system for quenching. The results are expressed in dpm and are corrected for nonspecific absorption of ^{3}H-Z. The standard error of the mean never exceeds 10%. When microscopic enumeration of ^{3}H-Z particles in phagocytes is performed, the assay is carried out on monolayers plated on coverslips as described.[19]

IgG Coating of ^{3}H-Z. To detect phagocytosis via the Fc receptor, ^{3}H-Z are coated with specific antibodies. Rabbit anti-yeast sera are obtained by injection iv of 5×10^9 yeast cells (*Saccharomyces cerevisiae,* fixed with 2% glutaraldehyde) three times a week for 2 weeks. Sera are collected at the fourth week after immunization and absorbed four times on *E. coli.* The agglutination titer of the sera (for yeast cells) should be about 1 : 500. Addition of 0.1 M α-methyl-D-mannoside reduces the agglutination titer to 1 : 2, suggesting that the antisera are mainly directed against the yeast cell wall mannan. ^{3}H-Z (2×10^7 particles) is suspended in 1 ml of PBS containing the specified antiserum concentration. The suspension is incubated for 30 min at 37°, and the particles are then washed three times in PBS. Control particles are incubated with normal rabbit serum.

[18] Z. Bar-Shavit, A. Raz, and R. Goldman, *Eur. J. Immunol.* **9,** 385 (1979).
[19] Z. Bar-Shavit and R. Goldman, *Exp. Cell Res.* **99,** 221 (1976).

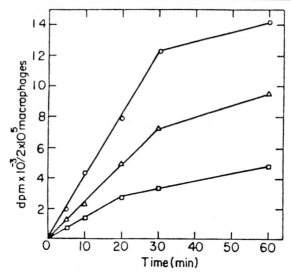

Fig. 1. Time course of phagocytosis of ³H-Z. The phagocytosis assay of thioglycolate-elicited macrophages (2 × 10⁵ per well) was carried out in a 24-well plate using three concentrations of particles. □—□ (1.5 × 10⁶), △—△ (3.75 × 10⁶), ○—○ (7.5 × 10⁶) ³H-Z (preparation II) in 0.5 ml of PBS. Each time point was assessed in a separate plate.

The association of radioactivity with the cells monolayers is time and dose dependent (Fig. 1). Figure 2 demonstrates the correlation between radioactivity and actual number of particles per cell. Figure 3 shows that the uptake of ³H-Z is linearly dependent on the number of phagocytes per well. Figure 4 shows the detection of Fc receptors by precoating the [³H]zymosan with antimannan antiserum.

The uptake of [³H]zymosan particles obeys all the accepted criteria for a phagocytic assay. The assay shows saturation kinetics with respect to time and dose of particles and is linearly dependent on the number of phagocytes. It is totally inhibited by sodium azide and shows a good correlation between assessment of uptake via radioactivity and direct enumeration of particles within cells using the light microscope.

The use of [³H]zymosan particles offers several advantages over the use of other particles described in the literature.

1. The possibility of obtaining particles with high specific activity enables the assessment of the phagocytic response of a small number of phagocytes. We have demonstrated that good, reproducible results can be obtained with 10–25 × 10³ phagocytes depending on the phagocyte used. The labeling procedure can be manipulated so that particles of a higher

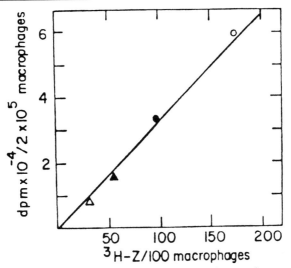

FIG. 2. Correlation of ³H-Z uptake as assessed by direct microscopic count (abscissa) and by radioactive analysis (ordinate). The phagocytosis assay was carried out in a 24-well plate. Thioglycolate-elicited macrophages (2 × 10⁵) were plated either directly on the well bottom or on 13-mm cover glasses placed in the wells (for microscopic examination). △, ▲, ●, and ○ denote 0.5 × 10⁶, 1 × 10⁶, 2 × 10⁶, and 4 × 10⁶ ³H-Z (preparation I) per well in 0.5 ml PBS.

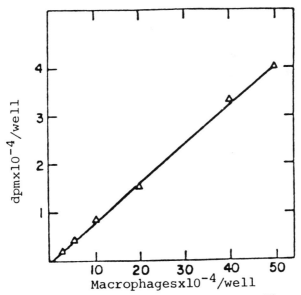

FIG. 3. Dependence of phagocytosis on the number of phagocytes. The specified number of macrophages was plated onto wells in 24-well plates. The phagocytic response was assessed using 5 × 10⁶ ³H-Z (preparation I) per well in 0.5 ml PBS.

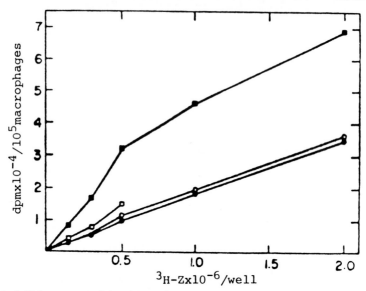

Fig. 4. Enhancement of the phagocytic response by precoating of ^{3}H-Z with a specific antimannan antiserum. Thioglycolate-elicited macrophages (10^5 per well) were plated onto 96-well plates and incubated with the specified number of ^{3}H-labeled Z (in 0.2 ml PBS) treated as described in Materials and Methods. ●—●, ^{3}H-Z; ○—○, ^{3}H-Z treated with normal rabbit serum (1 : 10 dilution); □—□, ■—■, ^{3}H-Z treated with antimannan serum diluted 1 : 20 and 1 : 10, respectively.

specific activity can readily be obtained, if needed. The sensitivity of this method is higher compared to other methods.[20-24]

2. The particles are stable and can be used for a period of at least 6 months. The assessment of phagocytosis by measuring the uptake of ^{51}Cr-labeled erythrocytes[23] depends on a constant supply of fresh sheep red blood cells and on labeling of the erythrocytes prior to each experiment.

3. ^{3}H-Z particles can be used for detection and assessment of Fc and C3b receptors on both phagocytic cells and on nonphagocytic cells. In many clinical tests, subpopulations of lymphoid cells are screened and diagnosed according to these parameters.

4. The method enables a very rapid manipulation of large numbers of samples. It is possible to use either the 24-well or the 96-well microtiter

[20] J. Roberts and J. H. Quastel, *Biochem. J.* **89,** 150 (1963).
[21] R. J. Downey and B. F. Diedrich, *Exp. Cell Res.* **50,** 483 (1968).
[22] R. H. Michell, S. J. Pancake, J. Noseworthy, and M. L. Karnovsky, *J. Cell Biol.* **40,** 216 (1969).
[23] B. Mantovani, M. Rabinovitch, and V. Nossenzweig, *J. Exp. Med.* **135,** 780 (1972).
[24] T. P. Stossel, R. J. Mason, R. J. Hartwig, and M. Vaughan, *J. Clin. Invest.* **51,** 615 (1972).

plates interchangeably. This results in many advantages over using separate 35-mm tissue cultures plates; i.e., reduction in cost of plates, of media, and in the amount of various effector molecules the action of which is being studied.

The assay described utilizes adherent phagocyte monolayers incubated with a suspension of [³H]zymosan particles with no stirring or shaking of the plates. Each phagocyte encounters a number of particles which depends on the surface area of the well and the absolute number of the particles. Thus, for comparison of absolute uptake of particles by phagocytes adherent on wells differing in diameter, the surface area of the wells has to be taken into account. The ratio of areas of the wells in 24- and 96-well plates is 7.11. Therefore 5×10^6 ³H-Z per well in the 24-well plate correspond to 0.7×10^6 ³H-Z per well in a 96-well plate. In fact, under the experimental condition described in the legend to Fig. 4 about 1.5 and 1.0 ³H-Z particles are taken up per macrophage in the 96-well (Fig. 4) and 24-well plates (Fig. 3), respectively.

The microscopic evaluation of phagocytosis of ³H-Z by macrophage monolayers has shown that for particles of this size, there is a lag period between attachment and ingestion. At 20 min of interaction of macrophage monolayers with the particles, 30% of the associated ³H-Z is not yet internalized. Thus, it seems that this assay is not suitable for evaluation of initial rates, utilizing very short assay periods. This method is not suitable for the assay of initial rates of phagocytosis (1–5 min), since phagocytosis depends on the settling of particles at the cell–liquid interface. A thorough mixing of macrophages and particles is needed for quick assays and this can be obtained only in suspension cultures or under rotatory motion. We have found[25] that rotatory motion inhibits the interaction of yeast cells with macrophages. Suspension culture phagocytosis assays require large amounts of phagocytes. In the assays described in the literature[24] there was no evaluation of the actual amounts ingested and the question of external association is still open.

Binding Studies

Radiolabeled SP and NT (³H- or ¹²⁵I-SP and ³H-NT) are available from New England Nuclear or Amersham. Binding studies can be carried out using either cell suspensions or cell monolayers. Since the phagocytosis stimulation studies are performed in monolayers, similar conditions are often chosen for binding.

---

[25] Z. Bar-Shavit, Ph.D. Thesis, The Weizmann Institute of Science, Israel (1980).

Macrophages (7×10^5/well) are plated onto 24-well tissue culture plates. The cells are allowed to adhere for 2 hr, washed, and cultured in serum-containing medium for 24 hr. The monolayers are washed three times with PBS and incubated with either radiolabeled NT or SP in PBS (0.5 ml/well) for 30 min at 22°. The concentration of the labeled peptides in these experiments ranges from 10^{-10}–10^{-6} M. Nonspecific binding is assessed by parallel incubation of the monolayers with the radioactive peptide in the presence of 500-fold nonradioactive peptides. The incubation is terminated by aspirating the buffer from the cell monolayers and vigorous washing of the monolayers for five times with PBS. The monolayers are then lysed (by either detergent or 0.1 N NaOH) and the radioactivity is determined. Using this method, the existence of two classes of binding sites for NT can be shown on macrophages[3]: high-affinity (K_D = 0.9 nM, approximately 5000 sites/cell) and low-affinity (K_D = 28 nM, approximately 33,000 sites/cells) receptors. Macrophages have only one class of receptors for SP, which is similar to the low affinity sites of NT and to tuftsin receptors.[2]

[16] Opsonic Activity of Fibronectin

By Frank A. Blumenstock, Thomas M. Saba, Pina Cardarelli, and Claire H. Dayton

Plasma fibronectin is a plasma protein that has been documented to specifically bind to a variety of other macromolecules including collagen, actin, DNA, hyaluronic acid, fibrin, heparin,[1,2] as well as to microparticulates such as some bacteria,[1,2] immune complexes,[3,4] and artificial gelatinized particles.[5] A variety of studies have demonstrated that fibronectin can mediate monocyte or macrophage binding and endocytosis of microparticulates as well as soluble macromolecular substances. The methods described here are those which have been utilized in our laboratory to investigate this process.

[1] K. M. Yamada, *Annu. Rev. Biochem.* **52**, 761 (1983).
[2] D. F. Mosher, *Annu. Rev. Med.* **35**, 561 (1984).
[3] M. Vuento, M. Korkolainen, and V.-H. Stenman, *Biochem. J.* **205**, 303 (1982).
[4] A. Hantanen and J. Keski-Oja, *Clin. Exp. Immunol.* **53**, 233 (1983).
[5] T. M. Saba and E. Jaffe, *Am. J. Med.* **68**, 577 (1980).

Purification of Plasma Fibronectin

Gelatin–Sepharose Column

  Reagents

  CNBr-activated Sepharose 4B (Pharmacia)
  Gelatin (Nutritional Biochemical Company)
  0.2 M NaHCO$_3$ (fresh)
  1 M NaCl
  0.001 M HCl
  1 M Ethanolamine in 0.1 M NaHCO$_3$ (fresh) containing 0.5 M NaCl
  0.1 M Sodium acetate (pH 4.0) in 1 M NaCl
  0.1 M Sodium borate (pH 8.0) in 1 M NaCl
  1 M NaCl in 0.1 M phosphate (pH 7.4), 3 mM benzamidine-HCl and
    0.02% ME

Affinity Column Preparation. Purified plasma fibronectin will be isolated by affinity chromatography using a gelatin–Sepharose 4B affinity column. Column preparation consists of dissolving gelatin (Nutritional Biochemical Company) in boiling 1 M NaCl at a concentration of 4 mg/ml. The solution is allowed to cool and 12.5 ml of gelatin solution is added to 12.5 ml of freshly prepared 0.2 M NaHCO$_3$. This solution is added to 5 gm of cyanogen bromide (CNBr)-activated Sepharose (Pharmacia) which had previously been washed five times with 200 ml of 0.001 M HCl on a sintered glass filter. The washed CNBr-Sepharose is then allowed to react with 25 ml of the gelatin solution for 2 hr, with gentle agitation. The gelatin solution is removed from the gelatin–Sepharose by suction through the sintered glass filter. Following a wash with 0.1 M NaHCO$_3$ in 0.5 M NaCl, the gelatin–Sepharose is allowed to react with 25 ml of 1 M ethanolamine in 0.1 M NaHCO$_3$ and 0.5 M NaCl for 2 hr at room temperature. This allows for saturation of any remaining unreacted sites on the gelatin–Sepharose.

At the termination of the reaction with ethanolamine, the residual noncovalently bound protein is washed from the gel by three washing cycles, each of which consists of a 1-hr wash with 0.1 M acetate buffer (pH 4.0) in 1 M NaCl followed by a 1-hr wash with 0.1 M borate buffer (pH 8.0) in 1 M NaCl. Five grams of CNBr-activated Sepharose conjugated to gelatin yields enough gel to fill a 10.0-ml bed column. The column is then cast and the gelatin–Sepharose is first washed extensively with column wash buffer, 1 M NaCl, 0.1 M NaPO$_4$, 3 mM benzamidine, 0.02% ME, and 0.01% NaN$_3$ (pH 7.4). The column is then washed with 100 ml of 8 M urea in column wash buffer (pH 7.4). Following urea treatment, the column is reequilibrated with wash buffer. When not in use, the column is stored at 5° and contains 0.01% NaN$_3$. The capacity of the column is 80–100 ml of plasma and can be reused 10–15 times.

Isolation of Plasma Fibronectin

Reagents

BaCl$_2$

(NH$_4$)$_2$SO$_4$

1 M NaCl in 0.01 M phosphate (pH 7.4), 3 mM benzamidine-HCl, and 0.02% ME

1 M Urea in 0.01 M phosphate (pH 7.4), 3 mM benzamidine-HCl, (neutral pH, 0.02%)

4 M Urea in 0.01 M phosphate (pH 7.4), 3 mM benzamidine-HCl (0.02%)

Isolation of Plasma Fibronectin. Whole blood is collected from ether-anesthetized rats by inferior vena cava puncture using a 10-cm^3 plastic syringe containing 0.8 ml of citrate–phosphate–dextrose (CPD)[5a] anti-coagulated with 3 mM benzamidine-HCl. For every 10 ml of blood drawn, 1.5 ml of CPD is used. The blood is centrifuged at 5000 g for 20 min at 5° and the citrated plasma collected. The addition of 15 mg BaCl$_2$/ml of plasma allows for removal of the vitamin K-dependent coagulation proteins. This suspension is stirred gently for 30 min at 4° and centrifuged at 12,000 g for 10 min at 4°. Excess barium is removed from the supernatant by addition of 9.5 mg/ml (NH$_4$)$_2$SO$_4$ and the BaSO$_4$ removed by centrifugation. The plasma is applied to a gelatin–Sepharose column, and the fibronectin purified by an affinity procedure originally described by Engvall and Ruoslahti.[6]

Before application of plasma through the gelatin–Sepharose column, the column is extensively washed with column wash buffer until a stable baseline is recorded at 280 nm. Plasma is pumped through the column by a peristaltic pump at a flow rate of 25 ml/hr. This is followed by a 1-hr wash in column buffer until baseline returns to zero, followed by a 1-hr wash with 1 M urea.

Plasma fibronectin is eluted from the gelatin by pumping 4 M urea in 0.01 M sodium phosphate buffer, pH 7.4, containing 3 mM benzamidine through the column and 10-min interval fractions collected. A sharp peak in the 280 nm absorbance is an indication of the eluted protein. The

---

[5a] Abbreviations: CPD, citrate–phosphate–dextrose; ME, 2-mercaptoethanol; PBS, phosphate-buffered saline; RES, reticuloendothelial system; HBSS, Hanks' balanced salt solution; DMEM, Dulbecco's modified Eagle's medium; DIcI, 1-ethyl-3-(3-dimethylaminopropyl)carbodiimide; GBSS, Gey's balanced salt solution.

[6] E. Engvall and E. Ruoslahti, *Int. J. Cancer* **20**, 1 (1977).

fractions collected from this peak are pooled and the urea removed from the protein by overnight dialysis at 4° in 0.2 M sodium phosphate (pH 7.4) containing 0.02% ME. The dialyzed protein is repurified as described above, dialyzed vs PBS (pH 7.4) at 4° for 24 hr, and stored at −70°.

Assessment of Opsonic Activity of Fibronectin

The opsonic activity of fibronectin has been assessed by a variety of techniques. This section will describe several of them. All of these techniques utilize gelatinized target particles, and in most cases these particles are radioactively labeled, and macrophage uptake of the particles, as mediated by fibronectin, is assessed by radioactivity.

In Vitro Liver Slice Assay[7]

In this assay the ability of plasma, serum, or fibronectin to mediate the uptake of a radioactive gelatinized lipid emulsion by the phagocytic cells in rat liver slices is assessed. Thus, fresh rat liver slices are incubated with solutions containing fibronectin in the presence of a radiolabeled gelatinized lipid emulsion, under defined conditions. The liver slices are then evaluated for the uptake of radioactivity. The extent of uptake is then defined as the opsonic activity of the solutions tested. Electron microscopy has verified that the uptake of the lipid particulate by the liver slices is indeed phagocytic in nature and not adherent, as has been observed utilizing inert nonmetabolizable particulates.[8]

This assay suffers from the fact that it utilizes large amounts of plasma, serum, or fibronectin in the determination of opsonic activity. However, it is easily performed without extensive preparation and gives reliable results, although like other bioassays it will produce data with a high degree of variability.

Equipment

37° Dubonoff metabolic shaker bath (Thomas Scientific)
25-ml Erlenmeyer flasks, heavy walled
No. "0" rubber stoppers
Stadie–Riggs tissue slicer and blades (Thomas Scientific)
Waring blender with microblender attachment. The volume of the stainless steel blending chamber is 30 ml

[7] T. M. Saba, J. P. Filkins, and N. R. DiLuzio, RES, *J. Reticuloendothel. Soc.* **3**, 398 (1966).
[8] B. C. Dillon, J. E. Estes, T. M. Saba, F. A. Blumenstock, E. Cho, S. K. Lee, and E. P. Lewis, *Exp. Mol. Pathol.* **38**, 208 (1983).

Material and Reagents

Glycerol

Soya lecithin (Sigma)

[125]I-Labeled triolein (New England Nuclear, special order). This material is quite expensive, but 5 mCi will last approximately 6 months with normal usage. When using the ^{125}I-labeled triolein to prepare the stock lipid emulsion, one-third of the material is used to prepare the first 3 months' supply of stock emulsion. After 3 months, the rest of the ^{125}I-labeled triolein is used to prepare a second 3-month supply of the stock emulsion

Peanut oil (Planters)

Krebs–Ringer phosphate buffer.[9] Krebs–Ringer phosphate buffer is prepared from the following stock solutions:

  A. 0.154 M NaCl
  B. 0.154 M KCl
  C. 0.110 M CaCl$_2$
  D. 0.154 M MgSO$_4$
  E. 0.100 M Phosphate buffer (pH 7.4)

The working Krebs–Ringer phosphate buffer is prepared by combining the stock solutions in the following ratios:

  100 Parts solution A
  4 Parts solution B
  3 Parts solution C
  1 Part solution D
  20 Parts solution E

After mixing, this buffer should be kept in a glass-stoppered vessel in the cold. A precipitate may form which will redissolve at 37°. Therefore mix thoroughly prior to use

Heparin (Upjohn, 1000 U/ml)

Male rats (200–400 g)

95% O$_2$–5% CO$_2$

Isotonic saline

5% Dextrose in water

Gelatin (Nutritional Biochemical Company)

Preparation of Gelatinized RES Test Lipid Emulsion. Preparation of stock emulsion: A stock emulsion is prepared by high-speed emulsification of glycerol, ^{125}I-labeled triolein in a peanut oil diluent, and soya lecithin in a weight ratio of 10:10:1. The glycerol (10 g) is added to 1 g of lecithin in the microblender and emulsified for 10 min. The radioactive

---

[9] W. W. Umbreit, R. H. Burris, and J. E. Stanffer, *in* "Manometric Techniques," 4th ed., p. 132. Burgess, Minneapolis, Minnesota, 1964.

[125]I-labeled triolein plus peanut oil (10 g) is added dropwise over 10–15 min while continuing emulsification. The anhydrous stock emulsion is stored at 4° and can be used for up to 3 months.

Preparation of 1% gelatinized lipid emulsion for in vitro use: A gelatin solution of 0.1% (w/v) is prepared in 5% dextrose and water by adding the appropriate volume of near boiling 5% dextrose and water to the appropriate amount of gelatin. The mixture is stirred until the gelatin is in solution, and the solution is then allowed to cool to room temperature. The gelatin solution is then adjusted to a pH of 7.4 with concentrated NaOH. It is important not to exceed a pH of 7.5 because if the solution is back titrated with HCl and then used to gelatinize the lipid emulsion, very little phagocytic uptake of the emulsion occurs even in the presence of high concentrations of fibronectin. The correct amount of gelatin solution is then added to the anhydrous stock emulsion to equal a 1% (w/v) emulsion. Prior to use, this preparation is allowed to incubate at 37° for 30 min while being gently agitated to disperse the lipid evenly into the dextrose–gelatin solution.

Bioassay of opsonic activity: The liver slice bioassay is carried out in the following manner. The 25-ml Erlenmeyer flasks in a rack are cooled on ice. To the flasks, plasma (heparinized), serum, or fibronectin (in a buffer at physiological pH) is added and adjusted to a total volume of 3.0 ml with cold Krebs–Ringer phosphate. To each of the flasks, 0.1 ml of heparin (1000 U/ml) is added and the flasks gently agitated and kept on ice. Immediately prior to the preparation of liver slices, 0.2 ml of the gelatinized lipid emulsion is added to each flask.

Liver slices are prepared from adult male rats (200–400 g). The animals are ether anesthetized and a laparatomy performed to visualize the liver. Care must be taken not to cut through the diaphragm. The intestines are pushed to one side and the vena cava and the portal vein are cut with scissors and the animal allowed to bleed into the peritoneum. When the liver has blanched (2–3 min), the lobes of the liver are excised and placed in a 200-ml beaker containing 200 ml of ice cold Krebs–Ringer phosphate on ice. The removal of the lobes of the liver should take 2–3 cuts with scissors and the size of tissue obtained should be slightly larger than 25 mm in diameter. The liver tissue is transferred into another beaker on ice containing ice-cold Krebs–Ringer phosphate. Cooling the liver tissue as much as possible and as quickly as possible makes it easier to slice the liver with the Stadie–Riggs tissue slicer.

Preparing liver slices using the Stadie–Riggs tissue slicer quickly and reproducibly takes practice but can be achieved. Prior to use, the tissue slicer and the blade should be rinsed thoroughly with cold Krebs–Ringer phosphate. Having a plastic wash bottle full of cold Krebs–Ringer phosphate will help. The cold lobes of liver tissues are placed on the pedestal of the tissue slicer and sliced with a sawing movement rather than trying to push the slicer through the liver tissue. The first liver slice is discarded since one side consists of Glisson's capsule. Subsequent slices are placed in ice-cold Krebs–Ringer phosphate on ice. A large glass or plastic Petri dish works best. With practice, one can obtain 10 slices from each lobe of the liver. The slices are then placed into the 25-ml Erlenmeyer flasks. The flasks are then gassed with 95% O_2–5% CO_2 and capped with the rubber stoppers. The flasks are then placed into a metabolic shaker bath and gently agitated for 30 min at 37°. The length of time between obtaining the liver tissue and placing the flasks in the shaker bath should be no more than 20 min.

When setting up the assay flasks, each sample is assayed in triplicate and a buffer blank containing no serum or fibronectin is always included. Additionally, 0.2-ml aliquots of the test emulsion are placed in radioactivity-counting tubes for analysis, since this represents the challenging dose of test colloid. After the flasks have been incubated for 30 min at 37°, they are returned to an ice bath. The liver slices are removed from the flasks and washed twice with cold isotonic saline, blotted dry on filter paper, weighed, and placed in counting tubes for analysis of radioactivity. The liver slices should range between 200 and 400 mg. It has been found that using a Roller–Smith balance works best for rapidly weighing the tissues. Following analysis of the radioactivity using a gamma counter, the results are expressed as the percentage of the added dose of colloid that is present in the liver slice and normalized to a per 100 mg basis. That is,

$$\frac{\text{(Counts/min liver slice)}}{\text{(Counts/min in standard)}} = \frac{\text{percentage challenge dose}}{100 \text{ mg liver tissue}}$$
$$\text{(liver slice weight, in 100 mg)}$$

Table I demonstrates the results obtained in measuring the opsonic activity of various concentrations of purified human fibronectin utilizing the liver slice bioassay. In this assay, 1.0 ml of fibronectin solutions at the various fibronectin concentrations is added to 2.0 ml of Krebs–Ringer phosphate in the assay flasks.

Macrophage Monolayer Assay[10]

While the liver slice assay is useful in demonstrating the opsonic activity of plasma fibronectin, it suffers from the fact that it cannot be used to study the basic mechanisms whereby fibronectin mediates the attachment

[10] F. A. Blumenstock, T. M. Saba, E. Roccario, E. Cho, and J. E. Kaplan, *J. Reticuloendothel. Soc.* **30,** 61 (1981).

TABLE I
Opsonic Activity of Purified Human Plasma
Fibronectin Analyzed by Liver Slice Bioassay

Fibronectin added $(\mu g)^a$	Opsonic activity (percentage added dose/100 mg)
0	0.27 ± 0.01
100	0.92 ± 0.06
200	2.49 ± 0.23
400	4.44 ± 0.62
700	7.36 ± 1.09
1000	10.52 ± 1.53

[a] Added to a total incubation volume of 3.0 ml. All media were supplemented with heparin (100 U/flask). Incubation time was 30 min.

of gelatinized particles to macrophage cell surfaces. Furthermore, because of the large quantities of material (plasma, serum, fibronectin) required for the assay, it is important to develop an assay utilizing isolated macrophages to increase the sensitivity of the opsonic assay and to investigate basic mechanisms governing macrophage–fibronectin interactions. Gudewicz et al.[11] were the first to utilize isolated peritoneal cells to assess the opsonic activity of fibronectin. These investigators utilized a gelatinized latex particle, while the assay described here utilizes a gelatinized fixed sheep erythrocyte that is ^{51}Cr-labeled and can be used as a labeled test particle in an isotopic assay for plasma fibronectin. The principle of the assay is very similar to the liver slice assay except that elicited rat peritoneal macrophages are used. The assay tests the ability of fibronectin to moderate the attachment and uptake of the gelatinized target particle to the macrophage monolayer.

Equipment

Humidified CO_2 incubator (37°)
Oscillating shaker

Materials and Reagents

Hanks' balanced salt solution (HBSS)—Gibco
Dulbecco's modified Eagle's medium (DMEM)—Gibco
Penicillin–streptomycin (5000 U/ml—5000 μg/ml)—Gibco
Calf serum (heat inactivated)—Gibco

[11] P. W. Gudewicz, J. Molnar, M. Z. Lai, D. W. Beezhold, G. E. Siefring, Jr., R. B. Credo, and L. Lorand, *J. Cell Biol.* **87,** 427 (1980).

Trypan blue solution (0.4%)—Gibco
Tissue culture plate (24 wells)—Costar
Heparin (from beef lung) (1000 U/ml)—Upjohn
Gelatin—ICN Pharmaceuticals, Inc.
PBS—pH 7.2–7.4
    A. Dulbecco's PBS (10×),   100 ml—Gibco
    B. Distilled water,   899 ml
    C. 5 N NaOH,   1.0 ml

If the 10× Dulbecco's PBS is diluted to 1× with distilled water, and no NaOH added, the pH is 4.75. This PBS (pH 4.75) is utilized in the gelatinization of the fixed ^{51}Cr-labeled sheep erythrocytes
Formaldehyde solution (37%, w/w)
Sodium [^{51}Cr]chromate sterile—New England Nuclear
CPD
    A. Citric acid (monohydrate),   3.27 g
    B. Sodium citrate (dihydrate),   26.3 g
    C. Sodium biphosphate (NaH_2PO_4-
       anhydrous),   1.93 g
    D. Dextrose (anhydrous),   <u>23.2 g</u>
                    to 1000 ml with DH_2O

Working HBSS with sodium bicarbonate
    A. HBSS (10×),   100 ml
    B. 7.5% $NaHCO_3$,   4.65 ml
    C. DH_2O,   895.35 ml

    1-Ethyl-3-(3-dimethylaminopropyl)-carbodiimide [HCl (DICI)]—Sigma

Preparation of Gelatinized ^{51}Cr-Labeled Sheep Red Blood Cells.
1. Preparation: Fresh sheep blood (50 ml) is collected into 7.0 ml CPD and centrifuged at 3000 g (4°). The cells (~25 ml) are washed five times with isotonic saline by centrifugation (4°) and resuspended in 40 ml of PBS (pH 7.4) containing 500 μCi of $Na_2^{51}CrO_4$. The suspension is agitated very slowly at room temperature for 1 hr. Following the incubation with $^{51}CrO_4$, the suspension is washed a number of times with isotonic saline by centrifugation at 3000 g until the supernatant has minimal ^{51}Cr counts (100–500 cpm/ml).

The labeled sheep erythrocytes are then suspended in 200 ml of PBS (pH 7.4). A dialysis bag containing 50 ml of 37% formaldehyde is inserted into the vessel containing the erythrocyte suspension and gently agitated overnight at room temperature. *A magnetic stirrer should not be used.*

The entire vessel should be agitated. A plastic bottle of adequate volume works nicely. The next morning the dialysis bag is cut and the mixture is agitated an additional 2 hr at room temperature. After fixation, the suspension is centrifuged, washed with isotonic saline five times, and stored as a 50% suspension in PBS (pH 7.4) at 4°. The fixed cells can be used for at least 1 month.

2. *Gelatinizing the fixed labeled sheep erythrocytes:* The fixed cells are gelatinized on the day of use. Cells that are not freshly gelatinized do not respond well to the opsonic stimulus of fibronectin.

Gelatin is dissolved in PBS (pH 7.4) at a concentration of 2 mg/ml. This usually requires warming of the solution. Alternatively, the PBS can be heated and an appropriate volume of the hot PBS is then added to the gelatin and the suspension stirred until the gelatin is in solution. After allowing the gelatin solution to cool to room temperature, 20 ml of this solution is added to the suspension of labeled sheep erythrocytes. To prepare the cells for gelatinization, 2 ml of the stock erythrocyte suspension is removed, centrifuged, and the supernatant PBS removed. The cells are suspended in 20 ml of Dulbecco's PBS (pH 4.75), 400 mg of DICI is added immediately thereafter, and the mixture is combined with the 20 ml of gelatin solution and agitated for 2 hr at room temperature. Following gelatinization, the suspension is centrifuged for 10 min at 3000 g at 4°, the supernatant discarded, and the cells washed five times with isotonic saline. Following washing, the gelatinized, labeled, fixed sheep erythrocytes are suspended in DMEM to yield a hematocrit of 4%.

Preparation of Rat Peritoneal Macrophage Monolayers (see also this series, Vol. 108 [25]). Rat peritoneal macrophages are elicited by injecting 15 ml of a 1% sodium caseinate solution in PBS (pH 7.4) per 100 g body weight. The sodium caseinate solution is prepared by warming the PBS to 60° and then, while stirring slowly, adding the sodium caseinate. After the protein has dissolved, the solution is allowed to cool to room temperature prior to intraperitoneal injection into the ether-anesthetized rats.

Ninety-six hours after caseinate injection, the rats are placed in an etherized chamber until they are dead. Ten milliliters of PBS (pH 7.4) containing 50 U of heparin is injected into the peritoneal cavity and the fluid is collected from the peritoneum. This is most easily accomplished using a 50-cc sterile plastic syringe without a needle and aspirating the peritoneal fluid through an appropriately sized incision in the peritoneum. The fluid is filtered through a sterile funnel containing a small amount of cotton gauze. The cell suspension is centrifuged at 100 g for 5 min at 4°. Contaminating erythrocytes are removed by adding 5 ml of cold distilled water to the cell pellet, immediately vortexing for 10 sec, and immediately adding 15 ml of cold 1.2% NaCl. The macrophages are then washed twice with 20 ml of HBSS (pH 7.4 with $NaHCO_3$) by centrifugation at 100 g for 5 min at 4°. The cells are suspended to a concentration of 2 × 10⁶ cells/ml in DMEM containing 20% calf serum and penicillin/streptomycin at 100 U/ml. This cell suspension is transferred into each 10 × 17 mm well of the 24-well tissue culture plate and incubated for 2 hr at 37° in humidified 5% CO_2–95% air.

Bioassay for Fibronectin Opsonic Activity. After the cells have incubated for 2 hr, the DMEM is removed from the monolayer by careful aspiration. This will contain the nonadherent cells. The monolayer is rinsed once with 1 ml of DMEM (without calf serum or penicillin/streptomycin) and then 0.5 ml of DMEM (without calf serum or penicillin/streptomycin) is added. In each well are also added, in the following order: 10 μl heparin (1000 U/ml), 20 μl serum, plasma, or sample to be tested for opsonic activity, 100 μl of the gelatinized labeled fixed sheep erythrocyte suspension, and 370 μl DMEM (without calf serum and penicillin/streptomycin). The assay plates are incubated for 2 hr at 37° in a humidified 5% CO_2–95% air environment. Following incubation, the medium is aspirated from the monolayers, which are carefully washed three times with cold PBS (pH 7.4). The plates are air dried at room temperature and then the monolayers digested with 1.0 ml of 1.0 M NaOH for at least 30 min, but preferably overnight. Four wells containing monolayers in each plate are not used (i.e., containing 990 μl of DMEM and 10 μl of heparin) and are assayed for protein after NaOH digestion utilizing the method of Lowry[12] with bovine serum albumin as a standard. Each well should yield approximately 100 μg/protein. The NaOH digests to be analyzed for radioactivity are transferred to counting tubes and the wells washed once with 1 ml of isotonic saline with the wash being added to the counting tubes. To duplicate or triplicate counting tubes containing 2.0 ml of distilled water, 100 μl of the target cells used in the assay is added. The ⁵¹Cr radioactivity of the samples is then determined. Fibronectin-mediated uptake is determined as the percentage of the added target particles taken up in the presence of fibronectin (serum, plasma, or fibronectin sample) minus the percent taken up in the absence of any fibronectin (appropriate control). That is,

$$\frac{\text{Net percentage uptake}}{\text{Milligram cellular protein}} = \frac{\text{sample count} - \text{blank count}}{\left(\begin{array}{c}\text{counts in 100 }\mu\text{l} \\ \text{labeled erythrocytes}\end{array}\right)\left(\begin{array}{c}\text{milligram cellular} \\ \text{protein/well}\end{array}\right)}$$

Table II demonstrates the results obtained in measuring the opsonic activity of various concentrations of purified human fibronectin utilizing the rat peritoneal macrophage monolayer assay.

---

[12] O. Lowry, N. J. Rosebrough, A. L. Farr, and R. J. Randall, *J. Biol. Chem.* **193,** 265 (1951).

TABLE II
OPSONIC ACTIVITY OF PURIFIED HUMAN PLASMA FIBRONECTIN
ANALYZED BY PERITONEAL MACROPHAGE MONOLAYER ASSAY

Fibronectin added $(\mu g)^a$	Opsonic activity (net percentage added dose/2×10^6 cells)
10	6.17 ± 0.56
20	8.47 ± 0.65
40	11.33 ± 0.94
80	14.44 ± 0.95

^{a} Added to a total incubation volume of 1.0 ml. Heparin was added at a dose of 10 U/well. Incubation time was 2 hr at 37°.

Kupffer Cell Monolayer Assay (see also this series, Vol. 108 [26])

Cell populations of the RES are of current scientific interest, especially with respect to victims of trauma, burn, and postinjury sepsis.[13,14] Because Kupffer cells represent a major population of RE phagocytic cells,[15] it is most desirable to have an efficient method for isolating and culturing these cells.

In our laboratory, Kupffer cells are isolated from endotoxin-treated rats using a combined collagenase perfusion and pronase digestion. The Kupffer cell yield obtained from endotoxin-treated rats is much greater than that obtained from non-endotoxin-treated controls.[16] This may be due to local proliferation and/or recruitment from the monocyte pool. Additionally, activated Kupffer cells demonstrate an enhanced spreading capacity on plastic culture dishes. This phenomenon is also noted in inflammatory peritoneal macrophages.[17] Activated Kupffer cells have been shown to have increased phagocytic and pinocytic potential.[17,18] Furthermore, lipopolysaccharide is normally absorbed in the portal circulation, Kupffer cells having the capacity to phagocytize the endotoxin.[19]

In the following isolation procedure, thorough exsanguination of the livers with HBSS is a necessary step prior to collagenase perfusion. Incubation in a protease solution selectively destroys hepatocytes, which represent the major liver cell population.[20] Kupffer cells themselves are re-

[13] T. M. Saba, *Arch. Intern. Med.* **126**, 1031 (1970).
[14] J. E. Kaplan and T. M. Saba, *Am. J. Physiol.* **230**, 7 (1976).
[15] T. M. Saba, *Prog. Liver Dis.* **7**, 109 (1982).
[16] P. M. Cardarell, Ph.D. Thesis Dissertation, Albany Medical College, Albany, New York (1985).
[17] Z. A. Cohn, *J. Immunol.* **121**, 813 (1978).
[18] M. L. Karnovsky and J. K. Lazdins, *J. Immunol.* **121**, 809 (1978).
[19] D. P. Praaning-van Dalen, A. Brouwer, and D. L. Knook, *Gastroenterology* **81**, 1036 (1981).
[20] D. M. Mills and D. Zucker-Franklin, *Am. J. Pathol.* **54**, 147 (1969).

sistant to a number of proteases. One disadvantage of using pronase is possible ingestion of degraded parenchymal cells by Kupffer cells; however, using only collagenase treatment leads to mixed cell populations with flourishing hepatocytes. Further purification of nonparenchymal cells is accomplished by a metrizamide density gradient.[21] Metrizamide is nonionic, highly soluble, and has a high density due to its triiodobenzene ring. These properties make it suitable for separation of biological materials. Using these methods, it is possible to consistently produce a viable Kupffer cell population which is 90–95% pure.

Materials (all materials in direct contact with cells should be sterile)
1. Cell isolation
   Collagenase, type II (Sigma)
   Protease, type XIV (Sigma)
   Metrizamide, analytical grade (Accurate Chemical)
   DNase I (Sigma)
   Dulbecco's phosphate-buffered saline without Ca^{2+} and Mg^{2+}
   Trypan blue
   Gey's balanced salt solution (GBSS) (Gibco)
   HBSS without Ca^{2+} and Mg^{2+} (Gibco)
   Sterile NaCl (0.9%)
   Pentobarbital sodium injection (Nembutal) (50 mg/ml)
   Lipopolysaccharide *S. enteritidis* (5 mg/ml) (Difco)
   GBSS without NaCl and glucose
   Sterilization filter units—0.20 μm
   95% O_2–5% CO_2 gas mixture
2. Cell culture
   DMEM—low glucose (Gibco)
   Calf serum—heat inactivated (Gibco)
   Antibiotic–antimycotic solution (10,000 U penicillin; 10,000 μg/ml streptomycin; 25 μg/ml Fungizone) (Gibco)
   24-Well plastic tissue culture plates

Equipment (equipment in direct contact with cells should be sterile)
Laminar flow hood
Surgical pack for small animals
Glass beakers and bottles
Nylon gauze
Centrifuge
37° Shaking water bath

---

[21] D. L. Knook and E. C. Sleyster, *Exp. Cell Res.* **99**, 444 (1976).

Vented Erlenmeyer flask
Small intramedic polyethylene plastic cannulas (i.d. 0.58 mm, o.d. 0.965 mm) (Clay Adams)
Hemacytometer
Petri plate
1000-μl and 50-μl pipets
Peristaltic perfusion pump (Holter model RE161)
37° CO_2 incubator

Preparation of Solutions

1. 0.03% Collagenase[21a]
   -Dissolve 30 mg collagenase type II in 100 ml HBSS
   -Stir gently
   -Filter sterilize
   -Keep at 37° until use
2. 0.2% Pronase[21a]
   -Dissolve 300 mg protease type XIV in 150 ml GBSS
   -Stir until solution is homogeneous
   -Filter sterilize
   -Keep at 37°—use within 30 min
3. DNase I
   -Dilute to 1 mg/ml with double-distilled water
   -Can be stored in 1-ml aliquots at −20° indefinitely
4. GBSS without glucose and NaCl
   -Dissolve the following constituents per 1 liter distilled H_2O:
   KCl, 0.37 g
   $NaHCO_3$, 2.27 g
   $CaCl_2$, 0.17 g
   $MgCl_2 \cdot 6H_2O$, 0.21 g
   $Na_2HPO_4 \cdot 7H_2O$, 0.266 g
   KH_2PO_4, 0.03 g
   $MgSO_4 \cdot 7H_2O$, 0.07 g
   -pH to 7.4
   -Osmolarity should be between 290 and 310 mOsm/liter
   -Autoclave or filter sterilize
   -Check pH at least once a month
   -May be stored at 4° indefinitely
5. 30% Metrizamide gradient[21a]
   -Add 4.2 g metrizamide and 14 mg glucose to 14 ml GBSS without NaCl

[21a] Made fresh before each isolation.

-Stir until *all* substrate is dissolved—takes at least 10 min
-Filter sterilize
-Keep at 37° until used
(Note: metrizamide solution can be stored at 4° for up to 1 week; however, care must be taken not to expose the solution to light)
6. Culture medium
-Use sterile technique
-Add 20 ml calf serum and 2 ml penicillin/streptomycin to a final volume of 100 ml DMEM
-Place in 37° incubator 15 min before use

Procedure. Kupffer cell isolation: Kupffer cells are isolated using a modification of previously described methods.[21-23] For monolayer assays, two male Sprague-Dawley rats weighing 250–300 g each are given intraperitoneal injections of 100 μg *Salmonella enteritidis* endotoxin for three consecutive days prior to isolation. Animals are provided with food and water *ad libitum*. On the fourth day, the rats are anesthetized with nembutal (0.1 ml/kg) intraperitoneally. The surgical site is clipped and scrubbed with a bactericidal solution. Working in a sterile surgical area, a midline incision is made and the portal vein cannulated. Care must be taken to ensure that the cannula is not lodged in one lobe of the liver. After the cannula has been fastened in place, the liver is pump perfused with HBSS at a flow rate of 15–20 ml/min. At the onset of perfusion, the inferior vena cava is cut and the liver excised, taking care not to nick the bowel. When both lobes have been thoroughly perfused with HBSS, the collagenase solution is added. Light manual massage of poorly perfused areas will often improve circulation. Following the collagenase perfusion, the livers are rinsed briefly in sterile saline, the cannulas are removed, and the livers are placed on nylon gauze in a Petri dish above an Erlenmeyer flask. A small amount of pronase solution is added and the livers are lightly scraped with a spatula, thus causing them to dissociate. When the livers have been finely digested, they are channeled into the flask along with the remainder of the pronase solution. One milligram of DNase is added, and the flask with a vented top is attached to a gas mixture of 95% O_2–5% CO_2 and incubated in a 37° shaking water bath for 55 min. When gassing the solution, only enough pressure should be used to ripple the top of the solution. At the end of the incubation the solution is filtered through nylon gauze into four 50-ml sterile plastic centrifuge tubes. This and subsequent steps are carried out in a laminar flow hood to maintain sterility. The tubes are capped and centrifuged for 5 min at 350 g at 10°. The supernatant is removed and the pellets are resuspended in GBSS, consolidated

[22] A. C. Munthe-Kaas, T. Berg, P. O. Selgen, and R. Seljelid, *J. Exp. Med.* **141**, 1 (1975).
[23] R. Zahlten, H. Hagler, M. Nejtek, and C. J. Day, *Gastroenterology* **75**, 80 (1978).

TABLE III
OPSONIC ACTIVITY OF PURIFIED HUMAN PLASMA FIBRONECTIN
ANALYZED BY RAT KUPFFER CELL MONOLAYER ASSAY

Fibronectin added $(\mu g)^a$	Opsonic activity (net percentage added dose/2 × 10^6 cells)
10	12.59 ± 1.19
20	17.30 ± 1.51
30	16.52 ± 0.13
50	20.57 ± 1.43
80	27.67 ± 1.36

^{a} Added to a total incubation volume of 1.0 ml. Heparin was added at a dose of 10 U/well. Incubation time was 1 hr at 37°.

into two tubes, and centrifuged for another 5 min at the same speed. Resulting pellets are pooled and washed once more. The final pellet is resuspended thoroughly in a total volume of 10 ml GBSS, the metrizamide solution is added, and the cells are well resuspended. The solution is separated into four 15-ml sterile plastic centrifuge tubes and a 0.5-ml aliquot of GBSS is slowly layered on top of each tube. This is a critical step, and must be done with great care. The tubes are centrifuged for 15 min at 1400 g at 10°. A clear band of nonparenchymal cells should now be present at the top of each tube. These layers are collected, washed twice in GBSS, and centrifuged for 5 min at 350 g for each wash. The Kupffer cell pellet is thoroughly resuspended in a final volume of 10 ml culture medium. Using a sterile pipet, a 50-μl aliquot is added to 450 μl PBS, and 50 μl of this solution is resuspended in 50 μl trypan blue to be counted. Viable cells are counted using a hemacytometer. Cells are plated at 2.5 × 10^6/ml using the following dilution formula: Number of cells/grid × 1000 × 10 × 20 × 10. Cells are plated in 24-well plastic culture dishes and incubated for 22 hr. After one wash with DMEM, the cells are ready to be challenged, in the manner previously described utilizing peritoneal macrophages.

Table III demonstrates the results obtained in measuring the opsonic activity of various concentrations of purified human fibronectin utilizing the rat Kupffer cell monolayer assay.

Acknowledgments

Experimental studies were supported by Grant GM-21447 from the National Institutes of Health and a grant from USV/Armour Pharmaceuticals/Revlon Health Care Group, Tuckahoe, New York. Purified human fibronectin used in Tables I, II, and III was provided by USV/Armour Pharmaceuticals, Tuckahoe, New York. These data were previously reported in *Am. J. Med.* **80,** 229 (1986) and are used here with permission.

Section V

Enzymes and Metabolic Activity of Phagocytes

Editors' Note: The articles in this section deal with several aspects of the biochemistry of the Respiratory Burst. Other methods for the study of H_2O_2 production, oxygen consumption, superoxide production, oxidizing radical production, hexose monophosphate shunt activity, peroxidase, and degranulation of polymorphonucleates may be found in [3] of this volume. Methods for the study of other enzymes of the leukocyte metabolism can also be found in [3]. Other subjects relevant to the biochemistry of phagocytosis have been presented in other volumes of this series and will not be duplicated here. Thus, in Vol. 105 ("Oxygen Radicals in Biological Systems"), articles may be found on superoxide dismutase [9]–[11], glutathione peroxidase [12], catalase [13], and oxidative activities of phagocytic cells [45]–[52]. Methods for the study of phagocytic cell proteases and their inhibitors have been discussed in [42]–[44] and [54]–[56] of Vol. 80.

[17] Respiratory Burst during Phagocytosis: An Overview

By MANFRED L. KARNOVSKY and JOHN A. BADWEY

Granulocytes, some classes of macrophages, and natural killer cells release substantial amounts of superoxide (O_2^-) and hydrogen peroxide (H_2O_2) when their plasmalemma is perturbed. This response and the concomitant phenomena (e.g., increased oxygen consumption, hexose monophosphate shunt activity) have been termed the "respiratory burst." They result from the stimulation of an NAD(P)H oxidase in the plasmalemma. The perturbations induced by physiological processes (e.g., phagocytosis; antibody-mediated extracellular interactions) can be mimicked by certain soluble or "quasi-soluble" agents (e.g., *cis*-unsaturated fatty acids, phorbol esters[1-3]). Recently, use of the latter stimuli has been particularly helpful in detecting the probable involvement of phosphatidylinositol-specific phospholipase C and protein kinase C in the mechanism by which the oxidase is activated.[4-6]

Superoxide and H_2O_2, formed on the external surface of the plasmalemma and internal surface of the phagosome, may interact to form hydroxyl radical (OH·) and perhaps singlet oxygen ($^1O_2$). These products of oxygen reduction (i.e., O_2^-, H_2O_2, OH·) and excitation ($^1O_2$) constitute key components of the oxygen-dependent mechanisms by which foreign cells are destroyed by the leukocytes mentioned earlier.[1,3] Detection and quantification of cellular O_2^- and H_2O_2 release is facilitated by the use of enzymes specific for the substances mentioned (e.g., superoxide dismutase, catalase, peroxidase). Unfortunately, highly specific assays of that nature are not available for OH· and $^1O_2$, although the chemical and physical methods used are adequate.[7]

[1] J. A. Badwey and M. L. Karnovsky, *Annu. Rev. Biochem.* **49**, 695 (1980).

[2] B. M. Babior, *N. Engl. J. Med.* **298**, 659 (1978).

[3] S. J. Klebanoff and R. A. Clark, "The Neutrophil Function and Clinical Disorders." North-Holland Publ., Amsterdam, 1978.

[4] D. M. Helfman, B. D. Appelbaum, W. R. Vogler, and J. Kuo, *Biochem. Biophys. Res. Commun.* **111**, 847 (1983).

[5] J. A. Badwey, J. T. Curnutte, J. M. Robinson, C. B. Berde, M. J. Karnovsky, and M. L. Karnovsky, *J. Biol. Chem.* **259**, 7870 (1984).

[6] J. T. Curnutte, J. A. Badwey, J. M. Robinson, M. J. Karnovsky, and M. L. Karnovsky, *J. Biol. Chem.* **259**, 11851 (1984).

[7] J. A. Badwey, J. M. Robinson, M. J. Karnovsky, and M. L. Karnovsky, *in,* "Handbook of Experimental Immunology" (D. M. Weir, L. A. Herzenberg, C. C. Blackwell, and L. A. Herzenberg, eds.), Vol. 2, Ed. 4, Ch. 50, p. 50.1. Blackwell, Edinburgh, 1986 (in press).

The most thoroughly studied cidal system known to function in the phagosome consists of myeloperoxidase, H_2O_2, and chloride. This system has been shown *in vitro* to be effective in killing bacteria, yeast, mycoplasma, viruses, and tumor cells.[3] Although the detailed chemical basis for the effective killing activity of this system is not yet clear, studies indicate that hypochlorite[8] and chloramines[9] are likely to be involved.

The diversity of the oxygen-dependent cidal mechanisms is underscored by the fact that neutrophils deficient in peroxidase (an inherited condition) are capable of normal killing *in vitro* and individuals so affected are generally in good health.[2]

The complex nature of the NAD(P)H oxidase activity is now being revealed. This system consists of at least two components, a flavoprotein[10] and a *b*-cytochrome with a low midpoint potential.[11] Another inherited disease in humans (chronic granulomatous disease of childhood) has served to provide clues regarding the enzyme mechanisms involved in the "respiratory burst."[2,3,11] The steps in the activation of this oxidase, (e.g., putative phosphorylation, membrane fusion), between phosphoinositide turnover, protein kinase C activation, and O_2^- release, are unknown.

Apart from the cidal properties mentioned, "active oxygen species" produced by phagocytes have been implicated recently in a variety of processes. Examples include mutation of normal cells,[12] suppression of murine lymphocyte proliferation,[13] inactivation of leukotrienes,[14] and the production of a potent chemotactic factor in serum.[15] Cellular ramifications of the "respiratory burst" are thus likely to become of interest to a broad spectrum of biologists.

---

[8] J. E. Harrison and J. Schultz, *J. Biol. Chem.* **251**, 1371 (1976).

[9] S. J. Weiss, M. B. Lampert, and S. T. Test, *Science* **222**, 625 (1983).

[10] T. G. Gabig, *J. Biol. Chem.* **258**, 6352 (1983).

[11] A. W. Segal, A. R. Cross, R. C. Garcia, N. Borregaard, N. H. Valerius, J. F. Soothill, and O. T. G. Jones, *N. Engl. J. Med.* **308**, 245 (1983).

[12] S. A. Weitzman and T. P. Stossel, *Science* **212**, 546 (1981).

[13] Z. Metzger, J. T. Hoffeld, and J. J. Oppenheim, *J. Immunol.* **124**, 983 (1980).

[14] W. R. Henderson and S. J. Klebanoff, *Biochem. Biophys. Res. Commun.* **110**, 226 (1983).

[15] W. F. Petrone, D. K. English, K. Wong, and J. M. McCord, *Proc. Natl. Acad. Sci. U.S.A.* **77**, 1159 (1980).

[18] NADPH Oxidase from Polymorphonuclear Cells

By THEODORE G. GABIG and BRUCE A. LEFKER

The NADPH oxidase enzyme system from polymorphonuclear leukocytes is responsible for the sudden burst of oxygen consumption that follows exposure of these cells to certain activating agents. The enzyme system catalyzes the univalent reduction of molecular oxygen at the expense of pyridine nucleotide (NADPH) oxidation. The membrane-associated enzyme system has a vectorial orientation in the plasma membrane and/or phagosomal membrane of the cell. A variety of reduced oxygen species derived from the O_2 product of the oxidase system mediates microbicidal, cytotoxic, and cytolytic effects of these phagocytic cells. This enzymatic activity is postulated to be a multicomponent electron transport chain consisting of a flavoprotein, a unique low-potential cytochrome b_{-245} (referred to as cytochrome b_{559} in this chapter), and ubiquinone-50. The flavoprotein component (E-FAD) has been completely resolved from the cytochrome b_{559} and shown to be the initial electron acceptor from NADPH, whereas the cytochrome b_{559} is thought to be a terminal electron donor to oxygen. Ubiquinone-50 has been postulated to be part of the terminal oxidase segment of the chain; however, its electron carrier function in the oxidase has not been studied directly in relation to the other components. Oxidase activity is not expressed in unstimulated neutrophils or subcellular fractions prepared from them, and attempts to activate the oxidase in subcellular neutrophil preparations have been unsuccessful. Oxidase activation is a highly specific process controlled by complex mechanism(s) at the cellular level. Our laboratory approached the characterization of this oxidase system with the aid of strategic and tactical principles developed by other groups of investigators for the characterization of mitochondrial and microsomal membrane electron transport systems. Additionally, certain unique characteristics of the polymorphonuclear leukocyte NADPH oxidase enzyme system were taken into account in the development of the following scheme for isolation, assay, resolution, and partial characterization of this oxidase system.

Preparation of the NADPH Oxidase and Resolution into Components

Materials

Cytochrome c (horse heart, type VI), superoxide dismutase (bovine erythrocyte), NADPH (preweighed vials), flavin adenine dinucleotide, riboflavin, deoxycholic acid, cholic acid, and phorbol myristate acetate can be obtained from Sigma. Lymphoprep is available from Accurate Chemical Company, and helium (99.9999%) and a catalytic gas purifier (model 4506) from Matheson, Inc. Impurities in deoxycholic acid and cholic acid are absorbed with activated charcoal in hot 50% (v/v) ethanol solution.[1] The bile acids are recrystallized, then brought to 10% (w/v) solution, pH 8.4 (deoxycholate), or 20% (w/v) solution, pH 8.2 (cholate), by neutralization with potassium hydroxide. A Coomassie Blue R-250 dye binding protein assay kit can be obtained from Pierce, Inc.

Preparation and Activation of Polymorphonuclear Leukocytes

Polymorphonuclear leukocytes are prepared from fresh whole blood anticoagulated with acid–citrate–dextrose by dextran sedimentation followed by centrifugation over lymphoprep and hypotonic lysis of red blood cells[2] (see also this series, Vol. 108 [9]). The final cell preparation is >95% neutrophils, containing fewer than 1–2 platelets per leukocyte and <0.1% red blood cells; the remaining cells are mononuclear cells, eosinophils, and basophils. Neutrophils (6–9×10^8) are suspended in Dulbecco's phosphate-buffered saline containing 1 mM NaN$_3$ at a cell density of 10^8/ml. An equal volume of Dulbecco's phosphate-buffered saline containing 1 mM NaN$_3$ and 2 μg/ml phorbol myristate acetate is prewarmed to 37°. The cell suspension is warmed to 37° for 3 min, then mixed with the phorbol myristate acetate solution, and incubated for 20 min in a gently shaking metabolic incubator at 37°. The activation is terminated by cooling the neutrophil suspension to 0° in an ice bath. All subsequent steps are performed at 0–4° unless otherwise indicated. The neutrophils are sedimented by centrifugation at 800 g for 3 min. Cells are resuspended in 8.5 ml unbuffered 0.25 M sucrose by vigorous vortex mixing; remaining cell clumps are then disrupted by gentle hand homogenization in a Potter-Elvehjem homogenizer.

Neutrophil Fractionation

Cell suspension is disrupted by sonication for three 15-sec intervals at 20-W power with the standard probe tip of a Branson 200 sonifier set on pulsed power, 50% duty cycle. For different sonifiers, the power and time of sonification should be adjusted so that only 70–80% of the cells are completely disrupted (i.e., at least 20% of the cells should remain intact as assessed by phase-contrast microscopy so that unwanted disruption of primary and specific neutrophil granules is avoided). Unbroken cells and

[1] Y. Hatefi, this series, Vol. 53, p. 3.
[2] A. Boyum, *Scand. J. Clin. Lab. Invest.* **21**, 77 (1968).

TABLE I
PREPARATION OF THE PARTICULATE NADPH OXIDASE FROM
POLYMORPHONUCLEAR LEUKOCYTES

Preparation[a]	Total protein (mg)	Specific activity (mU/mg protein)	Yield (%)	Purification (-fold)
Postnuclear supernatant[b]	57.0	49	100	1
Supernatant I	37.4	57	76	1.2
Pellet II	4.7	403	68	8.2

[a] Identification of the preparations follows the nomenclature in Fig. 1.
[b] Starting material is approximately 5.9×10^8 polymorphonuclear leukocytes before stimulation and disruption.

nuclei are sedimented by centrifugation at 800 g for 5 min. Exactly 8.0 ml of the postnuclear supernatant is then placed in a 25-ml glass flask and stirred constantly with a magnetic stirrer while 3.32 g of sucrose is slowly added. The suspension should be stirred until all sucrose is dissolved, at which time the volume will have increased to approximately 10 ml. The suspension is then ultracentrifuged at 30,000 rpm in a Beckman Ti 75 fixed angle rotor for 30 min. The supernatant is completely decanted from the greenish-brown tightly packed pellet and diluted to approximately 50 ml by addition of ice-cold distilled H_2O. The diluted supernatant is then ultracentrifuged at 30,000 rpm in a Beckman Ti 75 rotor for 30 min. The partially translucent, faintly salmon-pink pellet of this centrifugation is enriched in NADPH-dependent O_2^--generating activity and can be stored at $-70°$ for up to 6 months without appreciable loss of enzymatic activity. The material at this stage of preparation is free of enzymatically or spectrally detectable myeloperoxidase and contains noncovalently bound FAD, ubiquinone-50, and a unique low-potential[3] cytochrome b_{559} (see below). This two-step subcellular fraction scheme for obtaining this particulate oxidase fraction contains minor modifications that results in slightly higher final specific activity compared to the original method.[4] A representative experiment illustrating the expected purification and yield during the fractionation is shown in Table I. Figure 1 illustrates the modified procedure. Increasing the time of stimulation with phorbol myristate acetate from 7 to 20 min is responsible for the major portion of increased specific oxidase activity as well as specific content of FAD, ubiquinone-50, and cytochrome b_{559}. Ultracentrifugation in the Beckman ultracentrifuge allows greater temperature control (compared to the original Sor-

[3] A. R. Cross, F. K. Higson, O. T. G. Jones, A. M. Harper, and A. W. Segal, *Biochem. J.* **204**, 479 (1982).
[4] T. G. Gabig, E. W. Schervish, and J. T. Santinga, *J. Biol. Chem.* **257**, 4114 (1982).

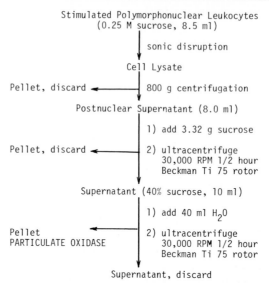

FIG. 1. Preparation of the particulate NADPH oxidase from stimulated polymorphonuclear leukocytes. Neutrophilic polymorphonuclear leukocytes are isolated from fresh human blood, stimulated, sonically disrupted, then fractionated as described in the text. The final pellet contains NADPH-dependent O_2^- generating activity in 60–70% yield and 7- to 8-fold purification.

vall high-speed centrifugation method) and results in fewer preparations with unexplained low specific activity of the oxidase. The extreme heat lability of the NADPH-dependent O_2^--generating activity has been noted by nearly every group of investigators working with it, which should serve to emphasize the importance of temperature control during all manipulations, especially after the stimulated neutrophils have been disrupted.

Resolution of the Particulate Oxidase into Flavoprotein, Ubiquinone-50, and Cytochrome b_{559} Components

All steps are performed at 0–4°. The particulate oxidase fraction from the preceding step is taken up in 50 mM Tris–HCl, pH 8.2, by gentle hand homogenization in a glass homogenizer with a loose-fitting Teflon pestle. The protein concentration, measured by the Coomassie Blue R-250 dye binding methods,[5] is adjusted to 4–5 mg/ml. The weight of the suspension is accurately determined, then exactly 55 μl of 20% (w/v) potassium cholate per gram of suspension is added while stirring continuously. The suspension is stirred for an additional 30 min and then ultra-

[5] M. M. Bradford, *Anal. Biochem.* **72**, 248 (1976).

centrifuged for 30 min at 30,000 rpm in a Beckman Ti 75 rotor. The resulting supernatant, containing the flavoprotein component of the oxidase in approximately 70% yield and 3-fold purification, is free of spectrally detectable cytochrome b_{559}, but is contaminated with approximately 10% of the ubiquinone-50 initially in the particulate oxidase starting material.

The faintly pink, translucent pellet from the preceding step is taken up in 1 M KCl, 50 mM Tris–HCl, pH 8.4, by hand homogenization and adjusted to a protein concentration of 4–5 mg/ml. Exactly 110 μl of 20% (w/v) potassium cholate per gram of suspension is added while the suspension is stirred. The suspension is stirred constantly for an additional 30 min, then ultracentrifuged for 30 min in a Beckman Ti 75 rotor at 30,000 rpm. The supernatant of this centrifugation contains approximately 80% of the ubiquinone-50, 30% of the flavoprotein, and 20–30% of the cytochrome b_{559} originally in the particulate oxidase starting material. The pellet of this centrifugation is translucent and distinctly pink, it contains the cytochrome b_{559} in approximately 70–80% yield and 5-fold purification and has no detectable flavoprotein. However, it is contaminated with approximately 10% of the ubiquinone-50 that was initially in the starting material. The cholate resolution scheme is shown in Fig. 2.

Characterization of the Particulate NADPH Oxidase and Resolved Components

Particulate Oxidase

The particulate oxidase fraction contains enzymatic activity that catalyzes the following reaction:

$$\text{NADPH} + 2O_2 \rightarrow \text{NADP}^+ + 2O_2^- + H^+ \tag{1}$$

Assay of this enzymatic activity is quantified as the rate of superoxide dismutase inhibitable ferricytochrome c reduction in the presence of NADPH in air-saturated buffer. Reagents are added to each of a matched pair of 1-cm path length spectrophotometric cuvettes as follows:

Reagents

0.1 M Potassium phosphate buffer, pH 7.0 (0.7 ml to each cuvette)
1 mM Ferricytochrome c solution in distilled H_2O (0.1 ml to each cuvette)
1 mM NADPH, prepared fresh daily (0.1 ml to each cuvette)
10% (w/v) Potassium deoxycholate, pH 8.4 (5 μl to each cuvette)
Superoxide dismutase, 3 mg/ml distilled H_2O (10 μl to reference cuvette only)

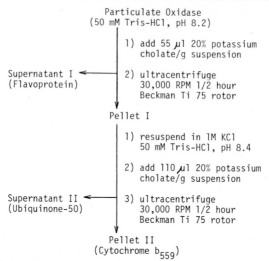

FIG. 2. Resolution of the particulate NADPH oxidase into flavoprotein, ubiquinone-50, and cytochrome b_{559} containing fractions. Details of the bile salt resolution are described in the text.

Distilled H_2O (10 μl to sample cuvette only)

Particulate oxidase suspension, diluted to 0.4–0.8 mg protein/ml in 25% (v/v) ethylene glycol prior to assay (0.1 ml to each cuvette)

The oxidase enzyme is kept on ice until immediately prior to assay; the remaining reagents are equilibrated at room temperature. Reactions, conducted at 25°, are initiated by the addition of 0.1 ml particulate oxidase suspension to each cuvette. The rate of absorbance change is recorded at 550 nm for 1–2 min with the superoxide dismutase-containing cuvette in the reference position and the remaining cuvette in the sample position of a split beam spectrophotometer. The rate of absorbance change should be linear for the initial 1–2 min of assay. The oxidase suspension should be appropriately diluted so that the initial rate does not exceed 0.1 A/min. Under these conditions, the initial rate of absorbance change is linear with respect to neutrophil oxidase protein in the assay and represents NADPH-dependent O_2^- product generation. A reduced minus oxidized millimolar extinction at 550 nm of 21.0 for cytochrome c is used to calculate the rate of superoxide generation:

$$(A/min)/21.0 = \mu mol\ O_2^-/min/ml\ of\ reaction\ mixture \qquad (2)$$

The specific activity is calculated from the amount of neutrophil oxidase protein in the 1-ml reaction mixture.

TABLE II[a]

STOICHIOMETRY OF PURPORTED ELECTRON
CARRIERS IN THE PARTICULATE NADPH OXIDASE

FAD	Ubiquinone-50	Cytochrome b_{559}
238 ± 10	203 ± 30	520 ± 40

[a] Measurements of FAD, ubiquinone-50, and cytochrome b_{559} performed as described in the text. Data given in pmol/mg protein, mean ± SE ($n = 3$).

The specific activity of the NADPH oxidase enzyme system during the steps of a typical oxidase preparation is shown in Table I along with the relative yield and purification. The 0.05% deoxycholate in the assay mixture increases the apparent specific activity by 2- to 2.5-fold in comparison to assay in the absence of detergent, a manifestation of enzyme latency. Higher concentrations of deoxycholate will cause progressive inhibition of oxidase activity. Certain cationic detergents (e.g., cetyltrimethylammonium bromide) strongly inhibit at all concentrations. The neutral detergent Triton X-100 induces dilipidation of the oxidase complex and dissociation of noncovalently bound FAD from the flavoprotein dehydrogenase component, and requires the presence of exogenous FAD and sonicated phosphatidylethanolamine in the assay for expression of enzyme activity.[6] This particulate oxidase preparation consumes NADPH and generates protons in the stoichiometry expected from Eq. (1) (i.e., NADPH : O_2 : H^+ of 1 : 2 : 1).[7] It contains a flavoprotein dehydrogenase (E-FAD), ubiquinone-50, and cytochrome b_{559} in a stoichiometric ratio of approximately 1 : 1 : 2,[8] as shown in Table II. For measurements of cytochrome b_{559}, difference absorption spectroscopy is used because the turbid preparation scatters light. These measurements, performed anaerobically, are described below. The content of noncovalently bound FAD in the particulate oxidase preparation is measured by standard methods[9] after trichloroacetic acid precipitation. The optical spectrum of the flavoprotein in the native particulate oxidase preparation cannot be reproducibly recognized because of the much greater extinction of the cytochrome b_{559} in the 430–560 nm region. Measurement of the fluorescence excitation and emission spectra of the resolved flavoprotein

[6] T. G. Gabig and B. M. Babior, J. Biol. Chem. **254**, 9070 (1979).
[7] T. G. Gabig, B. A. Lefker, P. J. Ossanna, and S. J. Weiss, J. Biol. Chem. **259**, 13166 (1984).
[8] T. G. Gabig and B. A. Lefker, Clin. Res. **32**, 368A (1984).
[9] H. Burch, this series, Vol. 3, p. 960.

fraction is described below. Ubiquinone-50 in the particulate oxidase fraction is measured by a standard reverse-phase HPLC method following extraction with light petroleum ether.[10]

Resolved Flavoprotein

The 30,000 rpm supernatant fraction of the 1% cholate extraction (Fig. 2) contains the flavoprotein component of the oxidase detectable in the optically clear preparation by fluorescence excitation and emission spectra.[11,12] The extinction of the flavoprotein is apparently too low for reliable measurement by its UV-visible absorption spectrum at this concentration. Coincident with the initial resolution of the flavoprotein by 1% cholate extraction, all detectable NADPH-dependent O_2^--generating activity is destroyed, and is not recovered in any of the resultant fractions shown in Fig. 2. The resolved flavoprotein-containing fraction has no NADPH oxidase activity in air-saturated buffer, but it has detectable diaphorase activity when assayed in the presence of 2,6-dichlorophenol or ferricyanide as artificial electron acceptors.[13] A modified Thunberg cuvette system for measurement of anaerobic absorption or fluorescence excitation/emission spectra is shown in Fig. 3. The cuvettes may be constructed from components (standard 1-cm pathlength quartz spectrophotometric or fluorimetric cuvettes, quartz–borosilicate graded glass seal, standard ground glass fittings and stopcocks available from Kontes, Vineland, NY) by a trained chemistry glassblower. Gas-tight syringes, obtainable from Hamilton Company, are modified to fit the ₮17/22 male fitting by epoxy cementing the sleeve of a ₮17/22 female fitting over the outer barrel of the Hamilton syringe. The cuvettes allow a sample placed in the main cuvette compartment to be cyclically evacuated and flushed with inert gas (purified nitrogen, argon, or helium) until the desired anaerobic conditions are achieved. A gas train described by Beinert et al.[14] in a previous volume of this series is utilized for this purpose. Up to two separate anaerobic additions in fixed volume can be made from the side bubble or sidearm of this closed system, or serial anaerobic additions of a third component can be made via the gas-tight Hamilton syringe. The sample compartment cover of the spectrophotometer or spectrofluorimeter is modified to accommodate the increased height of the cuvette system. This system was employed in our laboratory to demonstrate that the resolved flavoprotein was reduced by NADPH under anaerobic conditions as reflected by dis-

[10] A. I. Tavares, N. J. Johnson, and F. W. Hemming, Biochem. Soc. Trans. 5, 4771 (1977).
[11] T. G. Gabig, J. Biol. Chem. 258, 6352 (1983).
[12] T. G. Gabig and B. A. Lefker, J. Clin. Invest. 73, 701 (1984).
[13] T. G. Gabig and B. A. Lefker, Biochem. Biophys. Res. Commun. 118, 430 (1984).
[14] H. Beinert, W. H. Orme-Johnson, and G. Palmer, this series, Vol. 54, p. 111.

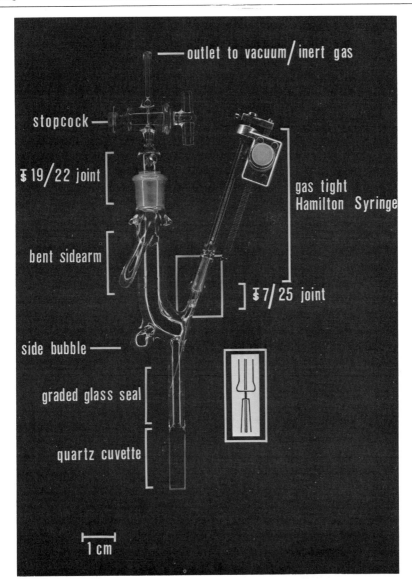

FIG. 3. Modified Thunberg cuvette system for anaerobic UV-VIS absorption and fluorescence spectroscopy. The cuvette system was constructed from commercially available components as described in the text. The outlet stopcock was connected to an anaerobic gas train system for equilibration with oxygen-free inert gas/vacuum via a three-way stopcock. The side bubble and/or sidearm were utilized to make anaerobic additions of one or two reagents (e.g., substrate, inhibitors) to the NADPH oxidase sample in fixed volume. The gas-tight Hamilton syringe was used to make serial 1-μl additions of a single reagent. Specific applications are described in the text.

appearance of its characteristic fluorescence excitation maximum at 450 nm.[13] Additional applications are discussed below.

Ubiquinone-50

Crawford and Schneider[15] were the first to demonstrate the ubiquinone content of neutrophils disproportionately high in comparison to mitochondrial content of these cells. They identified the quinone as ubiquinone-50 and provided evidence that it may be a component of the NADPH-dependent O_2^--generating oxidase on the basis of the ability of soluble Q analogs to stimulate the respiratory burst of intact neutrophils as well as augment the NADPH-dependent O_2^--generating activity of the subcellular oxidase.[16] The major portion of ubiquinone-50 from the native particulate oxidase, as prepared in our laboratory, is extracted with 1 M KCl, 2% cholate (supernatant II, Fig. 2). This component has not been characterized further in our laboratory.

Resolved Cytochrome b_{559}

The final pellet of the cholate resolution scheme (Fig. 2) contains the cytochrome b_{559} in particulate form. As noted above, this fraction has no NADPH-dependent O_2^--generating activity. In addition it has no detectable NADPH oxidase activity in air-saturated buffer, and no measurable NADPH diaphorase activity in the presence of artificial electron carriers (2,6-dichlorophenol indophenol or ferricyanide).[13] The cytochrome b_{559} in this fraction, as well as that in the initial particulate oxidase preparation, is quantified by its dithionite reduced minus oxidized difference absorption spectrum performed anaerobically in the modified Thunberg cuvette system. The millimolar extinction coefficient, based on its pyridine hemochrome, ($\varepsilon_{red-ox}^{559-540\,nm} = 21.6$) was published by Cross et al.[3] This is in close agreement with the extinction measured by titration with a known concentration of dithionite solution in our laboratory by the method of Burleigh et al.[17] The modified Thunberg cuvette system was also used to demonstrate that the cytochrome b_{559} in the native particulate oxidase (prior to cholate resolution) was reduced by NADPH under anaerobic conditions[4] whereas the resolved cytochrome b_{559} was not reduced under identical conditions.[13]

[15] D. R. Crawford and D. L. Schneider, *Biochem. Biophys. Res. Commun.* **99,** 1277 (1981).
[16] D. R. Crawford and D. L. Schneider, *J. Biol. Chem.* **257,** 6662 (1982).
[17] B. D. Burleigh, G. P. Foust, and C. H. Williams, *Anal. Biochem.* **27,** 536 (1969).

[19] NADH Oxidase from Guinea Pig Polymorphonuclear Leukocytes

By JOHN A. BADWEY and MANFRED L. KARNOVSKY

This oxidase from guinea pig polymorphonuclear leukocytes (PMN)[1] catalyzes a cyanide-insensitive production of H_2O_2 from NADH.[1a] The stoichiometry of the reaction is

$$NADH + O_2 + H^+ \rightarrow H_2O_2 + NAD^+$$

Superoxide (O_2^-) is produced as an intermediate during this reaction. For every 1.0 nmol of NADH oxidized only 0.3 nmol, at best, of O_2^- is produced.[2] This stoichiometry is consistent with two modes of electron release from the enzyme: a divalent pathway which results in the direct transfer of two electrons to O_2 to produce H_2O_2, and a univalent pathway which involves the transfer of single electrons to O_2 to produce O_2^-. The latter pathway produces H_2O_2 via dismutation of O_2^-.

$$E \cdot H_2 + O_2 \rightarrow E + H_2O_2 \qquad \text{(divalent pathway)}$$
$$E \cdot H_2 + O_2 \rightarrow E + 2O_2^- + 2H^+ \qquad \text{(univalent pathway)}$$
$$2O_2^- + 2H^+ \rightarrow H_2O_2 + O_2 \qquad \text{(dismutation reaction)}$$

NADH-oxidase is thought to function as a peripheral membrane protein in association with the cytoplasmic surface of the phagosomal membrane. Approximately 50% of this enzyme is membrane bound if the cells are homogenized in 349 mM sucrose. If the cells are homogenized in 154 mM KCl, or if the membrane-bound enzyme is exposed to these conditions, virtually all of the enzyme is recovered in the supernatant fraction.[3] Cytochemical studies have confirmed the expectation that H_2O_2 produced during phagocytosis by guinea pig neutrophils is associated with the phagosomal membrane and the plasmalemma.[4] The latter structure invaginates to produce the phagosome.

---

[1] Abbreviations: PMN, polymorphonuclear leukocytes; AIK, alkaline isotonic potassium medium.
[1a] R. H. Cagan and M. L. Karnovsky, Nature (London) 204, 255 (1962).
[2] J. A. Badwey and M. L. Karnovsky, J. Biol. Chem. 254, 11530 (1979).
[3] R. L. Baehner, N. Gilman, and M. L. Karnovsky, J. Clin. Invest. 49, 692 (1970).
[4] J. A. Badwey, J. T. Curnutte, J. M. Robinson, J. K. Lazdins, R. T. Briggs, M. J. Karnovsky, and M. L. Karnovsky, J. Cell. Physiol. 105, 541 (1980).

Guinea pig neutrophils contain high levels of an aldehyde oxidase which also produces O_2^- and H_2O_2.[5] Substrates for the aldehyde oxidase are formed during phagocytosis by an oxidative decarboxylation and deamination of amino acids, catalyzed by the myeloperoxidase–H_2O_2–Cl^- system.[6] This activity copurifies with NADH oxidase. Aldehyde oxidase exhibits several striking similarities, and also dissimilarities, compared to the NADH oxidase. The possibility exists that the NADH-oxidase and aldehyde oxidase activities are located at separate sites on a single, bifunctional enzyme.[5] The present chapter describes the assay, partial purification, and properties of NADH oxidase.

Assay Method

Principle. The oxidation of NADH is followed spectrophotometrically at 340 nm, taking 6.22×10^3 M^{-1} cm^{-1} as the extinction coefficient.[7] This reaction, of course, measures all NADH-oxidizing activities present in a given sample.

Reagents

N-2-Hydroxyethylpiperazine-N'-2-ethanesulfonic acid buffer (HEPES), 50 mM, pH 7.0 at 37°

NADH, 3.2 mM (prepare just prior to use in 50 mM HEPES, pH 7.0, and maintain on ice)

Procedure. The reaction mixture (1.0 ml) contains 0.16 mM NADH (50 μl) and 50 mM HEPES (0.94 ml), pH 7.0, plus a suitable amount of enzyme (10–20 μl) (see below) which is added last to initiate the reaction. Prior to the addition of enzyme, the reaction mixture in a 1-cm quartz cuvette is incubated at 37° in a thermostatted spectrophotometer for 7 min to ensure temperature equilibration. Samples containing enzyme are maintained on ice prior to assay.

Units. One unit of activity is the amount of enzyme oxidizing 1 μmol NADH/min under the above conditions. Specific activity is expressed as units of activity per milligram of protein.

Application to Crude Homogenates. Kinetic and chromatographic data indicate that the NADH-oxidase activity of the soluble fraction, and particularly that which results from the ammonium sulfate fractionation step described below, represents a single molecular species.[2]

[5] J. A. Badwey, J. M. Robinson, M. J. Karnovsky, and M. L. Karnovsky, *J. Biol. Chem.* **256**, 3479 (1981).

[6] S. K. Adeniyi-Jones and M. L. Karnovsky, *J. Clin. Invest.* **68**, 365 (1981).

[7] B. L. Horecker and A. L. Kornberg, *J. Biol. Chem.* **175**, 385 (1948).

Isolation of Polymorphonuclear Leukocytes

For this study guinea pigs of either sex, weighing 500 to 1000 g, and fed chow pellets and water *ad libitum,* are used. These animals are injected intraperitoneally with 20–30 ml of a sterile colloidal suspension of 12.0% (w/v) sodium caseinate in normal saline, and 18 hr later are killed by exposure to ether. The peritoneal cavities are opened, the cell suspension is removed with a pipet, and the peritonea are washed twice with ice-cold saline. The cell suspension and peritoneal washes are combined and filtered through nylon gauze. Cells are pelleted by centrifugation at 250 g for 10 min at 4°. Contaminating erythrocytes, when present, are lysed by suspending the cell pellets in ice-cold deionized-distilled water for 10 sec with mixing on a Vortex mixer before restoring isotonicity with concentrated saline [9.0% (w/v)]. This lysis procedure does not disrupt PMN. Cell pellets are pooled and washed twice with cold saline. The yields are variable, but normally 0.5 to 1.5 × 10⁹ cells (>90% PMN) can be obtained from each animal.

Partial Purification Procedure

All steps except the last (Sepharose 6B chromatography) are carried out on ice or in a cold room at 4°.

Step 1. Preparation of the Soluble Extract. Leukocytes are pelleted by centrifugation for 10 min at 250 g and suspended at 20% concentration (v/v) in ice-cold 154 mM KCl containing 0.32 mM KHCO₃ (alkaline isotonic potassium medium, AIK). The suspension is homogenized for 3 min utilizing a Teflon pestle in a 20-ml glass tube packed in ice. The homogenate is centrifuged at 250 g for 10 min and the supernatant is removed and saved. The cell pellet is resuspended in AIK medium and a second and third homogenization performed in identical fashion. The supernatants from the homogenizations are combined, centrifuged at 30,000 g for 15 min, and the resulting pellet is discarded.

Step 2. Ammonium Sulfate Fractionation. A cold saturated ammonium sulfate solution containing 154 mM KCl is slowly added to the supernatant over a period of 15 min to bring it to 30% saturation. This suspension is allowed to stand for 1 hr. The precipitate is removed by centrifugation at 25,000 g for 20 min and discarded, and the supernatant is brought to 56% ammonium sulfate saturation. After 1 hr, the resulting precipitate is removed by centrifugation as described above, taken up in a minimum volume of 5.0 mM HEPES, pH 7.0, and dialyzed against 3.0 liters of this buffer overnight. A large precipitate may form during dialysis which can be removed by centrifugation in a clinical centrifuge and discarded. The dialyzed preparation is centrifuged at 100,000 g for 1 hr to remove opalescent material.

At this stage, the NADH oxidase is free of contaminating enzymes that may affect the assay [e.g., NADH phosphodiesterase(s)] and is suitable for kinetic studies.[2]

Step 3. Sepharose 6B Chromatography. The enzyme solution is applied to a Sepharose 6B column (2.5 × 100 cm) equilibrated with 20 mM HEPES buffer, pH 7.0, containing 154 mM KCl. The column is run at a flow rate of ca. 10 ml/hr. The enzyme emerges as a single, sharp peak in the M_r 300,000 region. Peak fractions are concentrated and stored frozen at $-20°$.

These preparations generally represent a purification over the crude homogenate of 20- to 50-fold, and recoveries of 80 to 100% are routinely achieved.

Properties

Stability. The enzyme is stored at $-20°$ in small aliquots to prevent repeated freezing and thawing. No diminution in activity was observed upon storage in this fashion for periods up to 6 months.

Characteristics. NADH-oxidase exhibits normal Michaelis–Menten kinetics for NADH with a K_m of approximately 0.4 mM. The maximum velocity is 9.72 ± 1.66 SD ($n = 5$) nmol H_2O_2 produced/min/10^7 cells.[2] The pH optimum is 4.5 to 5.0[1] and the apparent molecular weight is 310,000.[2] The enzyme exhibits only slight activity with NADPH, and is inactive with a variety of compounds known to serve as substrates for various oxidases (e.g., xanthine, L-leucine, D-alanine).[2] Phosphate, nucleotides, and a variety of polyanions (e.g., citrate) inhibit the enzyme. Inhibition by ATP is competitive with the substrate NADH with a K_i of 20 μM. Chelation of the ATP with magnesium or other divalent cations relieves this inhibition.[2] Since phosphate is inhibitory, its use as a buffer with this enzyme must be avoided.

Distribution. NADH oxidase activity is present in monocytes from guinea pigs and mice, and macrophages from mice at levels greater than those observed in guinea pig PMN.[8] Whether the properties of the enzyme from mononuclear phagocytes are similar to those exhibited by the enzyme from the PMN remains to be determined.

Acknowledgment

Work reported from this laboratory was supported by a grant from the United States Public Health Service, N.I.H. Nos. AI-03260.

[8] S. R. Simmons and M. L. Karnovsky, *J. Exp. Med.* **138**, 44 (1973).

[20] Human Myeloperoxidase and Hemi-myeloperoxidase

By Patricia C. Andrews and Norman I. Krinsky

Introduction

Myeloperoxidase (MPO; donor: H_2O_2 oxidoreductase, EC 1.11.1.7)[1] is an enzyme found in the azurophilic granules of mammalian neutrophils[1a,2] and also identified in human monocytes.[3,4] Its function is to kill bacteria which have been phagocytosed by these cells.[5,6] This bactericidal activity comes from its ability to oxidize halides to their hypohalous acids.[7,8]

When iodide is included, iodination of proteins[5] and thyroxine production[9] are both observed. With chloride as the halide, peptide bonds are broken,[10] reactive chloramines and aldehydes are formed,[11–14] and free sulfhydryls are oxidized.[15] In a halide-dependent reaction, MPO can lead to the oxidation of methionine to its sulfoxide derivative.[16] This oxidation of methionine causes inactivation of the chemotactic stimuli, C5a[17,18] and

[1] Abbreviations: MPO, myeloperoxidase; CTAB, cetyltrimethylammonium bromide; DTT, dithiothreitol; TMB, 3,3′,5,5′-tetramethylbenzidine; PBS, phosphate-buffered saline; DMFA, N,N′-dimethylformamide; PMNL, polymorphonuclear leukocyte; SDS, sodium dodecyl sulfate.

[1a] K. Agner, Acta Physiol. Scand. 2, Suppl. 8, 1 (1941).

[2] J. Schultz, R. Corlin, F. Oddi, K. Kaminker, and W. Jones, Arch. Biochem. Biophys. 111, 73 (1965).

[3] A. Bos, R. Wever, and D. Roos, Biochim. Biophys. Acta 525, 37 (1978).

[4] R. I. Lehrer and M. J. Cline, J. Clin. Invest. 48, 1472 (1969).

[5] S. J. Klebanoff, J. Exp. Med. 126, 1063 (1967).

[6] S. J. Klebanoff, J. Bacteriol. 95, 2131 (1968).

[7] K. Agner, in "Structure and Function of Oxidation Reduction Enzymes" (A. Åkeson and A. Ehrenberg, eds.), p. 329. Pergamon, Oxford, 1972.

[8] J. E. Harrison and J. Schultz, J. Biol. Chem. 251, 1371 (1976).

[9] V. Stolc, Biochem. Biophys. Res. Commun. 45, 159 (1971).

[10] R. J. Selvaraj, B. B. Paul, R. R. Strauss, A. A. Jacobs, and A. J. Sbarra, Infect. Immun. 9, 255 (1974).

[11] J. M. Zgliczynski, T. Stelmaszynska, J. Domanski, and W. Ostrowski, Biochim. Biophys. Acta 235, 419 (1971).

[12] R. A. Johnson and F. D. Green, J. Org. Chem. 40, 2186 (1975).

[13] J. M. Zgliczynski, T. Stelmaszynska, W. Ostrowski, J. Naskalski, and J. Sznajd, Eur. J. Biochem. 4, 540 (1968).

[14] B. B. Paul, A. A. Jacobs, R. R. Strauss, and A. J. Sbarra, Infect. Immun. 2, 414 (1970).

[15] E. L. Thomas, Infect. Immun. 23, 522 (1979).

[16] M.-F. Tsan, J. Cell. Physiol. 111, 49 (1982).

[17] R. A. Clark and S. J. Klebanoff, J. Clin. Invest. 64, 913 (1979).

[18] R. A. Clark and S. Szot, J. Immunol. 128, 1507 (1982).

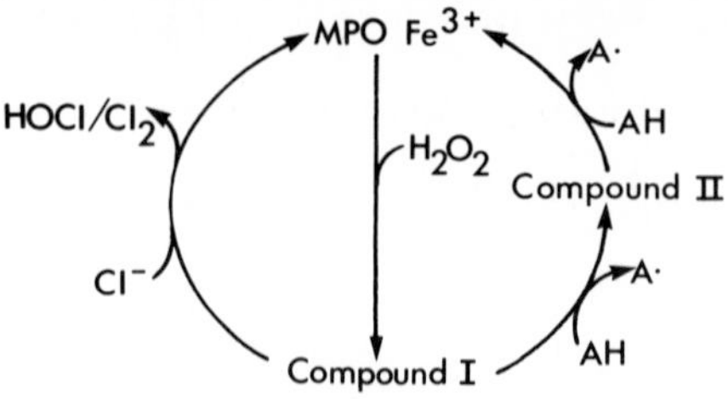

FIG. 1. Mechanism of MPO catalysis. (Reprinted with permission of CRC Press, Boca Raton, FL.)

formylmethionylleucylphenylalanine,[17–19] inactivation of the phagocytosis-stimulating activity of Fc and C3b opsonized yeast particles,[20] and of α_1-proteinase inhibitor.[21] Along with inactivating neutrophil NADPH oxidase,[22] these effects of HOCl production suggest a host regulatory role for MPO.

Assay for Myeloperoxidase

In our original assay system for MPO[23,24] we utilized a noncarcinogenic, nonmutagenic derivative of benzidine, 3,3',5,5'-tetramethylbenzidine (TMB) as substrate. The assay was designed to measure MPO activity in the presence or absence of halides. Since MPO oxidizes halides to hypohalous acids such as hypochlorous acid (HOCl), the assay was stoichiometric inasmuch as each H_2O_2 molecule produced either one molecule of HOCl or two oxidized dye molecules (Fig. 1). The HOCl produced will oxidize two dye molecules maintaining stoichiometry. Inasmuch as chloride can act as an allosteric inhibitor in a pH-dependent manner,[25] the assay reported below was carried out in the *absence* of any added halide. An improvement[26] of the original assay[25] resulted in an 8-fold increase in sensitivity and will be described below. The stock solutions used for this

[19] T. A. Lane and G. E. Lamkin, *Blood* **61**, 1203 (1983).
[20] O. Stendahl, B. I. Cable, C. Dahlgren, J. Hed, and L. Molin, *J. Clin. Invest.* **73**, 366 (1984).
[21] N. R. Matheson, P. S. Wong, and J. Travis, *Biochem. Biophys. Res. Commun.* **88**, 402 (1979).
[22] R. C. Jandl, J. André-Schwartz, L. Borges-Dubois, R. S. Kipnes, B. J. McMurrich, and B. M. Babior, *J. Clin. Invest.* **61**, 1176 (1978).
[23] P. C. Andrews and N. I. Krinsky, *J. Biol. Chem.* **256**, 4211 (1981).
[24] P. C. Andrews and N. I. Krinsky, *in* "Handbook of Methods for Oxy Radical Research" (R. A. Greenwald, ed.), p. 297. CRC Press, Boca Raton, FL, 1985.
[25] P. C. Andrews and N. I. Krinsky, *J. Biol. Chem.* **257**, 13240 (1982).
[26] K. Suzuki, H. Ota, S. Sasagawa, T. Sakatani, and T. Fujikura, *Anal. Biochem.* **132**, 345 (1983).

assay[26] consist of the following: phosphate-buffered saline (PBS), N,N'-dimethylformamide (DMFA), 15 mM H_2O_2, 20 mM TMB in DMFA, 200 mM phosphate buffer, pH 5.4. The reaction mixture consists of the following components:

	Volume (μl)	Final concentration	
		(mM)	(%)
PBS	200		40
DMFA	40		8
H_2O_2 (15 mM)	10	0.3	
Phosphate buffer (200 mM)	200	80	
TMB (20 mM) in DMFA	40	1.6	
Total volume	490		

The mixture is incubated at 37° for 2 min and the reaction is initiated by the addition of 10 μl of either MPO or a polymorphonuclear leukocyte (PMNL) supernatant containing approximately 1 μg MPO. At the end of 3 min, the reaction is stopped by the addition of 1.75 ml of 200 mM sodium acetate buffer (pH 3.0) and the mixture is placed on ice. One unit of MPO activity is defined as the amount of enzyme that will give a 655 nm absorbance change of 0.10/min. This assay has been reported to detect the MPO in the supernatant of a 2×10^6/ml suspension of PMNL stimulated with cytochalasin B (5 μg/ml) and fMet-Leu-Phe (10^{-6} M) for 15 min at 37°. As little as 10 μl of the PMNL supernatant from this preparation can be assayed for MPO activity.

Isolation

There are almost as many effective MPO isolation procedures as there are laboratories which study this enzyme. Sources of MPO have included pig,[27,28] cow,[28] guinea pig,[29] horse,[30] both normal[1,13,23,31–37] and leukemic

[27] T. Odajima and I. Yamazaki, *Biochim. Biophys. Acta* **284**, 360 (1972).
[28] V. N. Kokryakov, A. I. Borisov, S. V. Slepenkov, and S. N. Lyzlova, *Biokhimiya* **47**, 100 (1982).
[29] R. K. Desser, S. R. Himmelhoch, W. H. Evans, M. Januslea, M. Mage, and E. Shelton, *Arch. Biochem. Biophys.* **148**, 452 (1972).
[30] A. Dublin and J. Silberring, *Anal. Biochem.* **72**, 372 (1976).
[31] I. Olsson, T. Olofsson, and H. Odeberg, *Scand. J. Haematol.* **9**, 483 (1972).
[32] A. R. J. Bakkenist, R. Wever, T. Vulsma, H. Plat, and B. F. Van Gelder, *Biochim. Biophys. Acta* **524**, 45 (1978).

humans,[1,13,38–43] HL-60 cells,[44] dog,[38] and rat.[45,46] Of particular interest is the isolation of human MPO by Con A–Sepharose affinity chromatography and gel filtration with a yield of 114%.[33] While the 21-fold purification of MPO isolated by this method is promising, the methods used to demonstrate purity are equivocal and the actual purity of the preparation is difficult to assess. Our own method[23] was developed to avoid the use of proteolytic enzymes and NaCl during MPO isolation so that the intact enzyme could be used in kinetic studies examining the role of chloride in MPO catalysis. The procedures for the production of hemi-MPO apply to MPO isolated by any method that suits the investigator.

The isolation of white blood cells or purified neutrophils as starting material are standard methods for which the reader is referred to original articles for the best protocol for a given source (see this series, Vol. 108 [9]).

Purification of Native Human Myeloperoxidase

The isolation and purification of native MPO from human white blood cell pellets consist of detergent solubilization, ammonium sulfate precipitation, and gel filtration. The method described below results in an MPO preparation which is 89–93% pure as judged by *RZ* (comparison with crystalline canine MPO; *RZ* is the ratio of heme absorbance at 430 nm to protein absorbance at 280 nm and is used as a measure of purity) and native gel electrophoresis and has a yield of approximately 50% (Table I).

Detergent Solubilization of White Blood Cells

White blood cells are lysed by the addition of 0.5% cetyltrimethylammonium bromide (CTAB) (Sigma) in distilled water and homogenized

---

[33] D. P. Merrill, *Prep. Biochem.* **10,** 133 (1980).

[34] M. R. Andersen, C. L. Atkin, and H. J. Eyre, *Arch. Biochem. Biophys.* **214,** 273 (1982).

[35] W. M. Nauseef, R. K. Root, and H. L. Malech, *J. Clin. Invest.* **71,** 1297 (1983).

[36] R. L. Olsen and C. Little, *Biochem. J.* **209,** 781 (1983).

[37] S. O. Pember, R. Shapira, and J. M. Kinkade, Jr., *Arch. Biochem. Biophys.* **221,** 391 (1983).

[38] J. E. Harrison, S. Pabalan, and J. Schultz, *Biochim. Biophys. Acta* **493,** 247 (1977).

[39] K. Agner, *Acta Chem. Scand.* **12,** 89 (1958).

[40] J. Schultz and K. Kaminker, *Arch. Biochem. Biophys.* **96,** 465 (1962).

[41] N. Newton, D. B. Morell, L. Clarke, and P. S. Clezy, *Biochim. Biophys. Acta* **96,** 476 (1965).

[42] J. W. Naskalski, *Biochim. Biophys. Acta* **485,** 291 (1977).

[43] C. L. Atkin, M. R. Andersen, and H. J. Eyre, *Arch. Biochem. Biophys.* **214,** 284 (1982).

[44] M. Yamada, M. Mori, and T. Sugimura, *Biochemistry* **20,** 766 (1981).

[45] J. Schultz, A. Gordon, and H. Shay, *J. Am. Chem. Soc.* **79,** 1632 (1957).

[46] N. Newton, D. B. Morell, and I. Clarke, *Biochim. Biophys. Acta* **96,** 463 (1965).

TABLE I
PURIFICATION OF NATIVE MYELOPEROXIDASE

Preparation step	RZ	Specific activity (U/mg)	Purification factor	Yield[a] (%)
CTAB lysate	0.11	2.76	1.0	100
50% (NH₄)₂SO₄ supernatant	0.15	3.06	1.1	86
65% (NH₄)₂SO₄ supernatant	0.11	0.76		17
65% (NH₄)₂SO₄ pellet	0.26	15.4	5.6	72
Sephadex G-150, peak A	0.70	47.3	17.1	61
Peak A rechromatographed on Sephadex G-150	0.73	49.5	17.9	56

[a] Yield was determined as the percentage of peroxidase activity remaining at each step.

with a motor-driven Potter-Elvehjem glass and Teflon homogenizer. The use of CTAB (a cationic detergent) to lyse the cells has the advantages of precipitating cellular DNA, which is otherwise viscous and hygroscopic, and of preventing the association of MPO ($pI > 11$) with the DNA. In a typical preparation 400 ml of outdated white blood cells (containing on the order of 10^{10} cells) are lysed in 200 ml 0.5% CTAB. This solution is left for 18 hr at 4° before centrifugation at 15,000 g (4°) for 15 min. The green supernatant fraction containing the cellular lysate can be stored at $-20°$ at this point, if the experimenter wishes to pool lysates collected from several days' worth of white blood cell pellets.

Ammonium Sulfate Precipitation

All centrifugations are carried out at 15,000 g (4°) for 15 min. CTAB lysates are centrifuged to remove any particulate material which has formed due to storage conditions. The supernatant, while being stirred, is treated with solid ammonium sulfate (311 mg/ml) to yield a final concentration of 50% saturation. This solution is incubated on ice for 30 min and centrifuged to remove the precipitate. The supernatant, containing the MPO, is again treated with solid ammonium sulfate (100 mg/ml supernatant) to raise the concentration to 65% saturation, incubated on ice 30 min, and centrifuged to collect the MPO precipitate. The pellet is suspended in 50 mM Tris, pH 7.0, containing 0.5% CTAB. Again, at this point, the MPO preparation may be frozen for later use. For the typical preparation yielding 200 ml CTAB lysate, the 65% saturated ammonium sulfate pellet is resuspended in a 5-ml volume. We have noticed that the addition of ammonium sulfate to levels higher than 70% saturation causes the MPO precipitate to float on the ammonium sulfate solution. The activity of MPO does not appear to be affected, but it is more difficult to cleanly separate the precipitate from the solution.

Gel Filtration on Sephadex G-150

A Sephadex G-150, superfine grade, column (2.5 × 30 cm; about 150 ml volume) equilibrated and eluted with 50 mM Tris, pH 7.0, containing 0.5% CTAB (column buffer), is used in the final steps of purification. This column is developed at room temperature and we have found that chromatography against gravity (i.e., upside down) gives slightly better peak separation and a longer lifetime of the column before it must be recast. The addition of CTAB to the column buffer is required to prevent the MPO from sticking to the Sephadex beads at neutral pH. If the column is used at pH 4.5, no CTAB is required, suggesting that nonspecific charge effects are the cause of this problem. If azide is used as a bacteriostatic agent, three column volumes of azide-free column buffer should be passed through the column to remove this MPO inhibitor.

If the resuspended 65% saturated ammonium sulfate pellet has been frozen, it should be centrifuged at 15,000 g for 15 min to remove any precipitates which have formed. The supernatant (typically 5 ml) is then applied to the Sephadex G-150 column and 0.9-ml fractions are collected. The elution profile obtained is shown in Fig. 2. MPO elutes in the first peak (peak A), which appears about 15 ml after the void volume. This peak corresponds to an apparent M_r of 88,000 (compared to 173,000 by analytical ultracentrifugation), which is probably due to residual charge effects retarding the elution of the enzyme. The second eluting peak (peak B) contains some 430 nm absorbing material but is devoid of peroxidase activity. Peak A fractions are combined (typically 10 ml) and solid ammonium sulfate (430 mg/ml) is added to give a final concentration of 65% saturation. The solution is incubated at 4° for 30 min and centrifuged for 15 min at 15,000 g. The concentrated MPO pellet is resuspended in 4 ml column buffer and rechromatographed on the Sephadex G-150 column. On this column MPO elutes as a single peak. Fractions with RZ values of 0.68 and above are pooled, concentrated with 65% saturated ammonium sulfate, resuspended in 2 ml distilled H_2O, and dialyzed against distilled H_2O to remove CTAB and salts. The RZ, specific activity, purification factor, and yield at each step of this purification scheme are shown in Table I.

Production of Human Hemi-myeloperoxidase

As discussed below, MPO is a tetrameric enzyme with two pairs of subunits and contains one heme prosthetic group attached to each of the

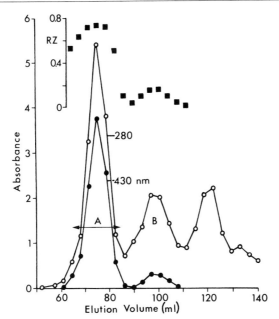

Fig. 2. Elution profile of first Sephadex G-150 column. The absorbance at 280 (○) and 430 (●) nm was determined for fractions (0.9 ml) collected from Sephadex G-150 chromatography of crude MPO. The RZ (A_{430}/A_{280}) (■) was also determined for each fraction. Peak A contained the MPO activity. Fractions with an $RZ > 0.4$ (arrows) were pooled for further purification. The void volume of this column was 58 ml. (Reprinted with permission of the American Society of Biological Chemists, Inc.)

two heavy subunits. The addition of dithiothreitol (DTT) to the enzyme results in the cleavage of a single disulfide bridge between the two heavy chains, forming a half-enzyme containing one heavy chain and one light chain, called hemi-MPO. If the free sulfhydryl is alkylated with iodoacetic acid or iodoacetimide, a stable preparation of hemi-MPO can be isolated.[23]

Native MPO (6–10 mg) in 2 ml 50 mM Tris, pH 7.0, containing 0.5% CTAB is incubated for 1 hr at room temperature with 200 μl 0.4 M DTT. This solution turns bright green and remains so until the alkylation step. At the end of the incubation period 200 μl of 1 M iodoacetimide or iodoacetic acid is added and the incubation continued for 1 hr on ice. During this incubation period, a white precipitate forms and is removed by centrifugation at 8000 g for 2 min in a Brinkmann microfuge. Control experiments have shown that this precipitate forms independently of the presence of protein. The supernatant is chromatographed on Sephadex G-150 under the same conditions used for the isolation of native MPO.

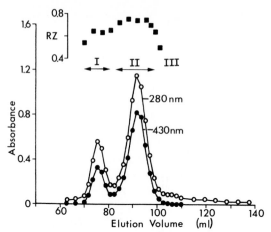

Fig. 3. Elution profile of hemi-MPO on Sephadex G-150. MPO which had been treated with dithiothreitol and iodoacetamide was chromatographed on Sephadex G-150. The absorbance at 280 (O) and 430 (●) nm and the *RZ* (■) of each fraction (0.9 ml) was determined. Arrows show which fractions were pooled for further analysis. The void volume of this column was 58 ml. (Reprinted with permission of the American Society of Biological Chemists, Inc.)

The elution profile of the hemi-MPO preparation is shown in Fig. 3. The first peak elutes (Fig. 3) in the same volume as native MPO and has been determined to be the unreduced native enzyme. The second peak to elute contains hemi-MPO, usually with a slightly increased *RZ* value and specific activity as compared with the native MPO preparation from which it is formed (Table II). This is probably due to the removal of contaminants which remain in the native MPO peak. The position of the hemi-MPO peak corresponds to an apparent M_r of 52,000. As in the case of the native enzyme, the true molecular weight is larger (73,000, as determined by analytical ultracentrifugation[23]). A third peak always appears as a small shoulder on the tail end of the hemi-MPO peak. This peak is devoid of 430

TABLE II
PURIFICATION OF HEMI-MYELOPEROXIDASE ON
SEPHADEX G-150

Fraction	RZ	U/mg	$M_r{}^a$
Native myeloperoxidase	0.73	49.5	88,000
Peak I (myeloperoxidase)	0.64	41.3	88,000
Peak II (hemi-myeloperoxidase)	0.74	51.2	52,000

[a] Determined from elution volume on Sephadex G-150.

nm-absorbing material and peroxidase activity. Its identity has not been determined.

The hemi-MPO fractions with RZ values greater than 0.68 are pooled and concentrated by precipitation with 65% saturated ammonium sulfate, as described for native MPO. The half-enzyme is resuspended in H_2O and dialyzed against distilled H_2O. Hemi-MPO, like native MPO, can be stored at $-20°$ with no adverse effects. This reduction, alkylation, and gel filtration procedure results in the conversion of 50–75% of native MPO into hemi-MPO.

Hemi-MPO has thus far been shown to differ from the native enzyme only in size. Thus, if hemi-MPO is either present in source cells, as may be the case for HL-60 cells, or is formed during isolation of native MPO (reducing agents should be *avoided* during isolation), it may not be detected. Hemi-MPO can easily be distinguished from MPO by SDS gel electrophoresis of unreduced samples. These samples should *not* be boiled as spurious protein bands at about 40,000 Da result.[47]

Physicochemical Properties of Myeloperoxidase

MPO appears to be synthesized during the myelocyte and promyelocyte stages of granulocyte maturation.[48] *In vitro* translation of HL-60 (human promyelocyte leukemia cell line) cellular mRNA and subsequent immunoprecipitation with MPO-specific antibodies shows that the nascent polypeptide chain has an M_r of 77,000.[48] HL-60 cells labeled for 30 min with [^{35}S]methionine, lysed, and the MPO isolated by immunoprecipitation show a major band on SDS gels (under reducing conditions) of 75,000 Da and a minor band at 77,000 Da, suggesting a rapid initial proteolytic processing of the nascent chain.[44,48,49] Other investigators, following a similar protocol with HL-60 cells, find a 90,000-Da polypeptide labeled during a 1-hr incubation with radioactive amino acids.[50] Since different laboratories have reported a wide range of molecular weights for MPO and its subunits, the values cited below reflect the SDS gel banding position of MPO subunits determined by different investigators and not the width of MPO bands on a single gel. During transport from the rough endoplasmic reticulum to azurophil granules, this polypeptide chain (77,000–90,000 Da) undergoes dimerization,[49] further proteolytic processing,[48–50] and the addition of carbohydrate[51] and heme prosthetic groups.[44,49] Fully processed MPO has two identical heavy subunits

[47] P. C. Andrews, C. Parnes, and N. I. Krinsky, *Arch. Biochem. Biophys.* **228**, 439 (1984).
[48] H. P. Koeffler, J. Ranyard, and M. Pertcheck, *Blood* **65**, 484 (1984).
[49] M. Yamada, *J. Biol. Chem.* **257**, 5980 (1982).
[50] I. Olsson, A. Persson, and K. Stromberg, *Biochem. J.* **223**, 911 (1984).
[51] R. L. Olson and C. Little, *Biochem. J.* **222**, 701 (1984).

(55,000–62,000 Da),[31,38] two identical light subunits (10,000–15,000 Da)[31,38] two heme prosthetic groups,[38,39] carbohydrate attached to the heavy subunits,[51] and a disulfide bridge between the two heavy subunits.[23]

When native MPO (M_r 135,000–150,000) is subjected to reduction under denaturing conditions, the four subunits are completely dissociated.[38,39,51] Under these conditions, the heme prosthetic group can be found associated with the heavy subunit, while the light subunit is free of heme. If MPO is boiled in SDS *without* reduction and gel filtered, heme can be found associated with the light subunit,[51] which may suggest a close proximity between the heme group and the light subunit in the native enzyme. If MPO is subjected to reduction and alkylation under nondenaturing conditions, the enzyme is split in half and a hemi-MPO composed of one heavy subunit, one light subunit, and presumably one heme group is formed.[23] Hemi-MPO exhibits the same catalytic activities, substrate regulation, and heme spectrum as the native enzyme.[23,47]

[21] Cytochrome b-245

By ANTHONY W. SEGAL, ANGELA M. HARPER, ANDREW R. CROSS, and OWEN T. G. JONES

Engulfment of particles by the phagocytic cells in blood and tissues (neutrophils, monocytes, macrophages, and eosinophils) is associated with a burst of nonmitochondrial respiration. This is important for the killing of bacteria[1] and the elevation of the pH within the phagocytic vacuole.[2] The oxidase system of these cells[3] has recently been identified as an electron transport chain containing a very low potential cytochrome b-245, which is probably the terminal oxidase itself (Fig. 1).

This electron transport chain is defective in the inherited syndrome of chronic granulomatous disease (CGD),[4,4a] which manifests as an unusual predisposition to bacterial infection. The commonest subgroup of this

---

[1] J. A. Badwey and M. L. Karnovsky, *Annu. Rev. Biochem.* **49**, 695 (1980).

[2] A. W. Segal, M. Geisow, R. Garcia, A. Harper, and R. Miller, *Nature (London)* **290**, 406 (1981).

[3] A. W. Segal and O. T. G. Jones, *Nature (London)* **276**, 515 (1978).

[4] Abbreviations: EDTA, ethylenediaminetetraacetic acid; PNS, postnuclear supernatant; DIFP, diisopropyl fluorophosphate; CGD, chronic granulomatous disease; MOPS, 3-[N-morpholino]propanesulfonic acid; PMA, phorbol myristate acetate.

[4a] A. W. Segal, A. R. Cross, R. C. Garcia, N. Borregaard, N. H. Valerius, J. F. Soothill, and O. T. G. Jones, *N. Engl. J. Med.* **308**, 245 (1983).

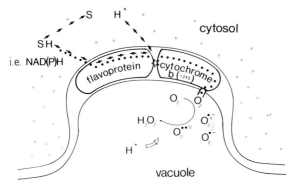

FIG. 1. Proposed model of cytochrome b-245 electron transport chain. This is thought to involve one (or possibly two) flavoproteins and the b cytochrome, which becomes located in the wall of the phagocytic vacuole. (Reproduced with permission from Ref. 13.)

disease is inherited in an X-linked manner and is caused by an absence of the cytochrome, whereas it is present and its properties appear to be normal in patients with an autosomal recessive inheritance, suggesting an abnormality of a proximal electron donor or activation system.

Most of the work on this oxidase has been performed on peripheral blood neutrophil granulocytes because they can be conveniently isolated in relatively large numbers and in a very pure state.

This chapter will describe the purification of neutrophilic leukocytes from peripheral blood, the isolation of their membranes which are enriched in the cytochrome, the preparation of a solubilized oxidase, and the purification of the cytochrome itself. The techniques for measuring the cytochrome in these various preparations will be described as will the methods used to determine some of its properties.

Purification of Neutrophils (see also this series, Vol. 108 [9])

Principle

Most methods depend on the facts that: (1) erythrocytes that have been induced to aggregate or form "rouleau" by the addition of agents such as dextran or methylcellulose sediment more rapidly than leukocytes; (2) neutrophils and eosinophils are denser than the other leukocytes from which they can be separated by centrifugation through discontinuous density gradients; and (3) contaminating erythrocytes can be selectively lysed under hypotonic conditions or by incubation with NH_4Cl.

Normal human blood is the optimal source of cells but samples from a single individual do not contain enough neutrophils for large preparations of membranes or for the purification of the cytochrome. Sources of larger numbers of human cells include buffy coat residues that are produced as a by-product of the purification of platelets, and blood from patients from chronic myeloid (granulocytic) leukemia. Buffy coat residues are usually very concentrated and should be diluted (3 : 1) with saline; they can then be processed like peripheral blood. To obtain large numbers of cells the leukocytes from patients with chronic myeloid leukemia can be removed from the circulation of the patient by leukophoresis, a process in which the peripheral blood is circulated through a zonal centrifuge and the band of leukocytes selectively withdrawn. This is usually performed therapeutically in subjects in whom the massive numbers of neoplastic cells produce problems of hyperviscosity in the circulation. Large numbers of cells can also be easily obtained from the blood of animals. Their isolation from porcine blood is described in this chapter.

Methods

Isolation of Granulocytes from Human Peripheral Blood. Blood is drawn into a plastic syringe containing heparin (preservative free, Paines and Byrne, Ltd., Greenford, Middlesex; 5IU/ml). Dextran (M_r 150,000–500,000; Pharmacia Fine Chemicals, Uppsala, Sweden) as a 10% (w/w) solution in normal saline (0.15 M, 0.9% NaCl, w/w) is added to give a final concentration of 1% and mixed by inversion of the syringe, which is then held vertically with the needle uppermost. For larger numbers of cells, buffy coat residues are mixed with normal saline in a ratio of 3 : 1, dextran is added as above at a final concentration of 1%, the solutions are mixed, and the erythrocytes are allowed to sediment in a measuring cylinder. After about 30 min the erythrocytes will have sedimented and the supernatant plasma, containing the leukocytes, is then layered onto the Ficoll/Hypaque (type FP, Sigma Chemical Company; 1.077) in a volume-to-volume ratio of 4 : 1.

After centrifugation at 400 g for 15 min at 20° the granulocytes and the residual erythrocytes will have passed through the Ficoll/Hypaque and formed a pellet. The plasma and cells above the pellet are removed and discarded. Erythrocytes are lysed by the addition of 25 ml of distilled water, which is pipetted up and down vigorously for 30 sec before isotonicity is restored by the addition of 25 ml of double-strength saline.

The neutrophils are then centrifuged at 200 g for 5 min at 20°, after which they should be visible as a greenish-yellow pellet. If there are any residual erythrocytes visible on the surface of the neutrophils the hypotonic lysis step is repeated. Finally the cells are washed in saline to remove traces of hemoglobin. The cells can be resuspended in any physiological medium. The more convenient are Hanks' balanced salt solution, RPMI-1640, modified Krebs–Ringer buffer (see the following section), or Na/K phosphate (50 mM)-buffered normal saline containing glucose (10 mm) and heparin (5 IU/ml).

Purification of Neutrophils from Pig Blood. Blood is collected during the normal slaughtering operations of an abattoir. The blood is drained from the jugular vein into a small bucket containing anticoagulant. Whole blood (about 4 liters) is taken into ethylenediaminetetraacetic acid (EDTA), (BDH Chemicals, Ltd., Poole, Dorset), dissolved in a minimum volume of normal saline so as to give a final concentration in the blood of 0.15% (w/v) to prevent coagulation. Catering gelatin (50 ml/liter, BDH) is added from a freshly prepared 25% (w/v) solution and the mixture allowed to stand for about 1 hr. The top layer, enriched in leukocytes, is sucked off and the cells collected by centrifugation at 400 g for 7 min at room temperature. The cells are washed once in normal saline containing heparin (5 IU/ml) and EDTA (1 mM). The cells are subjected to three cycles of osmotic shock in order to lyse any erythrocytes present: they are collected by centrifugation and suspended for 30 sec in 5 vol of 0.2% NaCl containing EDTA (1 mM) and heparin (5 IU/ml); isotonicity is restored by the addition of an equal volume of 1.6% NaCl containing EDTA and heparin. Finally, when free of hemoglobin, the leukocytes are suspended in normal saline containing EDTA and heparin. The suspension is layered onto an equal volume of Ficoll and centrifuged for 20 min at 400 g. The neutrophil pellet is collected, washed twice with normal saline containing EDTA and heparin, and finally suspended in 50 ml modified Krebs–Ringer buffer, pH 7.4, containing HEPES (50 mM), NaH$_2$PO$_4$ (1 mM), NaCl (100 mM), KCl (5 mM), MgCl$_2$ (1 mM), CaCl$_2$ (1 mM), and supplemented with 2 mM glucose and heparin (5 IU/ml). The yield of neutrophils is about 5×10^9 cells from 1 liter of blood.

Purification of Granulocytes from Patients with Chronic Granulocytic Leukemia. Leukopheresis samples from these patients are an excellent source of large quantities of human cells (100–200 ml of packed cells) which are usually fairly normal but can occasionally be slightly immature.

The granulocytes predominate to such an extent that it is unnecessary to separate the other leukocytes. All that is required is to remove the erythrocytes. This is done in peripheral blood samples as described above and in leukopheresis samples by pelleting the cells at 400 g for 10 min, resuspending the pellet in 1 liter NH$_4$Cl (150 mM), and stirring the suspension with a magnetic stirrer for 30 min at room temperature. The cells are pelleted at 400 g for 10 min and the incubation with NH$_4$Cl and centrifugation are repeated until all erythrocytes have been eliminated.

Notes

1. One of the major difficulties in the handling of neutrophils is their tendency to clump with each other and with other leukocytes. This clumping is promoted by cooling the blood to 4°. Therefore the blood and the solutions used in the separation procedures should not be cooled below room temperature. Once the cells have been purified and suspended in medium they can be cooled and stored at 4°.

2. Neutrophils are relatively short lived. Blood should usually be processed as soon as possible. The yields become very poor more than 24 hr after venesection.

3. There is no direct evidence for a viral cause of chronic myeloid leukemia. Blood samples from these patients are not subjected to special procedures in diagnostic laboratories. However, prudence dictates that sensible precautions (such as refraining from mouth pipetting) should be employed with these specimens.

Fractionation of Neutrophils to Purify Membranes and Specific Granules

To investigate the spectral and bioenergetic properties of cytochrome *b*-245 it is optimal to separate it from other chromophores such as myeloperoxidase and mitochondrial cytochromes. This is best done by fractionation on continuous or discontinuous gradients of sucrose and a technique for the latter will be described in detail.

Method

The cells are centrifuged at 400 *g* for 5 min at 4°, resuspended in ice-cold 8.6% sucrose, and again centrifuged as above. The pellet is transferred to a precooled Dounce homogenizer (15 ml; Kontes Glass, Vineland, NY) and 8.6% sucrose is added to give a final concentration of about 1×10^8 cells/ml. Disruption of the cells by vigorous homogenization is monitored by phase-contrast microscopy and is usually advanced after about 50–100 strokes of the tight-fitting B pestle. The homogenate is then centrifuged at 400 *g* for 10 min at 4° after which a clear line of demarcation should be apparent between the postnuclear supernatant (PNS) above and the nuclei and unbroken cells below (nuclear pellet). The PNS is decanted and, if necessary, the nuclear pellet can be rehomogenized in fresh 8.6% sucrose and the centrifugation and separation steps repeated. Some difficulty may be experienced in disrupting dilute cell suspensions. This can be facilitated by suspending the cells in hypotonic (5%) sucrose (to swell them) and then lysing the plasma membrane with a brief burst of sonication. Homogenization should then be completed with a few strokes of the Dounce homogenizer.

A discontinuous sucrose gradient is constructed by gently layering sucrose solutions in the centrifuge tube, starting with the most dense. This should be done just prior to use and they should be maintained at 4° to minimize diffusion. The design of the gradients can be altered to optimize separation of the requisite organelle. In general, equal volumes of the various sucrose solutions and of the PNS are used. The concentrations of sucrose at which enrichment of the various subcellular particles takes place are as follows:

Cellular constituent	Position of enrichment in sucrose gradient
Cytosol	Above 15% (w/w)
Membranes	At interface between 15 and 34%
Mitochondria	At interface between 34 and 38%
Specific granules	At interface between 38 and 43%
Azurophil granules	At interface between 43 and 60%

Centrifugation is carried out in a swinging bucket rotor (e.g., Sorvall AH 627) at 4°. Centrifugation time to equilibrium is about 10^8 g-min, but adequate separations can be achieved in about one-tenth of the time. The different bands are then aspirated from above with a syringe and needle.

Note

All sucrose solutions contain EDTA (1 mM, pH 7.4) and heparin (15 IU/ml) to minimize clumping. Accuracy of the concentrations of all solutions should be checked with a refractometer.

Activation of Neutrophils and Preparation of Soluble Oxidase[5]

The neutrophil suspension (50 ml) in modified Krebs–Ringer buffer is made 1 mM with diisopropyl fluorophosphate (DIFP; Sigma Chemical Company) at 0°. After 5 min the cells are washed twice with the modified Krebs–Ringer buffer, then incubated at 37° in an approximately 100-ml vol and stimulated by the addition of 1 μg/ml phorbol myristate acetate (Sigma Chemical Company; 1 mg/ml in dimethyl sulfoxide). After 5 min the temperature is rapidly reduced to 4° by the addition of chips of frozen modified Krebs–Ringer buffer and the cells are harvested by centrifugation. The cells are resuspended in disruption buffer, which has the following composition: sucrose (8.6%, w/v), MgSO₄ (4 mM), NaN₃ (1 mM), CaCl₂ (1 mM), deoxyribonuclease 1 (10 μg/ml), silicone antifoam (0.01%, w/v; British Drug Houses) and phenylmethylsulfonyl fluoride (5 μg/ml).

[5] A. R. Cross, J. F. Parkinson, and O. T. G. Jones, *Biochem. J.* **223**, 337 (1984).

Chymostatin, leupeptin, antipain, pepstatin, and 7-amino-1-chloro-3-L-tosylamidoheptan-2-one (all supplied by Sigma) are added at 1 μg/ml each. The suspension is disrupted by nitrogen cavitation after stirring at 500 psi for 20 min in a nitrogen bomb (Parr Instrument Company) at 0°. The suspension is centrifuged at 100,000 g for 30 min. The pellet is suspended in 20 mM glycine buffer, pH 8.0, containing 25% (v/v) glycerol, 1 mM MgSO$_4$, 0.5 mM CaCl$_2$, 1 mM NaN$_3$, 0.25% (v/v) Lubrol PX (Sigma), 0.25% (w/v) sodium deoxycholate (BHD), and the same mixture of proteinase inhibitors as in the disruption buffer. The suspension is sonicated three times for 20 sec at 100 W, with 1-min cooling between treatments, in a Branson sonifier and then centrifuged at 100,000 g. The clear supernatant contains about 40% of the activity of the PNS.

Purification of Cytochrome b-245[6]

Homogenization of Cells and Preparation of Organelle Pellets

Although approximately $3-5 \times 10^{11}$ neutrophils, derived from chronic granulocytic leukemic patients, are generally used, the methods can be scaled down for smaller numbers of cells. The cells are pretreated with DIFP (2 mM), homogenized in 400 ml 8.6% sucrose containing EDTA and heparin (see above) using a Polytron homogenizer (Kinematica, Luzern, Switzerland), and a PNS is prepared as described above. The pellet is rehomogenized and a second, and subsequently a third, PNS is removed. Homogenization and centrifugation are carried out at 4° and the proteinase inhibitors, phenylmethylsulfonyl fluoride (Sigma Chemical Company; 1 mM) and Trasylol (Bayer, Newbury, UK; 100 U/ml), are present throughout. The combined PNS's are made to 1.5 liters with 8.6% sucrose (containing 1 mM EDTA and 5 IU/μl heparin) and centrifuged [69,900 g (av), 45 min at 4°] in 70-ml tubes using a Sorvall A-641 rotor. The supernatant is discarded and the centrifuge tubes are refilled with PNS without removing the pellet until all the PNS has been processed. The pelleted material is then resuspended using the Polytron homogenizer in 8.6% sucrose, an aliquot is taken for protein determinations, and the suspensions centrifuged again.

Detergent Wash and Extraction

The pellets are first washed with cholate and the cytochrome is then solubilized with Triton N-101. At the detergent-protein ratios used so-

[6] A. M. Harper, M. J. Dunne, and A. W. Segal, *Biochem. J.* **219**, 519 (1984).

dium cholate does not remove cytochrome b-245, but serves to enhance the specific content of the cytochrome within the Triton N-101 extract. In particular, the Triton N-101 extract is virtually free of myeloperoxidase, the major chromophore found in human neutrophils. This greatly facilitates the quantitation of the cytochrome.

The pelleted material is resuspended at a concentration of 15 mg protein/ml in buffer A supplemented with sodium cholate (5 mg/ml), as indicated in Fig. 2. Buffer A has the following composition: Tris–acetate, pH 7.4 (100 mM), KCl (0.1 M), glycerol (20%, v/v), dithiothreitol (1 mM), EDTA (1 mM), and butylated hydroxytoluene (30 μM). The organelles are immediately centrifuged and the wash with sodium cholate is repeated. The pellets are then washed once in an equal volume of buffer without cholate before solubilization of the cytochrome b-245 with Triton N-101 (Sigma) in buffer A (for details see Fig. 2). The combined Triton N-101 extracts contain 60–65% of the cytochrome b-245 with a specific content of 600–700 pmol of cytochrome b-245/mg of protein.

Chromatography of Triton N-101 Solubilized Cytochrome b-245

The solubilized cytochrome b-245 is contaminated with approximately 10% mitochondrial derived cytochromes b. These are removed by passing the extract through a column of n-aminooctyl-Sepharose under the conditions shown in Fig. 2. The obtained eluate is then loaded onto a column of heparin–agarose connected in tandem with the first column. More than 90% of the cytochrome b-245 is absorbed to the heparin–agarose. The second column is then disconnected, washed, and approximately 60–80% of the cytochrome is eluted with a gradient of 0.10–0.25 M NaCl in buffer A containing 10 mg/ml Triton N-101. The fractions containing cytochrome b-245 at a specific concentration greater than 10–11 nmol/mg of protein (generally eluted between 0.18 and 0.22 M NaCl) are pooled. The combined fractions contain about 30% of the absorbed cytochrome b-245. Approximately 100–150 nmol of cytochrome b-245, purified about 150-fold, can be prepared from $(3–5) \times 10^{11}$ cells within 3–4 days. The recovery is approximately 15%.

Spectrophotometric Measurement of Cytochrome b

The Identification of Cytochrome b-245 by Difference Spectroscopy

Reduction by Dithionite. In order to determine the presence or absence of cytochrome b-245 in neutrophils it is most convenient to measure the absorption spectrum of the solution before and after reduction of the

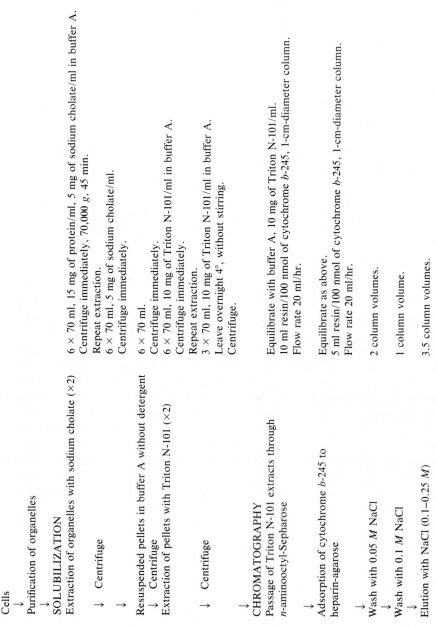

Cells
→
Purification of organelles
→

SOLUBILIZATION
Extraction of organelles with sodium cholate (×2)   6 × 70 ml, 15 mg of protein/ml, 5 mg of sodium cholate/ml in buffer A.
                                                     Centrifuge immediately, 70,000 g, 45 min.
                                                     Repeat extraction.
→   Centrifuge   6 × 70 ml, 5 mg of sodium cholate/ml.
                                                     Centrifuge immediately.

Resuspended pellets in buffer A without detergent   6 × 70 ml.
→   Centrifuge   Centrifuge immediately.
Extraction of pellets with Triton N-101 (×2)   6 × 70 ml, 10 mg of Triton N-101/ml in buffer A.
                                                     Centrifuge immediately.
                                                     Repeat extraction.
→   Centrifuge   3 × 70 ml, 10 mg of Triton N-101/ml in buffer A.
                                                     Leave overnight 4°, without stirring.
                                                     Centrifuge.
→

CHROMATOGRAPHY
Passage of Triton N-101 extracts through   Equilibrate with buffer A, 10 mg of Triton N-101/ml.
n-aminooctyl-Sepharose   10 ml resin/100 nmol of cytochrome b-245, 1-cm-diameter column.
                                                     Flow rate 20 ml/hr.
→
Adsorption of cytochrome b-245 to   Equilibrate as above.
heparin-agarose   5 ml resin/100 nmol of cytochrome b-245, 1-cm-diameter column.
                                                     Flow rate 20 ml/hr.
→
Wash with 0.05 M NaCl   2 column volumes.
→
Wash with 0.1 M NaCl   1 column volume.
→
Elution with NaCl (0.1–0.25 M)   3.5 column volumes.

FIG. 2. Flow scheme for the isolation of cytochrome b-245 from human neutrophils. (Reprinted with permission from Ref. 6.)

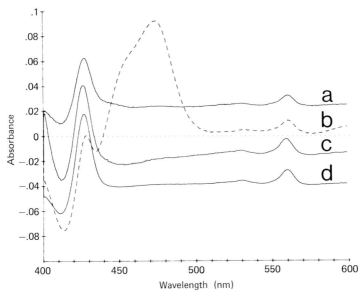

FIG. 3. Reduced-minus-oxidized difference spectra of (a) intact human neutrophils (4 × 10⁷ cells; 4.3 mg protein/ml); (b) sample in spectrum (a) after the addition of detergent (Triton X-100, 1%); (c) a preparation of solubilized oxidase from pig neutrophils (578 pmol of cytochrome *b* and 1.5 mg protein/ml); and (d) a purified preparation of human cytochrome *b*-245 (520 pmol and 35 μg protein/ml).

cytochrome. This can be done in a single operation by measuring the difference spectrum. Cells should be present at a concentration of about 1 × 10⁷/ml in a medium such as Krebs–Ringer or RPMI without colored pH indicator. The cells are aliquoted into two cuvettes and a baseline comparison is measured from 400 to 600 nm in a dual-beam spectrophotometer. An instrument with a sample position designed for use with turbid suspensions is best suited to such crude preparations. A few crumbs of dithionite are then added to one of the cuvettes, the sample is mixed, and the spectral change induced by reduction of the cytochrome are recorded. A typical spectrum of cytochrome *b*-245, with absorption maxima at 559, 529, and 428 nm, is shown in Fig. 3. Neutrophils contain few mitochondria and very little cytochrome P-450 or cytochrome b_5, therefore from this simple spectrum a good estimate of the amount of cytochrome *b*-245 present can be obtained. The height of the peak at 559 nm above the trough at 540 nm is measured and the following extinction coefficient[7] is used:

[7] A. R. Cross, F. K. Higson, O. T. G. Jones, A. M. Harper, and A. W. Segal, *Biochem. J.* **204**, 479 (1982).

$$E_{559-540 \text{ nm}} = 21.6 \text{ cm}^{-1} \text{ m}M^{-1}$$

A sample of neutrophils containing 2×10^7 cells/ml would give $A_{559-540 \text{ nm}}$ of 0.002, which corresponds to 100 pmol cytochrome b-245/ml.

If the difference spectrum is distorted by aggregation or other effects leading to sloping baselines or spectra, the cytochrome b-245 content can be calculated from the reduced − oxidized spectrum by drawing a line to join the two troughs in the spectrum at about 540 and 570 nm and measuring the height of the peak at 559 nm above this line.

Cytochrome b-245 is located on the surface of the cell in the plasma membrane as well as in internal organelles. Thus, reduction of the cytochrome is usually incomplete when sodium dithionite is added to suspensions of intact neutrophils. In addition, aggregation of intact neutrophils may cause severe interference by significant increases in light scattering. For accurate quantitations these problems can be overcome by disrupting the cells with detergent (such as 0.1% Triton X-100) or by any of the common techniques for homogenization (including gentle sonication). However, disruption of the cells exposes intracellular myeloperoxidase to reduction by the dithionite and this may distort the spectrum and impair quantitation (Fig. 3). Accurate differentiation of this cytochrome from other hemoproteins can only be achieved by measurement of the peak at 559 nm that is reduced at low potential (see below).

Stimulated Autologous Reduction. Cytochrome b-245 is present in the cells of some patients with CGD, but it does not function.[4] This situation can be identified by stimulating the oxidase system under anaerobic conditions. In normal cells the cytochrome is reduced by electron flow down the chain, but under anaerobic conditions it remains in the reduced state, because it is unable to transfer its electrons to oxygen, the natural acceptor. The following test allows one to distinguish CGD cells from normal cells. Cell suspensions are prepared as above. Ideally, the cells are placed in anaerobic (Thunberg) cuvettes gassed with argon or nitrogen, a baseline is determined, and the test sample is stimulated by the addition of 10 μg/ml of PMA from a side arm, after which the spectral changes are recorded. However, for most qualitative studies it is adequate to suspend the cells in plastic cuvettes covered with paraffin or plastic film and gently gassed with N_2. The PMA can be added by injection through the covering film, since the relatively small amounts of residual oxygen are rapidly removed by the initial stages of the respiratory burst itself. Difference spectroscopy is performed as described above. In cells with a normal electron transport chain the reduction of the cytochrome is observed over the next 10–20 min, whereas in CGD cells no reduction of the cytochrome takes place.

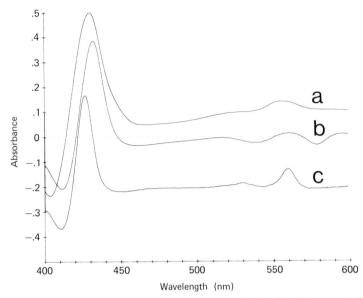

FIG. 4. Reduced − oxidized spectra of (a) purified rat mitochondria (4 mg protein/ml); (b) human hemoglobin (4.45 nmol and 71 μg protein/ml; equivalent to a 1-in-2000 dilution of blood); (c) pure cytochrome b-245 (3.47 nmol and 230 μg protein/ml).

Notes

1. The reduced − oxidized spectrum of cytochrome b-245 is fairly similar to that of hemoglobin and the mitochondrial cytochromes (Fig. 4). It is thus essential that preparations of neutrophils be free of hemoglobin when attempts are made to quantitate the cytochrome by simple difference spectroscopy. This technique cannot be used with cells that are relatively rich in mitochondria or endoplasmic reticulum.

2. Cell metabolism can change the pH in the extracellular medium. Suspending media should therefore be free of indicators, the color changes of which can cause erroneous results.

Oxidation–Reduction Midpoint Potential Titration and Quantitation of the Low-Potential Cytochrome b Component of Cells

The most striking property of cytochrome b-245 is its very low oxidation–reduction midpoint potential ($E_{m\,7.0}$) of −245 mV.[8] This is about 200 mV lower than any of the b-type cytochromes commonly found in mammalian cells. It is close, however, to the $E_{m\,7.0}$ of cytochrome P-450. Thus, difference spectra can be obtained that differentiate cytochrome b-245

[8] A. R. Cross, O. T. G. Jones, A. M. Harper, and A. W. Segal, *Biochem. J.* **194**, 559 (1981).

from other b- and c-type cytochromes as well as from hemoglobin and myeloperoxidase in contaminated preparations of neutrophils, mixed cell populations, or crude cell homogenates. Although these measurements can be conducted using homogenates of whole cells (which might be necessary when relatively few cells can be collected from unusual patients), they are best performed using preparations of purified membranes or cellular organelles (see above). (*Note:* The $E_{m\,7.0}$ values for the cytochrome obtained from detergent extracts may differ from those obtained with intact cells or membranes as a result of conformational changes caused by the solubilization.)

Method. The test sample is suspended in a solution of MOPS (50 mM) and KCl (100 mM), pH 7.0, at a concentration of about 2–3 mg protein/ml. The suspension is transferred to a cuvette fitted with a platinum electrode, a calomel reference electrode, and ports which permit gassing with scrubbed O_2-free nitrogen or scrubbed O_2-free argon. Another port, sealed with suba-seal caps, permits injection of oxidant and reductant solutions into the cuvette. The cuvette is stirred either magnetically or with a top drive stirrer. The scrubbing solution (Fieser's solution), used to remove O_2 from the gases, contains 200 g KOH, 20 g sodium anthraquinone 2-sulfonate, and 150 g sodium dithionite per liter. All connections should be air impermeable, using butyl rubber or soft copper tubing. The description of such a cuvette assembly is given by P. L. Dutton in this series.[9] The following mediators are added to the suspension in order to accelerate reaching oxidation–reduction equilibrium between the measuring electrodes and the membrane suspension: phenazine methosulfate, phenazine ethosulfate, 2-hydroxy-1,4-naphthoquinone, anthraquinone 2,6-disulfonate, 3,6-diaminodurene, and duroquinone, all at 12.5 μM; pyocyanine at 6 μM; and anthraquinone (10 μl of a saturated solution in ethanol added to 4 ml of membrane suspension).

The redox cuvette is placed in the sample position of a scanning spectrophotometer and a sample of the same membrane suspension is placed in the reference cuvette. The oxidation–reduction potential (E_h) in the sample cuvette is lowered by injecting microliter aliquots of sodium dithionite solution (around 0.1% in 100 mM MOPS buffer, pH 7.0) and is elevated by injecting a potassium ferricyanide solution (10 mM). The E_h is systematically lowered from about 0 to −300 mV and then returned to 0 mV and the difference spectra are recorded from 500 to 600 nm at approximately 10-mV intervals (see Fig. 5). Cytochrome b-245 is measured as the difference in absorbance at 559 − 540 nm as the E_h decreases from −200 to −300 mV. Differences due to other, higher potential, cyto-

[9] P. L. Dutton, this series, Vol. 54, p. 420.

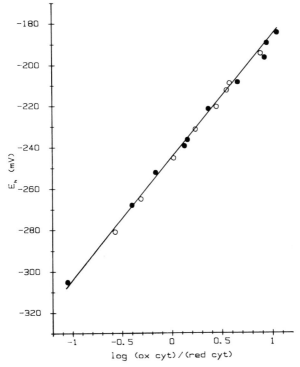

Fɪɢ. 5. Potentiometric titration of the cytochrome *b*-245 in the membranes (2.1 mg protein/ml) prepared from human neutrophils. ●, Reductive titration with sodium dithionite; ○, oxidative titration with potassium ferricyanide.

chromes are not measured and do not interfere with the determination of cytochrome *b*-245.

Note. The equilibration of cytochrome *b*-245 with the electrode is slow and patience is required. The E_h should be held in the region of -200 to -300 mV for about 5 min before a spectrum is recorded.

The Reaction of Cytochrome b-245 with CO

Cytochrome *b*-245 has a relatively low affinity for CO ($K_D = 1.18$ m*M*).[7] It is therefore important to saturate the medium with CO in order to observe the binding of CO to cytochrome *b*-245.

The suspension of material containing the cytochrome is reduced with a few crystals of the dithionite, divided between two matched cuvettes and a baseline recorded in the spectrophotometer. The cuvette in the sample position is removed and carbon monoxide is bubbled through the cuvette by means of a Pasteur pipet attached to a cylinder gas supply line.

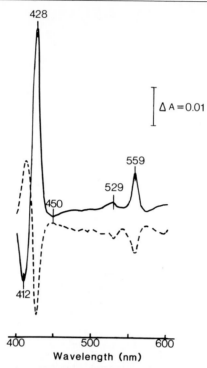

FIG. 6. Dithionite reduced minus oxidized difference spectrum (———); dithionite reduced plus CO, minus dithionite reduced difference spectrum (----). The purity of the cytochrome was 13.8 nmol/mg protein; its concentration was 0.46 μM.

The gas flow should be regulated to introduce 1 to 2 bubbles per second through the sample, which should be kept continuously in the dark to avoid photodissociation of the cytochrome b–CO complex. The gas should be passed for several minutes to ensure that the liquid phase is saturated with CO. When this has been achieved the difference spectrum of the (reduced + CO) − reduced sample is obtained. The formation of the cytochrome b–CO complex is indicated by the appearance of troughs at 559 and 430 nm and of a peak at 419 nm (Fig. 6). The extent of binding will depend on the CO concentration in solution and, since this is temperature dependent, the quantity of cytochrome complexed to CO will be in the region of 35% at 20° and 60% at 4°. The proportion may be estimated using the $E_{575-557\,nm} = 21.6$ mM^{-1} cm^{-1}.[10]

Note. A number of other compounds which bind CO and may interfere with the assay may be present in samples. These include myeloperoxidase, mitochondrial cytochrome oxidase, and hemoglobin.

[10] B. Chance, *in* "Haematin Enzymes" (J. E. Falk, R. Lemberg, and R. K. Montan, eds.), Vol. II, p. 433. Pergamon, Oxford, 1961.

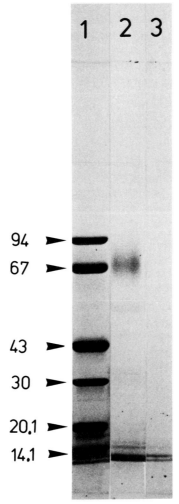

FIG. 7. SDS–polyacrylamide gel electrophoresis of cytochrome b-245 eluted from heparin agarose. Tracks: (1) M_r markers (values × 10^3 are indicated at the right of the figure); (2) peak eluted fraction of cytochrome b-245 from heparin agarose (24 μg, 0.48 nmol); (3) buffer and proteinase inhibitors. Electrophoresis was carried out in a 1-mm thick, 5–15% acrylamide gradient slab gel, which was water cooled, at a voltage of 150 V until the tracking dye, bromophenol blue, reached the bottom of the stacking gel, and then at 250 V until completion.

Electrophoresis of Cytochrome b-245

Sodium Dodecyl Sulfate/Polyacrylamide Gel Electrophoresis (SDS/PAGE)

This cytochrome runs as a broad band with an apparent M_r of 68,000–78,000 (Fig. 7) because it is a heavily glycosylated glycoprotein. Removal of the carbohydrate moieties with endoglycosidase F causes the apoprotein to migrate as a single, compact band with an apparent M_r of 55,000.[11] Aggregation to form high M_r (>100,000) material is marked if the samples are boiled for too long (>3 min) or the ionic strength of the samples is too high (>0.2 M). Samples are therefore routinely prepared for electrophoresis by first dialyzing against buffer containing Tris acetate, pH 7.4 (10 mM), glycerol (20%), dithiothreitol (0.1 mM), EDTA (1 mM), and butylated hydroxytoluene (30 mM). Before loading, samples are boiled for 2 min in SDS (2%) EDTA (20 mM), and dithiothreitol (100 mM). Without pretreatment of the cells with DIFP there are a large number of other protein-staining bands visible on the gels with M_r values of <20,000. Where gels are to be stained for heme,[12] dithiothreitol is specifically omitted from the samples and they are not boiled. Even so there is a marked tendency for the heme to dissociate from the protein during electrophoresis and the cytochrome b-245 stains only weakly.

Nondenaturing Electrophoresis and Isoelectric Focusing

Although the cytochrome itself has an apparent molecular weight of above 75,000–90,000, when solubilized in buffer containing Triton N-101 it forms a complex structure with an apparent molecular weight of several hundreds of thousands as determined by gel exclusion chromatography. This large complex structure has precluded successful nondenaturing electrophoresis and isoelectric focusing.

[11] A. M. Harper, M. F. Chaplin, and A. W. Segal, *Biochem. J.* **277**, 783 (1985).
[12] P. E. Thomas, D. Ryan, and W. Levin, *Anal. Biochem.* **75**, 168 (1976).
[13] A. W. Segal, *in* "Advances in Host Defense Mechanisms" (J. I. Gallin and A. S. Fauci, eds.), Vol. III, p. 121. Raven Press, New York, 1983.

[22] Measurement of Hydrogen Peroxide Production by Phagocytes Using Homovanillic Acid and Horseradish Peroxidase

By Marco Baggiolini, Walter Ruch, and Philip H. Cooper

The products of the respiratory burst of phagocytes are superoxide and H$_2$O$_2$.[1] Both may be taken as a measure of the response of these cells to stimulation with particles (e.g., bacteria) and a variety of solutes like phorbol myristate acetate (PMA)[1a] or chemotactic molecules. H$_2$O$_2$ is a compound of the peroxidase-dependent microbicidal system which was described originally in neutrophils by Klebanoff[2] and is considered an important factor in cytolysis and parasite killing mediated by mononuclear phagocytes.[3]

We describe here a method for the determination of H$_2$O$_2$ produced by neutrophils, monocytes, and macrophages.[3a] The assay is based on the oxidation of homovanillic acid (HVA) into a fluorescent dimer, which is catalyzed by horseradish peroxidase and depends on the H$_2$O$_2$ generated by the cells.[4] The principle of this reaction was described by Guilbault *et al.*[5] and its use with neutrophils was briefly reported by Rossi *et al.*[6] (see also this volume [3]).

Reagents

Saline Medium. Hanks' balanced salt solution (HBSS) as described originally by Hanks and Wallace[7] is used. A phosphate-buffered saline (PBS) solution, containing 137 mM NaCl, 2.7 mM KCl, 8.1 mM Na$_2$HPO$_4$, 1.5 mM KH$_2$PO$_4$, 0.9 mM CaCl$_2$, 0.5 mM MgCl$_2$, and 1 g/liter D-glucose[8] can be used alternatively for neutrophils.

[1] B. M. Babior, *N. Engl. J. Med.* **298**, 659 (1978).

[1a] Abbreviations: HBSS, Hanks' balanced salt solution; PBS, phosphate-buffered saline; HVA, homovanillic acid; EDTA, ethylenediaminetetraacetic acid; PMA, phorbol myristate acetate; DMSO, dimethyl sulfoxide.

[2] S. J. Klebanoff, *J. Exp. Med.* **126**, 1063 (1967).

[3] C. F. Nathan, S. C. Silverstein, and Z. A. Cohn, *J. Exp. Med.* **149**, 100 (1979).

[3a] The material presented in this chapter has been published in part in Ruch *et al.*[4]

[4] W. Ruch, P. H. Cooper, and M. Baggiolini, *J. Immunol. Methods* **63**, 347 (1983).

[5] G. G. Guilbault, P. J. Brignac, and M. Juneau, *Anal. Chem.* **40**, 1256 (1968).

[6] F. Rossi, G. Zabucchi, P. Dri, P. Bellavite, and G. Berton, *Adv. Exp. Med. Biol.* **121A**, 54 (1980).

[7] J. H. Hanks and R. E. Wallace, *Proc. Soc. Exp. Biol. Med.* **71**, 196 (1949).

[8] B. Dewald, U. Bretz, and M. Baggiolini, *J. Clin. Invest.* **70**, 518 (1982).

H₂O₂ Standard. A 40 m*M* stock solution is prepared by diluting perhydrol (hydrogen peroxide) (E. Merck AG, Darmstadt, FRG) 1 : 200 in distilled water. This solution is kept at 4°. Further dilutions are made in HBSS prior to use, and the exact concentration is determined by spectrophotometry at 240 nm where the molar extinction coefficient of H_2O_2 is 40.[9]

HVA-Peroxidase. A solution of 0.4 m*M* homovanillic acid puriss (Fluka AG, Buchs, Switzerland) containing 4 U/ml of horseradish peroxidase, type II (Sigma Chemical Corp., St. Louis, MO) is prepared in HBSS. The peroxidase units are those defined in the Sigma catalog.

Two stop solutions are used. One of pH 12.0, a 100 m*M* glycine-NaOH buffer containing 25 m*M* EDTA, and one of pH 10.4, a 50 m*M* glycine-NaOH buffer containing 5 m*M* EDTA. EDTA is necessary to chelate Ca^{2+} and Mg^{2+}, which would otherwise precipitate as hydroxide.

Zymosan from Saccharomyces cerevisiae (Biosynth AG, Zürich, Switzerland). Zymosan is boiled in isotonic NaCl for 1 hr, washed three times by centrifugation (1000 *g*, 5 min), resuspended in HBSS (50 mg/ml), and autoclaved at 120°. Portions of 1 ml of the sterile suspensions are stored in a refrigerator.

Opsonization of Zymosan. Opsonization is carried out immediately prior to use by incubating 1 vol of the above suspension with 4 vol of autologous serum for 30 min at 37°, and then washing three times with HBSS.

Phorbol Myristate Acetate (Consolidated Col., St. Louis, MO). Phorbol myristate acetate is stored at −70° as a stock solution of 2 mg/ml in dimethyl sulfoxide (DMSO). Immediately prior to use this solution is diluted to the desired concentration with HBSS.

Cells

Human neutrophils are purified from anticoagulated blood,[8] suspended in isotonic NaCl (10^7–10^8 cells/ml), and kept at 6–10° until use (see this series, Vol. 108 [9]). Storage in ice should be avoided because of more rapid loss of viability.

Mononuclear phagocytes may be used as a suspension or as monolayers in small plastic culture wells. In both cases the medium is HBSS. Mouse bone marrow-derived macrophages are obtained by proliferation in Teflon bag cultures.[10,11] They are harvested from the bags, centrifuged at 200 *g* for 10 min at room temperature, and resuspended at the desired

[9] R. F. Beers and I. W. Sizer, *J. Biol. Chem.* **195,** 133 (1952).
[10] C. Neumann and C. Sorg, *Eur. J. Immunol.* **10,** 834 (1980).
[11] P. H. Cooper, P. Mayer, and M. Baggiolini, *J. Immunol.* **137,** 913 (1984).

density in medium 199 containing 15% heat-inactivated horse serum and 2 mM glutamine for seeding culture plates, or in HBSS for direct use. Cultures are established in 24-well plates (Linbro Company, Inc., Hamden, CN) by distributing 0.5×10^6 cells/well in 1 ml of medium.[11] After 2 hr at 37° the cultures are washed three times with warm HBSS to remove nonadherent cells, and the adherent cells are cultured further[11] or used for the H_2O_2 assay.

Human monocytes are purified from buffy coats of donor blood by centrifugal elutriation[12,13] and used as a suspension in HBSS or as adherent cultures in flat-bottom 96-well microtiter plates. Cultures are established by distributing $0.02-0.2 \times 10^6$ cells/well in 0.2 ml of DMEM containing 3% heat-inactivated human ABO serum. After 1 hr at 37°, the cultures are washed twice with HBSS to remove nonadherent cells, and the adherent cells are cultured further in the medium indicated or used for the H_2O_2 assay.

Assay Procedure

Two-Milliliter Assay

One milliliter of HBSS and 0.5 ml of HVA-peroxidase solution are prewarmed at 37°. Samples of the cell suspension (0.25 ml containing $0.2-1.0 \times 10^6$ cells) are then added, and the reaction is started shortly thereafter by the addition of the stimulus (e.g., zymosan, 4–16 mg/ml, or PMA, $10^{-9}-10^{-6}$ M) in 0.25 ml HBSS. Nonstimulated controls are treated identically, except that 0.25 ml of HBSS is added instead of stimulus. Incubation is carried out at 37° for the desired time (5–30 min for neutrophils, up to a few hours for mononuclear phagocytes) and then stopped by the addition of 0.25 ml of the pH 12 stop solution. The tubes are then centrifuged at 1200 g for 10 min at room temperature and the fluorescence of the clear supernatant is measured.

Similar conditions are adopted with adherent macrophages in 24-well (16-mm-diameter) plates. The culture medium is aspirated and the adherent cells (0.5×10^6 per well) are overlayed with 1.25 ml prewarmed HBSS. Prewarmed HVA-peroxidase solution (0.5 ml) is then added, and the reaction is started by the addition of the stimulus in 0.25 ml prewarmed HBSS or HBSS alone for the controls. The plates are covered and kept at 37° in a CO_2 incubator for the desired time. The reaction is stopped with 0.25 ml of pH 12 stop solution. The reaction mixture (which may contain detached cells and debris) is transferred to small plastic

[12] K. J. Clemetson, J. L. McGregor, R. P. McEver, Y. V. Jaques, D. F. Bainton, W. Domzig, and M. Baggiolini, *J. Exp. Med.* **161**, 972 (1985).
[13] This series, Vol. 108 [20].

tubes, centrifuged as above, and the fluorescence of the clear supernatant is measured.

Standard and blank assays are performed in the absence of cells and either in the presence or absence of the stimulus. In tubes, the cell sample is substituted by 0.25 ml of a standard solution (containing 0.5 to 10 nmol H_2O_2) and 0.25 ml of HBSS, respectively. In culture wells, the desired standard concentrations of H_2O_2 are incorporated in the 1.25 ml of HBSS added initially, HBSS alone being used for the blanks.

Microassay

These conditions were established for microcultures of human monocytes and monocyte-derived macrophages in 96-well plates (see under "Cells"). The plates are centrifuged (300 g for 5 min), the medium is discarded, and the cells are overlayed with 0.1 ml of prewarmed HVA-peroxidase solution. The reaction is started with the stimulus in 0.1 ml of prewarmed HBSS (a plate shaker may be used for short mixing). The plates are then covered and kept at 37° in a CO_2 incubator for the desired time. At the end of incubation, the plates are centrifuged as above and 0.150 ml of the reaction mixture is transferred to tubes containing 1.0 ml of pH 10.4 stop solution. Standards and blanks are performed in wells without cells, containing 0.1 ml of HVA-peroxidase solution and 0.1 ml of H_2O_2 standard HBSS, respectively.

Fluorimetry

The conditions are those described by Guilbault et al.[5]: λ_{max}, excitation = 312 nm, λ_{max}, emission = 420 nm. Our studies have shown that the fluorophore is stable at pH 10–11 for at least 2 hr at room temperature and in daylight.[4] It was reported, however, that ultraviolet light increases the fluorescence of HVA due to nonspecific oxidation.[14] For precaution, it is therefore advisable to keep the samples in the dark if they cannot be measured immediately.

Typical Results

The conditions for the H_2O_2 assay procedure proposed here have been established through systematic variation of the composition of the reaction mixture, the incubation time, and the pH of the fluorochrome solution.[4] One unit of peroxidase per milliliter and 100 μM HVA were found to be sufficient for the detection of H_2O_2 even if supplied at once at levels

[14] H. E. Hirsch and M. E. Parks, Anal. Biochem. **122**, 79 (1982).

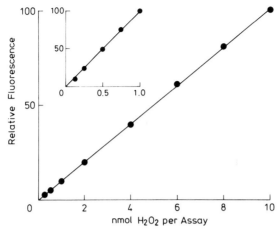

FIG. 1. Dependence of fluorescence intensity from the amount of H_2O_2 added to the reaction mixture described under "Two-Milliliter Assay." The reaction was stopped 30 min after addition of H_2O_2. (Reproduced with permission from Ruch *et al.*[4])

which were higher than those normally produced by neutrophils or mononuclear phagocytes (10 nmol/assay). Under these conditions, the reaction is rapid, reaching completion in less than 5 min. The fluorescence intensity of the HVA-oxidation product increases sharply between pH 6 and 9 and reaches a plateau approximately at pH 10. For this reason, a stop solution is used which brings the pH of the reaction mixture to 10.2–10.4.

A curve relating standard amounts of H_2O_2 added to the assay mixture and intensity of fluorescence is shown in Fig. 1. The relationship is linear over the test range of 0.1 to 10 nmol H_2O_2/assay, the correlation coefficient of the regression line calculated for the experimental points being better than 0.999.

Assessment of H_2O_2 produced by phagocytes was also highly reproducible. The variation between samples from the same preparation of cells which were assayed in parallel was less than 4%. The table shows the results obtained over a period of several months with different preparations of various types of phagocytes stimulated with PMA or zymosan.

Comments

Superoxide dismutase, which accelerates the conversion of superoxide to H_2O_2, and sodium azide, which inhibits catalase and other oxidoreductases possibly competing with the H_2O_2 detection system, may be added to the mixture to increase H_2O_2 recovery. Such additions, however, are not always necessary, and their effect should be explored before

H₂O₂ PRODUCTION BY STIMULATED PHAGOCYTES OF DIFFERENT ORIGIN

Cells	Stimulus[a]	Time (min)	Nanomoles H₂O₂ produced/ 10 μg DNA (mean ± SD)	Number of experiments[b]
Human neutrophils (in suspension)	PMA	10	13.4 ± 3.7	12
Human monocytes (adherent)	PMA	120	26.7 ± 9.7[c]	8[d]
	Zymosan	120	16.9 ± 3.3[c]	9[d]
Mouse peritoneal macrophages (adherent)	PMA	120	1.9 ± 0.9	3
	Zymosan	120	7.1 ± 1.9	9
Mouse bone marrow (in suspension)	Zymosan	120	2.3 ± 0.8	45

[a] Zymosan: 1 mg/ml final concentration; PMA: 10^{-8} M.
[b] Each experiment was carried out in duplicate or triplicate with a different cell preparation.
[c] Nanomoles $H_2O_2/10^6$ cells.
[d] Microassay.

adopting them. In our hands, the yield of H_2O_2 following stimulation of human or bovine neutrophils with opsonized particles was 30 to 50% higher in the presence of 0.08 mg/ml superoxide dismutase (Sigma Corp.) and 1 mM sodium azide. Using the microassay described, the H_2O_2 yield from human monocytes stimulated with PMA or zymosan increased by 30 to 70%, depending on the donor, in the presence of 1 mM azide. Under these conditions, however, azide had no effect on H_2O_2 recovery from monocyte-derived macrophages. At the concentrations indicated, sodium azide and superoxide dismutase did not affect the determination of standard samples of H_2O_2 by the HVA horseradish peroxidase system.

A number of alternative organic compounds may be considered as hydrogen donors. p-Hydroxyphenylacetic acid[5] and 3-(p-hydroxyphenyl)propionic acid[15] yield oxidation products with high fluorescence intensity. In the detection system described, however, these hydrogen donors must be used in much higher concentrations (10 and 400 times that of HVA, respectively), and are therefore more likely to exert cytotoxic effects. We have compared the sensitivity of H_2O_2 detection in standard samples with HVA and the two other hydroxyphenyl derivatives mentioned above and found no significant difference.

[15] K. Zaitsu and Y. Ohkura, Anal. Biochem. 109, 109 (1980).

[23] Assay of the Extracellular Hydrogen Peroxide Pool Generated by Phagocytes

By SAMUEL T. TEST and STEPHEN J. WEISS

The stimulation of human neutrophils with either soluble or particulate stimuli results in the activation of a plasma membrane-associated NADPH oxidase and the generation of $O_2^{\cdot-}$.[1] The $O_2^{\cdot-}$ then dismutates either spontaneously or enzymatically to form H_2O_2, which rapidly enters into an equilibrium among interconnected pools in the phagocytic vacuole, the cytoplasm, and the extracellular space.[2] The accurate quantitation of H_2O_2 released into the extracellular space requires a method which does not perturb the H_2O_2 concentration in any of its pools. Previous assays have required substrates which either scavenge H_2O_2 as it is released into the extracellular space or alter its usual routes of catabolism.[3] The development of the hydrogen peroxide-sensing platinum anode[4] has permitted the design of an assay for the continuous measurement of extracellular H_2O_2 release by intact cells.[2,5,6] When used to measure H_2O_2 release by triggered phagocytes, the polarographic method results in the consumption of insignificant amounts of H_2O_2 for analysis, thus allowing quantitation of the extracellular H_2O_2 pool without altering its size.

Isolation of Human Neutrophils

Neutrophils are isolated from the anticoagulated peripheral venous blood (heparin or EDTA) of normal human volunteers by a modification of the method originally described by Bøyum[7] (see also this series, Vol. 108 [9]). This method involves Ficoll-Hypaque density centrifugation followed by dextran sedimentation.[8] Contaminating erythrocytes are removed by hypotonic lysis in distilled water.[8] The neutrophils are suspended in Dulbecco's phosphate-buffered saline (pH 7.4; Grand Island Biological Company, Grand Island, NY, or M. A. Bioproducts, Walkers-

---

[1] S. J. Weiss, *in* "Handbook of Inflammation" (P. A. Ward, ed.), Vol. 4, p. 37. Am. Elsevier, New York, 1983.

[2] S. T. Test and S. J. Weiss, *J. Biol. Chem.* **259**, 399 (1984).

[3] S. J. Klebanoff, *Ann. Intern. Med.* **93**, 480 (1980).

[4] L. C. Clark, Jr., this series, Vol. 56, p. 448.

[5] C. B. Schroy and J. E. Biaglow, *Biochem. Pharmacol.* **30**, 3201 (1981).

[6] M. E. Varnes, S. W. Tuttle, and J. E. Biaglow, *Adv. Exp. Med. Biol.* **159**, 49 (1983).

[7] A. Bøyum, *Scand. J. Clin. Lab. Invest.* **21**, Suppl. 97, 31 (1968).

[8] S. J. Weiss, P. K. Rustagi, and A. F. LoBuglio, *J. Exp. Med.* **147**, 316 (1978).

ville, MD) supplemented with 1 mg/ml of glucose. The final cell concentration can be determined by counting an aliquot of the cell preparation on a Coulter counter or a hemacytometer.

Measurement of Extracellular H_2O_2 Generation by Neutrophils (see also this volume [3])

Reagents

Dulbecco's phosphate-buffered saline, pH 7.4, supplemented with 1 mg/ml glucose

Reagent H_2O_2, 30% (any reagent grade H_2O_2 may be used but should be stabilizer free)

Neutrophils, 6×10^6 cells/ml

Phorbol myristate acetate (Consolidated Midland Corp., Forrester, NY)

A stock preparation of 1 mg/ml is prepared in dimethyl sulfoxide and stored in 0.5-ml aliquots at $-70°$. One such aliquot may be kept at $-20°$ and used as a stock for daily experiments without significant loss of activity if immediately refrozen after thawing and use. Final preparations are made in Dulbecco's buffer. PMA[8a] is not stable in aqueous solutions and should be used promptly after dilution

Opsonized zymosan

Commercially obtained zymosan (Sigma Chemical Company, St. Louis, MO; Worthington Biochemicals, Freehold, NJ) may be opsonized by preparing a suspension at 50 mg/ml in buffer, adding 3 vol of freshly prepared human serum and incubating for 30 min at 37° in a water bath. The opsonized zymosan is then washed three times and suspended in Dulbecco's buffer at the desired concentration

Sodium azide

Other neutrophil triggering agents (i.e., C5a, N-formyl-Met-Leu-Phe, etc.) or additions may be used also and should be prepared or diluted in Dulbecco's buffer

H_2O_2-Detecting System (see also this volume [3])

H_2O_2 release is monitored continuously with a YSI model 25 oxidase meter fitted with a YSI 2510 oxidase probe (Yellow Springs Instrument Company, Yellow Springs, OH). The electrode is covered with a single layer of polycarbonate membrane (0.015-μm pore size; Nuclepore Corp.,

[8a] Abbreviation: PMA, phorbol myristate acetate.

Pleasanton, CA). If more than one electrode is used, it is important that the shiny side of the membrane face the same direction (usually outward) on each electrode. The oxidase meter can be set in either of two modes. When operated in the nanoamp mode, the probe current is displayed directly in nanoamps (nA) up to a maximum of 100 nA. When operated in the variable sensitivity mode, the instrument can be calibrated with known amounts of reagent H$_2$O$_2$ to read desired concentrations of H$_2$O$_2$ at full scale with probe currents of approximately 10–110 nA.[9] The latter mode is preferred for cell experiments. The current output may be recorded continuously by attaching the oxidase meter to an appropriate recorder. In order to allow simultaneous recording from three or four probe-meter assemblies, a flip-flop timer (model 342; Automatic Timing and Controls Company, King of Prussia, PA) may be interposed between two of the oxidase meters and one of the recorder inputs on a dual channel recorder.

The prepared probe is plugged into the oxidase meter, immersed in Dulbecco's buffer, and allowed to stabilize for approximately 3 hr. If experiments are to be performed at 37°, the buffer should be at this temperature during probe stabilization. When the probe current reaches steady state, the probe is stabilized. It is best to leave the entire system in continuous operation, with the probe immersed in buffer at room temperature overnight, for optimal stability of the probe current.[9] Membrane changes are best accomplished at the end of any day's experiments, so that the probe may restabilize overnight after being unplugged.

When used in the variable sensitivity mode, the meters are calibrated prior to each experiment with dilutions of reagent H$_2$O$_2$ calculated on the basis of an extinction coefficient at 230 nm of 81 M^{-1} cm^{-1}.[10] The probe should detect H$_2$O$_2$ concentrations as low as 0.25 μM and consistently give linear response curves from 0 to 100 μM with correlation coefficients $\geq$0.99.

Assay Conditions

Buffer preparations containing neutrophils are placed in siliconized cuvettes and incubated at 37° with constant stirring in a YSI model 5301 bath stirrer assembly (the cuvettes are supplied with this apparatus) connected to a constant temperature water circulator (one model is the Haake Model FE; Haake Instruments, Inc., Rochelle Park, NJ). For accurate measurement of H$_2$O$_2$ release, proper stirring of the mixtures is essential. The magnetic stirring bars should be large enough to provide adequate

[9] Instruction Manual, YSI Model 25 Oxidase Meter and YSI 2510 Oxidase Probe, YSI Co., Yellow Springs, Ohio.
[10] J. W. T. Homan-Muller, R. S. Weening, and D. Roos, *J. Lab. Clin. Med.* **85**, 198 (1975).

mixing, but small enough so as not to touch the oxidase probe or damage the neutrophil suspension. Stirring bars measuring 3×10 mm are ideal for the assay conditions described below. The final reaction volume may be varied, but measurements may be readily accomplished in a final volume of 4 ml. The probes should be allowed to stabilize in Dulbecco's buffer at $37°$ for at least 3–4 hr before taking experimental measurements. Bicarbonate buffers should not be used because proper control of the pH is difficult to maintain during the course of the assay. Following stabilization the electrodes are calibrated with reagent H_2O_2, the cuvettes are rinsed or replaced with clean cuvettes, the probes are rinsed free of H_2O_2 with double-distilled H_2O, and reimmersed in buffer at $37°$.

The recordings obtained in a typical assay of extracellular H_2O_2 release by triggered neutrophils are shown in Fig. 1. For experiments without sodium azide, the meter is calibrated so that full scale (100 nA) is equivalent to 25 μM H_2O_2. Dulbecco's buffer (3.7 ml) is placed in the cuvette and equilibrated to $37°$. A 200-μl aliquot of neutrophils (stock concentration 6×10^6 cells/ml) is added to give a final cell concentration of 3×10^5 cells/ml and this is followed by 100 μl of the stimulating agent. Results are shown for PMA at a final concentration of 25 ng/ml or serum-opsonized zymosan at a final concentration of 1.25 mg/ml as the stimuli. Extracellular H_2O_2 is detected within approximately 30–40 sec following the addition of either of these stimuli. The addition of exogenous catalase will rapidly consume extracellular H_2O_2 and should return the probe reading to baseline. H_2O_2 concentration determined by the electrode response may be confirmed by removing an aliquot from the experimental mixture and determining the H_2O_2 concentration by an independent assay technique (e.g., the potassium thiocyanate-ferrous ammonium sulfate method of Thurman et al.[11]). In experiments in which sodium azide is added to prevent H_2O_2 catabolism by neutrophil catalase or myeloperoxidase,[1] the meter is calibrated so that full scale is equivalent to 75 or 100 μM and the azide is added so as to give a final concentration of 1 mM (or whatever the desired concentration).

Following each experimental measurement the probe should be removed from the buffer solution (without unplugging it) and rinsed thoroughly with a dilute solution of HOCl (100–200 μM) to destroy any enzyme activity or substances on the probe or membrane surface which might interfere with further measurements. This is followed by several rinses with double-distilled H_2O, after which the probe is immersed in buffer at $37°$. The system should then be recalibrated with reagent H_2O_2. The probe current should remain steady for several minutes following

---

[11] R. G. Thurman, H. G. Ley, and R. Scholz, Eur. J. Biochem. **25**, 420 (1972).

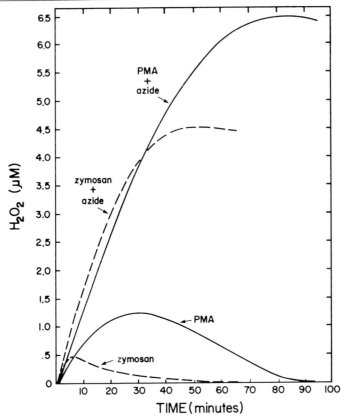

FIG. 1. Extracellular H_2O_2 concentration surrounding triggered neutrophils. Neutrophils (3×10^5/ml) were stimulated with either PMA (25 ng/ml), or serum-opsonized zymosan (1.25 mg/ml), and the H_2O_2 concentration was measured continuously. Measurements were made simultaneously from paired cell preparations with or without sodium azide (1 mM). Time is in minutes from addition or stimulus. (From Test and Weiss,[2] reproduced with permission.)

addition of a known amount of H_2O_2. If there is loss of probe current, indicating persistence of H_2O_2-catabolizing substances on the probe or membrane, the probe should be cleaned again as above and the calibration repeated before proceeding to the next experimental assay.

Other additions, including both intact cells and/or chemical substances, may be used with the polarographic technique.[2,5,6] Care should be taken in the preparation of cellular additions to avoid the release of cell-associated H_2O_2-catabolizing substances (i.e., catalase, myeloperoxidase, glutathione peroxidase) into the suspending buffer and it should be noted that erythrocyte contamination in the neutrophil preparation will lower the concentration of the extracellular H_2O_2 pool. Any additions to the reaction mixture should first be analyzed under cell-free conditions for their ability to interfere with the assay by inducing a current at the anode or by autooxidizing in solution to generate H_2O_2. Substances known to have such effects include ascorbate, uric acid, glutathione, cysteamine, xanthine, and NADPH.[6] In addition, we have found that human serum and methionine have a similar effect. The effects of these substances may be minimized by placing an ultrafine cellulose acetate membrane, which will exclude substances of $M_r > 150–200$, between the probe and the polycarbonate membrane.[9] Some substances (i.e., taurine) induce an initial response, after which a steady state is reached. In this instance, the new steady state may be used as the baseline for quantitation of H_2O_2 release by triggered neutrophils. Chemical additions should also be checked for their ability to consume reagent H_2O_2 in a cell-free system before use in assays involving intact cells. Ferricytochrome c is one such substance which interferes with the assay by consuming free H_2O_2.[2]

Calculation of Data

Maximum H_2O_2 release can be attained readily from the curve obtained when assays are performed with continuous recording. This value may be expressed either as a concentration or converted to total nanomoles of H_2O_2 released. In order to allow comparisons of total H_2O_2 generation over a period of time, data may also be expressed as the integral of the H_2O_2 concentration (c) with respect to time (t) or ct. These values are calculated by measuring the area under the recorded curve of H_2O_2 generation. Although heme-enzyme inhibitors (e.g., azide, cyanide, aminotriazole) cause a large increase in the extracellular H_2O_2 concentration (see Fig. 1), caution should be exercised in expressing these results as the "total" amount of H_2O_2 generated by the triggered neutrophils. Heme-enzyme inhibitors do not block glutathione peroxidase activity and may also artifactually increase H_2O_2 generation by interfering with the termination of the respiratory burst.[2]

Acknowledgments

This work was supported by National Institutes of Health Grants 5 R01 HL-28024 and R01 AI 21301. Samuel T. Test was supported by a Research Fellowship from the American Heart Association of Michigan.

[24] Microassays for Superoxide and Hydrogen Peroxide Production and Nitroblue Tetrazolium Reduction Using an Enzyme Immunoassay Microplate Reader

By EDGAR PICK

Macrophages of various tissue origin produce copious amounts of superoxide (O_2^-) and hydrogen peroxide (H_2O_2) when adequately stimulated.[1] This process is accompanied by a marked increment in oxygen uptake and increased utilization of glucose via the hexose monophosphate shunt (HMPS). The coordinated sequence of reactions is known as the "oxidative" or "respiratory" burst. The primary reaction of the oxidative burst is the one-electron reduction of oxygen to O_2^- through the action of a cyanide-resistant NADPH oxidase that is dormant in resting cells.[2] H_2O_2 is generated by either spontaneous or enzymatic dismutation of O_2^-, a reaction in which one molecule of O_2^- is oxidized by another. Macrophages utilize oxygen radicals produced in the course of the oxidative burst as cytotoxic agents for the destruction of phagocytosed pathogenic microorganisms and for damaging extracellular targets.

As opposed to the neutrophil that possesses a rather constant oxygen radical-generating capacity, the ability of macrophages to produce O_2^- and H_2O_2 is variable and determined by tissue origin, stage of differentiation, and the influence of modulatory agents. Factors such as bacterial lipopolysaccharide or cell wall peptidoglycan and, most notably, a T cell-derived lymphokine that has been identified as interferon-γ, augment the ability of macrophages to produce oxygen radicals in response to membrane stimuli.[3,4] This ability is well correlated with an enhanced microbicidal capacity directed against a number of pathogenic bacteria, fungi, and protozoa, a phenomenon known as macrophage activation. There is considerable experimental evidence showing that increased oxygen radical production by macrophages is an adequate biochemical correlate for at least some forms of macrophage activation.

We are describing three assays for the quantitation of the cyanide-resistant oxidative metabolism of macrophages that are in current use in our laboratory. The techniques, as described, were developed for use

---

[1] J. A. Badwey and M. L. Karnovsky, *Annu. Rev. Biochem.* **49,** 695 (1980).

[2] P. Bellavite, G. Berton, P. Dri, and M. R. Soranzo, *J. Reticuloendothel. Soc.* **29,** 47 (1981).

[3] C. F. Nathan, *Fed. Proc., Fed. Am. Soc. Exp. Biol.* **41,** 2206 (1982).

[4] E. Pick, *Clin. Immunol. Newsl.* (in press).

METHODS IN ENZYMOLOGY, VOL. 132

with guinea pig macrophages but they were applied successfully to human and mouse macrophages and to human neutrophils and eosinophils, as well.

Principle Common to All Methods

The three methods described in this section have in common the measurement of O_2^- and H_2O_2 production and nitroblue tetrazolium (NBT)[4a] reduction by cells cultured in 96-well microplates with the aid of an enzyme immunoassay microplate reader fitted with appropriate filters for the photometric determination of the respective reaction products.

Apparatus

The instrument used is an 8-channel vertical light path filter photometer that permits sequential reading of absorbances through individual wells of 96-well flat-bottom microplates. Such instruments were originally designed for the reading of enzyme-linked immunosorbent assays (ELISAs) in microtitration plates. We have successfully used the "Titertek Multiskan" (Cat. No. 78-503-00) and the dual wavelength version "Titertek Multiskan MC" (Cat. No. 78-530-00), both manufactured by Flow Laboratories. Other authors report using the "MicroELISA Automatic Reader" manufactured by Dynatech (Cat. No. AM 120) and similar equipment made by other manufacturers should be equally satisfactory. Interface software packages are available for most types of equipment for connection to microcomputers.

Special interference filters are required for the performance of the O_2^- (550 nm), H_2O_2 (600 nm), and NBT reduction (550 nm) assays. These are not routinely provided with equipment originally intended for reading ELISAs performed in microplates. However, the filters can be obtained from the manufacturers of ELISA readers and we had no difficulty in acquiring such filters by special order.

Isolation of Macrophages

The following is a description of the procedure used in the author's laboratory for work with guinea pig macrophages (see also this series, Vol. 108 [25] and [26]).

[4a] Abbreviations: NBT, nitroblue tetrazolium; ELISA, enzyme-linked immunosorbent assay; PEC, peritoneal exudate cells; BSS, Hanks' balanced salt solution; DMEM, Dulbecco's modified Eagle's medium; MEM, Eagle's minimum essential medium; FCS, fetal calf serum; SOD, superoxide dismutase; DMSO, dimethyl sulfoxide; PMA, phorbol myristate acetate; Con A, concanavalin A; WGA, wheat germ agglutinin; HRPO, horseradish peroxidase.

Peritoneal exudate cells (PEC) are obtained by rinsing the peritoneal cavity with a large volume (100–150 ml) of ice-cold Earle's or Hanks' balanced salt solution (BSS). The cells are washed three times with BSS by centrifugation (350 g, 10 min at 4°) and resuspension and brought to a concentration of 2.5 × 10^6 cells/ml in Dulbecco's modified Eagle's medium (DMEM) or Eagle's minimum essential medium (MEM) supplemented or not with 2–15% fetal calf serum (FCS), heat inactivated at 56° for 30 min. The cell suspension is dispensed into the wells of 96-well flat-bottom tissue culture microplates (Microtest III with lid, No. 3072, Falcon or Nunclon Microwell, No. 167008, Nunc), 100 μl/well using a repeating dispenser (Multipette 4780, Eppendorf). Macrophages are allowed to adhere to the bottom of the wells by incubation of the plates for 1–2 hr at 37° in 95% air–5% CO_2. Nonadherent cells (erythrocytes and lymphocytes) are removed by shaking the plates for 30 sec on a platform shaker (Titertek plate shaker, Flow Laboratories) and rinsing the wells three times with 0.1-ml volumes of phenol red-free BSS. The BSS used for rinsing is dispensed simultaneously to 12 wells with the aid of a 12-channel Titertek pipet (Flow Laboratories). Automatic microplate washing devices intended for ELISAs were not found satisfactory for this purpose. This procedure leads to uniform macrophage monolayers being formed on the bottom of the wells corresponding to 10–15 μg of cell protein per well. From this step on the cells can be used for the assay of O_2^-, H_2O_2, or NBT reduction (indicating the level of oxygen radical production of nonactivated macrophages or macrophages activated by *in vivo* procedures) or subjected to activating treatments *in vitro*.

Treatment of Macrophages with Lymphokine

As an example of an *in vitro* activating procedure we describe the treatment of macrophages with lymphokine-containing culture supernatants.[5,6] The macrophage monolayers in the microplate wells are covered with 100-μl amounts of either DMEM or different dilutions (1/8–1/64) of supernatant from concanavalin A-stimulated guinea pig lymphocyte cultures[7] or of a mock supernatant derived from cultures of unstimulated lymphocytes. FCS is added to all media at concentrations varying from 2 to 15% as dictated by the minimal amount that ensures maximal cell viability in culture. Macrophages are normally cultured for 72 hr in 95% air–5% CO_2 with medium replacement by fresh material every 24 hr. For

---

[5] M. Freund and E. Pick, *in* "Thymic Hormones and Lymphokines" (A. L. Goldstein, ed.), p. 335. Plenum, New York, 1984.

[6] M. Freund and E. Pick, *Immunology* **54**, 35 (1985).

[7] E. Pick and P. Kotkes, *J. Immunol. Methods* **14**, 141 (1977).

this purpose, we use careful vacuum suction of the spent medium, one vertical row at a time, followed by the immediate addition of fresh medium with the use of a repeating dispenser. Care is taken to avoid dislodging the monolayer by excessive suction and to prevent cell damage by "drying out" of the wells. At the completion of the lymphokine exposure, O_2^- and H_2O_2 production or NBT reduction by macrophages is measured directly in the microplates used for culture, with the cells *in situ*.

Microassay of O_2^- Production (see also this volume [3])

Principle

The assay is based on the reduction of ferricytochrome *c* by O_2^-, the specificity of reduction being controlled by its inhibition by superoxide dismutase.[8]

Reagents

Earle's or Hanks' BSS without phenol red
0.15 *M* NaCl in distilled water (saline)
Ferricytochrome *c* from horse heart, type III or VI (Sigma Chemical Company). Dissolve just prior to the assay in BSS to a concentration of 160 μM
Superoxide dismutase (SOD), from bovine blood, 3000 U/mg protein (Sigma). Dissolve in saline as a stock solution of 1–5 mg/ml and store frozen at $-20°$
Dimethyl sulfoxide (DMSO; Sigma)

Stimulants of an Oxidative Burst

This list contains agents found by us to elicit O_2^- and H_2O_2 production in guinea pig peritoneal macrophages.[9] The stimulants to be used and their relative potencies in eliciting an oxidative burst vary with animal species, tissue origin of the cells, and, in the case of exudative macrophages, with the nature of the eliciting material.

Phorbol myristate acetate (PMA; Sigma) is dissolved in DMSO at a concentration of 0.2 m*M* and kept frozen at $-70°$ in the dark in glass vials. Dilute to desired concentration in the assay medium immediately before addition to the microplates. Final concentration of DMSO in the assay should not exceed 0.1%.

Zymosan (Sigma) is opsonized with fresh serum as follows: 100 mg is suspended in 10 ml saline and heated at $100°$ in a boiling water bath for 1 hr with occasional mixing. The particles are washed three times in saline

[8] B. M. Babior, R. S. Kipnes, and J. T. Curnutte, *J. Clin. Invest.* **52**, 741 (1973).
[9] E. Pick and Y. Keisari, *Cell. Immunol.* **59**, 301 (1981).

by centrifugation at 2000 g for 5 min and incubated with 6 ml fresh serum for 20 min at 37° in a shaking water bath. The material is now washed twice with saline and resuspended in saline to 50 mg/ml. Opsonized zymosan can be stored frozen at −70°.

Concanavalin A (Con A; type IV, Sigma) and wheat germ agglutinin (WGA; Sigma) are dissolved in water to a concentration of 5 mg/ml and stored frozen in small aliquots at −20°. Dilute in assay buffer to the desired concentration.

A 10 mM stock solution of A23187 (Calbiochem-Behring Corp.) is prepared in DMSO and stored at −70° in the dark. It is diluted in the assay buffer briefly prior to addition to the microwells, making sure that final DMSO concentration does not exceed 0.1%.

N-Formyl-L-methionyl-L-leucyl-L-phenylalanine   (formyl-Met-Leu-Phe; Sigma) is dissolved in DMSO to 1 mM and the stock solution stored frozen at −70°.

Phospholipase C (type I, from *Clostridium welchii*, 20 U/mg protein; Sigma) is dissolved in saline to a concentration of 1 mg protein/ml; this stock is kept frozen at −20°.

Sodium fluoride (E. Merck) is freshly dissolved in water to a concentration of 1 M prior to addition to the assay buffer.

Digitonin (Sigma) is dissolved in absolute ethanol to a concentration of 0.5–1 mg/ml on the day of the experiment and diluted in the assay buffer making sure that the final concentration of ethanol does not exceed 0.5%.

When using guinea pig peritoneal macrophages, the stimulating agents are added to the cells at the following final concentrations: PMA, 20–200 nM; opsonized zymosan, 0.5 mg/ml; Con A, 50 μg/ml; WGA, 50 μg/ml; A23187, 2 μM, formyl-Met-Leu-Phe, 1 μM, phospholipase C, 0.1–0.2 U/ml; sodium fluoride, 10 mM, and digitonin, 2.5 μg/ml.

Procedure

The procedure described here is in essence as reported by Pick and Mizel.[10] The ELISA reader should be fitted with a 550 nm interference filter. Immediately after preparation of the fresh macrophage monolayer or at different time intervals of exposure to lymphokine *in vitro*, the cells are covered with 100 μl/well of a 160 μM solution of ferricytochrome c in phenol red-free BSS containing the stimulants. For best reproducibility, stimulants are added to batches of the cytochrome c solution before its addition to the wells. To one vertical row (normally row No. 1 in the plate) add cytochrome c containing 300 U/ml of SOD. This row serves as

[10] E. Pick and D. Mizel, *J. Immunol. Methods* **46**, 211 (1981).

reference (blank) and the absorbances read on the other wells in the plate are expressed as Δ absorbance in relation to the blanking wells.

To other wells in the plate, add the stimulant-containing cytochrome *c,* preferably in vertically oriented groups of wells. To one group of wells add cytochrome *c* without stimulant, as an indicator of the basal level of O_2^- production. Cover the plates with lids and place in a 37° humidified incubator gassed with 95% air–5% CO_2. At desired time intervals, the plates are transferred to the ELISA reader; the instrument is blanked on the row of wells containing cytochrome *c* + SOD and instructed to sequentially read the absorbances of all wells at 550 nm in reference to the blank. With most modern automatic equipment, reading of one plate with print-out of absorbance values takes about 1 min, permitting repeated reading of the same plate and the recording of approximate reaction kinetics. In this form of the assay, the blanking procedure assures automatic deduction from the absorbance values read on experimental wells of absorbance due to the plate material, the cells themselves, and to cytochrome *c* reduction caused by a material other than O_2^- (SOD resistant). This reference value can be stored in the memory of the ELISA reader and samples in subsequent microplates can be read relative to it without the need of performing the blanking procedure anew. If there is reason to believe that the stimulant might affect cell morphology or spatial orientation (such as flattening, elongation, or aggregation) or reduce cytochrome *c* or induce O_2^--independent cytochrome *c* reduction by the cells, it is recommended that for each stimulant, one group of wells containing cells, cytochrome *c,* stimulant, and SOD should serve as reference for a second group of wells containing only cells, cytochrome *c,* and stimulant. The absorbances read in this manner are closest to true SOD-inhibitable cytochrome *c* reduction values that are recorded using a double-beam spectrophotometer. The absorbance at 550 nm values are converted to nanomoles of O_2^- based on the extinction coefficient of (reduced minus oxidized) cytochrome c[11]:

$$\Delta E_{550 \text{ nm}} = 21 \times 10^3 \ M^{-1} \ cm^{-1}$$

One mole of O_2^- reduces 1 mol of cytochrome *c*. Since the vertical light path passing through 100 μl cytochrome *c* added to the wells of a Microtest III plate (Falcon) is 3 mm, we can calculate that (nmol O_2^- per well) = (absorbance at 550 nm × 15.87). This formula should be slightly adjusted when a different type of microplate is used that might result in a light path of a different length for a 100 μl/well content.

[11] V. Massey, *Biochim. Biophys. Acta* **34,** 255 (1959).

The length of time for which O_2^- production occurs at a linear rate depends on the type of macrophage, on the cell density, and on the nature of the stimulus.[12] When determination of detailed kinetics is not essential, we limit ourselves to a 60- or 90-min end-point reading but it should be understood that recalculation of data as nanomoles O_2^-/minute based on end-point readings is not legitimate.

We find that the most accurate way of relating O_2^- production to cell number is to express results as nanomoles O_2^- per milligram macrophage protein per time interval. Protein concentration is determined on 8–24 wells containing macrophages at the cell concentration used for the O_2^- assay but not carried through the assay. These wells are covered with 50 μl/well of 1 N NaOH and the plate is incubated overnight at 37° in a water vapor-saturated incubator. The following day the NaOH digest from all wells is pooled after repeated suction–ejection pipetting with a 100-μl automatic pipet and protein concentration determined by the method of Lowry et al.[13] using bovine serum albumin as protein standard. Alternatively, results can be expressed as nanomoles O_2^- per 10^6 cells per time interval. Adherent macrophages can be counted conveniently after detergent lysis and staining of nuclei, as recently described.[14]

Comments

Many variations are possible in the arrangement of samples in the plate and in the choice of the reference wells.

In most later model ELISA readers, blanking on a group of eight vertical wells or on a single well are both possible. We routinely use blanking on eight wells but single well blanking can be used when the supply of cells is limited.

We do not routinely check whether cytochrome c reduction by stimulated macrophages is SOD inhibitable because there is no evidence for cytochrome c reduction by materials other than O_2^- present on the surface of or released by *intact* macrophages. On the other hand, one has to make sure that the stimulants themselves or any other agent that is being studied by addition to the assay buffer are not reducing cytochrome c.

Some versions of ELISA readers possess the capability of dual-wavelength absorbance measurement (such as the Titertek Multiskan MC, Flow). This makes it possible to assay O_2^- production by the difference in absorbance of cytochrome c at 550 and 540 nm.[15] Since both readings are

[12] R. B. Johnston, Jr., this series, Vol. 105, p. 365.
[13] O. H. Lowry, N. J. Rosebrough, A. L. Farr, and R. J. Randall, *J. Biol. Chem.* **193**, 265 (1951).
[14] A. Nakagawara and C. F. Nathan, *J. Immunol. Methods* **56**, 261 (1983).
[15] A. Nakagawara and S. Minakami, *Biochem. Biophys. Res. Commun.* **64**, 760 (1975).

performed on each individual well, nonspecific absorbance due to plate material and cells is automatically eliminated. Blanking of the apparatus can, therefore, be effected on air, empty wells, or the assay buffer. Specificity of cytochrome c reduction can be controlled by measuring the absorbance at 550 nm minus absorbance at 540 nm value in the absence and presence of SOD. For calculation of nanomoles O_2^- values, the initiators of the method[15] recommend the use of the following molar extinction coefficient of (reduced minus oxidized) cytochrome c: $E_{550-540\ nm} = 19.1 \times 10^3\ M^{-1}\ cm^{-1}$; this can be adapted for use with 100-μl reaction volumes and a 3-mm light path.

One limitation of the method, as described, is that O_2^- production is not interrupted during measurement of the absorbance values in the ELISA reader. With stimulants that induce high rates of O_2^- production and when several plates are to be read at what theoretically should be a single time interval, the introduction of some error is unavoidable. Also, when repeated readings of the same plate are being performed, the decrease in temperature from 37° to room temperature during plate reading might disturb the kinetics of O_2^- production.

When a single fixed time reading has to be performed, the reaction can be stopped by rapidly adding 30 U SOD per well (in 10 μl) using a 12-channel dispensing pipet and mixing the content of the wells for 10 sec on the microplate shaker.

Microassay of H₂O₂ Production (see also this volume [3])

Principle

The assay is based on the horseradish peroxidase-dependent oxidation of phenol red by H_2O_2 leading to the formation of a compound that, at an alkaline pH, exhibits increased absorbance at 600 nm.[16]

Reagents

Earle's or Hanks' BSS without phenol red

Phenol sulfonphthalein (phenol red), sodium salt (Sigma), is dissolved in distilled water at 10 g/liter (28 mM) and this stock solution is stored at 4°

Horseradish peroxidase (HRPO; type II, salt-free powder, 150–200 purpurogallin units/mg; Sigma) is dissolved in 0.05 M potassium phosphate buffer, pH 7.0, to a concentration of 1000 U/ml. This stock solution is kept frozen at −20° in small aliquots sufficient for the performance of single experiments

[16] E. Pick and Y. Keisari, *J. Immunol. Methods* **38,** 161 (1980).

H_2O_2 (30%; E. Merck): A 10 mM solution is prepared in distilled water and stored at 4° in the dark. The exact molar concentration of the H_2O_2 solution is to be determined by its absorption at 230 nm using an extinction coefficient of 81 M^{-1} cm^{-1} [17]

NaOH is dissolved in distilled water to a concentration of 1 N

The assay solution is prepared on the day of the experiment and can be used for at least 24 hr. For 10 ml of complete assay reagent mix 0.2 ml phenol red stock solution (final concentration is 0.2 g/liter or 0.56 mM), 0.2 ml HRPO stock solution (final concentration is 20 U/ml), and 9.6 ml BSS.

Stimulants of an Oxidative Burst

These are prepared as and used at the concentrations indicated for the O_2^- assay.

Procedure

The procedure described here is in essence identical to the original method published by us.[10] The ELISA reader should be fitted with a 600-nm interference filter. The principal modification is the replacement of the original phosphate-buffered saline–glucose in the assay solution with either Earle's or Hanks' BSS.

Macrophage cultures in microplates are established as for the O_2^- assay and the cells are covered with 100 μl/well amounts of the phenol red-HRPO solution containing the stimulants. As in the O_2^- assay, best reproducibility is achieved when stimulants are added to batches of the assay solution before its addition to the wells. To one vertical row of wells, add 110 μl/well of assay solution made alkaline by adding 1 ml 1 N NaOH to 10 ml phenol red-HRPO reagent. This group of wells serves as reference and the blanking procedure of the ELISA reader is performed on these wells (or a single well if the apparatus has a single well blanking option). The plates are covered with lids and incubated for the desired length of time (60–90 min) at 37° in a humidified incubator in 95% air–5% CO_2. The reaction is interrupted and the plates prepared for absorbance reading by adding 10 μl/well of 1 N NaOH with a repeating dispenser. This step brings pH to the alkaline range and induces immediate cell death with consequent interruption of H_2O_2 production. No shaking of the plate is required. After a 3-min equilibration interval, the plate is now transferred to the ELISA reader. The instrument is blanked on the row of wells that were covered from the beginning with NaOH-containing assay solution and absorbances of the experimental wells read at 600 nm against the

---

[17] J. W. T. Homan-Muller, R. S. Weening, and D. Roos, *J. Lab. Clin. Med.* **85**, 198 (1975).

reference wells. Blanking, as described, automatically deducts from absorbances read in experimental wells any contribution originating from the plate material, the cell mass itself, and the absorbance contributed by the phenol red-HRPO solution in the unoxdized state. Because macrophages in monolayer do not absorb light significantly at 600 nm, a row of wells without macrophages and containing 110 μl/well of alkalinized assay solution is also quite adequate for blanking.

For expression of results as nanomoles H_2O_2 per well, standard curves are prepared using dilutions of H_2O_2 in the complete assay solution from 1 to 100 μM; 100-μl amounts of the mixtures are dispensed into the wells of a microplate (normally one vertical row of eight wells for each concentration of H_2O_2) and 10 μl of 1 N NaOH is added to each well. The plates are read in the ELISA reader using a row of wells containing 100 μl of assay solution without H_2O_2, made alkaline with NaOH, as reference. A typical standard curve is linear in the 0–90 μM H_2O_2 range corresponding to a span of absorbance values from 0 to 0.54. By using strictly the same reagent stocks for preparing the assay solution used for the standard curve and for measurement of H_2O_2 produced by cells, we found it practical to use a formula based on the extinction coefficient of oxidized minus native phenol red at 600 nm. We are currently using the following formula,

$$\text{Nanomoles } H_2O_2 \text{ per well} = (\text{absorbance at 600 nm})16.7$$

(based on $\Delta E_{600 \text{ nm}} = 19.8 \times 10^3 \, M^{-1} \, cm^{-1}$).

Results are finally expressed as nanomoles H_2O_2 per milligram macrophage protein (or per 10^6 cells) per time interval. Protein content or cell numbers are determined as described for the O_2^- assay.

Comments

The H_2O_2 microassay is comparable in reproducibility, speed, and simplicity with the O_2^- assay. It is, however, more difficult to apply to kinetic experiments because reading of results involves lysis of cells by NaOH. However, the small number of cells and reagent quantities required for the performance of the assay make the preparation of a separate set of wells for each time interval a realistic alternative.

The phenol red microassay expresses a linear relationship between absorbance at 600 nm and H_2O_2 concentration up to 90 μM H_2O_2 (= 9 nmol H_2O_2 per well). When using high cell numbers, prolonged incubation, strong stimulants, or cells that have been primed for an enhanced oxidative burst there is a risk of reaching H_2O_2 concentrations that are outside the linear range of the assay. The simplest remedy for this is to shorten the incubation time or reduce the cell number.

The concentration of HRPO used in the assay (20 U/ml) is in excess of the saturating amount as established using H_2O_2 solutions of various concentrations. On some occasions, however, the use of lesser amounts of HRPO (2–6 U/ml) for the assay of H_2O_2 produced by macrophages resulted in incomplete detection of H_2O_2.[6] We suggested that this phenomenon is explained by considerable H_2O_2 generation by macrophages taking place in intracellular vacuoles and by the fact that the phenol red-HRPO assay is capable of detecting it because of the ability of HRPO and phenol red to penetrate the vacuoles.[6]

Microassay of NBT Reduction

Principle

The assay represents a modification of the quantitative NBT reduction assay[18] for use with cells cultured in 96-well microplates. The amount of reduced NBT (precipitated formazan) is measured directly in the cells present in the wells, without prior solubilization, with the aid of an ELISA reader fitted with a 550-nm filter.

Reagents

Earle's or Hanks' BSS without phenol red
Nitroblue tetrazolium (NBT; grade III, crystalline; Sigma)
Iodoacetamide (crystalline; Sigma)

Stimulants of an Oxidative Burst

These are prepared as and used at the concentrations indicated for the O_2^- assay by cytochrome c reduction.

Procedure

The procedure described here is in essence identical to the original method published by us.[19]

Macrophage cultures in microplates are established as for the cytochrome c reduction assay. To the cells we add 100 μl/well of a 1 mg/ml solution of NBT in phenol red-free BSS containing the stimulants. The NBT solution is prepared on the day of the assay and stimulants are added to batches of the solution before it is dispensed to the wells. One vertical row of eight wells serves as reference, and in these wells the cells are preincubated for 10 min at 37° with 100 μl/well of a 10 mM solution of iodoacetamide in BSS. When the NBT solution is added to the other wells in the plate, the iodoacetamide solution in the reference wells is removed

[18] R. L. Baehner and D. G. Nathan, *N. Engl. J. Med.* **278**, 971 (1968).
[19] E. Pick, J. Charon, and D. Mizel, *J. Reticuloendothel. Soc.* **30**, 581 (1981).

by suction and replaced by a solution containing both 1 mg/ml NBT and 10 mM iodoacetamide. The plates are now covered with lids and placed in a humidified incubator at 37° in 95% air–5% CO_2 for the desired time intervals. The amount of formazan accumulating in cells is quantitated in an ELISA reader fitted with a 550-nm filter after performing the blanking procedure on the wells containing iodoacetamide-pretreated cells. NBT reduction occurs exclusively in the cells and can be followed visually by watching the progressive darkening of the macrophage monolayer. The taking of repeated readings is recommended. The frequency of these is dependent on the type of cell examined and the nature and concentration of the stimulants (see Comments). Results are expressed as absorbance per milligram macrophage protein per time interval. Cell protein concentrations are determined as described for the cytochrome c reduction assay.

A permanent record of NBT reduction can be obtained with ease by removing the NBT solution and fixing the cells in the wells by the addition of 100 μl/well of absolute methanol for 10 min followed by rinsing in distilled water and drying.

Comments

The reference wells in this procedure contain cells of which the oxidative burst is inhibited by the sulfhydryl reagent, iodoacetamide. This inhibitor has been chosen because it neither reacts with NBT nor influences macrophage morphology or surface adherence.

NBT reduction is an irreversible process and, depending on the stimulant, a plateau of absorbance is reached after specific time intervals. Prolonging incubation beyond this time will generate artifactual results because the unstimulated (control) cells also reduce NBT and will accumulate increasing amounts of formazan leading to an apparent decrease in the difference in NBT reduction between stimulated and control cells. This is the rationale for the recommendation to make repeated readings on the same plate (intervals of 20–30 min are used by us for guinea pig peritoneal macrophages).

The choice of 550 nm for reading the absorbance of intracellular formazan is based on the reported use of this wavelength for the turbidimetric measurement of formazan in aqueous media.[20] Also, native NBT shows minimal absorbance at 550 nm. Indeed, results obtained by reading plates as described (without removal of nonreduced NBT) are identical to those found by reading plates after removal of NBT-containing medium from the wells.

[20] A. W. Segal and T. J. Peters, *Lancet* **1**, 1361 (1976).

The test described here measures only intracellular NBT reduction. No reduced NBT can be detected in cell-free supernatants taken from wells containing cells with massive intracellular formazan deposits.

Results derived by using the NBT microassay are not to be equalled with those derived by the cytochrome c reduction assay. The reasons for this discrepancy are not completely understood; among the commonly suggested explanations are reduction of NBT by O_2^--independent reactions (as indicated by incomplete inhibition by SOD) and the ability of NBT and cytochrome c to detect different pools of O_2^- (intracellular and extracellular, respectively). Of special interest is the finding that the relative potencies of various stimulants in eliciting NBT reduction and cytochrome c reduction by the same cells were not correlated.[19] This is best illustrated by the fact that, in guinea pig macrophages, formyl-Met-Leu-Phe and A23187 are relatively weak stimulants when assayed by extracellular cytochrome c reduction but belong to the most potent activators of NBT reduction.[19]

Pending the satisfactory elucidation of the biochemical basis of NBT reduction by stimulated phagocytes it is advisable to look upon NBT reduction as an empirical method and not a legitimate alternative assay for O_2^- generation.

Comments Relating to All ELISA Reader-Based Assays of the Oxidative Burst of Macrophages

The three microassays described above have a number of advantages over earlier methods of measuring oxygen radical production by phagocytes.

The tests require small numbers of cells and small amounts of assay reagents. The minimal and optimal cell number has to be determined for each cell type and the 10–15 μg cell protein/well that we use with guinea pig macrophages is only a rough indicator.

The most significant advantage of all three microassays is their simplicity and speed that, together with the small numbers of cells required, permit the performance of a large number of assays and numerous replicates for each experimental situation. In addition, the cytochrome c reduction and NBT reduction assays are appropriate for making time kinetic measurements. The microassays, therefore, combine the advantages of the discontinuous (fixed time) method[12] with those of the continuous assay.[21]

The simplicity of the methodics together with the capacity of assaying numerous samples from a relatively small pool of cells allows the perfor-

[21] M. Markert, P. C. Andrews, and B. M. Babior, this series, Vol. 105, p. 358.

mance of O_2^- and H_2O_2 measurements and NBT reduction assays in parallel. This capability is especially valuable when one is interested in studying the effect of a variety of activating agents or inhibitory drugs or numerous concentrations of the same agent on the same batch of cells. Also, when measurements are to be performed on cells maintained in culture, the fact that the assays are executed with cells *in situ* in microplates that also serve as culture vessels provides an additional element of simplicity.

Another benefit derived from the use of an ELISA reader for the recording of absorbance values is the ability to link up with microcomputers offering a variety of sophisticated data analysis modes.

There seems to be no limitation to the type of cells that can be investigated either in the surface-adherent state or in suspension. From the time that the microassays were first described for use with guinea pig macrophages, they were applied to human monocytes,[22–24] peritoneal macrophages,[23] neutrophils[22,25] and eosinophils[26]; and to mouse macrophages[27–29] and rat neutrophils.[30] Recently, a modified O_2^- microassay was successfully applied to the measurement of O_2^- production by membrane-bound NADPH oxidase derived from a cell homogenate.[31]

We found the microassays especially valuable in the assessment of the oxidative metabolism of lymphokine-activated macrophages. It is highly recommended to assay both O_2^- and H_2O_2 production and, if possible, NBT reduction because the information gained by using a single method does not reflect all the changes in oxygen radical production associated with macrophage activation. Thus, it is a common finding that activated macrophages produce disproportionately larger amounts of H_2O_2 than O_2^-.[6,32] Such a situation is, probably, the result of the fact that a large proportion of the O_2^- is more accessible to dismutation than to the extra-

[22] B. J. Goodwin and J. B. Weinberg, *J. Clin. Invest.* **70**, 699 (1982).
[23] J. B. Weinberg and A. F. Haney, *JNCI, J. Natl. Cancer Inst.* **70**, 1005 (1983).
[24] H. C. Stevenson, E. Bonvini, T. Favilla, P. Miller, Y. Akiyama, T. Hoffman, R. Oldham, and D. Kanapa, *J. Leuk. Biol.* **36**, 521 (1984).
[25] D. E. van Epps and D. E. Chenoweth, *J. Immunol.* **132**, 2862 (1984).
[26] L. Prin, J. Charon, M. Capron, P. Gosset, H. Taelman, A. B. Tonnel, and A. Capron, *Clin. Exp. Immunol.* **57**, 735 (1984).
[27] A. K. Sharp and M. J. Colston, *Eur. J. Immunol.* **14**, 102 (1984).
[28] C. P. Sung, C. K. Mirabelli, and A. M. Badger, *J. Rheumatol.* **11**, 153 (1984).
[29] B. P. Barna, S. D. Deodhar, S. Gautam, B. Yen-Lieberman, and D. Robert, *Cancer Res.* **44**, 305 (1984).
[30] K. O. Fennell, H. R. Creamer, W. L. Gabler, A. C. Brown, and W. W. Bullock, *Inflammation (N.Y.)* **8**, 287 (1984).
[31] Y. Kobayashi, M. Asada, and T. Osawa, *J. Biochem. (Tokyo)* **95**, 1775 (1984).
[32] A. Nakagawara, N. M. de Santis, N. Nogueira, and C. F. Nathan, *J. Clin. Invest.* **70**, 1042 (1982).

cellular cytochrome c reagent.[33] When H_2O_2 is generated by enzymatic dismutation, treating macrophages with an SOD inhibitor such as diethyl dithiocarbamate (for Cu,Zn-SOD)[33] or sodium nitroprusside (for all types of SOD)[34] will result in a switch from H_2O_2 to O_2^- liberation.

The value of the NBT reduction assay is that it is the only one to provide information on intracellular O_2^- production. Its use in the past was limited by the semiquantitative nature of the microscopic assay and the cumbersomeness of earlier quantitative methods.

The use of multiple oxidative burst stimulants in monitoring oxygen radical production by activated macrophages is recommended and can be managed with relative ease by employing the microassays. Activated macrophages might exhibit enhanced O_2^- or H_2O_2 production in response to some stimulants but not to others. An illustration of this is the enhanced O_2^- production of lymphokine-activated guinea pig macrophages in response to WGA, formyl-Met-Leu-Phe, and phospholipase C, but not to a group of five other stimulants that includes the commonly used PMA and opsonized zymosan.[5] This variability is, most likely, the expression of distinct transductional pathways being responsible for the activation of the O_2^--forming enzyme by different stimulants[35–37] and suggests that macrophage activation leads to changes in the receptor–enzyme coupling mechanism and not in the actual O_2^- forming enzyme.

Acknowledgments

The work reported in this chapter was supported by grants from the U.S.–Israel Binational Science Foundation (Nos. 1505 and 2730), the Arpad Plesch Research Foundation, and the Deutsches Krebsforschungszentrum, Heidelberg. We thank Mrs. Patricia Bar-On for expert secretarial assistance.

[33] S. Tsunawaki and C. F. Nathan, *J. Biol. Chem.* **259**, 4305 (1984).
[34] M. Freund and E. Pick, unpublished observation (1985).
[35] L. C. McPhail, P. Henson, and R. B. Johnston, Jr., *J. Clin. Invest.* **67**, 710 (1981).
[36] L. C. McPhail and R. Snyderman, *J. Clin. Invest.* **72**, 192 (1983).
[37] Y. Bromberg and E. Pick, *Cell. Immunol.* **79**, 240 (1983).

[25] Determination of Adenylate Cyclase and Guanylate Cyclase Activities in Cells of the Immune System

By Chandra K. Mittal

Adenylate cyclase and guanylate cyclase catalyze the formation of adenosine 3',5'-monophosphate (cyclic AMP) and guanosine 3',5'-monophosphate (cyclic GMP), respectively, from ATP and GTP in the presence of a metal.

$$\text{ATP} \xrightarrow[\text{Mn}^{2+}/\text{Mg}^{2+}]{\text{adenylate cyclase}} \text{cyclic AMP} + \text{PP}_i \text{ (pyrophosphate)}$$

$$\text{GTP} \xrightarrow[\text{Mn}^{2+}/\text{Mg}^{2+}]{\text{guanylate cyclase}} \text{cyclic GMP} + \text{PP}_i \text{ (pyrophosphate)}$$

With the exception of testes, mammalian adenylate cyclase is exclusively membrane bound and guanylate cyclase is distributed in particulate and cytosolic compartments. Although the two cyclases represent two entirely independent and distinct biological entities,[1] there is a structural analogy between the substrates (ATP and GTP) and the products (cyclic AMP and cyclic GMP) of the reactions they catalyze. Hence, the methods developed to assay adenylate cyclase and guanylate cyclase activities are also parallel but not identical. This chapter focuses on two highly sensitive assay methods applicable to both enzyme systems. The radiometric assay using [α-^{32}P]ATP or [α-^{32}P]GTP is extremely rapid, whereas the second method employing nonlabeled substrate and radioimmunoassay (RIA)[1a] of the product is relatively slow. It, however, offers the most sensitive, convenient, and least expensive method to measure cyclic nucleotides in adenylate and guanylate cyclase incubations. Both procedures have been successfully employed to study cyclic nucleotide metabolism in polymorphonuclear leukocytes, macrophages, lymphocytes, basophils, and mast cells. The method to be chosen is based on the level of enzyme activity to be investigated. This investigator has utilized the latter method for several years to study guanylate cyclase and adenylate cyclase activities[1] in various mammalian systems including polymorphonuclear leukocytes (PMNs).

---

[1] C. K. Mittal and F. Murad, in "Handbook of Experimental Pharmacology" (J. A. Nathanson and J. W. Kebabian, eds), Vol. 58, Part I, p. 225. Springer-Verlag, Berlin and New York, 1982.

[1a] Abbreviations: PMNs, polymorphonucleates; DTT, dithiothreitol; MIX, 1-methyl-3-isobutylxanthine; TCA, trichloroacetic acid; RIA, radioimmunoassay.

Preparation of Adenylate Cyclase and Guanylate Cyclase

PMNs, macrophages, or lymphocytes are sonicated with Polytron at 0° in 0.25 M sucrose containing 20 mM Tris–HCl (pH 8.0), 1 mM EDTA, and 1 mM dithiothreitol (DTT). For preparation of these cells see this series, Vol. 108 [9], [25], [26], [27], [28], and [29]. Homogenates can be either directly used as the source of the enzymes or, based on need, can be further fractionated to prepare crude/purified membranes and cell-free supernatants by standard centrifugation techniques. Methods are also available to purify adenylate cyclase and guanylate cyclase from mammalian tissues. Recently homogeneous preparations of guanylate cyclase from rat liver and lung cytosolic fractions[2,3] have been reported. For an elaborate discussion on the purification of guanylate cyclases, readers are referred to these reports.

Radiometric Assays

Principle. After incubation with the cyclase the excess substrate [α-[32P]ATP or [α-[32P]GTP is separated from the product cyclic[[32P]AMP or cyclic[[32P]GMP. This procedure[4] combines chromatography of nucleotides on Dowex-50 columns[5] and neutral alumina (aluminum oxide).[6] The amount of labeled cyclic nucleotide formed is then determined.

Adenylate Cyclase

Reagents

ATP, 100 mM
[α-[32P]ATP (10–50 Ci/mmol)
Cyclic AMP, 20 mM
Cyclic [8-[3H]AMP (5–10 Ci/mmol)
Creatine phosphate, 150 mM in 10 mM Tris–HCl, pH 7.5, containing creatine phosphokinase (600 U/ml)
1-Methyl-3-isobutylxanthine, 4 mM
Trichloroacetic acid, 30%
Dithiothreitol, 20 mM
Tris–HCl, 0.6 M, pH 7.5
$MgCl_2$, 0.2 M

[2] J. M. Baughler, C. K. Mittal, and F. Murad, *Proc. Natl. Acad. Sci. U.S.A.* **76**, 219 (1979).
[3] D. L. Garbers, *J. Biol. Chem.* **254**, 240 (1979).
[4] Y. Salomon, C. Londos, and M. Rodbell, *Anal. Biochem.* **58**, 541 (1974).
[5] G. Krishna, B. Weiss, and B. B. Brodie, *J. Pharmacol. Exp. Ther.* **163**, 379 (1968).
[6] A. A. White and T. V. Zenser, *Anal. Biochem.* **41**, 372 (1971).

Dowex-50 columns: Dowex AG 50W-X4 (200–400 mesh) from BioRad Labs is thoroughly washed in deionized water. Two milliliters of 50% resin (v/v) in water is poured over 0.5 × 15 cm columns with sintered glass bottoms. The columns are washed with 2 ml of 1 N HCl followed by water before use. The columns can be stored after acid treatment at room temperature. The resin in the column can be recycled for use by washing with 1 N HCl. Occasionally columns are washed with 1 N NaOH to dissolve the deposited protein followed by 1 N HCl treatment

Neutral alumina columns: Dry neutral alumina WM-3 (0.6 g) is put at the bottom of a glass wool-supported 0.5 × 15 cm column. The resin is washed with 10 ml of 1 M imidazole–HCl, pH 7.3, followed by 20–25 ml of 0.1 M imidazole buffer, pH 7.3, prior to use. These columns are also reusable

Procedure. Typically adenylate cyclase activity is determined in 12 × 75 mm disposable culture tubes in a total volume of 100 μl containing 5 μl of 0.6 M Tris–HCl, pH 7.5, 5 μl of 0.2 M MgCl$_2$, 5 μl of 150 mM creatine phosphate containing 3 U of creatine phosphokinase, 5 μl 4.0 mM 1-methyl-3-isobutylxanthine (MIX), 5 μl of 20 mM cyclic AMP, 5 μl 20 mM DTT, 5 μl of 10 mM ATP, and 10 μl [α-^{32}P]ATP (2–4 × 10^6 cpm). Other desired modulators, such as GTP, sodium fluoride, catecholamines, or other receptor-selective drugs, are added at the desired concentrations and/or glass-distilled water is added to obtain the total volume of 100 μl. Alternatively, with the exception of radioactive ATP and the modulators, other reaction ingredients can be mixed together in equal volumes and then 35 μl of this reaction cocktail is pipetted into the tubes. The reaction is initiated by adding the enzyme preparation, containing 15–20 μg membrane protein, to the prewarmed reaction mixture. After 10–15 min of incubation at 30°, the reaction is terminated by adding 25 μl of 30% ice-cold trichloroacetic acid (TCA). To a zero-time blank the TCA is added before the membrane preparation. A no-enzyme blank is prepared by substituting the membranes with 10 mM Tris–HCl, pH 7.5, otherwise treated as a regular incubation. To each assay tube 50 μl of cyclic[^{3}H]AMP (approximately 20,000 cpm) is added in order to monitor the recovery of cyclic[^{32}P]AMP formed during the enzyme incubation.

The isolation of labeled cyclic AMP is achieved by the procedure of Salomon *et al.*[4] To each incubation tube 0.8 ml of water is added, the contents are mixed, and poured over a column (0.5 × 15 cm) containing 1 ml Dowex-50 resin in water. The flow-through is discarded. The column is washed twice with 1 ml of water and the eluate is again rejected. This step removes most of the [α-^{32}P]ATP present in the reaction mixture. Labeled cyclic AMP is then eluted with 3 ml of water and collected in 13 × 100 mm test tubes. To each tube 0.2 ml of 1.5 M imidazole–HCl, pH 7.2, is added and the tube contents are poured onto 0.5-cm columns containing 0.6 g neutral alumina freshly prewashed with 8 ml of 0.1 M imidazole–HCl, pH 7.5. The flow-through is collected in a scintillation vial containing 14 ml of Aquasol. The column is washed with an additional 1 ml of 0.1 M imidazole buffer and the eluate is collected in the same vial.

Alternatively, if higher sensitivity is not needed, the collection of labeled cyclic AMP from Dowex-50 in tubes after [α-^{32}P]ATP removal can be avoided. Instead, the Dowex columns can be placed over a rack containing the alumina columns prewashed with 0.1 M imidazole–HCl, pH 7.3. Three milliliters of water is applied to the upper Dowex-50 column and the effluent is allowed to pass through the alumina column (lower) and flow-through is discarded. The alumina column rack is then placed over scintillation vials containing 15 ml of Aquasol and labeled cyclic AMP is eluted directly into the vials with 4 ml of 0.1 M imidazole buffer, pH 7.3. Vials are shaken vigorously and the radioactivity is determined.

Although no significant differences in the elution patterns have been observed in the past, it is always advisable that with each new batch of resin to be employed in assays the investigator must run a simple elution profile of labeled nucleotides under the exact experimental conditions, excluding the enzyme preparation. This will approximate the elution volume for a particular batch and help avoid the use of a defective resin. In view of the narrow resolution of Dowex-50 columns this experimental check will serve an extremely useful purpose.

Guanylate Cyclase

The method to determine guanylate cyclase activity described below employs a chromatographic procedure similar to that of adenylate cyclase but independently developed by Krishna and Krishnan.[7] It has been successfully employed by investigators.

Reagents

Tris–HCl, 1 M, pH 7.5
GTP, 100 mM
MgCl$_2$, 80 mM, or MnCl$_2$, 80 mM
Cyclic[8-^{3}H]GMP (5–15 Ci/mmol)
[α-^{32}P]GTP (10–50 Ci/mmol)
Theophylline, 40 mM, or 1-methyl-3-isobutylxanthine, 2 mM
Cyclic GMP, 20 mM
Creatine phosphate, 150 mM
Creatine phosphokinase

---

[7] G. Krishna and N. Krishnan, *J. Cyclic Nucleotide Res.* **1**, 293 (1975).

Procedure. The reaction mixture for a typical guanylate cyclase assay consists of 5 μl 1 M Tris–HCl, pH 7.5, 5 μl 80 mM MnCl$_2$ or MgCl$_2$, 25 μl 40 mM theophylline or 2.0 mM 1-methyl-3-isobutylxanthine, 5 μl 150 mM creatine phosphate containing 20 μg creatine phosphokinase (120–135 U/mg protein), 5 μl 20 mM cyclic GMP, 10 μl 10 mM GTP, 10 μl [α-^{32}P]GTP (containing 2–4 $\times$ 10^6 cpm), and various modulators (if desired) or H$_2$O. In a total volume of 100 μl the reaction is initiated by the addition of a particulate or cytosolic enzyme preparation (up to 50 μg protein). After 10–15 min of incubation at 37°, the reaction is terminated by adding 20 μl of 40% ice-cold TCA. Cyclic [8-^{3}H]GMP (10 μl) containing about 20,000 cpm is added to each assay tube. Samples are then transferred to an ice-water bath for about 15–20 min. To purify cyclic GMP, the tube contents are diluted with 0.5 ml of water and applied to 0.5 $\times$ 15 cm columns containing 1 ml Dowex-50 H$^+$ resin. The effluent is discarded. The column is washed with 0.5 ml water. Cyclic GMP is then eluted with 1 ml water and further purified on an alumina column like cyclic AMP in the adenylate cyclase assay as described above.

Nonradiometric Assays

Principle.[8] Nonradioactive nucleoside triphosphates (ATP or GTP) are incubated with the enzyme preparations and the produced cyclic nucleotides (cyclic AMP or cyclic GMP) are quantitated by radioimmunoassay (RIA). The sensitivity of the assay can be enhanced 50- to 100-fold by acetylation[9] of the cyclic nucleotides. This increases the sensitivity to the femtomole (10^{-15} M) range.

Guanylate Cyclase

Reagents (generally the same as described for the radiometric assay)
Sodium acetate buffer, 1 M, pH 4.0
Triethylamine
Acetic anhydride
Procedure. The reaction cocktail (100 μl total volume) consists of the same ingredients as those used for the radiometric assay for guanylate cyclase, except that the inclusion of unlabeled cyclic GMP is omitted and nonradioactive GTP is substituted for [α-^{32}P]GTP as the substrate. In this laboratory reactions are generally initiated by the addition of 10 μl of 10 mM GTP and incubations are carried out for 10 min at 37°. Reactions are stopped by adding 0.9 ml of ice-cold 50 mM sodium acetate buffer (pH 4.0) and placing the tube rack in a hot water bath (90°) for 3 min. Use of

[8] C. K. Mittal and F. Murad, *J. Biol. Chem.* **252**, 3136 (1977).
[9] J. F. Harper and G. Brooker, *J. Cyclic Nucleotide Res.* **1**, 207 (1975).

sodium acetate as a stopping solution has proved to be satisfactory for the radioimmunoassay. In addition, with this buffer system at pH 4.0 any nonenzymatic formation of cyclic GMP is avoided during the heating step.[10] Tubes are transferred to an ice-water bath to complete the denaturation of proteins. Nonenzyme and boiled enzyme controls are concurrently run with each assay. The heated tube contents are centrifuged at 2000 g for 15 min at room temperature to precipitate the denatured protein and the clear supernatant fractions are utilized for the determination of cyclic GMP. The radioimmunoassay for cGMP is described elsewhere in this volume.

Under basal conditions the rates of the guanylate cyclase reactions are very low when Mg^{2+} is used as the cofactor. The amount of cyclic GMP formed is about 10–15% of that formed with Mn^{2+}. In such cases the cyclic GMP should be acetylated in order to increase the sensitivity of the assay. The anti-cyclic nucleotide antibody to be employed in the RIA must be checked for cross-reactivity with GTP (because of its relatively high concentrations) as well as for any interference in the antigen–antibody reaction by constituents of the cyclase reaction mixture. The presence of interfering substances such as high concentrations of chelators, ions, and enzyme modifiers can invalidate the cyclic nucleotide measurements. Generally this can be overcome by diluting the sample. However, any persistent interference will necessitate the purification of the cyclic nucleotide. This can be achieved by chromatography on a Dowex-1 (AG1-X8) column as described elsewhere in this volume [26].

Adenylate Cyclase

The procedure described for the assay of guanylate cyclase is also applicable to determine adenylate cyclase activity. The reaction mixtures contain the same ingredients as those described for the radiometric assay of adenylate cyclase with two exceptions. Unlabeled cyclic AMP is completely excluded from the reaction mixture and nonradioactive ATP is employed as the substrate instead of [α-^{32}P]ATP.

Characterization of Product Formed by Adenylate and
   Guanylate Cyclase

Before undertaking detailed investigations on adenylate cyclase and guanylate cyclase activities it is essential to determine the authenticity of the product and the validity of the procedures employed to measure these products. This is especially important for the radiometric assays where the authenticity of the ^{32}P-labeled cyclic nucleotide formed is solely based

---

[10] H. Kimura and F. Murad, *J. Biol. Chem.* **249**, 329 (1974).

on the cochromatographic pattern of the product with 8-^{3}H-labeled cyclic nucleotide. This should be verified by independent criteria as outlined by the following procedures.

Neutral alumina eluate containing radiolabeled cyclic AMP or cyclic GMP is lyophilized and reconstituted in 200 μl of 50 mM Tris–HCl, pH 7.5; 100 μl of this cyclic nucleotide solution is incubated with 100 μl of 8.0 mM MgCl$_2$ and 10 μg of a commercial preparation of beef heart phosphodiesterase for 10 min at 37°. The remaining 100 μl is incubated as a control without the enzyme. The reaction is stopped by heating the tubes in a boiling water bath. Tube contents are cooled, diluted with water to 1 ml, and applied to 0.5 × 6 cm Dowex-1 chloride columns. After washing with H$_2$O the columns are sequentially eluted with HCl solutions of pH 2.7, 2.0, and 1.3 and the radioactivity is determined. Precise details of this procedure are given elsewhere in this volume [26]. Disappearance of the [32]P- and ^{3}H-labeled cyclic nucleotide peak after phosphodiesterase treatment will verify the existence of a cyclic nucleotide in the neutral alumina eluates.

Additional verification may be obtained by stopping the reaction, after phosphodiesterase incubation, by adding 0.8 ml 50 mM sodium acetate buffer (pH 4.0) followed by heating at 90° for 3 min. Then a RIA can be performed on the samples to determine the immunoreactivity of the cyclic nucleotide in neutral alumina eluates [26]. RIA being a competitive assay, a lack of dilution in the ^{125}I-labeled ligand binding to antibody in the enzyme-treated sample versus nontreated will further support cyclic nucleotide being the product of a cyclase reaction.

[26] Measurements of Cyclic Adenosine Monophosphate and Cyclic Guanosine Monophosphate Levels in Polymorphonuclear Leukocytes, Macrophages, and Lymphocytes

By Chandra K. Mittal

Intracellular levels of cyclic AMP and cyclic GMP are altered in response to a variety of stimuli,[1] including mitogens, calcium, neurohormones, autacoids, prostaglandins, oxygen radicals, nitro drugs, and nu-

[1] C. K. Mittal and F. Murad, in "Handbook of Experimental Pharmacology" (J. A. Nathanson and J. W. Kebabian, eds.), Vol. 58, Part I, p. 225. Springer-Verlag, Berlin and New York, 1982.

cleophilic substances. Some of these stimuli affect proliferation, phagocytosis, chemotaxis, and lysosomal enzyme release in neutrophils, macrophages, and lymphocytes. Although a variety of methods, including activation of cyclic nucleotide-specific protein kinases,[2,3] enzymatic conversion of cyclic nucleotides,[4] or protein-binding assays,[5,6] are available to determine cyclic nucleotide contents in tissues and cells, these are not the methods of choice for systems like neutrophils, macrophages, or lymphocytes because of their relatively low cyclic nucleotide levels. This chapter describes the radioimmunoassay (RIA) that offers the most sensitive device for cyclic nucleotide measurements in these types of cells. A procedure is also described to determine changes occurring in cyclic nucleotide levels by measuring the intracellular pool of radioactive cyclic nucleotides generated from a radiolabeled precursor.

Extraction of Cyclic Nucleotides from Cells

Polymorphonucleates, lymphocytes, and macrophages are prepared using techniques described in this series, Vol. 108 [9], [25], [26], [27], and [28].

Before intracellular levels of cyclic nucleotides can be determined at steady state in a cell suspension, it is essential to abruptly terminate the metabolic reactions and extract the nucleotides. This is achieved by homogenizing the cells in 8–10% trichloroacetic acid $(TCA)^{6a}$ at 0°. To a typical 1-ml cell suspension, 0.5 ml of 30% ice-cold TCA is added. Cells are immediately sonicated with a Polytron and are allowed to stand for 15–20 min to complete deproteinization. To each sample 2000 cpm of ^3H or ^{14}C-labeled cyclic GMP or cyclic AMP or both (depending on the ones needed) are added as an internal standard to monitor the recovery of these nucleotides in the extraction and purification steps. Samples are centrifuged at 0° to separate denatured protein from the TCA extracts. The protein is dissolved in 1 N NaOH and determined by the procedure of Lowry et al.[7] using bovine serum albumin as standard. To the TCA ex-

[2] J. F. Kuo and P. Greengard, J. Biol. Chem. 245, 2493 (1970).

[3] J. F. Kuo, T. P. Lee, P. L. Reyes, K. G. Walton, T. E. Donnelly, Jr., and P. Greengard, J. Biol. Chem. 247, 16 (1972).

[4] G. Schultz, J. G. Hardman, K. Schultz, J. W. Davis, and E. W. Sutherland, Proc. Natl. Acad. Sci. U.S.A. 70, 1721 (1973).

[5] A. G. Gilman, Proc. Natl. Acad. Sci. U.S.A. 67, 305 (1970).

[6] F. Murad, V. Manganiello, and M. Vaughan, Proc. Natl. Acad. Sci. U.S.A. 68, 736 (1971).

[6a] Abbreviations: RIA, radioimmunoassay; PCA, perchloric acid; TCA, trichloroacetic acid; TME, tyrosine methyl ester; ScAMP, succinyl cyclic AMP.

[7] O. H. Lowry, N. J. Rosebrough, A. L. Farr, and R. J. Randall, J. Biol. Chem. 193, 265 (1951).

tract 80 μl of 1 N HCl is added to yield a final concentration of 0.05 N. This protonates the TCA so that its removal by diethyl ether is facilitated. TCA is removed from the acidified supernatant fraction by five successive extractions with 4 vol of water-saturated ether. Prior water saturation of ether is essential in order to prevent loss of the aqueous layer containing cyclic nucleotides. Trace amounts of ether left in the samples are removed by aspirating the vapors after heating the tubes in a water bath at 60°. TCA-free extracts are stored under acidic conditions to prevent hydrolysis. Following neutralization of the HCl with 80 μl of 1 N NaOH and the addition of 80 μl 1 M sodium acetate buffer, pH 4.0, these extracts can be directly used to assay for cyclic nucleotides. In case any interference is encountered in the assay, a further separation and purification of cyclic nucleotide is essential. These procedures are described below.

Perchloric acid (PCA), 2 M, can also be used instead of TCA to extract the cyclic nucleotides from cells. PCA is quantitatively removed by the addition of an equimolar concentration of KOH or $KHCO_3$. However, in our laboratory TCA is routinely employed for the extraction of nucleotides.

Separation and Purification of Cyclic Nucleotides

Whole-cell extracts, prepared from cultured cells, contain nucleotides, ions, and electrolytes from the medium (Krebs–Ringer, Hanks', etc.) and other additives. These substances can potentially interfere with the cyclic nucleotide quantitation procedures. Additionally, if there is cross-interference between the two cyclic nucleotides themselves in the assay, it is necessary to separate them. Separation of cyclic nucleotides and purification can be achieved simultaneously. Among the procedures developed for this purpose, the use of ion-exchange chromatography has gained wide acceptance. Both anion- and cation-exchange resins have proved useful. Over a period of several years this technique has yielded reproducible results regardless of the batch of resin or the origin of the biological sample.

Anion-Exchange Chromatography. Anion-exchange resin Dowex-1 chloride (AG 1-X2, 50–100 mesh) from BioRad Laboratories has been widely used for this purpose. Glass columns with a sintered glass bottom can be obtained commercially or, alternatively, the columns can be made from pieces of glass tubing with one end constricted. A plug of glass wool is inserted to retain the resin. The resin is packed in the column to a height of 6 cm. Uniformity in the column size is essential to assure uniform elution of the cyclic nucleotides. The flow rate of a 0.5 × 6 cm Dowex-1 chloride column is 2.5 to 3.0 ml/min. The resin is washed with water until the pH of the wash is approximately 5.0. The column is then equilibrated with 5 ml 0.02 M Tris–HCl buffer, pH 7.4. The acidified biological extract (containing 0.05 N HCl) is neutralized with NaOH and applied to the column. The flow-through is discarded and the resin is rinsed twice with 5 ml water. The column is then washed with 30 ml HCl, pH 2.7. Following this wash, first the cyclic AMP is eluted with 20 ml HCl, pH 2.0, and then cyclic GMP with 20 ml HCl, pH 1.3. The purified fractions containing individual cyclic nucleotides are lyophilized and reconstituted in 1 to 1.5 ml 50 mM sodium acetate buffer, pH 4.0, for the assay. The recovery of labeled (^{3}H, ^{14}C) cyclic AMP or cyclic GMP is approximately 60–80% using AG 1-X2 chloride columns.

Use of Dowex-1 formate (AG 1-X8, 200–400 mesh) as the anion-exchange resin for the purification of cyclic nucleotides is also common.[6] Briefly, the ether-extracted and neutralized (as above) sample is applied to a 0.5 × 3 cm column. The column is washed with 10 ml redistilled water. Cyclic AMP is eluted with 10 ml 2 N formic acid and cyclic GMP with 12 ml 4 N formic acid and the samples are lyophilized. This purification method results in a complete separation and a quantitative recovery (80–95%) of each of the cyclic nucleotides.

Cation-Exchange Chromatography. Alternatively, cyclic nucleotides in TCA extracts can also be separated on a cation-exchange resin. A 0.6 × 5 cm column of Dowex-50 H$^+$ (AG 50W-X8, 100–200 mesh) is conveniently prepared in a disposable Pasteur pipet. The resin is equilibrated with 0.1 N HCl. One milliliter of a TCA extract is directly applied onto the column. The resin is washed with 4 ml 0.1 N HCl followed by 1 ml water and the eluate is discarded. With an additional 3–4 ml of water, both cyclic AMP and cyclic GMP are eluted. These fractions are pooled, lyophilized, and reconstituted as described above. Unlike Dowex-1, chromatography on Dowex-50 does not require a prior removal of TCA from the biological extract and the recovery of cyclic nucleotides is 60–80%. Although this procedure does not significantly separate cyclic AMP from cyclic GMP, it does remove large pools of related nucleotides. Dowex-50 chromatography can be successfully applied to determine radiolabeled intracellular pools of an individual cyclic nucleotide.

In general, the samples purified with either of the chromatographic procedures are suitable for direct quantitation of cyclic nucleotides by radioimmunoassay, but not for the determination of intracellular pools of radiolabeled cyclic AMP or cyclic GMP generated from radioactive adenine or guanine. Usually, the purified fractions still contain some quantities of related radiolabeled mono- and dinucleotides, especially those chromatographed over Dowex-1.[6] The separation of cyclic nucleotides from the contaminating nucleotides is achieved by adding 100 μl each of

0.25 M ZnSO$_4$ and 0.25 M Ba(OH)$_2$ to 1 ml neutralized eluate from the ion-exchange columns.[8] The precipitate is mixed and centrifuged. The supernatant is transferred to scintillation vials for the determination of radioactivity of radiolabeled cyclic nucleotide.

Radioimmunoassay (RIA) of Cyclic Nucleotides

Principle. In this assay fixed amounts of radiolabeled ligand and anti-cyclic nucleotide specific antibody are incubated with nonradioactive cyclic nucleotide. Because of the competition between labeled and nonlabeled ligand for limited antibody sites, there is a decrease in the binding of radioactive ligand to the antibody proportional to the amount of unlabeled cyclic nucleotide present. The antigen–antibody complex is precipitated with alcohol and the radioactivity is determined.

Preparation of ^{125}I-Labeled Succinyl Cyclic Nucleotide Tyrosine Methyl Esters. The cyclic nucleotide derivative is iodinated with the method of Hunter and Greenwood as adapted by Steiner et al.[9] A stock solution of commercially available (Sigma) succinyl cyclic AMP tyrosine methyl ester (ScAMP-TME) or succinyl cyclic GMP tyrosine methyl ester (ScAMP-TME) is prepared in distilled water (2.0 mg/ml). The stock solution can be stored at $-70°$ for 4 years without any change. The working stock solutions are diluted to 0.2 mg/ml and stored in 100-μl aliquots at $-20°$. Commercially supplied Na^{125}I (0.5–1.0 mCi/10–20 μl) is placed in a small glass conical centrifuge tube. It is diluted with 20 μl of 0.5 M potassium phosphate buffer, pH 7.0. Succinylated derivative (20 μl) is added to the radioactive mixture. Iodination is initiated with the addition of 10 μl of chloramine-T (0.5 mg/ml in 0.5 M potassium phosphate, pH 7.0). The tube is capped, the contents are immediately mixed, and the reaction is allowed to continue for 60 sec. The reaction is stopped by adding 50 μl of sodium metabisulfite (5 mg/ml in water). *Caution:* The whole procedure should be carried out behind lead bricks in an efficiently vented fume hood, with the counter tops covered with plastic-backed absorbant sheets. Personnel must wear disposable latex gloves.

Iodinated cyclic nucleotide is purified by paper chromatography. The reaction mixture is streaked onto a Whatman 31 ET paper strip (2 × 40 cm). After drying the strip, descending paper chromatography is performed using 1-butanol:water:glacial acetic acid (12:5:3). The solvent front is allowed to run to about 30 cm. The paper is then allowed to dry and the strip is cut into 1-cm pieces. Each section of paper is inserted into a 1-cm^3 syringe and eluted with 1.5 ml of 50 mM sodium acetate, pH 4.0

[8] G. Krishna, B. Weiss, and B. B. Brodie, *J. Pharmacol. Exp. Ther.* **163**, 379 (1968).
[9] A. L. Steiner, C. W. Parker, and D. M. Kipnis, *J. Biol. Chem.* **247**, 1106 (1972).

(0.5 ml to be used three times). The radioactivity of each eluent is then determined. Three peaks of radioactivity are observed on the chromatogram. The second peak (with R_f 0.6) represents ^{125}I-labeled succinyl cyclic nucleotide tyrosine methyl ester. Before this ligand is employed in the RIA it is advisable to dilute a small aliquot from each eluent to yield 10,000–20,000 cpm/100 μl and to determine the immunoreactivity of the radioactive ligand. This is accomplished by incubating the ligand alone with the cyclic nucleotide specific antibody and determining the extent of specific binding as described below. Fractions exhibiting 35–60% specific binding are most suitable for use. An equal volume of 1-propanol is added to these fractions and the mixtures are stored at $-20°$. This is to avoid disintegration of the ligand by repeated freezing and thawing. Because of the ability of 1-propanol to denature and/or precipitate protein components of the RIA system, the minimal dilution of the stock ligand is recommended only to the extent that no more than 5% propanol is carried in the RIA incubations.

Acetylation of Cyclic Nucleotides. For cells and tissue samples with low amounts of cyclic nucleotides a higher sensitivity of the immunoassay is desirable and this can be achieved by acetylation.[10] This derivatization enhances the affinity of the cyclic nucleotide for antibody by 50- to 100-fold and makes their measurement possible in the femtomole (10^{-15} M) range. To 0.5 ml of a TCA-free acidic extract or ion-exchange chromatographed biological sample, first 10 μl of triethylamine and then immediately 5 μl of acetic anhydride are added. The time lapse between these two additions should not exceed 4–5 sec in order to protect the cyclic nucleotide from hydrolysis at the alkaline pH of triethylamine. This can be conveniently accomplished by acetylating four to five samples at a time. At the end the pH of the biological sample is about 5.0. The standard solution of cyclic nucleotide is diluted to concentrations ranging from 1 to 1000 fmol/100 μl as described below and these standard solutions are also acetylated. Standards and unknowns are always freshly acetylated together. They gradually get deacetylated with time. In our laboratory insignificant changes have been observed in samples stored at $-20°$ for 10 days. However, in such cases standards should also be age matched.

Immunoassay Procedure. The assay at the femtomole level is essentially identical to that at the picomole level, except that in the latter case standards and unknowns are not acetylated. Molar extinction coefficients of 14,100 (258 nm) and 12,900 (254 nm) are used to determine the concentrations of the cyclic AMP and cyclic GMP standard solutions, respectively. Stock solutions are prepared by dissolving a small amount of cyclic

---

[10] J. F. Harper and G. Brooker, *J. Cyclic Nucleotide Res.* **1**, 207 (1975).

nucleotide in 50 mM sodium acetate buffer, pH 4.0, and measuring the optical densities in a spectrophotometer. A concentrated stock solution is adjusted to 10 μM (1000 pmol/100 μl) and then further diluted to 1 μM, before working standards are prepared to individually yield 0.01, 0.02, 0.04, 0.08, 0.16, 0.32, 0.64, 1.25, 2.5, 5.0, and 10 pmol of cyclic nucleotide/100 μl. For the femtomole assay 100 pmol/ml stock solution is diluted to give 1, 2, 4, 8, 16, 32, 64, 125, 250, 500, and 1000 fmol/100 μl prior to acetylation. A blank containing no cyclic nucleotide is always included as a routine.

One-hundred microliters of each standard and up to 200 μl of an unknown sample are pipetted into 12 × 75 mm disposable culture tubes. The volume of the standards and unknowns is brought to 200 μl with 50 mM sodium acetate, pH 4.0. Tube racks are then transferred to trays containing ice water; 100 μl of ^{125}I-labeled ScAMP-TME or ^{125}I-labeled ScGMP-TME (containing 10,000–15,000 cpm) is added. Commercially available cyclic nucleotide specific antibody (New England Nuclear) is diluted in bovine serum albumin (0.1 mg/ml in 50 mM sodium acetate, pH 4.0). The dilution of antibody should be sufficient to bind 35–60% of the radioligand. To each tube 100 μl of the antibody solution is added, tubes are mixed, and 400 μl of the reaction mixture is incubated overnight (10–14 hr) at 4°; 50 μl of bovine γ-globulin (1% in water) is added to each tube, followed by 2.5 ml ice-cold 95% isopropanol to precipitate the radioactive antigen–antibody complex. The tube contents are vortexed and then centrifuged at 2500 rpm for 30 min at 0°. Isopropanol is decanted and the tubes are air dried in an inverted position at room temperature prior to determination of the radioactivity in a gamma-counter.

In this assay, the decrease in ^{125}I counts bears an inverse relationship to the amount of cyclic nucleotide in the unknown biological sample. The amount of cyclic nucleotide is then calculated by comparing the unknown counts with that of the standard curve. The final results are expressed as picomoles or femtomoles of cyclic nucleotides present per milligram protein or per cell number.

To assure reproducibility and to exercise prudent quality control, it is advisable that the standards be run in duplicate and the unknowns in triplicate in titrating volumes (e.g., 50, 100, 150 μl of biological sample). This will be adequate to detect any interference encountered in the assay due to extraneous factors.

Section VI

Cell-Mediated Cytotoxicity

[27] Cytolytic Activity Mediated by T Lymphocytes

By Howard D. Engers, K. Theodor Brunner, and
Jean-Charles Cerottini

Introduction

Cytolytic T lymphocytes (CTL)[1] are thought to play a major role in the immune defense system against virus infections and allografts,[1a] and they are most likely involved in immune responses to tumors.[2] Alloantigen-specific CTL effector cells have been isolated from the lymphoid organs of animals undergoing allograft rejection[3] and virus-specific CTL have been demonstrated in animals infected with several different strains of viruses.[4] Tumor-specific CTL effector populations have been obtained from animals in the process of rejecting syngeneic tumors.[5] Highly enriched CTL populations and clones have been generated *in vitro* in allogeneic mixed lymphocyte cultures (MLC) or in syngeneic mixed-lymphocyte-tumor-cell cultures (MLTC).[3] Such populations and clones derived from them have been shown to mediate effector function upon transfer *in vivo*.

There have been numerous assay systems proposed for the determination of the cytolytic potential of lymphoid cell populations, including cell counting, inhibition of antibody plaque formation, cloning efficiency, or release of radioactive label. Of these various methods, the ^{51}Cr release assay described in this section represents a simple, rapid, and quantitative means of assessing CTL-mediated cytolysis *in vitro*.[6] It can also be used to monitor target cell lysis mediated by effector cells other than CTL.[1] The actual mechanism of target cell lysis by CTL is still poorly under-

---

[1] Abbreviations: CTL, cytolytic T lymphocytes; MLC, mixed lymphocyte culture; MLTC, mixed lymphocyte-tumor-cell culture; DMEM, Dulbecco's modified Eagle's medium; FBS, fetal bovine serum; EDTA, ethylenediaminetetraacetic acid; LU, lytic unit.

[1a] J.-C. Cerottini and K. T. Brunner, *Adv. Immunol.* **18**, 67 (1974).

[2] A. Feffer and A. L. Goldstein, "The Potential Role of T Cells in Cancer Therapy." Raven Press, New York, 1982.

[3] H. D. Engers and H. R. MacDonald, *Contemp. Top. Immunobiol.* **5**, 145 (1976).

[4] J. W. Blasecki, "Mechanisms of Immunity to Virus-Induced Tumors." Dekker, New York, 1981.

[5] K. T. Brunner, H. R. MacDonald, and J.-C. Cerottini, *J. Exp. Med.* **154**, 362 (1981).

[6] J.-C. Cerottini and K. T. Brunner, *in* "*In Vitro* Methods in Cell-Mediated Immunity" (B. R. Bloom and P. R. Glade, eds.), p. 369. Academic Press, New York, 1971.

stood.[7] It involves irreversible changes in the permeability of the target cell surface membrane. In addition, recent experiments suggest that nuclear disintegration involving DNA fragmentation may represent an important stage in cell-mediated cytolysis of target cells.[8]

Mixed Lymphocyte Cultures

General Considerations

The generation of CTL activities *in vitro,* using MLC and syngeneic MLTC, has allowed a detailed analysis of the cell-mediated immune reactions that occur during a response to allografts or tumor growth.[3,9] The method given below represents a simple, reproducible tissue culture method for the generation of high levels of CTL activities in MLC. It should be noted that the choice of appropriate responding and stimulating cell numbers as well as culture vessel geometry is important for each experimental system.[10,11]

Material Equipment

Dulbecco's modified Eagle's medium (DMEM); fetal bovine serum (FBS), heat inactivated at 56° for 45 min; "MLC" medium, consisting of DMEM supplemented with 5% FBS, 5×10^{-5} M 2-mercaptoethanol, 10 mM HEPES buffer, and for certain systems with augmented levels of L-glutamine (800 mg/ml), folic acid (10 mg/liter), L-arginine–HCl (200 mg/liter) and L-asparagine (36 mg/liter) (final concentrations); culture vessels, either 30-ml plastic tissue culture flasks (Falcon 3013) or 24-well plates; loose-fitting Tenbroeck all-glass homogenizers, 15 ml (Bellco glass, Vineland, NJ); sterile centrifuge tubes; hemocytometer or Coulter counter; centrifuge; CO_2 incubator; source of gamma irradiation.

Cell Source and Preparation

The method described is suitable for the induction of CTL *in vitro* in MLC using any given source of lymphoid cells, e.g., spleen, lymph node, thymus, or peripheral blood. These cells may be taken from normal (primary response) or immune (secondary response) animals.[3]

Taking murine spleen as an example of a source of responding cells, proceed as follows: (1) Remove spleens aseptically, homogenize with 3–4

---

[7] E. Martz, *Contemp. Top. Immunobiol.* **7,** 301 (1977).

[8] J. H. Russel, *Immunol. Rev.* **72,** 97 (1983).

[9] F. Plata, J.-C. Cerottini, and K. T. Brunner. *Eur. J. Immunol.* **5,** 227 (1975).

[10] F. W. Fitch, H. D. Engers, H. R. MacDonald, J.-C. Cerottini, and K. T. Brunner, *J. Immunol.* **115,** 1688 (1975).

[11] A. Weiss and F. W. Fitch, *J. Immunol.* **119,** 510 (1977).

strokes in a loose-fitting Tenbroeck all-glass homogenizer with 5 ml DMEM, decant cell suspension with plunger in, add 5 ml DMEM, repeat the homogenization procedure, pool the cells, and centrifuge at 400 g for 5 min at room temperature. (2) Resuspend the cell pellet in 10 ml DMEM (maximum four spleens per tube) using a wide-bore Pasteur pipet, and allow clumps to settle for 15–20 min at room temperature. (3) Discard sediment, centrifuge supernatant, and wash cell pellet once by centrifugation, then resuspend in MLC medium. (4) Prepare stimulating spleen cells from the desired normal animals as described above, and irradiate with a gamma source (2500 rads). Viable cell concentrations are determined by a vital staining procedure, such as trypan blue dye exclusion.

Culture Conditions

Responding and stimulating cells are mixed together in the desired ratios in the appropriate culture vessel, depending on the number of cells cultured; e.g., 5×10^6 responding cells plus 5×10^6 stimulating cells in 2.0 ml of MLC medium in 24-well plates, or 25×10^6 responding cells plus 25×10^6 stimulating cells in 20 ml MLC medium in upright plastic culture flasks. The cell *ratio* and *total* cell number must be varied to determine the optimal culture conditions for each particular system in question, particularly when dealing with syngeneic MLTC cultures where the optimal responding to stimulating cell ratio may be as high as 100 : 1.[9] In addition, when using immune responding cells to induce secondary responses, subcellular particulate alloantigen preparations may be substituted for intact irradiated cells as a source of stimulating alloantigen.[3]

The cultures are then incubated in a 5% CO_2–air incubator for the optimal period (generally 5–7 days). Again, time-course studies are imperative to confirm the optimal culture conditions for any given system.

The cells remaining in culture are harvested by centrifugation, washed once, and the viable cell number adjusted to the required concentration. These cells can then be tested for their cytolytic potential using the desired assay system, e.g., the ^{51}Cr release assay discussed below. Cell recoveries can range from 10 to 50% of the initial viable cell input, depending on the culture conditions chosen.[10]

Comments

The generation of CTL activities *in vitro* in MLC represents a considerable improvement over the initially studied *in vivo* models. MLC or MLTC provide a versatile, simple, and reproducible means of obtaining enriched CTL responses *in vitro*.[3] The CTL populations thus obtained provide a convenient starting point for the derivation of cloned CTL lines which then can be maintained in culture for months or years by periodic exposure to the appropriate antigen in the presence of filler cells and a source of interleukin 2.[12,13] In addition, MLC and MLTC culture supernatants provide a convenient source of soluble mediators (lymphokines), which are known to play an important role in many cell-mediated immune reactions.[14]

Chromium Release Assay

General Considerations

Chromium-51 is introduced into the target cell as chromate ($^{51}CrO_4^{2-}$). It is a relatively nontoxic gamma emitter of convenient half-life (27.8 days), which is easily taken up by most cells by an energy-independent process. Inside the cell, chromate is reduced to chromic ion, thus maintaining a concentration gradient allowing very high levels of uptake. Although it has generally been assumed that ^{51}Cr is complexed to cytoplasmic proteins, evidence has been presented that chromium exists in the cell mainly as a small molecule, possibly as a coordination compound with a variety of substances like citrate.[15] In contrast to many other trace-labeled metabolites, it is not reutilized after it has been released from a cell. Its release usually reflects irreversible target cell damage (lysis). However, there is always a spontaneous release of isotope even in the absence of cell damage. The spontaneous release varies for different target cells. The assay is therefore limited by the availability of appropriate target cells, i.e., target cells which release a low proportion of label spontaneously, but which then release a high proportion of label when lysed. In general, long-term assays (> 24 hr) are not feasible, and best results are obtained in short-term (3–6 hr) assays.

The ^{51}Cr release assay has been carried out using many different target cell types, ranging from relatively sensitive lymphoid tumor cell lines (either ascites or tissue culture) and mitogen-induced lymphoblasts to less sensitive target cells such as freshly isolated lymphocytes, macrophages, or fibroblasts. In general, cells which grow firmly attached to plastic culture vessels are less sensitive to lysis than those grown as suspension cultures.

[12] H. R. MacDonald, R. P. Sekaly, O. Kanagawa, N. Thiernesse, C. Taswell, J.-C. Cerottini, A. Weiss, A. L. Glasebrook, H. D. Engers, K. T. Brunner, and C. Bron, *Immunobiology* **161**, 84 (1982).

[13] H. D. Engers, T. Lahaye, G. D. Sorenson, A. L. Glasebrook, C. Horvath, and K. T. Brunner, *J. Immunol.* **133**, 1664 (1984).

[14] A. Kelso, A. L. Glasebrook, O. Kanagawa, and K. T. Brunner, *J. Immunol.* **129**, 550 (1982).

[15] C. J. Sanderson, *Biol. Rev. Cambridge Philos. Soc.* **56**, 153 (1981).

Materials and Equipment

$Na_2{}^{51}CrO_4$, sterile, pyrogen free, 1 mCi/ml with a specific activity of 300 or more mCi/mg (ICR, Belgium); FBS, heat inactivated at 56° for 45 min; DMEM; Tris–phosphate buffer (Tris buffer): 8 g NaCl, 0.38 g KCl, 3 g Tris, 0.1 g anhydrous Na_2HPO_4/liter of H_2O, pH 7.4; round-bottom tubes or U-bottomed plastic microtiter plates; HEPES buffer solution, 1 *M* in DMEM; automatic pipet (eight channel if using microplates); hemocytometer or Coulter counter; CO_2 incubator or a gas-tight box to be placed in a 37° incubator; centrifuge (with adaptors to spin microplates); gamma counter.

Labeling of Target Cells and Assay Procedure

Cells to be used as targets are washed once with assay medium (DMEM supplemented with 5% FBS and 10 mM HEPES) and the supernatant aspirated to leave 0.1–0.2 ml of medium. The cell pellet (2–3 × 10⁶ cells/tube) is resuspended by tapping the tube, and 0.2 ml of Tris buffer–5% FBS added. Then 0.1–0.2 ml of $Na_2{}^{51}CrO_4$ solution is added, and the cells incubated in a CO_2 incubator at 37° for 45–60 min with occasional shaking. The labeled cells are then washed three times with assay medium, and the cell suspension adjusted to a final concentration of 1 × 10⁵ viable cells/ml.

Cells grown as monolayer cultures can be labeled with $Na_2{}^{51}CrO_4$ either *in situ* (using 20–50 μCi/ml for 1–2 hr) or in suspension as described above, following minimal trypsination or EDTA–PBS treatment at 4° to obtain a single cell suspension.

After labeling with ⁵¹Cr and washing, 10⁴ target cells are dispensed in 0.1-ml aliquots (using an eight-channel pipet) into U-bottomed microplate wells already containing varied numbers of lymphoid cells, also suspended in 0.1 ml. Lymphocyte-to-target cell ratios usually vary from 100 : 1 to 0.1 : 1, depending on the source and lytic activity of the effector cell population under investigation. The microplates are then incubated at 37° in a CO_2 incubator. If test tubes are utilized instead of microplates, they are gassed with 5% CO_2 in air, stoppered, and incubated in a water bath or an ordinary 37° incubator. At the end of the desired assay interval (generally 3–6 hr), the plates or tubes are centrifuged at 1000 rpm for 5 min and 0.1 ml of supernatant fluid is removed using an automatic pipet and transferred to tubes to be measured in a gamma counter. The ⁵¹Cr released can also be estimated using a liquid scintillation counter if a gamma counter is not available.[16]

[16] H. B. Herscowitz and T. W. McKillip, *J. Immunol. Methods* **4**, 253 (1974).

Spontaneous release values are determined by incubating target cells alone, or together with normal lymphocytes.

Maximal release is determined by freeze–thawing or detergent treatment of the target cells or, if available, the plateau value obtained with an excess of CTL.

Quantitation of Results

The percentage of specific ^{51}Cr release (specific lysis) is calculated for each lymphocyte-to-target cell ratio using the following formula:

Percentage specific ^{51}Cr release =

$$\left(\frac{\text{experimental release} - \text{spontaneous release}}{\text{maximal release} - \text{spontaneous release}} \right) 100$$

This method of calculation is adequate for conditions where spontaneous release is relatively low ($< 25\%$). In systems where spontaneous release values exceed this level, it is more appropriate to use the following formula:

Percentage specific ^{51}Cr release =

$$\left(\frac{\text{experimental release} - \text{spontaneous release}}{\text{total incorporated}} \right) 100$$

The specific ^{51}Cr release values obtained are then plotted as a function of the $\log_{10}$ of each lymphocyte-to-target cell ratio, using semilog graph paper.[6] The resulting curves are generally linear between 20 and 80% specific lysis, and parallel to each other when several experimental groups are tested. One lytic unit (LU) is arbitrarily defined as that number of lymphocytes required to yield 50% specific ^{51}Cr release in the chosen incubation time. Using this value, the number of LU per 10^6 recovered cells or per organ and/or culture can then be calculated.

Comments

A typical ^{51}Cr release assay result is presented in Fig. 1. In this example, two CTL clones with different specificities (derived by micromanipulation techniques and propagated in interleukin 2-containing medium)[12,13] were tested on their respective target cells. In Fig. 1A, the C57Bl/6 (H-2^b) CTL clone CHA-11, specific for H-2^d alloantigen, gives optimal lysis of P-815 (H-2^d) target cells but virtually no lysis when tested on MBL-2, a C57Bl/6 (H-2^b) Moloney leukemia virus-derived syngeneic tumor cell line. In the reciprocal experiment, a Moloney leukemia virus-specific C57Bl/6 CTL clone (CHM-14) shows efficient lysis of the relevant MBL-2 tumor cells but no lysis of the irrelevant P-815 target cells (Fig. 1B).

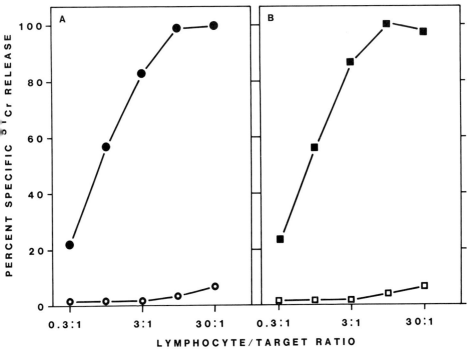

FIG. 1. Two different C57Bl/6 (H-2^b) CTL clones were derived by micromanipulation techniques[13] and tested for their ability to lyse the appropriate ^{51}Cr-labeled target cells. (A) shows the lytic activity of CTL clone CHA-11, specific for H—2^d alloantigens, on P-815 (H-2^d) target cells (●—●), and on syngeneic (H-2^b) Moloney leukemia virus-induced tumor target MBL-2 (○—○). (B) shows the lytic activity of CTL clone CHM-14, specific for Moloney leukemia virus-coded surface antigens, on P-815 (□—□) and MBL-2 (■—■) target cells.

When measured as a function of time, specific ^{51}Cr release proceeds linearly for several hours. This is mainly due to the fact that one CTL can lyse several target cells sequentially. It should be kept in mind that the rate of target cell lysis (hence, that of ^{51}Cr release) is a function of both the frequencies and efficiencies of CTL within the lymphoid population tested. Moreover, since CTL-mediated lysis is dependent upon close contact between CTL and target cells, the volume of the incubation mixture and the geometry of the incubation vessel can affect the velocity of target cell lysis. Finally, kinetic studies have shown that ^{51}Cr release is not concomitant to the time of the lethal hit (defined as an irreversible target cell damage) but may occur several minutes to a few hours later. Therefore, the percentage of specific ^{51}Cr release obtained at a given time point represents an underestimate of the percentage of target cells that actually are lethally damaged at this particular time point.

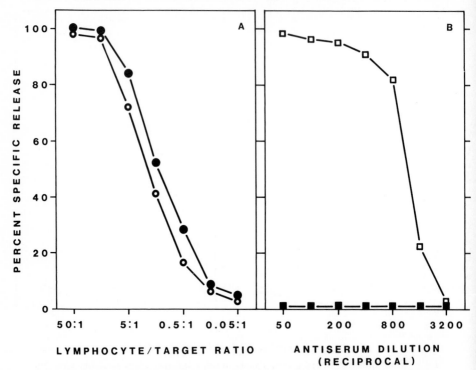

Fig. 2. A comparison of CTL-mediated cytolysis (A) with that induced by alloantibodies plus complement (B), using either ^{51}Cr as a cytoplasmic label (○—○; □—□) or ^{131}I-labeled IUdR as a nuclear label (●—●; ■—■). P-815 (H-2^d) target cells were labeled with either ^{51}Cr- or ^{131}I-labeled IUdR and then incubated with the C57Bl/6 anti-H-2^d CTL clone CHA-11 (A) or anti-H-2^d antibodies plus complement (B). After incubation for 90 min at 37°, the amount of radioactive label released in the supernatant either directly (^{51}Cr) or after treatment of target cells with 0.5% Triton X-100 (^{131}I) was determined.

While the ^{51}Cr release assay system described here represents the most widely used method for the determination of cell-mediated cytolytic activities, it is often useful to confirm the findings obtained with this assay by using an independent system to measure target cell lysis. In this context, it is noteworthy that a rapid DNA fragmentation process takes place in the nucleus of murine target cells exposed to CTL (but not to antibody and complement). Nuclear disintegration can be easily monitored by using target cells labeled with radioactive thymidine or iododeoxyuridine. As first demonstrated by Russel,[8] determination of the amount of radioactivity released after treatment of the target cells with a nonionic detergent provides a simple means to quantitate nuclear disintegration in such target cells (Fig. 2). The usefulness of this alternative method for measuring cell-mediated lysis is, however, limited since human target cells, in contrast to murine target cells, do not exhibit early DNA fragmentation upon lysis by CTL.

[28] Cytotoxicity by Natural Killer Cells: Analysis of Large Granular Lymphocytes

By John R. Ortaldo

Introduction

Natural killer cells (NK)[1] were discovered about 12 years ago[1a–5] during studies of cell-mediated cytotoxicity. Although investigators at that time expected to find specific cytotoxic activity of tumor-bearing individuals against autologous and allogeneic tumor cells, appreciable cytotoxicity was observed also with lymphocytes from normal individuals. Since then, studies on natural killer cells have expanded into a broad research area stimulated by the increasing indications that these cells may play a role in normal host natural resistance against cancer and infectious diseases.[6–18] NK are defined as effector cells with spontaneous cytotoxicity

[1] Abbreviations: NK, natural killer cells; LGL, large granular lymphocytes; ADCC, antibody-dependent cellular cytotoxicity; PBS, phosphate-buffered saline; FCS, fetal calf serum; MoAb, monoclonal antibody; IL-2, interleukin 2; LDA, limiting dilution analysis; PHA, phytohemagglutinin; IAC, immunoabsorbent column; LU, lytic units.

[1a] E. B. Rosenberg, R. B. Herberman, P. H. Levine, R. H. Halterman, J. L. McCoy, and J. R. Wunderlich, *Int. J. Cancer* **9**, 648 (1972).

[2] J. L. McCoy, R. B. Herberman, E. B. Rosenberg, F. C. Donnelly, P. H. Levine, and C. Alford, *Natl. Cancer Inst. Monogr.* **37**, 59 (1973).

[3] R. Kiessling, E. Klein, and H. Wigzell, *Eur. J. Immunol.* **5**, 112 (1975).

[4] R. Kiessling, G. Petranyi, G. Klein, and H. Wigzell, *Int. J. Cancer* **15**, 933 (1975).

[5] R. B. Herberman, M. E. Nunn, and D. H. Lavrin, *Int. J. Cancer* **16**, 216 (1975).

[6] R. K. Oldham, D. Siwarski, J. L. McCoy, E. J. Plata, and R. B. Herberman, *Natl. Cancer Inst. Monogr.* **37**, 49 (1973).

[7] G. Cudkowicz and P. S. Hochman, *Immunol. Rev.* **44**, 13 (1979).

[8] O. Haller, R. Kiessling, A. Örn, K. Kärre, K. Nilsson, and H. Wigzell, *Int. J. Cancer* **20**, 93 (1977).

[9] N. Hanna and R. Burton, *J. Immunol.* **127**, 1754 (1981).

[10] N. Hanna and I. J. Fidler, *JNCI, J. Natl. Cancer Inst.* **63**, 801 (1980).

[11] R. B. Herberman, *Adv. Cancer Res.* **19**, 207 (1974).

[12] R. B. Herberman, ed., "Natural Cell-Mediated Immunity against Tumors." Academic Press, New York, 1980.

against target cells, which lack the properties of classical macrophages, granulocytes, or cytotoxic T cells. The observed cytotoxicity does not show restriction related to the major histocompatibility complex. It should be emphasized that NK are not the sole effector cells of natural immunity but rather have been recognized as one of several types of cells which play an important role in natural resistance. In addition to NK cells, macrophages, natural antibodies, and polymorphonuclear leukocytes are other major effectors involved in natural resistance.[15,19–30] Until recently, the cells responsible for NK activity have been defined only in a negative way or by *in vitro* functional activity distinguishing them from typical T cells or B cells or macrophages. However, it is now possible to isolate populations of cells highly enriched for NK activity and for cells with a characteristic morphology of large granular lymphocytes (LGLs).[31–33] They comprise only about 5% of peripheral blood leukocytes and 1–3% of total mononuclear cells. LGL have been found in most species tested,[15,32] including human, mouse, hamster, rat, chicken, guinea pigs, and miniature swine. LGL contain azurophilic granules in their cyto-

[13] R. B. Herberman and H. T. Holden, *Adv. Cancer Res.* **27**, 305 (1978).

[14] R. B. Herberman, ed., "NK Cells and Other Natural Effector Cells." Academic Press, New York, 1982.

[15] R. B. Herberman and J. R. Ortaldo, *Science* **214**, 24 (1981).

[16] E. Ojo and H. Wigzell, *Scand. J. Immunol.* **7**, 297 (1978).

[17] G. Petrayni, R. Kiessling, S. Povey, G. Klein, E. Herzenberg, and H. Wigzell, *Immunogenetics* **3**, 15 (1976).

[18] S. Pollack, S. Heppner, R. J. Brawn, and K. Nelson, *Int. J. Cancer* **9**, 316 (1972).

[19] D. A. Chow, M. I. Greene, and A. H. Greenberg, *Int. J. Cancer* **23**, 788 (1979).

[20] S. A. Eccles and P. Alexander, *Nature (London)* **250**, 667 (1974).

[21] R. Evans, *Transplantation* **14**, 468 (1972).

[22] A. Tagliabue, A. Mantovani, M. Kilgallen, R. B. Herberman, and J. L. McCoy, *J. Immunol.* **122**, 2363 (1979).

[23] R. Evans, *J. Reticuloendothel. Soc.* **26**, 427 (1979).

[24] I. J. Fidler, *Cancer Res.* **34**, 1074 (1974).

[25] J. B. Hibbs, Jr., H. A. Chapman, Jr., and J. B. Weinberg, *J. Reticuloendothel. Soc.* **24**, 549 (1978).

[26] S. Korec, in "Natural Cell-Mediated Immunity against Tumors" (R. B. Herberman, ed.), p. 1301. Academic Press, New York, 1980.

[27] M. O. Landazuri, E. Kedar, and J. L. Fahey, *J. Natl. Cancer Inst. (U.S.)* **52**, 147 (1974).

[28] M. H. Levy and E. F. Wheelock, *Adv. Cancer Res.* **20**, 131 (1974).

[29] A. Mantovani, T. R. Jerrells, J. H. Dean, and R. B. Herberman, *Int. J. Cancer* **23**, 18 (1979).

[30] G. W. Wood and G. Y. Gillespie, *Int. J. Cancer* **16**, 1022 (1973).

[31] T. Timonen, C. W. Reynolds, J. R. Ortaldo, and R. B. Herberman, *J. Immunol. Methods* **41**, 269 (1982).

[32] C. Reynolds, T. Timonen, and R. B. Herberman, *J. Immunol.* **127**, 282 (1981).

[33] K. Kumagi, K. Itoh, R. Suzuki, S. Hinuma, and F. Saitoh, *J. Immunol.* **127**, 185 (1981).

plasm, and highly purified LGL can be isolated by discontinuous Percoll gradient centrifugation. LGL are generally nonphagocytic, nonadherent cells that possess receptors for the third component of complement as well as receptors for the Fc portion of immunoglobulin.[15,31–33] It is the latter receptor that allows the binding of antibody-coated targets to LGL cells and mediates the phenomenon of ADCC, a function previously attributed to the killer lymphocyte.[32] Therefore, LGL can mediate both forms of cytotoxicity. In the present chapter, the main focus will be on the purification of LGL cells and their subpopulations with the use of a variety of classical separation procedures, as well as monoclonal antibodies. The cytotoxic and noncytotoxic functions of LGL will be analyzed. Previously, NK cells have been primarily defined by their ability to lyse target cells *in vitro*. It is necessary to purify these cells in order to adequately assess their ability to mediate cytolysis as well as other noncytolytic functions and to characterize them in regard to their biological role *in vitro* and *in vivo*.

Procedures

Preparation of Cells

Mononuclear cells are separated by Ficoll-Isopaque one-step gradient centrifugation (Bøyum, 1968) (this series, Vol. 108 [9]) and suspended in growth medium (RPMI 1640 supplemented with 10% FCS, 0.29 mg/ml glutamine, and antibiotics) (Biofluids, Rockville, MD). The cells of the monocyte–macrophage series, as well as B cells, are depleted by sequentially incubating mononuclear cells on plastic flasks (60 min at 37°) and passage over nylon wool columns (Julius *et al.*, 1973) (this series, Vol. 108 [28]). To fractionate the nonadherent mononuclear cells in discontinuous Percoll density gradients, seven different concentrations of Percoll in growth medium are adjusted to physiological osmolarities using 10×-PBS (Ca^{2+} and Mg^{2+} free) (Gibco, NY) and distilled water, respectively (Table I).

Mononuclear cells (50–80 × 10^6) are layered on top of the gradient and the test tube is centrifuged at 550 × g for 30 min at room temperature. The resulting seven bands are collected from the top with a Pasteur pipet, washed twice in growth medium, and the cells counted. The yields of cells from the original input population consistently exceeds 80%.

Measurement of Osmolarity and Density of Percoll Solutions

All solutions for the fractionation of mononuclear cells on Percoll are prepared immediately before use. This is necessary because of the insta-

TABLE I
PREPARATION OF DISCONTINUOUS GRADIENTS FOR SEPARATION OF LGL

	Percoll $(\mu l)^a$				Density $(g/ml)^b$		
Fraction	Human	Rat	Mouse	Volume $(ml)^c$	Human	Rat	Mouse
1	2550	2650	2310	2.5	1.053	1.058	1.048
2	2700	2850	2850	2.5	1.060	1.063	1.063
3	2850	3050	3120	2.5	1.063	1.069	1.072
4	3000	3150	3390	1.5	1.068	1.073	1.079
5	3150	3300	3660	1.5	1.073	1.077	ND
6	3300	3450	3930	1.5	1.077	1.080	ND
7	4000	4000	4200	1.5	>1.080	>1.080	ND

[a] Percoll adjusted to 285 and 290 mOsm/kg H_2O with 10X-PBS in human, mouse, and rat experiments, respectively) was mixed with growth medium (RPMI 1640 with 10% FCS, 0.29 mg/ml glutamine and antibiotics, also adjusted to same osmolarities with distilled water) to the final volume of 6000 μl.
[b] Approximate density measured with density marker beads (Pharmacia Chemical, Uppsala, Sweden). ND, Not determined.
[c] Volume of each fraction in a 15-ml Falcon 2095 centrifuge tube.

bility of the osmolarity of the Percoll solutions when stored.[33] Low osmolarity of the Percoll is corrected to the appropriate level with 10× concentrated PBS, whereas high osmolarity is adjusted with stock Percoll solution (~10 mOsm). The osmolarity is measured with an osmometer based on the freezing point principle (Osmette A, Precision Instruments, Sudbury, MA). Usually about 8% (v/v) 10×-PBS in Percoll produces an osmolarity of approximately 290 mOsm/kg H_2O. The final adjustment to 285 mOsm/kg H_2O and 290 mOsm/kg H_2O in human and rat experiments, respectively, is performed adding stepwise 0.1% (v/v) 10×-PBS to Percoll to increase, or 1% (v/v) Percoll to lower the osmolarity. The tonicity of medium is adjusted similarly with 10×-PBS to increase or with distilled water to lower the osmolarity to the appropriate level. The proper osmolarity is essential to obtain consistent and reproducible separations.[31–33] Density marker beads (Pharmacia Chemicals, Uppsala, Sweden) are run through the gradients for the measurement of the density of each fraction.

Morphologic Analysis

Cell smears for the morphologic analysis are prepared by centrifuging for 7 min 0.2 × 10⁶ cells in 0.2 ml of growth medium on slides at 900 rpm with a cytocentrifuge (Shandon Southern Instr., Bewickley, PA). After air drying, the cells are fixed in methanol for 10 min, air dried again, and

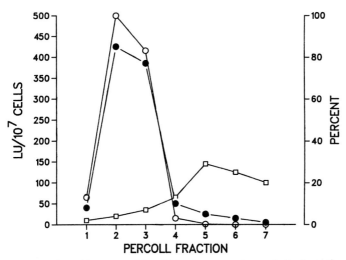

FIG. 1. Distribution of human peripheral blood mononuclear cells in discontinuous density gradient centrifugation, when osmolarity is 285 mOsm/kg H_2O. ○, Cytotoxicity on lytic units/10^7 cells; ●, frequency of LGL (in percent); □, percentage distribution of cells.

then stained for 25 min with 10% aqueous Giemsa at pH 7.4. The slides are examined under oil immersion microscopy.

Depletion of High-Affinity E-Rosette-Forming Cells from LGL-Enriched Low-Density Percoll Fractions (for more details see this series, Vol. 108 [7])

In order to remove contaminating T cells from low-density LGL-rich fractions, rosette depletion is performed. FCS, lymphocytes, and sheep erythrocytes are added in equal volumes with a lymphocyte : erythrocyte ratio of 1 : 150. The cells are centrifuged at 150 g for 5 min, and thereafter incubated 45–60 min in a water bath at 29°. The pellets are gently resuspended with a Pasteur pipet and the suspension layered on a top of a Ficoll-Isopaque (Litton-Bionetics, Rockville, MD) gradient, which is then centrifuged at 550 g for 30 min at room temperature. The cells at the interphase are collected, washed twice in growth medium, and counted. The recovery of cells from the original low-density Percoll fractions is >50%, and the purity of LGL consistently exceeds 90%.

Figure 1 demonstrates a typical pattern of results (using nonadherent mononuclear lymphocytes as a starting population) following isolation on a Percoll gradient. A high level of purification of both cells with morphology of LGL and cells with cytotoxic activity (against NK-susceptible targets) is obtained in the low-density fractions of Percoll. The majority of

TABLE II
ENRICHMENT OF HUMAN LGL BY DEPLETION OF HIGH-AFFINITY
E-ROSETTE-FORMING CELLS FROM LOW-DENSITY PERCOLL FRACTIONS

Fraction[a]	LU/10[7] cells	Percentage recovery of cells	Percentage LGL
Input	38	100	11
2 and 3	570	13	82
2 and 3[-]	678	58[b]	96
2 and 3[+]	28	20[b]	8

[a] Input: plastic adherent, nylon wool-treated peripheral blood mononuclear cells; 2 and 3: cells from low-density fractions of Percoll gradients; 2 and 3[-]: low-density cells depleted of high-affinity E-rosette-forming cells; 2 and 3[+]: high-affinity E-rosette-forming cells from the low-density fractions.
[b] Refers to percentage recovery relative to fraction 2 and 3.

the lymphoid cells are high-density cells, and sediment in the subsequent fractions (fractions 4–6). The low-density Percoll fractions contain 70–85% LGL, the remainder being other cell types, especially (or *predominantly*) activated T cells. Since the initial purification of mononuclear cells includes plastic adherence and passage through nylon wool, very few monocytes or B cells are present. Since T cells are known to have high-affinity receptors for sheep erythrocytes (E) and form rosettes at 29° as well as at 4°, whereas NK cells are known to form rosettes only at 4°,[15] this fraction must be further purified by depletion of cells forming rosettes at 29°. As shown in Table II, depletion of the high-affinity rosette-forming cells results in a considerable enrichment of LGL which is paralleled by an increase in the lytic activity per 10[7] cells. The major contaminants remaining in the high-density Percoll fractions after the rosetting are 1–2% of monocytes and cells morphologically similar to LGL but without granules (termed large agranular lymphocytes) and 0–1% blasts. It is critical that highly purified LGL be used as the starting population. Although the combination of various separation techniques is somewhat tedious, it provides highly enriched effector cell populations and allows for the use of monoclonal antibody for further analysis of the subpopulation of LGL.

Treatment with Monoclonal Antibody

The MoAb used in these studies are OKT3 (reactive with Pan T antigen), OKT8 (suppressor/cytotoxic T cell), OKT10 (thymocytes, activated T cells, NK cells), OKT11 (receptor for sheep red blood cells), OKM1 (granulocytes, monocytes, and LGL); all MoAb of the OK series are obtainable from Ortho Corporation, Raritan, NJ; also used are αHLA-DR (clone L243) (purchased from Becton Dickinson, Mountain View, CA), B73.1 (reactive with Fc receptor on LGL and neutrophils[34]), HNK-1 monoclonal antibody (LGL and some T lymphocytes[35]), and BRL-Mϕ (monocytes) (purchased from Bethesda Research Labs, Inc., Gaithersburg, MD). Rabbit anti-mouse brain serum (referred to as anti-Thy 1) is obtainable from Flow Labs, Rockville, MD, and used at the dilution of 1 : 100.

Treatment with MoAb is performed as follows: LGL (5×10^6) are treated with 20 μl of MoAb at the concentration of 1 μg/ml in 0.1 ml of PBS + 1% FCS (GIBCO, Grand Island, NY) for 45 min at 4°. At the end of the incubation, the cells are washed twice with PBS.

Most human LGL react with the following type of antibodies: (1) several monoclonal antibodies, B73.1, 3G8, and Leu-11, reactive with the Fc receptors for IgG on LGL and granulocytes, (2) rabbit anti-serum to the glycolipid, asialo-GM$_1$, which also reacts with monocytes and granulocytes, (3) OKT10, which also reacts with thymocytes and activated lymphocytes,[15,36] and (4) OKM1, which also reacts with monocytes, macrophages, polymorphonuclear leukocytes, and platelets.[15,36] (It appears to recognize receptor for C3bi.) Generally, the removal of cells bearing any of these markers, either by treatment with complement or by negative immunoaffinity procedures, results in the depletion of most of the detectable NK activity. In addition, LGL can be characterized by their lack of expression of some cell surface markers (e.g., the pan-T cell antigens such as Leu-1, OKT3, helper T antigens as defined by OKT4 or Leu-3, and the immature T cell antigens that are expressed on thymocytes, such as OKT6). Human LGL also do not express surface antigens detected by a number of monocyte-specific reagents, such as Leu-M1 or Mo-2.[15,37] In contrast, the pattern that is featured by all LGL is rather heterogeneous with respect to defined markers. Human LGL react to a variable extent with monoclonal antibodies directed against (1) the sheep erythrocyte receptor (Lyt 3, OKT11, Leu-5,[15,36]), (2) 3A1 (present on most T cells and on about 50–60% LGL), (3) HNK 1 (present on about 40–60% of LGL), (4) OKT8 (present on the cytotoxic suppressor T lymphocyte and on about 10–30% of LGL), and (5) 25% of the LGL react with HLA-Dr determinants.[15,36]

[34] B. Perussia, S. Starr, S. Abraham, V. Fannina, and G. Trinchieri, *J. Immunol.* **130**, 2133 (1983).

[35] T. Abo and C. M. Bolck, *J. Immunol.* **127**, 1024 (1981).

[36] J. R. Ortaldo, S. O. Sharrow, T. Timonen, and R. B. Herberman, *J. Immunol.* **127**, 2401 (1981).

[37] J. R. Ortaldo and R. B. Herberman, *Annu. Rev. Immunol.* **2**, 359 (1984).

Similar heterogeneity in phenotype has been seen in rodent (mouse and rat) NK cells.[15,37]

In summary, NK cells have a characteristic phenotype, for example, most human NK cells and LGL can be described as OKT3⁻, OKT4⁻, T10⁺, B73⁺, OKM1⁺, and Leu-M1⁻ cells. Thus, these cells have a unique but definable phenotype that sets them apart from other lymphoid cells.

Separation of LGL Subsets

LGL subsets are separated on an anti-F(ab')$_2$ immunoadsorbent column, following the method of Schlossman and Hudson[38] as modified by Nelson *et al.*[39] For further details see also this series, Vol. 108 [15]. Cyanogen bromide-activated Sepharose 4B, purchased from Pharmacia, is covalently coupled with affinity purified F(ab')$_2$ goat anti-mouse IgG, available from Cappel, West Chester, PA or Boehringer Mannheim, Indianapolis, IN, at 10 mg/15 g of gel. Binding efficiency of the protein ranges from 40 to 70%, as determined by measuring the reduction in the optical density at 280 nm of the protein solution after the coupling.

Immunoaffinity columns are prepared as previously described.[39] Briefly, anti-F(ab')$_2$–Sepharose 4B gel is placed in the barrel of a 10-ml plastic syringe, tightly fitted at the bottom with a sintered plastic disk (Bel-Art Product, Pequannock, NJ). Usually 3 ml of packed gel is used to separate 5×10^6 of cells. Before use, the gel is washed with RPMI 1640 + 10% FCS and incubated at 37° for 30 min, to allow settling of the beads. MoAb-treated LGL are resuspended in 1 ml of medium containing 1 mM EDTA solution (to prevent nonspecific binding) and carefully put onto the column. EDTA-containing medium is carefully added on top of the gel and the nonadherent cells are collected in 10–15 ml, at a slow flow rate (12 drops/min). Adherent cells are removed from the gel by mechanical agitation with a Pasteur pipet and collected by washing the gel at a rapid flow rate (1 ml/min). The cells are then washed in medium and incubated overnight at 37° before being tested for cytotoxicity. Analysis of the isolated cells shows that positive fractions contain <2% "inappropriate" cells, whereas negative fractions contained up to 10%.

Separation of LGL by Electronic Cell Sorting. Subsets of LGL may be isolated by flow cytometry. MoAb-treated LGL are incubated with a fluorescein-conjugated, affinity purified F(ab')$_2$ goat anti-mouse IgG (Cappel-Worthington, Cochranville, PA) at 50 μg/ml, for 40 min, at 4°. Cells are then washed once and analyzed for fluorescence. Positive and negative cells are separated on an Ortho cytofluorograph 50 H (Ortho Diagnos-

[38] S. F. Schlossman and L. Hudson, *J. Immunol.* **110**, 313 (1973).
[39] D. L. Nelson, B. M. Bundy, H. E. Pitchon, R. M. Blaese, and W. Strober, *J. Immunol.* **117**, 1472 (1976).

TABLE III
COMPARISON OF CYTOTOXIC ACTIVITY OF LGL SUBSETS SEPARATED BY TWO
DIFFERENT METHODS

Effector cells[a]	Immunoadsorbent column		Flow cytometry	
	LU/10^7 cells[b]	Percentage of control LGL	LU/10^7 cells[b]	Percentage of control LGL
Control LGL	270	100	150	100
B73.1$^-$	7	2	23	15
B73.1$^+$	444	164	211	140
OKM1$^-$	7	2	3	2
OKM1$^+$	948	350	181	120
OKT10$^-$	74	27	20	13
OKT10$^+$	265	98	135	90

[a] LGL were treated with MoAb and separated on immunoaffinity columns or by flow cytometry into positive and negative fractions. Control LGL were not pretreated or were passed through the immunoaffinity columns without previous treatment with MoAb. The two methods of separation were performed on different days.

[b] LU were calculated at 30% specific lysis.

tics Systems, Westwood, MA). For further details, see this series, Vol. 108 [19]. Reanalysis of the isolated populations reveals that contamination in both positive and negative fractions of "inappropriate" cells is less than 2%.

NK Activity in Different LGL Subsets. Using enriched LGL, NK activity of the various subsets can be tested. The results shown in Table III are from a representative experiment. Most of the activities are recovered in B73$^+$, OKM1$^+$, and OKT10$^+$ fractions. Since fractionation by cell sorting has been the standard method for obtaining cell subsets with low levels of contamination of inappropriate cells, it is important to examine the reactivity of NK subsets separated with Sepharose 4B columns and compare them to those on the cell sorter. It should be pointed out that separation by cell sorter, although providing high levels of purity, generally is inefficient for obtaining large numbers of cells. In contrast, the use of Sepharose 4B columns or other panning procedures results in high levels of cell recovery, with similar contamination with appropriate cells. Table III illustrates the patterns of reactivity and separation seen with immunoabsorbent columns using B73.1, OKM1, and OKT10 and with flow cytometry. Similar results have been obtained using panning procedure, where the antibodies are bound to plastic Petri dishes (this series, Vol. 108 [11] and [12]).

Complement-Dependent Lysis. MoAb-treated LGL are incubated with a 1 : 100 dilution of rabbit antimouse Ig (Sera-Lab, Westbury, NY) for 30 min at 4°. Cells are then washed and resuspended in 1 ml of RPMI 1640 containing 10% FCS and rabbit complement (1 : 10 dilution) (Cedarlane, Ontario, Canada), and placed in a 37° water bath for 1 hr with occasional shaking. Cells are washed twice with medium and viable cells are counted by trypan blue exclusion (see this series, Vol. 108 [6]). Viability of cells treated with complement alone ranges from 70 to 90%.

Cell Lines. Tumor cell lines (available from American Type Culture Collection) used as targets in the cytotoxicity assay are the following: K562, Daudi, and Alab. The target cell for antibody-dependent cellular cytotoxity is RL♂1, a murine T lymphoma,[13] coated with a 1 : 100 dilution of a rabbit anti-mouse brain serum (Flow Labs, Rockville, MD). All cell lines are maintained in RPMI 1640 + 10% FCS, 2 mM glutamine, and 50 μg/ml gentamycin and are subcultured three times a week.

Cytotoxic Activity of LGL Subsets against a Panel of NK-Susceptible Targets. In most studies on the cytotoxic reactivity of minor subpopulations of cells, K562 cells are used as target cells. Since LGL subsets may vary in their ability to kill different target cells, the various subsets should be tested against a wider panel of NK-susceptible target cells (Table IV). However, OKT10⁻ cells, which generally possess low activity against K562, have appreciable activity against Daudi cells. From these results one would conclude that general NK phenotype demonstrating reactivity to tumor targets is restricted to populations bearing the OKT10, OKM1, OKT11, and B73 markers. Differences, however, exist, especially when NK is measured against virus-infected or parasitic targets.

Cytotoxicity Assay (see also this volume [27])

Target cells are labeled with 100 μl ⁵¹Cr (New England Nuclear) for 1 hr at 37° and washed extensively before use. Target cells (5 × 10³ in 0.1 ml) and effector cells in 0.1 ml of medium are plated in microtiter plates (Linbro) at several effector-to-target ratios (25 : 1, 12 : 1, 6 : 1, 3 : 1). After 4 hr of incubation at 37°, the supernatants are harvested and the radioactivity determined in a gamma scintillation counter. Cytotoxicity is determined by the amount of ⁵¹Cr released from dead target cells. Three replicates for each experimental group are used and the percentage of specific lysis is calculated as mean ± SD according to the formula:

Percentage specific lysis =

$$\left(\frac{\text{cpm in experimental wells} - \text{cpm in wells with target cells alone}}{\text{cpm incorporated in target cells}} \right) 100$$

TABLE IV
CYTOTOXIC ACTIVITY OF DIFFERENT LGL SUBSETS
AGAINST NK-SUSCEPTIBLE TARGETS

Effector cells[b]	Percentage specific lysis[a]			
	K562	ADCC	Alab	Daudi
Unfractionated				
LGL	42	22	26	32
OKT10				
−	11	7	4	24
+	47	13	33	34
OKM1				
−	3	2	8	9
+	46	23	28	40
OKT11				
−	10	8	4	6
+	45	13	25	43
OKT8				
−	41	16	13	28
+	36	14	21	36
αDR				
−	35	13	16	29
+	22	13	28	11
B73.1				
−	5	1	7	8
+	51	32	20	29

[a] Effector cells were tested against the various targets
in a standard 4-hr ^{51}Cr-release assay at 20:1 effector
to target (E:T) ratio.
[b] LGL were treated with MoAb and separated on IAC
into positive and negative fractions. Control LGL
were passed through the IAC without any pretreatment.

A 6% increase in isotope release above baseline is statistically significant
at $p < 0.05$ (Student's t test).

Limiting Dilution Analysis (LDA)

For LDA experiments, culture medium is RPMI 1640, supplemented
with 10% human AB serum (M. A. Bioproducts, Walkersville, MD), 0.1
mM nonessential amino acids (GIBCO, Grand Island Biological Company, Grand Island, NY), 2 mM sodium pyruvate (GIBCO), 10 mM
HEPES (GIBCO), 2 mM glutamine (NIH media unit), 50 μg/ml gentamycin (Sigma, St. Louis, MO), 0.2 mM 2-mercaptoethanol buffered solution

(Sigma), and supplemented with interleukin 2 (IL-2) (this series, Vol. 116 [38]) [50% (v/v) supernatant of the gibbon lymphosarcoma cell line MLA 144, constitutively producing IL-2[40]]. LDA are performed as previously described.[41] Responder cells (ranging from 10,000 to 10 cells/well, 12 replicates/concentration) are dispensed in 0.1 ml of complete medium containing IL-2 into round-bottomed, 96-well plates (Linbro). Autologous irradiated (3000 rad) mononuclear cells (2 × 10^5/well) are added in 0.1 ml of the same medium, together with 0.1 μg/ml of phytohemagglutinin (PHA) (Burroughs-Wellcome, Welcome, England). Each well receives a total volume of 0.2 ml. Plates are cultured for 7 days for determination of proliferative precursor frequency and 8 days for cytotoxic precursor frequency. These incubation times are optimal for such studies.[41]

Proliferation is assessed by uptake of [^{3}H]thymidine (1 μCi/well) (specific activity, 6.7 mCi/mmol) (New England Nuclear) during the last 18 hr of incubation. The cells are then harvested onto glass fiber disks with an automatic harvester, and radioactivity is measured in a liquid scintillation counter.

For the determination of the cytotoxic progenitor frequency, plates are washed with medium without IL-2, and the cells are incubated in complete medium alone for 6–8 hr before the cytotoxicity assay, to reduce possible nonspecific lysis.[15] Medium (0.1 ml) is replaced with ^{51}Cr-labeled target cells (5000 cells/well). Supernatants are harvested after 4 hr of incubation, as indicated above (see cytotoxicity assay).

Uptake of [^{3}H]thymidine, or release of ^{51}Cr from targets, in wells (24 replicates) containing autologous feeder cells with medium and IL-2, serves as background control. Test wells are considered positive when values exceeded the mean control value by three SD. By linear regression analysis of the number of responders plated against the log of the percentage of negative wells, a minimal estimate of frequencies is calculated as the number of cells plated which gave 37% negative wells. The significance of differences between lines for different responders within the same experiment is calculated as described by Taswell.[42]

Progenitors of LGL

The LDA technique provides a minimal estimate of the frequency of IL-2-dependent progenitors against K562 or other target cells as well as the frequency of the proliferative progenitors. It should be noted, how-

---

[40] H. Rabin, R. F. Hopkins, F. W. Ruscetti, R. H. Newbauer, R. L. Brown, and T. G. Kawakami, *J. Immunol.* **127**, 1852 (1981).
[41] B. M. Vose and G. D. Bonnard, *J. Immunol.* **130**, 687 (1983).
[42] C. Taswell, *J. Immunol.* **126**, 1615 (1981).

TABLE V

COMPARISON OF THE PHENOTYPE OF FRESH NK CELLS AND OF IL-2-DEPENDENT
PROGENITORS PROLIFERATIVE AND OF CYTOTOXIC CELLS

Sample	OKT3	OKT8	OKT10	OKT11	OKM1	B73	HNK1
Proliferative progenitor	−	−	−	+	−	−	−/+[a]
Cytotoxic progenitor	−	−	+	+	−/+	−	−/+
NK (K562) fresh	−	−	+	−/+	+	+	−/+

[a] −/+ indicates activity found in both negative "−" and positive "+" population.

ever, that the limiting dilution assay used in such studies[39] identifies the most common progenitors, those which grow the fastest and kill the best.

Subsets of LGL were prepared by flow cytometry with a variety of monoclonal antibodies and a summary of the results is shown in Table V. The progenitors of the cells with NK activity using the LDA were predominantly found to be with the OKT11+ subset of LGL. Cells that were OKT11− had low frequencies of both proliferative and cytotoxic progenitors. OKM1+ cells had a high frequency of proliferative progenitors; however, both OKM1+ and OKM1− cells had an appreciable portion of cytotoxic progenitors. Cells that were B73− had a high frequency of proliferative progenitors and an appreciable frequency of cytotoxic progenitors. Thus, the expression of B73 on NK cells and their IL-2-dependent progenitors is quite different. This conclusion was further supported by the finding that B73+ cells, although highly cytotoxic, responded poorly in limiting dilution assays. The cytotoxic progenitors were also found to be mainly within the OKT10+ and OKT8− subset; however, these subsets had very poor proliferative progenitors in the limiting dilution assay. Depletion of HNK1+ or DR+ cells did not affect the generation of either proliferative or cytotoxic progenitors.

[29] Cytotoxicity by Macrophages and Monocytes

By ROEL A. DE WEGER, BERT A. RUNHAAR, and WILLEM DEN OTTER

Introduction

Macrophage or monocyte cytotoxicity (briefly macrophage cytotoxicity) can be induced in various ways *in vitro* or *in vivo*. This cytotoxicity can be divided in cytostatic and cytolytic cytotoxicity.[1-3] For the expression of both forms of cytotoxicity direct contact of the target cell with the macrophages is required. Although in various systems it has been shown that the cytotoxicity is due to a soluble factor released by the macrophage, direct contact of the target cell with the macrophage is required in order to initiate the release of the factor.[4,5] The factor itself is difficult to detect as it is rapidly inactivated by serum components,[6] and so routinely macrophage cytotoxicity is detected in assays in which macrophages and target cells are in close contact with each other.

Whether macrophages are cytostatic or cytolytic depends on the type of activation and the same macrophage population can express both activities.[2,3] Therefore the choice of the cytotoxicity assay depends on the type of cytotoxic activity that has to be measured. Cytolytic activities are generally detected by assays that measure the release of radiolabel from the target cells. These labels can be cytoplasmic labels such as $Na^{51}CrO_4$, [^3H]proline,[7,8] or Indium-111 oxine,[9] or DNA labels such as [^3H]thymidine[8] or [^{125}I]iododeoxyuridine.[7] Cytostatic activities can be detected by assays that measure the incorporation of radiolabels in DNA such as [^3H]thymidine and [^{125}I]iododeoxyuridine.[10,11] The overall effect of cytostatic and cytolytic macrophage activity can be studied by visual cell counting of nonadherent target cells. We have selected several of these

---

[1] P. Pucetti and H. T. Holden, *Int. J. Cancer* **23**, 123 (1979).
[2] M. W. Campbell, M. M. Sholley, and G. A. Miller, *Cell. Immunol.* **50**, 153 (1980).
[3] S. K. Chapes and S. Haskill, *Cell. Immunol.* **76**, 49 (1983).
[4] B. M. Eggen, O. Bakke, and J. Hammerström, *Scand. J. Immunol.* **18**, 13 (1983).
[5] W. J. Johnson, C. C. Shisnant, and D. O. Adams, *J. Immunol.* **127**, 1787 (1981).
[6] D. O. Adams, *J. Immunol.* **124**, 286 (1980).
[7] C. G. Brooks, *J. Immunol. Methods* **22**, 23 (1978).
[8] R. Keller and R. Keist, *Br. J. Cancer* **37**, 1073 (1978).
[9] R. H. Wiltrout, D. Taramelli, and H. T. Holden, *J. Immunol. Methods* **43**, 319 (1981).
[10] H. G. Opitz, D. Nierhammer, H. Lemke, H. D. Flad, and R. Huget, *Cell. Immunol.* **16**, 379 (1975).
[11] R. Evans and C. G. Booth, *Cell. Immunol.* **26**, 120 (1976).

assays that will be described in detail, together with pitfalls that one might encounter.

Methods for Determining Macrophage/Monocyte Cytotoxicity

General Considerations

Effector Cells. Cytotoxicity of murine peritoneal macrophages is usually tested in confluent macrophage monolayers in flat-bottom wells. Depending on the method used for the detection of cytotoxicity 24 ($\varnothing$ 16 mm)-, 48 ($\varnothing$ 9 mm)-, or 96 ($\varnothing$ 6 mm)-well culture plates (e.g., Costar) can be used. To obtain confluent monolayers, respectively, 8×10^5, 4×10^5, and $1-3 \times 10^5$ macrophages per well are seeded in these culture plates. Macrophages obtained from the murine peritoneal cavity (see this series, Vol. 108 [25]) can be tested after activation *in vivo* (e.g., after immunization with tumor cells[12]) or *in vitro* (e.g., with lymphokines and lipopolysaccharide[13]). This macrophage cytotoxicity is preferably measured within several hours after harvesting the *in vivo* activated macrophages or after *in vitro* activation, as the cytotoxicity tends to diminish during culture. In some cases it has been shown that this reduction of cytotoxicity is due to the production of prostaglandins by activated macrophages.[14]

Murine peritoneal macrophages can be washed thoroughly 2–3 hr after culturing to remove nonadherent cells (see this series, Vol. 108 [27]). Macrophages that have been activated *in vitro* are even more strongly adherent than normal macrophages and can be washed vigorously before testing of the cytotoxicity against tumor cells. To be sure that only macrophage cytotoxicity is measured the washed macrophage monolayers can be treated with an anti-T cell serum [e.g., anti-Thy-1 (Olac, England)] or anti-NK-cell serum (Becton Dickinson, USA) in the presence of complement in order to remove T cells or NK cells that stick on the monolayer.[15,16] After incubation with these antisera the monolayers are washed and the target cells can be added to the monolayers.

For most cytotoxicity experiments in murine systems, it is essential to use "clean" mice, as many conventionally reared animals have spontaneous cytotoxic macrophages, which will interfere with the induction of

[12] W. Den Otter, H. F. J. Dullens, H. Van Loveren, and E. Pels, *in* "The Macrophage and Cancer" (K. James, B. M. McBride, and A. Stuart, eds.), p. 300. Econoprint, Edinburgh, 1977.

[13] M. S. Meltzer and L. P. Ruco, *Adv. Exp. Med. Biol.* **121B**, 381 (1980).

[14] S. M. Taffet and S. W. Russell, *J. Immunol.* **126**, 424 (1981).

[15] R. A. de Weger, E. Pels, and W. Den Otter, *Immunology* **47**, 541 (1982).

[16] R. Keller, T. Bächi, and K. Okumura, *Exp. Cell Biol.* **51**, 158 (1983).

cytotoxicity. This spontaneous cytotoxicity seems to be due to subclinical infections (for details, see Ref. 17).

Human monocytes are relatively weakly adherent and cannot be washed as vigorously as murine macrophages. Monocytes are often obtained from blood leukocytes by adherence. The contaminating lymphocytes within these monocyte suspensions contain a relative high concentration of NK cells, indicating that NK cells adhere to monocyte monolayers.[18-20] This will influence the measured cytotoxicity. Therefore pure monocyte suspensions should be used for the cytotoxicity tests. The most pure monocyte suspensions have been obtained by counterflow elutriation (see this series, Vol. 108 [20]). To test the cytotoxicity of human monocytes the target cells can be added directly to the more than 90% pure monocyte suspension or after a 24-hr preculture of the monocytes.

Target Cells. Although it was generally accepted that activated macrophages could kill all types of tumor cells, recently resistant tumor target cells have been described, especially resistance to the cytotoxicity of human monocytes (Table I).[21] It is accepted, however, that activated macrophages will only kill transformed target cells, whereas normal cells are not killed although some growth inhibition may occur.[22-24] When the cytotoxicity of macrophages or monocytes has to be tested against a target cell of unknown susceptibility, it is useful to test a susceptible target cell as well, to measure whether the macrophages actually have cytotoxic capacity. The macrophage cytotoxicity can be measured at effector cell : target cell (E : T) ratios from 20 : 1 up to 1 : 1. Generally an E : T of 10 : 1 is used. It is important to use target cells at a density that ensures a normal proliferation rate (e.g., $1-3 \times 10^5$ cells/ml). Proliferation of the target cell is not a requirement for macrophage cytolysis. However, when the tumor cell concentration is too low or too high a reduction in cell growth will occur, resulting in a decrease of the calculated cytotoxicity.

Media. Macrophage cytotoxicity can be tested in many types of media such as RPMI, and Eagles' minimum essential medium or Fischer's medium. Generally the medium is supplemented with 10% fetal bovine se-

[17] E. Pels and W. Den Otter, *Br. J. Cancer* **40**, 856 (1979).
[18] B. Freundlich, G. Trinchieri, B. Perussia, and R. B. Zurier, *J. Immunol.* **132** (3), 1255 (1984).
[19] D. A. Horwitz, A. C. Bakke, W. Abo, and K. Nishiya, *J. Immunol.* **132**, 2370 (1984).
[20] B. M. Eggen, *Acta Pathol. Microbiol. Immunol. Scand., Sect. C* **90C**, 123 (1982).
[21] A. Mantovani, T. R. Jerrels, J. H. Dean, and R. B. Herberman, *Int. J. Cancer* **23**, 18 (1979).
[22] A. Mantovani, A. Tagliabue, J. H. Dean, T. R. Jerrelsand, and R. B. Herberman, *Int. J. Cancer* **23**, 28 (1979).
[23] W. T. Piessens, W. Churchill, and J. R. David, *J. Immunol.* **114**, 293 (1975).
[24] M. DeBray-Sachs, C. Boitard, R. Assan, and J. Hamburger, *Transplant. Proc.* **13**, 1111 (1981).

Target cell line	Percentage specific release ($[^3H]Thy)^a$
SL2 (murine lymphosarcoma)^b	71 ± 4
TU5 (murine SV40 transformed renal epithelial line)^c	13 ± 4
K562 (human erythroleukemia line)^d	4 ± 2

^{a} Mean of three experiments performed in triplicate.
^{b} H. F. J. Dullens, J. Milgers, C. D. H. Van Basten, R. A. De Weger, E. De Heer, and W. Den Otter, *Leuk. Res.* **6,** 425 (1980).
^{c} A. Mantovani, Z. Bar-Shavit, G. Peri, N. Polentarutti, C. Bordignon, C. Sessa, and C. Mangioni, *Clin. Exp. Immunol.* **39,** 776 (1980).
^{d} R. B. Herberman and M. T. Holden, *Adv. Cancer Res.* **27,** 305 (1978).

rum in murine systems, whereas for human monocytes 20% pooled human AB, Rh$^+$ serum is required. Often, addition of serum is not an absolute requirement for the measurement of cytotoxicity, but in serum-free media (e.g., Neuman-Tytell) the cytotoxicity is lower than in serum-containing media. In most media and sera lipopolysaccharide (LPS)^{24a} is present as a contaminant. This LPS can enhance the cytotoxicity significantly. LPS is even a requirement for the induction of lymphokine-induced cytotoxicity.[12] Levels of LPS can be measured by the *Limulus* amebocyte lysate test (Associates of Cape Cod, Inc., Woods Hole, MA).

Incubation Period. Macrophage cytotoxicity is often considered to be a rather slow process, which means that a significant cytotoxicity can be measured only after about 15 hr. Still, in some cases a more rapid effect of the cytotoxicity can be measured depending on the activation method, the target cell, and the detection method used (Fig. 1). In general, a 24-hr incubation period is advisable. Longer incubation periods have the disadvantage that tumor cell growth is reduced due to exhaustion of the medium or too high a cell concentration. This is observed often as an increase in spontaneous release of radiolabel in the controls which will result in a reduced calculated cytotoxicity (Fig. 1).

Cytotoxicity Tests

Visual Cell Counting

Materials

Target cells: Nonadherent tumor cells
Culture plates: Flat-bottom wells (24 or 48 wells, Costar, England)

^{24a} Abbreviations: LPS, lipopolysaccharide; SDS, sodium dodecyl sulfate; $[^3H]Thy$, $[^3H]thy$-midine.

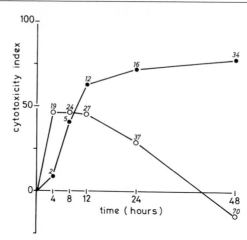

FIG. 1. Cytotoxicity of human monocytes. Influence of the incubation time on the calculated cytotoxicity index. The cytotoxicity of monocytes was determined against SL2 tumor cells in RPMI supplemented with 20% AB serum. Human monocytes (more than 90% pure) were obtained by counterflow elutriation. ○, Cytotoxicity index determined with the ^{51}Cr release assay; ●, cytotoxicity index determined with the [^{3}H]thymidine release assay. Spontaneous release as percentage of the maximal release is given in italic numbers at each measure point. A representative experiment is shown (mean of experiment in triplicate).

Trypan blue (Fluka, Switzerland)
Indian ink (1/10 diluted; Talens, the Netherlands)
Hemocytometer
Phase-contrast microscope

Procedure. Target cells ($1.5–2 \times 10^5$ cells/ml) are added to the macrophage monolayers (E : T = 10 : 1) and incubated for 24 hr at 37°. After incubation 10 μl Indian ink (1/10 diluted) is added and the culture is incubated for another 30 min at 37°. Subsequently, the culture fluid is vigorously mixed in the well with a Pasteur pipet. Although nonadherent target cells are used, a vigorous mixing of the cultures is required as the tumor cells can strongly adhere to the macrophage surface.[25,26] After mixing 50 μl of trypan blue is added and this suspension is immediately counted in a hemocytometer. The Indian ink particles are taken up by the macrophages; this facilitates the distinction between tumor cells and macrophages. In order to further distinguish between macrophages/monocytes and tumor cells it is also possible to incubate the hemocytometer in a moist incubator at 37° for 30 min. The macrophages/monocytes adhere

[25] T. A. Hamilton and M. Fishman, *J. Immunol.* **127**, 1702 (1981).
[26] S. D. Somers, J. P. Martin, and D. O. Adams, *J. Immunol.* **131**, 2086 (1983).

TABLE II
CYTOLYTIC AND CYTOSTATIC ACTIVITY OF MURINE MACROPHAGES[a]

Effector cells	[³H]Thymidine release[b] [cpm (spec. rel.)]	[³H]Thymidine incorporation[c] [cpm (GI)]	Visual cell counting[d] [cell number (CI)]
None	1,366	119,719	99
Control macrophages	1,595 (2%)	107,994 (10%)	93 (6%)
Activated macrophages	5,927 (52%)	3,532 (96%)	40 (60%)

[a] SL2 tumor cells were incubated 24 hr without macrophages or with murine peritoneal macrophages which were activated with SL2-sensitized lymphocytes and tumor cells (activated macrophages) or macrophages incubated with normal lymphocytes and tumor cells (control macrophages). The ratio tumor cells : macrophages is 1 : 10 [R. A. De Weger, E. Pels, and W. Den Otter, *Immunology* **47**, 541 (1982)]. Given in parentheses are the calculated specific release (spec. rel.), growth inhibition (GI), or cytotoxicity index (CI) of the control and activated macrophages compared with tumor cells cultured alone. A representative experiment is the mean of an experiment performed in triplicate.
[b] [³H]Thy-labeled tumor cells (max. release, 12,207 cpm) were incubated with the macrophages.
[c] Nonlabeled tumor cells were incubated with the macrophages in medium containing [³H]Thy (165,616 cpm/well).
[d] Tumor cells were incubated with the macrophages and after 24 hr the number of tumor cells is determined in a hemocytometer.

and spread on the glass surface and can easily be distinguished from the nonadherent tumor cells by phase-contrast microscopy. Trypan blue is added to distinguish between viable (nonstained) and dead (blue) target cells. To obtain reliable results the controls are counted first. In the controls at least 80 cells are counted. This determines the area in the hemocytometer (number of squares) that has to be counted. Subsequently, the same area is counted with cells from the test. The cytotoxicity index (CI) is determined with the following formula:

$$CI = [(N - T)/N]100$$

in which N is the number of viable tumor cells in the control and T is the number of viable tumor cells in the test.

As controls, cultures of tumor cells without macrophages (e.g., to determine the spontaneous macrophage cytotoxicity) or cultures from tumor cells with control macrophages are used. The spontaneous cytotoxicity of control murine macrophages should range between +10 and −10%. Otherwise the macrophages are inhibiting or stimulating tumor growth. Examples of experiments are given in Tables II and III.

TABLE III
CYTOTOXICITY OF HUMAN MONOCYTES[a]

Effector cells	Incubation period (hours)	^{51}Cr release[b] [cpm (percentage spec. rel.)]	[^{3}H]Thymidine release[c] [cpm (percentage spec. rel.)]	Visual cell counting[d] [cell number (CI)]
None	4	261	159	88
Monocytes	4	302 (5%)	365 (4%)	36 (59%)
None	24	431	849	84
Monocytes	24	330 (−17%)	3826 (69%)	10 (88%)

[a] Cytotoxicity of monocytes was detected against SL2 tumor cells in RPMI supplemented with 20% human AB serum. Human monocytes (more than 90% pure) were separated by counter flow elutriation. A representative experiment is the mean of an experiment in triplicate.
[b] Maximal release: 1014 cpm.
[c] Maximal release: 5187 cpm.
[d] CI, Cytotoxicity index.

[³H]Thymidine Release

Materials

Target cells: Adherent or nonadherent tumor cells
[³H]Thymidine ([³H]Thy; Amersham, England)
Sodium dodecyl sulfate (SDS; Serva, West Germany)
Culture plates: flat-bottom wells (96 wells)

Procedure. Target cells (2×10^5 cells/ml) are cultured in medium with serum and 0.2–0.5 μCi of [³H]Thy/ml for 24 hr at 37°. The concentration of [³H]Thy used depends on the incorporation rate of the target cells. Subsequently, the cells are washed with medium (three times) to remove free radioisotope and resuspended at a concentration of 2×10^5 cells/ml. Fifty microliters of this suspension is mixed with SDS (final concentration 1%). Radioactivity in the supernatant of this mixture is used as the maximal release value. To the macrophage monolayers (in 100 μl medium) 100 μl of the target cell suspension is added (E : T = 10 : 1). After incubation (routinely 24 hr) the culture plate is centrifuged and 100 μl of the supernatant is collected from each well and the radioactivity is measured.

The specific [³H]Thy release is calculated according to the formula:

Percentage specific release =

$$\left(\frac{\text{cpm in test} - \text{cpm spontaneous release}}{\text{cpm maximal release} - \text{cpm spontaneous release}} \right) 100\%$$

Representative experiments are shown in Tables II and III.

^{51}Cr *Release* (see also this volume [27])

Materials

Targets: Adherent or nonadherent tumor cells
Na^{51}CrO$_4$ (Amersham, England)
Sodium dodecyl sulfate (SDS)
Culture plates: Flat-bottom wells (96 wells)

Procedure. Target cells are centrifuged and the pellet is resuspended in medium containing Na^{51}CrO$_4$ (100–200 μCi/2 $\times$ 10^6 cells). This suspension is cultured for 1 hr at 37°. Subsequently the cells are washed twice with medium, incubated 15 min at 37°, and washed again to remove free radiolabel.

The incubation procedure to test the cytotoxicity is the same as described for [^{3}H]Thy release, but the routine incubation period is 4 hr as the spontaneous ^{51}Cr release of most murine tumor target cells (1–2% label per hour) is too high to obtain a reliable measurement after 24 hr (see Table III and Fig. 1).

[^{3}H]Thymidine Incorporation

Materials

Target cells: Adherent and nonadherent tumor cells
[^{3}H]Thymidine ([^{3}H]Thy)
Sodium dodecyl sulfate (SDS)
Culture plates: Flat-bottom wells (96 wells)

Procedure. Tumor cells (1 $\times$ 10^5 cells/ml) are incubated with the macrophage monolayers (E : T = 10 : 1) in medium containing serum and 0.2 μCi/ml [^{3}H]Thy. After 24 hr the tumor cells are harvested with an automated cell harvester. The cells are washed and the incorporated [^{3}H]Thy activity is determined by liquid scintillation counting. Before measuring the incorporated [^{3}H]Thy the tumor cells are lysed in 1% SDS. The growth inhibition (*GI*) is expressed as:

$$GI = \left(\frac{\text{cpm in control} - \text{cpm in test}}{\text{cpm in control}} \right) 100$$

Controls are tumor cells grown in the absence of macrophages. The *GI* of nonactivated macrophages is calculated as well, and compared with the *GI* of activated macrophages (Table II). When the macrophages are cytolytic, the [^{3}H]Thy incorporation is reduced. This is, however, not entirely due to an active growth inhibition of the macrophages but is at least in part due to the reduced number of tumor cells incorporating [^{3}H]Thy.

Growth inhibition assay should therefore be performed with noncytolytic macrophages.

Concluding Remarks

Visual cell counting has the advantage that it measures the overall cytolytic and cytostatic activity of the macrophage population. Besides, by cell counting the "condition" of the tumor cells can be observed and the proliferation of the target cells can be assessed. A disadvantage of the visual cell counting assay is that it can be used only for nonadherent cells. (For adherent target cells a different and more complicated method has been described by Weinberg and Hibbs.[27])

In the cell-counting method, sufficient cells per well have to be counted to make the tests reliable (at least 80 cells of the controls). Under those conditions, the method is highly reproducible when performed properly. Cell counting, however, is laborious compared with the radiolabel methods. The advantages of the latter methods are therefore that it is easier to handle large numbers of samples and that one can use relatively small samples. Besides, visual cell counting is often unreliable to detect NK-cell, T-cell, or K-cell cytotoxicity, as the distinction between target cells and effector cells is usually difficult.

At the beginning of a series of experiments it is advisable to perform cell-counting assays together with radiolabel release or radiolabel incorporation assays in order to be sure what is being measured by the radiolabel experiments and to test the effect of the label on the target cell susceptibility.[28,29] For example, mitomycin C-treated tumor cells can still incorporate [³H]thymidine although there is no increase in cell number, thereby invalidating [³H]thymidine incorporation data as a measure for cell multiplication (R. D. W., unpublished observations). Another pitfall is that the spontaneous release of [³H]thymidine from rapidly growing tumor cells is often higher than from slowly growing tumors. This is shown by the results with visual cell counting (unpublished observations). This may influence the cytotoxicity calculated from the results of the [³H]thymidine release assay.

The method of incorporation of radiolabels to measure cytostatic macrophage activity has to be used with care as well, as it is known that macrophages can produce cold nucleotides that compete with the incorporation of radioactive nucleotides. This has been shown for [³H]thymi-

---

[27] J. B. Weinberg and J. B. Hibbs, in "Manual of Macrophage Methodology" (H. B. Herscowitz, H. T. Holden, J. A. Bellant, and A. Ghaffar, eds.), p. 345. Dekker, New York, 1981.

[28] R. Evans and D. M. Eidler, JNCI, J. Natl. Cancer Inst. 71, 983 (1983).

[29] I. J. Fidler, J. Natl. Cancer Inst. (U.S.) 45, 773 (1970).

dine and [^{125}I]iododeoxyuridine.[10,11] To study the release of these nucleotides, the supernatants of macrophage cultures have to be tested for their activity to reduce the incorporation of radioactive nucleotides into tumor cells and the test has to be compared with data from the visual cell counting assay.

[30] Cytolytic T-Cell Clones and Hybridomas

By HARALD VON BOEHMER and WERNER HAAS

Introduction

Cloned T-cell populations are used for the biochemical analysis of antigen receptors, lymphokine receptors, secreted lymphokines, and their genes. T-cell clones can also be used to study the antigen-dependent regulation of expression of lymphokine receptors as well as lymphokine secretion.

Two types of CTL clones can be distinguished according to their growth requirements: CTL-A clones, which can be maintained in culture by regular addition of stimulator cells as well as IL-2[1] and CTL-B clones, which can be grown in IL-2-containing media in the absence of stimulator or feeder cells. We now have found a third type of CTL clone which can be stimulated to grow without addition of exogenous IL-2. In this chapter we describe the methodology used to obtain the various types of CTL clones and the methods used to generate cytolytic T-cell hybridomas.

Materials

BSS

Cells are washed in BSS which is prepared by dissolving in 1 liter H_2O: 0.01 g phenol red, 0.14 g $CaCl_2 \cdot 2H_2O$, 8.00 g NaCl, 0.40 g KCl, 0.20 g $MgSO_4 \cdot 7H_2O$, 0.06 g KH_2PO_4, 0.24 g $Na_2HPO_4 \cdot 2H_2O$.

Ca^{2+}- and Mg^{2+}-Free Medium

Medium used to detach adherent cells is prepared by dissolving in 5 liters H_2O: 0.05 g phenol red, 41.00 g NaCl, 1.00 g KCl, 14.43 g

---

[1] Abbreviations: BSS, balanced salt solution; CTL, cytolytic T lymphocytes; IL-2, interleukin 2; CS medium, culture medium containing supernatants of concanavalin A-stimulated spleen cells; Con A, concanavalin A; PEG, polyethylene glycol; DMSO, dimethyl sulfoxide; IFN, interferon; IMDM, Iscove's modified Dulbecco's medium.

$Na_2HPO_4 \cdot 12H_2O$, 1.00 g KH_2PO_4, 1.45 g EDTA Titriplex III (Merck, Darmstad, West Germany).

GKN Solution

GKN solution, which is used to dissolve polyethylene glycol, is prepared by dissolving in 1 liter H_2O: 0.40 g KCl, 8.00 g NaCl, 1.77 g $Na_2HPO_4 \cdot 2H_2O$, 0.69 g $H_2PO_4 \cdot H_2O$.

PEG Solution

Polyethylene glycol solution is prepared freshly before each fusion. A 40% PEG solution is prepared by mixing: 2.20 ml GKN, 0.80 ml H_2O, 0.04 ml 0.1 N NaOH, 2 g PEG (MW 4000; Merck).

The mixture is boiled for 10 min and then kept at 37°.

Normal Culture Medium

Normal culture medium is prepared by dissolving in 900 ml H_2O, 17.670 g IMDM (1-liter package, Gibco Europe, Paisley, Scotland), 3.024 g $NaHCO_3$, 0.500 ml 0.1 M 2-mercaptoethanol (Merck-Schuchhardt, Hohenbrunn, FRG, No. 805740), L-thioglycerol (1%, v/v) (Sigma, St. Louis, MO) (can be used instead of mercaptoethanol), 10 ml H_2O containing 10^5 U penicillin and 100 mg streptomycin (KC Biological, Lenexa, KS). The mixture is brought to 1 liter with H_2O. Finally 100 ml fetal calf serum is added. The medium is filtered and stored at $-20°$ in 200-ml aliquots.

IL-2-Containing Media (see also this series, Vol. 116 [38])

IL-2-containing media are prepared as described by Schreier and Tees.[1a] Lewis rat spleen cells (5×10^6/ml) are incubated in normal culture medium containing 4 μg/ml concanavalin A. Supernatants are harvested after 48 hr, filtered, and stored at $-20°$. Usually 4–5 liters of supernatant are obtained from 50 spleens. The supernatants are tested for growth-promoting activity with an IL-2-dependent cell line CTL-L16 (see Smith[2]) and diluted according to the IL-2 titer (usually 1 : 5) with normal culture medium. This medium containing crude supernatants is called CS medium. Partially purified IL-2 is prepared from rat spleen cell cultures containing mitogen and 0.025% bovine serum albumin instead of serum. The supernatant is harvested after 48 hr, proteins precipitated with 80% saturated ammonium sulfate, dialyzed, and fractionated on Sephadex

[1a] M. H. Schreier and R. Tees, *Int. Arch. Allergy Appl. Immunol.* **61**, 227 (1980).
[2] K. A. Smith, *Contemp. Top. Immunobiol.* **11** (1980).

G-100.[3] Appropriate aliquots of each fraction are tested for IL-2 activity and pooled accordingly. The partially purified IL-2 is added to normal culture medium to give a final concentration of 5% which induced optimal proliferation of CTL-L16 cells.

Recently, we have supplemented normal culture medium with IL-2 of human origin obtained from a gene cloned in *Escherichia coli* (Biogen, Geneva, Switzerland).

Selection Medium

CS medium containing hypoxanthine (10^{-4} M), aminopterin (4 × 10^{-7}), thymidine (1.6 × 10^{-5} M), and ouabain (10^{-3} M) is used to select hybrids.

Methods

In Vivo Immunization

In mixed leukocyte culture strong cytolytic T-cell responses to non-MHC antigens are usually obtained after *in vivo* immunization only. To induce *in vivo* T-cell responses to the weak male-specific transplantation antigen HY, female mice are immunized with 10^7 male spleen cells ip and the spleens are removed 14 days later. To induce *in vivo* CTL responses to hapten-coupled cells, mice are injected with 10^7 hapten-coupled spleen cells ip or iv (iv-injected cells were previously irradiated with 2000 R). For details see Haas and von Boehmer.[4]

Generation of CTL in Bulk Cultures

In primary mixed leukocyte culture, 10^7 spleen or lymph node cells are cultured together with 10^7 X-irradiated (2000 R) stimulator spleen cells in 5 ml culture medium in small plastic flasks (Falcon 3024F). The flasks are put in an upright position in a 37° incubator with a humidified atmosphere containing 6% CO_2. Restimulation is carried out 10–14 days after initiation of the cultures. Surviving responder cells (1 × 10^6) are cultured with 10^7 X-irradiated stimulator spleen cells in 5 ml culture medium as described for primary cultures. Further restimulations are carried out at 7-day intervals. In some cases the cultures are supplemented with partially purified IL-2.

[3] M. H. Schreier, N. N. Iscove, R. Tees, L. Aarden, and H. von Boehmer, *Immunol. Rev.* **51,** 315 (1980).
[4] W. Haas and H. von Boehmer, *Eur. J. Immunol.* **14,** 1069 (1984).

Isolation of CTL Clones

After one to five restimulations, cells are cloned by micromanipulation. Viable cells from bulk cultures are purified by centrifugation over Ficoll (d 1.090) and diluted to <10³ cells/ml. Single cells are picked in drawn-out 50-μl microtubes (Richard-Allan, Medical Industries). The microtubes are drawn out with a pipet puller (F. S. Hockman, Scientific Instruments, Media, PA). Single cells are transferred into wells of flat-bottom microtiter plates containing 10⁶ X-irradiated stimulator cells per well and partially purified IL-2 at a final concentration of 5%. After 7–10 days the content of wells in which cell growth is observed is transferred in Costar plates 3524 containing 10⁷ X-irradiated stimulator cells in 2 ml medium supplemented with partially purified IL-2.

Detachment of Adherent Cells

Some CTL clones, in particular type 3 clone (see the section, "CTL Clones"), adhere rather strongly to plastic surfaces. To detach them, the culture medium is aspirated and cold Ca^{2+}- and Mg^{2+}-free medium is added. Most of the Ca^{2+}- and Mg^{2+}-free medium is carefully aspirated after 1 min and 5 min later the cells are suspended in fresh culture medium.

Generation of Cytolytic T-Cell Hybridomas

Cytolytic hybrids are obtained by fusing murine CTL clones with murine or rat thymoma cells.[5–6a]

Selective Systems

Cytolytic T-cell hybridomas are generated by fusing ouabain-resistant and thioguanine-resistant (HGPRT⁻) BW5147 thymoma cells (American Type Culture Collection, Rockville, MD) with CTL clones. The hybrids are selected in CS medium containing HAT, which eliminates the HGPRT⁻ tumor cells, and 10^{-3} M ouabain, which eliminates the CTL. Hybrids selected in medium without IL-2 are noncytolytic.[5]

[5] M. Nabholz, M. Cianfriglia, O. Acuto, A. Conzelmann, W. Haas, H. von Boehmer, H. R. MacDonald, H. Pohlit, and J. P. Johnson, *Nature (London)* **287,** 437 (1980).

[5a] A. Conzelmann, A. Silva, M. Cianfriglia, C. Tougne, R. P. Sekaly, and M. Nabholz, *J. Exp. Med.* **156,** 1335 (1982).

[6] W. Haas and P. Kisielow, *Eur. J. Immunol.* **15,** 755 (1985).

[6a] M. Nabholz, M. Cianfriglia, A. Oreste, A. Conzelmann, A. Weiss, W. Haas, and H. von Boehmer, in "Research Monographs in Immunology" (G. J. Hämmerling, U. Hämmerling, and J. F. Kearney, eds.), Vol. 3. Elsevier North-Holland Publ., Amsterdam, 1982.

The generation of variant cell lines resistant to purine and pyrimidine analogous and to ouabain as well as the advantage of using such variants are described in detail elsewhere.[7-11]

Fusion. Cell fusion is performed according to the procedure described by Galfré *et al.*[12]

Hybrid Selection. Two days after the fusion, 100 μl selection medium is added to each well. The selection medium is CS medium containing hypoxanthine (10^{-4} M), aminopterin (4×10^{-7} M), thymidine (1.6×10^{-5} M), and ouabain (1×10^{-3} M). Deoxycytidine is added to a final concentration of $2 \times 10^{-5} M$ (Conzelmann *et al.*[13]) to avoid the cytostatic effects of thymidine. This precaution does not seem to be essential since all our hybrids were obtained in the absence of this drug. Four to 5 days after the fusion 150-μl supernatants are removed from each well and replaced by 150 μl fresh selection medium. Cell growth can be observed usually in some wells 10 days but in some cases as late as 30 days after the fusion. Usually cell growth is observed in 10 to 40% of the cultures which have received 3.6×10^4 fused cells/well. Confluent monolayers in microculture wells are detached with Ca^{2+}- and Mg^{2+}-free medium (see "Materials") and the cells are transferred into larger vessels of 24-well Costar trays (Costar 3524) containing CS medium with hypoxanthine (10^{-4} M) and thymidine (1.6×10^{-5} M) to overcome the residual toxic effects of aminopterin. After 3–4 days confluent monolayers are detached and transferred to 10 ml drug-free CS medium in small plastic flasks.

Isolation of Cytolytic Clones

Virtually all cultures obtained from fusions of type 3 CTL clones (see below) with BW5147 express specific cytolytic activity but the cytolytic activity of most hybrids is lower than that of the parental CTL clone. Cultures with high specific cytolytic activity are cloned in CS medium. Each well of a microtiter plate receives on average one hybrid cell and 2 ×

[7] R. E. Giles and F. H. Ruddle, *in* "Tissue Culture: Methods and Applications" (P. F. Kruse, Jr., and M. K. Patterson, eds.), p. 475. Academic Press, New York, 1973.

[8] R. M. Baker, D. M. Brunette, R. Makovitz, C. H. Thompson, G. F. Whitmore, L. Siminovitch, and J. E. Till, *Cell* **1,** 9 (1974).

[9] B. B. Beezley and N. H. Ruddle, *J. Immunol. Methods* **52,** 269 (1982).

[10] W. N. Choy, T. V. Gapalakrishnan, and J. W. Littlefield, *in* "Technics in Somatic Cell Genetics" (J. W. Shay, ed.), p. 11. Plenum, New York, 1982.

[11] J. Morrow, *in* "Technics in Somatic Cell Genetics" (J. W. Shay, ed.), p. 1. Plenum, New York, 1982.

[12] G. Galfré, S. C. Howe, C. Milstein, G. W. Butcher, and J. C. Howard, *Nature (London)* **266,** 550 (1977).

[13] A. Conzelmann, P. Corthésy, M. Cianfriglia, A. Silva, and M. Nabholz, *Nature (London)* **298,** 170 (1982).

10^4 X-irradiated (2000 R) peritoneal cells in 200 μl CS medium. The peritoneal cells are obtained by flushing the peritoneal cavity of syngeneic or allogeneic mice with BSS (see also this series, Vol. 108 [25]). Clones which have reached confluence are transferred to Costar trays (Costar 3524), expanded in CS medium, and tested for cytolytic activity. Clones with high cytolytic activity are maintained in CS medium and usually recloned several times to select clones with stable expression of cytolytic activity. Recloning is usually done in CS medium in the absence of any feeder cells.

Identification of Hybrids

The cells grown in selection medium are likely to be hybrids since cell growth in cultures containing the mixture of unfused parental cells in selection medium is not observed. Hybrids are distinguishable from the parental cells by the following observations:

1. CS^- variants (see below) can be obtained regularly from fused cells. We have never obtained CS^- variants from type 3 CTL clones although one such variant clone has been described.[14]

2. In comparison with the CTL clones, hybrids usually grow somewhat faster and to higher cell densities (up to 2 × 10^6 cells/ml).

3. The DNA content of hybrids is higher than that of each parental cell. The cellular DNA content is determined according to Taylor and Milthorpe[15] using the DNA binding dye propidium iodide and a fluorescence-activated cell sorter.

4. The hybrids have more chromosomes than each of the parental lines but usually fewer than the summation of both parental lines.

5. The hybrids express markers characteristic for both parental cell lines. The thymoma BW5147 is derived from AKR mice which express Thy-1.1. Since most other mouse strains express Thy-1.2, hybrids can be identified by serology. In many cases hybrids can be identified also with antisera against major histocompatibility antigens. Alternative methods to identify hybrid cells are described in detail elsewhere.[9,16]

Functional Tests

Proliferation Assay. Antigen-induced proliferation is measured by culturing 5 × 10^4 T cells with 10^6 X-irradiated stimulator spleen cells (which

[14] W. R. Benjamin, P. S. Steeg, and J. J. Farrar, *Proc. Natl. Acad. Sci. U.S.A.* **79**, 5379 (1982).

[15] I. W. Taylor and B. K. Milthorpe, *J. Histochem. Cytochem.* **28**, 1224 (1980).

[16] N. H. Ruddle, *in* "Lymphokines" (M. Feldmann and M. H. Schreier, eds.), Vol. 1, p. 49. Academic Press, New York, 1982.

in some cases are depleted of T cells) in 0.2 ml culture medium in flat-bottom microtiter wells in the presence or absence of exogenous IL-2; 2 μCi [^{3}H]thymidine is added per well after 48 hr of culture and the radioactivity is determined 24 hr later.

Assay for Cytolytic T Cells. This assay is performed as described previously.[17] See also this volume [27].

Detection and Elimination of Mycoplasma

Mycoplasma infection is known to induce alterations of various types in cell lines.[18] Among these are alterations of karyotypes, cloning efficiency, and the capacity of cells to give rise to hybrids.[11]

Test for Mycoplasma Infection. A modification of the method described by Fogh and Fogh[19] is used. Costar wells (Costar 3524) receive 4 × 10^4 HeLa cells in 1 ml culture medium and 1 ml of supernatant from cell lines or media to be tested. After incubation for 2–3 days, the medium is aspirated and 750 μl 0.6% sodium citrate is added. Then 250 μl H$_2$O is added dropwise to induce hypotonic swelling of the cells followed, after 5 min, by the addition of 1 ml of freshly prepared Carnoy's fixative (25% glacial acetic acid and 75% ethanol). The fixative is removed after 10 min and 500 μl fresh fixative is added. After 10 min the fixative is removed and samples are dried. To stain DNA, a 0.01% solution of quinacrine dihydrochloride is added. After 10 min, the wells are dried and the base of each well is cut off, mounted on a slide, and observed under a fluorescence microscope at 100× magnification. Mycoplasmas are readily detectable mainly associated with the surface of the cells.[18]

Elimination of Mycoplasma.[20] Cells are cloned in CS medium containing the antibiotics lincomycin (Gibco) and tylosin (Flow Laboratories). Both drugs are used at the maximum concentration tolerated by the cells (25–50 μg/ml). No mycoplasma is detected in clones derived from such cultures within a period of up to 4 months.

Freezing of CTL Clones[21]

Most CTL clones and hybridomas survive freezing and thawing very well, while some type 3 CTL clones (see below) survive poorly. Soon

---

[17] H. Pohlit, W. Haas, and H. von Boehmer, *Eur. J. Immunol.* **9**, 81 (1979).
[18] J. Fogh, *in* "Contamination in Tissue Culture" (J. Fogh, ed.), p. 66. Academic Press, New York, 1973.
[19] J. Fogh and H. Fogh, *Proc. Soc. Exp. Biol. Med.* **117**, 899 (1964).
[20] L. Schimmelpfeng, U. Langenberg, and J. H. Peters, *Nature (London)* **285**, 661 (1980).
[21] S. P. Leibo and P. Mazur, *in* "Methods in Mammalian Reproduction" (J. C. Daniel, Jr., ed.), p. 151. Academic Press, New York, 1978.

after recloning and functional analysis, clones are expanded in larger flasks (Falcon 3013F), harvested, and suspended in CS medium at 5×10^7 cells/ml. Two-milliliter screw-top plastic tubes (ST 506, Sterilin IG, Zürich), sterilized at 140° for 40 min, receive 100 μl cell suspension (5×10^6 cells). The tubes are kept in ice water before addition of cold BSS containing 28.4% DMSO. The tubes are then immersed in precooled ethanol, which is then cooled at a rate of $-1°/2$ min in a cooling chamber (Multi-Cool FTS-Systems, Inc.). At -6 to $-10°$ ice formation is induced by grasping each tube with a forceps precooled in liquid nitrogen. The samples are then immediately reimmersed into the ethanol. The seeding for each sample requires 1–2 sec. At $-60°$ samples are transferred to liquid nitrogen.

To thaw, samples are transferred to a beaker precooled in liquid nitrogen and allowed to warm in air. After 15–20 min, the samples are almost thawed and transferred to ice water. Completely thawed samples are transferred to a 37° water bath and prewarmed CS medium (30°) is added as follows: 5 times 20 μl, 6 times 50 μl, 4 times 100 μl, 2 times 500 μl over 2–3 min. Samples are mixed gently after each addition. The cell suspension is then transferred to 20 ml warm CS medium and distributed in Costar trays (Costar 3524). Each of 10 wells receives 2 ml cell suspension (2.5×10^5 cells originally frozen) and 2×10^5 irradiated (2000 R) peritoneal cells in 100 μl CS medium. With some clones growth is obvious after 2–3 days in all wells. Other clones are more difficult to recover, growth being observed only after 5–10 days and often in only some of the wells.

Applications

In Vitro Selection of Antigen-Specific Cytolytic T Cells

In mixed leukocyte cultures heterogeneous cell populations are stimulated. In addition to CTL, T helper cells are induced and factors are produced which may activate and/or expand so-called "bystander" B and T cells.[22] Nevertheless, a significant enrichment of antigen-specific T cells can be obtained by stimulation of primed or unprimed cells with X-irradiated spleen cells expressing the appropriate antigens. Approximately 20–30% of T cell clones which are isolated 10–12 days after the primary in vitro stimulation with a cloning efficiency of 10–12% express specificity for stimulator cell antigens. Further enrichment is obtained by repeated in vitro stimulation with appropriate X-irradiated stimulator spleen cells.[23,24] The observation that specific T cells can be selected by such procedures

[22] H. von Boehmer, J. Immunol. 112, 70 (1974).
[23] H. von Boehmer and W. Haas, Immunol. Rev. 54, 27 (1981).
[24] R. T. Hünig and M. J. Bevan, J. Exp. Med. 155, 111 (1982).

indicates that in heterogeneous populations of recently activated T cells there is a growth advantage for cells stimulated with antigen. This preferential growth is observed also with cultures supplemented with IL-2. Recently it has been shown that antigenic stimulation increases expression of receptors for IL-2.[25]

CTL Clones

Type 1 CTL Clones. Cells selected in bulk culture are cloned by micromanipulation with cloning efficiencies ranging from 10–30%. In a typical case male-specific CTL from female (B6 × CBA) F$_1$ hybrid mice immunized with male CBA spleen cells were cloned. Seven days after the immunization the female cells were restimulated in culture with male CBA spleen cells and cloned 10 days later. Most (>80%) of the clones were cytolytic and could be divided into two types according to their growth requirements: between 20 and 40% could be induced to grow by T cell-depleted spleen cells in the absence of exogenous IL-2. These type 1 CTL clones are restricted by class I MHC antigens (they lyse class II negative targets) but induction of their proliferation requires special stimulator cells, most of which express class II antigens.[26,27] This phenotype of CTL clones is not stable in culture and invariably changes after a few weeks into another phenotype (type 2).

Type 2 CTL Clones. Type 2 CTL clones require antigen and exogenous IL-2 to grow. Antigenic stimulation increases the number of IL-2 receptors expressed per cell. If antigen is not provided the cells do not proliferate in response to IL-2 and die after 8 to 10 days of culture. In contrast to type 1 clones, type 2 clones can be maintained in culture for several months without any detectable change in their ability to proliferate or to lyse target cells.

Type 3 CTL Clones. Type 3 CTL clones are derived from variants which eventually overgrow other cells. These cells express IL-2 receptors constitutively—hence their growth advantage in IL-2-containing media—and do not require any stimulator or feeder cells. Type 3 CTL clones show chromosomal abnormalities[28] and probably have gone one step toward malignant transformation. The specific lytic activity is often maintained in type 3 CTL clones and therefore they are useful for the analysis of antigen-specific receptors and their genes.

[25] D. A. Cantrell and K. A. Smith, *J. Exp. Med.* **158**, 1895 (1983).
[26] H. von Boehmer, P. Kisielow, W. Leiserson, and W. Haas, *J. Immunol.* **133**, 59 (1984).
[27] H. von Boehmer and K. Turton, *Eur. J. Immunol.* **13**, 176 (1983).
[28] J. P. Johnson, M. Cianfriglia, A. L. Glasebrook, and M. Nabholz, *in* "Isolation, Characterization and Utilization of T Lymphocyte Clones" (C. G. Fathman and F. W. Fitch, eds.). Academic Press, New York, 1982.

Cytolytic T-Cell Hybridomas

T-Cell Hybridomas Which Express Cytolytic Activity Constitutively. These clones are obtained by fusing BW5147 cells (American Type Culture Collection, Rockville, MD) with type 3 CTL clones. The hybrids obtained in these crosses resemble the parental CTL clones in that they are adherent, express IL-2 receptors and cytolytic activity constitutively, and require exogenous IL-2.[5,6] In our experience the great majority of hybrids selected in CS medium express specific cytolytic activity when tested about 3 weeks after the fusion. The cytolytic activity of most of these hybrids is weaker than that of the parental CTL line but stable cytolytic hybridomas are obtained after recloning for two to three times.

The IL-2-dependent (CS$^+$) hybrids give rise to IL-2-independent variants (CS$^-$).[5a,6] The great majority of the CS$^-$ variants resemble the parental tumor cells in that they are not adherent, do not express cytolytic activity, and do not require exogenous IL-2 to grow. CS$^-$ variants do not express IL-2 receptors and grow in the presence of anti-IL-2 receptor antibodies (Haas and Kisielow, 1984). CS medium does not induce expression of cytolytic activity in CS$^-$ variants. The rate of production of CS$^-$ variants in different CS$^+$ hybrids varies from 1 per 2×10^5 cell divisions to 0 per 4×10^8 cell divisions.[5a,6]

Cytolytic activity and IL-2 dependence are lost frequently in CS$^+$ hybrid clones producing CS$^-$ variants at a high rate. The reason for the concomitant loss of cytolytic activity and IL-2 dependence is not clear. It may be that the binding of IL-2 to IL-2 receptors—which is found to be sufficient to induce cytolytic activity in type 2 CTL clones[29]—is an absolute requirement for expression of cytolytic activity. However, we found one CS$^-$ variant clone which resembles the parental CTL clone rather than the parental tumor cells in that it is adherent and expresses specific cytolytic activity. This cytolytic CS$^-$ variant grows in the presence of anti-IL-2 receptor antibodies and does not express detectable levels of IL-2 receptors. We have fused a thioguanine-resistant (HGPRT$^-$) clone derived from the cytolytic CS$^-$ hybrid again to type 3 CTL clones. From these crosses hybrids are obtained which express cytolytic activity with both parental specificities and which are, like the parental type 3 CTL clone, IL-2 dependent. All CS$^-$ variants selected from these hybrids retain the expression of specific cytolytic activity.

Interferon Production by Cytolytic Hybrids and Noncytolytic Variants. All cytolytic hybrids tested so far produce γ-IFN. Low amounts of γ-IFN are secreted constitutively by the cytolytic CS$^-$ variant as well as by cytolytic CS$^+$ hybrids grown in long-term cultures for several months.

[29] L. Lefrançois, J. R. Klein, V. Paetkau, and M. J. Bevan, *J. Immunol.* **132,** 1845 (1984).

Large amounts of γ-IFN are found in supernatants of all hybrids cultured for 24 hr in the presence of 10 μg/ml Con A or of appropriate stimulator cells.[6] IFN is also produced by noncytolytic CS⁻ variants although to a lower degree than by the cytolytic hybrids. The observation that the noncytolytic hybrids produce IFN in response to antigen demonstrates that the lack of expression of cytolytic activity in the CS⁻ variants is not due to a lack of antigen receptor expression. CS⁻ variants are therefore a convenient source of T cells for antigen receptor analysis.

Acknowledgments

We thank Dr. Pawel Kisielow for performing the interferon assays and for reading the manuscript which will be published in a similar form in a book entitled *Methods in Immunology*, edited by I. Lefkovits and B. Pernis and published by Academic Press.

Section VII

Assays for Cytotoxic Activity

[31] Quantitative Cytotoxic Assay Using a Coulter Counter

By YOSHITAKA NAGAI, MACHIKO IWAMOTO, and
TOSHIHARU MATSUMURA

Cytotoxicity has been assayed either by determining the number of cells that become unable to exclude a dye, such as trypan blue, or by determining the amount of radioisotope released from preradiolabeled cells. The dye exclusion method is tedious and time consuming, but it is useful for a few samples. An unavoidable delay when many samples are handled causes an increase in the number of stained cells. Moreover, the presence of cell aggregates or erythrocytes causes interference. The radioisotope method is suitable to handle many samples, but it requires expensive instruments and reagents. Sometimes, spontaneous release of radioisotope complicates the analysis.

It has been known that damaged cells are more susceptible to protease treatment than intact cells. To quantitate complement-mediated immune cytolysis of lymphosarcoma cells and mouse spleen cells, Stewart and Goldstein[1] combined the procedure of protease digestion with that of electronic counting. In this method, cells are first treated with a protease to completely digest damaged cells and subsequently with a detergent to lyse intact cells as well as erythrocytes. The number of residual nuclei from intact cells is determined electronically. We modified and extended the method of Stewart and Goldstein. This assay[2] is rapid and reproducible and gives results comparable with those given by a dye-exclusion method. As our objective was to develop a rapid and quantitative method for the determination of antibody-mediated cytotoxicity, we describe examples using thymocytes as target cells.

This cytotoxicity assay, when appropriately modified, is applicable to the determination of viability of tissue culture cells. Therefore, examples using L-929 mouse fibroblasts and FM3A mouse mammary carcinoma cells are also described.

Application to Antibody-Mediated Cytotoxicity

Reagents

The medium used in all procedures is Eagle's MEM, adjusted to pH 7.2 with 7.5% NaHCO₃. Pronase E (Kakenkagaku Company, Tokyo,

---

[1] C. C. Stewart and S. Goldstein, *J. Lab. Clin. Med.* **84**, 425 (1974).
[2] M. Iwamoto and Y. Nagai, *Jpn. J. Exp. Med.* **51**, 109 (1981).

Japan)[3] is dissolved in phosphate-buffered saline (PBS),[3a] pH 7.2, at a concentration of 3 mg/ml and stored at $-20°$ in 2- to 3-ml aliquots. Pronase (purchased from Calbiochem) can be used as well as Pronase E. The frozen stock solution is thawed just before use and is filtered through a 0.45-μm millipore filter for removal of interfering precipitates. Isoton II and Zap-Oglobin are purchased from Coulter Diagnostics (Hialeah, FL). Zap-Oglobin is a hemolysing agent for diagnostic use, which lyses erythrocytes and the membranes of nucleated cells to liberate the nuclei of leukocytes for counting. Trypan blue is dissolved in PBS at a concentration of 0.2%. Lyophilized guinea pig serum, dissolved in doubly distilled water and diluted 1:4 with medium, is used as complement after absorption with agar.[4] Although any medium appropriate for the culture of target cells may be used, addition of 10 mM HEPES and 1% BSA will prevent cell damage.

Target Cells

Thymus of 5- to 8-week-old male C3H/He mouse is removed and teased apart with sharp forceps in medium. The cell suspension is pipetted repeatedly and then filtered through a stainless steel screen (200 mesh). For further details, see this series, Vol. 108 [6]. To obtain a 100% dead cell suspension, a portion of the thymocyte suspension is incubated at 56° for 30 min. The untreated portion is used as 100% viable cells. The two cell suspensions are diluted with medium to 5×10^6 cells/ml, equal volumes of the two suspensions are mixed and used as 50% dead cells. Viability of these suspensions, measured with trypan blue dye exclusion, gave a value 2–7% dead cells for viable cells, 45–53% dead cells for 50% dead cells, and 100% dead cells for the 100% dead cell preparation.

Coulter Counter Setting

In our experiments, a Model ZBI Coulter counter was used with a 100-μm orifice aperture tube. The common settings of the counter were as follows: manometer, 0.5 ml; matching switch, 20 K; gain trim, 0. The threshold settings for thymocyte nuclei were A (1/amplification), 1/2; I (1/aperture current), 0.354; T_L (lower threshold), 15; T_U (upper threshold), 60 (nuclear threshold). The settings for intact thymocytes were A, 1; I, 1/2; T_L, 15; T_U, 60 (cellular threshold). The K value of our Coulter counter was measured as 11.5, so that the volume difference per each threshold was 2.04 μm^3 for nuclear threshold and 5.75 μm^3 for cellular threshold.

[3] The name Pronase E was changed to Actinase E recently.
[3a] Abbreviations: PBS, phosphate-buffered saline; BSA, bovine serum albumin.
[4] A. Cohen and M. Schlesinger, *Transplantation* **10**, 130 (1970).

Analysis of Volume Frequency Distribution

In order to obtain a precise counting of the target cells and an effective rejection of debris, the most important adjustment is the threshold setting. The volume distribution pattern of thymocyte nuclei is first examined to determine the threshold setting. Thymocyte suspensions diluted with Isoton II with or without Zap-Oglobin (three drops/10 ml of counting solution) are analyzed for volume frequency distribution. Figure 1A illustrates the volume distribution of a thymocyte suspension treated with Zap-Oglobin, and it is obtained from counting the number of particles in each threshold at a setting of $A = 1/2$ and $I = 0.354$, which corresponds to 2.04 μm^3 per each threshold. In Fig. 1A, two populations of particles are seen. The first distribution, showing a marked decrease from a threshold of 1 to 10, is due to debris. The second distribution, giving a peak at a threshold of about 30, is due to thymocyte nuclei. The mean volume of the nuclei was calculated to be 67 μm^3. According to this distribution pattern, the particles which are counted between threshold 15 and threshold 60 are measured as thymocyte nuclei. On the other hand, the thymocyte suspension without Zap-Oglobin shows a similar distribution pattern at a setting $A = 1$ and $I = 1/2$, and the mean volume is 178 μm^3 (Fig. 1B).

The volume distribution pattern of erythrocytes (mean volume 77 μm^3) overlaps with that of thymocyte nuclei; erythrocytes, however, are completely solubilized by addition of Zap-Oglobin (Table I). Therefore, contaminating erythrocytes do not interfere in the cytotoxicity assay.

Recovery of Cell Counts after Treatment with Zap-Oglobin and Pronase E

Addition of Zap-Oglobin after incubation with Pronase E results in the liberation of nuclei from the remaining viable cells. The effects of Zap-Oglobin and protease treatment on viable cells and dead cells are shown in Table I.

Cells (5×10^5) of each suspension in 0.3 ml of medium containing 1% BSA are incubated with 0.3 ml of PBS or 0.3 ml of 3 mg/ml of Pronase E at 37°. After 30 min, 0.5 ml of the mixture is added to a counting vial containing 10 ml of Isoton II and is counted at cellular threshold (values under column headings 1 and 3, in Table I). Then, three drops (0.1 ml) of Zap-Oglobin are added to the vial and the sample is counted once more at nuclear threshold (values under column headings 2 and 4, in Table I). For the calculation of cell recovery after treatment with Zap-Oglobin and Pronase E, the complete recovery is set at the counts with Zap-Oglobin but without Pronase E treatment.

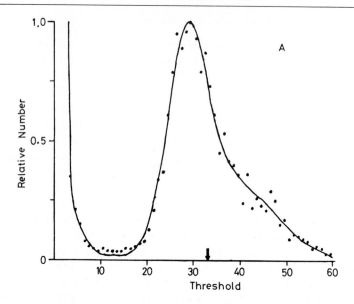

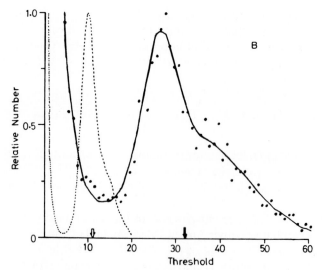

Fig. 1. (A) Volume frequency distribution patterns of thymocyte nuclei. Sample treated with Isoton II and Zap-Oglobin. The mean volume of thymocyte nuclei (◆) is 67 μm^3. $A = 0.5, I = 0.354, 2.04$ μm^3 per each threshold. (B) Volume frequency distribution patterns of thymocytes (solid line), sample treated with Isoton II only, and thymocyte nuclei (broken line). The mean volume of thymocytes (◆) is 178 μm^3 and that of nuclei (◇) is 67 μm^3. $A = 1, I = 0.5, 5.75$ μm^3 per each threshold.

TABLE I

RECOVERY OF CELL COUNTS AFTER TREATMENT WITH ZAP-OGLOBIN AND PRONASE E

Cell preparations	Number of particles/ml ($\times 10^{-4}$)[a]				Percentage viability (dye exclusion)
Pronase E[b]:	1	2	3	4	
Zap-Oglobin[c]:	–	–	+	+	
	–	+	–	+	
Viable cells	4.50 ± 0.04 (100)	4.50 ± 0.02 (100)	4.20 ± 0.04 (93.3)	4.26 ± 0.08 (94.7)	97.2
50% Dead cells	3.42 ± 0.01 (104.0)	3.29 ± 0.04 (100)	2.20 ± 0.03 (66.9)	1.94 ± 0.06 (58.2)	52.2
100% Dead cells	3.47 ± 0.10 (108.1)	3.21 ± 0.07 (100)	0.59 ± 0.03 (18.4)	0.03 ± 0.01 (0.93)	0.0
Erythrocytes	4.67 ± 0.14 (100)	0.02 ± 0.01 (0.43)		0.03 ± 0.01 (0.93)	

[a] Mean ± SD for triplicate samples. Percentage recovery is shown in parentheses. See text for counter settings.

[b] +, Incubation with 0.3 ml of 3 mg/ml Pronase E at 37° for 30 min. –, Incubation with 0.3 ml of PBS at 37° for 30 min.

[c] +, Presence of three drops of Zap-Oglobin in counting solution; –, without Zap-Oglobin.

TABLE II
DOSE RESPONSE OF PRONASE E TREATMENT

Pronase E concentration (mg/ml)	Percentage decrease from Pronase E-free control		
	Viable cells	50% Dead cells	100% Dead cells
0	0	0	0
1	0	39.5	89.5
2	1.0	43.2	88.5
3	0.4	48.5	96.1
Trypan blue[a]	5.0	50.0	100

[a] Percentage of dead cells estimated by the trypan blue dye exclusion method without Pronase E treatment.

In the case of a viable cell suspension, there is no difference in the counts for cells and nuclei, suggesting that the effect of protease is negligible (the percentage viability by dye exclusion is 97.2% and the percentage recovery with Pronase-Zap treatment is 94.7%). However, protease decreases the count of dead cells by about 80%. Addition of Zap-Oglobin causes the complete destruction of dead cells, yielding background level counts.

When erythrocytes are present together with the target cells, they are completely solubilized with Zap-Oglobin, and so have no effect on the count of the target cells.

Effect of the Amount of Pronase E and of the Incubation Time

The dose dependence of the effect of Pronase E on viable and dead cells is shown in Table II. Satisfactory results are obtained with Pronase E at 3 mg/ml. With this amount the observed lysis is 0.4% for viable cells, 48.5% for 50% dead cells, and 96.1% for 100% dead cells. Only partial digestion is observed at lower protease concentrations, and 5 mg/ml Pronase E has the same effect as 3 mg/ml Pronase E.

The percentage of cells digested at 37° as a function of time does not change appreciably. After incubation at 37° for 30 min, only 5% of the viable cells are lost, but the percentage lysis of 100% dead cells is 98.5%, and that of 50% dead cells is 52.4%. Similar results are obtained after incubation for 60 min.

TABLE III
REPRODUCIBILITY OF PRONASE E AND ZAP-OGLOBIN PROCEDURE

Well number	Viable cells			50% Dead cells			100% Dead cells		
	Count displayed[a]		Number of cells/ml[b] (10^{-4})	Count displayed		Number of cells/ml (10^{-4})	Count displayed		Number of cells/ml (10^{-4})
1	16,286	16,305	3.40	7,593	7,407	1.50	128	106	0.02
2	16,588	16,673	3.48	7,711	7,823	1.55	160	128	0.03
3	16,758	16,782	3.52	7,875	8,035	1.59	132	112	0.02
4	16,399	16,490	3.43	7,991	7,892	1.59	129	135	0.03
5	16,524	16,621	3.45	7,630	7,536	1.51	125	160	0.03
Average			3.46			1.55			0.03
SD[c]			0.05			0.04			0.01

[a] Number of particles in 0.5 ml of counting solution at a setting of $A = 0.5$, $I = 0.354$, $T_L = 15$, $T_U = 60$.

[b] Sum of two displayed counts indicating cells/1 ml of counting solution. If the displayed count is over 10,000, the value is corrected by the available correction table of the Coulter counter.

[c] Standard deviation.

Reproducibility of the Method

To test the reproducibility of this method, five samples of each cell suspension are treated with 3 mg/ml of Pronase E at 37° for 30 min, then counted with Zap-Oglobin at the nuclear threshold. The results listed in Table III demonstrate the high reproducibility of this method. Furthermore, the counts of each cell suspension are stable for at least 3 hr,[2] which gives adequate time for measuring many samples at one time.

Assay Procedure of Antibody-Mediated Cytotoxicity

Target cells (5×10^5 thymocytes in 0.1 ml of medium) are added to a plastic tube containing 0.1 ml of diluted antiserum and 0.1 ml of complement. Normal serum is present instead of antiserum in the controls. Each reaction mixture is incubated for 1 hr at 37° in a humidified atmosphere of 5% CO_2–95% air. After incubation, 0.3 ml of Pronase E solution is added to each tube and the mixtures are further incubated for 30 min. At the end of the incubation, 0.5 ml of the final mixture from each tube is transferred to a counting vial containing 10 ml of Isoton II and three drops (0.1 ml) of Zap-Oglobin. After mixing, the nuclei in the solution are counted in a Coulter counter at nuclear threshold. Counts displayed in the counter window are the number of particles in 0.5 ml of counting solution. Counts/ 1 ml of the solution are calculated as the sum of the counts determined by

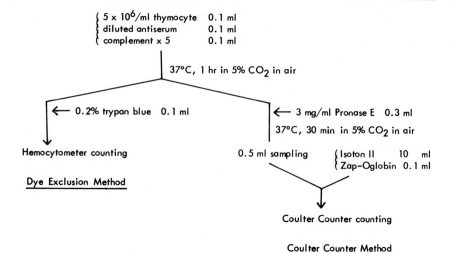

FIG. 2. Procedures for the dye-exclusion and Coulter counter methods.

counting the same sample twice. Results are expressed as follows:

$$\text{Percentage lysis (percentage decrease)} = \left(1 - \frac{\text{counts in sample}}{\text{counts in control}}\right) 100$$

All tests are performed at least in triplicate. The procedures are summarized in Fig. 2, together with the dye-exclusion method. When a titration of anti-mouse brain-associated theta (θ) serum with C3H/He mouse thymocyte is carried out by these two procedures, the titration curves obtained are in good agreement.[2]

In this assay, the complement solution supplies an adequate amount of proteins to protect the target cells from harmful digestion. BSA (1%) may be used if less or no complement is present in the mixture.

The cell number determined with the Coulter counter is linear at cell concentrations from 1×10^3 to 1×10^5 cells/ml.

Application to Tissue Culture Cells

For this purpose the method is modified in two major points: Dispase is used instead of Pronase, and the Zap-Oglobin treatment is omitted. Thus, what is determined by this method is the number of cells with membranes that are resistant to dispase treatment.[5,6] Dispase (available

[5] T. Matsumura, R. Konoshi, and Y. Nagai, *In Vitro* **18**, 510 (1982).
[6] T. Matsumura and M. Nagata, *Tissue Cult.* **9**, 348 (1983) (in Japanese).

from Boehringer-Mannheim) is a neutral protease purified from *Bacillus polymyxa*. It is active and stable during incubation in a tissue culture medium containing fetal bovine serum.[7,8]

Cell Preparations

Cells, e.g., L-929 mouse fibroblasts in monolayer, and FM3A mouse mammary carcinoma cells in suspension, are cultivated in MEM supplemented with 5% (for FM3A cells) or 10% (for L cells) inactivated fetal bovine serum. A monodisperse suspension of live FM3A cells is obtained directly from the mother culture. L-cell monolayers are suspended either by mechanical dislodging, or by trypsinization, or by a dispase treatment using 1000 U (Anson's hemoglobin lysing unit)[7,8] dispase/ml in the growth medium. The cells are centrifuged and the pelleted cells are resuspended by pipetting. This cell suspension is called a live cell suspension. An aliquot of this suspension is frozen and thawed twice by alternatively putting it into a liquid nitrogen bath and a water bath. A dead cell suspension is thus obtained. The number of dye-excluding cells in both cell suspensions is determined by the use of erythrosin dye.[9] Aliquots of the live cell suspension and of the dead cell suspension are mixed, so that the mixed suspension contains 20×10^4 to 25×10^4 cells/ml with the ratio of dye-excluding cells to total cells varying from less than 0.1% to more than 90%.

Dispase Treatment

One milliliter of each of these mixed cell suspensions and 1 ml growth medium containing 2000 U/ml dispase are added to glass vials with 4.5 cm² bottom surface area (No. 986542, Wheaton Scientific). The vials are incubated for 30 min at 37° in a humidified atmosphere containing 5% CO_2. After the incubation, 16 ml PBS is added to each vial, and the vials are kept in a ice-cold tray until electronic counting.

Electronic Counting

Cells with a volume of more than 1.0×10^3 μm^3 are enumerated as viable cells. In our counter, the lower threshold value, the upper threshold value, the aperture current value, and the amplification value were set at 21, 100, 4, and 2, respectively.

[7] T. Matsumura, T. Yamanaka, S. Hashizume, Y. Irie, and K. Nitta, *Jpn. J. Exp. Med.* **45**, 377 (1975).

[8] T. Matsumura, K. Nitta, M. Yoshikawa, T. Takaoka, and H. Katsuta, *Jpn. J. Exp. Med.* **45**, 383 (1975).

[9] H. J. Phillips and R. V. Andrews, *Exp. Cell Res.* **16**, 678 (1959).

Reliability and Comments

Optimal time of incubation with dispase is from 15 to 60 min. An electronic count value for a dead cell suspension, which contains less than 0.1% dye-excluding cells, is less than 5% (for L cells) or 2% (for FM3A cells) of that for the live cell suspension from which the dead cell suspension was prepared. The electronic count value and the percentage of dye-excluding cells are proportional from 0.1% to more than 90% dye-excluding cells. For L cells, since the background value is relatively high, the following procedure is adopted. A vial containing the mixed cell suspension is incubated for a period sufficient for the live cells to attach to the bottom of the vials. The medium is then removed by suction and a 1-ml aliquot of 1000 U/ml dispase solution is dispensed into each vial for a further 30 min incubation. The vials are placed for a few seconds onto a Vortex mixer to complete cell detachment. Then the cell suspensions are diluted with PBS and subjected to electronic counting as above.[5]

Although the electronic counting method is simple and rapid, it also has a number of drawbacks. The number of viable cells determined by the present method and that determined by a dye-exclusion test, although they are proportional, differ considerably. The electronic count value may sometimes be only 70% of the value obtained by dye exclusion. This is because the electronic count value depends on the volume of the cells, which is affected by the culture conditions. Consequently, this method is less sensitive than a dye-exclusion test; the difference is at least a factor of 10. It is recommended that the electronic counting method be calibrated every time with a dye-exclusion test. The Coulter counter method is not generally applicable to any kind of tissue culture cells. It can be used for suspended cells and cells completely dissociable with dispase (e.g., L cells), but not for cells forming aggregates upon dispase treatment (e.g., HeLa cells).[8] The optimal concentration of dispase should be examined when other lines of cells are used, because treatment with high concentrations may sometimes lead to severe injury to the live cells.

[32] Continuous Assay for Cytolytic Activity

By STEVE J. MCFAUL and JOHANNES EVERSE

The most commonly used assays for cellular damage are trypan blue exclusion,[1] ^{51}Cr release,[2-6] and monitoring the leakage of cellular constituents such as enzymes or ATP.[7-10] Trypan blue exclusion is based on the passive diffusion of trypan blue, a high-molecular-weight dye, across damaged cell membranes; healthy cells exclude the dye. The damaged cells acquire a blue appearance in a light microscope field and, thus, can be counted among healthy cells. Chromium-51 release is based on the leakage of ^{51}Cr from cells which were previously allowed to passively accumulate the isotope. Background leakage from healthy cells is subtracted from the total leakage in order to quantitate the number of damaged cells. Enzyme activities and/or ATP levels are measured by appropriate methods in the assay supernatant following the removal of cells by centrifugation. The extent of cellular damage is directly proportional to the level of these constituents.

Each of these methods shares a common disadvantage: they are static measurements that require the removal of aliquots from the assay medium at timed intervals. In addition, trypan blue exclusion is quite insensitive, and none of the above methods detects actual cell lysis. An additional disadvantage of measuring ^{51}Cr release is the inconvenience of working with radioactive isotopes.

Such measurements cannot provide detailed information concerning the progress of the cytolytic activity, unless a large number of samples are taken at specified time intervals. It would be extremely cumbersome if this would have to be done for each assay. Most cytotoxic assays are therefore done by performing an end-point analysis after the time at which the cytotoxic activity ceased has been established by analyzing a standard

[1] E. A. Boyse, L. J. Old, and I. Chouroulinkov, *Methods Med. Res.* **10**, 39 (1964).

[2] D. O. Adams, W. J. Johnson, E. Fiorito, and C. F. Nathan, *J. Immunol.* **127**, 1973 (1981).

[3] E. C. Jong and S. J. Klebanoff, *J. Immunol.* **124**, 1949 (1980).

[4] R. A. Clark, S. J. Klebanoff, A. B. Einstein, and A. Fefer, *Blood* **15**, 161 (1975).

[5] W. R. Henderson, E. Y. Chi, E. C. Jong, and S. J. Klebanoff, *J. Exp. Med.* **153**, 520 (1981).

[6] R. Canty and J. Wunderlich, *J. Natl. Cancer Inst. (U.S.)* **45**, 761 (1970).

[7] H. J. Sips and M. N. Hamers, *Infect. Immun.* **31**, 11 (1981).

[8] L. M. Aledort, R. Weed, and S. Troup, *Anal. Biochem.* **17**, 268 (1966).

[9] E. Beutler and C. K. Mathoi, *Blood* **30**, 311 (1967).

[10] G. J. Brewer and R. D. Powell, *J. Lab. Clin. Med.* **67**, 726 (1964).

assay. However, performing an end-point analysis on a standard assay to which, e.g., an inhibitor has been added only provides information as to the number of cells lysed at the time of the assay. Whether cell lysis occurs at a slower rate but eventually reaches the same level as that obtained without the inhibitor, or simply less cells are lysed before the cytotoxic activity ceases, or possibly a lag time is created in the lytic profile, are questions that cannot be answered from an end-point analysis.

Questions such as these can readily be answered by using an assay in which the toxic action can be continuously monitored. In this section we describe such an assay in which erythrocytes are used as the target cells. The method, however, could be adapted to utilize other target cells as well.

The continuous cytolytic assay described here is an adaptation of the lysozome assay developed by Shugar[11] in 1952. It is based on the fact that the turbidity of a suspension of erythrocytes, as measured by light scattering, decreases considerably upon lysis of the cells. Changes in light scattering can be measured conveniently and continuously in a spectrophotometer, using a wavelength at which none of the components of the system has any appreciable absorbance.

The observed changes in absorbance allows one to calculate rates of the lytic reaction as well as other kinetic parameters. The method allows one to ascertain that the end of the lytic activity has been reached before the assay is terminated. Furthermore, the effects of various agents on the cytolytic profile can be readily observed and these effects can be expressed in a quantitative manner, as illustrated below.

In the following detailed description of the method the lactoperoxidase–H_2O_2–iodide system is used as the cytolytic agent. However, other toxic substances, the action of which leads to erythrocyte lysis, could be used under the same conditions.

Reagents

Freshly drawn heparinized blood is centrifuged at 2000 g for 3 min and the packed erythrocytes are washed twice with phosphate-buffered saline (PBS).[11a] The erythrocytes are then suspended in PBS at a concentration of 2 × 10^7 cells/ml. The suspension is kept in ice until use

Buffer: 9.5 mM sodium phosphate, 140 mM NaCl, pH 7.2 (PBS)

Lactoperoxidase (Sigma Chemical Company, 100 U/mg), 10 μg/ml, in PBS

Hydrogen peroxide, 1 mM in PBS

Potassium iodide, 250 μM in PBS

[11] D. Shugar, *Biochim. Biophys. Acta* **8**, 302 (1952).
[11a] Abbreviations: PBS, phosphate-buffered saline; RBC, red blood cells.

Cytolytic Assay

The following amounts are placed in a polystyrene 1-ml cuvette (Fisher Scientific):

660 μl Buffer
100 μl Erythrocyte suspension
100 μl Lactoperoxidase solution
100 μl KI solution

After mixing the reagents the cuvette is placed in the spectrophotometer, the cell housing of which is maintained at 37°. After incubation for a few minutes to allow for temperature equilibration, 40 μl of the hydrogen peroxide solution is added and the change in absorbance is monitored at 600 nm. The reference cuvette contains all ingredients, except erythrocytes.

Note: If the cytotoxic assay takes a relatively long time to reach its end-point, errors may be introduced in the observed optical density changes due to some settling of the erythrocytes and some spontaneous lysis that may occur. These errors can be mostly eliminated by having erythrocytes also present in the reference cuvette (and omitting H_2O_2 or lactoperoxidase instead). This presents a technical difficulty, however, because at the onset of the reaction the difference in turbidity between the assay cuvette and the reference cuvette is zero, which the spectrophotometer will record as zero absorbance. Upon the initiation of cell lysis the light passing through the assay cuvette will increase relative to that passing through the reference cuvette, which will be indicated as a negative absorbance. This difficulty can be overcome by reversing the arrangement of the cuvettes, i.e., the sample cuvette is placed in the reference beam and the reference cuvette is placed in the sample beam of the spectrophotometer. This arrangement will invert the observed change in optical density, i.e., an increase in optical density, rather than a decrease, will be registered as cells lyse. An example of this is presented in Fig. 1, which shows the profile of a cytolytic assay using the peroxidase system.

The wavelength that should be used to determine the lytic activity may vary dependent on the cell type used as well as on the spectral properties of the toxic agents to be tested. The optimal wavelength can be determined experimentally by recording the absorbance of the reaction mixture between 400 and 800 nm before any cell lysis occurs and again after cell lysis is complete. Figure 2 shows the results obtained with erythrocytes and lactoperoxidase. It is clear from the figure that the largest difference between lysed and intact cells is observed between 600 and 800 nm while the lysate has little absorbance in this region. Any wavelength between

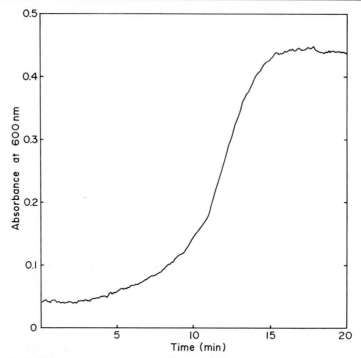

FIG. 1. Turbidometric profile of peroxidase-mediated cytolysis of erythrocytes. Assay cuvettes contained PBS, 40 μM H_2O_2, 25 μM KI, 1 μg lactoperoxidase, and 1 × 10⁶ erythrocytes in a 1-ml total volume. (Reprinted from McFaul *et al.*, *Proc. Soc. Exp. Biol. Med.* **179,** 331–337, 1985, by permission.)

600 and 800 nm can therefore be appropriately used to measure erythrocyte lysis (600 nm was arbitrarily chosen in the procedure described above).

Calculations

If the change in absorbance is linear, the rate of cell lysis can be calculated from its slope using the equation:

$$\Delta(RBC)/\Delta t(min) = antilog[(\log \Delta A_{600} + b)/m(t_2 - t_1)]$$

where the rate is expressed as the number of cells lysed per minute, and m and b are the slope and intercept, respectively, of the relationship between erythrocyte density and the absorbance at 600 nm, described by the following equation:

$$\log A_{600} = m \log[RBC] - b$$

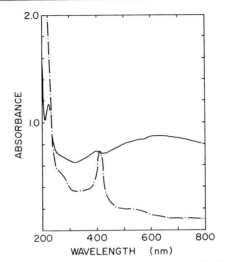

FIG. 2. Wavelength profile of a human erythrocyte suspension in PBS before (—) and after (−·−) lysis of the cells with lactoperoxidase. (Reprinted from McFaul *et al.*, *Proc. Soc. Exp. Biol. Med.* **179**, 331–337, 1985, by permission.)

This relationship is shown in Fig. 3. The relationship is linear between 4×10^5 and 1×10^7 erythrocytes/ml, with a correlation coefficient of 0.999 as determined with a Beckman model 24 spectrophotometer. However, since the observed absorbance due to light scattering can vary

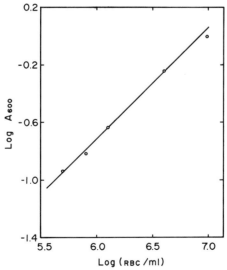

FIG. 3. Standard curve showing the relationship between erythrocyte concentration and the absorbance change at 600 nm.

somewhat among different spectrophotometers, it is suggested that a standard curve of log A_{600} vs log[RBC] be constructed and the appropriate values for m and b be determined before using this assay.

Calculations of a large number of rates may be done most efficiently and reproducibly by selecting a calculating constant ΔA_{600} and Δt required to achieve this change in absorbance. Under this approach (log ΔA_{600} + b)/m is a constant, which facilitates the calculations.

Comparison of the Continuous Assay with Other Assays

The results shown in Fig. 1 indicate that the cytolytic activity of the lactoperoxidase–H_2O_2–I^- system toward erythrocytes proceeds with an initial lag time of several minutes, followed by the actual lysis of the cells. In order to use the continuous assay with confidence for the determination of cytolytic activity it is important to demonstrate that similar results are obtained using conventional methods. This was done for the cytolytic system under discussion and the results are shown in Fig. 4, in which the rate of hemoglobin release, total protein release, ATP release, and ^{51}Cr release are measured under the conditions described in Fig. 1. The data show that the curve obtained with the continuous assay is very similar to the curves obtained by measuring the release of cellular components. The rate of ^{51}Cr release, however, was found to be somewhat slower.

Additional Comments

It should be noted that the order in which the individual components are added to the cuvette may have a significant effect on the results. For instance, when lactoperoxidase, H_2O_2, and iodide are mixed in the absence of target cells, the cytolytic mixture loses its cytolytic activity with a half-life of about 35 sec. Thus, if the erythrocytes are added last, a significantly lower cytotoxicity may be observed than when hydrogen peroxide is added last. A few simple experiments will usually indicate whether such a problem exists.

Erythrocytes as the target cells are ideal in the continuous assay, because they do not contain intracellular organelles. Intracellular organelles do cause light scattering after cells are lysed. Therefore, the difference in optical density observed between intact cells and lysed cells is less when the cells contain nuclei, mitochondria, etc., than when the target cells do not contain such organelles.

Erythrocytes that are stored at 4° for a few days are more susceptible to lysis than freshly drawn cells, indicating that the membrane becomes more fragile upon storage. It is therefore important, when a comparison of the cytotoxic activity of various compounds is to be carried out, to do so

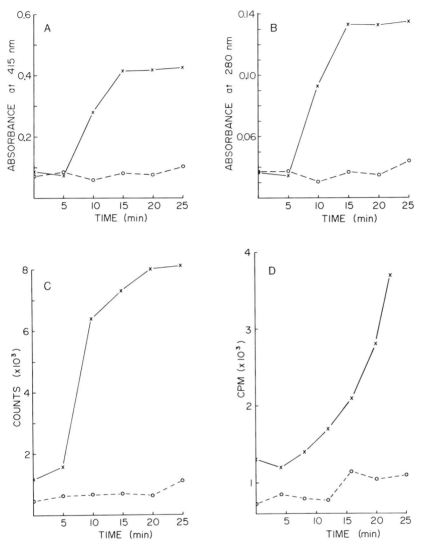

FIG. 4. Release of cellular components from human erythrocytes upon treatment with the lactoperoxidase–H_2O_2–I^- system. The reaction mixtures were constituted as in Fig. 1. At the specified time intervals the cells were removed by centrifugation and the supernatant was analyzed for its hemoglobin content (A), protein content (B), and ATP content using a luciferase kit (C). The release of ^{51}Cr from cells previously incubated with this isotope is shown in D. ×, Complete reaction resulting in cell lysis; ○, lactoperoxidase omitted from reaction mixture. (Reprinted from McFaul *et al.*, *Proc. Soc. Exp. Biol. Med.* **179**, 331–337, 1985, by permission.)

with the same batch of erythrocytes. Differences in sensitivity toward cytolytic agents are also observed with erythrocytes harvested from different animals. A "standard" assay should therefore be performed at relatively frequent intervals (at least once a day) in order to assure day-to-day reproducibility. It is especially important to maintain the proper osmolarity of the assay solution. Slightly hypotonic solutions will observably contribute to cell fragility and increase the lytic rate, whereas hypertonic solutions may inhibit cell lysis.

Finally, the rate of cell lysis is strongly dependent on the temperature of the reaction mixture. Therefore, obtaining reproducible results requires that the temperature of the assay be the same in each experiment. This can be achieved by preequilibrating all buffers and stable solutions at or slightly above the desired assay temperature and using as small a volume as is practical of any solution that is at a significantly different temperature, e.g., is kept on ice. Equilibrating the assay solution in the spectrophotometer for several minutes and starting the reaction by adding a preequilibrated solution greatly enhances the reproducibility of the experiments.

Acknowledgments

This work was supported in part by Grant D-676 of the Robert A. Welch Foundation and by Grant CA-32715 of the National Cancer Institute, National Institutes of Health.

[33] Measurement of Phagocytosis and Cell-Mediated Cytotoxicity by Chemiluminescence

By Knox Van Dyke, Michael R. Van Scott, and Vincent Castranova

Introduction

Chemiluminescence (CL)[1] is the production of light generated from chemical sources.[1a] Generally the light so produced is in the visible range (400–600 nm). However, light can also be in the ultraviolet or infrared range. Bioluminescence is the production of light from biological sources.

[1] Abbreviations: CL, chemiluminescence; LSC, liquid scintillation counter; FMLP, n-formylmethionylleucylphenylalanine; PMA, phorbol myristate acetate; PAF, platelet-activating factor; DMSO, dimethyl sulfoxide.
[1a] K. Van Dyke, ed., "Bioluminescence and Chemiluminescence: Instruments and Applications," Vol. 1, p. 1. CRC Press, Boca Raton, Florida, 1983.

Ultimately the bioluminescent source stems from chemical entities and can be considered a branch of chemiluminescence.

Cellular CL is usually considered the production of light from single cells. Cell CL is generally descriptive of the light produced from activated phagocytic cells such as granulocytes, monocytes, and alveolar or peritoneal macrophages. However, depending on the stimulant, nonphagocytic cells such as erythrocytes, platelets, and lymphocytes can also produce light. Also there is a variety of marine bacteria and fungi that may produce light. In addition to single cells producing CL, cells can be organized into organs which produce luminescence. This response is described as whole organ CL.[2]

Allen and co-workers are generally credited with the recognition of cellular CL as it relates to phagocytic cells.[3] Although initial studies were limited to monocytes and polymorphonuclear leukocytes, later reports have shown that macrophages could also produce cellular CL.[4]

In the early phases of phagocytic CL only the "natural light" resulting from the interaction of reactive cellular products with particles was measured. However, measurement of natural light requires relatively large numbers of phagocytic cells. The problem of low sensitivity can be overcome by the use of "bystander" molecules, i.e., chemicals which are activated by reactive cellular products to produce light more efficiently.[5] This increased sensitivity allows the use of fewer cells to monitor CL and, thus, allows research with neonates, patients with certain leukemias and neutropenias, and those with immune disorders.[6,7] Initially, only luminol was used as the photofluor but now the use of both luminol and lucigenin are commonplace.

The production of cellular CL from phagocytic cells seems to follow a general course.[8] Membrane receptors of phagocytes can be stimulated by either particles or certain soluble stimulants, e.g., formylated peptides, phorbol myristate acetate, or concanavalin A. Such receptor binding results in transmembrane ionic movements and depolarization of the

[2] A. Boveris, E. Cadeñas, and B. Chance, *Fed. Proc., Fed. Am. Soc. Exp. Biol.* **40**, 195 (1981).

[3] R. C. Allen, R. C. Stjernhölm, and R. H. Steele, *Biochem. Biophys. Res. Commun.* **47**, 679 (1972).

[4] P. R. Miles, P. Lee, M. A. Trush, and K. Van Dyke, *Life Sci.* **20**, 165 (1977).

[5] R. C. Allen and L. D. Loose, *Biochem. Biophys. Res. Commun.* **69**, 245 (1976).

[6] M. E. Wilson, M. A. Trush, and K. Van Dyke, *J. Immunol. Methods* **23**, 315 (1978).

[7] P. Stevens, D. J. Winston, L. S. Wong, and K. Van Dyke, *Infect. Immun.* **22**, 41 (1978).

[8] K. Van Dyke, C. Van Dyke, D. Peden, M. Matamoros, V. Castranova, and G. S. Jones, *in* "Bioluminescence and Chemiluminescence" (M. Deluca and W. D. McElroy, eds.), p. 45. Academic Press, New York, 1981.

membrane potential.[9] This depolarization may trigger a variety of cellular biochemical events which result in activation of the respiratory burst, lysosomal enzyme release, lipid metabolism, and phagocytosis. Ionophores for Ca^{2+} or Na^+, which do not bind to receptors but induce ionic currents which directly depolarize the cells, also stimulate phagocytes.[10] This general activation of phagocytic activity can be monitored by measuring the generation of cellular CL.

Methods

Instrumentation for Detection of Cellular CL

In our previous review[11] we noted that the phenomena of cellular CL could be followed in an LSC or in less expensive photometers. In addition, we presented the rationale for use of luminol, presentation of data, and assessment of opsonophagocytic dysfunctions. However, these LSC systems are not specifically designed for CL studies. LSC systems generally cannot be easily temperature controlled at 37° and are not adapted for rapid sampling. Such flexibility is important since the metabolic reactions which result in CL are very temperature-dependent and may occur within a few seconds. In addition, photometers are not sufficiently sensitive. Thus, large numbers of cells and high doses of stimulant are necessary to measure cellular CL. In fact, they are often too insensitive to measure natural light. Lastly, neither of these systems are normally equipped with data-handling capability. Recently, sensitive state-of-the-art luminometers with or without data-handling capability have been introduced. These instruments are specifically designed to measure CL. The systems are temperature controlled and capable of monitoring both "natural" and "luminol-augmented" CL.

There are two basic approaches that can be used to follow cellular CL: (1) A single-well, temperature-controlled detector can be used to follow CL. This system can monitor CL from a single sample. Multiple samples can be run by changing samples manually. The detector can be attached to real-time data reduction via computer. Such a system is the less expensive and can handle low to intermediate numbers of samples. (2) A sample switching single-well, temperature-controlled detector can be used to follow CL from a number of samples. Both the detection chamber and the belt assembly are temperature controlled. This system can be attached to real-time data collection to record and analyze the results.

[9] G. S. Jones, K. Van Dyke, and V. Castranova, *J. Cell. Physiol.* **106**, 75 (1981).
[10] V. Castranova, G. S. Jones, R. M. Phillips, D. Peden, and K. Van Dyke, *Biochim. Biophys. Acta* **645**, 49 (1981).
[11] M. A. Trush, M. E. Wilson, and K. Van Dyke, this series, Vol. 57, p. 462.

There are three major manufacturers of equipment designed specifically to analyze cellular CL. They are Berthold (Wildbad, West Germany), LKB, Wallac (Turku, Finland), and Packard Instrument Company (Downer's Grove, IL). These instruments and their characteristics have been described.[1a] Briefly, Berthold makes three different instruments: (1) a six-chamber system (LB9505) in which six separate analyses can be accomplished simultaneously, (2) a moving-chain multiple sampler (LB950) capable of handling 400 samples, and (3) a single-well counter (9500T). All three systems can be linked via computer to give continuous display, printout, and data analysis. LKB makes a 25-sample changer system that can be linked to computer data-handling systems. Packard makes a 48-sample discrete sample system (6500 Model) which controls temperature and mixing of samples. The display system can be linked via computer for real-time data analysis. Choice among these instruments depends on the number of samples, the applications desired, and the availability of funds. However, these instruments are clearly superior to LSC, which are not built specifically for measurement of cellular CL. Another major advantage is that smaller tubes and, therefore, fewer cells can be used and that vials do not have to be dark adapted. In contrast to using an LSC, one may work efficiently in a well-lighted room using these new systems.

Measurement of Cellular CL

Many studies of cellular CL with phagocytes have been accomplished with partially isolated or isolated phagocytes.[11] Isolation of cells is time consuming. In addition, biological activity may be affected by the isolation procedure. For these reasons, the use of whole blood has been introduced.[12–15] Although erythrocytes tend to quench some of the CL from whole blood, a variety of approaches can be taken to lessen or eliminate this problem. The best solution to the problem is to use highly diluted blood to eliminate erythrocyte quenching of light.[16,17]

Whole Blood CL. Recently, methods have been devised to screen for defects in the phagocytic cells of patients using whole blood samples

[12] H. Fischer, T. Kato, H. Wokalek, M. Ernst, H. Eggert, and E. T. Rietschel, in "Bioluminescence and Chemiluminescence" (M. DeLuca and W. D. McElroy, eds.), p. 617. Academic Press, New York, 1981.

[13] H. Faden and N. Maciejewski, *J. Reticuloendothel. Soc.* **30**, 219 (1981).

[14] T. Tono-Oka, N. Ueno, T. Matsumoto, M. Ohkawa, and S. Matsumoto, *Clin. Immunol. Immunopathol.* **26**, 66 (1983).

[15] B. Descamps-Latscha, A. T. Nguyen, R. M. Golub, and M. N. Feuillet-Fieux, *Ann. Immunol.* **133**, 20 (1982).

[16] R. C. Allen and B. A. Pruitt, *Arch. Surg. (Chicago)* **117**, 133 (1982).

[17] G. Bruchelt and K. H. Schmidt, *J. Clin. Chem. Clin. Biochem.* **22**, 1 (1983).

rather than preparations of isolated granulocytes. Defects which can be monitored by whole blood CL may result from the following causes:

1. Genetic errors producing either cellular or hormonal defects (e.g., chronic granulomatous disease or myeloperoxidase deficiency)
2. Defects resulting from immunosuppressive drugs administered for organ transplants, cancer chemotherapy, autoimmune diseases, or treatment for severe rheumatoid arthritis (methotrexate)
3. Immunodeficiency states resulting from human T-cell leukemia virus III (HTLV-III), which causes acquired immunodeficiency syndrome (AIDS)
4. Extensive burns which affect the complement system via humoral rather than cellular changes

These defects (either cellular or humoral) can be detected by whole blood chemiluminescence using a series of soluble or particulate stimulants. The soluble stimulants (purchasable from Sigma Chemical Company) include PAF, FMLP, Ca^{2+} ionophore (A23187), and PMA. Typical whole blood CL data are shown in Fig. 1A–C. Detection of cellular defects using this system for chronic granulomatous disease would yield CL equivalent to resting cell level defects and myeloperoxidase deficiency would produce only 5% of the CL from normal cells.[18]

The assay of luminol-dependent whole blood chemiluminescence can be accomplished under a variety of conditions. The simplest of the conditions is the following:

1. Dilute heparinized whole blood 1 : 10 with HEPES-buffered solution (400 μl)
2. Add 50 μl of 1 × 10^{-3} M luminol
3. Add 50 μl of stimulant

However, the blood may be diluted 1 : 100 and CL assayed as follows:

1. Dilute heparinized whole blood 1 : 100 with HEPES-buffered solution (400 μl)
2. Add 50 μl of 1 × 10^{-5} M luminol
3. Add 50 μl of stimulant

The CL reaction is measured for 10 min at 37°. The concentration of the reagents used in whole blood is the following:

1. HEPES buffer—145 mM NaCl, 5 mM KCl, 10 mM HEPES, 1 mM $CaCl_2$, 1 mM $MgCl_2$, and 5 mM glucose (pH 7.4)

---

[18] L. DeChatelet and P. Shirley, *Clin. Chem.* (*Winston-Salem, N.C.*) **27,** 1739 (1981).

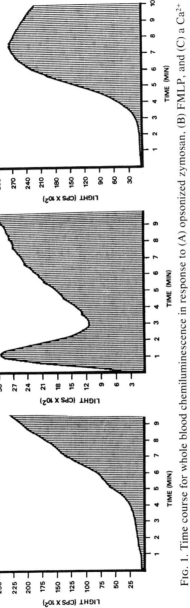

FIG. 1. Time course for whole blood chemiluminescence in response to (A) opsonized zymosan, (B) FMLP, and (C) a Ca^{2+} ionophore (A23187).

2. Stimulants
   a. Zymosan—1 mg of zymosan suspended in 1 ml of HEPES-buffered solution and sonicated for 15 sec
   b. PAF is added to bovine serum albumin (2.5 mg/ml in Hanks' balanced salt solution) at a concentration of 10^{-3} M
   c. FMLP—dissolved in 1 ml of DMSO and diluted with HEPES-buffered medium to 10^{-4} M
   d. PMA—dissolved in 1 ml of DMSO and diluted with HEPES-buffered medium to 10^{-4} M
   e. A23187—dissolved in 1 ml of DMSO and diluted with HEPES-buffered medium to 10^{-4} M

An alternative to the use of luminol as a chemiluminescent bystander molecule is the use of lucigenin (Sigma Chemical Company) at the same concentrations as luminol. Lucigenin displays different kinetics of light production than luminol and is thought to reflect the production of superoxide $(\cdot O_2)^-$ better than luminol.[16] Apparently, luminol is an indicator of peroxidase activity (either hydrogen peroxide or lipid peroxides).

The whole blood assay is best used when screening for defects of the cellular or humoral system. Both erythrocytes and serum can partially quench the production of cellular CL but the measurement of light emission can still be used to clearly establish whether a defect exists. Comparison of samples using blood from a single individual is valid, but comparison between samples from different individuals is not. The reasons for this are that the number of light-producing cells could vary between individuals and that the quenching of light, regardless of the mechanism, could vary between blood from different individuals. However, whole blood CL is an effective and quick screening method for detection of the presence or absence of cellular or humoral defects. It was not meant to be a quantitative measurement of cellular CL per se. That is best accomplished using isolated cell preparations.

Whole blood luminol-dependent CL can also be used to follow the kinetics of opsonization in whole blood (Fig. 1A). When particles are coated with immunoglobulins and/or complement, they can be recognized by human neutrophil and/or other granulocytic cells. If bacteria or yeast cell walls (zymosan particles) are added to whole blood, the particles become opsonized and light is generated as the particles are engulfed. Since some bacteria are not opsonized by a given individual, defects of opsonization are readily detected.[7] The more the blood is diluted the less the light is quenched. However, opsonization kinetics are more difficult to study when the opsonins are highly diluted.

Measurement of CL from Isolated Phagocytic Cells

Phagocytic cells that produce CL when stimulated by particles (either opsonized or nonopsonized) or soluble stimulants are granulocytes (neutrophils, eosinophils, and basophils), monocytes, and alveolar or peritoneal macrophages. Although activation of blood phagocytes can be investigated by monitoring CL in whole blood or other heterogeneous cellular preparations, it is often necessary to use purified preparations of phagocytes in order to study dose–response relationships of stimulants or inhibitors of CL and to detail mechanisms involved in stimulus–activation coupling. In addition, cellular purification is often required for studies involving phagocytes from tissues other than blood. We have outlined methods for purification of phagocytic cells from a variety of tissues in another publication ("Cellular Chemiluminescence," CRC Press, 1986) (see also this series, Vol. 108, [25], [26], [27], [28], [29]).

While a fairly wide variety of conditions can be utilized, the following considerations are important. Cells must be isolated using techniques which do not damage or activate the phagocytic cells. Second, cells must be counted microscopically with a hemocytometer or electronically with a cell counter (Coulter Instrument, Hialeah, FL). In general visual observations are helpful from a qualitative point of view, but electronic cell counting is easier and more precise. Last, use of phosphate-based buffers should be avoided since phosphate interacts with calcium, resulting in precipitation of the ionized calcium. For this reason, buffers such as HEPES are recommended.

There are several important considerations with regard to the stimulant used with the different types of phagocytes. Indeed, all types of phagocytes do not respond to all soluble stimulants. There is also a difference in the reaction to particles. Alveolar macrophages do not require opsonization (coating of particles with antibody and/or complement) to phagocytize and hence to produce CL.[19] On the other hand, blood granulocytes, monocytes, peritoneal macrophages, and leukocytes do require serum factors for proper recognition of particles.[11] Particles such as zymosan or bacteria can be opsonized by adding 1 mg of zymosan or 10^9 colony-forming units of bacteria to 1 ml of HEPES-buffered medium. This suspension is mixed in a 1 : 1 ratio with serum and incubated at 37° for 30 min. It is important to draw the blood in heparin rather than EDTA or citrate because opsonization requires ionic calcium to activate alternate complement pathways and EDTA or citrate would chelate calcium. Serum is obtained by simply centrifuging the blood at 10,000 g for 1 min at 2° to pellet the blood cells.

---

[19] T. D. Sweeney, V. Castranova, L. Bowman, and P. R. Miles, *Exp. Lung Res.* **2**, 85 (1981).

The CL reaction for phagocytic cells can be accomplished under the following conditions: (1) Add 10^6 cells to 400 μl of HEPES-buffered medium (10^7 cells should be added when less active macrophages are used); (2) add 50 μl of 10^{-6} M luminol or lucigenin (bystander molecules); (3) the reaction is initiated by adding 50 μl of soluble stimulant or particles to the cell suspension. CL is then measured at 37°.

The addition of luminol or lucigenin as a bystander molecule greatly increases the sensitivity of detection of CL. This affords two advantages: (1) fewer cells are required to monitor CL and (2) differences in resting (unstimulated) CL between different cellular preparations can be determined, giving an indication of the state of activity of the cells *in vivo*. However, the "natural" CL system, i.e., measurement of CL without luminol or lucigenin, is less prone to artifact, such as direct reaction of additives with the bystander molecule to either quench or initiate CL. Such artifacts are of concern when screening substances for their toxicity to phagocytes or their ability to alter cellular secretion of reactive products.

As with whole blood, CL from isolated phagocytes can be expressed in multiple ways, including slope, peak height, and total CL. Generally, we have found total CL, i.e., integration of the response under the CL curve, to be the most useful means of data presentation. However, peak height data are readily available. The control value where buffer containing no stimulant is added to the cell preparation should be subtracted from all data before presentation. In cases where luminol or lucigenin are not added, the blank values are low and almost unmeasureable.

Use of CL to Monitor Viability of Phagocytic Cells

Since phagocytic cells play an important role in natural defense mechanisms, studies concerning the toxicity of various compounds on these cells are of great importance. In toxicological investigation one could monitor any of a number of physiological parameters as an indication of cellular viability. A list of possible parameters is as follows:

1. Membrane integrity as monitored by trypan blue exclusion
2. Cellular metabolism as monitored by oxygen consumption
3. Phagocytic rate as monitored by the uptake of particles
4. Lysosomal enzyme activity
5. Lysosomal enzyme release
6. Release of reactive forms of oxygen as monitored by superoxide secretion, hydrogen peroxide release, or CL

Recent studies indicate that these parameters display a wide range of susceptibilities to toxic agents.[20,21] Trypan blue exclusion is a relatively insensitive test for toxicity, i.e., many cellular functions are inhibited by doses of toxins too low and for exposure times too short to cause leakiness of the plasma membranes of phagocytes. Lysosomal enzyme activity and lysosomal stability are also fairly insensitive. Phagocytic rate and oxygen consumption are sensitive tests for toxicity. However, inhibition of particulate-induced release of reactive forms of oxygen is the most sensitive indicator of cellular toxicity as judged by exposure time and dose.

Measurement of superoxide anion release is made by monitoring the reduction of cytochrome c at 550 nm while hydrogen peroxide release is measured fluorometrically at an excitation wavelength of 350 nm and an emission of 460 nm using scopoletin as the fluorescent probe. Both of these assays depend on changes in either the absorbance or fluorescence of an indicator. Therefore, reaction of a toxic agent with these indicators would result in artifactual results. Indeed, metallic ions and organic dusts such as cotton have been shown to cause artifacts in this assay.[22,23] In such cases, measurement of CL without the use of luminol has been found to be a useful alternative to the cytochrome c or scopoletin assays for monitoring the release of reactive forms of oxygen by phagocytes.

Summary

The generation of CL by phagocytes has been shown to be a valuable tool for monitoring the activity of phagocytic cells. It has been used to investigate mechanisms by which stimulants or inhibitors act to affect the function of phagocytic cells. CL has also been used as an assay system to detect various disease states. Last, it has been shown to be a very sensitive assay system for determining the toxicity of environmental or occupational agents on phagocytic cells.

[20] V. Castranova, L. Bowman, P. R. Miles, and M. J. Reasor, *Am. J. Ind. Med.* **1**, 349 (1980).
[21] V. Castranova, L. Bowman, J. R. Wright, H. Colby, and P. R. Miles, *J. Toxicol. Environ. Health* **13**, 845 (1984).
[22] V. Castranova, L. Bowman, M. J. Reasor, and P. R. Miles, *Toxicol. Appl. Pharmacol.* **53**, 14 (1980).
[23] P. R. Miles, M. J. Reasor, C. A. Glance, and V. Castranova, *Proc.—Cotton Dust Res. Conf., Beltwide Cotton Prod. Res. Conf., 8th, 1984*, p. 121 (1984).

[34] Activated Macrophage-Mediated Cytotoxicity: Use of the *in Vitro* Cytotoxicity Assay for Study of Bioenergetic and Biochemical Changes That Develop in Tumor Target Cells

By JOHN B. HIBBS, JR., and READ R. TAINTOR

Cytotoxic activated macrophages do not cause rapid or global metabolic disorganization in tumor target cells. The metabolic dysfunction induced by cytotoxic activated macrophages in tumor target cells develops slowly, is quite discrete, and in many cases is fully reversible. For example, cytotoxic activated macrophages inhibit DNA synthesis and mitochondrial respiration while certain other metabolic pathways, such as glycolysis, remain functional.[1] Recent experiments suggest that this reproducible pattern of metabolic perturbation may be due to cytotoxic activated macrophage-induced iron loss from tumor target cells.[2] This may result in inhibition of certain enzymes that require iron for catalytic activity (i.e., NADH : ubiquinone oxidoreductase, succinate : ubiquinone oxidoreductase, etc.[3]) and, as a result, block metabolic pathways with enzymatic activity vulnerable to inhibition by depletion of intracellular iron.

Target cells of cytotoxic activated macrophages slowly develop metabolic inhibition and the metabolic inhibition in many tumor cell lines is completely reversible.[1-3] Since it is possible to remove target cells from monolayers of cytotoxic activated macrophages, the target cells can be used for bioenergetic or biochemical analysis.

In addition, methods (preferably quantitative) are needed in each experiment to show that target cells cocultivated with activated macrophages develop characteristic and reproducible metabolic inhibition. This serves as a baseline for identifying the presence of cytotoxic effect and is necessary for interpretation of the significance of bioenergetic or biochemical changes that are being measured. An investigational approach such as this can provide information leading to a mechanistic understanding of the role of macrophages in regulation of cell proliferation.

[1] D. L. Granger, R. R. Taintor, J. L. Cook, and J. B. Hibbs, Jr., *J. Clin. Invest.* **65**, 357 (1980).
[2] J. B. Hibbs, Jr., R. R. Taintor, and Z. Vavrin, *Biochem. Biophys. Res. Commun.* **123**, 716 (1984).
[3] D. L. Granger and A. L. Lehninger, *J. Cell Biol.* **95**, 527 (1982).

Macrophage Activation

In order for macrophages to induce a cytotoxic effect on target cells, they must undergo functional change.[4] This functional change is described as activation. However, the term activation is not precise and also is used to describe the acquisition or loss of functional capabilities by macrophages that are unrelated to the expression of cytotoxic activity.[5-7] Here we describe methods for measuring activated macrophage-mediated cytotoxicity for eukaryotic target cells. This cytotoxic effect is immunologically nonspecific at the effector level and occurs without specific presensitization to target cell antigens. Macrophages expressing this function we term cytotoxic activated macrophages with the caveat that the type of cytotoxic effect observed with many lines of transformed target cells is cytostasis rather than cytolysis. Indeed, activated macrophage-induced metabolic changes in target cells may be more appropriately considered in the more general context of cell growth regulation rather than in the narrower context of cytotoxicity.

Macrophage activation is a process of differentiation resulting from a complex interaction of environmental chemical signals that either increase or decrease the probability of expression of macrophage-mediated cytotoxicity for target cells.[8] The following description of factors that influence the level of macrophage activation, both *in vivo* and *in vitro*, provides information helpful for planning and interpreting studies of macrophage cytotoxic activity *in vitro*.

Macrophages representing different functional states (within a continuum of macrophage differentiation) can be isolated from appropriate groups of mice and can also be generated *in vitro* by exposing the macrophages to appropriate stimuli.[8-11]

1. Resident macrophages are obtained from the unstimulated peritoneal cavities of normal mice (also termed normal macrophages). They often do not respond effectively *in vitro* to chemical signals that induce the expression of tumor cell killing.

[4] J. B. Hibbs, Jr., H. A. Chapman, Jr., and J. B. Weinberg, *in* "Mononuclear Phagocytes Functional Aspects" (R. Van Furth, ed.), p. 1681. Martinus Nijhoff, The Hague, 1980.
[5] R. J. North, *J. Immunol.* **121,** 806 (1978).
[6] R. Takemura and Z. Werb, *Am. J. Physiol.* **246,** C1 (1984).
[7] H. A. Chapman, Jr., Z. Vavrin, and J. B. Hibbs, Jr., *Cell* **28,** 653 (1982).
[8] J. B. Hibbs, Jr., R. R. Taintor, H. A. Chapman, Jr., and J. B. Weinberg, *Science* **197,** 279 (1977).
[9] S. W. Russell, W. F. Doe, and A. T. McIntosh, *J. Exp. Med.* **146,** 1511 (1977).
[10] J. B. Weinberg, H. A. Chapman, Jr., and J. B. Hibbs, Jr., *J. Immunol.* **121,** 72 (1978).
[11] L. P. Ruco and M. S. Meltzer, *J. Immunol.* **121,** 2035 (1978).

2. Stimulated macrophages, i.e., macrophages which have entered early stages of the activation response, are obtained from mice injected intraperitoneally (ip) with 10% peptone, thioglycolate broth, or other sterile nonimmunogenic inflammatory stimulants (also termed elicited or inflammatory macrophages). Stimulated macrophages are not cytotoxic for transformed cells but become cytotoxic when exposed to large amounts of bacterial lipopolysaccharide (LPS)[11a] or highly pure recombinant γ-interferon or lymphocyte supernatants with macrophage-activating factor (MAF) activity plus picogram or nanogram amounts of LPS (20 ng/ml or less).[8-12] The MAF activity in lymphokine preparations may be identical to γ-interferon.[12]

3. Noncytotoxic activated macrophages are obtained from the peritoneal cavity of mice with chronic *Mycobacterium bovis*, strain Bacillus Calmette–Guérin (BCG), or toxoplasma infection (also termed nontumoricidal activated macrophages or primed macrophages). Many other agents such as *Corynebacterium parvum* or pyran copolymer also can be used to induce noncytotoxic activated macrophages *in vivo*.[13,14] In addition, noncytotoxic activated macrophages are produced *in vitro* by exposure of stimulated macrophages to MAF-containing lymphocyte supernatants or low concentrations of γ-interferon. Noncytotoxic activated macrophages have a markedly lowered threshold for expression of cytotoxic effect because treatment of the macrophages with picogram or nanogram (20 ng/ml or less) quantities of LPS induces cytotoxic effect for target cells.[8-11] Naturally occurring infections in a mouse colony can be as effective as experimentally produced BCG or toxoplasma infection in causing *in vivo* acquisition of the noncytotoxic or cytotoxic activated state by the peritoneal macrophages. A mouse colony infection should be suspected when peritoneal macrophages from normal mice or mice inoculated with a nonimmunologic sterile inflammatory agent such as peptone are cytotoxic when exposed to very small quantities of LPS *in vitro*.

4. Cytotoxic activated macrophages (also termed tumoricidal activated macrophages) are fully differentiated cells capable of mediating target cell cytotoxicity. At times, when removed from the peritoneal cavity of mice infected with BCG or toxoplasma, activated macrophages

---

[11a] Abbreviations: LPS, lipopolysaccharide; MAF, macrophage-activating factor; DMEM, Dulbecco's modified minimal essential medium; DMEM–G, DMEM without glucose; DMEM+G, DMEM containing 20 mM glucose; CS, calf serum; FBS, fetal bovine serum; BCG, Bacillus Calmette–Guérin; [^{3}H]TdR, [^{3}H]thymidine; TCA, trichloroacetic acid.

[12] L. P. Svedersky, C. V. Benton, W. H. Berger, E. Rinderknecht, R. N. Harkins, and M. A. Palladino, *J. Exp. Med.* **159**, 812 (1984).

[13] J. L. Krahenbuhl, L. H. Lambert, Jr., and J. S. Remington, *Immunology* **31**, 837 (1976).

[14] A. M. Kaplan, P. S. Morahan, and W. Regelson, *J. Natl. Cancer Inst. (U.S.)* **52**, 1919 (1974).

function as cytotoxic effector cells without apparent exposure to additional differentiation signals *in vitro* rather than as noncytotoxic activated macrophages. This inconsistency is often due to LPS contamination of the tissue culture medium which can be detected and measured with the *Limulus* amebocyte (available from Sigma Chemical Company) lysate assay.[10,15] Polymyxin B, (Sigma Chemical Company), which binds to LPS and neutralizes its activity, can also be used to detect LPS contamination of the tissue culture environment. Strong evidence for LPS contamination exists if activated macrophages are cytotoxic for tumor cells in the absence of polymyxin B and noncytotoxic in the presence of 10–25 µg/ml of polymyxin B.[10] It is also likely that other more physiologic explanations exist for expression of cytotoxicity by activated macrophages without further exposure to differentiation signals *in vitro*. This may include the relative intensity of biochemical signals inducing macrophage differentiation to the cytotoxic activated state *in vivo* as well as the net effect of biochemical signals that could either enhance or inhibit differentiation of activated macrophages before removal of the macrophages from the mouse peritoneal cavity.[16] The possible role of *in vivo* factors that determine whether macrophages function as noncytotoxic activated macrophages or cytotoxic activated macrophages *in vitro* in the absence of LPS has not been explored.

The above discussion is relevant for modulation of mouse macrophage differentiation in strains of mice sensitive to LPS.[10] Macrophages from other species respond in a similar way to similar differentiation signals but the role of LPS may be less important in the *in vitro* assay depending on the sensitivity of that species to LPS.[17,18]

Cocultivation Procedures

Culture Medium

Dulbecco's modified minimal essential medium (DMEM) is prepared with all components except glucose and pyruvate. To this 20 m*M* HEPES buffer, 100 U/ml penicillin, and 100 µg/ml streptomycin are added. This glucose-free medium is designated DMEM–G. Where noted, glucose is added to DMEM–G to a final concentration of 20 m*M* (DMEM+G). Culture medium is supplemented with calf serum (CS) or fetal bovine serum (FBS) (HyClone Laboratories, Sterile Systems, Inc.), depending

[15] J. P. Levin, A. Tomasulo, and R. S. Oser, *J. Lab. Clin. Med.* **75**, 903 (1970).
[16] H. A. Chapman, Jr., and J. B. Hibbs, Jr., *Science* **197**, 282 (1977).
[17] J. B. Weinberg and A. F. Haney, *JNCI, J. Natl. Cancer Inst.* **70**, 1005 (1983).
[18] D. J. Cameron and W. H. Churchill, *J. Immunol.* **124**, 708 (1980).

on the target cell requirements. When glucose-free medium is needed, CS and FBS are dialyzed (MW 12,000, cut-off dialysis tubing) against phosphate-buffered saline (PBS), pH 7.4.

Many commonly used tissue culture medium formulations such as medium 199, minimum essential medium, and DMEM (low glucose) contain 5.6 mM glucose. If medium with 5.6 mM glucose is used for the cocultivation assay, artifactual lysis of the target cells with the nonlytic phenotype such as L1210 or L10 cells may occur. This is a result of rapid glucose depletion from the culture medium during the cocultivation by activated macrophage effector cells and by the target cells. The target cells consume glucose at an accelerated rate during the cocultivation period because the relatively inefficient glycolytic pathway becomes the sole source of ATP once cytotoxic activated macrophages have induced inhibition of electron transport in target cell mitochondria.[1,3] We supplement the cocultivation medium to a total of 20 mM glucose to avoid artifactual target cell lysis due to glucose depletion during the assay.

Target Cells. The methylcholanthrene-induced murine (DBA/2) lymphoblastic leukemia cell line (L1210),[19] and the diethylnitrosamine-induced guinea pig (strain 2) hepatoma cell line (L10),[20,21] are maintained in suspension culture by serial *in vitro* culture in DMEM +20 mM glucose (total) plus 5% FBS (L1210) or 5% CS (L10). Other nonadherent cell lines, as well as adherent cell lines, can be used for the type of studies outlined in this chapter.

Preparation of Cytotoxic Activated Macrophage Effector Cells. Mice, 6 or more weeks of age, are infected ip with 5×10^6–1×10^7 colony-forming units of the Pasteur strain BCG (TMC 1011; originally obtained from the Trudeau Institute), 17 to 22 days before harvesting the peritoneal exudate cells. Three to 5 days before removing the peritoneal cells, the mice also are inoculated with 1.0 ml of 10% peptone ip. Other methods can be used with equal effectiveness to produce *in vivo* macrophage activation for expression of cytotoxicity. Establishing chronic infection with *Toxoplasma gondii*,[22] or ip inoculation of nonliving agents such as *C. parvum* vaccine or pyran copolymer are examples.[13,14] When mouse peritoneal exudate cells are used as the source of activated macrophages for

[19] L. W. Law, T. B. Dunn, P. J. Boyle, and J. H. Miller, *J. Natl. Cancer Inst. (U.S.)* **10,** 179 (1949).

[20] B. Zbar, H. T. Wepsic, H. J. Rapp, T. Borsos, B. S. Kronman, and W. H. Churchill, Jr., *J. Natl. Cancer Inst. (U.S.)* **43,** 833 (1969).

[21] B. Zbar, H. T. Wepsic, H. J. Rapp, J. Whang-Peng, and T. Borsos, *J. Natl. Cancer Inst. (U.S.)* **43,** 821 (1969).

[22] J. B. Hibbs, Jr., L. H. Lambert, Jr., and J. S. Remington, *Nature (London), New Biol.* **235,** 48 (1972).

in vitro studies, the agent (living or nonliving) used to induce *in vivo* activation should also be administered intraperitoneally.[23] In addition, macrophage activation for expression of cytotoxicity can be effected *in vitro* by exposing stimulated macrophages to lymphokine-rich supernatants or purified preparations of γ-interferon.[8,12,24,25]

Macrophage–Tumor Cell Cocultures. Mice are anesthetised and then exsanguinated by decapitation. The ventral surface is soaked with 70% ethanol. Using two sterile forceps, the abdominal skin is pulled simultaneously in cephalad and caudad directions, which exposes the ventral surface of the parietal peritoneum. Five milliliters of DMEM + G is injected into the peritoneal cavity using a syringe with a 22-gauge needle. The parietal peritoneum is then moistened with a small amount of tissue culture medium and massaged with sterile forceps for 5–10 sec. Next, the parietal peritoneum is held up with forceps, and a small opening is cut using sterile scissors. Peritoneal cells are obtained by removing the culture medium from the peritoneal cavity through this small opening with a sterile exudate or Pasteur pipet. The suspension of peritoneal cells is placed in a 50-ml polypropylene centrifuge tube (Falcon) and maintained at 4° on ice. The usual yield of peritoneal exudate cells from a BCG-infected mouse is 15–30 × 10⁶. The peritoneal exudate cells are washed by centrifugation (180 g for 5 min) and counted in a hemocytometer or Coulter counter (model ZB1). Activated macrophage monolayers are prepared by adding 90 × 10⁶ peritoneal exudate cells from BCG-infected mice to Costar 3100 (80 mm diameter) tissue culture dishes in 20 ml DMEM + G and adhered for 1 hr at 37° in a 5% CO_2 atmosphere. Nonadherent peritoneal exudate cells are then removed by washing three times with DMEM + G. This results in confluent monolayers of adherent mononuclear cells with very few nonadherent cells remaining. Log-phase L1210 cells (8 × 10⁶) or log phase L10 cells (4 × 10⁶) are added to the activated macrophage monolayers in DMEM + G + 5–20 ng/ml LPS with 5% FBS (L1210) or 5% CS (L10). LPS is added to the culture medium to ensure that all monolayers of activated macrophages will express cytotoxicity for the tumor target cells.[8–11] Final medium volume is 30 ml. Although larger diameter Petri dishes are most useful for preparing a large number of target cells for bioenergetic or biochemical analysis, tissue culture chambers of other sizes can be used for the cocultivation. For example, to Costar 3506 (35 mm diameter) tissue culture chambers, 17.5 × 10⁶ peritoneal cells, 1.6 × 10⁶ L1210 cells, or 8 × 10⁵ L10 cells, and 6 ml of culture medium are added. To Costar 3524 (16 mm diameter) tissue

[23] J. B. Hibbs, Jr., *Transplantation* **19**, 81 (1975).
[24] W. F. Piessens, W. H. Churchill, Jr., and J. R. David, *J. Immunol.* **114**, 293 (1975).
[25] L. P. Ruco and M. S. Meltzer, *Cell. Immunol.* **32**, 203 (1977).

culture chambers, 3.5×10^6 peritoneal cells, 3.0×10^5 L1210 cells, or 1.5×10^5 L10 cells, and 1.2 ml of culture medium are added. Cocultures are incubated at 37° in a 95% air and 5% CO_2 humidified atmosphere for 24 hr. Kinetic analysis of the development of target cell metabolic perturbations can be made by varying the length of the cocultivation period.

Separation of Target Cells from Contaminating Macrophages after Cocultivation

Percoll gradients are prepared by a method previously described.[26] Briefly, mix seven parts Percoll (Pharmacia) with six parts of double-strength Hanks' balanced salt solution, pH 7.4. Add 13 ml of this mixture to a 15-ml round-bottomed polycarbonate centrifuge tube (1648 International Equipment Company) and centrifuge at 21,000 g_{av} for 40 min at 20° in a model J21B refrigerated centrifuge (Beckman) using a JA20 rotor (34° fixed angle). Cool gradients in a refrigerator at 4° until needed for cell separation. To remove target cells from the activated macrophage monolayer, attach one end of a short piece of Tygon tubing (25–30 mm length and 6 mm i.d.) to a 30-ml lock tip syringe (Pharmaseal) and the other end to a 5-ml plastic pipette (Falcon). Using this pipetter, aspirate and express the culture medium 10 times vigorously to detach target cells adherent to the macrophage monolayer. Add culture medium containing the separated cells plus a 10-ml rinse to a 50-ml polypropylene centrifuge tube. Centrifuge at 180 g for 5 min and resuspend in 0.5–1.0 ml of DMEM + G. At this step, L1210 cell and L10 cell viability should be greater than 90% as measured by trypan blue exclusion.

To separate the resuspended cells that were removed from the macrophage monolayer from contaminating macrophages, carefully layer these cells in 0.5–1.0 ml of DMEM + G on top of preformed continuous Percoll gradients (see above) and centrifuge at 1000 g for 20 min at 4° in a model PR-6 refrigerated centrifuge (International) in a swinging bucket rotor. Viable L1210 and L10 cells band above the macrophages and the few nonviable cells band above the viable L1210 and L10 cells. The viable target cell band is harvested with a Pasteur pipet, washed three times with DMEM + G by centrifugation (180 g for 5 min), counted, and tested for viability by trypan blue exclusion. The target cells can then be used for bioenergetic or biochemical analysis.

Using continuous Percoll gradients, the yield of L1210 cells is 80–85% and L10 cells is 85–90% of the number initially added to the activated macrophage monolayer. Following Percoll separation, L1210 cells con-

---

[26] F. Gmelig-Meyling and T. A. Waldmann, *J. Immunol. Methods* **33**, 1 (1980).

tain 10–18% macrophages and L10 cells 5–10% macrophages. Further purification of the target cells can be achieved by repeating the Percoll separation procedure. Control target cells, either cultured alone or with stimulated or resident macrophages, are run on Percoll gradients in an identical fashion to those cocultivated with activated macrophages.

Determination of Cytotoxicity

Two phenotypic responses to activated macrophage-induced cytotoxicity exist.[1,27–29] Both phenotypes develop inhibition of mitochondrial respiration and inhibition of DNA replication, but their ultimate fate in response to cocultivation with activated macrophages is different. The nonlytic phenotype responds with prolonged cytostasis but eventually recovers if glucose is available for glycolytic ATP production. Cells with lytic phenotype progress to cytolysis in the presence or absence of glucose. Target cells with the nonlytic phenotype are most useful for studying bioenergetic and biochemical effects induced by cytotoxic activated macrophages, because tumor cells with this phenotype do not undergo significant lysis during cocultivation. As a result, evidence of activated macrophage-induced cytotoxicity cannot be detected by the usual methods of prelabeling target cells with chromium-51 (see this volume [27]) or [3H]thymidine ([3H]TdR) (see this volume [29]) and subsequent determination of lysis by measurement of release of the isotope label into the culture medium during the cocultivation. However, predictable metabolic changes do occur in target cells which do not undergo lysis (such as L1210 and L10 cells) and these metabolic changes can be used to document evidence of activated macrophage-induced cytotoxicity. Cytotoxic activated macrophage-induced changes in the target cells that can be used to detect cytotoxicity include lysis of the target cells in glucose-free medium[1]; release of iron-59 from prelabeled target cells[2]; and inhibition of [3H]TdR uptake by the target cells.[13] It is often useful to cocultivate the target cells to be used for biochemical analysis in 80-mm diameter tissue culture chambers and the target cells used for documentation of the activated macrophage-induced cytotoxic effect in smaller diameter 16-mm tissue culture chambers.

[27] J. L. Cook, J. B. Hibbs, Jr., and A. M. Lewis, Jr., *Proc. Natl. Acad. Sci. U.S.A.* **77,** 6773 (1980).

[28] J. L. Cook, J. B. Hibbs, Jr., and A. M. Lewis, Jr., *Int. J. Cancer* **30,** 795 (1982).

[29] J. B. Hibbs, Jr. and D. L. Granger, *in* "Self Defense Mechanisms—Role of Macrophages" (D. Mizuno, Z. A. Cohn, K. Takeya, and N. Ishida), p. 319. Univ. of Tokyo Press, Tokyo, 1982.

Methods for Documenting the Cytotoxic Effect of Activated Macrophages for Tumor Target Cells

Measurement of Iron-59 (^{59}Fe) Release. L1210 or L10 cells ($\cong 10^5$/ml), are labeled in DMEM + G plus 5% FBS (L1210) or 5% CS (L10) for 24 hr with 1 μCi/ml ^{59}Fe citrate (New England Nuclear).[2,30] Cells are then washed four times with DMEM + G plus 1% CS by centrifugation (180 *g* for 5 min), counted, and total ^{59}Fe label incorporated into the target cells determined. A known number of labeled target cells is pelleted and lysed with 100 μl 1 *N* KOH at 60°for 30 min. Four milliliters Aquasol II is then added to the 100-μl sample followed by 100 μl 1 *N* HCl and the radioactivity is determined in a scintillation counter. The remaining labeled target cells are added to activated macrophage monolayers, stimulated macrophage monolayers, resident macrophage monolayers, or control chambers and incubated for the length of time desired. To determine release from target cells cocultivated with activated macrophages, or otherwise experimentally manipulated (experimental release), or from target cell cultures alone (spontaneous release), medium is sampled after the desired interval of cocultivation or cultivation. A 100-μl aliquot of centrifuged supernatant (180 *g* for 5 min) is transferred to 4 ml of Aquasol II and the radioactivity is determined using scintillation counting. The measurements are corrected for quench, and the percentage specific release calculated by the following formula:

Percentage specific release =

$$\left(\frac{\text{experimental release} - \text{spontaneous release}}{\text{total release} - \text{spontaneous release}} \right) 100$$

Specific release of ^{59}Fe also can be conveniently quantitated using a gamma counter.

Inhibition of Mitochondrial Respiration. A large number of target cells ($4–8 \times 10^6$) is needed to directly measure O_2 consumption with a Clark oxygen electrode. As a result, a simple alternative method capable of detecting inhibition of mitochondrial respiration in a relatively small number of target cells was devised.[1] The technique used is based on the fact that target cells acquire a strict requirement for glucose, or other hexose (mannose or fructose), capable of being utilized by the glycolytic pathway, in order to remain viable after cocultivation with activated macrophages (see Fig. 1). Transformed cells with the nonlytic phenotype develop inhibition of mitochondrial respiration and inhibition of DNA

---

[30] Iron-55 can be substituted for ^{59}Fe in experiments measuring iron release from target cells of activated macrophages.

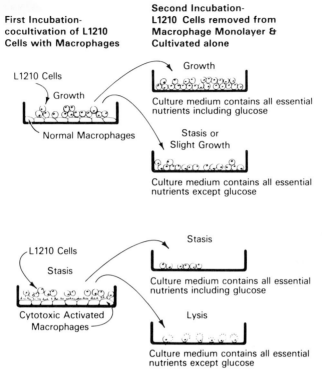

FIG. 1. Diagrammatic representation of the two-step *in vitro* culture system used to demonstrate inhibition of mitochondrial respiration in target cells of cytotoxic activated macrophages. The scheme shows the glucose requirement for survival of transformed cells with the nonlytic phenotype (L1210 cells) after a period of cocultivation with cytotoxic activated macrophages. L1210 cells do not develop a similar requirement for glucose after a period of cocultivation with normal (resident) or stimulated macrophages. See Granger *et al.*[1] for experimental details.

synthesis and eventually recover after a variable period of cytostasis if glucose is available for glycolytic ATP production.[1,3,29] Inhibition of mitochondrial respiration in target cells with the nonlytic phenotype can be detected by removing them from the activated macrophage monolayers and culturing them in medium without glucose. If inhibition of mitochondrial respiration has developed, cells of the nonlytic phenotype will survive in culture medium with glucose but undergo cytolysis in culture medium without glucose. This technique cannot be used to detect inhibition of mitochondrial respiration in tumor cells with the lytic phenotype because they progress to cytolysis in the presence or absence of glucose.

To determine evidence of inhibition of mitochondrial respiration, following a period of cocultivation (first incubation) the target cells are separated from the macrophage monolayer using the pipetter described above. An aliquot of medium containing approximately 10^5 target cells (most of the target cells remain available for bioenergetic or biochemical analysis) is harvested, added to a Falcon 2054 centrifuge tube, together with cold PBS (4°), to a total volume of 5 ml and centrifuged at 180 g for 5 min. The supernatant is removed down to the cell pellet with gentle suction. The cell pellet is resuspended in 5 ml of cold PBS and the centrifugation is repeated. The supernatant is removed again with gentle suction. DMEM–G (0.2 ml) plus 2% dialyzed FBS (L1210) or 2% dialyzed CS (L10) is added and the cell pellet is resuspended. A 50-μl aliquot of this cell suspension (~2.5 × 10^4 cells) is added to flat-bottom Falcon 3072 microtest chambers. Then 50 μl of DMEM–G or DMEM + G plus 2% appropriate dialyzed serum is added and the cells are incubated for an additional 18–24 hr (second incubation). Tumor cells with the nonlytic phenotype such as L1210 and L10 cells that undergo activated macrophage-induced inhibition of mitochondrial respiration lose viability in glucose-free medium but survive and eventually resume proliferation in medium supplemented with glucose.[1,2] Control L10 and L1210 cells survive in DMEM–G.[1,2,31] Measurement of viable and nonviable cells by trypan blue exclusion in DMEM + G and DMEM–G after a second incubation is a simple and reliable method for detecting target cells that have undergone activated macrophage-induced inhibition of mitochondrial respiration. This correlates directly with measurement of O_2 consumption using a Clark O_2 electrode.[1]

Measurement of Inhibition of [³H]TdR Uptake. Uptake of [³H]TdR can be used to measure DNA synthesis, but if macrophages are present in the same tissue culture chamber as the target cells during the labeling period, significant artifactual inhibition of [³H]TdR uptake may occur.[32] This is due to the fact that macrophages lack thymidine kinase and secrete large amounts of thymidine into the culture medium.[33] The macrophage-derived thymidine competes with the added [³H]TdR, causing artificially

[31] The nutrient requirement of our strain of L1210 cells in DMEM–G has recently changed. To obtain >90% viability of normal L1210 cells in glucose free medium we supplement DMEM–G (as well as DMEM + G) used during the second incubation with 1 mM sodium acetate. Sodium acetate does not change the response of cytotoxic activated macrophage-injured L1210 cells to culture in DMEM–G.

[32] H. G. Opitz, D. Niethammer, R. C. Jackson, H. Lemke, R. Huget, and H. S. Flad, *Cell. Immunol.* **18,** 70 (1975).

[33] J. J. Stadecker and E. R. Unanue, *J. Immunol.* **123,** 568 (1979).

low levels of [³H]TdR incorporation. However, if the target cells are removed from the macrophage monolayers at the termination of the co-cultivation period and immediately pulse labeled with [³H]TdR the effects of competition with macrophage derived thymidine can be avoided.

Target cells are removed from the macrophage monolayers as described above.[34] An aliquot of approximately 10⁵ target cells is centrifuged (180 g for 5 min) and resuspended in DMEM + G plus 5% serum containing 0.5 μCi/ml [³H]TdR. The target cells and appropriate control cells are labeled for the desired interval of time and processed as previously described.[13]

Measurement of [³H]TdR Release. Evaluation of release of [³H]TdR can be used to document lysis of target cells with the lytic phenotype or absence of lysis of cells with the nonlytic phenotype such as L1210 and L10 cells.[35] Target cells ($\cong 10^5$/ml) are labeled in DMEM + G plus 5% FBS (L1210) or 5% CS (L10) for 24 hr with 0.2 μCi/ml [³H]TdR and processed as previously described.[1] (Details of these methods have been given in [29] of this volume and will not be duplicated here.)

Concluding Remarks

In this chapter we describe a method for the preparation, for bioenergetic or biochemical analysis, of large numbers of target cells that have been cocultivated with cytotoxic activated macrophages. Following the cocultivation period, the target cells are removed from the macrophage monolayers and separated from contaminating macrophages on continuous Percoll gradients. The target cells can then be used for measurement of changes (e.g., enzyme activity, O₂ consumption), caused by the effector macrophages during the cocultivation. The kinetics of development of these changes by target cells can be studied by varying the length of the cocultivation period.

The target cells that we use to study activated macrophage-induced metabolic changes do not undergo cytolysis. Therefore, techniques to document metabolic changes are needed as an indicator that a cytotoxic effect has occurred. Several methods for documenting activated macrophage induced metabolic effects are described (measurement of ⁵⁹Fe re-

---

[34] The Percoll separation step can be omitted. [³H]TdR uptake by L1210 cells and L10 cells that are removed from monolayers of cytotoxic activated macrophages after 24 hr of cocultivation is the same with and without Percoll separation.

[35] A 48-hr cocultivation period with cytotoxic activated macrophages is usually required to get significant release of [³H]TdR from tumor target cells with the lytic transformed cell phenotype [see J. L. Cook, J. B. Hibbs, Jr., and A. M. Lewis, Jr., *Proc. Natl. Acad. Sci. U.S.A.* **77**, 6773 (1980)].

lease, lysis of target cells in culture medium without glucose, and measurement of inhibition of [³H]TdR uptake).

Acknowledgments

This work was supported by the Veterans Administration. We are grateful to G. Shaw and T. Childs for typing the manuscript.

[35] Intracellular Killing of Bacteria and Fungi

By LINDA L. MOORE and JAMES R. HUMBERT

Introduction

Much interest has been generated in recent years regarding the complex metabolic events which lead to ultimate microbial killing within phagocytic cells; several other chapters in this volume address that issue in detail ([17] through [26]). Measurement of bactericidal activity remains the essential physiological tool that links the clinical expression of phagocytic defects with their biochemical and molecular etiology; for this reason, bactericidal activity measurements are often used as a diagnostic screening test in clinical medicine, or as a logical source of reference when one experiments with inhibitors or enhancers of metabolic steps which should influence bacterial killing in phagocytes.

Among the numerous assays pertaining to bactericidal activity, we have chosen to describe three variants which should be applicable to most clinical or experimental laboratory situations. Of the three methods described here, method A can easily be adapted to any bacterial strain by using modifications proposed by Leijh et al.[1] which are briefly discussed later in this chapter. Method B is readily applicable to the study of fungi as well as bacteria and uses minuscule amounts of blood (see also this volume [3]). These three methods can be profitably applied to the study of any type of phagocytic cells, although they have been developed mostly for investigations with neutrophils or monocytes.

Separation techniques for neutrophils and monocytes have been described elsewhere in this series (Vol. 108 [9]) and will not be duplicated here. See also Casale and Kaliner[2] and Bøyum.[3]

---

[1] P. C. Leijh, M. T. van den Barselaar, I. Dubbeldeman-Rempt, and R. van Furth, Eur. J. Immunol. 10, 750 (1980).

[2] I. B. Casale and M. Kaliner, Immunol. Methods 55, 347 (1982).

[3] A. Bøyum, Scand. J. Clin. Lab. Invest. 21, Suppl. 97, 9 (1968).

Methods

Killing Assay Using Bacterial Colony Counts (Method A)

Principle. This method is our modification of a method first described by Tan,[4] whereby patient and control polymorphonuclear leukocytes are incubated with a test organism (in this case, *Staphylococcus aureus* 502A) for a selected time interval. The percentage of bacteria killed and the number of ingested but live bacteria can be determined with the use of lysostaphin and trypsin. Although this method is specific for determining only staphylococcal killing, bactericidal and fungicidal methods applicable to other species are presented later in this chapter. This method can be used profitably with other phagocytic cells.

Equipment

37°, 5% CO_2 incubator or a shaking water bath (37°)
Centrifuge
Vortex mixer

Supplies/Reagents

Hanks' balanced salt solution (HBSS, GIBCO, Grand Island, NY)[4a]
  Trypticase soy agar (TSA, BBL Microbiology Systems, Cockeysville, MD) is prepared as a 4% (w/v) solution with deionized, distilled water, is autoclaved and poured into 100 × 15 mm Petri dishes (Scientific Products, McGaw Park, IL)
Trypticase soy broth (TSB, BBL Microbiology Systems, Cockeysville, MD) is prepared as a 3% solution (w/v) in deionized, distilled water, and is autoclaved in bacterial culture tubes (15–20 ml size)
Eighteen-hour trypticase soy broth culture of *Staphylococcus aureus* 502A (obtainable from the authors of this article at the Buffalo Children's Hospital)
Lysostaphin (Sigma Chemical) at a final concentration of 500 U/ml
Trypsin (Microbiological Associates) at a final concentration of 2.5%
Serum from a type AB donor (frozen in 0.5 ml aliquots)
Opsonin solution (0.5 ml AB serum added to 2.5 ml sterile HBSS)
Patient and control polymorphonuclear leukocytes (to separate PMNs from peripheral blood see refs. 2 and 3)

[4] J. S. Tan, C. Watanakunakorn, and J. P. Phair, *J. Lab. Clin. Med.* **78,** 316 (1971).
[4a] Abbreviations: TSA, trypticase soy agar; HBBS, Hanks' balanced salt solution; TSB, trypticase soy broth; TVB, total viable bacteria; IB, intracellular bacteria; EB, extracellular bacteria; AO, acridine orange.

Procedures

1. Prepare suspensions of separated patient and control leukocytes to equal 10^7 cells/ml in sterile HBSS (see above).

2. Centrifuge, wash, and resuspend the 18-hr *S. aureus* culture to obtain a suspension of 10^8 bacteria/ml in sterile HBSS.

3. To 0.1 ml adjusted *S. aureus* suspension in capped plastic tubes, add 0.4 ml opsonin solution and 0.5 ml leukocytes or HBSS (for a leukocyte-free control). Immediately take 0.1 ml of suspension from each mixture and serially dilute this amount into sterile H_2O to achieve a final dilution of 1 : 10,000. Plate 0.1 ml of this 1 : 10,000 dilution onto TSA plates and incubate overnight at 37°. The resultant colony-forming units represent an estimation of bacterial numbers at time 0.

4. Incubate the leukocyte and control tubes (prepared in step 3) at 37° for 2 hr. At the end of this incubation period divide the mixture in each tube into two portions, add the reagents indicated, and proceed as shown in the following scheme.

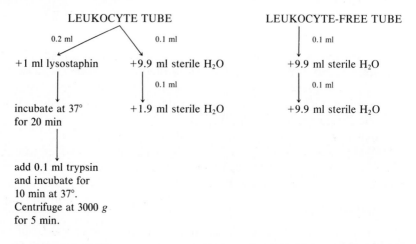

5. Plate 0.1 ml from each water tube onto TSA plates and incubate at 37° overnight. Colony-forming units represent total viable bacteria (TVB).

6. Resuspend the pellets obtained in the lysostaphin/trypsin tubes into 10 ml sterile H_2O. Vortex the tubes and transfer 0.1 ml into 1.9 ml sterile H_2O. Plate 0.1 ml onto TSA plates. Incubate at 37° overnight. Colony-forming units represent the number of viable intracellular bacteria (IB).

7. Interpretation: At the end of an overnight incubation the colonies on each plate must be counted. Calculations for total viable bacteria

(TVB), intracellular bacteria (IB), and extracellular bacteria (EB) are as follows: For leukocyte-free mixtures,

$$\text{Percentage TVB} = \left(\frac{\text{colony count at 120 min}}{\text{colony count at time 0}}\right) 100$$

(This serves as a check for bacterial growth in serum-containing buffer.) For leukocyte mixtures,

$$\begin{array}{c}\text{Percentage}\\\text{TVB}\end{array} = \left[\frac{\text{colony count at 120 min} \times 0.2 \text{ (a dilution factor)}}{\text{colony count at time 0}}\right] 100$$

$$\text{Percentage IB} = \left[\frac{\text{colony count (IB) at 120 min}}{\text{colony count at time 0}}\right] 100$$

Percentage EB = percentage TVB − percentage IB

Expected Results. When *S. aureus* 502A is used with human neutrophils, the expected results are as follows (2 SD above mean), as derived from 21 adult values in our laboratory: percentage TVB (120 min) $<15.7\%$; percentage IB (120 min) $<5.6\%$; percentage EB (120 min) $<10.6\%$.

Assessment of Microbial Killing by a Fluorochrome Assay[5] (Method B)

Principle. After whole blood is allowed to clot on coverslips, monolayers of adherent cells are incubated with standardized suspensions of bacteria or fungi. After incubation the monolayers are briefly stained with dilute acridine orange (AO) and examined under ultraviolet light. Dead organisms fluoresce red, while live organisms fluoresce green. Killing capacity is measured by comparing the numbers of ingested dead organisms to the total numbers of organisms ingested. When whole blood is used, cell monolayers consist mostly of neutrophils; the method can be adapted for other phagocytic cells by overlaying the coverslips with an appropriate cell suspension.

Equipment

37° Humidified CO_2 incubator

Petroff Hauser chamber (a bacterial counting chamber, VWR Scientific, Pittsburgh, PA)

UV fluorescence microscope with a halogen light source and 100× oil immersion lens)

[5] C. G. Pantazis and W. T. Knicker, *J. Reticuloendothel. Soc.* **26,** 155 (1979).

Stage micrometer
Ocular micrometer

Supplies and Reagents

18-mm Circular glass coverslips
100 × 15-mm Petri dishes
HBSS (GIBCO, Grand Island, NY)
Normal autologous or homologous human serum
Acridine orange (J. T. Baker)
McCoy's 5A medium (Colorado Serum Company)
A broth culture of the target yeast or bacteria

Procedures

Monolayer preparation:

1. Whole blood (0.2 ml) (without anticoagulant) is dripped onto a coverslip and allowed to spread onto most of the surface.
2. The coverslip is placed onto a saline-moistened tissue or a circle of filter paper cut to fit the bottom of a Petri dish.
3. The Petri dish is covered and placed into a 37°, 5% CO_2 incubator for 30 min.
4. The coverslip is rinsed with warmed (37°) HBSS.

Bacterial challenge:

1. A suspension of bacteria or yeast (in McCoy's medium with 10% serum) is laid over the monolayer (at a predetermined concentration of bacteria per adherent cell).
2. The coverslip is again incubated in a 37°, humidified 5% CO_2 incubator for 30–90 min. Appropriate incubation times can be chosen.
3. After incubation, the coverslip is rinsed with warmed (37°) HBSS.

The number of adherent cells can be estimated by air drying one or more coverslips, staining with Wright's stain, and counting 10–20 random fields. The average number of adherent cells per coverslip can be determined by computing the radius of one field (measure field with ocular and stage micrometers), the number of possible fields on that coverslip, and the total area of the coverslip.

Assessment of Phagocytosis and Killing

1. The coverslip is transferred to dishes containing AO (at a dilution of 1 : 10,000) and allowed to stain for 45–60 sec.

2. The coverslip is drained and mounted, wet side down, onto a glass slide and sealed with clear nail polish.
3. The preparation is examined under UV fluorescence (100× oil immersion objective) using a halogen light source.
4. The total number of red and green intracellular organisms in 100 leukocytes are counted. Killing is assessed by relating the average number of red (dead) intracellular organisms to the total number of red and green organisms ingested.

Expected Results. In our laboratory when *Candida albicans* is used as the test organism with human neutrophils in 24 experiments, killing at 30 min is 36% ± 7.6 (2 SD). When tested at 60 min killing is 63% ± 12.2 (2 SD).[6]

Differentiation between Attached and Ingested Organisms. Extracellular fluorescence can be quenched by adding crystal violet (0.5 mg/ml in 0.15 M NaCl) to the AO-stained and counted preparations.[7] Crystal violet does not penetrate phagocytic cells, thus any fluorescence seen can be attributed to intracellular organisms (see also this volume [6]).

Killing Assay Using a Temperature-Sensitive Mutant as a Target Organism (Method C)

Principle. A novel assay developed by Hooke[8] confronts the problem of bacterial growth during the course of the assay. The target organism used in this assay is an *Escherichia coli* mutant (Easter strain, E/2/64) which does not grow at the temperature of the assay (37°) but retains its colony-forming ability when subsequently held at 25°. The procedures used for generating the strain (mutagenesis, segregation, enrichment, and isolation) were originally described by Oeschger and Berlyn[9] and will be presented briefly here. *E. coli* strain Easter (American Type Culture Collection) is grown to the logarithmic phase at 37° and is then incubated with *N*-methyl-*N'*-nitrosoguanidine, a DNA mutagen. The mutagenized cells are washed, diluted, and incubated overnight at 22° in order to reach the logarithmic growth phase. The cultures are then shifted to a temperature of 35° and incubated with penicillin (final concentration of 5000 U/ml). The cells are then washed and grown overnight to logarithmic phase. D-Cycloserine, at a final concentration of 2 mM, is incubated with the cells for 3 hr at 35°, after which time they are washed and returned to 22°.

[6] J. M. Bernstein, L. L. Moore, and P. Ogra, *in* "Recent Advances in Otitis Media with Effusion" (D. Lim, ed.), p. 129. B. C. Decker, Inc., Philadelphia, Pennsylvania, 1982.
[7] J. Hed, *FEMS Microbiol. Lett.* **1,** 357 (1977).
[8] A. M. Hooke, M. P. Oeschger, B. J. Zeligs, and J. A. Bellanti, *Infect. Immun.* **20,** 406 (1978).
[9] M. P. Oeschger and M. K. Berlyn, *Mol. Gen. Genet.* **134,** 77 (1974).

Portions of the culture are then plated at 22° and colonies which develop are replica plated at 35°. Bacteria which grow only at 22° are considered temperature-sensitive mutants.

Equipment

37° Incubator
End-over-end rotating tube rack

Supplies and Reagents

G agar (9) is composed of 1% tryptone, 0.1% yeast extract, 0.25% NaCl, and 1.5% agar
Krebs–Ringer phosphate buffer
Pooled fresh human serum
Polymorphonuclear leukocytes or macrophages
E. coli strain Easter[9] or any other similarly behaving temperature-sensitive mutant organism which survives at 37° but replicates only at a lower temperature

Procedures

1. Peripheral blood leukocytes are washed and resuspended to the desired density in Krebs–Ringer phosphate buffer containing 8% pooled fresh human serum.
2. Equal volumes of phagocytic cells and bacterial cell suspensions are mixed to a final cell: bacterial ratio of 1 : 4 in 1 ml. A leukocyte-free bacterial suspension serves as the control.
3. The suspensions are incubated at 37° with end-over-end rotation.
4. Portions are removed at 0 time, 1, and 2 hr. These aliquots are placed into distilled water (which lyses the phagocytes). Serial dilutions of the lysates are plated and incubated at 25°. The remaining viable bacteria (extracellular and intracellular together) are represented by colony-forming units.

Comments

The three methods described above should be useful in most situations when a careful assessment of bactericidal activity by neutrophils or other phagocytes is required. They do have specific advantages and limitations, however, which will be discussed further. A problem common to all measurements of bactericidal activity concerns the distinction (or rather the lack of it) between intracellular bactericidal activity and the preceding step, phagocytic uptake. This problem is not easily circumvented. At worst, researchers can rely on a separate, but specific phagocytic assay to document the *absence* of a phagocytic defect; in such cases, conclusions concerning bactericidal activity could be profitably drawn from comparisons with normal control cells. In other situations, when phagocytosis is increased or decreased over normal values, interpretation of bactericidal activity becomes more complex and indeed of questionable validity. Of the three methods described, the fluorochrome method provides a practical means of minimizing the fluctuations in phagocytosis, since serial monitoring by direct visualization provides a satisfactory, if not perfect, impression of phagocytic vs bactericidal activity. The pH must be rigorously maintained, however, as fluorescence of organisms will turn to an uninterpretable brown if the pH is allowed to rise. This method has the additional advantage of requiring very small amounts of peripheral blood, and thus can be used easily for studies in infants or in small laboratory animals. Other modifications of bactericidal assays have been used to increase their specificity; Leijh et al.,[1] for instance, propose to use a very short initial exposure of bacteria to phagocytes (3 min), in the presence of phenylbutazone (2 mg/ml); 2 washings at 4° remove the extracellular bacteria and the drug (which temporarily paralyzes the bactericidal activity by interfering with H_2O_2 production in the pentose phosphate pathway); then initial (zero value) samples are removed and cultured, and the cell/bacteria interaction is allowed to proceed as described for the first method described above.

Another common problem concerns the complication of bacterial growth during the assays, which tends to blur results and renders precise kinetic studies of bactericidal activity almost impossible. The third method presented resolves that issue elegantly, by resorting to a mutant bacterial species which multiplies only at a temperature much lower than that of the bactericidal assay.

Finally, while we have presented standardized versions of three tests, many variants of these assays exist and should be encouraged for specific purposes. Particularly, the use of high bacteria/cell ratios (100 : 1, 1000 : 1) should be considered if *maximum* killing capacity is of interest to the investigator; temperature dependence (20 vs 37, 40, or 42°) should also be considered if appropriate with the ecology of pathogens under study, or if the investigator is interested in reproducing *in vitro* the temperature conditions occurring in specific clinical situations.[10]

Acknowledgments

    This chapter reports work supported in part by the Association for Research of Childhood Cancer. The expert secretarial assistance of Deborah Machinski is gratefully acknowledged.

[10] C. J. van Oss, D. R. Absolom, L. L. Moore, B. H. Park, and J. R. Humbert, *J. Reticuloendothel. Soc.* **27,** 561 (1980).

Section VIII

Cytotoxic Activity of Macrophages

[36] Induction of Specific Macrophage Cytotoxicity

By ROEL A. DE WEGER and WILLEM DEN OTTER

Macrophage cytotoxicity can be induced in various ways. Bacterial infections *in vivo* or bacterial products [e.g., lipopolysaccharide (LPS),[1] muramyl dipeptide] can induce nonspecific macrophage cytotoxicity[1a–3] *in vivo* and *in vitro*. This nonspecific cytotoxicity can also be induced by lymphokines (macrophage-activating factor; MAF) and LPS or interferon.[4,5] These lymphokines are produced by lymphocytes after stimulation by mitogens or soluble antigens.[5,6]

Specific macrophage cytotoxicity was first described by Evans and Alexander[7,8] in murine tumor systems. They showed that specific macrophage cytotoxicity was due to a specific T-cell factor, specific macrophage-arming factor (SMAF). The factor adheres to the macrophages, thus providing them with a specific receptor for tumor cells. Others also have described SMAF or SMAF-like factors in tumor systems[9–11] and in parasite systems.[12,13]

We have shown that the induction of lymphocytes with the capacity to produce SMAF depends on immunization with high doses of intact tumor cells and that *in vitro* the lymphocytes are triggered only by intact cells.[14] This suggests that the lymphocytes react directly with the tumor cells (no antigen presentation by macrophages). SMAF is an antigen-specific factor that is absorbed from the supernatant only by the specific tumor cells or

---

[1] Abbreviations: LPS, lipopolysaccharide; SMAF, specific macrophage-arming factor; PBS, phosphate-buffered saline; FBS, fetal bovine serum; MAF, macrophage-activating factor.

[1a] J. B. Hibbs, *J. Natl. Cancer Inst. (U.S.)* **53**, 1487 (1974).

[2] D. Taramelli and L. Varesio, *J. Immunol.* **127**, 58 (1981).

[3] S. Nagao, T. Miki, and A. Tanaka, *Microbiol. Immunol.* **25**, 41 (1981).

[4] D. Taramelli, M. B. Bagley, H. T. Holden, and L. Varesio, *Adv. Exp. Med. Biol.* **155**, 487 (1982).

[5] L. P. Svedersky, C. V. Benton, W. H. Berger, E. Rinderknecht, R. N. Harkins, and M. A. Palladino, *J. Exp. Med.* **159**, 812 (1984).

[6] L. P. Ruco and M. S. Meltzer, *J. Immunol.* **119**, 889 (1977).

[7] R. Evans and P. Alexander, *Transplantation* **12**, 227 (1971).

[8] R. Evans and P. Alexander, *Nature (London)* **236**, 168 (1972).

[9] M. L. Kripke, M. B. Budmen, and I. J. Fidler, *Cell. Immunol.* **30**, 341 (1977).

[10] M. L. Lohmann-Matthes, B. Kolb, and H. G. Meerpohl, *Cell. Immunol.* **41**, 231 (1978).

[11] M. Horowitz and B. F. Argyris, *Immunol. Commun.* **6**, 297 (1977).

[12] K. K. Sethi and H. Brandis, *Z. Immunitaetsforsch.* **154**, 226 (1978).

[13] T. Shirahata and K. Shimize, *Microbiol. Immunol.* **23**, 17 (1979).

[14] R. A. DeWeger, E. Pels, and W. Den Otter, *Immunology* **47**, 541 (1982).

cells sharing antigens of the specific tumor cells. The adherence of SMAF to macrophages is not antigen specific and not MHC restricted. Macrophages armed by SMAF express a specific cytotoxicity.[14-16] Macrophages can also bind other specific T-cell factors such as specific T-suppressor factors, after which the macrophages produce nonspecific suppressor factors.[17,18] So, it seems that macrophages can function in an antigen-specific manner during an immune response by binding specific T-cell factors. This is of particular importance during tumor growth, as during tumor growth in several experimental tumor systems specific T-cell factors or antigen-specific macrophage functions have been detected.[18-21]

Thus, specific macrophage cytotoxicity provides a bioassay for the presence of the specific T-cell factor SMAF.

Methods for Measuring Specific Macrophage Cytotoxicity

Immunization and Collection of Lymphocytes

In allogeneic murine tumor systems mice are injected subcutaneously on the chest with a high dose (10^7 cells) of allogeneic tumor cells in 0.1 ml phosphate-buffered saline (PBS). Murine tumor systems used are C57BL (H-2^b) mice immunized with SL2 (DBA/2, H-2^d) lymphosarcoma cells or DBA/2 (H-2^d) mice immunized with EL4 (C57BL, H-2^b) lymphosarcoma cells (available from American Type Culture Collection). The mice develop large tumors between days 5 and 10 that are rapidly rejected after 10 days.[14] Between days 6 and 8 the draining lymph nodes and the spleens are collected. The organs are squeezed through a metal sieve and the resulting single-cell suspension is washed and filtered through glass wool to remove dead cells. The lymphocytes are suspended in medium (Fischer's medium or MEM) supplemented with 10% FBS (2×10^6) cells/ml). Before use the lymphocytes are cultured for 2 hr at 37° in glass flasks to remove adherent cells.

In syngeneic murine-tumor systems mice are injected with 10^7 irradiated (5000 rad) syngeneic tumor cells intraperitoneally (ip) at day 0 and day 10. At day 20 the mice receive a third ip injection with 5×10^6

[15] E. Pels, R. A. De Weger, and W. Den Otter, *Immunobiology* **16**, 84 (1984).
[16] E. Pels, R. A. De Weger, and W. Den Otter, *Int. Arch. Allergy Appl. Immunol.* **74**, 140 (1984).
[17] W. Ptak, M. Zembala, G. L. Asherson, and J. Marcinkiewicz, *Int. Arch. Allergy Appl. Immunol.* **65**, 121 (1981).
[18] T. A. Koppi and W. J. Halliday, *Cell. Immunol.* **76**, 29 (1983).
[19] J. A. Bluestone and C. Lopez, *Cancer Invest.* **1**, 5 (1983).
[20] Y. Yamamura, *Int. J. Cancer* **19**, 717 (1977).
[21] Q. W. Ye, M. B. Mokyr, J. M. Pyle, and S. Dray, *Cancer Immunol. Immunother.* **16**, 162 (1984).

nonirradiated tumor cells. At day 27 peritoneal exudate cells are collected from mice that do not develop an ip tumor (more than 90% of the mice) after this immunization.[22] Syngeneic murine-tumor systems used are DBA/2 mice immunized with SL2 lymphosarcoma cells and BALB/c mice immunized with MPC11 plasmacytoma cells (available from American Type Culture Collection). The peritoneal exudate cells are suspended in Fischer's medium containing 10% FBS in a cell concentration of 5×10^6 lymphocytes/ml. This cell suspension is cultured 2 hr at 37° in a glass flask to remove adherent cells. The cell suspension is then transferred to a new glass flask and cultured for another 2 hr. After this procedure the cell suspension contains more than 95% lymphocytes. Subsequently the cells are passed over a glass wool filter to remove dead cells and the cells are resuspended in a concentration of 2×10^6 lymphocytes/ml in medium containing 10% FBS. After the immunization procedure described for the syngeneic system, spleen or lymph node lymphocytes produce far less SMAF than the peritoneal lymphocytes. The lymphocytes obtained in allogeneic and syngeneic systems are used in the same way for testing the SMAF production and for the induction of specific macrophage cytotoxicity. Lymphocytes from the same strain of mice not immunized with tumor cells are used as controls.[14,15]

Macrophages (see also this series, Vol. 108 [25] and [27])

Specific macrophage cytotoxicity can be studied *in vitro* with resident peritoneal macrophages or elicited exudate macrophages. Routinely, we prefer to use nonelicited resident peritoneal macrophages. The macrophages should not express spontaneous cytotoxicity as this significantly hampers the detection of SMAF.[23,24] Macrophages are seeded into flat-bottom culture wells [8×10^5 macrophages in 24-well plates ($\varnothing$ 16 mm) or 3×10^5 macrophages in 96-well plates ($\varnothing$ 6 mm)] directly after harvesting from the peritoneal cavity. After 2–3 hr the macrophage monolayers are washed to remove nonadherent cells. Although there is no genetic restriction, we routinely use macrophages of the same strain of mice as the SMAF-producing lymphocytes.

Production and Detection of SMAF

Lymphocytes (2×10^6 cells/ml) are mixed with the specific tumor cells (2×10^5 cells/ml) used for immunization at a lymphocyte: tumor cell ratio

[22] H. van Loveren, J. W. De Groot, and W. Den Otter, *Proc. Soc. Exp. Biol. Med.* **167,** 207 (1981).
[23] E. Pels and W. Den Otter, *Br. J. Cancer* **40,** 856 (1979).
[24] E. Pels and W. Den Otter, *IRCS Med. Sci. Libr. Compend.,* **4,** 385 (1976).

TABLE I
SPECIFICITY OF THE CYTOTOXICITY OF ARMED MACROPHAGES IN AN
ALLOGENEIC SYSTEM[a]

Tumor target	Strain of origin	H-2 type	Cytotoxicity index[b]	
			SMAF armed	Lymphocyte armed
SL2	DBA/2	d	48 ± 6	64 ± 3
MOPC 195	BALB/c	d	34 ± 10	59 ± 7
MPC 11	BALB/c	d	32 ± 7	54 ± 7
EL 4	C57BL	b	22 ± 6[c]	30 ± 9[c]
TLX-9	C57BL	b	23 ± 7[c]	33 ± 7[c]
AK-SL2	AKR	k	25 ± 12[c]	32 ± 8[c]

[a] Sensitized lymphocytes obtained from C57BL (H-2[b]) mice immunized with SL2 (H-2[d]) tumor cells, were used to produce SMAF (SMAF armed) or to arm macrophages directly (lymphocyte armed).
[b] Macrophage cytotoxicity was determined by cell counting. Mean ± SEM of four experiments performed in triplicate.
[c] Significantly different from the cytotoxicity against SL2 cells ($p < 0.05$).

of 20:1, in Fischer's medium containing 10% FBS and 5×10^{-5} M 2-mercaptoethanol. The cells are cultured for 24 hr at 37°. After incubation the cells are centrifuged and the supernatant collected. The supernatant can be tested directly or stored at −20°. In the unfractionated supernatant SMAF is unstable at room temperature.[16]

The supernatant (0.6 ml/8 × 10[5] macrophages) is added to macrophage monolayers (cultured for 2 hr) and incubated for 4 hr at 37°. Subsequently, the monolayers are washed and the cytotoxicity of the armed macrophages against tumor cells is measured (see [34], this volume). Routinely an effector:target cell ratio of 10:1 is used and cytotoxicity is determined by cell counting. Results of a typical experiment in an allogeneic system, C57BL mice (H-2[b]) immunized with SL2 (H-2[d]) tumor cells, are shown in Table I.

Direct Arming of Macrophages by Lymphocytes

An alternative method to induce specific macrophage cytotoxicity is to incubate the tumor-sensitized lymphocytes and tumor cells directly with a macrophage monolayer, in a lymphocyte:macrophage:tumor cell ratio of 20:15:1 [e.g., 0.3 ml tumor cells (2 × 10[5] cells/ml) and 0.6 ml lymphocytes (2 × 10[6] cells/ml) are added to a well of a 24-well culture plate, containing 9 × 10[5] macrophages (monolayer)]. Macrophage monolayers

TABLE II
SPECIFICITY OF LYMPHOCYTE ARMED MACROPHAGES IN SYNGENEIC SYSTEMS[a]

Tumor target	Strain of origin	H-2 type	Cytotoxicity index[b]	
			BALB/c vs MPC 11	DBA/2 vs SL2
SL2	DBA/2	d	22 ± 4^c	60 ± 7
L5178Y	DBA/2	d	NT	27 ± 9^d
MOPC 195	BALB/c	d	19 ± 9^c	22 ± 9^d
MPC 11	BALB/c	d	44 ± 11	20 ± 5^d
EL4	C57BL	b	NT	19 ± 3^d

[a] Sensitized lymphocytes obtained from BALB/c mice immunized with the syngeneic MPC 11 tumor cells (BALB/c vs MPC 11) or from DBA/2 mice immunized with the syngeneic SL2 tumor cells (DBA/2 vs SL2) were used to arm macrophages.
[b] Cytotoxicity was determined by cell counting. Mean $\pm$ SEM of four experiments performed in triplicate. NT, Not tested.
[c] Cytotoxicity significantly different from cytotoxicity against MPC 11 cells ($p < 0.05$).
[d] Cytotoxicity significantly different from cytotoxicity against SL2 cells ($p < 0.05$).

can be used 2–3 hr after seeding or after overnight culture. The macrophages are cultured with lymphocytes and tumor cells for 24 hr. After incubation, the macrophage monolayer is washed vigorously to remove the nonadherent lymphocytes and tumor cells. The washed macrophages are then challenged with tumor cells to measure the cytotoxicity. Cytotoxicity is measured in an effector–target cell ratio of 10 : 1 in a 24-hr assay as described elsewhere (see [28], this volume). Cytotoxicity can be determined by cell counting or by a [^{3}H]thymidine release assay. Results of experiments in allogeneic and in syngeneic systems are summarized in Tables I and II.

Comments

In the allogeneic system the armed macrophages express the highest cytotoxicity against the target cells bearing the same H-2 as the sensitizing tumor cells. For example, when the SMAF-producing lymphocytes are obtained from C57BL (H-2^b) mice immunized with SL2 cells (H-2^d), the highest cytotoxicity is expressed against H-2^d-bearing tumor cells (Table I). In the DBA/2 (H-2^d) versus EL4 (H-2^b) tumor cell system the highest cytotoxicity is induced against H-2^b-bearing tumor cells.[16] In the syngeneic system the armed macrophages express the highest cytotoxicity against the sensitizing tumor cells (Table II), and so is not H-2 directed.

Both in the allogeneic and in the syngeneic systems cytotoxicity is also expressed against other nonspecific tumor targets; this cytotoxicity is only about 50% of the specific cytotoxicity.[16]

In control experiments for SMAF production, macrophages are incubated with supernatants of cultures of normal lymphocytes and tumor cells. In control experiments for macrophages armed directly by lymphocytes, the macrophages are incubated with (1) normal lymphocytes with or without tumor cells or (2) with sensitized lymphocytes without tumor cells. None of these controls should express an increased cytotoxicity compared with macrophages cultured for 24 hr in medium alone.[14,15] To establish whether specific T-cell cytotoxicity instead of specific macrophage cytotoxicity is measured, the monolayers can be treated before measuring the cytotoxicity with anti-Thy-1 serum in the presence of complement.[14,15]

To determine whether specific antibodies are involved, the armed macrophages can be treated with anti-murine immunoglobulin antibodies[14,15] before testing of the cytotoxicity.

Concluding Remarks

The results obtained with macrophages armed by SMAF and with macrophages armed directly by lymphocytes are running in parallel when tested together.[14,15] The arming of macrophages by direct incubation of the lymphocytes with the macrophages results in a higher cytotoxicity although the nonspecific cytotoxicity is also somewhat higher than the nonspecific cytotoxicity induced by SMAF (Table I). We prefer to use the "lymphocyte-armed macrophages" for studying the specific macrophage cytotoxicity. To study the factor itself we use the SMAF armed macrophage bioassay.

The arming of macrophages by the T-cell factor SMAF[16] is clearly distinct from the induction of nonspecific macrophage cytotoxicity by the lymphokine macrophage activating factor (MAF), by several criteria:

1. The induction of macrophage cytotoxicity by MAF is LPS dependent, whereas the induction of cytotoxicity by SMAF is not (R.D.W., unpublished observations).

2. MAF renders resident macrophages only slightly cytotoxic and acts preferentially on exudate macrophages,[6] whereas SMAF renders both types of macrophages equally cytotoxic.

3. Macrophages incubated with MAF and LPS express a higher phagocytic and lysosomal enzyme activity, whereas these activities are not enhanced in armed macrophages (unpublished observations).

[37] Activation of Macrophages with Oxidative Enzymes

By Doris L. Lefkowitz, Stanley S. Lefkowitz, Ru-Qi Wei, and Johannes Everse

Introduction and Background

The role of macrophages in host defenses is becoming increasingly recognized. Activation of these cells results in their ability to kill both neoplastic cells and obligate intracellular parasites. The phenomenon of activation of macrophages has been extensively reviewed recently.[1,2]

Activation of murine macrophages to tumoricidal activity results from macrophages interacting with lymphokines through a defined sequence of events.[3] Accompanying this activation are significant changes in cellular metabolism including generation and release of reactive oxygen intermediates. The capacity to generate increased amounts of reactive oxygen intermediates is a consistent biochemical marker of activation. It has been shown that increased amounts of macrophage activator produce a linear increase in the rate of superoxide production. In addition to the changes in oxygen intermediates, there are alterations in intracellular components, changes in membrane properties and increased secretion of various substances.

A number of different substances have been shown to possess the ability to activate macrophages. In addition to lymphokines including interferon, substances such as lipopolysaccharides and muramyl dipeptide, are capable of causing macrophage stimulation and activation. We have found that certain oxidative enzymes, i.e., peroxidases, are also capable of stimulating and activating macrophages.

Peroxidases are widely distributed in nature. High levels are found in certain plants and fruits. The enzyme from horseradish is probably one of the best characterized of the plant enzymes. Peroxidases are also found in many mammalian tissues. It is well known that polymorphonuclear leukocytes contain a high level of peroxidase activity, amounting to as much as 5% of their total protein content. High levels are also found in eosinophils and monocytes. At the present time the biological function(s) of peroxidases is (are) still unknown, especially those of the plant enzymes. *In vitro,* peroxidases are known to catalyze the dehydrogenation of a variety

---

[1] D. O. Adams and M. G. Hanna, Jr., *Contemp. Top. Immunobiol.* **13,** 1 (1984).

[2] D. O. Adams and T. A. Hamilton, *Annu. Rev. Immunol.* **2,** 283 (1984).

[3] M. S. Meltzer, M. Occhionero, and L. P. Ruco, *Fed. Proc., Fed. Am. Soc. Exp. Biol.* **41,** 2198 (1982).

of hydroxylated aromatic compounds using H_2O_2 as their second substrate.

In recent years several papers have reported that certain peroxidases such as myeloperoxidase and lactoperoxidase possess cytotoxic activity toward bacteria and mammalian tumor cells. Thus, myeloperoxidase in conjunction with H_2O_2 and a halide ion forms a powerful cytotoxic "complex."[4,5] Other peroxidases are capable of forming similar cytotoxic complexes.[6] *In vitro,* these complexes have been shown to be cytotoxic to bacteria, fungi, and various mammalian cells, indicating that their cytotoxic activity is not specific with respect to their target. Recent experiments from this laboratory indicated, however, that *in vivo* horseradish peroxidase and lactoperoxidase specifically attacked tumor cells.[7]

It has recently been shown that exogenous horseradish peroxidase can enhance the ability of macrophages to kill intracellular parasites.[8] This ability is lost when the cells are treated with peroxidase inhibitors. Others have shown that eosinophil peroxidase-coated tumor cells are 32 times more sensitive to lysis than normal tumor cells when incubated with macrophages.[9] Subsequently, we demonstrated that macrophages can be stimulated to the tumoricidal state by various peroxidases.[10] This activation is affected by inhibitors of the peroxidase reaction, suggesting that a product of the peroxidase may be the actual macrophage activator.

In this chapter we outline the methods used for activating macrophages with peroxidases. Activation can be measured in a variety of different ways, several of which are described in other chapters of this volume. In this chapter we present two assays that are routinely used in our laboratory and that are modifications of several published methods.[11,12] The first one employs a target cell cytostasis assay and the second determines the increase in the "oxidative burst" as measured by the increase in superoxide production, using the chemiluminescence of luminol as the indicator.

[4] S. J. Klebanoff, *J. Clin. Invest.* **46,** 1078 (1967).
[5] S. J. Klebanoff and D. C. Smith, *Gynecol. Invest.* **1,** 21 (1970).
[6] K. E. Everse and J. Everse, *in* "Air Pollution—Physiological Effects" (J. J. McGrath and C. D. Barnes, eds.), p. 1. Academic Press, New York, 1982.
[7] K. E. Everse, H. Lin, E. L. Stuyt, J. C. Brady, F. Buddingh, and J. Everse, *Br. J. Cancer* **51,** 743 (1985).
[8] Y. Buchmuller and J. Mauel, *J. Reticuloendothel. Soc.* **29,** 181 (1981).
[9] C. F. Nathan and S. J. Klebanoff, *J. Exp. Med.* **155,** 1291 (1982).
[10] R. Q. Wei, S. S. Lefkowitz, D. L. Lefkowitz, and J. Everse, *Proc. Soc. Exp. Biol. Med.,* in press.
[11] N. O. Olsson, A. Leclerc, J. F. Jeannin, and F. Martin, *Ann. Immunol.* **133,** 245 (1982).
[12] J. B. Weinberg, H. A. Chapman, and J. B. Hibbs, Jr., *J. Immunol.* **121,** 72 (1978).

Microassay for Target Cell Toxicity

The microassay described here is different from other cytotoxicity assays in that it does not require the presence of a radioactive marker. A fixed number of target cells are exposed to activated macrophages for a period of 48 hr and at the end of this time the cells are stained with a dye. The amount of stain present in wells containing target cells and activated macrophages is then compared with the amount of stain present in wells containing target cells with nonactivated macrophages.

Obviously, this type of assay cannot distinguish whether the activated macrophages exert a cytostatic or a cytolytic effect on the target cell. However, whether cytostasis or cytolysis is observed appears to be dependent on the type of target cell used rather than reflecting differences in the degree of macrophage activation. Therefore, if the purpose of the experiment is solely to determine the state of macrophage activation, this assay is well suited and any problems related to the use of radioactive isotopes can be avoided.

Mouse peritoneal macrophages are used in this assay because these cells can be obtained in large amounts following induction. In addition, inbred mouse strains are readily available, assuring reproducibility of data. Finally, much of the available literature, and methodology, refers to murine macrophages as a model of macrophage activation.

The target cells selected for this assay are 3T12 cells (American Type Culture Collection, Rockville, MD). These cells are maintained in Dulbecco's minimal essential medium (MEM)[12a] containing 10% fetal calf serum (FCS).

Macrophage Collection (see also this series, Vol. 108 [25] and [26])

1. Macrophages are obtained from 8- to 16-week-old C57BL/6 mice (Jackson Laboratories).
2. Mice are injected ip with 1 ml thioglycolate broth using a 25-gauge needle.
3. Four days later the mice are sacrificed by cervical dislocation.
4. With their ventral side up, the skin is swabbed with 70% alcohol, cut along the midline, and removed.
5. The animal is injected ip with 8 ml of cold phosphate-buffered saline (PBS), pH 7.2, containing 10 U heparin/ml.

[12a] Abbreviations: PBS, phosphate-buffered saline; MEM, Dulbecco's minimal essential medium; FCS, fetal calf serum; LPS, lipopolysaccharide; BSA, bovine serum albumin; HRP, horseradish peroxidase; AT, 3-aminotyrosine.

6. The abdomen is gently massaged for 1 min and the fluid withdrawn (PBS may be slowly withdrawn and reinjected several times to dislodge cells).
7. Cells are then centrifuged at 150 g for 10 min at 5° and resuspended in the appropriate media.

It is important to remove the peritoneal cells with a minimum of erythrocyte contamination. This is because erythrocytes interfere with both the macrophage activity and the chemiluminescence. The yield of peritoneal exudate cells obtained varies depending on conditions. The number obtained from mice injected with thioglycolate broth varies between 10–15 × 10^6, whereas the number obtained from noninduced mice is approximately 2 × 10^6. Of this number approximately 75 to 80% are macrophages, 20 to 25% lymphocytes, and 1 to 2% granulocytes. After attachment more than 99% are mononuclear phagocytes. The morphology of induced cells is different from that of the resident cells. The former are considerably larger and will spread with extended processes whereas the latter are rounded and considerably smaller.

Assay Procedure

1. Peritoneal cells, collected as described above, are suspended in Dulbecco's minimal essential medium (MEM) without serum.
2. Cells are washed twice in MEM without serum using centrifugation at 150 g for 10 min and then resuspended in MEM at 10^6 cells/ml.
3. One hundred microliters containing 10^5 cells is added to each well of a flat-bottom tissue culture plate (Falcon 3040).
4. After 2 hr incubation at 37°, under 5% CO_2, the nonadherent cells are removed by gentle aspiration of the medium.
5. The cell monolayer is washed twice by adding 150 μl of medium and aspirated as above.
6. Cells are fed with 250 μl MEM containing 2% fetal calf serum (FCS) (Sterile Systems, Logan, UT) and incubated for 3 days.
7. After incubation the medium is removed and 50 μl of fresh medium with FCS are added to each well.
8. An additional 50 μl either of "activating factors" or control medium may be added to each well.
9. One hundred microliters of medium containing 6 × 10^3 3T12 target cells are added to each well and incubated at 37° under 5% CO_2.
10. After 48 hr incubation, the medium is aspirated and the cells washed with MEM containing 2% FCS four times to remove nonadherent cells.

11. The number of residual target cells are assessed by one of two methods:

A. Microscopic visual counting

   1. Cells are fixed with absolute methanol and stained with Giemsa.
   2. The number of target cells is determined by counting the number of 3T12 cells in at least four microscopic fields of each well using 450× magnification.

B. Dye uptake method measured either in a spectrophotometer or an ELISA reader

   1. Adherent cells are fixed with 0.1 ml of 10% formalin in PBS.
   2. The cell monolayers are washed using 0.01 M borate buffer at pH 8.4 then stained for 30 min with 0.5% methylene blue in borate buffer.
   3. The cells are washed four times, by aspiration with the borate buffer, and then allowed to dry at room temperature.
   4. Each well is filled with 200 μl of 0.1 N HCl which elutes the dye from the residual cells.
   5. The absorbance is read at 664 nm either directly in each well on an ELISA photometer such as the Dynatech Minireader II, Alexandria, VA, or with a spectrophotometer. For reading with a spectrophotometer the dye solution is quantitatively transferred into a 3-ml cuvette containing 2 ml deionized water.

Each experiment consists of a series of evaluations using different peroxidase dilutions. An average of at least eight wells are made for each experiment, which is carried out in triplicate. The following controls should be present in each experiment: macrophages only; macrophages with 3T12 cells, but no other additions; 3T12 cells with activator but without macrophages; and 3T12 cells only. Control points are routinely determined in triplicate and the results are averaged. A sample experimental layout is shown in Fig. 1. The percentage cytotoxicity is calculated from the average value obtained from the combination of activated macrophages and 3T12 cells with the average values obtained from the wells containing macrophages and 3T12 cells without activators, using the formula:

Percentage cytotoxicity = 100% − (experimental/control)100%

Figure 2 illustrates typical results obtained with the dye-binding method using the ELISA reader.

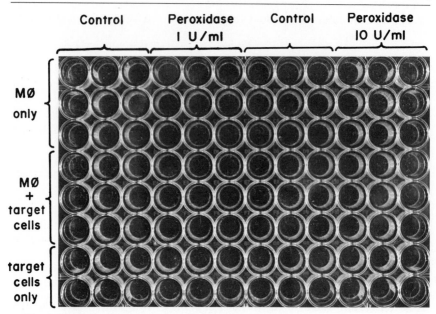

FIG. 1. Example of the experimental layout of a macrophage (MØ) activation experiment in a 96-well plate.

Comments

The ratio of effector cells to target cells may be varied from 30 : 1 to 10 : 1. The former ratio results in considerably more cell killing but differences between the amount of dye taken up by the macrophages alone and by the combination of macrophages plus target cells is less and therefore more difficult to interpret. A lower ratio, e.g., 15 : 1, although resulting in less killing, is preferable because of the greater differences in dye uptake.

Before the target cells are added, the macrophages are incubated from 2 hr to 3 days. The reason is to allow induced macrophages to stabilize their level of activity prior to the addition of activators. Using this procedure, the population of adherent cells consists of >99% macrophages.

It has been reported by other investigators[11,12] that macrophages do not stain well with methylene blue. However, the number of macrophages present in this assay is at least 10 to 30 times the number of target cells; the macrophages can therefore account for a significant portion of the total staining. Another source of errors can be the fact that active, cytotoxic macrophages frequently detach from the surface.[9] The staining method nevertheless proved useful in assessing the number of residual target cells. Excellent correlation is usually obtained between this method

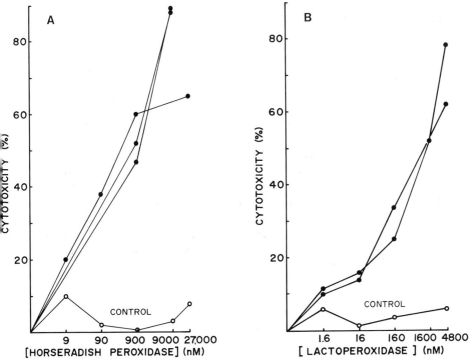

FIG. 2. Cytotoxicity of macrophages against 3T12 cells. Macrophages are stimulated by horseradish peroxidase (A) or lactoperoxidase (B). Control values represent the toxicity of the peroxidases to 3T12 cells in the absence of macrophages.

and by counting the residual target cells under the microscope (average value of the number of target cells counted in each of four fields).

It is important that scrupulous care be taken to minimize the presence of lipopolysaccharide (LPS) in reagents, glassware, etc. Trace amounts of LPS, which are frequently present, may significantly alter results due to its macrophage-activating properties. The difficulty in completely excluding the presence of LPS in solutions is well recognized. Minimizing contamination can be done by utilization of freshly prepared solutions, maintenance of sterility, etc. Removal of LPS contamination can be implemented using Detoxi-Gel, Pierce Chemical Company, Rockford, IL. It is strongly recommended that reagent solutions be tested for the absence of LPS. In order to ascertain that valid results are obtained with unknown compounds it is advisable to do a control assay using a known macrophage activator, such as LPS or murine interferon (α and b).[10] We found that removal of the activating agent, after exposure to the macrophages for 20 hr and prior to the addition of the target cells, does not yield different results from experiments in which the activator remains present during the incubation with the target cells.

Determination of Macrophage Activation by Chemiluminescence (see also this volume [33])

One of the functions of phagocytic cells is to ingest and kill microorganisms. It has been shown that these cells emit light while ingesting particles.[13-15] Light arises from the production of high-energy compounds produced during bactericidal activity. This spontaneous luminescence can be used as an assay of phagocytic function or as a sensitive means to quantitate microbicidal metabolic activity. However, these measurements require large numbers of cells and a very sensitive light detector, such as a scintillation counter.

Phagocytosis by macrophages is also accompanied by a large increase in the production of superoxide ions, which are released into the medium. Superoxide spontaneously dismutates to O_2 and H_2O_2 within a few seconds:

$$2O_2^- + 2H^+ \rightarrow H_2O_2 + O_2$$

The hydrogen peroxide formed can be measured rapidly and quantitatively with the use of luminol (5-amino-2,3-dihydro-1,4-phthalazine-dione):

| luminol | hydrogen peroxide | 3-amino-phthalic acid |

As indicated in the equation, each molecule of H_2O_2 in the presence of excess luminol produces one quantum of light. The photons produced can be determined with the use of a scintillation counter in the out-of-coinci-

[13] R. C. Allen, R. L. Stjernholm, and R. H. Steel, *Biochem. Biophys. Res. Commun.* **47,** 679 (1972).
[14] R. C. Allen, *in* "Chemical and Biological Generation of Excited States" (W. Adam and G. Cilento, eds.), p. 309. Academic Press, New York, 1982.
[15] C. S. F. Easmon, P. J. Cole, A. J. Williams, and M. Hastings, *Immunology* **41,** 67 (1980).

dence mode or with the use of a photon counter. For reasons of convenience and simplicity of operation the use of a photon counter is usually preferred over that of a scintillation counter.

Thus, increases in phagocytic activity of macrophages in response to the addition of a stimulant can readily be measured by determining the increase in chemiluminescence in the presence of luminol. The increase in chemiluminescence upon stimulation of macrophages with an appropriate stimulus is rapid and occurs within a minute after the addition of the stimulating agent. The luminol method therefore provides an easy and rapid way of measuring the activation of macrophages. The method is also extremely sensitive if a sensitive light meter (e.g., a photon counter) is used. Moreover, a continuous measurement over a given period of time can be made if the counter is connected to a suitable chart recorder.

In this chapter we will present the chemiluminescence method for measuring the increase in phagocytic activity of macrophages stimulated with a peroxidase, since we have used the method extensively for this purpose. However, with appropriate modifications the method could also be used to measure the effects of other stimulatory agents.

Preparation of Reagents

Zymosan Solution. The following procedure is to be carried out under aseptic conditions. Suspend 10.0 mg zymosan (Sigma Chemical Company, St. Louis, MO)/ml H_2O and boil for 30 to 45 min. After cooling, centrifuge at 250 g for 10 min and resuspend the pellet in equal amounts of PBS and guinea pig complement (Cappel Labs, West Chester, PA) (1 : 1). After incubation at room temperature for 30 min, centrifuge and resuspend the pellet in equal amounts of complement and PBS as before. Incubate at room temperature for an additional 30 min, then centrifuge and discard supernatant. Add 10× vol of PBS, centrifuge, and repeat with another 10× vol. Resuspend as 1 mg zymosan/ml PBS and freeze aliquots at −70°.

Luminol Solution. One milligram each of luminol and crystallized globulin-free bovine serum albumin (BSA) (Sigma) are mixed per milliliter of PBS. The mixture is stirred for 10 min, then filtered using a 0.2-μm filter. The filtrate is aliquoted and stored at 4° in a covered container to prevent exposure to light.

Assay Procedure

1. Suspend peritoneal cells in MEM without phenol (Auto POW, Flow Laboratories, Rockville, MD) containing 1% albumin.
2. Centrifuge cells at 150 g for 10 min, wash two more times with MEM as above.

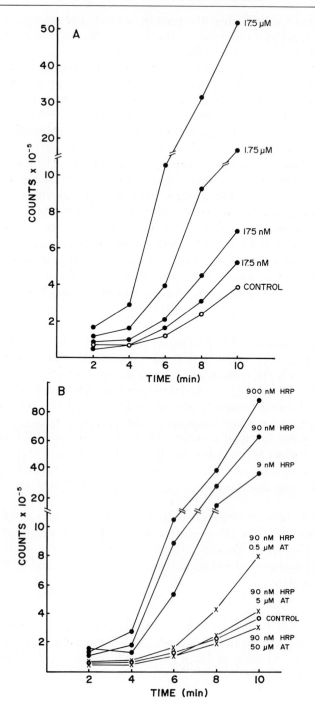

3. Resuspend cells in 2 ml MEM, count, aliquot, and adjust cell concentration to 1×10^6/ml.
4. Add 100 μl cells to each 8 × 50 mm tube (Evergreen Scientific, Los Angeles, CA).
5a. For peritoneal cells assayed in suspension proceed to Step 8.
5b. For cell monolayers incubate 30 min at 37° under 5% CO_2.
6. Discard supernatant and wash monolayer in each tube 3× with 200 μl of medium, then add 100 μl fresh medium and return to incubator for 30 min.
7. Just prior to assay, discard medium and add 100 μl of fresh medium which is maintained in the incubator under 5% CO_2.
8. Each tube containing either cell suspension or monolayer contains the following:
   a. 100 μl Fresh medium
   b. An additional 100 μl medium with or without "activator"
   c. 100 μl Zymosan
   d. 30 μl Luminol
9. Place in luminometer (Turner Designs, Mountain View, CA, model 20e) at 37° and program for 10 2-min readings.
10. Results are recorded and plotted as time vs light emission (counts).

In general, data are collected at 2-min intervals and plotted against time. Figure 3A and B illustrates sample data indicating a dose-related effect of peroxidases on the respiratory burst. It is also apparent that 3-aminotyrosine, an inhibitor of peroxidase, readily interferes with the chemiluminescence of macrophages stimulated by horseradish peroxidase.

Comments

A number of possible pitfalls are present in this procedure. It is important to maintain sterility throughout these studies using scrupulously clean glassware, solutions, etc., to prevent the introduction of bacteria and other particles which would interfere with and/or enhance macrophage responses.

Erythrocyte contamination of peritoneal exudate cells interferes in general with the reactivity of macrophages. Extensive red cell contamina-

---

FIG. 3. (A) Chemiluminescence of macrophages exposed to various concentrations of microperoxidase. (B) Chemiluminescence of macrophages exposed to various concentrations of horseradish peroxidase in the absence and presence of 3-aminotyrosine.

tion results in macrophage activation which will ultimately result in decreased reactivity to specific activators.

Another factor which can play havoc with the procedure is variation in cell numbers. Not all of the peritoneal exudate cells are mononuclear. This is especially important using cell suspensions which may contain neutrophils. The latter may be more reactive than the mononuclear cells.

A second problem in the "suspension system" is mononuclear cell adherence to the wall of the culture vessel. With continual agitation at 0°, some attachment of the macrophages still occurs, ultimately resulting in reduced numbers available for the assay. Using attached peritoneal exudate cells, "contaminating cells," are reduced. As with other assays, it is not always possible to obtain a uniform number of cells in each tube, requiring the necessity to run a number of controls to minimize variability.

Suspended cells lose some of their activity after several hours. Even those which are attached can be used for approximately 5 hr before reactivity begins to decline. Other factors influencing their reactivity include pH, temperature, contaminants. It is important to maintain all reagents at constant temperature. It is also essential to monitor the pH of the medium since the pH indicator phenol red is not used. In addition, small amounts of serum containing trace amounts of hemoglobin can stimulate the reaction.

Among the advantages of these methods is extreme sensitivity. It is known that sensitivity of chemiluminescence can exceed 10^{-12} M.[16] In addition, the reaction is extremely rapid and effects on macrophage function can be determined in minutes. This procedure can also be used to measure the production of high-energy oxygen intermediates. Using appropriate inhibitors, one can study the production of certain oxidative intermediates which function in various biochemical pathways. The procedure also lends itself to comparative studies between various activators particularly for the identification of the portion of a molecule which is the "activating moiety." It is clear that the use of chemiluminescence represents a valuable adjunct to studying many diverse reactions which are triggered in the process we call macrophage activation.

Acknowledgment

This work was supported by Grant CA 32715 of the National Cancer Institute, U.S. Public Health Service.

[16] T. P. Whitehead, L. J. Kricka, T. J. N. Carter, and G. H. G. Thorpe, *Clin. Chem.* (*Winston-Salem, N.C.*) **25,** 1531 (1979).

[38] Assay of a Cytocidal Protein Excreted by Activated Macrophages

By BETH-ELLEN DRYSDALE, CHARLES M. ZACHARCHUK,
MASAKO OKAJIMA, and HYUN S. SHIN

Macrophages, when activated by a variety of agents, become cytotoxic for nucleated target cells. One way macrophages mediate the killing of the target cells is through the release of a soluble factor designated as macrophage cytotoxic factor (Mφ-CF).[1,1a] When cells which adhere to the surface of culture vessels are used as targets, the ability of Mφ-CF to kill the target cells can be measured by determining the number of dead cells that detach from the culture vessel. Another assay method involves the use of radiolabeled thymidine or its analog, iododeoxyuridine, since target cells damaged by Mφ-CF cease to incorporate these DNA precursors. These two basic assay methods are the subject of this chapter. There are, however, other assay techniques that can be used. For example, dead and live cells can be enumerated using a Coulter counter,[2] or the extent of target cell damage can be measured by the release of tritiated thymidine from prelabeled target cells.[3]

Production of Mφ-CF

Medium containing Mφ-CF can be generated by culturing mouse, rat, guinea pig, or rabbit macrophages in the presence of an activator. The crude Mφ-CF preparations are stable at $-70°$ for several months. The molecular weights of mouse, guinea pig, and rabbit Mφ-CF are approximately 45,000 as measured by gel filtration. The production of Mφ-CF from these three species is described below.

Reagents

Culture Media. Mouse, guinea pig, and rabbit Mφ-CF are made in serum-free medium consisting of RPMI-1640 supplemented with 100 U penicillin/ml, 100 g of streptomycin/ml, and 2 mM glutamine (RPMI-PS). During the plating of the mouse and guinea pig macrophages, fetal calf

[1] Abbreviations: Mφ-CF, macrophage cytotoxic factor; RPMI-PS, RPMI supplemented with penicillin and streptomycin; TCM, tissue culture medium; LPS, lipopolysaccharide; BCG, Bacillus Calmette–Guérin; PEC, peritoneal exudate cells; IUdR, 5-iodo-2'-deoxyuridine.

[1a] B.-E. Drysdale, C. M. Zacharchuk, and H. S. Shin, *J. Immunol.* **131**, 2362 (1983).

[2] M. K. Gately and M. M. Mayer, *J. Immunol.* **112**, 168 (1974)

[3] D. N. Mannel, M. S. Meltzer, and S. E. Mergenhagen, *Infect. Immun.* **28**, 204 (1980).

METHODS IN ENZYMOLOGY, VOL. 132

serum is added to the RPMI-PS such that the final concentration of serum is 10%. The medium containing serum is referred to as tissue culture medium (TCM).

Activating Agents. The lipopolysaccharide (LPS) used is prepared by the Westphal method[4] from *Escherichia coli* strain 0127 : B8 or 0111 : B4 (Difco Laboratories, Detroit, MI). LPS is dissolved to a final concentration of 0.2 mg/ml in 10 mM phosphate-buffered saline, pH 7.2, containing 0.15 M sodium chloride, 100 U penicillin/ml, 100 g of streptomycin/ml, and 4% fetal bovine serum. The solution is aliquoted, and stored at $-20°$. Stock solutions (10 mM) of the calcium ionophore, A23187 (Sigma, St. Louis, MO, or Calbiochem, La Jolla, CA) are prepared in dimethyl sulfoxide. The solvent alone, at the concentrations used, has no effect on Mϕ-CF generation.

Agents Used to Elicit Macrophages. Mouse and guinea pig peritoneal exudates containing macrophages are elicited with hydrolyzed starch (Connaught, Toronto, Canada); starch is mixed with saline (3% w/v) and heated to 90° with constant stirring for approximately 2.5 hr. This homogeneous suspension is then aliquoted and autoclaved. Once cooled it is refrigerated, and left at 4° for at least 2 weeks before use. While the 3% starch is most frequently used in our laboratory, other agents such as thioglycolate can be used to elicit mouse or guinea pig macrophages capable of making Mϕ-CF. Fluid thioglycolate medium, USP (Difco Laboratories) is prepared by adding 29.5 g of thioglycolate to 1 liter of water. The medium is dissolved by boiling for approximately 1 min. It is aliquoted, autoclaved, and then stored at 4°. To elicit rabbit alveolar macrophages, heat-killed Bacillus Calmette–Guérin (BCG) (ITR, Biomedical Research, Chicago, IL) is used.[5] The BCG is ground and then sonicated in mineral oil at a concentration of 1 mg/ml.

Procedure (see also this series, Vol. 108 [25], [27])

Murine peritoneal exudate cells (PEC) are obtained from 6- to 8-week-old male or female C3HeB/FeJ mice (Jackson Laboratory, Bar Harbor, ME) 3 days after an intraperitoneal (ip) injection of 3 ml of 3% starch. The cells are harvested by peritoneal lavage with 6 ml of ice-cold Hanks' balanced salt solution containing antibiotics. Guinea pig PEC are obtained from 300- to 500-g male Hartley guinea pigs (Buckberg Lab Animals, Tomkins Cove, NY) in the same way except that 20 ml of starch are injected ip and the volume of fluid used for lavage is 100 ml.

---

[4] O. Westphal and K. Jann, *Methods Carbohydr. Chem.* **5,** 83 (1965).
[5] Q. N. Myrvik, E. S. Leake, and S. Oshima, *J. Immunol.* **89,** 745 (1962).

Following lavage, tubes containing the fluid are placed on ice, and cell clumps and tissue debris are allowed to settle out of the cell suspension for 5 min. The suspension is then transferred to another tube and the cells are washed once by centrifugation. After the wash the cells are counted. The yield averages 1×10^7 per mouse and 2.5×10^7 per guinea pig. Greater than 90% of the cells are viable, and stained preparations show that 60–70% of the cells are macrophages by morphology. The cells are resuspended in TCM to a final concentration of 3×10^6/ml, and are then plated in plastic culture flasks by adding a sufficient volume of the cell suspension such that there are 10^6 PEC/cm^2 surface area. After allowing the cells to adhere for 2.5 hr at 37° in a humidified 5% CO_2 environment, the medium and nonadherent cells are aspirated, and the monolayers are washed twice with RPMI-PS. The volume of each wash is equal to the volume of the cell suspension initially plated in the flask. Differential counts of stained monolayers prepared in this manner show that ≥97% of the cells are macrophages.[1] After the second wash, a volume of RPMI-PS corresponding to the initial volume of the cell suspension is placed in the flask. An activator is then added in a small volume, i.e., no greater than 1/10 the total volume in the flask. When LPS is used to activate mouse cells, the final concentration in the flask is 5 to 10 ng/ml while a concentration of 10 μg/ml is used to activate the guinea pig cells. For two of the three species tested, mouse and guinea pig, the calcium ionophore A23187 at concentrations ranging from 5 to 10 μM will also activate the macrophages to produce Mφ-CF. While the ionophore alone is not as good an activator as LPS, the combination of LPS and A23187 can be used because these two agents act in a synergistic manner on mouse and guinea pig cells. After the addition of the activator(s), the macrophage monolayers are incubated for 6 to 8 hr at 37° in a humidified, 5% CO_2 environment.

To prepare rabbit Mφ-CF, alveolar macrophages are obtained from 2.5- to 3.5-kg female New Zealand and white rabbits (Bunnyville Farms, Littlestown, PA) that had been injected in the marginal ear vein 3 to 4 weeks earlier with 0.3 to 0.4 ml of the BCG preparation. The rabbits are sacrificed and the lungs removed[6,7] as follows. (For further details, see this series, Vol. 108 [25].) A horizontal incision is made in the mid neck of the rabbit and the trachea is exposed by blunt and sharp dissection. The trachea is then dissected free of its surrounding structures and cut just below the level of the thyroid cartilages. The thorax is then entered through an incision at the level of the xyphoid process. With the trachea

[6] J. J. Windle, H. S. Shin, and J. F. Morrow, J. Immunol. **132**, 1317 (1984).
[7] Q. N. Myrvik, E. S. Leake, and B. Fariss, J. Immunol. **86**, 128 (1961).

free and the chest open one can dissect the cardiopulmonary complex free from its vascular and chest wall attachments. Care is taken during dissection to leave the trachea intact and not to puncture the lungs. The cells are harvested by repeated pulmonary lavage with five to six washes of 50 ml of ice-cold Hanks' balanced salt solution containing antibiotics. The fluid is injected through the trachea, the lung is massaged gently, and the lavage fluid is poured into a collection tube. Cell clumps and debris are allowed to settle for 5 min, the cell suspension is transferred to other tubes, and the cells are washed three times. After washing, the rabbit cells are counted and their concentration is adjusted to 2×10^6/ml. The viability is at least 90% and the yield is approximately 10^9 cells per rabbit. Since differential counts show that these cell preparations are at least 90% macrophages by morphology, they are not further purified. LPS is added at a final concentration of 10 μg/ml. The cells are plated in plastic culture flasks at 3.5×10^5 cells/cm^2 surface area. The flasks are then incubated at 37° in a 5% CO$_2$–95% air atmosphere for 6 to 8 hr. At this point, the Mϕ-CF from different animal species is processed in the same way. After the 6- to 8-hr incubation, the medium is collected, centrifuged to remove cells and debris, and filtered through a 0.45-μm filter. This material may be stored at $-70°$, but to have a stable standard it is concentrated 20- to 30-fold by filtration in an Amicon apparatus with a 10,000-Da cutoff membrane and dialyzed against 10 mM phosphate-buffered saline, pH 7.2. The concentrated, dialyzed material is aliquoted and stored at $-70°$.

In comparing the activity of the three different preparations, prior to concentration, the rabbit material has the highest titer, ranging from 6000 to 8000 U/ml (the unit is defined as described below). Guinea pig Mϕ-CF has 1500 to 2500 U/ml while the mouse Mϕ-CF has a titer of 400 to 800 U/ml. At present it is not clear whether the titers of Mϕ-CF from these three sources differ because of the amount of Mϕ-CF produced or because of differences in functional activity.

Assay of Mϕ-CF Activity

Reagents and Cell Lines

Culture Medium. TCM is used to maintain all of the target cell lines and is also used in the Mϕ-CF assays.

Cell Lines. For Mϕ-CF assays, an adherent murine fibroblast line, L-929 (available from the American Type Culture Collection, Rockville, MD), is used most frequently. This cell line is very sensitive to the action of Mϕ-CF and adheres to culture vessels, making it an excellent target for the first assay described below. This line is maintained in TCM and is passaged by trypsinization.

Nonadherent cell lines such as the C3H lymphosarcoma, 6C3HED (available from Jackson Laboratory, Bar Harbor, ME), and the DBA/2 mastocytoma, P815 (available from the American Type Culture Collection, Rockville, MD), can also be used as targets as described below. These cell lines are also maintained in TCM.

Other Reagents. To harvest the adherent L-929 target cells, a solution of 0.25% trypsin (w/v) in Hanks' balanced salt solution without calcium or magnesium (Gibco, Grand Island, NY) is used. Actinomycin D (Cosmegen, Merck Sharpe and Dohme, West Point, PA) is often used to increase the sensitivity of the L-929 cells to Mφ-CF. The contents of each vial of actinomycin are dissolved in 1.1 ml of sterile distilled water according to the manufacturer's directions. The stock solution is stored in 50-μl aliquots at $-20°$. Crystal violet is used to stain viable adherent cells. It is prepared as a 10% stock solution in absolute ethanol (w/v), and it is diluted to 0.2% in distilled water before use. If nonadherent cells are used as targets, [5-^{125}I]iodo-2'-deoxyuridine ([^{125}I]IUdR) with a specific activity of 5 Ci/mg is used in determining cell viability.

Procedure

The killing of the L-929 cell line is most often used to measure the cytotoxic activity of Mφ-CF. This method is similar to that described by Ruff and Gifford.[8] Serial 2- or 3-fold dilutions (0.1 ml final volume) of test samples are made in 96-well microtiter plates with a multichannel pipet. L-929 cells are harvested from confluent cultures by treating the monolayers with the trypsin solution for 3–4 min at 37°. The cells are washed once in TCM and counted. The concentration of L-929 cells is adjusted to 5×10^5/ml in TCM. Actinomycin D is added at a final concentration of 1 μg/ml (100 μl of stock/25 ml of cell suspension). Using the multichannel pipet, 0.1 ml of the L-929 cell suspension is added to each of the microtiter wells. The cells are incubated with Mφ-CF at 37° in a humidified 5% CO_2 environment for 18 hr, the supernatants are discarded, and the remaining viable adherent cells are stained by immersing the microtiter plates into a beaker of crystal violet (0.2% in 2% ethanol) for 10 min. Microtiter plates are thoroughly rinsed with tap water and dried. When dry, 0.1 ml of 1% sodium lauryl sulfate is added to each well to solubilize the stained cells. The absorbance of each well is read at 570 nm with a model MR580 micro-ELISA autoreader (Dynatech, Alexandria, VA). Percentage cytotoxicity is defined as the relative absorbance (*Abs*) of sample compared to control (TCM only) wells:

$$\text{Percentage cytotoxicity} = [1 - (Abs_{\text{Sample}}/Abs_{\text{Control}})]\ 100$$

[8] M. R. Ruff and G. E. Gifford, *J. Immunol.* **125**, 1671 (1980).

A unit of cytotoxicity is arbitrarily defined as the reciprocal dilution of sample required to reduce the absorbance by 50%. Cytotoxic activity is plotted as percentage cytotoxicity versus log sample dilution to linearize the dose–response curve. A least-squares analysis of the data points makes it possible to determine the reciprocal dilution of sample at 50% cytotoxicity. The sample activity is expressed in units/milliliter. It should be noted that the presence of actinomycin D makes the assay substantially more sensitive. If actinomycin D is omitted, the titer of a standard lot of Mφ-CF tested on L-929 cells would be decreased by at least 50-fold.

A modification of this assay can be used to examine the effect of Mφ-CF preparations on nonadherent target cells such as 6C3HED lymphosarcoma or P815 mastocytoma cells as well as adherent cells. The cell types to be studied are harvested, counted, and suspended at 2.5×10^5/ml. They are not treated with actinomycin D but are added directly to microtiter plates containing the samples diluted as described above. The cells are incubated with Mφ-CF for 2 days at 37° in a humidified 5% CO_2 environment. Instead of staining with crystal violet to quantitate the number of viable cells remaining, the viability of the target cells is assessed by their ability to incorporate [^{125}I]IUdR. A 50-μl aliquot (containing 0.05 μCi) is added to each well. The plates are incubated for 2 hr after which uptake is terminated by the addition of 50 μl of thymidine (1 mg/ml) and the plates are put on ice. The contents of the wells are collected on filter strips using a cell harvester in order to separate incorporated label from free label. The strips are presaturated with the thymidine solution to minimize binding of free radiolabel. The contents of the wells are aspirated, and the wells are washed with 0.2 ml of 0.2 N NaOH to remove all of the cells. Filter strips are cut and the radioactivity is determined.

The suppression of target cell growth is expressed as percentage cytotoxicity, which is defined as follows:

$$\text{Percentage cytotoxicity} = [1 - (CPM_{\text{TC} + \text{Mφ-CF}}/CPM_{\text{TC}})]\ 100$$

where mean counts incorporated by target cells in the presence of Mφ-CF, $CPM_{\text{TC} + \text{Mφ-CF}}$, are divided by the counts incorporated by target cells in TCM alone, CPM_{TC}. Background counts obtained from wells containing only medium are insignificant, and therefore no correction for background is necessary. Units of Mφ-CF activity can be calculated as described above. The main advantage of the [^{125}I]IUdR incorporation assay is not the sensitivity, but the ability to study the action of Mφ-CF on nonadherent as well as adherent target cell lines.

Comments (see this series, Vol. 116 [33] and [34])

Cytotoxic factors other than Mφ-CF can be assayed using the same procedures described above. Lymphotoxin, a cytotoxic factor from activated lymphoid cells, and tumor necrosis factor, a cytotoxic protein present in the serum of BCG-infected animals subjected to a lethal dose of LPS, are both measurable by these methods. Interestingly, recent studies indicate that these cytotoxic factors are similar, if not identical to, Mφ-CF.[9]

Acknowledgments

We thank Dr. R. F. Siliciano, Dr. J. H. Peters, and Ms. D. Berry for their help in the preparation of this manuscript.

[9] C. M. Zacharchuk, B.-E. Drysdale, and H. S. Shin, *Proc. Natl. Acad. Sci. U.S.A.* **80**, 6341 (1983).

[39] Assays Detecting the Antibody-Dependent and Independent Binding and Cytolysis of Tumor Cells by Murine Macrophages

By WILLIAM J. JOHNSON and DOLPH O. ADAMS

I. General Comments

The host wide system of macrophages (termed the mononuclear phagocyte system, or MPS)[1] is a population of cells which are readily modulated by environmental influences to change profoundly their physiology and functional capacities.[1a] This process, which has generally been termed activation, represents a complex web of potential developments.[1a] Therefore, some critical issues in analysis of the MPS include the following: What are the various states of activation? How can they be distinguished? What signals regulate these states of activation? How are these signals transduced in macrophages to alter function? What specific

[1] Abbreviations: MPS, mononuclear phagocyte system; LPS, lipopolysaccharide; HBSS, Hanks' balanced salt solution; PEC, peritoneal exudate cells; SDS, sodium dodecyl sulfate; FCS, fetal calf serum; EDTA, ethylenediamine tetraacetic acid; EMEM, Eagle's minimum essential medium; ADCC, antibody-dependent cellular cytotoxicity; BCG, Bacillus Calmette–Guérin; TG, thioglycolate.

[1a] D. O. Adams and T. A. Hamilton, *Annu. Rev. Immunol.* **2**, 283 (1984).

changes in cellular physiology and function are necessary for a given functional alteration? How can various subpopulations of macrophages be distinguished from one another? To all of these questions, a useful approach has been to develop libraries of objective, quantitative markers of altered physiology and function.[1a] Assays for the capture and lysis of tumor cells have proved to be one of the most valuable markers for addressing such questions.[2,3] In addition, such assays permit delineation of both mechanisms and regulation of kill by macrophages, the effect of various drugs and pharmacologic agents on macrophage competence for kill, and the functional state of macrophages within tumors and how this relates to the ultimate fate of neoplasms.[1a-3]

The assays described in this chapter quantify the lytic interactions of murine macrophages and tumor cells in culture, and allow studies on the responsiveness of macrophage populations to signals which regulate the binding and subsequent kill of tumor cells. In order to obtain a reliable indication of the interactions between these two cell types, several variables must be controlled as strictly as possible.

A. Animal Health

A major variable in assays of macrophage function is the overall cleanliness and health of the animals used. Mice obtained from contaminated animal colonies often have macrophages, which are hyperresponsive to regulatory signals or even capable of spontaneous tumor cytolysis.[4-6] Although bacterial, fungal, and parasitic infections are *generally* not a common problem, chronic and often inapparent infections with viruses are frequently encountered; murine hepatitis virus, LDH virus, and Sendai virus are common sources of problems. Animals should be obtained only from reliable vendors, maintained in extremely clean facilities, and checked frequently for parasitic bacteria and viral infections.

B. Endotoxin Contamination

Contamination of media, serum, and reagents with LPS has emerged as a major consideration in studies on murine macrophages, particularly

[2] D. O. Adams and P. Marino, *Contemp. Hematol./Oncol.* **3**, 69 (1984).
[3] W. J. Johnson, S. D. Somers, and D. O. Adams, *in* "Contemporary Topics in Immunobiology" (D. O. Adams and M. G. Hanna, eds.), p. 127. Plenum, New York, 1984.
[4] M. S. Meltzer, *in* "Methods for Studying Mononuclear Phagocytes" (D. O. Adams, P. J. Edelson, and H. S. Koren, eds.), p. 133. Academic Press, New York, 1981.
[5] M. S. Meltzer, *Cell. Immunol.* **22**, 176 (1976).
[6] W. J. Johnson and E. Balish, *J. Reticuloendothel. Soc.* **28**, 55 (1980).

when cytolytic functions are considered.[7-9] Because endotoxin at low concentrations (i.e., the picogram to nanogram range) can significantly affect tumoricidal activity of macrophages alone or in combination with various lymphokine preparations, resolution of this problem is essential. Despite increasing awareness of this problem, many reagents (particularly those containing protein) are contaminated with sufficient endotoxin to affect macrophage function.[7-9] In general, one must screen tissue culture media and serum and select lots that are extremely low in endotoxin (i.e., ≤ 0.01 ng/ml). When the use of contaminated reagents cannot be avoided, decontamination can be attempted as a number of procedures for the detoxification and determination of very minute amounts of endotoxin have been described.[7-9]

The procedures which have proved useful for murine mononuclear phagocytes may not be adequate for human cells, since human cells are exquisitely sensitive to endotoxin (i.e., they may respond to less than 10 pg of endotoxin).[10]

C. Cell Culture Conditions

The fundamental principles of culturing macrophages have recently been reviewed.[11] Under most circumstances, basal medium supplemented with fetal calf serum (FCS) will suffice for culture of macrophages and of tumor cells. We routinely use Eagle's minimum essential medium (EMEM) supplemented with 10% FCS (Hyclone), 25,000 U/100 ml penicillin, 12.5 mg/100 ml streptomycin, and 29.2 mg/100 ml fresh L-glutamine (gentamycin is toxic to some tumor cell lines, such as the MCA-I fibrosarcoma used in our studies). Special precautions may need to be taken for serum. We routinely use unfiltered serum, which has been shipped to us frozen directly on dry ice and which has been frozen and thawed *only one time* after being aliquoted. The serum is used immediately on the day of removal from the freezer, and any leftover serum is degraded to use in routine tissue culture. Medium for each experiment should be made fresh daily, and particular emphasis should be placed on making sure that the glutamine is potent. In some experiments, it may be useful to supplement the medium further with nonessential amino acids, ascorbic acid, sodium pyruvate, and selected vitamins.

[7] J. B. Weinberg, *in* "Methods for Studying Mononuclear Phagocytes" (D. O. Adams, P. J. Edelson, and H. S. Koren, eds.), p. 139. Academic Press, New York, 1981.

[8] J. B. Weinberg, H. A. Chapman, and J. B. Hibbs, *J. Immunol.* **121,** 72 (1978).

[9] J. L. Pace, S. M. Taffet, and S. W. Russell, *J. Reticuloendothel. Soc.* **30,** 15 (1981).

[10] M. J. Pabst, H. B. Hedegard, and R. B. Johnston, *J. Immunol.* **128,** 123 (1982).

[11] D. O. Adams, this series, Vol. 58, p. 494.

An essential aspect of assaying cytolytic function is developing the habit of inspecting all cultures by inverted phase microscopy daily. This is of particular importance for long-term assays (more than 1 day). Overall health of the cultures, as well as an estimation of the degree of tumor cytolysis, can be obtained.

In general, macrophages grow best in culture densities that allow reasonable contact with one another but enough room to spread. We have found a density of 250,000–300,000 adherent macrophages/cm^2 to be optimal. Since different populations of macrophages adhere to different degrees after plating [i.e., 40–50% of resident macrophages adhere vs 85–90% of thioglycolate (TG)-elicited macrophages], this must be taken into account in setting up the cultures. A variety of methods exist to quantify cell density, including direct counting, quantification of cell protein, or quantification of DNA.[12-14]

Care in the maintenance of tumor cells to be used in assays of cytolytic function is important for two reasons. First, cultures that are confluent or overconfluent often do not incorporate radioactive labels (especially nuclear labels such as [^{3}H]thymidine) as well as subconfluent cultures. In addition, the spontaneous release of radioactive label from unhealthy cultures will often be much higher than normal, which makes interpretation of data difficult. Second, mycoplasmal contamination can affect tumoricidal assays in many different ways.[15,16] Cultures of tumor cells thus should be maintained carefully and checked for mycoplasma regularly (see this volume [32]).

D. Interpretation of Data

The assays described in this chapter can be usefully interpreted, if certain fundamentals are kept clearly in mind. First, the purity of the effector population must be clearly established and rigorously documented. For example, small numbers of specifically sensitive T cells (even as low as less than 1% of the adherent population) can contribute profoundly to lysis over 48 hr if the animals have been previously immunized against the tumor target in question.[17] Likewise, the precise conditions in the culture need to be rigorously defined. For example, the presence of any unwanted antibody needs to be rigorously excluded. Second,

[12] D. O. Adams, in "Method for Studying Mononuclear Phagocytes" (D. O. Adams, P. J. Edelson, and H. S. Koren, eds.), p. 325. Academic Press, New York, 1981.

[13] D. O. Adams, in "Methods for Studying Mononuclear Phagocytes" (D. O. Adams, P. J. Edelson, and H. S. Koren, eds.), p. 331. Academic Press, New York, 1981.

[14] P. J. Edelson, in "Methods for Studying Mononuclear Phagocytes" (D. O. Adams, P. J. Edelson, and H. S. Koren, eds.), p. 337. Academic Press, New York, 1981.

[15] J. N. Dietz and B. C. Cole, *Infect. Immun.* **37**, 811 (1982).

[16] J. Lowenstein, S. Rottem, and R. Gailly, *Cell. Immunol.* **77**, 290 (1983).

[17] D. O. Adams, unpublished (1980).

the number of effectors present must be also rigorously quantified. When comparing various populations of macrophages, the final number of adherent mononuclear phagocytes in the monolayers should be quantified, so that functional comparisons are made between equal numbers of macrophages.[12] Third, each assay requires use of appropriate technique and specificity controls. For technique controls, a known effective population of macrophages should be tested in all experiments as a positive control. For negative controls, one should conduct the assay in the presence of an irrelevant antibody or of macrophages which are not competent to perform the function in question. Fourth, consideration should be given to the events actually transpiring in the targets. The procedures in this chapter, aside from binding, deal strictly with lysis of tumor cells. In many experimental systems, stasis of tumor cells can also be observed and this possibility needs to be considered in interpretation of results.[18,19] Last, the end result being measured needs to be verified. As noted above, one should always inspect the cultures to verify microscopically and to assess roughly the extent of target lysis or binding. In setting up an assay involving radioactive label one further needs to determine that all of the radioactive label can be accounted for. That is, in a given experiment, the material present in several sets of wells should be subdivided into (1) label in adherent targets; (2) label in floating targets; and (3) label in cell-free supernatant. The total of these three should equal the total label in the wells (i.e., in a total detergent lysate). One should also be cautious of high degrees of spontaneous or background release, since it is difficult to assess the biological significance of lysis when background release is high.

II. Assays for Binding and Cytolysis of Nonadherent Targets Labeled with Chromium-51

A. Introduction

Short-term assays for lysis are well described.[20–22] Chromium-51 (^{51}Cr) is a common radiolabel for use in assays of cytotoxic function, although ^{111}In is also used. The many advantages of the former isotope

[18] A. C. Allison, *Immunol. Rev.* **40**, 3 (1978).

[19] M. C. Gynongyossy, F. A. Liabeu, and P. Goldstein, *Cell. Immunol.* **45**, 1 (1979).

[20] A. T. Taniyama and H. T. Holden, *in* "Manual of Macrophage Methodology" (H. B. Herscowitz, H. T. Holden, J. A. Bellanti, and A. Ghaffar, eds.), p. 323. Dekker, New York, 1981.

[21] R. H. Wiltrout, D. Taramelli, and H. T. Holden, *in* "Manual of Macrophage Methodology" (H. B. Herscowitz, H. T. Holden, J. A. Bellanti, and A. Ghaffar, eds.), p. 337. Dekker, New York, 1981.

[22] S. W. Russell, *in* "Methods for Studying Mononuclear Phagocytes" (D. O. Adams, P. J. Edelson, and H. S. Koren, eds.), p. 793. Academic Press, New York, 1981.

include its ease of use, low cost, and reproducibility. The main disadvantage is a spontaneous leakage of label that can approach 1.5–2% of the total label per hour. Thus, this radiolabel is useful only for assays of 18 hr or less duration. We have used the release of ^{51}Cr successfully to quantify binding and cytolysis.[23–27]

B. Preparation and Plating of Macrophages (see also this series, Vol. 108 [25] and [27])

Both binding and cytolysis assays are performed in 96-well plates (i.e., 6-mm wells) with flat bottoms (Linbro). The macrophages are routinely harvested from the peritoneal cavities of mice by lavaging with Hanks' balanced salt solution (HBSS) supplemented with 10 U/ml of heparin. The peritoneal exudate cells (PEC) are collected in polypropylene tubes, and aliquots are taken for both total and differential cell counts. The PEC are then washed twice in HBSS and resuspended to a constant number of macrophages per milliliter, based on the total and differential cell counts. Macrophages are then added to the wells and the plates are shaken. The shaking, which is done just before the trays are slid into the incubator, consists of three vigorous side-to-side and three vigorous front-to-back movements. The purpose of this procedure is to distribute the cells, be they targets or macrophages, evenly over the bottom of the wells. The exudate cells are then allowed to adhere for 1 to 2 hr, whereupon nonadherent cells are washed away with HBSS. The use of HBSS is preferred over phosphate-buffered saline, because of the glucose in HBSS. The washes are performed by pipetting 0.2 ml of HBSS into each well, vigorously swirling the fluid with the pipet, and removing it by aspiration. We generally wash macrophage monolayers a minimum of three times. The monolayers should be washed in all directions very vigorously, and the thoroughness of the resultant wash should be routinely and consistently monitored by microscopy. The washing step is critical to all of these assays in terms of achieving the desired number and purity of effectors as well as appropriately bound targets.

C. Labeling of Targets (see also this volume [27])

The P815 mastocytoma target (American Type Culture Collection, Rockville, MD) is generally used in these assays, although other non-

[23] D. O. Adams and P. A. Marino, J. Immunol. **126,** 981 (1981).

[24] W. J. Johnson, P. A. Marino, R. D. Schreiber, and D. O. Adams, J. Immunol. **13,** 1038 (1983).

[25] P. A. Marino and D. O. Adams, Cell. Immunol. **54,** 11 (1980).

[26] P. A. Marino and D. O. Adams, Cell. Immunol. **54,** 26 (1980).

[27] S. D. Somers, P. A. Mastin, and D. O. Adams, J. Immunol. **141,** 2086 (1983).

adherent cells such as lymphomas, leukemias, or carcinomas can serve equally well. The P815 cell line retains chromium well after labeling and is susceptible to killing by activated macrophages over a relatively short period of time (~16 hr).

The target cells (generally 10×10^6 total or more if needed) are taken from the cultures where the cell density is approximately 5×10^5 cells/ml. The cells are centrifuged at 200 g for 5 min, and 0.5 ml of medium and serum plus 400 μCi $^{51}CrO_4$ (specific activity of 50–400 μCi/mg) are added to the pellet, and the cells are *gently* resuspended. Incubation at 37° for 45 min to 1 hr is sufficient for significant uptake and retention of the ^{51}Cr label. In order to minimize the spontaneous leakage of label from the tumor cells, the cells are allowed to "leak" for 1 hr. The cell pellet is washed with 10–20 ml of medium plus serum, then 10 ml of medium plus serum is added and the cells resuspended and allowed to incubate another 1 hr at 37°. The cells are then centrifuged at 200 g for 5 min, resuspended in medium plus serum, counted in a hemocytometer, and adjusted to the appropriate cell density (depending on the assay).

D. Performing the Assays (see also this volume [27])

The basic assay can be used to quantify cell–cell interaction (i.e., target binding) (for basic reviews of binding, see Adams *et al.*[2,3]). To assess the binding of tumor cells to macrophages, 0.1 ml of the labeled target cell suspension containing 5×10^5 cells/ml is added to the 6-mm wells containing macrophages.[25–27] An additional 0.1 ml of medium is added to the 6-mm wells containing macrophages. An additional 0.1 ml of medium is added to each well, except for the maximal release controls, which receive 0.1 ml of 0.25% sodium dodecyl sulfate (SDS). The cultures are then incubated undisturbed for 1 hr at 37° under 5% CO_2. Unbound tumor cells are removed by 3–4 vigorous washings of each well with HBSS plus 5% FCS in a bidirectional manner. Specifically, the wells are washed, the plate rotated 90°, each well is washed, the plate is rotated 90°, etc. Washes are best done with a pipet that can eject the wash fluid with force (e.g., Becton Dickinson Biopet). If a multitip pipet is used, only the four central channels should be used to ensure a uniform wash. The controls for the performance of these assays include spontaneous adherence of targets to plastic alone. It is important to note that the total binding of tumor cells to activated macrophages comprises two distinct compartments: selective and nonselective binding.[27] the nonselective compartment can be estimated by adding the tumor cells to nonactivated macrophages; this can be further estimated for activated macrophages by previous trypsinization of the macrophages (0.1 g% trypsin in HBSS for 20 min at 37°). Again, verification of binding results by actual microscopic examination of the plates is vital. The remaining adherent macrophages and bound tumor cells are then lysed in 0.2 ml of 0.25% SDS; and an aliquot removed for assessment of bound radioactivity. Binding is calculated by the relationship,

$$\text{Number of targets bound} = \left(\frac{\text{cpm bound to macrophages}}{\text{total cpm added}}\right)(5 \times 10^4)$$

If the number of either macrophages or tumor cells per well is too high, difficulties can arise in harvesting due to the large amount of DNA released after cell lysis, producing a viscous solution from which it is difficult to remove aliquots reproducibly.

To assess the cytolysis of tumor targets by macrophages, the following modifications are made.[24,25] Tumor targets (4×10^4) are added to wells containing macrophages at a density of $1-3 \times 10^5$ adherent cells/cm^2. These cultures, as well as tumor cells cultured in medium in the absence of macrophages plus maximal release controls as described above, are incubated for 16 hr at 37° under 5% CO_2. The assay is terminated by removing an aliquot from each well (carefully remove 0.1 ml of cell-free supernatant) and transferring the aliquots to tubes for determination of radioactivity. The percentage cytolysis is calculated using the relationship,

$$\text{Percentage net cytolysis} = \left(\frac{\text{cpm from well with macrophages} - \text{cpm from wells without macrophages}}{\text{total cpm}}\right)100\%$$

The spontaneous release of ^{51}Cr from healthy P815 targets is routinely 18–25% over the course of the assay and should never be more than 2% per hour (i.e., 32% over a 16-hr assay period). The following results tabulated below are taken from an actual experiment and are typical of those seen using this assay.

Group	cpm	Percentage cytolysis
P815 alone—maximal release	2634 ± 68	—
P815 alone—spontaneous release	642 ± 58	— (24% spontaneous release)
Macrophages cultured with 10 ng/ml LPS plus P815	598 ± 47	0
Macrophages cultured with 10 U/ml γ-interferon plus P815	601 ± 50	1
Macrophages cultured with both interferon and LPS plus P815	2082 ± 64	55

III. Assay for Quantitation of Cytotoxicity by Release of [³H]Thymidine from Tumor Targets (see also this volume [29])

A. *Introduction*

Long-term assays for quantitative cytolysis by macrophages are well established.[28-31] The use of a nuclear label avoids the potential pitfall of a high spontaneous release of cytoplasmic label (such as ^{51}Cr); use of [³H]thymidine also does not affect target cell proliferation as has been reported for [^{125}I]uridine.[32-34] It has been our experience that adherent cell lines such as the MCA or 1023 fibrosarcomas release only 1–4% of the incorporated label over the 48-hr assay period, if proper attention is paid to culture conditions. There are at least two disadvantages to the use of this assay rather than the ^{51}Cr-release assay. First, nuclear labels such as [³H]thymidine are released from dead cells more slowly than cytoplasmic labels such as ^{51}Cr.[35] This assay can require 48 hr for determination of cytolysis as opposed to 16–18 hr. Second, ³H is a beta emitter, so detection of released label requires the use of scintillation fluids. In our experience, macrophages capable of killing ^{51}Cr–P815 targets are also capable of killing [³H]thymidine–MCA targets. Recent evidence, however, has implied that results obtained in these two assays may diverge in some instances.[35,36]

B. *Plating of Macrophages*

The major difference in the preparation of macrophages for this assay compared to the ^{51}Cr–P815 assay is that they are plated in 24-well rather than 96-well plates. The appropriate density for the plated macrophages remains 250,000 adherent macrophages/cm² (a range of 0.05 to 3.0 × 10⁵). To ensure a uniform distribution of cells in the wells, the plates must be shaken gently several times at right angles before they are placed in the incubator for the macrophage adherence step. This helps avoid uneven

[28] M. S. Meltzer, *in* "Manual of Macrophage Methodology" (H. B. Herscowitz, H. T. Holden, J. A. Bellanti, and A. Ghaffar, eds.), p. 329. Dekker, New York, 1981.

[29] R. B. Herberman, *in* "Methods for Studying Mononuclear Phagocytes" (D. O. Adams, P. J. Edelson, and H. S. Koren, eds.), p. 801. Academic Press, New York, 1981.

[30] M. S. Meltzer, *in* "Methods for Studying Mononuclear Phagocytes" (D. O. Adams, P. J. Edelson, and H. S. Koren, eds.), p. 785. Academic Press, New York, 1981.

[31] D. O. Adams, *J. Immunol.* **124**, 286 (1980).

[32] D. O. Adams, K. J. Kao, R. Farb, and S. V. Pizzo, *J. Immunol.* **124**, 293 (1980).

[33] B. P. Le Mevel, R. K. Oldham, S. A. Wells, and R. B. Herberman, *J. Natl. Cancer Inst.* (*U.S.*) **51**, 1511 (1963).

[34] R. C. Seeger, S. A. Rayner, and J. J. T. Owen, *Int. J. Cancer* **13**, 697 (1974).

[35] M. S. Meltzer, personal communication (1985).

[36] J. Fidler, personal communication (1985).

distribution of cells around the periphery of the wells, due to the vortex effects created by walking.

C. Labeling of Targets

The preparation of adherent tumor targets for this assay is accomplished in two different ways. In one, the targets are prelabeled in 75-cm^2 tissue culture flasks and added to the macrophage monolayers in the 24-well plates. Alternatively, the tumor cells are seeded in the empty wells of 24-well plates and labeled overnight; macrophages are then added the next day. Both procedures yield similar results in terms of the amount of cytolysis observed. It should be emphasized that radioactive labels, particularly over long periods of time, can be toxic to the target cells. The total amount of label incorporated in the entire population of targets in the well should thus be ~5000 cpm.

In order to obtain prelabeled tumor targets, the tumor targets are grown to subconfluence in 75-cm^2 plastic tissue culture flasks. [^{3}H]Thymidine at 0.05 μCi/ml (specific activity of 2 Ci/mmol) of culture fluid is then added and the cells are allowed to label overnight. The MCA and 1023 murine fibrosarcomas both incorporate [^{3}H]thymidine during this time period. The use of slower growing cell lines may require longer labeling periods. After the labeling period, the medium is removed and the cell monolayer is rinsed five times with HBSS. Five milliliters of 0.25% trypsin–EDTA is then added. When the cells have detached, they are monodispersed with a pipet and transferred to a 50-ml polypropylene tube in medium plus 10% FCS. The trypsin is removed by washing twice in medium plus 10% FCS. A cell count is performed, and the tumor cells resuspended to a concentration of 40,000 cells/ml. The cells are now ready to add to the macrophage monolayers.

For labeling the cells in the wells of 24-well plates, subconfluent target cells are trypsinized as above, and resuspended to 60,000 cells/ml. [^{3}H]Thymidine is added at a final concentration of 0.05 μCi/ml. The cells are mixed, and 0.5 ml of target–cell suspension is added to each well of a 24-well plate. The paltes are shaken and placed in a 37°, 5% CO_2 incubator for 18–24 hr. After incubation, the wells are washed three times with HBSS.

D. Performing the Assay

Macrophages are added to the wells at a concentration of ~1 × 10^6/ml (or less depending on the number desired) in EMEM supplemented with 10% FCS, 25 mM HEPES buffer, ascorbic acid (5 mg/100 ml), penicillin, streptomycin, glutamine, and nonessential amino acids. This medium is better suited to the long-term (48-hr) assay. Nonadherent cells are washed off in the usual way after a 2-hr adherence. The plates are then incubated for an additional 48 hr.

At the end of this incubation period, 0.5 ml of supernatant from each well is carefully removed and transferred to scintillation vials containing 0.2 ml of 2.5% SDS. The control wells (i.e., tumor cells alone) are treated with 0.5 ml of 2.5% SDS, and 0.75 ml of this supernatant is transferred to scintillation vials. Ten milliliters of a suitable scintillation cocktail (such as Aquasol II, New England Nuclear) is added, and the radioactivity is determined in a scintillation spectrometer. The resultant cytotoxicity is calculated using the same mathematical relationship as used for the ^{51}Cr-release assay.

IV. Short-Term Assay for the Determination of the Binding and Cytolysis of Antibody-Coated Lymphoma Targets Labeled with [51]Cr (ADCC)

A. *Introduction*

Recently, the antibody-dependent cellular cytotoxicity (ADCC) reaction has been divided into two fundamental forms: a rapid and a slow variant.[37,38] Assays are presented for both here. The assays to be described for quantification of ADCC are very similar to those already described for antibody-independent binding and cytolysis of tumor cells. The major differences lie in the target cell of choice and the necessity of a suitable antibody preparation.

Assays for rapid ADCC are well described.[39–41] We have found the HSB target line, a human T cell lymphoma, to be sensitive to macrophage-mediated ADCC and suitable for this assay.[42] We have prepared antisera to this tumor cell line by repeated intraperitoneal injections (1 × 10^6 cells/0.2 ml HBSS per mouse per week) for 6–10 weeks. Two weeks prior to tumor cell injection, the mice are immunized with an intraperito-

[37] D. O. Adams, M. S. Cohen, and H. S. Koren, *in* "Macrophage Mediated Antibody Dependent Cellular Cytotoxicity" (H. S. Koren, ed.), p. 43. Dekker, New York, 1983.

[38] S. D. Somers, W. J. Johnson, and D. O. Adams, *in* "Cancer Immunology: Innovative Approaches to Therapy" (R. Herberman, ed.), p. 69. Martinus Nijhoff, Boston, 1986.

[39] H. S. Koren and D. G. Fischer, *in* "Methods for Studying Mononuclear Phagocytes" (D. O. Adams, P. J. Edelson, and H. S. Koren, eds.), p. 813. Academic Press, New York, 1981.

[40] H. S. Koren, *in* "Methods for Studying Mononuclear Phagocytes" (D. O. Adams, P. J. Edelson, and H. S. Koren, eds.), p. 315. Academic Press, New York, 1981.

[41] C. F. Nathan, L. H. Brukner, G. Kaplan, J. Unkeless, and Z. A. Cohn, *J. Exp. Med.* **152**, 183 (1980).

[42] W. J. Johnson, D. P. Bolognesi, and D. O. Adams, *Cell. Immunol.* **83**, 170 (1984).

neal injection of 1×10^7 viable BCG organisms to facilitate the antibody response. One week after the last tumor cell injection, blood is collected. The serum is separated and stored at $-70°$ until use. The anti-HSB serum alone is not capable of causing tumor cell lysis in the absence of complement. Each batch of antiserum should be checked for the optimal concentrations to be used in each assay.

B. Preparation and Plating of Macrophages

The macrophages are prepared as described in Section II,B. As a routine, several effector : target ratios are examined in each assay so that different densities of macrophages are prepared. Macrophages elicited by casein (100 μg/mouse) or activated by *Propionibacterium acnes* (Burroughs Wellcome Company, 700 μg/mouse) or BCG are most effective in this assay, while those elicited by thioglycolate broth are much less effective.[37,42]

C. Labeling of Targets

The HSB targets are labeled in the same manner described in Section II,C. The HSB lymphoma is maintained in culture in RPMI 1640 medium, supplemented with 10% FCS, penicillin, streptomycin, and glutamine. These cells label best when taken from cultures of approximately 1 to 1.5 $\times 10^6$ cells/ml.

D. Performing the Assay

The medium used for the ADCC assay is RPMI 1640 plus 10% FCS. To assess the antibody-dependent binding of tumor cells to macrophages, 0.1 ml of the labeled target cell suspension (containing 40,000 ^{51}Cr-labeled HSB lymphoma cells) is added to the microtiter wells containing adherent macrophages.[42] An additional 0.1 ml antisera diluted to an optimal concentration in RPMI 1640 plus 10% FCS is then added. Controls include medium alone or an irrelevant antibody diluted to a similar concentration, as well as the addition of 0.1 ml of 0.25% SDS for the tumor cell maximal release control. The plates are then centrifuged briefly (50 g for 30 sec) to optimize cell contact in this short-term assay (centrifugation does not alter the extent of either binding or cytolysis in the assay). The plates are placed at 37° under 5% CO_2 for 30 min. Each well is then washed vigorously four times (twice in each direction) as described in Section II,D. The remaining macrophages and bound tumor cells are then lysed with 0.2 ml of 0.25% SDS, and an aliquot is removed for assessment of bound radioactivity. The number of antibody-coated targets bound is calculated as described in Section II,D.

To assess the cytolysis of antibody-coated tumor targets, the plates are placed at 37° under 5% CO_2 for 5 hr after brief centrifugation.[42] The assay is terminated by removing an aliquot from each well and determining the released radioactivity in a gamma counter. Again, the degree of cytolysis is determined by the mathematical relationship described in Section II,D.

V. Long-Term Assay for the Determination of Cytolysis of Antibody-Coated Tumor Cells Labeled with [³H]Thymidine

Assays for slow ADCC have also been described.[43] The assay for slow ADCC is performed in the same way as above with two exceptions.[44] First, adherent colorectal carcinoma cells labeled with [³H]thymidine are used in place of ⁵¹Cr-labeled HSB lymphoma targets. Second, the assay requires 48 hr for completion because of the use of this combination of label and target cells. Unlike the direct lysis of [³H]thymidine-labeled fibrosarcoma targets described in Section III, this assay works well in 96-well microtiter plates.

The antibodies used are frequently reactive with human colon carcinoma target cells, but other targets are also suitable.[45] Antibodies of several isotypes (e.g., IgG_1, IgG_{2a}, IgG_{2b}, and IgG_3) are capable of directing this type of cytolysis.[38,44] We have most often used a monoclonal antibody against the 1116 human colorectal cancer cells of the IgG_{2a} isotype.[44]

A. Preparation and Plating of Macrophages

Murine peritoneal macrophages are harvested from animals as described in previous sections. Our experience has been that thioglycolate broth (Difco) (1 ml of 4% solution/mouse)- and pyran copolymer (100 μg/mouse)-elicited macrophages are most efficient at mediating lysis in this system. By contrast, BCG or *Propionibacterium acnes*-activated mouse macrophages are not nearly as effective in mediating such lysis.[44]

B. Labeling of Targets

The SW-1116 human colorectal carcinoma (available from American Type Culture Collection, Rockville, MD) is frequently used in these assays, though the assay is suitable for testing many other lines.[44] The cell

[43] P. Ralph, *in* "Macrophage-Mediated Antibody-Dependent Cellular Cytotoxicity" (H. S. Koren, ed.), p. 71. Dekker, New York, 1983.

[44] W. J. Johnson, Z. Steplewski, T. J. Matthews, T. A. Hamilton, H. Koprowski, and D. O. Adams, *J. Immunol.* **136**, 4704 (1986).

[45] M. Z. Herlyn, Z. Steplewski, D. Herlyn, and H. Koprowski, *Proc. Natl. Acad. Sci. U.S.A.* **76**, 1438 (1979).

line 1116 is maintained by *in vitro* passage in EMEM–10% FCS. The 1116 targets are prelabeled in 75-cm^2 flasks as described in Section III,C. Two major differences from that procedure must be noted. First, the amount of label added per 75-cm^2 flask of cells is higher than that described for the MCA and 1023 fibrosarcomas. A total of 6 μCi [^{3}H]thymidine is added per flask. Second, labeling is allowed to occur for at least 36 hr as the SW-1116 line grows slowly as compared to 1023 or MCA targets. The length of the labeling period as well as the amount of [^{3}H]thymidine used do not result in an increased spontaneous release of label from the cells (i.e., <5%).

C. Performing the Assay

The assay is performed in the same way as the short-term ADCC assay described in Section IV,D, with the following differences: The microtiter plates are not centrifuged after the addition of targets and antibody, as the length of time required for the assay is more than sufficient to result in optimal cell–cell contact. The plates are incubated at 37° under 5% CO_2 for 48 hr. Aliquots are then removed for scintillation counting as described previously. Calculations for the resultant degree of tumor cytolysis are as described previously.

VI. Concluding Remarks

The assays described above are common assays, or variations on common assays, used to quantify interactions between tumor cells and macrophages. With repeated experience as well as the proper attention to variables known to affect the assays these can become highly reliable and reproducible assays. Their many applications include the assessment of spontaneous tumoricidal capacities of different macrophage populations, the determination of drug effects on macrophage tumoricidal capacities, investigating the roles of regulatory molecules on tumoricidal function, as well as the potential of different antibody preparations or isotypes for immunotherapeutic purposes. With more research into the mechanisms of tumor lysis as well as the mechanisms by which macrophages can be activated to recognize and kill tumor cells, these assays will assume even more importance.

[40] Preparation and Characterization of Chloramines

By EDWIN L. THOMAS, MATTHEW B. GRISHAM, and

M. MARGARET JEFFERSON

Interest in chloramines has increased as a result of recent studies indicating that these oxidizing agents are produced by neutrophils, monocytes, and perhaps also eosinophils and other leukocytes.[1-7] Chloramines represent an important class of leukocyte oxidants, and contribute to oxidative microbicidal, cytotoxic, and cytolytic activities,[6-11] chemical modification of regulatory substances,[5,12] and uptake and metabolism of nitrogen compounds.[8,13]

Chloramines are products of the reaction of hypochlorous acid (HOCl) or other chlorinating agents with primary and secondary amines. Mono-, di-, and trichloramine are also accepted names for derivatives of ammonia (NH_4^+). "Chloramine" has also been used to denote derivatives of amides, guanidino compounds, and sulfonamides. More accurate nomenclature specifies the position and number of Cl atoms and the structure of the nitrogen moiety, e.g., mono- and di-N-chloramines, RNHCl and RNCl$_2$; mono-N-chloramide, RCONHCl. All the derivatives contain covalent nitrogen–chlorine (N—Cl) bonds and are described collectively as N—Cl derivatives.[9]

Production of HOCl by leukocytes is the result of peroxidase-catalyzed oxidation of chloride (Cl$^-$) by hydrogen peroxide (H_2O_2). HOCl (pK_a 7.5) is in equilibrium with hypochlorite (OCl$^-$) and chlorine (Cl$_2$).

$$HOCl \rightleftharpoons OCl^- + H^+$$
$$HOCl + Cl^- + H^+ \rightleftharpoons Cl_2 + H_2O$$

[1] E. L. Thomas, M. M. Jefferson, and M. B. Grisham, *Biochemistry* **24**, 6299 (1982).

[2] S. J. Weiss, R. Klein, A. Slivka, and M. Wei, *J. Clin. Invest.* **70**, 598 (1982).

[3] E. L. Thomas, M. B. Grisham, and M. M. Jefferson, *J. Clin. Invest.* **72**, 441 (1983).

[4] M. Lampert and S. J. Weiss, *Blood* **62**, 645 (1983).

[5] S. J. Weiss, M. B. Lampert, and S. T. Test, *Science* **222**, 625 (1983).

[6] M. B. Grisham, M. M. Jefferson, and E. L. Thomas, *J. Biol. Chem.* **259**, 6757 (1984).

[7] M. B. Grisham, M. M. Jefferson, D. F. Melton, and E. L. Thomas, *J. Biol. Chem.* **259**, 10404 (1984).

[8] E. L. Thomas, M. B. Grisham, D. F. Melton, and M. M. Jefferson, *J. Biol. Chem.* **260**, 3321 (1985).

[9] E. L. Thomas, *Infect. Immun.* **23**, 522 (1979).

[10] E. L. Thomas, *Infect. Immun.* **25**, 110 (1979).

[11] E. L. Thomas, M. B. Grisham, and M. M. Jefferson, this volume [41].

[12] M. K. Rao and A. L. Sagone, Jr., *Infect. Immun.* **43**, 846 (1974).

[13] E. L. Thomas, M. J. Jefferson, and M. B. Grisham, unpublished results (1976–1984).

Under physiologic conditions of pH and Cl^- concentration, the predominant form is a mixture of HOCl and OCl^- ($HOCl/OCl^-$). HOCl and N—Cl derivatives may be considered to contain the oxidized Cl atom, represented as Cl^+. NH_2Cl, RNHCl, HOCl, OCl^-, and Cl_2 each contain one Cl^+ or two oxidizing equivalents and will oxidize 2 mol of a sulfhydryl compound (R'SH) to the disulfide (R'SSR'). $RNCl_2$ contains two Cl^+ so that each $RNCl_2$ oxidizes four R'SH.

$$RNHCl + 2R'SH \rightarrow RNH_3^+ + Cl^- + R'SSR'$$
$$RNCl_2 + 4R'SH \rightarrow RNH_3^+ + 2Cl^- + H^+ + 2R'SSR'$$

HOCl and N—Cl derivatives also react with many aromatic and unsaturated compounds to yield carbon–chlorine (C—Cl) derivatives. Unlike N—Cl derivatives, the C—Cl derivatives are not oxidizing agents, though some have mutagenic or other toxic activities.

The most abundant N-moieties available for reaction with HOCl in biological systems are primary amino groups (e.g., taurine, α-amino acids, polyamines, amino sugars, phosphatidylethanolamine, lysine residues, and protein amino termini). The combined concentration of primary amino groups in serum or the cytosol of neutrophils is 0.05 or 0.15 M, respectively.[13] NH_4^+ rarely achieves millimolar levels in plasma, though locally high concentrations may occur. For example, NH_4^+ and other weak bases may be concentrated in the acidified phagolysosome compartment of leukocytes. Other reactive N-moieties such as the secondary amino groups of spermine and spermidine and the guanidino group of arginine are present at lower concentrations. Sulfonamides do not occur naturally, but are of interest because they are used as antimicrobial agents and have effects on neutrophil functions.[3,14]

Amides (peptide bonds, asparagine and glutamine, acetylated amino sugars) are prevalent in biological materials, but chloramides hydrolyze readily to yield HOCl. Chloramides are obtained when water is excluded such as in organic solvents, but in aqueous media it is difficult to obtain evidence for reaction of HOCl with amides.[15] Indirect evidence for reaction of HOCl with peptide bonds is provided by HOCl-dependent conversion of proteins to low-molecular-weight fragments.[9,10,16] Fragmentation may be due to oxidation and rearrangement at aromatic amino acid residues,[17] as observed during oxidation by iodine (I_2). However, peptide

[14] R. I. Lehrer, *J. Clin. Invest.* **50,** 2498 (1971).

[15] W. E. Pereira, Y. Hoyano, R. E. Summons, V. A. Bacon, and A. M. Duffield, *Biochim. Biophys. Acta* **313,** 170 (1973).

[16] R. W. R. Baker, *Biochem. J.* **41,** 337 (1947).

[17] M. Morrison and G. R. Schonbaum, *Annu. Rev. Biochem.* **45,** 861 (1976).

bonds are also lost during the reaction of HOCl with polylysine[9] and polyserine,[13] suggesting that unstable chloramides are formed.

Leukocyte peroxidases catalyze oxidation of bromide (Br^-), iodide (I^-), and thiocyanate (SCN^-) as well as Cl^-. The concentration of Cl^- is usually higher than that of other ions, but I^- and SCN^- compete effectively with Cl^- for oxidation, and these ions are concentrated at a few sites *in vivo*. Moreover, HOCl and N—Cl derivatives oxidize I^- to I_2. Therefore, production of oxidants containing I^+ or SCN^+ moieties must be considered when characterizing oxidants produced by leukocytes. The predominant forms at physiologic pH are I_2 and $HOSCN/OSCN^-$.[17,18]

Preparation of Chloramines

Reaction of HOCl with N-Compounds. Small volumes of 10–100 mM OCl^- in dilute base are added (e.g., 4 additions 1 min apart) to a 0.1–10 mM solution of the N-compound at 4°. Commercially available $NaOCl$ solutions diluted with 0.1 M NaOH are stable for at least a week at 4°. Sulfate or Cl^- salts of N-compounds may be used, with phosphate buffer to control pH. Br^- salts, other salts with readily oxidized anions, and buffers with reactive N-moieties such as Tris and "Good" buffers[19] are not suitable.

One mole of NH_2Cl or most $RNHCl$ derivatives per mole of OCl^- is obtained when the pH is 5 or higher and the molar ratio of OCl^- to N-moieties ($OCl^-:N$) is 1:10. $RNHCl$ free of $RNCl_2$ or excess RNH_3^+ can often be obtained at a $OCl^-:N$ ratio of 1:1 at high pH (0.1 M NaOH). After adding OCl^-, the solution is adjusted to the desired pH. Preparation at high pH is required to obtain $RNHCl$ free of $RNCl_2$ with putrescine and other polyamines.

Preparation at pH 5 with high Cl^- (0.1 M or higher) favors $RNCl_2$ formation. Quantitative formation of $RNCl_2$ free of $RNHCl$ or RNH_3^+ is obtained at a $OCl^-:N$ ratio of 2:1. When the $OCl^-:N$ ratio is less than 2:1, but greater than 1:10, a mixture of RNH_3^+, $RNHCl$, and $RNCl_2$ is obtained. The relative amounts of $RNHCl$ and $RNCl_2$ depend on pH, Cl^- concentration,[20] and structure of the R moiety. The order of addition is also critical; adding amines into $HOCl/OCl^-$ solutions favors $RNCl_2$ formation.[1]

[18] E. L. Thomas, *Biochemistry* **20**, 3273 (1981).
[19] N. E. Good, G. D. Winget, W. Winter, T. N. Connolly, S. Izawa, and R. M. M. Singh, *Biochemistry* **5**, 467 (1966).
[20] J. C. Morris, *in* "Principles and Applications of Water Chemistry" (S. D. Faust and J. V. Hunter, eds.), p. 23. Wiley, New York, 1967.

NH_2Cl is stable, but $NHCl_2$ and NCl_3 decompose instantly.[21] CH_3NHCl is less stable than either NH_2Cl or most RCH_2NHCl derivatives. RCH_2NHCl and RCH_2NCl_2 derivatives such as TauNHCl and TauNCl_2 and $(RCH_2)_2NCl$ derivatives such as diethylamine-chloramine undergo less than 5% decomposition per hour at pH 7 at 37°. With some amines such as Tris, the RNHCl derivative is stable but $RNCl_2$ decomposes rapidly. Derivatives with oxidizable groups on the α-carbon undergo reactions in which Cl^+ is reduced and another portion of the molecule is oxidized. Such reactions result in loss of oxidizing equivalents and oxidative toxicity. Examples are the decomposition of RNHCl and $RNCl_2$ derivatives of α-amino acids to yield aldehydes and nitriles.[15,22–24] Chloroguanidino derivatives decompose over a period of hours at 37°, and RSO_2NHCl derivatives are relatively stable.

N—Cl derivatives are used within minutes of preparation to avoid decomposition or rearrangement. However, most are stable at 4° for hours or days at the appropriate pH (e.g., 8–10 for NH_2Cl or RNHCl and pH 5 for $RNCl_2$). In the absence of amines, $RNCl_2$ decomposes at high pH to yield a mixture of RNHCl and products without oxidizing activity.[1] When amines are present, $RNCl_2$ chlorinates the amines.[1,25] In the absence of amines, RNHCl dismutates at low pH to yield $RNCl_2$. Therefore, the RNHCl:$RNCl_2$ ratio may slowly increase at high pH and decrease at low pH.

Chlorination of amines by $RNCl_2$ and dismutation of RNHCl represent a form of exchange of Cl^+. Similarly, RNHCl undergoes exchange with RNH_3^+.

$$RNCl_2 + R'NH_3^+ \rightleftharpoons RNHCl + R'NHCl + H^+$$
$$RNHCl + R'NH_3^+ \rightleftharpoons RNH_3^+ + R'NHCl$$

These exchange reactions are much slower than the reaction of HOCl with RNH_2 or RNHCl, and may require minutes or hours to come to equilibrium. Nevertheless, exchange influences results of experiments *in vitro*[6–8] and may be important *in vivo*. Moreover, the composition of solutions of N—Cl derivatives can change, particularly when derivatives are subjected to extremes of pH or high concentration of N-compounds, and such changes can be significant when attempting to characterize the derivatives.

[21] E. Colton and M. M. Jones, *J. Chem. Educ.* **32**, 485 (1955).
[22] B. B. Paul, A. A. Jacobs, R. R. Strauss, and A. J. Sbarra, *Infect. Immun.* **2**, 414 (1970).
[23] J. M. Zgliczynski, T. Stelmaszynska, J. Domanski, and W. Ostrowski, *Biochim. Biophys. Acta* **235**, 419 (1971).
[24] T. Stelmaszynska and J. M. Zgliczynski, *Eur. J. Biochem.* **92**, 301 (1978).
[25] E. T. Gray, Jr., D. W. Margerum, and R. P. Huffman, *ACS Symp. Ser.* **82**, 264 (1978).

R-Labeled Derivatives. Derivatives radiolabeled in the R moiety are prepared by adding OCl^- to solutions of labeled N-compounds. Specific activity of the N—Cl derivative is that of the N-compound. For example, when RNHCl is prepared by adding 0.1 mM OCl^- to 1 mM RNH_3^+, 10% of the label is in the 0.1 mM RNHCl and 90% in the 0.9 mM RNH_3^+. Preparation and use of such derivatives presents no special problems, though the presence of excess labeled N-compound, reduction of the derivative to yield the labeled N-compound, exchange of Cl^+ with an unlabeled N-compound, or decomposition of the derivative to yield labeled products may complicate interpretation of results.

Cl-Labeled Derivatives. N—^{36}Cl derivatives are prepared by adding $O^{36}Cl^-$ to solutions of N-compounds. Label is introduced into OCl^- by allowing OCl^- to equilibrate with $^{36}Cl^-$. The major problem encountered in preparation and use of $N^{36}Cl$ derivatives is the low and uncertain specific activity. ^{36}Cl has a long half-life and thus a low specific activity, and commercially available NaOCl solutions contain high Cl^- (up to 1 M), which dilutes the specific activity of $^{36}Cl^-$. Moreover, when $O^{36}Cl^-$ is added to the N-compounds in solutions containing unlabeled Cl^-, the specific activity may be diluted, depending on the relative rates of reaction of $HO^{36}Cl$ with N-compounds and Cl^-. When $N^{36}Cl$ derivatives are added to solutions containing Cl^-, the specific activity may be further diluted, depending on rates of hydrolysis of the derivatives.

Results shown below illustrate one of many possible ways to prepare $N^{36}Cl$ derivatives of known specific activity. An OCl^- solution with low Cl^- is prepared, the OCl^- is allowed to equilibrate with $^{36}Cl^-$, and specific activity of $O^{36}Cl^-$ is determined after separating $HO^{36}Cl$ from $^{36}Cl^-$. The $N^{36}Cl$ derivatives are prepared and used in media without additional Cl^-. It is assumed that specific activity of $^{36}Cl^+$ in the $N^{36}Cl$ derivative is the same as that in $O^{36}Cl^-$.

As shown below, use of a Cl^--free medium is not essential when $NH_2^{36}Cl$ or $RNH^{36}Cl$ is prepared by adding $O^{36}Cl^-$ to a 10-fold excess of N-compound at pH 7. The reaction of $HO^{36}Cl$ with the N-compound is so fast that there is no equilibration of $HO^{36}Cl$ with unlabeled Cl^- in the medium. However, the label is partially diluted by Cl^- when $RN^{36}Cl_2$ is prepared by adding 2 mol $O^{36}Cl^-$ to 1 mol RNH_3^+.

After the derivatives are prepared, use of a Cl^--free medium is not essential for derivatives with strong N—Cl bonds such as $NH_2^{36}Cl$, $RNH^{36}Cl$, or $RN^{36}Cl_2$. The $^{36}Cl^+$ moiety in these derivatives does not equilibrate with unlabeled Cl^- at pH 7. Either these derivatives do not hydrolyze to yield $HO^{36}Cl$, or hydrolysis is so slow as to be undetectable.

Ethyl acetate (spectrophotometric grade; Fisher Chemical Company, Fair Lawn, NJ) is washed for 30 min with 1 vol of 0.1 mg/ml sodium borohydride in a Cl⁻-free medium consisting of 67 mM Na$_2$SO$_4$–32 mM phosphate buffer, pH 7. The solvent is washed once with 0.1 M NaOH and repeatedly with the Cl⁻-free medium, and allowed to stand undisturbed for 1 day. Equal volumes of washed ethyl acetate and 4–6% NaOCl (Fisher Chemical Company, Fair Lawn, NJ) are mixed and the aqueous phase made 32 mM in H$_3$PO$_4$ to extract HOCl into the organic (upper) phase. A portion of the organic phase is mixed with an equal volume of 0.1 M KOH to extract OCl⁻ into the aqueous phase.

A portion (1 ml) of the 30–50 mM OCl⁻ solution is supplemented with 50 μl (2.5 μCi) of 0.1–0.3 M H^{36}Cl (Amersham Corp., Arlington Heights, IL). After 1 hr at 4°, a portion of the O^{36}Cl⁻ solution is added to Cl⁻-free medium to obtain a mixture of HO^{36}Cl/O^{36}Cl⁻ and ^{36}Cl⁻. Equal portions are also added to Cl⁻-free medium containing small amounts of taurine, to obtain mixtures of HO^{36}Cl/O^{36}Cl⁻, ^{36}Cl⁻, and TauN^{36}Cl$_2$ (Fig. 1). After 15 min at 25°, the mixtures are extracted with equal volumes of washed ethyl acetate, portions of the organic phase are made 2 mM in dithiothreitol, and the extracted label is measured with a liquid scintillation spectrometer. Under these conditions, 80% of the oxidizing equivalents of HOCl/OCl⁻ is extracted, whereas Cl⁻ and TauNCl$_2$[25a] are not. Concentration of ^{36}Cl⁺ is calculated as two times the concentration of taurine that is required to block extraction. Specific activity of ^{36}Cl⁺ (or the O^{36}Cl⁻ solution) is calculated as $Y/0.8B$, where Y is label extracted in the absence of taurine, and B is ^{36}Cl⁺ concentration. Concentration of ^{36}Cl⁻ can also be calculated as $B(0.8A - Y)/Y$, where A is total label. For the preparation in Fig. 1, A is 166 nCi/ml or ca. 4 × 10^5 dpm/ml. Calculated values are 0.99 mM ^{36}Cl⁺ and 0.59 mM ^{36}Cl⁻, and specific activity of ^{36}Cl⁺ is 0.105 Ci/mol.

Figure 2A shows results of adding O^{36}Cl⁻ to solutions containing unlabeled Cl⁻ and then extracting. The amount of extracted label decreases with increasing Cl⁻ as the result of dilution of the label in HO^{36}Cl by Cl⁻. The curve drawn through the points is calculated from the equation, $Y = ABD/(B + C + X)$, where Y is extracted label, A is total label, B is Cl⁺ concentration, D is the fraction extracted ($D = 0.8$ for HOCl/OCl⁻), C is Cl⁻ concentration in the absence of added Cl⁻, and X is concentration of added unlabeled Cl⁻. Figure 2A also shows results of adding O^{36}Cl⁻ to solutions with unlabeled Cl⁻, adding excess NH$_4$⁺, and extracting label as NH$_2$^{36}Cl. The curve is calculated from the same equation with $D = 0.6$. The results indicate complete equilibration of label in HO^{36}Cl/O^{36}Cl⁻ with unlabeled Cl⁻, and also indicate that at these low Cl⁻ concentrations (<10

[25a] Abbreviations: TauNHCl and TauNCl$_2$, taurine monochloramine and dichloramine; Nbs, 5-thio-2-nitrobenzoic acid; Nbs$_2$, 5,5′-dithiobis(2-nitrobenzoic acid).

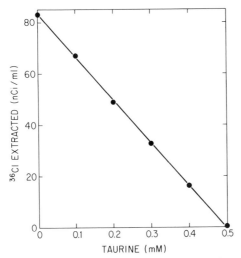

FIG. 1. Determination of $^{36}Cl^+$ concentration and specific activity. Portions (0.04 ml) of an $O^{36}Cl^-$ solution (ca. 25 mM) were added to 0.96-ml portions of Cl^--free medium containing varying concentrations of taurine. Taurine concentrations shown are final concentrations. After 15 min at 25°, 1 ml ethyl acetate was added and the amount of label extracted into the organic phase was measured.

mM), label is extracted as $HO^{36}Cl$ rather than $^{36}Cl_2$. In this experiment, the sum of the $HO^{36}Cl/O^{36}Cl^-$ and $^{36}Cl^-$ is less than 2 mM, so that 2 mM unlabeled Cl^- lowers the extraction of label by more than 50%.

Figure 2B shows results of adding $O^{36}Cl^-$ to N-compounds in solutions containing varying amounts of Cl^-. Unlabeled Cl^- has no effect on extraction of label in the form of $NH_2{}^{36}Cl$ or the $RNH^{36}Cl$ derivative of ethanolamine ($D = 0.33$). Therefore, there is no equilibration of label in $HO^{36}Cl/O^{36}Cl^-$ with Cl^- during the reaction of 1 mM $HO^{36}Cl$ with 10 mM N-compound in the presence of up to 0.15 M Cl^-. However, when 1 mM $O^{36}Cl^-$ is added to 0.5 mM ethanolamine to obtain 0.5 mM $RNH^{36}Cl_2$ ($D = 0.89$), the specific activity of $^{36}Cl^+$ is partially diluted by Cl^-. Figure 2C shows results of adding $O^{36}Cl^-$ to N-compounds without Cl^-, then adding unlabeled Cl^- and extracting. Cl^- has no effect on extraction of label in the form of $NH_2{}^{36}Cl^-$, $RNH^{36}Cl$, or $RN^{36}Cl_2$, indicating that the $^{36}Cl^+$ moieties do not equilibrate with Cl^-.

Enzymatic Preparation. Use of the myeloperoxidase–H_2O_2–Cl^- system as an enzymatic HOCl-producing system for preparation of N—Cl derivatives has no advantage over direct use of OCl^-, but is of interest for what it may reveal about myeloperoxidase activity *in vivo*. Adding H_2O_2 in small increments to a mixture of 0.1–0.2 M Cl^-, the N-compound at a 10-fold excess over H_2O_2, and catalytic amounts of purified myeloperox-

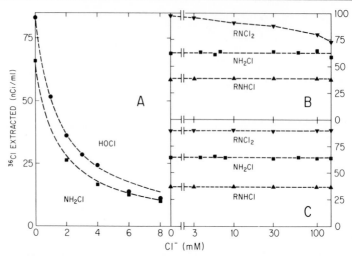

FIG. 2. Effect of unlabeled Cl^- on specific activity of $^{36}Cl^+$. (A) Portions (0.04 ml) of 25 mM $O^{36}Cl^-$ were added at 25° to 0.67 M Na_2SO_4–32 mM phosphate buffer, pH 7, containing varying concentrations of Cl^-. After 15 min, the mixtures received no further additions (●) or were made 10 mM in NH_4^+ (■). Final volumes were 1 ml, and concentrations shown are final concentrations. Mixtures were extracted with 1 ml ethyl acetate, and label in the organic phase was measured. (B) As in (A), except that the OCl^- was added to solutions containing varying concentrations of Cl^- and 0.5 mM ethanolamine (▼), 10 mM NH_4^+ (■), or 10 mM ethanolamine (▲). (C) As in (B), except that the OCl^- was added to ethanolamine or NH_4^+ in Cl^--free medium, varying concentrations of Cl^- were added, and the mixtures were extracted after 1 hr.

idase (0.1 μM) results in formation of 1 mol NH_2Cl or $RNHCl/mol$ of H_2O_2. Therefore, one $HOCl$ is produced per H_2O_2. $RNCl_2$ is formed when RNH_3^+ concentration is low, particularly at low pH, but quantitative formation of $RNCl_2$ is not obtained at a 2:1 ratio of H_2O_2 to RNH_3^+. When RNH_3^+ concentration is low, $HOCl$ reacts with and inactivates myeloperoxidase[26] and also reacts with H_2O_2,[27] resulting in loss of oxidizing equivalents of both $HOCl$ and H_2O_2.

$$HOCl + H_2O_2 \rightarrow H_2O + Cl^- + H^+ + O_2$$

As a result of these reactions, $HOCl/OCl^-$ does not accumulate during myeloperoxidase-catalyzed oxidation of Cl^-.[9,10,28]

Myeloperoxidase can be used to catalyze oxidation of $^{36}Cl^-$ to $HO^{36}Cl$ for preparation of N—^{36}Cl derivatives, but oxidation of one Cl^- per H_2O_2

[26] J. M. Zgliczynski, in "The Reticuloendothelial System" (A. J. Sbarra and R. R. Strauss, eds.), Vol. 2, p. 255. Plenum, New York, 1980.
[27] A. M. Held, D. J. Halko, and J. K. Hurst, J. Am. Chem. Soc. **100,** 5732 (1978).
[28] J. E. Harrison and J. Schultz, J. Biol. Chem. **251,** 1371 (1976).

TABLE I
MOLAR EXTINCTION COEFFICIENTS (E_M)

Functional group	N—Cl derivative	Wavelength (nm)						
		205	207	242	252	260	291	300
NH$_3$	NH$_2$Cl	25	50	429[a]	383	248	30	24
Primary amino	RNHCl	136	123	379	429[a]	379	63	36
Primary amino	RNCl$_2$	2720[a]	2675	540	190	171	327	370[a]
Secondary amino	RRNCl	362	294	178	259	296[a]	134	81
Guanidino	RNHC(NH)NHCl	1700	3000[a]	400	300	200	150	150
HOCl/OCl$^-$	pH 7	234	180	73	64	68	142[a]	134
	pH 5	195	146	97	67	52	35	30

[a] λ_{max} values. Parent N-compounds were NH$_4^+$, taurine (primary amine), diethylamine (secondary amine), and guanidinoacetic acid.

is obtained only at high Cl$^-$ concentrations. In experiments with purified myeloperoxidase or intact leukocytes, the presence of 0.1–0.2 M ^{36}Cl$^-$ makes it difficult to detect 1–100 μM N—^{36}Cl or C—^{36}Cl derivatives. Extensive washing with Cl$^-$-containing solutions may be required to displace noncovalently bound ^{36}Cl$^-$ from proteins or cells. Other methods for separating ^{36}Cl$^-$ from labeled derivatives are gel chromatography,[29] precipitation of ^{36}Cl$^-$ with mercuric salts,[30] and precipitation of ^{36}Cl-labeled macromolecules with organic solvents[30] or 5% (w/v) trichloroacetic acid.[11]

Identification and Quantitation

Absorption Spectra. N—Cl derivatives have characteristic though weak absorbance in the ultraviolet. NH$_2$Cl, RNHCl, RNCl$_2$, and RRNCl are identified and quantified from their differing λ_{max} and E_M values (Table I). When total Cl$^+$ is at least 0.1 mM and preferably 1 mM or higher, the composition of mixtures of these derivatives is determined from absorbance at the λ_{max} wavelengths using the E_M values and simultaneous equations. Table I gives E_M values to three significant figures to facilitate such calculations, but the precision of measurement is not as great as implied. In particular, E_M values for chloroguanidino derivatives are probably lower than the true values because these derivatives decompose and/or rearrange.

It is not possible to distinguish between similar derivatives. The struc-

[29] S. T. Test, M. B. Lampert, P. J. Ossanna, J. G. Thoene, and S. J. Weiss, *J. Clin. Invest.* **74,** 1341 (1984).
[30] J. M. Zgliczynski and T. Stelmaszynska, *Eur. J. Biochem.* **56,** 157 (1975).

ture of the R moiety does not influence λ_{max} or E_M of RNHCl or RNCl$_2$. Different RRNCl derivatives may have different λ_{max} values in the 260–270 nm range, but differences are small and E_M values are similar.

E_M and λ_{max} of N—Cl derivatives are independent of pH from 5 to 8, whereas the spectrum of HOCl/OCl$^-$ changes as the HOCl:OCl$^-$ ratio changes. Decreasing absorbance at 291 nm as pH is lowered suggests the presence of unreacted HOCl/OCl$^-$. For example, when OCl$^-$ is added to amides the spectrum is pH dependent and identical to that of HOCl/OCl$^-$.

When N—Cl derivatives are present in a mixture of proteins or other absorbing species, the spectrum of the N—Cl derivatives is obtained by measuring only that portion of the absorbance that is eliminated by a mild reducing agent.[3] Absorbance of a portion of the sample is measured before and after adding small volumes of 0.1 M sodium hydrosulfite (dithionite) in N$_2$-gassed water. Dithionite reduces N—Cl derivatives rapidly and stoichiometrically, so that absorbance decreases until all the N—Cl derivatives are reduced. Absorbance at 200–230 nm increases sharply as excess dithionite is added, because absorbance of dithionite is at least 10 times that of oxidized forms of dithionite. With these results as a guide, another portion of the sample is treated with an amount of dithionite just sufficient to reduce the N—Cl derivatives, and the spectrum of N—Cl derivatives is obtained by measuring the difference in absorbance between a nonreduced portion of the sample and the dithionite-reduced portion.

Determination of Cl$^+$ Using Nbs. N—Cl derivatives rapidly oxidize the sulfhydryl compound Nbs to the disulfide compound Nbs$_2$. Excess Nbs is added, and the amount of Cl$^+$ is calculated as one-half the difference between the amount of Nbs added and the amount remaining.[9,10] E_M of Nbs at 412 nm is 13,600, whereas Nbs$_2$ absorbance is usually negligible.[31] Because two Nbs are oxidized per Cl$^+$, the effective E_M is 27,200 and N—Cl derivatives at 5–10 μM are reliably quantified.

Nbs is prepared by reducing Nbs$_2$ (Sigma Chemical Company, St. Louis, MO). Complete reduction of Nbs$_2$ is obtained by adding excess sodium borohydride and then storing the solution overnight at 4° to permit decomposition of borohydride.[32] However, complete reduction is not essential because Nbs$_2$ does not interfere. A more rapid method is to add 4 μl of 2-mercaptoethanol to 100 ml of 1 mM Nbs$_2$ in phosphate buffer, pH 7, or phosphate-buffered saline, pH 7.4, to yield 0.6 mM Nbs. Solutions stored at 4° under N$_2$ are usable for days or weeks, though Nbs is slowly lost by air oxidation. Rapid loss of Nbs may indicate contamination by

[31] G. L. Ellman, *Arch. Biochem. Biophys.* **82**, 70 (1959).
[32] T. M. Aune and E. L. Thomas, *Eur. J. Biochem.* **80**, 209 (1977).

ferric ion (Fe^{3+}) or other metal ions. If Nbs concentration falls below convenient levels, it is better to prepare a fresh solution than to add more reductant, so as to avoid accumulation of oxidation products.

With each group of samples, a control is performed to determine Nbs concentration. The control has a volume, pH, buffering capacity, and ionic strength similar to that of the samples. Since Nbs absorbance is pH dependent, the buffering capacity of the Nbs solution is high, or additional buffer is added to ensure that absorbance of all samples is measured at the same pH, preferably 7 to 7.4. To avoid alkaline hydrolysis of Nbs_2, Nbs is not used at a pH above 8. To avoid adding a large volume of Nbs and thus a contribution by Nbs_2 absorbance, Cl^+ concentration is adjusted to 0.3 mM or less by dilution. When measuring HOCl/OCl^-, 10 mM taurine is added to convert HOCl/OCl^- to TauNHCl, preventing attack of HOCl on Nbs_2. Oxidation of Nbs is usually complete within the time required to mix. Absorbance of the control and samples is measured promptly to avoid further absorbance changes. Cl^+ concentration (μM) is calculated as $(BCE - AFG)/.0272AD$, where A is volume of Nbs added to the control, B is final volume of the control (including A), C is absorbance of the control, D is volume of the sample, E is volume of Nbs added to the sample, F is final volume of the sample (including E), and G is absorbance of the sample.

N—Cl derivatives accumulate in the extracellular medium when stimulated leukocytes are incubated *in vitro*. Nbs is used to measure the amount that has accumulated, but can't be used in a continuous assay for production of N—Cl derivatives, because Nbs is oxidized by the H_2O_2 and HOCl produced by leukocytes. To measure N—Cl derivatives in the medium,[3] the control and the incubation mixture containing leukocytes (both 0.5 ml) are placed on ice, and 1 μl of 15 mg/ml catalase, 0.1–0.5 ml Nbs in cold phosphate-buffered saline, and 2 ml of cold phosphate-buffered saline with 1 mM $MgSO_4$ are added. Following centrifugation (7000 g, 10 min, 4°), absorbance of the supernatant at 412 nm is measured. Alternatively, cells are removed by centrifugation, and catalase, Nbs, and buffer are added to a portion of the supernatant.[7]

When measuring oxidants in samples containing erythrocytes, hemoglobin released into the medium by a few lysed cells may contribute to absorbance.[8] To correct for hemoglobin, G is measured, and then a small volume (H) of 1 mM TauNHCl is added to a portion (I) of the supernatant to oxidize excess Nbs, and 412 nm absorbance (J) is measured again. Cl^+ concentration is calculated substituting K for G, where $K = (GI - HJ - IJ)/I$.

The Nbs method[32] was originally developed to measure HOSCN/$OSCN^-$, and can be used to measure oxidants containing Cl^+, Br^+, I^+, or

TABLE II

OXIDANT : REDUCTANT (O : R) RATIO[a]

	Oxidant					
Reductant	HOCl/OCl⁻	NH₂Cl	RNHCl	RNCl₂	I₂	HOSCN/OSCN⁻
Thiourea, $NH_2C(S)NH_2$	3 : 1	3 : 1	3 : 1	3 : 1	3 : 1	3 : 1
Thiocyanate, SCN^-	3 : 1	3 : 1	3 : 1	3 : 1	1 : 2	0[b]
Hypotaurine, RSO_2^-	1 : 1	1 : 1	1 : 1	1 : 1	1 : 1	0
Thiodipropionate, RSR	2 : 1	1 : 1	1 : 1	1 : 1	1 : 1	0
Dimethyl sulfoxide, RRSO	1 : 1	0	0	0	0	0
Phenol, C_6H_5OH	10 : 1	0	0	1 : 1,0	3 : 1	0
Alanine, $CH_3CH(NH_2)CO_2H$	2 : 1	1 : 1	0	1 : 1,0	0	0

[a] Oxidants at 0.3 mM (or 0.15 mM RNCl₂) were incubated at 37° with reductants at concentrations up to 0.6 mM at pH 7. After 1 hr, the concentration of oxidant remaining was measured with Nbs. RNHCl and RNCl₂ indicate TauNHCl and TauNCl₂. The O : R ratio is the negative slope of the plot of oxidant remaining vs initial reductant concentration, and is the highest molar ratio of oxidant to reductant at which complete reduction was obtained. RNCl₂ was treated as 2 Cl⁺.
[b] Zero indicates no reaction detected.

SCN⁺ moieties. The method provides no qualitative information; additional methods are required to identify the oxidant.

Selective Reduction. Differences in reactivity are used to identify oxidants (Table II). Portions of the oxidant solution are combined with varying amounts of reductant at pH 7. After 1 hr at 37°, the amount of oxidant remaining is measured with excess Nbs. Many compounds can be used as selective reductants; the only requirements are that the compound not oxidize Nbs, reduce Nbs₂, or otherwise interfere with measurements of Nbs absorbance.

All of the oxidants listed in Table II oxidize thiourea. Thiourea at 0.1 mM is sufficient to completely reduce 0.3 mM oxidant, so that the oxidant : reductant (O : R) ratio is 3 : 1. All the oxidants except HOSCN/ OSCN⁻ oxidize SCN⁻ and oxidize the sulfinic acid compound hypotaurine to the sulfonic acid compound taurine.

$$RNHCl + R'SO_2^- + H_2O \rightarrow RNH_3^+ + Cl^- + R'SO_3^-$$

Therefore, oxidation of thiourea but not SCN⁻ or hypotaurine indicates HOSCN/OSCN⁻. I₂ oxidizes SCN⁻, hypotaurine, phenol, and thioether (RSR) compounds such as thiodipropionic acid or methionine, which serves to distinguish between I₂ and HOSCN/OSCN⁻.

Oxidants containing I⁺ or Cl⁺ are also distinguished by differing reactivities: (1) I₂ reacts slowly with SCN⁻ so that the O : R ratio measured after 1 hr at 37° is low compared to that obtained with HOCl/OCl⁻ or

N—Cl derivatives. (2) I_2 reacts with phenol with an O : R ratio lower than that of $HOCl/OCl^-$ and higher than that of N—Cl derivatives. (3) HOCl/OCl^- but not I_2 is reduced by dimethyl sulfoxide.

HOCl oxidizes thioethers (RSR) or sulfoxides (RRSO) to sulfones ($RRSO_2$). Therefore, thiodipropionate and dimethyl sulfoxide reduce HOCl with O : R ratios of 2 : 1 and 1 : 1, respectively. In contrast, N—Cl derivatives oxidize thioethers to sulfoxides, and do not oxidize sulfoxides.

$$2HOCl + RSR \rightarrow 2H^+ + 2Cl^- + RRSO_2$$
$$HOCl + RRSO \rightarrow H^+ + Cl^- + RRSO_2$$
$$R'NHCl + RSR \rightarrow R'NH_3^+ + Cl^- + RRSO$$

The 1 : 1 stoichiometry of reduction by thiodipropionic acid, the absence of reduction by dimethyl sulfoxide, and the low O : R ratio with phenol rule out $HOCl/OCl^-$ or I_2 and confirm the presence of N—Cl derivatives.

Phenolic compounds (e.g., tyrosine residues) are good reducing agents for HOCl. The high O : R ratio probably indicates not only chlorination of the phenol ring, but also loss of ring structure and aromatic character. In contrast, NH_2Cl and TauNHCl are poor chlorinating agents for phenol. Reduction of $TauNCl_2$ is biphasic. Half the Cl^+ is lost at low phenol concentrations, corresponding to conversion of $TauNCl_2$ to TauNHCl and chlorination of phenol.

$$RNCl_2 + C_6H_5OH \rightarrow RNHCl + C_6ClH_4OH$$

Alanine reduces HOCl or NH_2Cl with an O : R ratio of 2 : 1 or 1 : 1, corresponding to formation and decomposition of alanine–dichloramine or -monochloramine. Alanine is a poor reductant for TauNHCl, indicating a slow rate of exchange of Cl^+. In contrast, reduction of $TauNCl_2$ is biphasic. Half the Cl^+ is lost at low alanine concentrations, corresponding to conversion of $TauNCl_2$ to TauNHCl and formation and decomposition of alanine–monochloramine. The other half of the Cl^+ is lost at much higher alanine concentrations, as expected from the reduction of TauNHCl. Alanine at 30–60 mM is required for complete reduction of 0.3 mM TauNHCl or 0.15 mM $TauNCl_2$ within 1 hr.

These observations indicate that chlorinating activity is in the order HOCl > $RNCl_2$ > NH_2Cl > RNHCl, though only half the Cl^+ moieties of $RNCl_2$ participate in rapid chlorination. Chlorinating activity of the N—Cl derivatives is not the result of hydrolysis to yield HOCl. Dimethyl sulfoxide or H_2O_2 does not reduce N—Cl derivatives, whereas these agents reduce HOCl. If the N—Cl derivatives were in equilibrium with HOCl, then reduction of HOCl would result in complete loss of Cl^+.

Dimethyl sulfoxide or H_2O_2 also does not reduce RSO_2NHCl. However, when 2 mol OCl^- are added to 1 mol RSO_2NH_2 in an attempt to prepare RSO_2NCl_2, the resulting solution acts as a mixture of 1 mol RSO_2NHCl and 1 mol $HOCl/OCl^-$. Therefore, RNH_3^+ and RSO_2NH_2 differ in that RSO_2NH_2 does not form a stable dichlorinated derivative.

Iodometry. HOCl and N—Cl derivatives oxidize I^- to yield one I_2 per Cl^+, and I_2 is quantified by titration with thiosulfate or from absorbance of I_2, I_3^-, or the starch–I_2 complex. Iodometry is a standard and sensitive method for measuring oxidants in chlorinated water, but is less reliable for solutions containing substantial amounts of biological materials. Iodometry has been used to detect accumulation of N—Cl derivatives in the extracellular medium following incubation of stimulated leukocytes *in vitro.*[2] For this method, it is essential to remove leukocytes prior to adding I^-, because I_2 reacts rapidly with cells. Moreover, I_2 reacts rapidly with proteins and other released or secreted materials in the medium. Therefore, concentrations of N—Cl derivatives determined with I^- may be lower than those determined with Nbs.

Oxidation or Chlorination of Dyes. Sensitive methods have been developed to measure oxidants in chlorinated water using aromatic compounds that undergo large changes in absorbance or fluorescence upon chlorination.[33] Absorbance changes accompanying chlorination of monochlorodimedone[34] have been used to measure peroxidase-catalyzed oxidation of Cl^-. Similarly, oxidation of Cl^- by leukocytes results in changes in fluorescence of the pH indicator fluorescein.[35] Chlorination of 1,3,5-trimethoxybenzene by leukocytes yields a derivative that is measured by HPLC techniques.[36] HOCl is usually assumed to be the chlorinating agent, though N—Cl derivatives may also chlorinate these compounds. Quantitative chlorination is probably not obtained in experiments with leukocytes, because the added compound must compete with all other chlorinatable substances.[36]

Extraction. Relative hydrophilic or lipophilic character is evaluated from the partitioning of derivatives between an aqueous and organic phase.[3,10] The percentage extraction of lipophilic derivatives into ethyl acetate is higher than obtained with less polar solvents, and phase separation is rapid and complete even when the aqueous phase contains protein.

[33] J. D. Johnson, *in* "Water Chlorination: Environmental Impact and Health Effects" (R. L. Jolley, ed.), p. 37. Ann Arbor Sci. Publ., Ann Arbor, Michigan, 1978.

[34] L. P. Hager, D. R. Morris, F. S. Brown, and H. Eberwein, *J. Biol. Chem.* **241,** 1769 (1966).

[35] J. K. Hurst, J. M. Albrich, T. R. Green, H. Rosen, and S. Klebanoff, *J. Biol. Chem.* **259,** 4812 (1984).

[36] C. S. Foote, T. E. Goyne, and R. I. Lehrer, *Nature (London)* **301,** 715 (1983).

Nevertheless, extraction into 2-octanol is advantageous because partition coefficients can be compared with published values for many membrane-permeable agents.[37] Chloroform is used for spectrophotometric experiments because solubility of chloroform in the aqueous phase is low.

As described above, solvents are treated to remove oxidizing agents and acids, and to saturate the solvent with water so as to minimize volume changes during extraction.[18] Equal volumes of aqueous sample and organic solvent are mixed and briefly centrifuged to aid phase separation. Oxidizing equivalents in portions of the aqueous and organic phase are measured with Nbs. When a portion of the organic phase is mixed with Nbs and buffer, Nbs absorbance is measured in the aqueous phase. Results are reported as the percentage of total oxidizing equivalents extracted into the organic phase, or as a partition coefficient, which is the ratio of concentrations in the organic and aqueous phases. The percentage of NH_2Cl extracted into ethyl acetate, 2-octanol, and chloroform is 60, 35, and 26%. Partition coefficients are 1.5, 0.54, and 0.35.

It is important to measure the oxidant in both phases. Low oxidant concentration in the organic phase may indicate decomposition rather than low solubility in the solvent. For example, $RNCl_2$ derivatives of putrescine and other polyamines undergo rapid decomposition when solutions are mixed with miscible or immiscible solvents with dielectric constants lower than 21 (less polar than acetone). Therefore, extraction can't be used to evaluate lipophilicity of these derivatives.

NH_2Cl is extracted and then back extracted into an aqueous medium to purify the derivative for spectrophotometric identification.[6] The reference cuvette contains a portion of the aqueous medium that has been equilibrated with the solvent. The highest possible concentration of NH_2Cl is obtained when volumes of the sample, solvent, and back-extraction medium are the same. Using smaller volumes of solvent and medium does not yield higher concentrations and larger volumes result in dilution.

N—Cl moieties have low pK_a values so that pH does not influence lipophilicity. For example, pK_a values of 1–2 were estimated for $RNHCl$,[25] indicating that $RNCl^-$ or RNH_2Cl^+ are not present in significant quantities at pH 5–8. However, an ionizable group in addition to the N—Cl moiety can influence lipophilicity. For example, deprotonation of the imidazole moiety of histamine–chloramines increases lipophilicity so that the derivatives are concentrated by extraction from the aqueous phase at pH 8 into a small volume of solvent followed by back extraction into a smaller volume of aqueous medium at pH 5.

Volatilization. NH_2Cl has a low vapor pressure and is separated from

[37] A. Leo, C. Hansch, and D. Elkins, *Chem. Rev.* **71,** 525 (1971).

other agents by allowing NH_2Cl to volatilize and then collecting NH_2Cl from the vapor phase.[6,7] The solution containing NH_2Cl and a second solution for collecting NH_2Cl are placed in separate compartments of a closed container that permits communication between the solutions through the vapor phase. For example, the NH_2Cl solution is placed in the outer well of a Conway vapor-diffusion chamber, 0.14 M NaCl–15 mM phosphate buffer, pH 7, is placed in the inner well, and the chamber is closed and gently mixed at 37° for 1–2 hr. At equilibrium, concentrations of NH_2Cl in the two solutions are equal, so that the percentage of total NH_2Cl obtained in the collecting solution depends on the ratio of volumes of the two solutions. NH_2Cl is identified and quantified from the 242-nm absorbance in the collecting solution. The reference cuvette contains a solution equilibrated under similar conditions with a dithiothreitol-reduced portion of the NH_2Cl solution, to correct for absorbance of other volatile substances such as NH_3.

Ultrafiltration and Chromatography. N—Cl derivatives are separated according to molecular weight by ultrafiltration or gel-exclusion chromatography.[7,9,10,29] Derivatives are measured in filtrates or effluent fractions by the Nbs oxidation assay. Good recoveries of N—Cl derivatives are obtained from columns of Sepharose 4B or Sephadex G-200, G-100, or G-50 (Pharmacia Fine Chemicals, Piscataway, NJ) equilibrated and eluted with 0.1 M NaCl–20 mM phosphate buffer, pH 5–8. In contrast, HOCl/OCl⁻ reacts with the column material and is not eluted. Poor recoveries may be obtained with Sephadex G-25 or G-10, apparently due to the presence of readily oxidized substances.[29] Reductants are removed by passing 1–10 mM HOCl/OCl⁻ over the column until HOCl/OCl⁻ concentration in the effluent equals that in the eluant, and then washing extensively.[9] This pretreatment does not alter resolving properties of the column, but does permit HOCl/OCl⁻ to be eluted.

$N^{36}Cl$ derivatives are detected by chromatographing nonreduced and dithiothreitol-reduced portions of a sample and measuring ^{36}Cl-labeled products that are present only in the nonreduced portion. Prior to chromatography, excess unlabeled Cl⁻ is added to exchange with noncovalently bound $^{36}Cl⁻$. Label not released by reduction or exchange with Cl⁻ indicates $C^{36}Cl$ derivatives. Paper electrophoresis is used to separate certain low-molecular-weight N—Cl derivatives from parent N-compounds and Cl⁻.[38] Derivatives are detected by spraying with KI or from the ^{36}Cl label.

Derivatization. $RNCl_2$ derivatives undergo a reaction that appears analogous to chlorination, but in which the R or R—N moiety rather than

---

[38] T. Stelmaszynska and J. M. Zgliczynski, *Eur. J. Biochem.* **45,** 305 (1974).

Cl becomes incorporated into an acceptor molecule.[1] Measuring incorporation of the radiolabeled R moiety into proteins and other biological materials provides a sensitive method to detect $RNCl_2$ formation. For example, myeloperoxidase-dependent incorporation of label from [^{14}C]glucosamine into neutrophil components was observed[39] before the role of $RNCl_2$ in this process was discovered.[1] Similar incorporation has subsequently been observed with many labeled amines.[13] Excess dithiothreitol is added to reduce $RNCl_2$ or other N—Cl derivatives and then labeled macromolecules are separated from labeled RNH_3^+ by precipitation with acid or organic solvents, gel chromatography, or electrophoresis.[1,39] Incorporation of label from the R moiety of $RNCl_2$ is a slow reaction with a poor yield, and there is no direct quantitative relation between the amount of incorporation and the amount of $RNCl_2$. Nevertheless, results obtained with neutrophils indicate that the cells produce HOCl, that the added amines encounter HOCl, and that RNHCl as well as $RNCl_2$ is produced.

Acknowledgments

This research was supported by Grants AI 16795, DE 04235, CA 21765, and CA 09346 from the National Institutes of Health and by the American Lebanese Syrian Associated Charities.

[39] S. I. Bearman, G. A. Schwarting, E. H. Kolodny, and B. M. Babior, *J. Lab. Clin. Med.* **96,** 893 (1980).

[41] Cytotoxicity of Chloramines

By Edwin L. Thomas, Matthew B. Grisham, and
M. Margaret Jefferson

Binding of ligands to membrane-surface receptors of neutrophilic leukocytes stimulates the reduction of oxygen (O_2) to superoxide (O_2^-) and the secretion of cytoplasmic granule components including myeloperoxidase into the intracellular phagolysosome compartment and the extracellular medium. Spontaneous or enzyme-catalyzed dismutation of O_2^- yields hydrogen peroxide (H_2O_2). Myeloperoxidase catalyzes the oxidation of chloride (Cl^-) by H_2O_2 to yield hypochlorous acid (HOCl), which reacts rapidly with ammonia (NH_4^+), amines (RNH_3^+), and other nitrogen compounds to yield chloramines and other nitrogen–chlorine (N—Cl) derivatives. Because the reaction of HOCl with N-compounds is fast and the concentration of N-compounds is high, it is unlikely that HOCl acts directly as the toxic agent produced by the myeloperoxidase–H_2O_2–Cl^- system. Instead, toxicity is mediated by N—Cl derivatives.[1-6]

Stimulated neutrophils produce NH_2Cl, RNHCl, and $RNCl_2$,[3-12] and it is of interest to study damage done by these oxidants to the neutrophils and to target cells such as microorganisms, erythrocytes, tumor cells, and normal host tissues. When neutrophils are incubated with target cells, N—Cl derivatives react with both types of cells so that results are influenced by the neutrophil : target cell ratio and the reactivity of N—Cl derivatives toward neutrophil and target cell components. Studies of mixtures of neutrophils and target cells also limit the choice of methods for evaluating cytotoxicity. An alternative approach is to identify N—Cl derivatives produced by neutrophils, to prepare solutions of these derivatives,[7] and to test their cytotoxicity with target cells alone.

Methods for evaluating cytotoxicity of N—Cl derivatives consist of measuring oxidation, chlorination, or other chemical modifications of cell components, and measuring lysis or the failure of specific structural or functional components of cells. Since N—Cl derivatives react with many cell components, it is impractical to consider all possible methods. However, sulfhydryl groups are especially susceptible to oxidative attack and are usually present in high concentrations relative to other cell components. A good correlation is observed between oxidation of cell sulfhydryls and loss of cell functions or viability. Similarly, many methods can be used to evaluate loss of cell functions. N—Cl derivatives inactivate enzymes and transport systems, lyse erythrocytes and nucleated cells, and mutagenize and kill bacteria. Methods described below illustrate a few approaches that have proved useful.

Two precautions should be observed. First, when attempting to evaluate damage to intracellular components, it is essential to remove oxidants

[1] E. L. Thomas, *Infect. Immun.* **23**, 522 (1979).
[2] E. L. Thomas, *Infect. Immun.* **25**, 110 (1979).
[3] M. B. Grisham, M. M. Jefferson, and E. L. Thomas, *J. Biol. Chem.* **259**, 6757 (1984).
[4] E. L. Thomas, M. B. Grisham, and M. M. Jefferson, *J. Clin. Invest.* **72**, 441 (1983).
[5] M. B. Grisham, M. M. Jefferson, D. F. Melton, and E. L. Thomas, *J. Biol. Chem.* **259**, 10404 (1984).
[6] E. L. Thomas, M. B. Grisham, D. F. Melton, and M. M. Jefferson, *J. Biol. Chem.* **260**, 3321 (1985).
[7] E. L. Thomas, M. B. Grisham, and M. M. Jefferson, this volume [40].
[8] E. L. Thomas, M. M. Jefferson, and M. B. Grisham, *Biochemistry* **21**, 6299 (1982).
[9] S. J. Weiss, R. Klein, A. Slivka, and M. Wei, *J. Clin. Invest.* **70**, 598 (1982).
[10] M. B. Lampert and S. J. Weiss, *Blood* **62**, 645 (1983).
[11] S. J. Weiss, M. B. Lampert, and S. T. Test, *Science* **222**, 625 (1983).
[12] S. T. Test, M. B. Lampert, P. J. Ossanna, J. G. Thoene, and S. J. Weiss, *J. Clin. Invest.* **74**, 1341 (1984).

from the cell surface and extracellular medium. Dilution and repeated washing of cells in cold buffer is usually sufficient to remove oxidants. The Nbs oxidation assay[7] is used to determine whether all oxidants have been removed. Second, when evaluating cytotoxicity of HOCl or N—Cl derivatives, the medium should contain no N-compound that could be chlorinated by HOCl or which could undergo exchange with N—Cl derivatives. For example, if HOCl or N—Cl derivatives are added to bacteria in a culture medium, the presence of NH_4^+ and other nutrients is likely to influence results. Even when cells are thoroughly washed, the metabolic production of NH_4^+ and release of other N-compounds into the medium must be considered.

Neutrophils as Targets of Their Own Cytotoxic Activities. Stimulated neutrophils undergo a progressive myeloperoxidase-dependent inhibition of O_2 and glucose metabolism.[4,13] This inhibition results from the reaction of HOCl with NH_4^+ produced by the cells.[3] NH_2Cl is produced and inactivates neutrophil components.[14]

Neutrophil preparations should be free of erythrocytes, because NH_2Cl and taurine–chloramines interact specifically with these cells,[3,6] Lysis of erythrocytes with water or high concentrations of NH_4^+ is avoided, because these methods may contaminate neutrophils with adsorbed erythrocyte enzymes such as catalase and superoxide dismutase and with NH_4^+. To avoid introducing iodide (I^-), metrizoate or diatrizoate is not used in preparing neutrophils. The following procedure[3] yields neutrophils free of erythrocytes, though small numbers of eosinophils and basophils may be present.

Whole blood (50 ml) is collected in acid–citrate–dextrose (ACD) anticoagulant solution. Cells are washed by dilution and centrifugation (1000 g, 10 min, 25°) in Cl^- medium, consisting of 0.14 M NaCl–15 mM potassium phosphate, pH 7.4. All steps below are in this medium. Cells are suspended to 50 ml in 0.1% (w/v) gelatin, mixed with 50 ml 3% (w/v) dextran (5 × 10^5 molecular weight; Sigma Chemical Company, St. Louis, MO), and allowed to settle for 30 min at 37°. Cells in the supernatant fraction are collected by centrifugation, suspended to 18 ml in 40% Percoll (Pharmacia Fine Chemicals, Piscataway, NJ), and 9-ml portions are layered on discontinuous gradients of 9 ml each of 50, 61, and 70% Percoll and 8 ml of 80% Percoll. After centrifugation (250 g, 30 min, 25°), cells at 61 : 70% and 70 : 80% interfaces are pooled, washed, and suspended to 18 ml, and 9-ml portions are layered on gradients of 9 ml of 70% and 8 ml of

[13] R. C. Jandl, J. André-Schwartz, L. Borges-DuBois, R. S. Kipnis, B. J. McMurrich, and B. M. Babior, *J. Clin. Invest.* **61**, 1176 (1978).
[14] E. L. Thomas, M. M. Jefferson, and M. B. Grisham, unpublished results (1976–1984).

80% Percoll. After centrifugation as detailed above, cells at the interfaces are pooled, washed, and suspended to 2×10^7/ml.

To measure O_2 uptake, neutrophils at 4×10^6/ml are incubated in 2.5 ml of Cl^- medium–1 mM MgSO$_4$–10 mM glucose. O_2 uptake is initiated by adding 5 μl of 40 μM PMA (P-L Biochemicals, Milwaukee, WI) in dimethyl sulfoxide (spectrophotometric grade; Aldrich Chemical Company, Milwaukee, WI). O_2 concentration is measured in a stirred chamber at 37° with a Clarke-type O_2 electrode (YSI, Yellow Springs, OH; model 53). Rates of O_2 uptake (μM/min) are calculated assuming a concentration of 0.2 mM O_2 for an air-saturated suspension at 37°.

To measure glucose metabolism, neutrophils at 4×10^6/ml are incubated in 0.5 ml Cl^- medium–1 mM MgSO$_4$–80 nM PMA–0.5 mM D-[U-^{14}C]glucose (Amersham Corp., Arlington Heights, IL)[14a] for 1 hr at 37° in closed, siliconized 15-ml glass tubes with continuous mixing to maintain aeration. Mixtures are centrifuged (7000 g, 10 min, 4°), and the cells and supernatants are separated. Cell-associated label (A; primarily glycogen) is determined by resuspending cells in 5 ml of cold Cl^- medium–1 mM MgSO$_4$, centrifuging again, resuspending to 0.5 ml in Cl^- medium, and measuring radioactivity in a 0.2-ml portion using a liquid scintillation spectrometer. Nonvolatile label in the supernatant (B; lactate and glucose) is determined by adding a 0.2-ml portion to 1 ml of 0.2 M HCl, drying under a stream of N_2, redissolving in 0.2 ml water, and measuring radioactivity. As a control, an identical incubation mixture is held at 4° for 1 hr, centrifuged, and nonvolatile label in the supernatant (C) is determined. Volatile label (D; primarily CO_2) is calculated as $C - B - A$. Lactate and glucose in the supernatant are separated by thin-layer chromatography of 50-μl portions on silica gel G plates in ethanol: NH$_4$OH : water, 20 : 1 : 4. Areas corresponding to lactate (R_f 0.7) and glucose (R_f 0.4) are scraped from the plates and radioactivity in lactate (E) and glucose (F) are measured. Label in lactate (G) is calculated as $EB/(E + F)$. Metabolized label (H) is calculated as $A + D + G$. Results are reported as the concentration of glucose metabolized (mM), 0.5 H/C, and the percentage of metabolized label in cell-associated forms, 100 A/H, in volatile forms, 100 D/H, and in lactate, 100 G/H.

Inhibition of O_2 or glucose metabolism is increased by adding superoxide dismutase (0.1 mg/ml), myeloperoxidase (0.1 μM), or NH$_4^+$ (up to 10 mM). Inhibition is decreased or blocked by adding catalase (30 μg/ml) or horseradish peroxidase–0.2 mM scopoletin as a trap for H_2O_2, by adding 30–100 μM 4,4′-diaminodiphenyl sulfone (dapsone, Sigma) as an inhibitor

---

[14a] Abbreviations: Nbs, 5-thio-2-nitrobenzoic acid; Nbs$_2$, 5,5′-dithiobis(2-nitrobenzoic acid); PMA, phorbol 12-myristate 13-acetate; GSH, reduced glutathione; EDTA, ethylenediaminetetraacetic acid; Hb, hemoglobin.

of myeloperoxidase, 10 mM taurine as a trap for HOCl, or 0.5 mM dithiothreitol to reduce HOCl and N—Cl derivatives, or in a Cl⁻-free medium[15] (67 mM Na$_2$SO$_4$–1 mM MgSO$_4$–32 mM potassium phosphate, pH 7.4).

Incubation of neutrophils or other cells with lipophilic N—Cl derivatives results in loss of sulfhydryl components, as a result of oxidation and perhaps also increased permeability of cell membranes and/or lysis. Loss of sulfhydryl components is determined by measuring total cell sulfhydryl content relative to untreated control cells. The method described below[16] measures the sum of protein sulfhydryl groups and low-molecular-weight sulfhydryl compounds such as GSH. More precisely, the method measures substances that reduce Nbs$_2$[17] and thus may include ascorbate and other reductants. On the other hand, aromatic sulfhydryl compounds such as ergothioneine do not reduce Nbs$_2$.[14] The method is applicable to neutrophils, other nucleated cells, and bacteria, but not erythrocytes.

Washed pellets (4 × 10^6 nucleated cells, or 1 × 10^9 bacteria) are suspended in 0.65 ml water, and 0.35 ml 0.1 M Tris–Cl (pH 8), 0.15 ml 10% (w/v) sodium dodecyl sulfate ("specially pure"; BDH, Poole, England), 10 μl of 1 M potassium EDTA (pH 7), and 30 μl of 3 mM Nbs$_2$ are added sequentially with thorough mixing. The air space above the samples is displaced with N$_2$, the tubes are closed, and the mixtures are incubated 30 min at 37°, placed on ice for 5 min, and centrifuged (12,000 g, 10 min, 4°). Sulfhydryl content (nanomoles) is calculated as $A(B - C)/D$, where A is volume (1.19 ml), B and C are 412 nm absorbance of the supernatant and reagent blank, and D is E μM of Nbs (0.0136).

The following method is used to measure acid-soluble sulfhydryl compounds of neutrophils and other nucleated cells, but not erythrocytes or bacteria. As mentioned above, reductants other than GSH are included but aromatic sulfhydryls are not. The washed pellet (8 × 10^6 cells) is suspended in 0.4 ml of 0.2% (w/v) Triton X-100 (Boehringer Mannheim, Indianapolis, IN). After 5 min at 25°, 0.4 ml of 5% (w/v) sulfosalicylic acid is added, and the mixture is centrifuged (12,000 g, 15 min, 4°). A portion (0.7 ml) of the clarified supernatant is neutralized with 10 μl of 10 M NaOH, added to 0.19 ml of 0.84 M Tris–Cl (pH 8)–52 mM potassium EDTA (pH 7)–0.3 mM Nbs$_2$, and incubated 30 min at 37° under N$_2$. Sulfhydryl content (nanomoles) is calculated as $0.8A(B - C)/0.7D$, where A is volume (0.9 ml), B and C are 412 nm absorbance of the sample and reagent blank, and D is 0.0136.

[15] R. A. Clark, S. J. Klebanoff, A. B. Einstein, and A. Fefer, *Blood* **45,** 161 (1975).
[16] E. L. Thomas, *J. Bacteriol.* **157,** 240 (1984).
[17] G. L. Ellman, *Arch. Biochem. Biophys.* **82,** 70 (1959).

Erythrocytes as Targets. Incubation of stimulated neutrophils with erythrocytes results in oxidation of erythrocyte Hb from oxyHb to the inactive metHb form. More than one mechanism contributes to Hb oxidation depending on experimental conditions. However, NH_2Cl appears to be the principal oxidant in the system described below.[4] Advantages of the neutrophil–erythrocyte system are that the target molecule (Hb) is present only in the target cell, that the high concentration and strong absorbance of Hb provide sensitive measures of both lysis and oxidation, and that NH_2Cl and HOCl differ in their attack on erythrocytes, which permits discrimination between these oxidants. NH_2Cl is more effective as an oxidant for Hb, but less effective as a lytic agent.

Whole blood (5 ml) is washed twice by centrifugation (1000 g, 10 min, 25°) in 50 ml Cl^- medium. Cells are suspended in 10 ml of 0.154 M NaCl adjusted to pH 7 with NaOH, and passed over a column (5-ml bed volume) prepared from 0.5 g each of microcrystalline cellulose and α-cellulose (Sigma).[18] Effluent (10 ml) is collected after erythrocytes begin to be eluted. Erythrocytes are collected by centrifugation, washed twice in 25 ml cold Cl^- medium, and suspended to 2×10^9/ml.

Neutrophils (4×10^6/ml) and erythrocytes (2×10^7/ml) are incubated in 1 ml of Cl^- medium–10 mM glucose–1 $MgSO_4$–80 nM PMA in closed siliconized 15-ml glass tubes at 37° with continuous mixing. Incubation mixtures are diluted with 0.5 ml cold Cl^- medium–1 mM $MgSO_4$ and centrifuged (7000 g, 10 min, 4°). The supernatant and pellet are separated, and the pellet is resuspended in 0.6 ml water, incubated 5 min at 25°, diluted with 0.6 ml of 0.2 M Tris–Cl (pH 7.4), and centrifuged (20,000 g, 2 min, 25°) to yield a clarified lysate. Percentage lysis is calculated as $100AB/CD$, where A and C are Hb absorbance at 409 nm in the supernatant and the clarified lysate. B and D are volumes of the diluted incubation mixture (1.5 ml) and the resuspended pellet (1.25 ml), assuming a pellet volume of 50 μl. Oxidation of Hb is determined by measuring absorbance of the supernatant and clarified lysate at 560, 576, and 630 nm in self-masking cuvettes, and calculating concentrations assuming E mM of 8.5, 15.3, and 0.2 for oxyheme ($Fe^{2+}O_2$); 12.7, 9.8, and 1 for deoxyheme (Fe^{2+}); and 4.3, 4.45, and 3.63 for metheme (Fe^{3+}).[19]

$$\text{Oxyheme } (\mu M) = (-83.6)A + (112.3)B + (-38.6)C$$
$$\text{Deoxyheme } (\mu M) = (146.9)A + (-80.6)B + (-75.2)C$$
$$\text{Metheme } (\mu M) = (-35.8)A + (16)B + (298.3)C$$

A, B, and C are absorbance at 560, 576, and 630 nm.

[18] E. Beutler, C. West, and K.-G. Blume, *J. Lab. Clin. Med.* **89**, 1080 (1977).
[19] A. Riggs, this series, Vol. 76, p. 5.

Incubation of erythrocytes with N—Cl derivatives results in oxidation of acid-soluble sulfhydryl compounds (primarily GSH), protein sulfhydryl groups (primarily cysteine residues of Hb), and heme moieties of Hb. Depletion of ATP and lysis may also be observed.[6]

To measure GSH,[20] washed pellets (1 × 10⁹ erythrocytes) are suspended in 0.9 ml water and incubated 5 min at 25°. The lysate is deproteinized by adding 0.4 ml of 5.2 M NaCl–7 mM EDTA–0.3 M H$_3$PO$_4$, incubating 5 min at 25°, and centrifuging (12,000 g, 10 min, 4°). A portion (0.6 ml) of the supernatant is neutralized with 0.15 ml of 2.4 M K$_2$HPO$_4$, 50 μl of 3 mM Nbs$_2$ in 0.2 M phosphate buffer (pH 7) is added, and the mixture is incubated 5 min at 37°. GSH (nanomoles) is calculated as $1.3A(B - C)/0.6D$, where A is volume (0.8 ml), B and C are 412 nm absorbance of the sample and reagent blank, and D is 0.0136.

To measure total sulfhydryls,[21] washed pellets (4 × 10⁷ erythrocytes) are suspended in 0.95 ml water, and 50 μl of 3 mM Nbs$_2$ is added. After 3 min at 37°, 1 ml each of ethanol and chloroform are added and mixed thoroughly, and 412 nm absorbance of the upper layer is measured. Sulfhydryl content (nanomoles) is calculated as $AE(B - C)/D$, where A is the volume before dilution (1 ml), E is a factor (1.354) that corrects for both dilution and the effect of the solvents on Nbs absorbance, B and C are 412 nm absorbance of the sample and reagent blank, and D is 0.0136.

Depletion of ATP may indicate greater inactivation of ATP-producing processes than of ATP-utilizing processes, increased membrane permeability and/or lysis, or direct chemical attack on ATP. When erythrocytes are incubated with high levels of HOCl or certain N—Cl derivatives, most of the ATP lost from the cells is recovered in the medium,[14] indicating that ATP-degrading enzymes are inactivated and that there is little attack on ATP. The method[6] below appears suitable for measuring ATP in the medium and the cellular (particulate) fraction for all cells studied.

Incubation mixtures are made 5 mM in dithiothreitol and centrifuged (12,000 g, 10 min, 4°). The pellet (derived from 2 × 10⁷ erythrocytes, 1 × 10⁶ neutrophils or other nucleated cells, or 1 × 10⁹ bacteria) is suspended in 1 ml of 0.15 M NaCl–1 mM MgSO$_4$. Portions (0.2 ml) of the supernatant and resuspended pellet are pipetted directly into 3 ml of 20 mM Tris-Cl (pH 7.5) in a boiling water bath, heated 5 min, and cooled to 25°. Portions (0.3 ml) are added to 0.15 ml of 20 mM Tris–Cl (pH 7.5) containing 1.5 mg/ml luciferase–luciferin reagent ("Firelight"; Turner Designs, Mountain View, CA), 10 mM MgSO$_4$, and 1 mM dithiothreitol in siliconized flat-bottom 1.5-cm diameter glass vials. After a 15-sec delay for mixing, inte-

[20] E. Beutler, O. Duron, and B. M. Kelley, *J. Lab. Clin. Med.* **61**, 882 (1963).
[21] N. S. Kosower, E. M. Kosower, and R. L. Koppel, *Eur. J. Biochem.* **77**, 529 (1977).

grated luminescence is measured for 1 min with a photometer (Turner Designs, model 20). ATP concentration is calculated from a standard curve prepared with ATP in 0.02 M Tris–Cl, pH 7.5.

Erythrocyte lysis is evaluated from release of Hb or of ^{51}Cr-labeled cell components as described below. High oxidant concentrations cause bleaching of heme absorbance and conversion of Hb to insoluble forms, so that both methods underestimate cytotoxicity.

Tumor Cells as Targets. Incubation of CCRF-CEM cells[22] (ATCC CCL 119; American Type Culture Collection, Rockville, MD) with increasing concentrations of certain N—Cl derivatives results in oxidation of sulfhydryls, ATP depletion, and lysis.[5,14] In contrast to results obtained with erythrocytes, NH$_2$Cl is at least as effective as HOCl as a lytic agent for CEM cells.

To evaluate lysis, intracellular components are labeled with [^{51}Cr]-chromate, and release of label in soluble forms is measured.[23,24] The method is also applicable to erythrocytes and perhaps neutrophils.[25] The ^{51}Cr method has already been described elsewhere in this volume [27] and will not be duplicated here.

Bacteria as Targets. Incubation of *Escherichia coli* with HOCl or N—Cl derivatives results in oxidation of cell sulfhydryl components, and chlorination of cell components to yield N—Cl and C—Cl derivatives.[1,14] Inhibition of O$_2$ uptake, loss of ATP, and loss of viability are correlated with loss of sulfhydryl components rather than chlorination.

Cells of the ML strain of *E. coli* (American Type Culture Collection) are used because they do not aggregate in the presence of cationic leukocyte proteins such as myeloperoxidase or lysozyme.[26] However, these bacteria are readily killed by serum and have not been successfully opsonized for studies of phagocytosis and intracellular killing. Cells are grown aerobically as a suspension to early stationary phase (16 hr) in a culture medium[27] containing 40 mM K$_2$HPO$_4$, 22 mM KH$_2$PO$_4$, 2 mM sodium citrate, 7.5 mM (NH$_4$)$_2$SO$_4$, 5 mM Tris–Cl (pH 6.6), and 1% (w/v) disodium succinate. Cells are washed twice by centrifugation (2000 g, 20 min, 4°) in 0.15 M NaCl and adjusted to 8 × 10^8 cells/ml or an absorbance of 1 at 600 nm. Incubation mixtures contain cells at 1 × 10^8/ml in 0.12 M NaCl–1 mM MgSO$_4$–50 mM, pH 6.6, potassium phosphate at 37°.

[22] G. E. Foley, H. Lazarus, S. Farber, B. G. Uzman, B. A. Boone, and R. E. McCarthy, *Cancer (Philadelphia)* **18,** 522 (1965).
[23] A. Slivka, A. F. LoBuglio, and S. J. Weiss, *Blood* **55,** 347 (1980).
[24] R. A. Clark and S. J. Klebanoff, *J. Exp. Med.* **141,** 1442 (1975).
[25] M.-F. Tsan, *J. Cell. Physiol.* **105,** 327 (1980).
[26] H. R. Kaback, this series, Vol. 22, p. 99.
[27] B. D. Davis and E. S. Mingioli, *J. Bacteriol.* **60,** 17 (1950).

During the incubation of cells with oxidants, most of the N—Cl derivatives of cell-surface components are shed into the medium and are measured following centrifugation (12,000 g, 10 min, 4°) by the Nbs oxidation assay.[7] Cells are resuspended in cold 0.15 M NaCl and centrifuged again, and sulfhydryl content is measured as described above.

Incorporation of label from $HO^{36}Cl$ or $N^{36}Cl$ derivatives into $C^{36}Cl$ derivatives of cell components is measured by diluting incubation mixtures (1 ml) with 20 μl of 0.1 M dithiothreitol, 0.1 ml of 1 M NaCl, and 1 ml of cold 10% (w/v) trichloroacetic acid, filtering through 0.45 μm nitrocellulose filters (Millipore Corp., Bedford, MA), and washing the filters with cold 5% trichloroacetic acid. Filters are dissolved in scintillation fluid and radioactivity is measured in a liquid scintillation spectrometer.

To measure O_2 uptake, cells are washed in cold 0.12 M NaCl–1 mM $MgSO_4$–50 mM potassium phosphate buffer (pH 7), and resuspended to 2×10^8 cells/ml in this medium with 1% disodium succinate. O_2 uptake is measured as described above. To measure viability, incubation mixtures are diluted with 0.15 M NaCl–0.5 mM dithiothreitol, serial dilutions are prepared in the culture medium without succinate, and 1-ml portions are plated on culture medium–1% (w/v) glucose–2% (w/v) agar. Colonies are counted after 1–2 days at 37°.

Among other methods that have been applied to bacteria are measurements of mutagenesis,[28,29] attack on iron-containing cell components resulting in release of iron in soluble forms,[30] release of cell-surface and periplasmic components into the medium,[1] and loss of ability to retain potassium (K^+).[29] Loss of K^+ may indicate increased cytoplasmic membrane permeability, or the inability to maintain a K^+ gradient, which requires a functional electron-transport chain.

Acknowledgments

This research was supported by Grants AI 16795, DE 04235, CA 21765, and CA 09346 from the National Institutes of Health and by the American Lebanese Syrian Associated Charities.

[28] K. L. Shih and J. Lederberg, *Science* **192**, 1141 (1976).
[29] C. N. Haas and R. S. Engelbrecht, *J. Water Pollut. Control Fed.* **52**, 1976 (1980).
[30] H. Rosen and S. J. Klebanoff, *J. Biol. Chem.* **257**, 13731 (1982).

[42] Liposome Encapsulation of Muramyl Peptides for Activation of Macrophage Cytotoxic Properties

By RAJIV NAYAR, ALAN J. SCHROIT, and ISAIAH J. FIDLER

A variety of *in vitro* assays have been utilized to elucidate the mechanism(s) by which macrophages can destroy susceptible target cells. These include inhibition of target cell growth (cytostasis),[1,2] cleared zones of tumor cell monolayers,[3] release of radioactive labels,[4] cinemicrographic analysis,[5] and sequential scanning and transmission electron microscopy.[6] Unfortunately, it has been difficult to compare the results from different investigations because of the multiplicity of variables involved in different assay systems, including but not limited to differences in macrophage sources and tumor cell type.

Recently, there has been a great expansion in the use of lipid vesicles as a drug delivery system.[7] The efficiency of macrophage activation by liposomes containing muramyl peptides far surpasses that produced by free (unencapsulated) muramyl peptides added to the culture medium.[8] Dose–response measurements have indicated that liposome-encapsulated muramyl dipeptide (MDP)[8a] renders macrophages tumoricidal at concentrations up to 4000 times lower than free MDP.[8] In addition, macrophages that have become refractory to reactivation by free MDP can be reactivated by liposome-encapsulated MDP. Activation via liposomes does not appear to require participation of cell-surface receptors but is believed to result from the interaction of MDP with intracellular target sites.[9] Be-

[1] R. Evans and P. Alexander, *Nature (London)* **236,** 168 (1972).

[2] R. Keller, *J. Exp. Med.* **138,** 625 (1973).

[3] J. B. Hibbs, Jr., *J. Natl. Cancer Inst. (U.S.)* **53,** 1487 (1974).

[4] K. C. Norburg and I. J. Fidler, *J. Immunol. Methods* **7,** 109 (1975).

[5] M. S. Meltzer, R. W. Tucker, and A. C. Brever, *Cell. Immunol.* **17,** 30 (1975).

[6] C. Bucana, L. C. Hoyer, B. Hobbs, S. Breesman, M. McDaniel, and M. G. Hanna, Jr., *Cancer Res.* **36,** 4444 (1976).

[7] G. Poste, D. Papahadjopoulos, and W. J. Vail, *in* "Uses of Liposomes in Biology and Medicine" (G. Gregoriadis and A. Allison, eds.), p. 101. Wiley, London, 1979.

[8] S. Sone and I. J. Fidler, *Cell. Immunol.* **57,** 42 (1981).

[8a] Abbreviations: MDP, muramyl dipeptide; MTP-PE, muramyl tripeptide phosphatidylethanolamine; IUdR, iododeoxyuridine; CHEM, Eagle's minimum essential medium supplemented with vitamins, pyruvate, nonessential amino acids, and L-glutamine; FCS, fetal calf serum; HBSS, Hanks' balanced salt solution; ip, intraperitoneally; MLV, multilamellar vesicles; SUV, small unilamellar vesicles; REV, reverse-phase evaporation vesicles; PS, phosphatidylserine; PC, phosphatidylcholine.

[9] I. J. Fidler, A. Raz, W. E. Fogler, L. C. Hoyer, and G. Poste, *Cancer Res.* **41,** 495 (1981).

cause of the limitations of using hydrophilic muramyl peptides *in vivo*,[10] a lipophilic muramyl dipeptide derivative, muramyl tripeptide phosphatidylethanolamine (MTP-PE), incorporated into the hydrophobic core of liposomes has been employed for the efficient activation of macrophages *in vivo* as well as under defined *in vitro* conditions.[11,12]

In this chapter, we describe a microcytotoxicity assay that measures the tumoricidal activity of macrophages activated by liposome-encapsulated muramyl peptides against tumor target cells prelabeled with [^{125}I]iododeoxyuridine ([^{125}I]IUdR). In addition, we discuss the advantages and disadvantages of hydrophilic and lipophilic muramyl peptide derivatives for activation of macrophage cytotoxic properties and the techniques employed to generate appropriate phospholipid vesicles to achieve macrophage activation. Finally, the procedure to assess macrophage activation is described.

Materials

Collection of Mouse Peritoneal Exudate Cells (PEC)

Thioglycolate (BBL, Cockeysville, MD)

Eagle's minimum essential medium supplemented with vitamin solution, sodium pyruvate, nonessential amino acids, and L-glutamine (CHEM) (Flow Laboratories, Rockville, MD)

Fetal calf serum (FCS)

Ca^{2+},Mg^{2+}-free Hanks' balanced salt solution (HBSS)

Wescodyne, iodine solution (West Chemical Products, New York, NY)

70% Ethanol

50-ml centrifuge tube, polypropylene (Falcon)

Micro Test II plates, 96-well, 38-mm^2 surface area (Falcon Plastics, Oxnard, CA)

Preparation of Liposomes

Chromatographically pure lipids: egg phosphatidylcholine, bovine brain phosphatidylserine (Avanti Biochemicals, Birmingham, AL)

Rotary vacuum evaporator (Buchi Rotavapor R or equivalent)

Nitrogen cylinder equipped with reducing valve

Organic solvents: chloroform, methanol

---

[10] M. Parant, F. Parant, L. Chedid, A. Yapo, J. F. Petit, and E. Lederer, *Int. J. Immunopharmacol.* **1**, 35 (1979).

[11] I. J. Fidler, S. Sone, W. E. Fogler, D. Smith, D. G. Braun, L. Tarcsay, R. H. Gisler, and A. J. Schroit, *J. Biol. Response Modif.* **1**, 43 (1982).

[12] A. J. Schroit, E. Galligioni, and I. J. Fidler, *Cell Biol.* **47**, 87 (1983).

Activation of Macrophages

Macrophages from desired species and body sites
CHEM plus 5% FCS
HBSS (Ca^{2+},Mg^{2+} free)
N-Acetylmuramyl-L-alanyl-D-isoglutamine (muramyl dipeptide, MDP)
(Ciba-Geigy, Ltd., Basel, Switzerland)
[N-Acetylmuramyl-L-alanyl-D-isoglutamine-L-alanyl-2-(1′,2′-dipalmi-
toyl)]phosphatidylethanolamine (muramyl tripeptide phosphatidyl-
ethanolamine, MTP-PE) (Ciba-Geigy, Ltd., Basel, Switzerland)

Assay of Macrophage Tumoricidal Activity

Macrophage population exposed to liposomes containing muramyl pep-
tides and control macrophages exposed to liposomes without mura-
myl peptides
Eagle's CMEM plus 5% FCS
[^{125}I]IUdR-labeled target cells (neoplastic and nonneoplastic)
0.1 N NaOH
Cotton swabs
Gamma counter

Procedures (see also this series, Vol. 108 [25] and [27])

Preparation of Mouse Peritoneal Macrophages

The methods for isolation of macrophages and for the *in vitro* culture
of monocytes and macrophages have already been outlined in this volume
([8] and [9]). In order to minimize the variables in the cytotoxicity assay,
isolation of mouse peritoneal macrophages is briefly described below.

Specific pathogen-free mice of strain B6C3F1 (NCI-Frederick Cancer
Research Facility's Animal Production Area) are injected ip with endotoxin-free thioglycolate (1 ml/10 g body wt) and killed by chloroform
euthanasia 4 days later. The mice are first rinsed in Wescodyne, and then
in 70% ethanol and pinned to a dissecting board. The abdominal skin is
reflected and 5–8 ml sterile Ca^{2+},Mg^{2+}-free HBSS is injected ip (25-gauge
needle).

Following gentle massage of the abdomen, the peritoneal fluid is aspirated (20-gauge needle), and the cell suspension is centrifuged (250 g, 7
min). The pellet is resuspended in serum-free Eagle's CMEM at a concentration of 5 × 10^5 macrophages/ml. The cells are plated into a 96-well
Micro Test II plates at a concentration of 10^5 cells/well (0.2 ml). Nonadherent cells (generally 10% of the total cell population) are removed by

PROPERTIES OF DIFFERENT LIPID VESICLES

Type	Size (diameter) (nm)	Trap volume (μl/μmol lipid)	Lamellarity
MLVs	50–250	2–2.5	Oligolamellar[12]
SUVs	25	<1	Unilamellar[13]
REVs	200	2–4	Oligolamellar[13]

washing with CMEM 60 min after the initial plating. For *in vitro* activation, macrophages are incubated with either empty liposomes or liposomes containing muramyl peptides in CMEM containing 5% FCS for 24 hr at 37° in a humidified atmosphere containing 5% CO_2.

Preparation of Liposomes Containing Muramyl Peptides

Types of Liposomes. Activation of macrophages by liposomes containing muramyl peptides can be achieved using three different types of liposomes: multilamellar vesicles (MLVs), small unilamellar vesicles (SUVs), and large unilamellar/oligolamellar vesicles produced by reverse-phase evaporation (REVs). The salient points in preparing different vesicles has been covered in great detail elsewhere.[13] The table summarizes some of the properties of the three types of vesicles described above.

MLVs form spontaneously when the amphipathic molecules such as phospholipids are dispersed in aqueous solutions.[3] The standard method for generating MLVs simply involves hydrating a dry lipid film on a glass test tube or a lyophilized phospholipid mixture by vortex mixing above the gel-to-liquid crystalline transition temperature of the lipid.[14] MLV liposomes are highly heterogeneous in size (50–250 nm in diameter) and oligolamellar. They have a relatively low encapsulated aqueous volume (2–2.5 μl/μmol lipid) and are obviously ideal candidates for entrapping lipophilic compounds.

Extensive sonication of MLVs results in the formation of smaller unilamellar vesicles (SUVs).[15] These vesicles have several disadvantages and are not recommended for this assay. Some of the disadvantages of using SUVs include the high radius of curvature of SUVs, which can lead to an asymmetric distribution of phospholipids between the inner and outer monolayers, and the drastic limitation in terms of the encapsulation of aqueous space per mole lipid: encapsulation efficiency is usually in the

---

[13] F. Szoka, Jr., and D. Papahadjopoulos, *Annu. Rev. Biophys. Bioeng.* **9**, 467 (1980).
[14] A. D. Bangham, M. M. Standish, and J. C. Watkins, *J. Mol. Biol.* **13**, 238 (1965).
[15] L. Saunders, J. Perrin, and D. B. Gammack, *J. Pharm. Pharmacol.* **14**, 567 (1962).

range of 0.1–1% and depends on the lipid composition. The most important attribute of SUVs is that they form a small homogeneous population of vesicles ~25 nm in diameter.

The reverse-phase evaporation vesicles (REVs) are formed from water-in-oil emulsions of phospholipids and buffer in an excess organic phase, followed by removal of the organic phase (diethyl ether, isopropyl ether, or mixtures of chloroform and isopropyl ether) under reduced pressure.[16,17] The REVs formed by this technique are a mixture of large unilamellar and oligolamellar vesicles and have an advantage of encapsulating a high percentage of the initial aqueous phase. The principal disadvantages of this method are (1) exposure of the material to be encapsulated to organic solvents and the required short periods of sonication (conditions that may denature unstable compounds) and (2) limited solubility of certain lipids (e.g., cholesterol, phosphatidylserine) in ether or ethanol, which requires further modification of techniques.

In the macrophage cytotoxicity assay outlined in this chapter, monolayer cultures are employed in the activation protocol. The effect of liposome structure and lipid composition on the activation of the tumoricidal properties of macrophages by liposomes containing muramyl peptides has been extensively investigated in order to identify the optimal type of liposomes for delivery of the macrophage-activating agents (MDP and MTP-PE) to macrophages.[12,18] The inclusion of phosphatidylserine (PS) in phosphatidylcholine (PC) liposomes at the molar ratio of 7(PC) : 3(PS) enhanced phagocytosis over that of pure PC liposomes.[18] Although incorporation of PS brought about significant enhancement in liposome permeability in the presence of serum, PS, however, was a prerequisite for the delivery of liposomes to the lungs following iv injection into mice.[18] With regard to the optimum type of liposomes, MLVs were found to be superior to LUVs or REVs containing equal amounts of lipid and entrapped muramyl peptides for the tumoricidal activation of macrophages on monolayer cultures.[12,18] This superiority can probably be attributed to a slow sustained release of liposome-encapsulated immunomodulating agents in phagocytosed macrophages.

Hydrophilic versus Lipophilic Muramyl Peptides for Activation of Cytotoxic Properties in Macrophages. Hydrophilic MDP has been shown to influence several macrophage functions *in vitro,* but no comparable effects have been observed *in vivo.* This is probably because of the half-life of MDP *in vivo* (<30 min), which is insufficient to render macrophages

[16] F. Szoka, Jr., and D. Papahadjopoulos, *Proc. Natl. Acad. Sci. U.S.A.* **75,** 4194 (1978).
[17] F. Olson, C. A. Hunt, F. C. Szoka, Jr., W. J. Vail, and D. Papahadjopoulos, *Biochim. Biophys. Acta* **557,** 9 (1979).
[18] A. J. Schroit and I. J. Fidler, *Cancer Res.* **42,** 161 (1982).

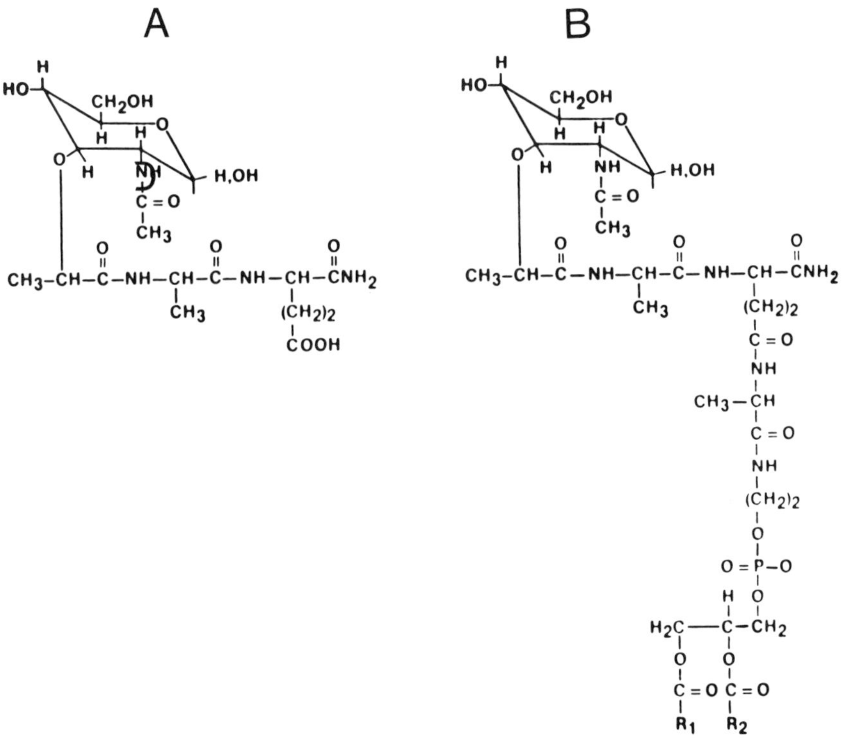

A

B

HO—

CH_2OH

H H O

O H NH — H,OH

C = O

CH₃

O O O

CH₃–CH–C–NH–CH–C–NH–CH–CNH₂

CH₃ (CH₂)₂

COOH

HO—

CH_2OH

H H O

O H NH — H,OH

C = O

CH₃

O O O

CH₃–CH–C–NH–CH–C–NH–CH–CNH₂

CH₃ (CH₂)₂

C = O

NH

CH₃–CH

C = O

NH

(CH₂)₂

O

O = P–O

H O

H₂C—C–CH₂

O O

C = O C = O

R₁ R₂

FIG. 1. Structure of (A) muramyl dipeptide and (B) muramyl tripeptide phosphatidyl-ethanolamine (R₁ and R₂ are fatty acid side chains).

tumoricidal.[10] The use of liposomes as carrier vehicles for MDP *in vitro* and *in vivo* has been reported as an efficient mechanism by which tumoricidal properties in macrophages can be activated. For example, delivery of liposome-encapsulated MDP to macrophages *in vivo* and *in vitro* activated macrophages for up to 3 days. In addition to rendering macrophages tumoricidal, it is equally important to sustain the tumoricidal state of macrophages beyond 3 days. As mentioned above, although inclusion of PS in liposomes is necessary for selective delivery to macrophages and increased localization in the lung, low-molecular-weight solutes such as MDP readily leak out of such liposomes. Clearly, the use of lipophilic derivatives of MDP would, therefore, increase the efficacy of liposomes for activating the tumoricidal properties of macrophages.

The lipophilic MDP derivative used in our laboratory is MTP-PE. The structures of MDP and MTP-PE are illustrated in Fig. 1. MTP-PE has several clear advantages over its water-soluble counterpart. The two acyl chains of MTP-PE ensure incorporation of the molecule into the hydrophobic core of the phospholipid bilayers of MLVs and thereby override any of the leakage problems associated with the smaller water-soluble MDP molecules. This property has been shown to lead to consistent and reproducible activation of macrophages. Furthermore, the degradation of MTP-PE incorporated into the liposomal phospholipid bilayer appears to be relatively slow and inefficient, which maintains tumoricidal activity in macrophages for longer periods of time (5 days) than do liposomes containing MDP. Finally, on a molar basis, MTP-PE is approximately 2.5 times more efficient than MDP in eliciting comparable levels of macrophage tumoricidal activity. Similar results have also been observed for *in vivo* and *in situ* activation of alveolar macrophages.

Liposome Preparation. MLVs are the liposomes of choice to achieve tumoricidal activation of macrophages. The phospholipid composition required for optimal phagocytosis of the MLVs is egg PC : bovine brain PS at the mole ratio of 7 : 3. The two types of MLVs employed are those encapsulated with the water-soluble MDP and those with the lipophilic MTP-PE incorporated directly into the phospholipid bilayers of the MLVs. Methods for preparing both types of liposomes are briefly described below.

Encapsulation of MDP within MLV is accomplished as follows. The appropriate amounts of phospholipids (typically 5–10 μmol) are mixed in chloroform, dried by rotary evaporation to a thin film, and placed under vacuum for 2 hr to remove any residual solvents. The lipid is hydrated in Ca^{2+},Mg^{2+}-free HBSS containing 100 μg MDP/ml by mechanical agitation on a vortex mixer. It should be noted that all reagents and media must be free of endotoxins. Absence of endotoxin contamination is ascertained by the *Limulus* amebocyte lysate assay (detection limit <0.1 ng/ml; Associate of Cape Cod, Inc.). The unencapsulated MDP is removed by centrifuging the MLVs at 30,000 g for 10 min. The MLV pellet is resuspended in HBSS and centrifuged once more to remove any remaining free MDP. For the *in vitro* cytotoxicity assay the MLV pellet is resuspended in an appropriate volume of CMEM containing 5% FCS to give a final lipid concentration of 500 μM. Liposomes are added to the macrophage cultures at a concentration of 100 nmol phospholipid/10^5 cells/0.2 ml. This dose has been found to be nontoxic to the macrophages and the target cells employed in the assay. For *in vivo* experiments the MLVs are resuspended in Ca^{2+},Mg^{2+}-free HBSS at 12.5 mM. Typically, 2.5 μmol phospholipid (0.2 ml) can be injected into the tail vein of a 20-g mouse using a 27-gauge needle.

Liposomes containing the lipophilic muramyl dipeptide are prepared as follows. MTP-PE is dissolved in methanol : chloroform (1 : 2, v/v) and added to the phospholipids in chloroform. The sample is dried to a thin film by rotary evaporation, and the MLVs are prepared as described above. More than 99% of the MTP-PE is vesicle associated and therefore the need for centrifuging the MLVs is eliminated. Typically, 4 μg MTP-PE/μmol phospholipid is incorporated for the *in vitro* and *in vivo* experiments.

In order to prevent lipid degradation, it is best to prepare liposomes on an "as needed" basis. Unfortunately, relatively little work has been done concerning the stability of liposomes during long-term storage. If liposomes need to be stored, it is best to store them in sterile conditions under nitrogen at 4°.[12]

In Vitro Activation of Macrophages by Liposome-Encapsulated Muramyl Peptides

Macrophages are incubated in medium alone, medium containing free MDP, or liposome-encapsulated MDP or MTP-PE for a period of 24 hr. Other controls can include liposomes containing entrapped HBSS and free MDP, so-called empty liposomes. After the 24-hr incubation period, the cultures are washed twice with warm CMEM containing 5% FCS, and [125I]IUdR-labeled target cells are added as described below.

In Vitro Assay of Macrophage-Mediated Cytotoxicity

Macrophage-mediated cytotoxicity is assessed by a radioactive release assay, which is illustrated in Fig. 2. Target cells in their exponential growth phase are incubated for 24 hr in medium supplemented with 0.2 μCi/ml [125I]IUdR (specific activity 200 mCi/mol; New England Nuclear, Boston, MA). The cells are then washed four times with warm HBSS to remove unincorporated radiolabel, harvested by a short trypsinization period (0.25% trypsin and 0.02% EDTA for 1 min at 37°), and resuspended in CMEM containing 5% FCS. Target cells (10^4) are added to each well (38 mm²) to obtain an initial macrophage-to-target cell ratio of 10:1. At this population density, normal (untreated) macrophages are not cytotoxic to neoplastic cells, whereas activated macrophages are. By 8 hr after seeding, no significant differences should be detected in the plating efficiency of [125I]-labeled target cells to the plastic or to macrophage monolayers (normal or activated). Radiolabeled target cells are also plated alone as an additional control group. The cultures are washed 24 hr after the addition of the tumor cells to remove all the nonadherent target cells and refed with fresh CMEM containing 5% FCS. This procedure permits an accurate determination of actual macrophage-mediated lysis as opposed to spontaneous target cell lysis (loss of radiolabel), since macrophage killing proceeds 24 hr after plating and reaches a maximum by 72 hr of cocultivation. After 72 hr cocultivation, the cultures are washed three

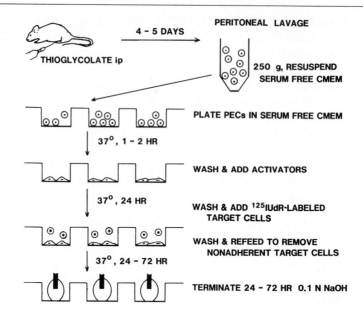

$$\% \text{ CYTOTOXICITY} = 100 \times \frac{\text{CPM OF TARGET CELLS} \atop \text{WITH NORMAL M}\Phi \quad - \quad \text{CPM OF TARGET CELLS} \atop \text{WITH TEST M}\Phi}{\text{CPM OF TARGET CELLS} \atop \text{WITH NORMAL M}\Phi}$$

FIG. 2. Preparation and assay of macrophage-mediated cytotoxicity.

times with HBSS, and the adherent (viable) cells lysed with 100 μl 0.1 N NaOH. The lysate is absorbed onto cotton swabs (Q-tips), and these are placed directly into 12 × 75 mm tubes and monitored for radioactivity in a gamma counter. The reproducibility of radiolabel recovery by the cotton swab method is good ($\pm 5\%$) and similar to that obtained by more conventional methods of collecting supernatants by fluid suction.

The percentage macrophage-mediated cytotoxicity is calculated as follows: Percentage cytotoxicity = 100 [(cpm in target cells and normal macrophages) − (cpm in target cells and test macrophages)/(cpm in target cells and normal macrophages)]. The results are analyzed for statistical significance by Student's two-tailed t-test.

Concluding Remarks

Muramyl peptides encapsulated into liposomes can render macrophages tumoricidal more efficiently than the free, unencapsulated agents. The amount of liposome-encapsulated MDP required for macrophage activation is ~4000 times lower than the amount of free MDP needed to render macrophages tumoricidal. finally, after the initial activation period, the liposome encapsulation of MDP (or MTP-PE) prolongs the duration of tumoricidal properties in the macrophages. This is probably because once endocytosed, MLVs function as "slow release depots" from which encapsulated material is released over an extended period of time.

[43] Intracellular Parasitism of Macrophages in Leishmaniasis: In Vitro Systems and Their Applications

By K.-P. CHANG, CAROL ANN NACY, and RICHARD D. PEARSON

The trypanosomatid protozoa of the genus *Leishmania* cause leishmaniases—chronic, often debilitating diseases of man and other animals. The females of the phlebotomine sandfly serve as the natural vector. In the transmission cycle, there are two stages of the parasites: the vector stage (promastigote) lives extracellularly in the digestive tract of the fly mostly as a motile cell with a long flagellum; and the mammalian stage (amastigote) is nonmotile and lives intracellularly in mononuclear phagocytes of the susceptible animals. The capacity of leishmanias to live in such diverse environments entails their evolutionary adaptation in the past, and indicates the present complexity of their interactions with the host. The transition of the extracellular promastigote into the intracellular amastigote and the proliferation of the latter in macrophages culminate in the clinical pathology and symptoms associated with various forms of leishmaniases.

At risk of leishmanial infection are millions of humans on all continents except, perhaps, Australia.[1] Clinical manifestations of leishmaniases often vary with the individual hosts and with the leishmanial species involved.[2] The diseases may be manifested as self-healing simple cutaneous lesions, disseminating cutaneous nodules, facial disfiguring mucocutaneous infection, and the potentially fatal visceral form. Clinical entities intermediate between these forms also exist. Much needed are new approaches for effective chemotherapy and immunoprophylaxis of leishmaniases.

Equally important are a number of unresolved biological issues which

[1] Z. Zuckerman and R. Lainson, in "Parasitic Protozoa" (J. P. Kreier, ed.), Vol. 1, p. 58. Academic Press, New York, 1977.
[2] P. D. Marsden, *N. Engl. J. Med.* **300**, 350 (1979).

METHODS IN ENZYMOLOGY, VOL. 132

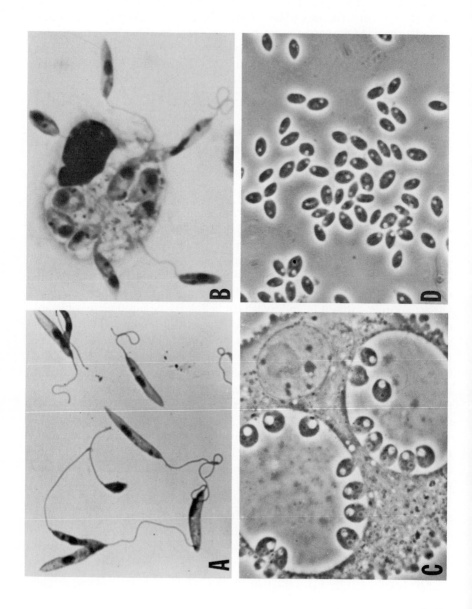

should be mentioned in any chapter concerning leishmanias. It has become increasingly evident in recent years that leishmanias and leishmaniases are much more complicated and diverse than previously thought.[3] There are some 20 named species/subspecies of *Leishmania*. Laboratory studies of these parasites and the field work of their interactions with man, reservoir animals, and sandfly vectors are far from complete. There are apparently still more leishmanias which have yet to be isolated and identified. In addition, sexual recombination may occur in leishmanias, as has been reported for the closely related *Trypanosoma*. It is important to remember these elements of disparity and uncertainty in our search for a common theme among leishmanias as a group in all areas under investigation, including the subject of this chapter.

A most striking feature common to all *Leishmania* species studied so far is their exclusive parasitization of phagolysosomal compartment of the mammalian macrophages.[4] This is the very intracellular site where macrophages kill invading microorganisms and dispose of unwanted materials. Figure 1 shows the appearance of leishmanias and their infection of macrophages *in vitro*. How leishmanias can take residence, differentiate, survive, and multiply in such an inhospitable environment has been a baffling question challenging all laboratory investigators.

Macrophages serve a dual role in leishmaniases. They are the host cells of the parasite and, at the same time, the immune effector or regulatory cells of the infected animals. One consequence of this, aside from its biological interest, is the convergence of investigators from diverse disci-

---

[3] K.-P. Chang, D. Fong, and R. S. Bray, *in* "Leishmaniasis" (K.-P. Chang and R. S. Bray, eds.), Chapter 1, p. 1. Elsevier Biomedical Press, Amsterdam, 1985.
[4] K.-P. Chang, *Int. Rev. Cytol., Suppl.* **14**, 267 (1983).

FIG. 1. (A) Leishmanial promastigotes grown in medium 199 plus 25 mM HEPES, pH 7.4, and 10% heat-inactivated fetal bovine serum. Promastigotes are slender cells of variable dimension, each with an anterior flagellum as long as or longer than the cell body. Each cell has a centrally located nucleus, above which is a kinetoplast or mitochondrial DNA network. Giemsa stain. (B) Early infection of a hamster peritoneal macrophage with leishmanial promastigotes for 30 min at 37°. Note that some promastigotes become adhered to the macrophage and others are phagocytized into the cell. Giemsa stain. (C) A mouse macrophage of the J774G8 line infected with promastigotes of *Leishmania mexicana amazonensis* (LV78) for 9 days. At this time, promastigotes have already differentiated into amastigotes in parasite-containing vacuoles, which are especially large in infection by this *Leishmania* species. Intracellular differentiation of leishmanias is marked by a reduction in the length of flagella and of the cell body. Phase-contrast microscopy of fresh material. (D) Amastigotes of *L. m. amazonensis* purified from infected macrophages as shown in (C) by cell breakage followed by centrifugation in Percoll gradient as described.[13] Phase-contrast microscopy of fresh material.

plines to the use of *Leishmania*–macrophage *in vitro* systems for different research objectives. Such *in vitro* systems have been used to study: (1) the cellular and molecular mechanisms of infection[4,5]; (2) microbicidal activities of macrophages under the influence of immune factors[6,7]; (3) efficacy of potential antileishmanial drugs[8,9]; and (4) maintenance and production of leishmanial amastigotes *in vitro*.[10]

We present in this chapter several *Leishmania*–macrophage *in vitro* systems which have been used with a certain measure of success in an attempt to answer some of the questions posed. Sufficient technical details are given in the hope that these systems can be exploited and improved to broaden the scope of investigation in these and other areas.

Acquisition, Cultivation, and Preparation of Leishmanias

There are many species and subspecies of *Leishmania* which one can obtain from individual investigators in the field. A large number of leishmanial isolates are available in a number of institutions, e.g., (1) Department of Parasitology, Division of Experimental Therapeutics, Walter Reed Army Institute of Research, Washington, DC; (2) Department of Medical Protozoology, Liverpool School of Tropical Medicine, Liverpool, UK; and (3) The Sandford F. Kuvin Center for the Study of Infectious and Tropical Diseases, Hebrew University Hadassah Medical School, Jerusalem, Israel. The U.S. federal regulations require a permit from the Centers for Disease Control for shipping leishmanias by mail. Leishmanias are classified as agents of moderate potential hazard at biosafety level II and should be treated accordingly.

The general methodology for the maintenance of leishmanias in culture and in laboratory animals has been recently reviewed.[11] With few exceptions, promastigotes of all species can be grown quite readily under axenic conditions in undefined or semidefined media. Some species can be adapted to grow in totally defined macromolecule-free media. However, most species lose their infectivity for macrophages *in vitro* or viru-

[5] R. D. Pearson, D. A. Wheeler, L. H. Harrison, and H. D. Kay, *Rev. Infect. Dis.* **5**, 907 (1983).
[6] J. Mauel, *Parasitology* **88**, 579 (1984).
[7] C. A. Nacy, C. N. Oster, S. L. James, and M. S. Meltzer, *Contemp. Top. Immunol.* **13**, 147 (1984).
[8] J. D. Berman and D. J. Wyler, *J. Infect. Dis.* **142**, 83 (1980).
[9] R. A. Neal and P. J. Mathews, *Trans. R. Soc. Trop. Med. Hyg.* **76**, 284 (1982).
[10] K.-P. Chang and W. R. Fish, *in* "*In Vitro* Cultivation of Protozoan Parasites" (J. B. Jensen, ed.), p. 111. CRC Press, Boca Raton, Florida, 1983.
[11] K.-P. Chang and L. D. Hendricks, *in* "Leishmaniasis" (K.-P. Chang and R. S. Bray, eds.), Chapter 13, p. 214. Elsevier Biomedical Press, Amsterdam, 1985.

lence for laboratory animals *in vivo* after repeated passages as promastigotes. There are several ways to circumvent these difficulties, e.g., cryopreservation of infective promastigotes and the use of amastigotes or freshly differentiated promastigotes. Most species of *Leishmania* can be passaged as amastigotes in laboratory animals. Detailed methodology for preparing the *Leishmania* species used by the authors of this chapter are given below.

Leishmania mexicana amazonensis (LV78, M1845)

This species is known to cause simple cutaneous, diffuse cutaneous, or, on rare occasion, mucocutaneous leishmaniasis of humans in South and Central America. The particular isolate (LV78) used by one of us (KPC) was originally isolated from a wild rat by Lainson and Shaw.[12]

Promastigotes are passaged weekly at $27 \pm 2°$ in medium 199 plus 10% heat-inactivated fetal bovine serum (HIFBS)[12a] (Characterized, Sterile System Company, Logan, Utah) and 25 mM HEPES, pH 7.4. One drop of the 1-week-old culture is transferred to 2 ml of the complete medium in a screw-capped plastic culture tube. To expand the culture, aliquots of the medium (4, 15, or 100 ml) are dispensed into tissue culture flasks (25, 75, or 175 cm^2) and inoculated with promastigotes. With a starting inoculum of 10^6 cells/ml, the culture reaches late-log or stationary phase of growth in about 3 days, giving a cell density of $5–10 \times 10^7$/ml.

Amastigotes of this species are passaged in BALB/c mice (25–30 g) under ether anesthesia by subcutaneous inoculation of 5×10^7 leishmanias/0.1 ml Hanks' balanced salt solution (HBSS)/mouse into the tail base. In several months, a histiocytoma develops and contains numerous parasitized macrophages. Closed lesions free from secondary infections may be obtained by caging mice individually. After prolonged infection, lesions invariably become necrotic, often with secondary bacterial infections. Dissemination of the parasites to other cutaneous and/or visceral sites may occur.

For the isolation of amastigotes, the histiocytoma of infected mice can be excised in its entirety and subjected to the isolation procedure developed for use with *in vitro* infected macrophages.[13] Each histiocytoma

---

[12] R. Killick-Kendrick, D. H. Molyneux, and R. W. Ashford, *Proc. R. Soc. London, Ser. B* **189**, 409 (1976).

[12a] Abbreviations: HIFBS, heat-inactivated fetal bovine serum; HBSS, Hanks' balanced salt solution; HO-MEM, modified minimal essential medium; PBS, phosphate-buffered saline; LK, lymphokines; Con A, concanavalin A; BCG, bacillus Calmette-Guerin; PPD, purified protein derivative of tuberculin.

[13] K.-P. Chang, *Science* **209**, 1240 (1980).

(with hairy skin removed) is ground in 10 ml phosphate-buffered saline (PBS), pH 7.5, containing 2 mM EDTA using a Ten Broeck-type glass homogenizer. The homogenate is allowed to stand for 5 min on ice, and the supernatant is then centrifuged at 3500 g for 5 min to sediment amastigotes and unbroken host cells. The pellet is resuspended in PBS–EDTA and the suspension is vortexed vigorously for 1 min to break infected macrophages. We then repeat the centrifugation at 3500 g for 5 min and the vortex step several times and/or passage the suspension through a 27-gauge needle. The suspension (10 ml) is centrifuged at 50 g for 1 min to remove the remaining intact host cells and at 3500 g for 5 min to pellet the amastigotes. We then resuspend the pellet in 5–7 ml 45% Percoll in PBS, introduce 0.5 ml 90% Percoll to the bottom of the tube with a Pasteur pipet, and centrifuge the tube at 3500 g for 30 min at 4°. After centrifugation, the top half of the 45% Percoll gradient containing broken host cells is discarded by aspiration. The layer of amastigotes at the interface of the gradients is collected and pooled. The preparation is washed with PBS–EDTA by centrifugation at 3500 g for 10 min and the pellet is resuspended in suitable medium for counting in a hemocytometer. Amastigotes isolated from *in vitro* infected macrophages are free from the contamination of host cells and cell debris. Those from the lesion (10^9–10^{10} amastigotes/lesion) are contaminated with a small number of host cells, e.g., lymphocytes and infected macrophages.

Leishmania donovani (Sudan Strain 3S)

This species is the agent of kala azar or human visceral leishmaniasis. Sudanese strains have been used widely by many U.S. investigators. The strains were originally isolated by Dr. L. Stauber in the Sudan and obtained from Dr. R. Herman, Rutgers University, New Brunswick, NJ, by one of us (RDP). The amastigotes have been maintained by serial passages in Syrian hamsters (*Mesocricetus auratus*) in the laboratory of RDP since 1978.

Amastigotes are isolated from spleens of animals infected 6–8 weeks previously. Hamsters are anesthetized and exsanguinated before splenectomy. Fulton and Joyner,[14] in 1949, used differential centrifugation of homogenized spleen for the isolation of *L. donovani* amastigotes. This has been adopted and modified for current use by different workers, and one version of these modifications used by one of us (RDP) is given below. Spleens are aseptically removed and ground in a tissue grinder (Tri-R Instruments, Rockville Center, NY) in sterile PBS. The homogenate is centrifuged at 150 g for 20 min to remove debris. The supernatant is then

[14] J. D. Fulton and L. D. Joyner, *Trans. R. Soc. Trop. Med. Hyg.* **43**, 273 (1949).

centrifuged at 1600 g for 20 min, and the pellet of amastigotes is resuspended and washed twice in PBS. Amastigotes can be applied to a column of CF-11 Whatman cellulose (Whatman, Inc., Clifton, NY) (sterilized by autoclaving) as described by Brazil.[15] The final pellet is suspended in a modification of the growth medium (HO-MEM) of Berens et al.[16] in which minimal essential medium (MEM) with HBSS (M. A. Bioproducts, Walkersville, MD) is used in place of MEM with Spinner salts, and hemin (20 μg/ml in 1 N NaOH) is added after, rather than before, filter sterilization. The concentration of amastigotes is determined by counting in a bacterial counting chamber (Petroff-Hausser).

Promastigotes are obtained by suspending amastigotes at 2–5 × 10⁵ cells/ml in 50-ml plastic tubes in 25-ml aliquots of modified HO-MEM, pH 7.2 to 7.4, with 10% HIFBS (Flow Laboratories, Rockville, MD). Penicillin, 100 U/ml, and gentamicin, 50 μg/ml, are added. Tubes are filled with 5% CO_2 in room air, sealed, and maintained at 26°. Promastigotes have also been used after one or two passages into fresh HO-MEM at 7- to 10-day intervals. At the time of study, promastigotes are concentrated by centrifugation at 1600 g for 15 min, washed twice in HBSS, and resuspended in medium 199.

Leishmania major (NIH Strain 173)

This species of *Leishmania* causes the simple cutaneous leishmaniasis or Oriental Sore of humans in the Old World. Most of these cutaneous lesions heal in months to years without treatment; only rarely is there systemic involvement or metastatic spread of organisms. The strain used by one of us (CAN) was isolated from a patient in Iran and given to the Laboratory of Parasitic Diseases, NIAID, NIH, by Dr. A. Ebrahimzadeh, Shapur University School of Medicine, Ahwaz, Iran.

BALB/cJ mice develop exceptionally large cutaneous lesions within 4 weeks of parasite inoculation, and are used for passage of amastigotes. Five to 10 × 10⁴ amastigotes are inoculated into the footpad tissue of an etherized animal, and the infected feet are used for amastigote harvest after 3 to 4 weeks. Feet are removed from a mouse killed by ether overdose, and are soaked in two successive 70% alcohol baths. Infected footpad tissue is disrupted by passage through a #50 mesh stainless steel screen into RPMI 1640 (GIBCO, Grand Island, NY) containing 10% HIFBS (Sterile System, Inc., Logan, UT) and 50 μg/ml gentamicin (MA Bioproducts, Rockville, MD) (assay medium). The resulting cell suspen-

---

[15] R. P. Brazil, *Ann. Trop. Med. Parasitol.* **72**, 579 (1978).
[16] R. L. Berens, R. Brun, and S. M. Krassner, *J. Parasitol.* **62**, 360 (1976).

sion is disrupted in a Ten Broeck tissue homogenizer to release amastigotes from infected macrophages, and centrifuged at 200 g for 10 min to remove cell debris. The supernatant is further centrifuged at 400 g for 10 min. The amastigote pellet is then suspended in a small volume of assay medium for counting. For quantitation, an aliquot of parasites is stained with 1% toluidine blue O and counted in a hemacytometer. Recovery of amastigotes ranges from 1 to 3×10^7 amastigotes per infected BALB/cJ foot.

Sources and Preparation of Macrophages

The outcome of leishmanial infection of macrophages *in vitro* varies not only with the parasite stage, species, and strain, but also with the tissue source and culture condition of the host cell. This may account in part for different results reported by investigators using similar populations of macrophages. In this section, we describe three different cell sources and culture systems in use in our laboratories.

J774G8 Macrophage Line. This cell line (J774) was originally derived from the peritoneal cavity of BALB/c mice.[17] Cells of this line are known to possess a number of characteristics typical of murine macrophages.[18] The clone, J774G8, was developed at the Rockefeller University, New York, NY, by Dr. J. Unkeless. Since 1979, these cells have been routinely grown at 35–37° in RPMI 1640 plus 10% HIFBS and 50 mM HEPES, pH 7.4, in 25-cm² tissue culture flasks. A regular incubator of constant temperature without the facility of humidification and 5% CO_2 atmosphere is adequate for routine culture of these cells in tightly closed flasks. Cells in culture spread out during the first 1–2 days. They subsequently become loosely adherent and can be easily dislodged for experimental use. Cell density reaches $2–3 \times 10^6$/ml in 3–4 days under the culture conditions described. We normally replace the spent medium for routine maintenance; cultures are split into several flasks after the addition of 2–3 vol of fresh medium for expansion.

Murine Peritoneal Macrophages. The use of specific pathogen-free mice is recommended to avoid subclinical viral and bacterial infections of animals that may influence macrophage functions.[19] Macrophages from most mouse strains can be used to analyze host–parasite interactions *in vitro;* however, certain mouse strains (C3H/HeJ, C57BL/10SnCr, A/J,

[17] P. Ralph, J. Prichard, and M. Cohn, *J. Immunol.* **114,** 898 (1975).
[18] P. S. Morahan, *J. Reticuloendothel. Soc.* **27,** 223 (1980).
[19] M. S. Meltzer, *in* "Methods for Studying Mononuclear Phagocytes" (D. O. Adams, P. J. Edelson, and H. S. Koren, eds.), p. 133. Academic Press, New York, 1981.

P/J, and BALB/cJ) have genetically controlled defects in macrophage activation and may be unsuitable for analysis of some immune reactions. The strain we use for analysis of macrophage activation is C3H/HeN, obtained from Sprague-Dawley, Frederick, MD.

   Resident peritoneal macrophages (see this series, Vol. 108 [25]): Cells are obtained from individual inbred mice by peritoneal lavage with 8–10 ml of assay medium supplemented with 10 U/ml sodium heparin (Abbott Laboratories, King of Prussia, PA).[20] Peritoneal fluids from 3 to 20 mice are pooled. Samples are removed for differential and total cell counts, and the remaining fluid is centrifuged at 250 g for 10 min at 4°. Differential cell counts are made on Wright-stained cell smears (Diff-Quick, Dade Diagnotics, Aquado, PR) prepared by cytocentrifugation (Cytospin centrifuge, Shandon Southern Instruments, Camberly, England). The differential of resident cell populations from healthy mice show 50 ± 5% macrophages, 45 ± 5% lymphocytes, 2 ± 3% polymorphonuclear leukocytes. Total cell counts are determined in a hemacytometer: yields are between 1 and 2 × 10⁶ macrophages per mouse. Washed cell suspensions are adjusted to 1 × 10⁶ macrophages/ml in assay medium without heparin and placed in sterile capped 12 × 75 mm polypropylene tubes (#2063, Falcon Plastics, Oxnard, CA) in 0.5-ml aliquots. Cell cultures are incubated for 2 hr at 37° in 5% CO_2 in moist air before exposure to parasites. Macrophages are exposed to amastigotes as unfractionated peritoneal cell cultures (5 × 10⁵ macrophages/0.5 ml).

   Inflammatory macrophages: Macrophages derived from circulating monocytes that migrate to sites of inflammation differ in many morphological and biochemical characteristics from differentiated resident peritoneal macrophages. We elicit inflammatory macrophage populations by intraperitoneal inoculation of 1 ml sterile inflammatory agents such as 1% starch (Difco), thioglycolate broth (Difco), phosphate-buffered saline (PBS), or FBS.[21] The procedure for obtaining these cells from mice is identical to that outlined above. Differentials for inflammatory macrophage populations vary with the stimulus and the duration of inflammation: Inflammatory agents that the macrophage cannot digest (starch, thioglycolate) induce the greatest influx of cells (yields of 5 to 10 × 10⁶/ mouse), and 80 to 95% of the population are macrophages; PBS and FBS are milder inflammatory stimuli, with correspondingly lower cell yields (2 to 3 × 10⁶/mouse), and little change in percentage of macrophages over

---

[20] C. A. Nacy and M. G. Pappas, *in* "Methods for Studying Mononuclear Phagocytes" (D. O. Adams, P. J. Edelson, and H. S. Koren, eds.), p. 745. Academic Press, New York, 1981.

[21] A. H. Fortier, D. L. Hoover, and C. A. Nacy, *Infect. Immun.* **38,** 1304 (1982).

resident cell populations (55 to 60% macrophages). Inflammatory cells can be further characterized by determining the percentage of cells that contain peroxidase-positive perinuclear granules (a cytochemical marker of immature macrophages: their percentage is low in resident cells and high in inflammatory macrophages) and 5'-nucleotidase (a cell membrane marker enzyme of differentiated cells: their percentage is high in resident cells and low in inflammatory populations).[22,23] The number of peroxidase-positive cells in resident populations is less than 2% and in inflammatory macrophage populations ranges from 15% (PBS-elicited cells at 24 hr) to 95% (starch-elicited cells at 24 hr).

Activated macrophages: In vivo activated macrophages can be obtained from mice injected with 1×10^6 colony-forming units of *Mycobacterium bovis* strain BCG (Phipps substrain, Trudeau Mycobacterial Collection, Saranac Lake, NY) or formalinized *Corynebacterium parvum,* 70 mg/kg (Wellcome Laboratories, Research Triangle Park, NC) intraperitoneally 10 days earlier. Cells are collected as before. Differentials are similar to those of resident cells, but $20 \pm 5\%$ of these activated macrophages are peroxidase positive. Macrophage yields are between 2 and 4×10^6/mouse. Alternatively, resident peritoneal macrophages can be activated *in vitro* by exposure to soluble products of antigen- or mitogen-stimulated lymphocytes (lymphokines, LK).[24] LK are produced by injecting C3H/HeN mice with 10^8 viable BCG intradermally at the base of the tail. Three to 6 weeks later, spleens are aseptically removed and passed through #50 mesh stainless steel sieves into assay medium. Spleen cells treated with 0.87% NH_4Cl lysis buffer (to lyse red blood cells) are centrifuged at 250 *g* for 10 min at 4° and resuspended to a concentration of 5×10^6 viable cells/ml in assay medium with 5% HIFBS. Aliquots of 20 ml spleen cell suspension with 50 to 100 μg/ml purified protein derivative of tuberculin (PPD, Connaught Medical Laboratories, Toronto, Canada) are incubated in upright 75-cm² plastic tissue culture flasks (#3023, Falcon Plastics) for 48 hr at 37°. LK-rich cell culture fluids are stored at 4° until used; shelf life of these LK are 2 to 4 months at 4°, 6 to 12 months at −20°. Spleen cells from untreated mice cultured with 5 μg/ml concanavalin A Sepharose beads (Con A, Sigma Chemicals, St. Louis, MO) are also used as an LK source. Residual Con A in the LK is removed by adsorption with 10 mg/ml Sephadex G-10 (Pharmacia Fine Chemicals) before storage. Control supernatants consist of the culture medium from spleen cells

[22] R. van Furth, J. G. Hirsch, and M. E. Fedorko, *J. Exp. Med.* **132,** 794 (1970).
[23] P. J. Edelson, *in* "Heterogeneity of Mononuclear Phagocytes" (O. Förster and M. Landy, eds.), p. 143. Academic Press, New York, 1981.
[24] C. A. Nacy, M. S. Meltzer, E. J. Leonard, and D. J. Wyler, *J. Immunol.* **127,** 2381 (1981).

of BCG-infected animals to which PPD is added after 48 hr of incubation, or of normal spleen cells to which 5 μg/ml Con A is added after the incubation period.

Human Peripheral Blood Monocyte-Derived Macrophages. These are obtained by a modification of Bøyum's technique[25-30] (see also this series, Vol. 108 [9]). In a typical study, 120 ml of blood is drawn from an adult donor via a 19-gauge butterfly needle. Ten milliliters is allowed to clot. The serum is removed and stored at −70°. The remaining 110 ml of blood is then either defibrinated to remove platelets and clotting factors or anti-coagulated with heparin, 10 U/ml of blood. Defibrination is accomplished by the method of Kay *et al.*,[31] in which eight wooden sticks are joined at one end and then placed into a 200-ml Erlenmeyer flask. The sticks spread out at the bottom of the flask and resemble the struts of an umbrella when positioned against the glass sides. The flask and sticks are autoclaved prior to use. Blood is added and the flask swirled manually for 10 min.

After defibrination, the blood is added to 165 ml of medium 199 (M. A. Bioproducts) and layered onto Ficoll–sodium diatrizoate (Histopaque: Sigma Chemical Company, St. Louis, MO). Approximately 30 ml of blood in medium 199 is carefully layered over 12.5 ml of Histopaque in each of nine 50-ml plastic tubes (Corning Glass Works, Corning, NY). After centrifugation at 400 g for 30 min, brake off, medium 199 and plasma remain at the top, Histopaque in the center, and an erythrocyte pellet at the bottom. The mononuclear cell layer is found at the interface between the top (medium 199–plasma) layer and the Histopaque. Mononuclear cells are removed, washed twice in HBSS (Grand Island Biological Company, Grand Island, NY), and resuspended in medium 199. The mononuclear cell layer contains 4–8 × 10⁷ cells, approximately 10–20% of which are monocytes.

Monocyte monolayers are prepared by placing 2.5×10^6/ml mononuclear cells in 0.5 ml medium 199 without serum onto 22-mm² glass coverslips (Corning Glass Works) positioned in 35 × 10 mm Petri dishes (Falcon, Oxnard, CA). Monocytes are allowed to adhere for 1 hr at 37° in 5% CO_2–95% room air. Alternatively, 15-mm diameter round coverslips

[25] A. Bøyum, *Scand. J. Clin. Lab. Invest., Suppl.* **97**, 77 (1968).

[26] R. T. Steigbigel, L. H. Lambert, Jr., and J. S. Remington, *J. Clin. Invest.* **53**, 131 (1974).

[27] R. D. Pearson, R. Romito, P. H. Symes, and J. Harcus, *Infect. Immun.* **32**, 1249 (1981).

[28] R. D. Pearson and R. T. Steigbigel, *J. Immunol.* **127**, 1438 (1981).

[29] R. D. Pearson, J. L. Harcus, D. M. Roberts, and G. R. Donowitz, *J. Immunol.* **131**, 1994 (1983).

[30] R. D. Pearson, J. A. Sullivan, D. Roberts, R. Romito, and G. L. Mandell, *Infect. Immun.* **40**, 411 (1983).

[31] H. D. Kay, H. T. Petrie, J. J. Burge, and L. W. Klassen. Submitted (1986).

(Bellco Glass, Inc., Vineland, NJ) or 13-mm^2 coverslips can be prepared for use in microtiter plates. After monocytes have adhered, nonadherent cells are gently washed away with warm medium 199 (37°). The adherent monocytes can then be used for study or allowed to convert to monocyte-derived macrophages by culturing monolayers in medium 199 with 13% fresh autologous serum, penicillin, 100 U/ml, and gentamicin, 50 μg/ml, at 37° in 5% CO_2–95% room air. Berman and Wyler[8] have also cultivated their monocytes in 10% autologous plasma. On day 6, the monocyte-derived macrophages were infected with amastigotes and resuspended in medium with 10% heat-inactivated fetal calf serum. Murray and co-workers[32,33] have cultured monocytes in 15–20% heat-inactivated pooled human serum. In general, the medium is changed at 48- to 72-hr intervals. Over 5–7 days, monocytes assume macrophage characteristics.[34,35] In our studies, each 22-mm^2 coverslip contains 1.5–5.0 × 10^5 adherent mononuclear cells. More than 98% of the adherent cells have strong nonspecific esterase activity, indicating that they are macrophages.[36,37]

Infection, Intracellular Growth of Amastigotes, and Applications

For the sake of simplicity, we shall limit our discussion to three *Leishmania*–macrophage combinations under specific experimental conditions used by the authors. The methods described are representative rather than exhaustive, but they are probably applicable to all host–parasite combinations with some modifications and cover the general principal methodology currently or previously used by others.[10]

Leishmania mexicana amazonensis–J774G8 Macrophage Continuous Culture System[13]

Infection. Macrophages of the J774G8 line and promastigotes of *L. mexicana amazonensis* (at late-log to stationary phase) grown as described are separately counted in a hemacytometer and harvested by centrifugation at 4° for 5 min at 500 and 3500 *g,* respectively. It is advisable to use leishmanias that are passaged for less than 1 year or about 260 generations as promastigotes. Both pellets in separate tubes are resuspended in RPMI 1640 plus 20% HIFBS and 50 m*M* HEPES, pH 7.4, to

[32] H. W. Murray and D. M. Cartelli, *J. Infect. Dis.* **72,** 32 (1983).
[33] H. W. Murray, B. Y. Rubin, and C. D. Rothermel, *J. Clin. Invest.* **72,** 1506 (1983).
[34] W. D. Johnson, Jr., B. Mei, and Z. A. Cohn, *J. Exp. Med.* **146,** 1613 (1977).
[35] A. Nakagawara, C. F. Nathan, and Z. A. Cohn, *J. Clin. Invest.* **68,** 1243 (1981).
[36] H. Braunsteiner and F. Schmalzl, *in* "Mononuclear Phagocytes" (R. van Furth, ed.), p. 62. Davis, Philadelphia, Pennsylvania, 1970.
[37] C. Y. Li, K. W. Lam, and L. Y. Yam, *J. Histochem. Cytochem.* **21,** 1 (1973).

2×10^6/ml for macrophages and 2×10^7/ml for promastigotes. The parasite pellet should be resuspended thoroughly by vigorous agitation. Then, the two suspensions are mixed at 1 : 1 ratio (macrophage-to-promastigote ratio 1 : 10), vortexed gently, and dispensed in 4-ml aliquots into 25-cm^2 tissue culture flasks. Thus, each flask contains 4×10^6 macrophages and 4×10^7 promastigotes. Separate pipetting of individual aliquots and rocking of the flasks immediately after seeding ensure even distribution of the cells. Culture flasks should be tightly closed and incubated at 32–35°. Medium is changed every 3–4 days. Spent medium can be simply aspirated off or centrifuged to recover the small number of floating cells. Culture should be split into two when the macrophage density reaches a number greater than 10^7 per flask. In continuous culture, medium is also renewed every 3–4 days during which time it is most convenient to remove floating macrophages together with the spent medium to allow the growth of amastigotes and macrophages in the adherent monolayer. Infected macrophages recovered from the spent medium can be replated in fresh culture flasks or used for the isolation of amastigotes. From 20 cultures (25-cm^2 TC flasks), 10^8–10^9 amastigotes are routinely obtained from the floating cells every 3–4 days.

Enumeration of Intracellular Amastigotes by Phase-Contrast Microscopy. Amastigotes of *L. mexicana* are comparatively large in size (2–4 μm in diameter) and produce large parasite-containing vacuoles in macrophages (Fig. 1C). These properties of the amastigotes make it possible to estimate the number of intracellular amastigotes by directly observing fresh infected culture under phase-contrast microscopy. The infected macrophages in a flask are completely resuspended by repeated pipetting and rinsing of the monolayer. The number of macrophages is counted in a hemacytometer. A 2- to 3-μl aliquot of the cell suspension is placed on a glass slide and covered with a 22-mm^2 coverslip. The amount of the culture fluid and the size of the slip are matched so that the macrophages are flattened, but not broken. Areas of view with flattened macrophages are selected and examined under a phase-contrast oil immersion lens for tallying at least 100 infected cells and the number of their intracellular amastigotes. The number of noninfected macrophages is counted simultaneously. The total number of amastigotes per culture is determined as: (the total number of macrophages) × (percentage infected cells) × (the average number of amastigotes per macrophage).

Intracellular Growth of Amastigotes. Table I presents a typical example of the infection. Leishmanias disappear from the culture at 37° or in the absence of macrophages. At 35°, there is a net gain of two to three amastigotes per promastigote in 7–10 days in the macrophage culture, resulting from intracellular differentiation and multiplication of the para-

TABLE I
MACROPHAGE- AND TEMPERATURE-DEPENDENT INTRACELLULAR GROWTH OF
Leishmania mexicana amazonensis IN J774G8 CELLS[a]

Culture periods (days)	Temperature (°C)	Number of J774G8 ($\times 10^{-6}$)	Percentage infection	Number of parasites per cell	Total number of leishmanias ($\times 10^{-6}$)
0	35	4	—	—	40[b]
3		6.8 ± 0.2	89 ± 2.2	9.6 ± 1.5	52.6 ± 3.4
7[c]		10.1 ± 1.8	74 ± 2.0	11.5 ± 1.5	91.2 ± 13
10		24.1 ± 2.7	65 ± 1.2	7.2 ± 0.8	112.4 ± 3.7
0	35	0	—	—	40[b]
3		0	—	—	50 ± 3.9
7		0	—	—	21 ± 1.8
10		0	—	—	1 ± 0.2
0	37	4	—	—	40[b]
3		8.4 ± 1.0	42 ± 1.8	5.3 ± 1.9	18.6 ± 2.1
7		12.7 ± 1.2	5 ± 0.8	1.4 ± 0.2	0.9 ± 0.2

[a] Promastigotes (4×10^7) and macrophages (4×10^6) are mixed at 10:1 ratio in 4 ml RPMI 1640 in 20% HIFBS and 50 mM HEPES and incubated in tightly closed 25-cm² TC flasks.

[b] Promastigotes grown at 27° in HEPES-buffered medium 199 plus 10% HIFBS and used as inoculum.

[c] Cultures were divided into equal portions in separate flasks. $N = 3$.

site. At this point, the culture can be sacrificed for the isolation of amasti-gotes by Percoll gradient centrifugation as described or can be continued indefinitely. Amastigotes isolated from animal lesions or infected macro-phages can also be used for infection. At a low macrophage-to-amastigote ratio of 1:1 or 1:2, a generation time of 24 hr has been observed for the rate of parasite multiplication in these macrophages.

Applications, Advantages, and Disadvantages. The *L. mexicana amazonensis*–J774G8 *in vitro* system has been used mainly for studying biochemical aspects of host–parasite interactions.[4] For example, this sys-tem has been used to demonstrate (1) changes in tubulin biosynthesis[38] and in antigenic proteins[39] during leishmanial differentiation; (2) post-transcriptional control of this developmentally regulated event for tubu-lin[40]; (3) the antileishmanial activity of formycin B,[41] and (4) the host–

[38] D. Fong and K.-P. Chang, *Proc. Natl. Acad. Sci. U.S.A.* **78,** 7624 (1981).
[39] K.-P. Chang and D. Fong, *Infect. Immun.* **36,** 430 (1982).
[40] M. Wallach, D. Fong, and K.-P. Chang, *Nature (London)* **299,** 650 (1982).
[41] D. Carson and K.-P. Chang, *Biochem. Biophys. Res. Commun.* **100,** 1377 (1981).

parasite metabolic interactions in heme biosynthesis.[42] These studies represent our efforts in understanding the biochemical mechanisms of intracellular survival, differentiation, and multiplication of leishmanias in macrophages.

The advantages of this system for these types of investigations are several, i.e., (1) the promastigotes of the *Leishmania* species and the J774 macrophages can be grown *in vitro* separately with ease as homogeneous cell populations; (2) sizable infected cultures can be established simply by combining the parasite and the macrophage in culture; (3) in a relatively short time period of 7–9 days, 10^9 amastigotes can be recovered with a high degree of purity from the infected culture by the simple isolation procedure described (this compares very favorably with *in vivo* infection which requires weeks and often months for the development of lesions adequate for the isolation of amastigotes); and (4) leishmanias in promastigote-to-amastigote and in amastigote-to-promastigote differentiation can easily be made available for investigation. This culture system, in essence, reproduces the entire life cycle of leishmanias *in vitro* without using laboratory animals.

The disadvantages of this system may include the following: (1) leishmanias which grow well in J774G8 cells are limited to several species from South America; (2) the macrophages are transformed cells which may not represent the normal host cells of leishmanias in animals; and (3) the system could be improved for overall efficiency in obtaining amastigotes at one time in the range of 10^{11}–10^{12} cells or protein in quantity needed for many types of biochemical experiments.

Leishmania donovani–Human Monocyte Monolayer System[27]

Infection. The interaction of *Leishmania donovani* with human monocytes or monocyte-derived macrophages can be assessed by incubating monolayers with promastigotes or amastigotes at a parasite-to-phagocyte ratio of 10 : 1 or 20 : 1 in 1 ml of medium 199 in the presence or absence of human serum. Parasite ingestion is usually assessed after 2 hr of incubation. To assess long-term intracellular parasite survival, infected monolayer preparations are washed after 2 hr and cultured in medium 199 which contains antibiotics and 13% fresh autologous human serum; this latter component is used to kill residual extracellular promastigotes and to maintain the macrophages.[28] Monolayers are then incubated at 37° and examined at 24, 48, 72, 120, or 168 hr.

[42] C. S. Chang and K.-P. Chang, *Mol. Biochem. Parasitol.* **16**, 267 (1985).

Enumeration of Intracellular Leishmanias by Fluorescent Microscopy. Monolayers are washed once in warm medium to remove extracellular parasites, and coverslips are then inverted onto a drop of acridine orange solution (Sigma Chemical Company, St. Louis, MO) at 10 μg/ml in PBS. Monolayers are examined by using a microscope equipped with epifluorescence and phase-contrast optics. In acridine orange-stained monolayers viewed by epifluorescence, the nucleus and kinetoplast of the parasite stain intensely green in contrast to the orange-staining granules and the larger, light green-staining nucleus of the monocyte. The percentage of infected monocytes and the number of parasites per 100 total monocytes is determined by examining ≥ 200 monocytes on each coded coverslip in duplicate. Acridine orange fluorescence allows easy detection of leishmanias in rounded, freshly isolated human monocytes and monocyte-derived macrophages.

Intracellular Growth of L. donovani in Human Macrophages. The majority of *L. donovani* promastigotes survive ingestion by oxidatively weak human monocyte-derived macrophages, as shown in Table II. The parasite assumes amastigote morphology and begins to multiply after approximately 48 hr. There is a net gain of 2.8 amastigotes per infecting promastigote over 7 days, resulting from intracellular multiplication of the parasite.

Application. It has been possible by using this and other *in vitro* systems to characterize many aspects of the interaction of leishmanias with human and rodent mononuclear phagocytes.[4,5,43–45] Recent evidence suggests that serum-independent attachment of *Leishmania* promastigotes to macrophages may be mediated by mononuclear phagocyte receptors for mannose present on the surface of promastigotes and also C3bi.[45a] The parasite is ingested by the macrophage and phagosome–lysosome fusion follows.

The survival of *L. donovani* within mononuclear phagocytes is dependent on multiple factors, including the capacity of the phagocyte to produce toxic oxygen intermediates, the magnitude of the phagocytosis-induced respiratory burst triggered by the parasite, and the susceptibility of the parasite in each stage of its life cycle to oxidative and other microbicidal mechanisms. Monocytes are well endowed with a variety of microbicidal mechanisms, but they lose much of their oxidative killing potency

[43] J. M. Blackwell and J. Alexander, *Trans. R. Soc. Trop. Med. Hyg.* **77**, 636 (1983).
[44] E. Handman, R. Ceredig, and G. F. Mitchell, *Aust. J. Exp. Biol. Med. Sci.* **57**, 9 (1979).
[45] J. Mauel, *in* "The Host Invader Interplay" (H. Van den Bossche, ed.), p. 165. Elsevier/ North-Holland Biomedical Press, Amsterdam, 1980.
[45a] J. M. Blackwell, R. A. B. Gzekowitz, M. B. Roberts, J. Y. Channon, R. B. Sim, and S. Gordon, *J. Exp. Med.* **162**, 324 (1985).

TABLE II
Association of *Leishmania donovani* with Human
Monocyte-Derived Macrophages[a]

Culture periods (days)	Percentage infection	Number of parasites per infected macrophages	Number of parasites per 100 total macrophages	Total number of *L. donovani* ($\times 10^{-5}$)
1	67.5 ± 5[b]	4.2 ± 0.7	297 ± 66	4.8 ± 0.8
2	66 ± 10	4.6 ± 0.7	333 ± 72	5.0 ± 0.2
3	76 ± 10	8.1 ± 1.6	682 ± 172	13.1 ± 2.7
5	67 ± 13	10.5 ± 2.5	787 ± 239	19.3 ± 6.6
7	79 ± 7	15.9 ± 3.2	1360 ± 351	17.9 ± 4.3

[a] Five-day-old human monocyte-derived macrophages were exposed to promastigotes at a 1:20 ratio in 1 ml for 2 hr at 37° in 95% room air–5% CO_2. Residual parasites were then removed. Monolayers were examined or resuspended in 2 ml of fresh medium with 10% fresh autologous serum and antibiotics and cultured for additional periods of time.[21]

[b] Mean ± SE ($N = 5$).

and all detectable lysosomal myeloperoxidase activity as they mature to macrophages. The magnitude of the oxidative response of these cells elicited by *L. donovani* and its susceptibility to toxic oxidants generated thereof vary with its stage: amastigotes trigger a smaller oxidative response when ingested by macrophages and are less susceptible to oxidants than promastigotes. Promastigotes are able to survive within oxidatively weak human monocyte-derived macrophages but the majority of promastigotes are killed by oxidatively potent monocytes or polymorphonuclear leukocytes[28,46] or murine peritoneal macrophages.[47] Whether promastigotes may convert to the more resistant amastigote stage within the sanctuary of oxidatively quiescent, cutaneous macrophages or cells of other lineages in natural infection is unknown.

Amastigotes are also readily ingested *in vitro* by monocytes or macrophages, survive, and multiply within them. Presumably, resolution of leishmanial infection in animals depends predominantly on cell-mediated immune mechanisms. Murray *et al.*[33] have shown that human monocyte-derived macrophages can be stimulated *in vitro* by lectin-induced lymphokines or γ-interferon to kill *L. donovani*. It is uncertain if the ability of amastigotes to infect monocytes or immature macrophages arriving at the site of infection may play a role in the pathogenesis of visceral leishmaniasis.

[46] K.-P. Chang, *Am. J. Trop. Med. Hyg.* **30**, 322 (1981).
[47] H. W. Murray, *J. Immunol.* **129**, 351 (1982).

Advantages and Disadvantages. There are several potential advantages of this type of *in vitro* system: (1) *in vitro* findings with primary culture of human mononuclear phagocytes may closely parallel human infection; (2) it is possible to follow the sequential events of parasite–phagocyte interactions with different parasite and phagocyte stages; and (3) the role of oxidative and other microbicidal mechanisms can be assessed. The disadvantages of the system include (1) the relatively limited number of monocytes available from human donors; (2) the potential transmission of hepatitis or even acquired immune deficiency syndrome (AIDS) to laboratory workers exposed to human blood; (3) the effects of glass or plastic adherence on macrophage function; and (4) the potential loss of some macrophage subpopulations from monolayer preparations over time.

Leishmania major–Murine Macrophage Nonadherent
  Culture System[20]

Infection. Amastigotes obtained from infected footpads of BALB/c mice (*L. major*) are adjusted to a concentration of 5×10^6/ml, and 0.1 ml of this amastigote suspension is added to each macrophage culture tube (a 1:1 amastigote:macrophage multiplicity of infection). Macrophages are exposed to parasites for 1 hr at 37°, and washed by low-speed centrifugation (50 g for 5 min); culture supernatants with uningested amastigotes are removed by aspiration and infected macrophages are resuspended in 0.5 ml assay medium and incubated at 37° for the remainder of the experiment.

Enumeration of Intracellular Amastigotes by Light Microscopy of Stained Cell Smears Prepared by Cytocentrifugation. At regular time intervals (1–72 hr), samples are removed to determine the number of macrophages/culture (by hemacytometer), and to prepare cell smears by cytocentrifugation (200 μl of 10^5–10^6 cells/ml for 7 min at 750 rpm) (Shandon Southern Instruments). The cell smears are stained with a modified Wright-Giemsa (Diff-Quick, Dade Diagnostics), and observed microscopically with oil immersion for cell differentials, percentage infected macrophages, and number of intracellular amastigotes/infected macrophage.

Intracellular Growth of Amastigotes. A comparison of the infection and replication of *L. major* in macrophages maintained as monolayers and in nonadherent culture is shown in Table III. Quantitative differences in the ability of murine peritoneal macrophages to support sustained replication of *L. major* amastigotes are associated with culture conditions of the host cell.[48] A number of physiologic and metabolic differences between

[48] C. A. Nacy and C. L. Diggs, *Infect. Immun.* **34,** 310 (1981).

TABLE III

Leishmania major Infection of Nonadherent and Adherent Macrophages

Time after infection (hr)	Nonadherent macrophages			Adherent macrophages	
	Percentage infected cells[a]	Parasites per macrophage[b]	Parasites per culture[c] (10^{-5})	Percentage infected cells	Parasites per macrophage
1	41 ± 1	1.4	3.2	31 ± 3	1.3
24	87 ± 2	2.1	9.1	95 ± 1	2.0
48	88 ± 3	3.2	14.4	94 ± 2	2.6
72	89 ± 2	6.8	29.0	90 ± 2	2.7
96	89 ± 2	9.6	51.3	89 ± 1	2.9
144	95 ± 5	12.5	49.8	84 ± 1	1.3
168	95 ± 3	24.1	95.7	55 ± 5	1.1
192	98 ± 2	25.0	56.3	41 ± 4	1.1

[a] Percentage infected macrophages was determined by microscopic examination of two cell smears each of duplicate samples at each time point (800 cells observed). Values are mean ± SEM.

[b] Number of intracellular amastigotes was observed for at least 100 infected macrophages and reported as mean number of parasites per infected cell.

[c] Total parasites per culture were calculated by multiplying the number of macrophages per culture by percentage infected macrophages and mean number of amastigotes per infected macrophage. The decrease in total parasites at 192 hr reflects a 50% drop in viability of macrophages maintained in culture for this time period; a similar decrease in viability is observed in uninfected macrophage cultures as well.[48]

adherent and nonadherent macrophages have been reported; at present, the physiological properties of nonadherent macrophages that contribute to increased survival and multiplication of *L. major* in this *in vitro* system are unknown.

After 1 hr exposure to amastigotes *in vitro,* similar numbers of macrophages are infected in resident and several different inflammatory cell populations. There is, however, a marked and significant predilection of the parasite to enter less differentiated peroxidase-positive macrophages.[21] Intracellular replication of the parasite (an increase in the total number of parasites over 72 hr) is not markedly different for the different cell populations.

In contrast, macrophages that are involved in *in vivo* immune reactions, such as cells obtained from BCG- or *Corynebacterium parvum*-treated mice, ingest 25–50% fewer amastigotes with 1 hr exposure to parasites than control untreated resident and inflammatory cells. These *in vivo* activated macrophages are also spontaneously cytotoxic to ingested amastigotes, and 95–100% of BCG- or *C. parvum*-treated macrophages are *Leishmania*-free by 72 hr in culture (Table IV).

TABLE IV

RESISTANCE TO INFECTION WITH, AND INTRACELLULAR DESTRUCTION OF, *Leishmania major*
AMASTIGOTES BY ACTIVATED MACROPHAGES

Source of macrophages	LK concentration	Percentage infected[a] macrophages at		Leishmanicidal[b] activity (%) at	
		1 hr	72 hr	1 hr	72 hr
BCG-treated mice	0	25	2	40	98
C. parvum-treated mice	0	27	5	35	91
Untreated mice	0	42	58	0	0
	1/6	22	10	48	78
	1/12	25	8	40	86
	1/24	27	12	35	79
	1/48	36	25	14	55
	1/96	43	46	0	20

[a] Resident macrophages are exposed to LK at different concentrations for a minimum of 4 hr at 37°,[59] washed by centrifugation at 200 g for 10 min, resuspended in assay medium, and infected with amastigotes of *L. major* for 1 hr. The percentage of macrophages with intracellular amastigotes is determined by microscopic examination of stained cell smears at 1 hr after infection. Results are expressed as mean percentage *Leishmania*-infected macrophages ± standard error of the mean (SEM) for 4–8 observations on 2–4 cultures (800–1600 macrophages observed). Resistance to infection is defined as the percentage decrease in infected macrophages in treated cultures compared to control cultures and is determined by the following formula:

$$\text{Resistance to infection} = \left(\frac{\text{percentage control macrophages} - \text{percentage infected treated macrophages}}{\text{percentage control macrophages}} \right) 100$$

[b] To induce the intracellular destruction of amastigotes independent of any LK effects on ingestion, macrophages are infected first, then treated with various concentrations of LK for 8–72 hr. The percentage of macrophages with intracellular amastigotes is determined by microscopic examination of stained cell smears at 72 hr after infection. Results are expressed as mean percentage *Leishmania*-infected macrophages ± standard error of the mean (SEM) for 4–8 observations on 2–4 cultures (800–1600 macrophages observed). Microbicidal activity is also defined as the percentage decrease in infected macrophages in treated cultures compared to control cultures, and is determined by the same formula as above.

Immunological Mediator-Induced Alterations in Intracellular Survival of Leishmania in Murine Macrophages. Treatment of resident peritoneal macrophages with LK *in vitro* induces both effects of *in vivo* activation, i.e., a decreased uptake of parasites at 1 hr and intracellular destruction of amastigotes at 72 hr. Table IV shows a representative dose–response for LK-induced macrophage resistance to infection with and intracellular destruction of *L. major* amastigotes. Prior exposure of macrophages to LK at $\frac{1}{6}-\frac{1}{24}$ dilutions for 4 hr at 37° reduces the infection of these cells by about 50%. A single pulse of LK administered at high concentrations 1 hr after infection of cells is sufficient to induce killing of 70–90% of amastigotes of *L. major* by macrophages. Elimination of intracellular *L. major* by macrophages is almost complete by 72 hr.

Comments. Although inflammatory macrophages are more responsive than resident cells for certain LK-induced tumoricidal and other effector activities, the cells that respond to LK for induction of intracellular destruction of amastigotes are resident macrophages.[49,50] Inflammatory macrophages are hyporesponsive to LK mediating leishmanicidal activity. This hyporesponsiveness is independent of the nature of the eliciting agent and increases in proportion to the number of immature cells that arrive at the peritoneal cavity during inflammation. Peripheral blood monocytes, the immediate precursors of inflammatory macrophages, also do not kill *L. major* after exposure to LK. The mechanism responsible for destruction of amastigotes of this strain of *Leishmania* may not be fully developed in immature cells.

The dramatic effects of LK on resident macrophage interactions with *Leishmania* amastigotes result from a cascade of activation reactions that depends on the nature of the target microorganism, the appropriate activating signals given in a precise sequence, and maturity of the effector cells.[7] The difference in capacity of activated murine resident cells to kill different *Leishmania* species may relate to a differential susceptibility of the respective amastigotes to the same killing mechanism, or to induction of different killing mechanisms effective for each strain. The Maria strain of *L. mexicana,* for instance, is unaffected by macrophage killing mechanisms induced by LK that eliminate intracellular *L. major* and *L. donovani.*[24,51–54] The activating factors that induce macrophage destruction of *L. major* and *L. donovani* have been only partially characterized:

[49] L. P. Ruco and M. S. Meltzer, *J. Immunol.* **120,** 1054 (1978).
[50] D. L. Hoover and C. A. Nacy, *J. Immunol.* **132,** 1487 (1984).
[51] P. Scott, D. Sacks, and A. Sher, *J. Immunol.* **131,** 966 (1983).
[52] K.-P. Chang and J. W. Chiao, *Proc. Natl. Acad. Sci. U.S.A.* **78,** 7083 (1981).
[53] C. G. Haidaris and P. F. Bonventre, *Infect. Immun.* **33,** 918 (1981).
[54] W. T. Hockmeyer, D. Walters, R. W. Gore, J. S. Williams, A. H. Fortier, and C. A. Nacy, *J. Immunol.* **132,** 3120 (1984).

one, and possibly a major activating agent in LK, is γ-interferon.[55] That there are other factors in LK that can induce intracellular killing has been demonstrated by determination of the size of LK that induce this effector function against amastigotes[24,52] and the nature of molecules that induce intracellular destruction of promastigotes[56] in murine macrophages. It is not yet clear whether these different factors activate distinct subpopulations of macrophages.

Applications. The advantage of this nonadherent culture system is that data can be expressed quantitatively, since the absolute number of macrophages can be easily determined. Application of this quantitative technique for short-term assays of 72 hr includes the following: comparative analysis of different *Leishmania* species,[51,54] analysis of macrophage function in different inbred mice,[57] genetic analysis of macrophage defects,[58] identification of macrophage subpopulations,[21,50] and determination of signals and steps involved in macrophage activation.[58,59] There are, however, several important considerations in selection of this assay: (1) The assay uses a mixed cell population. Interpretation of studies with peritoneal cells from *Leishmania*-infected animals that contain amastigote-reactive lymphocytes may be difficult; (2) the strain of *L. major* used in the studies outlined above is adapted for survival at 37°, and amastigotes convert to promastigotes in culture medium at temperatures less than 37°; and (3) macrophages obtained from the peritoneum of mice perform immunological functions maximally at 37°. For each degree change in culture temperature, there is a corresponding alteration in the capacity of these cells in response to LK.

Concluding Remarks

Leishmanial infection of macrophages represents a unique example of intracellular parasitism. Of particular interest is the dilemma in considering the role of macrophages in leishmaniasis: they are both the "benefactor" or the host cell from parasitological points of view and the "executioner" or the effector cell in the context of host immunity. In existence must be a delicate balance governed by intricate host–parasite interplays

[55] C. A. Nacy, W. T. Hockmeyer, W. R. Benjamin, S. L. James, and M. S. Meltzer, *Infect. Immun.* **40**, 820 (1983).
[56] E. Handman and A. W. Burgess, *J. Immunol.* **122**, 1134 (1979).
[57] C. A. Nacy, A. H. Fortier, M. G. Pappas, and R. R. Henry, *Cell. Immunol.* **77**, 298 (1983).
[58] C. A. Nacy, M. S. Meltzer, and A. H. Fortier, *J. Immunol.* **133**, 3344 (1984).
[59] C. N. Oster and C. A. Nacy, *J. Immunol.* **132**, 1494 (1984).

between the killing mechanisms of macrophages and the survival mechanisms of leishmanias. The *in vitro* systems presented here offer a way to analyze under controllable conditions some of the phenomena and factors involved in the balance of leishmania–macrophage interactions.

The phenomena of how leishmanias bind to macrophages (receptor mediated?) and how these parasites subsequently survive, differentiate, and multiply in phagolysosomes have been described in some detail and amply discussed on the basis of *in vitro* studies.[3–5,43,45] The biochemical basis of these cellular events in host–parasite interactions has just begun to receive attention. Of interest are the findings that membrane acid phosphatase of *L. donovani* inhibits superoxide production of phagocytes,[60] and that surface antigenic glycoproteins of *L. mexicana* are resistant to lysosomal enzyme degradation.[4] Although the molecular basis of these phenomena has yet to be defined, the observations support the contention that surface molecules of leishmanias may play an essential role in their intralysosomal survival in macrophages.[61,62]

Whatever the leishmanial mechanisms of intracellular survival may be, they apparently can be neutralized through the intervention of specific and nonspecific immune factors, as demonstrated *in vitro*. Before reaching their destination of macrophages, leishmanias can be intercepted and killed by the oxidatively active polymorphonuclear phagocytes[28,46] or by the lytic action of complement via classic[63] and/or alternative pathways.[64–66] Antibodies specific to the surface components of leishmanial promastigotes work against them in concert not only with complement but also with macrophages *in vitro*. Thus, appropriate monoclonal antibodies reactive to some *Leishmania* surface antigens effect complement-mediated immune lysis of the parasites[61] and reduce their infection of cultured macrophages.[61,67] In the latter case, perhaps, the binding of monoclonal antibodies to parasite surface antigens interferes with their functions critical for certain steps of intracellular parasitism. Either or both of these antibody-mediated antileishmanial mechanisms may ac-

---

[60] A. T. Remaley, D. B. Kuhns, R. E. Basford, R. H. Glew, and S. S. Kaplan, *J. Biol. Chem.* **259**, 11173 (1984).

[61] K.-P. Chang and D. Fong, *Ciba Found. Symp.* [N.S.] **99**, 113 (1983).

[62] D. M. Dwyer and M. Gottlieb, *J. Cell. Biochem.* **23**, 35 (1983).

[63] R. D. Pearson and R. T. Steigbigel, *J. Immunol.* **125**, 2195 (1980).

[64] D. M. Mosser and P. J. Edelson, *J. Immunol.* **132**, 1501 (1984).

[65] R. S. Bray, *J. Protozool.* **30**, 322 (1983).

[66] D. L. Hoover, M. Berger, M. H. Oppenheim, W. J. Hockmeyer, and M. S. Meltzer, *Infect. Immunol.* **47**, 247 (1985).

[67] E. Handman and R. E. Hocking, *Infect. Immun.* **37**, 28 (1982).

count for the finding that promastigote surface-specific monoclonal antibodies prevent the development of cutaneous lesions by the infection of mice with *L. mexicana*.[68] All this points to the feasibility of intervention in the early phase of leishmanial infection of macrophages as a strategy for the production of effective vaccines against leishmaniasis by, for example, the genetic engineering technology.

The resolution of leishmanial infection may depend ultimately on LK-mediated activation of macrophages to kill leishmanias in parasite-containing vacuoles, which are not readily accessible to humoral factors, such as antibodies and complement. Most striking is the *in vitro* inhibition of the uptake of leishmanias and intracellular killing of amastigotes by macrophages activated *in vitro* or *in vivo* by LK.[7] The antileishmanial activity of these macrophages for some *Leishmania* species may be closely related to the induction of oxidative microbicidal mechanisms in certain cell populations.[47] Other ill-defined mechanisms seem also to be involved.[52] In the area of macrophage activation, however, disparate findings with different *Leishmania* species and macrophage populations defies generalization of information obtained to date and cautions against extrapolation of *in vitro* data to *in vivo* situations. Characterization of LK that induce antileishmanial activities of macrophages, and mechanisms of the signal transduction for macrophage activation by these molecules are interesting subjects for investigation. Elucidation of these mechanisms may help explain the genetic and phenotypic variations observed with different *Leishmania* species in macrophage sensitivities to different LK signals in the induction of different leishmanicidal pathways.

[68] S. Anderson, J. R. David, and D. McMahon-Pratt, *J. Immunol.* **131,** 1616 (1983).

[44] Parasiticidal Activity of Macrophages against *Toxoplasma*

By Somesh D. Sharma, James R. Catterall, and Jack S. Remington

Toxoplasma gondii is a coccidian parasite capable of invading virtually every cell in the body. *Toxoplasma* have not been grown or sustained in a cell-free medium. Their ability to grow and multiply within cells, particularly macrophages, has provided an excellent model to assess the role cell-mediated immunity plays in the development of a protective response against this organism and the contribution of phagocytic cells, especially macrophages, as effectors of this resistance. A number of methods have been employed to examine the destruction of *T. gondii* within macrophages. These include visual observation of inhibition of multiplication or destruction of *Toxoplasma*,[1] uptake of nucleic acid precursors by multiplying *Toxoplasma* over and above that incorporated by the cell in which the organism is growing,[2] and determination of the number of parasites released from host cells by plaqueing techniques.[3] In most of the studies reported in the literature one of these assays has been utilized to assess the ability of macrophages from different animal compartments from humans as well as animal species to inhibit and/or kill *Toxoplasma*,[4] to elucidate the biochemical mechanisms responsible for the destruction of *Toxoplasma*,[5] and to assess the influence of lymphokine production and the ability of macrophages to respond to various lymphokines.[6]

Precautions

According to National Institutes of Health guidelines, *Toxoplasma* is a class 2 pathogen[7] and it is recommended that criteria to work with class 2 pathogens be followed. It is particularly important that immunosuppressed individuals and pregnant women with no previous exposure to this parasite avoid work with *Toxoplasma*. Safety glasses should be worn by all workers handling live *Toxoplasma*, contaminated needles should be discarded immediately, and live organisms should never come into contact with mucosal surfaces or breaks in the skin. It is recommended that serum from all personnel who work with this parasite be tested for antibody to *Toxoplasma* prior to working with the organism. Laboratory personnel who are accidentally inoculated with *Toxoplasma* and who do not have prior positive *Toxoplasma* serology should be referred to a physician immediately.

[1] J. Ruskin and J. S. Remington, *Antimicrob. Agents. Chemother.*, p. 474 (1968).

[2] E. R. Pfefferkorn and L. C. Pfefferkorn, *J. Protozool.* **24,** 449 (1977).

[3] V. L. Foley and J. S. Remington, *J. Bacteriol.* **98,** 1 (1969).

[4] R. McLeod, K. G. Bensch, S. M. Smith, and J. S. Remington, *Cell. Immunol.* **54,** 330 (1980); F. W. Ryning and J. S. Remington, *Infect. Immun.* **18,** 746 (1977).

[5] H. W. Murray and Z. A. Cohn, *J. Exp. Med.* **150,** 938 (1979); C. B. Wilson, V. Tsai, and J. S. Remington, *J. Exp. Med.* **151,** 328 (1980).

[6] S. E. Anderson, Jr., and J. S. Remington, *J. Exp. Med.* **139,** 1154 (1974).

[7] "Classification of Etiologic Agents on the Basis of Hazard," 4th ed. U.S. Department of Health, Education and Welfare, Public Health Service, Center for Disease Control, Office of Biosafety, Atlanta, Georgia, 1979.

Materials and Reagents

Materials

RPMI 1640, penicillin (pen),[7a] streptomycin (strep), fetal calf serum (FCS),[8] Hanks' balanced salt solution (HBSS), Dulbecco's phosphate-buffered saline (Grand Island Biological Company)

L929 cells (American Type Culture Collection)

Tissue culture flasks, dishes, pipets, etc. (Corning)

Petri dishes (#1029; Falcon Plastics)

Polycarbonate filters and holders (3 μm; 47 mm diameter); (Nucleopore)

Heparin (Rikker Laboratories)

Ficoll-Hypaque (Pharmacia)

Labtek slides (Labtek Industries)

Glass coverslips (Corning)

Acridine orange (Sigma)

Aminoacridine hydrochloride (Sigma)

Giemsa stain (Harelco)

Trichloroacetic acid (TCA) (Baker)

Sodium dodecyl sulfate (SDS) (Sigma)

Uracil (Sigma)

[5,6-^{3}H]Uracil, specific activity 48 Ci/mmol (New England Nuclear or Amersham)

Glass filters (Gelman Instrument)

Scintillation fluid (Beckman or New England Nuclear)

Feeding tube, size 5 French (Pharmaseal)

4-0 silk suture (Ethicon)

Reagents

PBS: 26.8 g Na_2HPO_4 (solution A), 13.8 g $NaH_2PO_4 \cdot H_2O$ (solution B), 171 g of NaCl per liter of H_2O (solution C); mix 71.25 ml of solution A with 28.75 ml of solution B and 50.0 ml of solution C. Adjust pH to 7.2 with HCl and dilute to 1 liter

Saline: 0.85 g NaCl per 100 ml of H_2O

Coverslip rinsing solution: 56 g KOH/1000 ml of H_2O

Complete medium: RPMI 1640 + 10% FCS + 100 U pen/ml + 100 g strep/ml. Filter sterilize with a 0.45-μm Millipore or equivalent filter

---

[7a] Abbreviations: pen, penicillin: strep, streptomycin; FCS, fetal calf serum; HBSS, Hanks' balanced salt solution; TCA, trichloroacetic acid; SDS, sodium dodecyl sulfate; PBS, phosphate-buffered saline; BCG, Bacillus Calmette–Guérin.

[8] Individual lots of FCS should be tested for their ability to support the formation of macrophage monolayers.

Acridine orange: 1 mg/ml of H_2O
Aminoacridine fixative: 0.4 g of aminoacridine hydrochloride in 50% ethanol
Giemsa stain: 7.415 g of original azure blend in 1000 ml of methanol
TCA solution: 49 g TCA, 10 g SDS, and 0.1 g of cold uracil/1000 ml of H_2O
TCA alone: 49 g of TCA/1000 ml of H_2O

Toxoplasma Strains

The highly virulent RH strain of *Toxoplasma* is most commonly used in our laboratory to assess the inhibition of growth and killing of this organism by macrophages, but less virulent strains have been utilized. These strains can be obtained from investigators working with the parasite.

Maintenance of *Toxoplasma*

The RH strain of *Toxoplasma* can be maintained in tissue culture or by serial intraperitoneal passage in mice. *Toxoplasma* obtained from either source have been successfully employed and the choice of one method over the other is largely dictated by individual preferences and the availability of tissue culture or animal facilities.

Maintenance in Tissue Culture. L929 cells are grown in RPMI 1640 containing 10% FCS, 100 U/ml pen, and 100 g/ml strep in 25-cm^2 plastic tissue culture flasks in a humidified atmosphere containing 5% CO_2. Cell cultures are split 1 to 3 when they are confluent by decanting the medium and scraping the adherent cells with a rubber policeman. An alternative procedure currently used in our laboratory takes advantage of the fact that L929 cells grow on, but do not attach tightly to, nontissue culture Petri dishes. They can be dislodged from the surface simply by vigorous pipetting and then transferred to tissue culture vessels for subsequent use. To maintain *Toxoplasma*, confluent monolayers in complete medium are infected with approximately one tachyzoite of the RH strain per 100 fibroblasts. After a 48-hr incubation period, tachyzoites released due to the lysis of the L929 cells are collected by decanting the medium and then centrifuged at 350–500 g in a tabletop centrifuge. To obtain adequate quantities of parasites for an experiment, confluent cultures are infected with 2–3 *Toxoplasma* per L929 cell. The supernatants containing freshly released parasites can be collected and used for infecting macrophages.

Maintenance in Vivo. Although any mouse strain can be used for passage of *Toxoplasma,* for economic reasons our laboratory routinely utilizes outbred Swiss Webster female mice weighing approximately 18–20 g. For *in vivo* passage, mice which had been inoculated 3 days earlier with *Toxoplasma* tachyzoites are sacrificed by inhalation of CO_2. The external abdominal wall is cleansed with 70% alcohol and incised with the aid of tissue forceps to expose the peritoneum. Sterile saline solution (1 ml) is injected into the peritoneal cavity with a syringe attached to an 18- or 22-gauge needle and approximately 1.5 ml of the peritoneal fluid–saline solution is withdrawn with a 5-ml syringe fitted with a 22-gauge needle. This solution is vigorously mixed by aspirating up and down with a syringe attached to a 27-gauge needle to release the tachyzoites from peritoneal cells. The number of *Toxoplasma* in this suspension are counted using a Nebauer hemacytometer under a 40× objective of a light microscope. For passage, a 0.2-ml aliquot containing 4×10^6 organisms is injected intraperitoneally into each mouse. When organisms are required for an experiment, *Toxoplasma* are passaged as described above except that the inoculum size is increased to 4×10^7 tachyzoites. Forty-eight hours after such an inoculation each mouse yields approximately 2×10^7 tachyzoites following filtration with a polycarbonate filter to remove peritoneal cells.

Methods

The sequence of steps involved in assessing destruction of *Toxoplasma* by normal or activated macrophages from any source or species is illustrated in Fig. 1.

Step 1A—Preparation of Peritoneal Macrophages

To obtain peritoneal macrophages, kill the animal by CO_2 inhalation. Flood the abdominal wall with 70% alcohol and expose the peritoneum. Using a syringe and 18- or 22-gauge needle, inject 4–5 ml sterile saline into the peritoneal cavity of a mouse or 50 ml into the peritoneal cavity of a rat.[9] Massage the peritoneal cavity gently and aspirate the fluid back using the syringe employed for injection.

Step 1B—Preparation of Alveolar Macrophages (see also this series, Vol. 108 [25])

Anesthetize the animal by CO_2 inhalation followed by cervical cord transection. Open the abdominal cavity and exsanguinate the animal by

---

[9] Avoid puncturing the gut or liver.

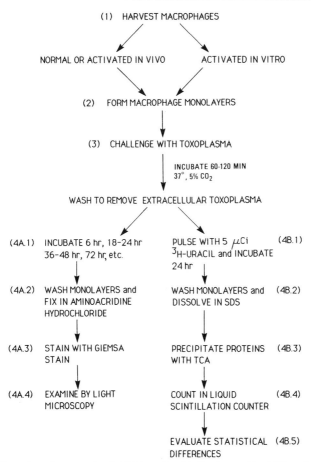

FIG. 1. Flow chart for assessing inhibition of multiplication and/or killing of *Toxoplasma* by macrophages.

cutting the aorta and inferior vena cava. Open the thoracic cavity from below and remove the anterior thoracic wall. Flush blood from the pulmonary vasculature with 20 ml (rat) or 3 ml (mouse) of PBS injected via the right ventricle, having first prevented backflow by placing artery forceps across the right atrium and inferior vena cava. Expose the trachea in the neck, incise its anterior wall, and introduce a catheter.[10] After securing the catheter with 4-0 silk, attach a syringe to its proximal end and lavage

[10] A suitable catheter for the rat trachea is a 5-French gauge feeding tube cut to a length of 5 cm. For the mouse, use the same kind of tube, but with a blunted 20-gauge needle inserted into its distal (pulmonary) end.

the lung 10–12 times, using a volume of approximately 0.8 ml in the mouse and 10 ml in the rat for each lavage.[11] In our laboratory, this procedure yields $1-3 \times 10^5$ cells in the mouse and $2-8 \times 10^6$ cells in the rat.

Step 1C—Preparation of Human Alveolar Macrophages

Human alveolar macrophages are obtained from patients undergoing fiber optic bronchoscopy, a relatively safe procedure with few contraindications.[12] Using a syringe and three-way stopcock, instill a 50-ml aliquot of sterile saline through the suction channel of the bronchoscope and aspirate immediately. The return volume averages 20–30 ml. Repeat up to a maximum of five times until 50–100 ml of fluid has been collected. If bronchial mucus is present, strain the fluid through one layer of surgical gauze prior to centrifugation and washing. In healthy nonsmokers, this procedure yields $2-10 \times 10^6$ cells comprising >85% macrophages.[13] If bronchial biopsies or brushings are being taken, they should be done after the lavage to prevent contamination with blood. It is also important that the tip of the bronchoscope be wedged tightly into a peripheral airway throughout the procedure—otherwise the lavage fluid will contain many cells from the large airways.

Step 1D—Preparation of Human Monocytes

Obtain 100–150 ml of heparinized blood from healthy volunteers. Dilute the blood with an equal volume of saline and layer 40 ml on top of a 10-ml cushion of Ficoll-Hypaque ($d = 0.077$) in 50-ml polycarbonate tubes. Centrifuge at 500–750 g in a Beckman TJ6 or equivalent centrifuge for 30 min at room temperature. The mononuclear cells will form an opaque ring at the Ficoll-Hypaque–plasma interface. Harvest mononuclear cells with the aid of a sterile Pasteur pipet. Wash the cells twice in HBSS solution by centrifugation at 350 and 200 g, respectively. Resuspend the cells in complete medium at the desired concentration. Use these cells within 1 hr to determine their ability to inhibit or kill Toxoplasma.

To obtain monocyte-derived macrophages, seed wells of a 24-well tissue culture dish containing glass coverslips with $1-2 \times 10^6$ monocytes in 1 ml of complete medium. Allow the monocytes to adhere to the cov-

---

[11] The composition of the lavage fluid is not critical but optimal yield is achieved by excluding divalent cations. We routinely use Dulbecco's PBS without Ca^{2+} or Mg^{2+}.

[12] J. F. Landa, "Diagnostic Techniques in Pulmonary Diseases," Part II, p. 655. Marcel Dekker, New York, 1981.

[13] G. W. Hunninghake, J. E. Godek, O. Kawanani, V. J. Ferrans, and R. G. Crystal, Am. J. Pathol. 97, 149 (1979).

erslips for 2–4 hr at 37° in a humidified atmosphere containing 5% CO_2. Aspirate the supernatant and wash the monolayer twice with warm HBSS to remove nonadherent cells. Reincubate the monolayers with 2 ml fresh complete medium and change the medium every 3 days thereafter until the cells are to be utilized in experiments. In our experience a minimum of 1 5- to 8-day incubation period is required before the monocyte develops the morphologic criteria of macrophages.

Macrophage Activation (see also this volume [43]). Macrophages can be activated *in vivo* by administration of *Corynebacterium parvum*,[14] BCG,[15] *Toxoplasma*,[1] or a wide variety of other immunomodulating agents. *In vitro* activation is generally achieved by incubating normal macrophages with lymphokines,[16] interferon, or interferon inducers.[17]

Formation of Monolayers

Step 2. Keep the harvested cells in tubes on ice. Centrifuge at 225 g for 10 min, wash twice in HBSS or PBS, recentrifuge, and resuspend in complete medium. Prepare macrophage monolayers in two- or four-chambered Labtek slides or on 15-mm glass coverslips[18] that have been placed in wells of 24-well tissue culture trays. Add 2×10^6 cells (1×10^6 cells if four-chamber Labtek slides are to be used) in a volume of 1.0 ml of complete medium to a single chamber or to the well of the tissue culture plate. Incubate cultures for 1–2 hr at 37° in an atmosphere containing 5% CO_2. At the end of this incubation period wash the wells vigorously with the aid of a repeating Cornwall syringe and a Pasteur pipet attached to a vacuum source.[19] Immediately after washing, and prior to moving to the next well, add 0.5 ml of saline to avoid drying of the monolayer and damaging the cells.

Challenge with Toxoplasma

Step 3. Obtain tachyzoites from tissue culture or from mice and wash twice with complete medium by centrifugation at 225 g for 15 min at 4°.

[14] J. E. Swartzberg, J. L. Krahenbuhl, and J. S. Remington, *Infect. Immun.* **12**, 1037 (1975).
[15] L. P. Ruco and M. S. Meltzer, *J. Immunol.* **120**, 1054 (1978).
[16] J. R. David, *Fed. Proc., Fed. Am. Soc. Exp. Biol.* **34**, 1730 (1975).
[17] M. Rabinovitch and S. I. Hamburg, "Mononuclear Phagocytes," Part II, p. 1515. Mosby, St. Louis, 1980.
[18] Coverslips are first cleaned by soaking in 1 M KOH solution overnight, rinsing exhaustively with cold deionized H_2O followed by rinsing in boiling double distilled H_2O. The coverslips are then sterilized by dry heat.
[19] It is recommended that the Pasteur pipet be held toward one end of the chamber or well and the syringe at the other end. This results in a whirlpool that removes nonadherent cells efficiently.

After the second wash, resuspend tachyzoites in complete medium. Take an aliquot and add 0.1 ml of acridine orange to a final concentration of 5 μg/ml and determine the viability of tachyzoites. Live tachyzoites contain bright orange lysosomes against a dark green cytoplasm whereas dead ones show light orange–green cytoplasmic fluorescence and no lysosomal staining. Adjust concentration to 2×10^6 viable tachyzoites/ml.

Aspirate the saline solution from wells or chambers and add 1 ml of tachyzoite suspension to each chamber or well, making certain that the entire monolayer is evenly covered. Incubate the cultures again in 5% CO_2 for 1 hr at 37°. At the end of the incubation, wash the monolayers at least five times as described above to remove extracellular *Toxoplasma*.

Step 4A.1. Add 1 ml of complete medium to all but one slide[20] and reincubate the remaining cultures for 6, 18–24, 36–48, and 72 hr.

Determination of Toxicity by Staining

Step 4A.2. At the end of each incubation period wash the wells or chambers with saline and fix in aminoacridine fixative for at least 1 hr at 4° (better results are obtained with an overnight incubation in the fixative).

Step 4A.3. Remove the coverslips from their wells and mount them with permount face up (i.e., the side containing the monolayer) on microscope slides. Stain the coverslips and/or the Labtek slides with Giemsa stain for 10 min and wash the excess stain under a gentle stream of H_2O from a faucet. Mount coverslips on stained slides with the aid of permount.

Step 4A.4. Using a 40× objective, determine the number of *Toxoplasma*/100 macrophages and the number of *Toxoplasma*/vacuole in each infected macrophage. It is relatively easy to identify and count the *Toxoplasma* (Fig. 2), but the following guidelines must be observed when assessing infected monolayers to obtain meaningful results:

1. Scan the entire slide to ensure that the fields counted are representative of the entire experiment. At least 200 cells should be counted for infection rates >20% and 500–1000 cells for infection rates <20%.

2. Only monolayers of similar cell density should be compared. A major decrease in cell density may imply loss of infected as well as uninfected adherent cells from an unhealthy monolayer. Assess cell density by counting the number of cells in 10 fields with the 40× objective.

3. Count only *Toxplasma* which have normal morphology. Dead organisms appear shrunken, with no clear distinction between nucleus and

[20] Process this slide according to the method described in Step 4A.2; this corresponds to 0 hr.

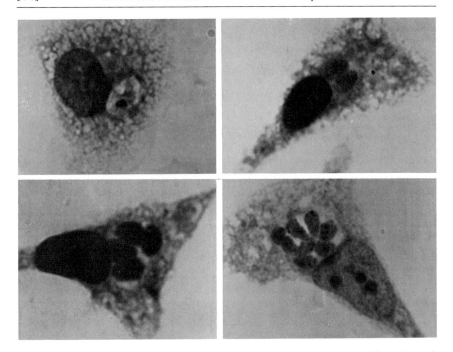

Fig. 2. *Toxoplasma gondii* within murine peritoneal macrophages, fixed in aminoacri-dine–ethanol and stained with Wright–Giemsa solution (×400). The macrophage mono-layers were exposed to tachyzoites for 1 hr, extracellular organisms were removed by washing, and medium was readded. The photographs were taken immediately after washing off the extracellular organisms (upper right-hand panel), and 8 hr (upper left) and at 18 hr (both lower panels).

cytoplasm, and are eventually digested by normal and activated macro-phages.

4. Uninfected cells must also be taken into account, for their number may vary as the experiment proceeds. Thus, viable *Toxoplasma* released from cells unable to kill them may infect a cell that was previously unin-fected, whereas a cell which normally kills the organism may revert from an infected to an uninfected state. For this reason, results are usually expressed as the number of *Toxoplasma*/100 macrophages rather than the number of *Toxoplasma*/100 infected macrophages.[21] As an increase in the number of *Toxoplasma*/100 macrophages could be due either to multipli-cation or to new infection, the number of *Toxoplasma*/vacuole should also be estimated, for this reflects intracellular multiplication.

[21] M. Chincilla and J. K. Frenkel, *Infect. Immun.* **19**, 999 (1978).

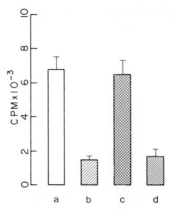

FIG. 3. Uptake of [³H]uracil by *Toxoplasma* in normal macrophages (a, c) and activated macrophages (b, d). Results of two separate experiments are presented and show that *Toxoplasma* continue to multiply in normal macrophages and take up more [³H]uracil in comparison to *Toxoplasma* that fail to multiply or are killed within activated macrophages.

Determination of Toxicity Using Radioactivity

Step 4B.1. Add 1 ml of complete medium to triplicate wells or chambers. Then add 10 μCi of [³H]uracil in a 10-μl volume. Include three wells of uninfected monolayers of each kind and three wells of tachyzoites alone as controls. Incubate the cultures in 5% CO_2 for 24 hr at 37°.

Step 4B.2. Wash the wells or chambers with saline as described above, then add 1 ml of 0.3 N TCA solution containing 1% SDS and 100 μg/ml cold uracil. Allow the precipitation to occur for at least 60 min at 4°. It is preferable to let it stand overnight.

Step 4B.3. Maintain precipitates on ice while collecting them on glass filters. Wash the filters five times with cold 0.3 N TCA. Rinse three times with 95% alcohol. Dry filters under an infrared lamp or in an oven at 60° for 1 hr.

Step 4B.4. Transfer the dried filters to vials containing scintillation fluid compatible with aqueous samples and determine radioactivity in a liquid scintillation spectrometer.

Step 4B.5. Express results as counts/minute (cpm) or difference in cpm between infected and uninfected cultures (Fig. 3). Evaluate statistical differences utilizing Student's t test.

Comments

Microscopic observation of intracellular *T. gondii* allows direct visualization of the organism. The radiometric method is quicker and more objective but does not distinguish between dead organisms and nonreplicating viable organisms. Thus, the methods are complementary and, ideally, should be used in parallel.

Acknowledgments

This work was supported by Grant Number AI 04717 from the National Institutes of Health. James R. Catterall is the recipient of an MRC (UK) Travelling Fellowship and a Francis S. North Foundation Fellowship.

[45] Parasiticidal Activity of Macrophages against *Plasmodium*

By Hannah Lustig Shear and Christian F. Ockenhouse

Introduction

Malaria, a protozoan parasitic disease caused by members of the genus *Plasmodium,* is still one of the major infectious diseases in the world today. The life cycle of this parasite alternates between a sexual cycle in the mosquito vector and two asexual cycles, first in the host's liver and then in the erythrocytes. The actual symptoms of malaria, the shaking chills, fever, and sweating, are caused by the cycle in the erythrocytes in which parasites periodically burst out of the red blood cells and invade fresh erythrocytes. A large part of the immune response is directed against this phase of the life cycle.

Numerous studies in both humans and experimental hosts indicate that acquired immunity to malaria is slow to develop, requires repeated exposure to the parasite, and wanes if the host is no longer exposed.[1] Although passively transferred immune serum provides some protection against malaria,[2] there are studies indicating that antibody alone is insufficient for protection[3] and that immunity to rechallenge may be cell mediated.[4]

The early studies of Taliaferro and Cannon[5,6] suggested that tissue macrophages play a role in immunity to malaria since parasitized erythro-

[1] J. H. L. Playfair, *Br. Med. Bull.* **38,** 153 (1982).
[2] S. Cohn, I. A. McGregor, and S. P. Carrington, *Nature (London)* **192,** 733 (1961).
[3] W. Cantrell, E. E. Elko, and B. M. Hopff, *Exp. Parasitol.* **28,** 291 (1970).
[4] J. L. Green and W. O. Weidanz, *Nature (London)* **290,** 143 (1981).
[5] W. H. Taliaferro and P. R. Cannon, *J. Infect. Dis.* **59,** 72 (1936).
[6] P. R. Cannon and W. H. Taliaferro, *J. Prev. Med.* **5,** 39 (1931).

cytes were observed within the phagocytic vacuoles of spleen macrophages of monkeys and birds infected with malaria. Other studies showed that human monocytes phagocytize *P. falciparum*[7,8] and that rodent macrophages take up *P. berghei*[9] and *P. yoelii.*[10] However, some investigators have had difficulty in demonstrating phagocytosis of parasitized erythrocytes *in vitro*,[11,12] and in some studies the levels of ingestion are low. In addition, it has been suggested[13] that phagocytosis cannot account for the clearance of a malarial infection and the appearance of "crisis forms" which are degenerating intraerythrocytic parasites. Since infections and agents which activate macrophages have a protective effect on experimental malaria infections,[14,15] it was suggested that activated macrophages secrete soluble mediators which are toxic to the parasites.

Early studies of Langhorne *et al.*[16] indicated that incubation of spleen cells from infected monkeys with parasitized erythrocytes reduced their ability to multiply. Later, Taverne *et al.*[17] demonstrated the killing of *P. yoelii* by cells of the monocyte–macrophage series. Data from this study suggested that fresh blood monocytes or activated rodent peritoneal cells were more effective killers than normal cells. However, phagocytosis could not be ruled out as the mechanism of killing in these studies. We, therefore, developed methods to detect the ability of rodent peritoneal macrophages and human monocytes to kill intraerythrocytic parasites *in vitro.*

In this chapter we describe methods for activating macrophages[9] which enhances the phagocytosis and killing of intraerythrocytic parasites. We also describe phagocytic and cytotoxicity assays using agents that are infectious for both humans and rodents.

Sources of Macrophages

Mice. Female SW mice, 4–6 weeks old, are obtained from Taconic Farms, Inc., Germantown, NY. Mice are sacrificed by anesthetizing with ether, followed by bleeding from the axillary vein.

[7] A. Vernes, *Lancet* **2**, 1297 (1980).

[8] A. Celada, A. Cruchard, and L. H. Perrin, *Clin. Exp. Immunol.* **47**, 635 (1982).

[9] H. L. Shear, R. S. Nussenzweig, and C. Bianco, *J. Exp. Med.* **149**, 1288 (1979).

[10] C. E. Tosta and N. Wedderburn, *Clin. Exp. Immunol.* **42**, 114 (1980).

[11] J. Chow and J. P. Kreier, *Exp. Parasitol.* **31**, 13 (1972).

[12] S. Trubowitz and B. Maseck, *Science* **162**, 273 (1968).

[13] A. C. Allison and I. A. Clark, *Am. J. Trop. Med. Hyg.* **26**, 216 (1977).

[14] I. A. Clark, A. C. Allison, and F. E. Cox, *Nature (London)* **259**, 310 (1976).

[15] C. M. Rzepczyk, *Immunology* **46**, 261 (1982).

[16] J. Langhorne, G. A. Butcher, G. H. Mitchell, and S. Cohn, *in* "The Role of the Spleen in the Immunology of Parasitic Diseases." Schwabe, Basel, 1979.

[17] J. Taverne, H. M. Dockrell, and J. H. L. Playfair, *Parasite Immunol.* **4**, 77 (1982).

Peritoneal Cells[18] (see also this series, Vol. 108 [25]). All cell preparations are carried out under sterile conditions. The peritoneal cavities of the mice are washed with 4.0 ml sterile PBS.[18a] The cells are centrifuged at 500g, resuspended in D10 [Dulbecco's modified Eagle's medium (DMEM) containing 10% fetal calf serum] [Grand Island Biological Co. (GIBCO), Grand Island, NY], and counted. Cells are resuspended to 1.0 × 10^6/ml and 0.1 ml plated onto 13- or 15-mm round glass cover slips (Clay Adams, Div. Becton Dickinson, Parsippany, NJ). In some experiments, mice are injected intraperitoneally with 1.0 ml brewer thioglycolate medium (Difco, Inc., Detroit, MI) 4 days before harvesting the cells.

Spleen Cells.[9] Spleens are teased in PBS, pH 7.4, containing 0.01M EDTA and 1% glucose. The spleen cell suspension is washed once by centrifugation, resuspended to 5 × 10^6 cells/ml in DMEM containing 20% fetal calf serum, 100 U/ml penicillin, and 100 mg/ml streptomycin (GIBCO). Cells (0.1 ml) are pipetted onto 13-mm round glass cover slips, incubated at 37° in 5% CO$_2$, 100% humidity for 45 min, and rinsed in DMEM. Macrophages are defined as adherent cells that ingest sheep erythrocytes sensitized with rabbit IgG. Lymphocytes, platelets, erythrocytic cells, and dendritic cells are also observed on the cover slips. About 40% of the adherent cell population of the spleen from normal mice is constituted by macrophages. In the spleen cultures of malaria-infected animals, this percentage increases to 67% on day 7 and to 75% on days 14–21. These macrophages often contain malaria pigment.

Human Monocytes and Monocyte-Derived Macrophages. Leukocyte concentrates (buffy coat cells) may be obtained from healthy volunteers. Leukocytes are separated from the red blood cells by centrifugation over Ficoll-Hypaque (Pharmacia Fine Chemicals, Piscataway, NJ) according to standard methods[18] (see also this series, Vol. 108 [9]). Cells are resuspended to 5 × 10^7/ml in RPMI 1640 (GIBCO) containing 15% human serum and penicillin (100 U/ml) and streptomycin (100 μg/ml) (GIBCO). As described for rodent macrophages, 0.1 ml of the cell suspension is pipetted onto glass coverslips, incubated for 1–2 hr at 37°, 7.5% CO$_2$, and rinsed. After washing, more than 95% of adherent cells are monocytes by morphologic criteria. The coverslips are placed into 16-mm, 24-well tissue culture clusters (Costar, Cambridge, MA) containing 0.5 ml *complete medium* [DMEM containing 15% human serum and penicillin (100 U/ml)

---

[18] P. Edelson and Z. A. Cohn, *in* "*In Vitro* Methods in Cell-Mediated and Tumor Immunity" (B. R. Bloom and J. R. David, eds.), p. 333. Academic Press, New York, 1976.

[18a] Abbreviations: PBS, phosphate-buffered saline; DMEM, Dulbecco's modified Eagle's medium; D10, DMEM containing 10% fetal calf serum; BCG, Bacillus Calmette–Guérin; LK, lymphokine; HK, heat killed; PE, parasitized erythrocytes; Con A, concanavalin A; PMA, phorbol myristate acetate; BSA, bovine serum albumin; EDTA, ethylenediaminetetraacetic acid.

and streptomycin (100 μg/ml)] and are either used immediately or cultured for 4 days, after which we call the cultures monocyte-derived macrophages.[19]

Macrophage Activation

Activation of Macrophages in Vivo. Macrophages are activated *in vivo* by infection of mice with either BCG (bacillus Calmette–Guérin) or with malaria parasites.[20]

BCG may be obtained from the Trudeau Institute, Saranac Lake, NY. Bacteria are mixed with 0.05% Tween 80 in equal volumes and sonicated three times with 3-sec bursts of an Astrason ultrasonic cleaner (Ultrasonic, Plainview, NY). The organisms are then resuspended to 10^8/ml in PBS and 0.2 ml injected/mouse intravenously. Three weeks after infection, mice are challenged intraperitoneally with 2×10^7 heat-killed BCG (see below).

Macrophages may also be activated by infection of mice with malaria, as described below.

Activation of Macrophages in Vitro. Lymphokines (LK) may be prepared from BCG-infected mice and from malaria-infected mice as follows.[20]

Heat-killed (HK) BCG is obtained by autoclaving a suspension of 5×10^6 viable BCG/ml at 15 lb/in.^{2} for 15 min. Malarial antigen is prepared from parasitized erythrocytes (PE) which are enriched by centrifugation of erythrocytes in DMEM at 750 g for 10 min. The upper layer is removed and usually contains >90% PE. A 40-μl sample of enriched PE pellet is lysed with 0.3 ml of 0.2% NaCl solution. Isotonicity is restored with the addition of an equal volume of 1.6% NaCl solution.

Unfractionated spleen cells (1.5×10^7/ml) from individual mice infected for 3 weeks with sonicated BCG or individual malaria-infected mice (30–60% parasitemia) are incubated with either 5×10^6 HK-BCG or 15 μl of lysed PE suspension/ml, respectively, for 48 hr at 37° in 7.5% CO_2. The cells are incubated in D10 containing 50 μM mercaptoethanol. Supernatants are collected and centrifuged at 500 g for 15 min, sterilized by filtration, and stored at −70°. Before use, supernatants are thawed and diluted 1 : 5 in D10. Mouse peritoneal macrophage monolayers are reincubated for up to 48 hr in 16-mm 24-well plates (Costar) in D10 plus spleen cell supernatants before the phagocytosis or cytotoxicity assay.

LK may also be prepared from concanavalin A (Con A)-stimulated human mononuclear cells[20] by cultivating unseparated mononuclear cells

[19] C. F. Ockenhouse, S. Schulman, and H. L. Shear, *J. Immunol.* **133,** 1601 (1984).
[20] C. F. Ockenhouse and H. L. Shear, *Infect. Immun.* **42,** 733 (1983).

(1×10^7 cells/ml) with Con A (15 μg/ml) in RPMI 1640 supplemented with 7.5% human serum, penicillin (100 U/ml), and streptomycin (100 μg/ml) and 50 μM mercaptoethanol, at 37°, 7.5% CO_2 (see this series, Vol. 116, [38]). After 48 hr the supernatants are collected, centrifuged at 700 g, sterilized by filtration through 0.45-μm amicon filters, and stored at 4 or −70°. Control LK are prepared by adding Con A to the cell suspension at the end of the incubation period.

Human macrophages are activated by incubation with Con A-induced LK (10% v/v) or with recombinant DNA-derived human γ-interferon (500 U/ml) (Genentech, Inc., South San Francisco, CA) for 72 hr, beginning on day 3 or 4 after the monocytes are separated from whole blood. Fresh LK are added with complete medium every 24 hr.

Maintenance of Parasites

Rodent malaria parasites are maintained by blood passage in mice. Parasitemias are determined on Giemsa-stained blood films. Blood is obtained in heparin, the number of red cells counted, and a fixed number of parasites injected into mice.

In most experiments, 10^4 infected erythrocytes of *Plasmodium berghei* (NK 65 strain) *P. yoelii* (nonlethal), or *P. chabaudi* (lethal) are injected intravenously.[20a] In some experiments, 10^6 *P. yoelii*-infected erythrocytes are injected intraperitoneally. Parasitized erythrocytes are purified by first spinning heparinized blood diluted in 15 ml DMEM at 750 g for 10 min. The dark upper portion of the pellet containing the parasitized erythrocytes is removed and resuspended in 2.0 ml PBS. The washed parasitized erythrocytes are then passed through a syringe containing 6 ml of CF-11 Whatman cellulose powder. The cellulose retains virtually all the leukocytes and most of the platelets and allows the erythrocytes and reticulocytes to pass through.[21]

P. falciparum parasites (clone A-2 of the FCR/3 Knob$^+$ strain) are cultured according to the method of Trager and Jensen[22] in RPMI 1640 (GIBCO) containing 10% human serum, *N*-trismethyl-2-aminoethanesulfonic acid (50 μg/ml), hypoxanthine (50 μg/ml), and glucose (2 mg/ml) (all from Sigma Chemical Company, St. Louis, MO) in human blood type A$^+$ erythrocytes. Parasites are synchronized by the sorbitol lysis method[23] as follows: 5-ml cultures of *P. falciparum* are centrifuged at 200 g for 5 min. The supernatant is discarded and the cells are resuspended in

[20a] These parasites are available from the author (Dr. H. L. Shear) upon request.
[21] E. Beutler, C. West, and K. G. Blume, *J. Lab. Clin. Med.* **88,** 328 (1976).
[22] W. Trager and J. B. Jensen, *Science* **193,** 673 (1976).
[23] C. Lambros and J. P. Vanderberg, *J. Parasitol.* **65,** 418 (1979).

2.5 ml of a 5% solution of D-sorbitol for 5 min at room temperature. The cells are resuspended in the same volume of D-sorbitol and centrifuged once again (200 g, 5 min). An equal volume of RPMI 1640 containing 10% human serum is added to the pellet and the cells are washed twice. Cultures are reestablished by addition of uninfected erythrocytes and medium to give a 12.5% hematocrit and a 0.1% parasitemia. The cultures are maintained in a Heraeus incubator (The Rupp and Bowman Company, Voorhees, NJ) at 37°. Before the cytotoxicity assay, human erythrocytes (A^+) are washed in RPMI 1640 and adjusted to 2% hematocrit at 1.5–2% parasitemia in complete culture medium.

Parasites are used at the late trophozoite or early schizont stage. The erythrocytic cycle of *P. falciparum* malaria is characterized by early trophozoites which have a ring form. In Giemsa-stained preparations, the cytoplasm is pale blue and there are 1–2 violet dots of chromatin (excellent drawings may be found in Katz et al.[23a]). The late trophozoites are larger and usually have a compact rather than ameboid cytoplasm. As the parasite matures to an early schizont or dividing form, numerous chromatin clumps are present in the cytoplasm. Mature schizonts contain 8–32 blue-stained bodies, each containing a violet dot of chromatin. At 48-hr intervals the schizonts rupture, releasing the newly formed merozoites which invade fresh erythrocytes giving rise to new ring forms (Fig. 1).

Phagocytic Assays

A procedure adapted from Bianco[24] is used. Macrophages plated on glass coverslips are rinsed in DMEM and overlayered with 0.1 ml of a suspension of parasitized erythrocytes (diluted to 0.5% or 2 × 10^8/ml). The coverslips are incubated for 45 min at 37°, 7.5% CO_2, 100% humidity. After the incubation, coverslips are rinsed in DMEM to assess binding to macrophages or in PBS diluted 1 : 5 with distilled water (which lyses the bound erythrocytes) to determine the number of erythrocytes ingested. Coverslips are then fixed in glutaraldehyde (1.25%, 10 min), rinsed in distilled water, and viewed by phase-contrast microscopy.

Cytotoxicity Assays against Intraerythrocytic Malaria Parasites

Cytotoxicity of Activated Macrophages against P. yoelii. The ability of macrophages to kill intraerythrocytic *P. yoelii* can be assessed in two

[23a] M. Katz, D. D. Despommier, and R. Gwadz, "Parasitic Diseases," Fig. 28.11. Springer-Verlag, Berlin and New York, 1982.

[24] C. Bianco, *in* "*In Vitro* Methods in Cell-Mediated and Tumor Immunity" (B. R. Bloom and J. R. David, eds.), p. 407. Academic Press, New York, 1976.

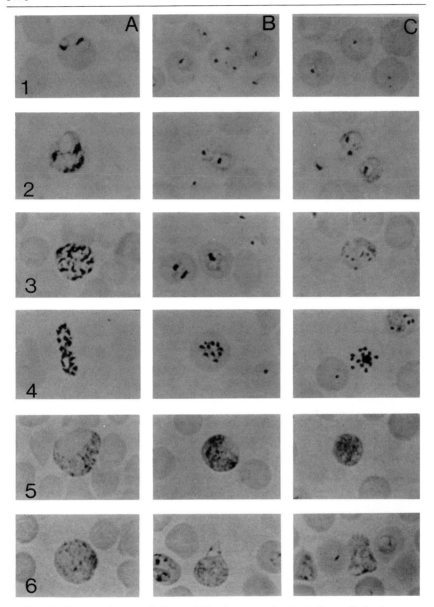

Fig. 1. Comparative morphology of blood stages of rodent plasmodia in the mouse (Giemsa stain) A1–6: *P. yoelii;* B1–6: *P. chabaudi;* C1–6: *P. petteri;* A1–C1: early ring stages; A2–C2: young trophozoites; A3–C3: segmenting late trophozoites; A4–C4: mature schizents; A5–C5: microgametocytes; A6–C6: macrogametocytes. From "Rodent Malaria" (R. Killick-Kendrick and W. Peters, eds.), p. 68. Academic Press, New York, 1978.

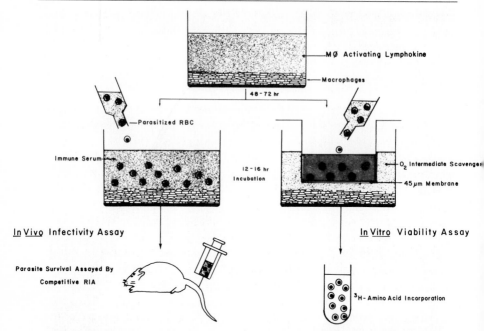

FIG. 2. Macrophage-mediated cytotoxicity assay. Macrophages are activated by incubation with macrophage-activating lymphokine. Parasites are added either directly onto the
macrophages (left) or in Adaps chambers (right). Parasite viability is measured either by
injection into mice followed by RIA for parasite antigen in the blood of injected mice (left) or
by their ability to incorporate ³H-labeled amino acids.

ways (Fig. 2). The first method, shown on the left of Fig. 2 (*In Vivo
Infectivity Assay*), is to incubate macrophages and infected erythrocytes
in direct contact and then inject the contents of the reaction into naive
mice. The level of parasitemia is then determined by a competitive RIA
which measures the inhibition of antibody binding to malarial antigen-
coated plates.

   The RIA, adapted from Avraham *et al.*[25] and Mackey *et al.*[26] is as
follows.[27] A 25-μl sample of blood is drawn from mice injected 3 days
previously with viable PE (10^2 to 10^6), or PE that survived the cytotoxicity assay. The blood sample is resuspended to 0.1 ml, and 50 μl of diluted
blood is placed into the wells of a 96-well microtiter plate. The plates are
frozen and thawed six times to disrupt parasite antigen. Anti-*P. yoelii*

[25] H. Avraham, D. T. Spira, and D. Sulitzeanu, *J. Parasitol.* **68,** 177 (1982).
[26] L. Mackey, L. Perrin, and P. H. Lambert, *Parasitology* **80,** 171 (1980).
[27] C. F. Ockenhouse and H. L. Shear, *J. Immunol.* **132,** 424 (1984).

hyperimmune serum (50 μl) at a titer of 1/8000 or 1/16,000 is added, and the plates are held overnight at 4°. (The hyperimmune serum is prepared by injecting ip into *P. yoelii*-treated mice 400 μg chloroquine in 0.5 ml for three consecutive days, followed by three weekly injections of 3×10^6 PE each. The mice are bled 1 week after the last injection.) Samples (30 μl) from the microtiter plates are placed in each of two *P. yoelii* antigen-coated wells and incubated for 6 hr at 4°. Malaria antigen-coated wells are prepared by resuspending 0.3 ml of packed PE in 100 ml PBS. Fifty-microliter samples are dispensed into each well of a disposable polyvinyl chloride 96-well plate (Dynatech Laboratories, Alexandria, VA). These plates are frozen and thawed six times, washed with 1% bovine serum albumin (BSA) in PBS, and incubated with 1% BSA for 30 min before the addition of the 30-μl samples of transferred antibody. After washing to remove unbound antibody, 30 μl of affinity-purified ^{125}I-labeled anti-mouse IgG (Kirkegaard and Perry Laboratories, Inc., Gaithersburg, MD) is added and incubated at room temperature for 1 hr. The anti-mouse IgG is labeled with 1 to 2 mCi ^{125}I in the presence of iodogen according to instructions by the manufacturer (Pierce Chemical Company, Rockford, IL). The wells are washed three times, dried, cut out, and the radioactivity is determined in a gamma counter. The results are expressed as the percentage of antibody binding inhibition for each sample compared with control wells. Controls include PBS, as well as normal frozen and thawed erythrocytes, incubated with normal and immune serum. The inhibition of antibody binding is directly proportional to parasites surviving the cytotoxicity assay. It is also possible to calculate the percentage surviving parasites for each sample from a standard curve obtained simultaneously with each assay from mice injected with viable parasites.

Using this method it has been established that macrophages activated *in vivo* and *in vitro* are able to reduce the number of viable parasites to a significantly greater degree than macrophages from normal mice. However, since this method does not allow one to differentiate between intracellular and extracellular killing, a method has been developed using chambers which separated the macrophages from the parasites by a Millipore filter,[20] shown on the right of Fig. 2 (*In Vitro* Viability Assay).

Preparation of the Chambers. The preparation of the chambers is critical because of potential problems with sterility and toxicity. The method used is adapted from Ullrich and Zolla-Pazner,[28] as follows: Adaps chambers (Adaps, Inc., Dedham, MA) are washed and dried. Millipore filters (0.45 μm) are then glued with a spare amount of silicone glue

---

[28] S. E. Ullrich and S. Zolla-Pazner, *J. Immunol. Methods* **39**, 147 (1980).

to the bottom of the chamber vessels, taking care not to get any glue inside the chamber. The chambers are dried for about 3 hr and then detoxified by incubating them in PBS containing 5% fetal calf serum for 48 hr, with two changes of the same buffer. Macrophages ($3-4.5 \times 10^6$) are plated on 13-mm glass coverslips which are then placed in 24-well tissue culture clusters (Costar). These serve as the bottom wells. The Adaps chambers are placed in the wells to form the upper chamber. Parasitized erythrocytes ($1-2 \times 10^7$) are pipetted into the upper chamber. Approximately 2 mm separates the macrophage monolayers from the parasites.

Assessment of Parasite Viability in Vitro. After incubation of the parasitized erythrocytes with macrophages for 12–16 hr as described above, the parasitized erythrocytes are resuspended in the chambers and transferred to 1.5-ml microfuge tubes. The cells are washed with PBS and assayed for viability by the synthesis of parasite-specific protein. For this, the cells are resuspended in 200 μl isoleucine-free RPMI medium to which 3.2 μCi/ml of [^{3}H]isoleucine is added. The tubes are incubated for 2 hr at room temperature on a shaker. The material is then washed and the radioactivity is determined in a liquid scintillation counter.

The results obtained with this assay confirm the findings with direct incubation of macrophages and parasitized erythrocytes; i.e., macrophages activated *in vivo* or *in vitro* are able to kill intraerythrocytic parasites. However, in this assay, the macrophages have to be triggered to secrete a cytotoxic factor by adding a phagocytic stimulus (for example, 10^6 PE) to the bottom well. Further, when the phagocytic stimulus consists of parasitized erythrocytes, hyperimmune serum enhances the killing (Table I).

Cytotoxicity of LK-Activated Macrophages against P. falciparum. Mononuclear leukocytes, monocyte-derived macrophages or LK-activated, monocyte-derived macrophages are assayed for their ability to kill *P. falciparum* as described in Ref. 21.

Medium is removed from each of the wells containing the effector cell population to be tested and is replaced with 0.5 ml trophozoite–schizont-infected *P. falciparum* erythrocytes (1.5 to 2% parasitemia, 2% hematocrit) resuspended in RPMI 1640 complete medium. The effector and target cells are then incubated under the conditions for culturing *P. falciparum* described above. After 24 hr the cultures are replenished with fresh medium. During this time, schizogony is completed, resulting in invasion of released merozoites into uninfected erythrocytes. At the conclusion of the incubation period (normally 48 hr) the erythrocytes are removed and parasitemias are determined by counting at least 1000 red blood cells on Giemsa-stained blood films. Control cultures consisting of parasitized

TABLE I
EFFECT OF HYPERIMMUNE SERUM ON THE KILLING OF *P. yoelii* PE BY
LK-ACTIVATED MACROPHAGES[a]

Macrophage	Phagocytic stimulus[b]	Antibody titer[c]	Percentage [³H]isoleucine incorporation compared with control[d,e]
−	−	−	100
+	−	−	96 ± 6
+	+	−	65 ± 3
+	+	10^2	50 ± 2
+	+	10^3	41 ± 6
+	+	10^4	64 ± 4
+	PMA (200 ng/ml)	−	62 ± 4

[a] Reprinted with permission from Ockenhouse and Shear.[27]

[b] Resident macrophages (4.5×10^6) were stimulated for 48 hr by an LK prepared from the spleens of *P. yoelii*-infected mice. Target PE (1×10^7) were suspended in Adaps chambers separated from macrophages by a 0.45-μm filter.

[c] Lysed PE served as the phagocytic stimulus to macrophages in bottom chamber of culture vessel.

[d] Hyperimmune serum (final dilution 10^{-2} to 10^{-4}) was added with lysed PE to macrophage monolayers.

[e] Results expressed as the mean cpm [³H]isoleucine incorporation ± SE of triplicate samples compared with control PE incubated without hyperimmune serum and in the absence of effector cells.

erythrocytes cultured in the absence of effector cells have an 8- to 10-fold increase in parasitemia after 48 hr. The results are expressed as the percentage inhibition of multiplication of parasites exposed to effector cells as compared to control cultures.

Identification of the Effector Molecule Involved in Cytotoxicity against Intraerythrocytic Malaria Parasites. LK-activated macrophages are incubated in the presence of a phagocytic stimulus (lysed PE, 10^7/ml), opsonized zymosan (0.5 mg/ml), or a membrane-triggering agent [phorbol myristate acetate (PMA), 500 ng/ml]. The degree of cytotoxicity against *P. yoelii* or *P. falciparum* in these cultures is compared with similar cultures to which various inhibitors of reactive oxygen metabolites are added (Table II). We found that catalase (500 μg or 1 mg/ml) and not superoxide dismutase (1 mg/ml), histidine (10 mM), or mannitol (50 mM) prevents parasite death. This suggests that H_2O_2 is toxic to *Plasmodia*.

TABLE II
PARTICIPATION OF REACTIVE OXYGEN INTERMEDIATES IN THE KILLING
OF *P. yoelii* PE BY LK-ACTIVATED MACROPHAGES[a]

Phagocytic stimulus[b]	Scavengers	Percentage [³H]isoleucine incorporation[c,d]
None	None	91.0 ± 6
PMA (500 ng/ml)	None	40.0 ± 7
Zymosan (0.5 mg/ml)	None	33.0 ± 9
PE (2×10^7)	None	31.0 ± 4
PE	Catalase, 500 μg/ml	66.0 ± 7
PE	Catalase, inactivated	39.0 ± 5
PE	SOD, 1 mg/ml	41.0 ± 4
PE	Histidine, 10 mM	23.0 ± 6
PE	Mannitol, 50 mM	37.0 ± 7
PE	Pretriggered with PMA; glucose depleted for 2 hr	86.0 ± 11

[a] Reprinted with permission from Ockenhouse and Shear.[27]
[b] Resident macrophages (4×10^6) were stimulated for 48 hr by an LK prepared from spleens of *P. yoelii*-infected mice. Target PE (1×10^7) were suspended in Adaps chambers separated from macrophages by 0.45-μm filter.
[c] Phagocytic stimulus was added to the macrophage monolayer incubated in the bottom chamber of the culture vessel. Only the target PE in the upper chamber were assayed for [³H]isoleucine incorporation.
[d] Results expressed as the mean cpm [³H]isoleucine incorporation $\pm$ SE of triplicate samples compared with control PE incubated in the absence of effector cells.

This is corroborated by the finding that macrophages depleted of H_2O_2 and cultured in the absence of glucose cannot kill *P. yoelii*-infected erythrocytes. In addition H_2O_2 generated enzymatically is toxic to *P. yoelii* and *P. falciparum*.[19,27]

[46] Parasiticidal Activity of Macrophages against Schistosomes

By STEPHANIE L. JAMES

Chronic infection with a variety of pathogens has been shown to stimulate enhanced resistance to subsequent infection by antigenically unrelated microorganisms. As first described in bacterial systems,[1] this type of nonspecific immunity is mediated by macrophages that are activated for enhanced effector function upon stimulation by mediator molecules (lymphokines) produced as a result of antigen-specific lymphocyte response to the initial infection.[2] These activated macrophages acquire potent microbicidal effector mechanisms capable of destroying diverse intracellular and extracellular targets, including bacteria, rickettsia, protozoal parasites, and tumor cells.[3] However, surely one of the most ambitious feats performed by activated macrophages is the killing of the invasive larval form of the helminthic parasite *Schistosoma mansoni,* a multicellular target many times their own size (Fig. 1). This phenomenon has been most extensively studied using murine effector cells. Macrophage activation to kill schistosomula can be achieved *in vivo* by treatment with the nonspecific activators *Mycobacterium bovis,* strain BCG, or *Corynebacterium parvum,*[4,5] as well as through specific immunization of the animal by primary schistosome infection[6] or vaccination with radiation-attenuated parasites.[7] As previously described in other systems,[2] activated cytocidal macrophages can be collected from these animals at the site of the original inoculation (usually the peritoneal cavity) or at the site of specific antigen challenge. *In vitro,* peritoneal macrophages from normal control mice can be activated to kill schistosomula by treatment with lymphokines obtained from mitogen-stimulated lymphocytes[8] or from specific antigen

---

[1] G. B. Mackaness, *J. Exp. Med.* **129,** 973 (1969).

[2] J. B. Hibbs, Jr., *in* "The Macrophage" (M. Fink, ed.), p. 83. Academic Press, New York, 1976.

[3] C. A. Nacy, C. N. Oster, S. L. James, and M. S. Meltzer, *Contemp. Top. Immunobiol.* **13,** 147 (1984).

[4] A. A. F. Mahmoud, P. A. Peters, R. H. Civil, and J. S. Remington, *J. Immunol.* **122,** 1655 (1979).

[5] S. L. James, E. J. Leonard, and M. S. Meltzer, *Cell. Immunol.* **74,** 86 (1982).

[6] S. L. James, E. Skamene, and M. S. Meltzer, *J. Immunol.* **131,** 948 (1983).

[7] S. L. James, P. C. Natovitz, W. L. Farrar, and E. J. Leonard, *Infect. Immun.* **44,** 569 (1984).

[8] D. T. Bout, M. Joseph, J. R. David, and A. R. Capron, *J. Immunol.* **127,** 1 (1981).

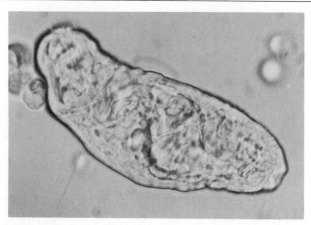

FIG. 1. Living schistosomulum in culture with nonactivated murine peritoneal macrophages. Note translucency and clarity of internal organ structure in viable larva. Effector-to-target size differential is obvious from comparison with cells at upper right and in background ($\times$1000).

stimulation of sensitized T lymphocytes.[5,7] In all cases examined to date, macrophages activated to kill schistosomula also demonstrate cytotoxic activity against tumor cell targets *in vitro*. In both instances, the cytotoxic effector mechanism would be expected to function extracellularly due to the size differential between macrophage and target. As for stimulation of macrophage-mediated tumor cell killing, γ-interferon has been implicated as an important macrophage-activating lymphokine for schistosomulum killing.[7] Moreover, when several inbred strains of mice, previously characterized as defective in various aspects of macrophage activation for tumor cell killing after *in vivo* treatment with bacterial agents, are utilized to compare the genetic control of induction of macrophage cytotoxicity, larvicidal activity develops concurrently with tumoricidal activity.[6] The evidence therefore suggests that regulation of macrophage effector function for cytotoxicity against extracellular helminth or tumor targets may be identical. Observations that certain inbred strains of mice with inherent defects in macrophage function also fail to develop resistance to challenge schistosome infection after primary exposure to radiation-attenuated[9] or viable[6] parasites indicate that activated macrophages may play an important role in resistance to this helminthic infection *in vivo*.

[9] S. L. James, R. Correa-Oliveira, and E. J. Leonard, *J. Immunol.* **133**, 1587 (1984).

Preparation of Schistosomula

The *S. mansoni* life cycle can be maintained in the laboratory by transmission through the *Biomphalaria glabrata* snail intermediate host and a rodent definitive host as described in detail elsewhere.[10] Cercariae, the infective form for the mammalian host, are collected by placing infected snails in a glass beaker containing sufficient spring water to cover and exposing them to fluorescent light at 32° for 1 to 2 hr. As this entire procedure represents a potentially biohazardous situation due to the presence of infective cercariae, laboratory personnel should be cautioned to avoid the area, and care should be taken to wear surgical gloves whenever handling the cercariae. Any skin site inadvertently exposed to cercariae should immediately be flooded with ethanol to kill the parasites.

Schistosomulum targets can be prepared by either of two techniques, with no apparent differences in susceptibility to macrophage-mediated killing *in vitro*. Skin-stage larvae are made by allowing *S. mansoni* cercariae to naturally penetrate an isolated preparation of rodent skin into tissue culture media.[11] Mechanically transformed schistosomula are prepared by artificially severing the tails of the cercariae from the bodies, then incubating the bodies in tissue culture media to allow the structural and metabolic changes consistent with transformation to the parasitic, physiologically adapted form.[12]

Method A: Skin-Stage Schistosomula

1. Tissue culture media should be prewarmed to 37°. Preferred media are either Earle's lactalbumin hydrolysate (Elac)[12a] or Eagle's minimum essential medium with Earle's salts (EMEM) supplemented with 100 U/ml penicillin, 100 μg/ml streptomycin, and 2% (v/v) heat-inactivated (56° for 1 hr) fetal bovine sera (FBS), pH adjusted to 7.2.

2. Mice or rats may be sacrificed by ether overdose; however, care should be taken not to allow contact of ether with the abdominal skin. The abdominal skin is shaved and washed with spring water. The shaved area is resected and removed, leaving the peritoneum intact. In the case of rat

[10] C. L. Lee and R. M. Lewert, *J. Infect. Dis.* **99**, 15 (1956).

[11] M. A. Stirewalt, D. R. Minnick, and W. A. Fregeau, *Trans. R. Soc. Trop. Med. Hyg.* **60**, 352 (1966).

[12] F. J. Ramalho-Pinto, G. Gazzinelli, R. E. Howells, T. A. Mota-Santos, E. A. Figueredo, and J. Pellegrino, *Exp. Parasitol.* **36**, 360 (1974).

[12a] Abbreviations: FBS, fetal bovine serum; EMEM, minimum essential medium with Earle's salts; PPD, purified protein derivative; EDTA, ethylenediaminetetraacetic acid; DMEM, Dulbecco's modified Eagle's medium; Elac: Earle's lactalbumin hydrolysate.

skin, it will be necessary to remove the residual fascia and fat from the underside of the skin by gentle rubbing with a cotton gauze sponge moistened in tissue culture media. Float isolated skin dermal side down on media within a Petri dish to prevent drying until ready to use.

3. A penetration apparatus, designed to create two chambers for holding water (top) and media (bottom) separated by the isolated skin, will be required.[13] This can take many forms; however, a convenient device can be designed from a glass O-ring joint (Corning No. 6780) consisting of members approximately 15 cm long and 2 cm i.d., of which one end has been sealed to a point. The sealed tube is filled with warm media. The previously prepared skin is cut to a convenient size to fit this tube and stretched epidermal side up over the top, taking care to avoid trapping any air between skin and media. A rubber O-ring is then placed over the skin, and the remaining member of the joint is placed on top and secured by a locking pinch clamp. The upper compartment is filled with spring water and the apparatus is checked for leakage.

4. Water containing cercariae is then added to the upper chamber, and the apparatus is placed in a water bath at 37° so that the lower chamber only is immersed. The upper chamber is blocked from any light by covering with foil, since cercariae are phototropic. The cercariae are allowed to penetrate and transform for approximately 3 hr.

5. After 3 hr, a small pellet of schistosomula should be visible at the tip of the lower chamber. The apparatus is then disassembled and the medium is aspirated down to approximately 3 cm from the pellet. The schistomula are resuspended and transferred to a sterile 15-ml centrifuge tube. They are then washed five times in sterile media containing penicillin and streptomycin.

Method B: Mechanically Transformed Schistosomula

1. A one-step density gradient is used for separation of cercarial bodies from tails.[14] Percoll (Pharmacia Fine Chemicals, Uppsala, Sweden), 24 ml, plus 4 ml of 10× concentrated MEM, 1 ml of antibiotic solution containing 10,000 U/ml penicillin and 10,000 g/ml streptomycin, and 1 ml of 1 M HEPES buffer are placed in a polypropylene 50-ml centrifuge tube. The mixture is diluted to 40 ml with sterile distilled water, mixed, and placed on ice.

2. The freshly collected cercariae suspension is placed into a 50-ml polypropylene centrifuge tube and cooled on ice for 30 min to inhibit cercarial movement in the water. The tube is then centrifuged for 2 min at

---

[13] J. A. Clegg and S. R. Smithers, *Int. J. Parasitol.* **2,** 79 (1972).
[14] J. K. Lazdins, M. J. Stein, J. R. David, and A. Sher, *Exp. Parasitol.* **53,** 39 (1982).

70 g to gently pellet the parasites. The water is aspirated and the pellet is resuspended in a final volume of cold EMEM not greater than 5 ml. The suspension is agitated in a vortex mixer for 45 sec up to 2 min, until cercarial tails are broken off by vortex shearing. The suspension is then layered over the cold density gradient.

3. The gradient is centrifuged at 300 g for 15 min in a refrigerated centrifuge at 6°. Refrigeration is necessary to inhibit the motility of tails and any residual cercariae, and thus to prevent disruption of the gradient.

4. Following centrifugation, the composition of the media–Percoll interface is examined. It should consist almost entirely of cercarial tails. A pellet of bodies and residual cercariae should be visible. The gradient is aspirated down to approximately 2 cm from the pellet, taking care to remove all of the interface material. The pellet is resuspended in 50 ml MEM and centrifuged 1 min at 300 g to dilute and remove the remaining Percoll. The media is aspirated and the parasites are resuspended in 20 ml Elac or MEM with 100 U/ml penicillin and 100 g/ml streptomycin. The mixture is incubated for 3 hr at 37° in a humidified 95% air, 5% CO_2 atmosphere with frequent agitation.

5. The larvae are washed in five changes of the sterile incubation media.

Activation of Mouse Peritoneal Macrophages

As with other assays for macrophage activity,[15] it is crucial to use healthy mice as the cell source in these experiments since many of the common viral infections can lead to spurious results. Also to be avoided is the use of strains with characterized defects in macrophage function, such as those with the *Lps* defect, or those derived from the A or P strains.[6] Macrophages can be activated either *in vivo* or *in vitro*.

Method A: Macrophage Activation in Vivo—Intraperitoneal Injection of the Stimulating Agent

Intraperitoneal (ip) injection of *Mycobacterium bovis,* strain BCG (Phipps substrain, TMC No. 1029, Trudeau Mycobacterial Collection, Saranac Lake, NY), at 1 × 10⁶ colony-forming units/mouse, will allow recovery of activated peritoneal macrophages from 7 to 35 days later. Killed *C. parvum* (Wellcome Research Laboratories, Beckenham, England) injected ip at a dose of 70 mg/kg will also provide activated peritoneal cells after 7 to 21 days.

[15] M. S. Meltzer, *in* "Manual of Macrophage Methodology" (H. Herscowitz, H. Holden, J. Bellanti, and A. Ghaffar, eds.), p. 329. Dekker, New York, 1981.

Schistosome-specific stimulation can be provided by ip injection of approximately 60 *S. mansoni* cercariae. While many of these larvae will mature normally and eventually come to reside in the portal venous system, some will remain as stunted worms within the peritoneal cavity and will thus provide continued antigenic stimulation for immune responses leading to macrophage activation. Larvicidal cells can be harvested from the peritoneal cavity at 6 to 12 weeks after infection.

Method B: Macrophage Activation in Vivo—Peritoneal Challenge of Immunized Mice

Activated macrophages can be recovered from the peritoneal cavities 16 to 24 hr after ip injection of 50–100 μg of purified protein derivative of tubercle bacilli (PPD) (Connaught Laboratories, Inc., Toronto, Canada) in mice immunized 3 to 6 weeks previously by intradermal inoculation with 5×10^6 colony-forming units of BCG.

In the schistosome system, activated peritoneal macrophages can be harvested 16 to 24 hr after ip injection of 250 μg soluble adult worm antigen in mice immunized 4 to 10 weeks before by percutaneous exposure to radiation-attenuated cercariae. The antigen is prepared using adult schistosomes, collected from the portal bloodstream of mice after 6 weeks of infection, that have been washed in buffered saline; the worms can be stored at $-20°$ to allow collection of sufficient quantities. The parasites are homogenized in buffered saline, and the homogenate is centrifuged at 100,000 g for 1 hr at $4°$. The supernatant fluid, containing soluble worm antigens, is removed, filter sterilized, and standardized according to protein content. For immunization of mice, cercariae are exposed to 50 Krad of gamma irradiation. The attenuated cercariae are then allowed to penetrate the shaved and washed abdominal skin or tail skin of anesthetized or restrained mice. Approximately 500 cercariae are used per animal during an exposure interval of 30 min to 1 hr.

Method C: Macrophage Activation in Vitro—Preparation of Lymphokine-Containing Supernatant Fluids

1. Lymphokines can also be prepared *in vitro* using either bacterial or schistosome-derived stimuli, and used to activate control macrophages *in vitro*. For this procedure, spleens are removed aseptically from normal mice or from animals immunized by treatment with BCG (5×10^6 units injected intradermally), infection with *S. mansoni* (60 cercariae injected ip), or vaccinated with irradiated cercariae (500 larvae allowed to penetrate the skin). The spleens are pressed through sterile 50-mesh sieves or teased into suspension with forceps in cold RPMI 1640 medium. Large

fragments of tissue are removed following gravity sedimentation. Single cell suspensions are centrifuged at 300 g for 10 min at 4°. The medium is aspirated and cell pellets are resuspended in red blood cell lysing buffer (0.16 M NH$_4$Cl, 0.01 M KHCO$_3$, 10^{-4} M sodium EDTA, pH 7.4) at a ratio of 0.5 ml buffer for each spleen equivalent. After 1–2 min, the lysing buffer is diluted with RPMI 1640 and the cells are again centrifuged. Pellets are resuspended in cold RPMI 1640 and the total leukocyte number determined by hemocytometer counts.

2. Spleen cells are resuspended at 10^7/ml in RPMI 1640 containing 10 mM HEPES buffer, 50 μg/ml gentamicin, and 5% FBS. The spleen cell suspensions are cultured with the appropriate stimulant for 48 hr at 37°, 5% CO_2. Cells from BCG-immune mice are treated with the specific antigen PPD at 50 μg/ml. Schistosome-immune cells are incubated with soluble adult worm antigens at 500 μg/ml. Normal cells can be stimulated to produce lymphokines by addition of the mitogen concanavalin A at 3 μg/ml. Supernatant fluids are collected and cleared of cells by centrifugation. In the case of mitogen-stimulated cells, the concanavalin A can be removed by addition of Sephadex G-10 (10 mg/ml) prior to centrifugation.[15] Since the antigens cannot be easily removed, appropriate controls consisting of antigen only or supernatant fluid from incubation of nonimmune cells with antigen can be included. Lymphokine-containing supernatants can be stored at $-20°$ for up to 6 months.

3. Peritoneal macrophages from normal mice are activated by incubation with lymphokine prior to exposing the cells to schistosomula. Macrophages are incubated at 37° with spleen cell supernatant fluids at dilutions of 1 : 10 or greater for 4 to 6 hr. Lymphokine can then be removed by centrifugation or allowed to remain present during the larvicidal assay.

Collection of Peritoneal Macrophages (see also this series, Vol. 108 [25])

Activated peritoneal cells are elicited as described above. Control macrophages from normal mice can be elicited by injection of 1 ml of 1.2% sodium caseinate in buffered saline 3 to 5 days prior to assay.

1. Mice are lightly anesthetized by brief exposure to ether and immediately decapitated, in order to exsanguinate the animal and minimize the possibility of internal bleeding causing red cell contamination of the macrophage preparation.

2. Abdominal skin is wetted with 70% ethanol and a small incision is made through the skin only, taking care not to clip the peritoneal wall, at the midpoint of the abdomen. The skin is then pulled back toward the head and tail to reveal the peritoneal area.

3. Using a 10-ml syringe equipped with a 19-gauge needle, 5 to 10 ml of cold harvest media [Dulbecco's modified Eagle's medium with 4.5 g of glucose/liter (DMEM) supplemented with 5 U heparin and 50 μg gentamicin/ml and 2% FBS] is injected into the peritoneal cavity. Without removing the needle, the medium is slowly withdrawn and the fluid is placed in polypropylene centrifuge tubes on ice.

4. Samples of the cells are counted by hemocytometer to determine the total number of leukocytes present. Other samples, containing approximately 10^5 cells in 150 μl of media with 10% FBS, are prepared for differential cell counting by centrifugation onto glass slides in a cytocentrifuge (Cytospin, Shandon Southern, Inc., Sewickley, PA). Slides are then stained with Diff-Quik and the percentages of macrophages, lymphocytes, and polymorphonuclear cells present are determined.

5. The peritoneal cells are then centrifuged at 300 g, 10 min at 4°, and the pellets are resuspended at 1.6×10^6 macrophages/ml in DMEM containing 10% FBS and gentamicin.

Larvicidal Assay

If the larvicidal assay is to be performed in suspension culture, the macrophage-containing cell populations are placed directly into polypropylene test tubes, to which the macrophages adhere poorly. Under circumstances in which the presence of contaminating cell types is undesirable, adherent cell monolayers can first be prepared. It should be noted that the use of suspension cultures has previously been found to maximize macrophage larvicidal activity,[16] apparently by allowing optimal conditions for effector cell–target contact.

Lymphokine-activated macrophages can efficiently kill high percentages of schistosomula in a non-antibody-dependent manner, i.e., via nonspecific effector mechanisms. However, when the activity of the macrophage population is low, as is the case when cells recovered from S. mansoni-infected mice are used, macrophage-mediated schistosomula killing can be enhanced by addition of sera from immunized (S. mansoni-infected or vaccinated) mice to the assay or by preopsonizing the larval targets, thus creating conditions of improved macrophage adherence to schistosomula.

1. Aliquots containing approximately 8×10^5 macrophages (0.5 ml vol) are placed into sterile 10 × 75 mm round-bottomed polypropylene test tubes (suspension culture) or into 16-mm diameter wells of sterile polystyrene tissue culture plates (monolayer culture). To allow the mac-

[16] S. L. James, A. Sher, J. K. Lazdins, and M. S. Meltzer, *J. Immunol.* **128,** 1535 (1982).

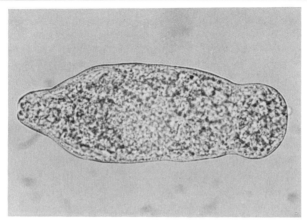

FIG. 2. Dead schistosomulum after 48-hr culture with lymphokine-activated macrophages. Note flattened and stiff appearance of larva, with internal granularity and loss of organ structure (×1000).

rophages to adhere to the plastic for monolayer cultures, the plate is incubated 2 hr at 37°. Nonadherent cells are then removed by washing three times with cold culture media, and the wells are filled with 0.5 ml of DMEM with 10% FBS and gentamicin. Duplicate samples are established for each assay condition.

2. Washed schistosomula are counted in three 50-μl aliquots under a dissecting or inverted tissue culture microscope. The larvae are resuspended in the above media at a concentration that will deliver approximately 80 schistosomula per tube or well, giving a final effector : target ratio of $10^4 : 1$, in a total volume of 1 ml of media. If lymphokine or immune sera are to be present in the assay, this volume must be taken into account. Standard procedure calls for the addition of cells in 0.5 ml and

TYPICAL LEVELS OF MACROPHAGE-MEDIATED SCHISTOSOMULUM
KILLING IN THE *in Vitro* LARVICIDAL ASSAY

Conditions	Larvicidal activity (%)
Macrophages from control mice	3
Same, opsonized schistosomula	5
Lymphokine-activated macrophages	100
BCG-activated macrophages	70
Same, opsonized schistosomula	100
S. mansoni-activated macrophages	20
Same, opsonized schistosomula	40

larvae in either 0.5 or 0.4 ml of media, with additional reagents added in the remaining 0.1 ml. The highest concentrations of lymphokine or immune sera used in these assays are therefore 10% of volume. In the case of antibody opsonization, the larvae may be pretreated in 10% immune sera for 30 min to 1 hr and then washed prior to addition of the parasites to the assay.

3. The schistosomula are incubated with the cells for 48 hr at 37°, 5% CO_2. Larval viability at the end of this time is evaluated microscopically at 100× magnification. For suspension cultures, the upper 0.9 ml of media is removed and the remaining 0.1 ml, containing settled larvae and cells, is placed onto a glass microscope slide, coverslipped, and examined with a standard light microscope using the 10× objective. For monolayer assays, individual wells are evaluated using an inverted tissue culture microscope with the 10× objective. Larval viability is determined on the basis of overall morphology and motility. While viable larvae appear translucent, with internal organ structures visible, and exhibit representative stretching and contracting movements, dead larvae appear opaque and granular, with loss of internal structure, and are nonmotile (Fig. 2). Macrophage-mediated killing is reported as larvicidal activity, determined as the mean percentage of recovered larvae which are nonviable in duplicate samples. Typical results[5] at 48 hr are shown in the table.

4. Results of the *in vitro* assay can be verified in an *in vivo* infectivity assay, in which the ability of larvae that have been cultured with activated macrophages to mature into adult worms is assessed.[16] After 24 to 48 hr of culture, the content of each sample is injected intravenously into a normal syngeneic mouse. Six weeks later, adult worms are perfused from the hepatic portal system[17] and counted. Decreased recovery of adult schistosomes should reflect increased *in vitro* larvicidal activity of macrophages.

[17] S. R. Smithers and R. J. Terry, *Parasitology* **55**, 695 (1965).

Author Index

Numbers in parentheses are footnote reference numbers and indicate that an author's work is referred to although the name is not cited in the text.

Subject Index

M

types of, 283
type-specific antibody as, 285
Opsonization
    aspecific, 302
    cellular defects, 314, 316–317
    defective, clinical consequences of,
        313–317
    definition, 283
    effective, measurement of, 318
    heat-labile, 302
    heat-stable, 302
    humoral defects, 314–316
    kinetics of, in whole blood, whole
        blood luminol-dependent Cl used
        to follow, 504
    of serum-resistant strains, 168
    of serum-susceptible strains, 168
    specific, 302
        vs. aspecific, role of, 288–290
Oriental Sore, 609
Osmolarity, effect on phagocytosis, 99
Osteoclast, exocytosis by, 9
Osteomyelitis
    S. marcescens, 293
    staphylococcal, opsonic antibody re-
        sponse in, 285
Oxacillin
    effect on microbe–phagocyte interac-
        tions, 309
    effect on microbial susceptibility to
        lysozyme, 310
Oxidation radicals
    detection, by measuring release of
        ethylene from 2-keto-4-methyl-
        thiobutyric acid, 157
    formation, measurement, 117–118, 157–
        158
Oxygen, solubility in water, at different
    temperatures, 149–150
Oxygen consumption, 353
    measurement, 147–150
    neutrophil
        manometric measurement, 112
        potentiometric measurement, 112
    test, 129
Oxygen electrode, 149–150, 518
Oxygen intermediates, production, chemi-
    luminescence used to measure, 548
Oxygen uptake, measurement
    manometric, 147–149
    with oxygen electrode, 149–150

P

P815 mastocytoma, cell labeling, 560–561
Penicillin
    effect on microbial susceptibility to
        lysozyme, 310
    effect on phagocytosis, 308–309
Percoll suspensions
    osmolality, measurement, 236
    pH, measurement, 236
    for phagocyte purification, preparation,
        235
    specific gravity, measurement, 236
Periodic acid–Schiff stain, 135
    for polysaccharides, 179–180
Peritoneal macrophages. See also Murine
        peritoneal macrophages
    preparation, 630
Peroxidases
    biological functions, 537
    distribution, 537
    leukocyte, reaction catalyzed, 571
    macrophage stimulation and activation,
        537
    reaction catalyzed, 537–538
Phagocyte particle interaction, 10–11
Phagocytes
    cellular CL, course, 499–500
    cryopreservation of, 234
    activation of, 9–10
    apical surface, removal of single recep-
        tor type from, 206–207
    differentiating between types, specific
        staining techniques for, 132–135,
        171–180
    functional heterogeneity, detection of,
        204
    heterogeneity response of, 217
    from human blood, 225–243
        changes in cell density during iso-
            pycnic centrifugation, 227
    lysing, by bacteria, 15
    metabolic alterations caused by phago-
        cytosis, studying, 220–221
    monolayers of, preparation, 215–216
    mononuclear, 5–6
        life span, 6
    movement, 7–8
    nonphagocytic functions of, 6–7
    nonprofessional, 6